PLANTS FOR MAN

SECOND EDITION

PLANTS FOR MAN

ROBERT W. SCHERY

Director, The Lawn Institute

Formerly Assistant Professor of Botany,
The Henry Shaw School of Botany,
Washington University

Research Associate, Missouri Botanical Garden

PRENTICE-HALL, INC. *Englewood Cliffs, New Jersey*

10   9   8   7   6   5   4   3   2   1

ISBN: 0-13-681254-6

Library of Congress Catalog Card Number 72-140

PRENTICE-HALL INTERNATIONAL, INC., *London*
PRENTICE-HALL OF AUSTRALIA, PTY. LTD., *Sydney*
PRENTICE-HALL OF CANADA, LTD., *Toronto*
PRENTICE-HALL OF INDIA PRIVATE LIMITED, *New Delhi*
PRENTICE-HALL OF JAPAN, INC., *Tokyo*

Printed in the United States of America

Contents

Preface

THE AIM OF THIS BOOK IS TO BRING TOGETHER GLEANINGS OF INFORMATION from an immense and scattered field related to useful plants. The hope is to portray them in an interesting, uncomplicated way, yet without sacrificing encyclopedic values of the book. Hopefully the book provides a "bird's eye view" of plant–man interdependency, and makes one aware of how fundamental plants are for man and his society (sometimes overlooked amidst the technological confusion of the modern world).

Of course technology is not foreign to the searching out, growing, harvesting, and processing of plants and their products. Agriculture and industrial processing are a part of economic botany. The book aims, however, more to create an awareness of the ubiquitous ways in which plants serve man than to provide technological or industrial detail. In an age when most people of the Western World are far removed from plants, unlike their farming ancestors, a person may be curious as to where the paper on which he writes comes from, or the shortening used in cooking his food, or the fiber that makes his clothing, or the drugs that cure his illnesses, and so on. The book should satisfy such curiosity.

The ebb and flow of plant usage changes with the times, and is much influenced by tangential factors such as labor costs, transportation, marketing systems, sponsored research, and even diplomacy. But the fact that the majority of the people on earth today still eke out a sustenance by simple gathering, planting, and harvesting, often as part of semi-literate societies, can't be overlooked. Before revising the first edition it was thought that many of the listings could be abandoned as obsolete, only to find upon closer scrutiny that man's dependence upon obscure plants, uncultivated sources, and new introductions has, if anything, increased. Not all species having a direct influence upon man's well-being these days can be listed, and in underdeveloped parts of the world many are probably still unrecorded. All plants are in a sense economic, because of their influence on the environment; but the abundance of those directly employed by man is indicated by the nearly 600 pages in the Uphof *Dictionary of Economic Plants*, where dozens of species are listed on almost every page.

The world cries out for conservation and more intelligent use of plants. Traditionally the illiterate food gatherer and the subsistence agriculturist have given scant thought to preserving life forms influential to their well being. Modern, technological civilization has been equally remiss in destroying habitat, polluting the environment, and overtaxing the world's resources. A review of the plants which support man may help the reader achieve better perspective on the conduct of human affairs. In ancient times the finding and growing of plants was essential for very survival. No less are they required to maintain a modern civilization, even though the majority of people, beneficiaries of plant usage, seldom stop to consider their dependency upon the green biosphere.

That plants are gathered, traded, and sold brings economics into the the picture: profit is what "makes the mare go." Paradoxically the "best" plant may not dominate or even be used, if others can be more conveniently grown, transported to market, processed, or have some other economic advantage. The discussions at the end of various chapters should make clear that a commanding market position today is no assurance that the same plant will necessarily dominate tomorrow. And no doubt there are still botanic gems to be found in the wild, which could serve mankind exceedingly well, bring fortune to many.

We hope that elaborating upon the relatively few plants of major economic importance in each of the categories represented by the various chapters will provide an understanding of that particular plant-based industry, and constitute a less tedious review than would be recitation of lengthy lists of plants that are lesser sources of these materials. Condensed listings of the less widely-used sources follow the major discussions, in order to give some breadth and background. Such listings are more for reference than for study, and in conjunction with the index should prove useful for providing a broad review and some understanding of what plant–man relationships "are all about".

PLANTS FOR MAN

PART I

Introduction

MAN-PLANT INTERDEPENDENCY HAS EXISTED SINCE HIS ADVENT. THE INTEREST in useful plants has shaped man and his civilizations.

CHAPTER 1

Man's Relationship with Plants

The Setting. MAN'S PLACE IN THE UNIVERSE is falling into better perspective as astronomy, geology, and other scientific disciplines marshal facts and inferences. Man on earth is most certainly "located on an insignificant mote in an incomprehensible vastness." Astronomers speculate about cosmic evolution that began some 10 billion years ago with a "big bang" of dense energy–matter that has been expanding and changing ever since. Primitive molecules compounded into more complex ones. From a cloud of them the sun condensed nearly 5 billion years ago, including its satellites among which is earth. More billions of years and rocks had solidified, light ones "floating" on the surface above heavier ones to become continents between water-filled basins, the oceans. It was a vastly different world than we know, barren and devoid of life. The atmosphere probably contained little oxygen but some hydrocarbon (such as methane). Volcanos, and the leftovers from creation, provided elements and simple compounds. Under the conditions of radiation and temperature prevailing the raw materials for life formed—linkages of carbon, hydrogen, nitrogen, and other elements. The first amino acids, nucleotides, and perhaps even simple proteins arose (molecules such as these have been fashioned in the laboratory abiogenetically). Even then the amino acids and nucleotides may have yielded primitive DNA, a beginning genetic code.

Evidence of primitive life is found in rocks 3 billion years old and older. There were wormlike creatures a billion years ago, and an "explosion" of life into many different forms a half-billion years later. Four hundred million years ago aquatic forms began to invade the land; 200 million years later there had evolved many relatively advanced plants and animals such as those of Figs. 1-1 and 1-2. The time scale for all of this is suggested by Calvin, Fig. 1-3, and Cloud, Fig. 1-4. This evolution was slow; if the 10 billion years since creation of the universe is likened to a calendar year, the first land life would not have arisen until December, and modern man would have appeared only 15 minutes before midnight on December 31. But, "to Him, Who commands time, all things are possible."

Advent of man and modern plants. Back in Mesozoic times, some 100 million years ago, the dominance of the mighty dinosaurs was challenged by an upstart but better-adapted class, the mammals. In the plant kingdom, too, there was an advent of tremendous moment—the first appearance in the world of the Angiosperms, or seed plants, predestined within a few brief geological ages to dominate the land areas of the earth.

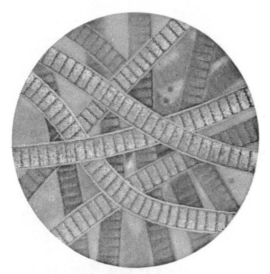

Two billion years ago.

Life began in the sea. Its actual beginnings and its earliest forms are lost, but this we know: some two billion years ago plants resembling today's algae were performing photosynthesis—using the energy of sunlight to make food. For the next billion and a half years, life remained in the water, evolving slowly.

420 million years ago.

Life invaded the land, and the first tree raised itself above the bare rock. Instead of roots, a tangle of specialized branches rested in slight depressions in the ground, amid moss and seaweed flung up by the sea. In time, the plants on the land would help decompose the bare rock and create soil.

370 million years ago.

Life on land became abundant and complex. Dense forests covered an earth that was uniformly hot and humid. The trees in these forests were giant ferns and club moss that grew as tall as 150 feet. The animals were insects and giant ancestors of the modern scorpion.

Figure 1-1 Early life and the evolution of forests, today's most abundant biomass. (Courtesy St. Regis Paper Co.).

300 million years ago.

For millions of years these conditions continued. Plant life was so abundant it frequently had no time to decay in the marshy soil, but piled up, layer upon layer. Increasing weight ultimately turned it into coal and oil. Reptiles appeared, but insects dominated—including dragon-flies with two-foot wingspreads.

200 million years ago.

The early dinosaurs appeared. With the ground hardening, plants found it more difficult to reproduce. So a new method of reproduction evolved: seeds, which could survive until the right conditions of soil and moisture started them growing. Among the first seed-producing plants were the conifers.

180 million years ago.

As the dinosaurs disappeared, parts of the earth began to experience seasonal changes in weather. A new kind of tree developed that could shed its leaves and remain dormant through long, cold winters. These were the deciduous trees, flowering plants representing the most advanced form of plant life on earth.

Fig. 1-1 cont'd

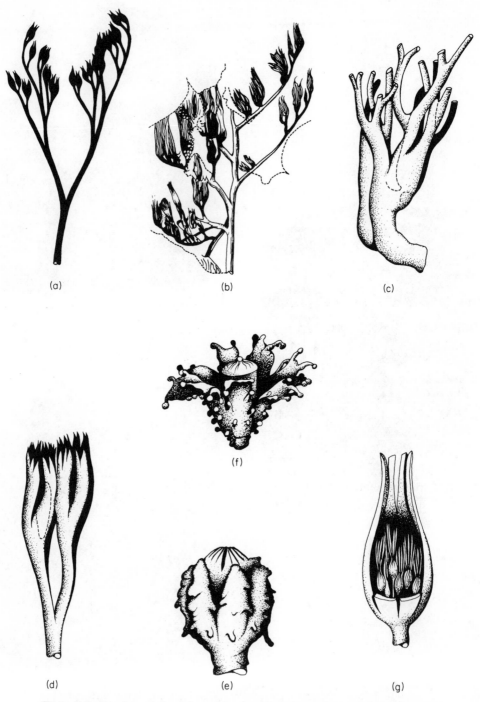

Figure 1-2   Drawings of the first seed-bearing plants, the pteridosperms; a and b show frond-like cupules, c–g, representative seed-bearing organs. (After Henry N. Andrews, *Science* **142**, 925–31 (1963).)

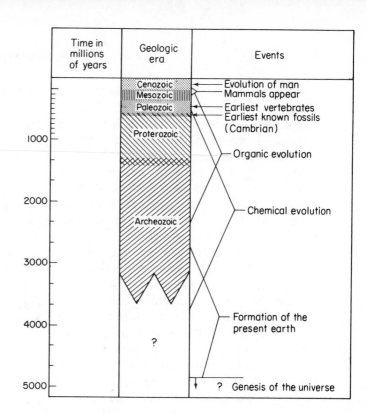

Time in millions of years	Geologic era	Events

Cenozoic — Evolution of man / Mammals appear
Mesozoic
Paleozoic — Earliest vertebrates / Earliest known fossils (Cambrian)

1000 — Proterozoic — Organic evolution

2000

Archeozoic — Chemical evolution

3000

4000 — Formation of the present earth

?

5000 — ? Genesis of the universe

Figure 1-3 Time scale for total evolution. (After Melvin Calvin, *AIBS Bulletin* **31**, Oct. 1962.)

Figure 1-4 Postulated main features of interacting biospheric, lithospheric, and atmospheric evolution on the primitive earth. (After Preston E. Cloud, Jr., *Science* **160**, 732 (May 17, 1968).)

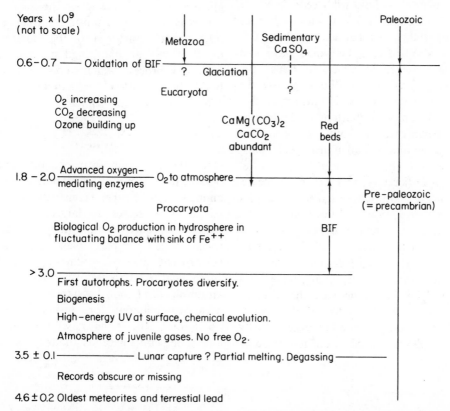

Years x 10^9 (not to scale)

Paleozoic

Metazoa   Sedimentary $CaSO_4$

0.6–0.7 —— Oxidation of BIF ——   ? Glaciation

Eucaryota   ?

O_2 increasing
CO_2 decreasing
Ozone building up

$CaMg(CO_3)_2$
$CaCO_2$
abundant   Red beds

1.8 – 2.0 $\dfrac{\text{Advanced oxygen-}}{\text{mediating enzymes}}$ — O_2 to atmosphere

Pre-paleozoic (= precambrian)

Procaryota

Biological O_2 production in hydrosphere in fluctuating balance with sink of Fe^{++} BIF

> 3.0

First autotrophs. Procaryotes diversify.

Biogenesis

High–energy UV at surface, chemical evolution.

Atmosphere of juvenile gases. No free O_2.

3.5 ± 0.1 ——————— Lunar capture ? Partial melting. Degassing ———

Records obscure or missing

4.6 ± 0.2 Oldest meteorites and terrestial lead

Mammals and Angiosperms jointly threaded their successful way through the Mesozoic and Cenozoic eras; their history is written in the fossil record. And finally we find man, the culminating mammal, dependent for his very existence, directly or indirectly, upon the seed plants; and many of the seed plants (e.g., maize and other cereals, most fruit and vegetable plants) dependent in turn upon man for their care and perpetuation.

Let us now shift our focus briefly to the geologically recent Pleistocene period, the "Ice Age," a mere million years or so preceding historical times. Man had evolved in the Old World, and experienced there four or five epochs of glaciation. During at least the later Pleistocene, he invaded the New World. He existed by gathering fruits, seeds, or succulent herbage of Angiosperm plants; or by killing game that in turn fed upon an extensive supply of Angiospermous forage. But whether partially sedentary and existing by the former means, or nomadic and existing by the latter, man was dependent even then upon plants and plant products, particularly the seed plants. Without wild seeds he would have had no gruels or flours; without fibrous barks or stems, no cordage or basketry. Without woody stems, he would have had inadequate shelter in which to house the simple family group; without grasslands, inadequate herds to yield meat. And— perhaps even more important to the history of the human race—without the appearance of herbaceous annuals, all of them seed plants, man could not have developed agriculture. Only with agriculture, this improved means of supporting multitudes of sedentary peoples in a given area, could civilization with its large cities and their necessary accouterment of libraries, schools, public buildings, and mechanical conveniences have come into being. From this early beginning, plants and plant products have remained the primary base upon which all modern civilization is built. In this book we are concerned chiefly with those plants utilized by our modern civi-

lization, but we cannot forget that all kinds of plants affect man and his existence, however minutely, and are therefore part and parcel of "economic botany" broadly considered.

Photosynthesis and the living world. All students of biology and elementary botany are cognizant of the fundamental importance of photosynthesis, the process whereby green plants utilize the energy of sunlight to build up complex organic molecules otherwise unavailable to most living organisms (see Table 1-1). Often quoted has been the phrase, "all flesh is grass"—meaning of course that growing plants (ordinarily surviving by photosynthetic activity) supply, directly or indirectly, the food upon which all animal life depends. This remarkable process, photosynthesis, depends upon the highly complex pigment–catalyst, chlorophyll. Chlorophyll, drawing upon the energy of sunlight, can cause inorganic molecules, carbon dioxide (CO_2) and water (H_2O), to become an organic molecule such as glucose ($C_6H_{12}O_6$), and at the same time can release to the atmosphere free oxygen (O_2), the necessary gas for the breathing of most of the animal kingdom. This most important process, which stores more than a billion billion kilocalories annually as food, can be epitomized chemically as:

$$6CO_2 + 6H_2O \xrightarrow[\text{sunlight energy}]{\text{chlorophyll and}} C_6H_{12}O_6 + 6O_2.$$

Once the transition from the inorganic to organic molecules is attained, both plants and animals can further construct by their metabolic processes the large and complex molecules needed to build plant and animal tissues. The complex plant products so constructed include cellulose, the basic component of wood and other fibers; starch, an omnipresent food substance, and often a necessary material for various industrial operations; oils, as those used for margarine and soap; proteinaceous foods; and resins, latices, tannins, and a host of other products. Little wonder, then, that so much stress is laid upon photosynthesis as a fundamental biological process, for without it the world would be a dreary place, uninhabited by any of the higher forms of life. Moreover, it would be physically different from the world we know now, with a deficiency of gaseous oxygen, and with accelerated erosion of land masses unprotected by vegetation and plant litter.

Decay and the living world. Have you ever stopped to consider what this world would be like in the absence of the very simple plants and their near relatives, the bacteria and other fungi? These simplest of plants, often lacking even true nuclei, seem to lie at the opposite end of the evolutionary sequence from the Angiosperms, or seed plants, whose importance we have already stressed. Yet they alone in the plant kingdom rival the seed plants in frequency and omnipresence, for bacteria are everywhere, in untold quantities. We may condemn bacteria and other fungi as the cause of many serious diseases, yet without them the earth would be a sorry place for higher organisms to live. There would be no decay or putrefaction, for enzymes secreted by the ever-present microorganisms break down or digest the complex molecules built up through the constructive process of photosynthesis. In so doing these microorganisms release for their own use the energy captured from sunlight by chlorophyll; in turn they can use this energy to elaborate the molecules needed for their metabolism. Any "by-products" of the enzymatic breakdown are eventually returned once again to the realm of the inorganic—to the air and the soil and the sea. Were there no decay, the surface of the earth might be littered to an unpredictable depth with the carcasses and remains of countless generations of animals and plants. Vital mineral elements would remain locked within this debris, unavailable to future generations of plants and animals. Thus an atom would likely be used but once, and soils would soon become seriously depleted. The mechanical inconvenience of undecomposed plant and animal remains would be beyond description. Certainly we are fortunate that the living world became a place of scavengers as well as of builders.

Since prehistoric times man has made direct use of microorganisms and their peculiar ability to supply specific catalytic enzymes. Thus yeast may have been the first among domesticated plants, used even by savage man to break down starch and the hard, horny seeds of Angiosperms into simpler, more palatable, and easily digestible food and drink. By essentially this same process we obtain today industrial alcohol, vinegar, alcoholic beverages, cheese, fibers by retting—to mention only a few of the industrial products enlisting the aid of microorganisms for their preparation.

The pace of change steps up. Explosive growth of technology during the latter half of the twentieth century threatens to remake in mere decades an earth's surface that evolution took eons to perfect. This technology has made possible many undreamed-of comforts, and it has created problems such as pollution which threaten civilization. Plant materials have been no more spared technological change than have nonliving parts of the environment. Plant substance is readily transformed by the chemist today, and substitutes are easily found for items which do not fit economical mass production well.

In technologically advanced countries plant materials that are costly, difficult to produce, or occur only in limited supply find little market. They are overwhelmed by standardized commodities inexpensively procurable in huge and constant supply. If a natural fiber such as ramie, or even "king cotton," cannot compete in the marketplace, plenty of substitutes are on hand to take over (as, indeed, has been the case with synthetic

Table 1-1

Energy from the Sun and Its Availability to Man Through Food (Ottawa Area)

	Cal/cm^2	Cal/cm^2	Horsepower-hours per acre (thousands)	Car mileage from energy per acre
1. Received at top of the atmosphere—annual total		228,000	14,400	5.5 million
2. Growing season total (17 April to 27 Oct.)		163,000	10,300	4.0 million
3. At ground level (incident on the crop)		109,000	6,900	2.6 million
4. Absorbed by crop (25% reflected)		82,000	5,200	2.0 million
5. Retained (after loss by back-radiation)		39,000	2,500	1.0 million
6. Available for heating plant tissue and air, and for photosynthesis (after loss due to transpiration)		5,500	350	134,000
7. Stored in plants (as digestible energy)				
Oats: grain		64	4.0	1,560
straw		57	3.6	1,390
Barley: grain		67	4.2	1,630
straw		53	3.3	1,290
Hay: (timothy and clover)		104	6.6	2,540
(alfalfa)		151	9.5	3,680
Corn: silage		183	11.5	4,470
8. Efficiency of animals for conversion of plant digestible energy to food energy available to humans:				
Milk	18%	23	1.5	560
Pork	16	20	1.3	500
Eggs	5	6.4	0.42	155
Fowl (meat)	4	5.1	0.33	120
Beef	2.8	3.5	0.22	85

Adapted from George W. Robertson, "Where does the Sunshine Go?" Greenhouse Garden Grass (Plant Research Institute), Research Branch, Canada Department of Agriculture, Vol. 6, No. 3, Autumn, 1967.

fibers such as nylon, the acrylics, and polyesters). Acetone, once produced by fermentation, is today less expensively synthesized from petroleum products. With millions of tons of economically produced soybean oil at hand, capable of being transformed by hydrogenation into suitable substitutes for other oils, industry need not concern itself overly much about minor sources. The trend is unmistakably toward huge businesslike operations built about relatively few plant species. Marginal plants or processes are soon squeezed out.

Already many plant products are mostly of historical interest, except perhaps in undeveloped countries where subsistence agriculture persists. Natural dyes are seldom used, and tannins extracted from plants are giving way to chemical tanning and synthetic leather. Rubber, resins, and other once-prominent exudates or extractions are being substituted by synthetics, usually derived from petroleum or minerals. Even wood, once man's fuel, shelter, and source of fiber, may be threatened. The modern alchemist can make plastics from waste wood (or other organic and mineral materials) which often have advantages over lumber. It is possible to process and flavor soybean meal to make it resemble beef, and various plant derivatives can be made scarcely distinguishable from dairy items for which they substitute.

We can expect those plants which fit mechanized agriculture well to prosper in the years ahead; but wild plants are likely to be more and more bypassed and forgotten. Yet wild plants may one day be needed. Indeed, mixed vegetation of many humble species (including oceanic plankton), is already vital for maintaining an atmosphere suited to advanced forms of life. For example, without absorption of carbon dioxide by green plants a heat-conserving "greenhouse effect" might raise air temperature enough to melt polar ice, raise the sea level, and severely alter the environment. Wild plants as well as weeds and crops are one of man's best buffers against extermination through his own pollution! But, as the natural environment becomes altered, certainly extinctions will occur and the reservoir of plant types become depleted. Fortunately, some natural reserves for endemic wildlife are being set aside, and germplasm banks created to retain for future breeding purposes cultivated varieties that have become obsolete. Concentration on greater quantities of stereotyped products, at the expense of the varietal wealth nature provides, may be good short-term economics, but inimical to long-term progress into an uncertain future.

SUGGESTED SUPPLEMENTARY REFERENCES

Baker, H., *Plants and Civilization*, Wadsworth Pub. Co., Belmont, Calif., 1965.

Bernal, J.D., *The Origin of Life*, World, Cleveland, 1967.

Calvin, Melvin, *Communication: From Molecules to Mars*, American Institute of Biological Science Bul. 29–44 (October, 1962).

Dodge, B.S., *Plants that Changed the World*, Little, Brown Co., Boston., 1959.

Edlin, H.L., *Man and Plants.*, Aldus Books, London, 1967.

Janick, Jules, R. W. Schery, F. W. Woods, and V. W. Ruttan, *Plant Science, An Introduction to World Crops*, Freeman, San Francisco, 1969.

Kenyon, D. H., and G. Steinman, *Biochemical Predestination*, McGraw-Hill, N. Y., 1969.

National Academy of Science, National Research Council, *Resources and Man*, Freeman, San Francisco, 1969.

Oparin, A. I., *The Chemical Origin of Life* (Am. Lecture Series Publ. 588) (originally in Russian, 1936), trans. Ann Synge, C. C. Thomas, Springfield, Ill., 1964.

————, *Genesis and Evolutionary Development of Life*, trans. E. Maass, Academic Press, N. Y., 1968.

Sauer, C. O. (ed. J. Leighly), *Land and Life, Univ. California Press*, Berkely, Calif., 1965.

Schultes, R. E., and A. F. Hill, *Plants and Human Affairs*, Lab. Manual for Biology 104, Botanical Museum of Harvard Univ., Cambridge, Mass., 1968.

Schwanitz, F., *The Origin of Cultivated Plants*, Harvard Univ. Press, Cambridge, Mass., 1966.

Thomas, William L., Jr. (ed.), *Man's Roll in Changing the Face of the Earth*, Univ. of Chicago Press, Chicago, 1956.

USDA Yearbooks, Washington, D.C., especially 1964 (*Farmer's World*), 1960 (*Power to Produce*), 1958 (*Land*), and 1951 (*Crops in Peace and War*).

Van Dyne, G. M., *The Ecosystem Concept in Natural Resource Management*, Academic Press, N. Y., 1969.

Wilsie, Carroll P. *Crop Adaptation and Distribution*, Freeman, San Francisco, 1961.

CHAPTER 2

Man's Economic Interest in Plants

IN CHAPTER 1 WE SURMISED HOW VERY EARLY the mammal called man may have existed. From the beginning of his history, he must have had an economic interest in the vegetation about him, even though he originally displayed little conscious thought concerning it. Soon, however, began the long process of man controlling or altering the environment for his use and comfort. Early records of primitive man show him in control of fire, for which he would need dry wood, a plant product. Undoubtedly he devised means (such as girdling) for killing woody vegetation by stripping away tree or shrub bark with a stone or other implement. Soon he devised tools and weapons that often demanded a wooden shaft or handle. Again, fibrous plant tissue was early recognized as valuable for binding and weaving; and of course certain plants had to be identified as suitable for food, others as unpalatable or even poisonous.

More than 10 millennia ago agriculture was born. Man learned to recognize specific plants and animals as especially valuable. He encouraged these, by planting or feeding and by discouraging their competitors. Thus domestication came about, and certain species of plants and animals lent themselves to man's use in exchange for protection. The efficacy of early man's attempts at do-

mestication cannot be denied. There exist today few economically important plants or animals that have been domesticated within historic times. With most domesticated plants the wild ancestral species are unknown or doubtfully known.

As nomadic groups abandoned chance gathering of wild food in favor of intentional (even if casual) sowing of seed, there no doubt occurred many independent beginnings to agriculture. Especially well documented by archeological findings and artifacts is the origin of crop cultivation in the Near East and in the Tehuacan Valley of Mexico. These two centers exemplify well the process of domestication. Vavilov's pioneering work concerning centers of domestication (Fig. 2-1) will be mentioned shortly, but for the moment let us imagine how the process might have occurred north of the Persian Gulf, and in south-central Mexico.

Early domestications. The eastern Mediterranean area from Greece to the Zagros Mountains in western Iran has been the locale for many early civilizations and much of recorded ancient history. The Fertile Crescent of the Tigris–Euphrates valley, with the Zagros mountains just to the east, afforded an unusually rich selection of habitats for diverse plants occupying varying ecological niches. That the western Zagros foothills were unusually well suited to early settlement is proven by the abundant archeological findings there. To the east were the inhospitable high mountains, which, however, could provide excellent summer habitat for nomadic bands. West of the foothills lay somewhat drier steppes, and finally the semiarid alluvial bottoms with rich soil, amenable to very productive agriculture once the art of irrigation was learned.

Apparently the settlements were fairly permanent in the foothills, but the tribes seasonally nomadic, relying upon the chase for wild game, collecting wild fruits and seeds, too. Evidence of wild alfalfa and primitive grains are repeatedly uncovered, as well as the bones of numerous animals and birds. Probably the migrating tribes would harvest an early crop on the steppes, another at higher elevations in the foothills a month or so later, and still others in the mountains during summer. Visualizing frequent migration through such a wealth of locations, indeed there would be remarkable opportunity for both chance and intentional selection, and for making introductions into other habitats.

The archeological record seems to indicate dependence upon wild food from about 40,000 to 10,000 B.C. Around 10,000 B.C. considerable transfer of species from native habitats to more convenient locations near the burgeoning settlements occurred. Wheat and barley, the earliest grains, were brought from the foothills to the valley and accorded some tillage and irrigation. Given this advantage, it is not hard to imagine aggressive annual grasses such as these adapting quickly. Their populations changed rapidly to forms favored by cultivation. Perhaps the dawn of argiculture involved nothing much more than such bringing of useful wild grain seeds from more distant locations to disturbed soils of

prehistoric camps, and noting how they prospered there. Naturally, the most useful types would be harvested (such as grain heads with nonshattering seeds). Sowing them into richer valley soils supplied with irrigation, and grubbing out competing vegetation, would complete the transformation to domesticated crop. A sequence of steps like this very well may have occurred with wheat in Mesopotamia. Barley may have been domesticated simply as an adventive weed impossible to eliminate from the wheat; rye and oats were apparently introduced into agriculture in this fashion, at a later date in southeastern Europe. Also introduced into cultivation in the Zagros foothills were several millets, field pea, lentil, vetch, chickpea, horse bean, and flax; the wine grape, olive, date, apple, pear, cherry, fig, and other horticultural crops.

In the New World, intensive archeological investigation has been especially rewarding in the Tehuacan Valley, some 150 miles south of Mexico City. Because of dryness, remains are especially well protected in the many caves and middens. Human coprolites often indicate exactly the food plants consumed. And the Tehuacan Valley seems to be at least one location where the modern type of maize was first brought into cultivation, along with beans, squash, and several other important food species.

As in the Fertile Crescent, there is advantageous juxtaposition of varied habitat, the dry valley being ringed by foothills and mountains with a wealth of ecological niches. Carbon-14 dating indicates that the valley was inhabited, and incipient agriculture underway, for several millennia prior to 7200 B.C. Artifacts are remarkably well preserved in the middens, shedding light even on such details as months of occupancy and weather conditions. Before 7200 B.C. the valley was only lightly populated, by small groups that changed camp from time to time and subsisted mainly upon wild plants and hunting. Pollen records show that the valley was grassland, and, although there were flint tools, nothing yet well adapted for cultivation of the soil.

During the next several thousand years up to 3400 B.C. the nomadic groups began to gather for longer periods at base camps. Crude plantings of such seeds as squash, chilis, and avocado were made with primitive digging sticks. Grain seems to have been gathered from the wild, although there were tools for its grinding. As time progressed, populations continued to increase, and to become more sedentary (although still disbursing seasonally for hunting and collecting plants from the wild).

In the two millennia after 3400 B.C. maize, pumpkin, certain beans, amaranth, zapote, and several squashes were brought into cultivation. Yet, wild plants and animals still accounted for about half of the food supply, to judge from feces. During the last 1,000 years B.C. peoples of the Tehuacan Valley became essentially full-time agriculturists. By then an early form of modern maize had been introduced, a mainstay that permitted sedentary habitations replete with permanent houses and various ceremonial structures. Irrigation was established, and additional food plants introduced (e.g., tomato, peanut, and lima bean). The wild type of maize was gathered until about 3000 B.C., brought to the valley for cultivation made possible by irrigation. By 1500 B.C. improved maize appeared. This important genetic advance may have resulted from natural hybridization of wild maize with teosinte (*Euchlaena mexicana*), which is itself perhaps a cross between an ancient maize and gamagrass (*Tripsacum dactyloides*). So useful was the improved maize that the wild form was abandoned and had become extinct before the time of Christ. In the Tehuacan Valley, as in the Near East, techniques which reduce need for migration in the search for

wild plants and animals were gradually adopted as they were proved useful through trial and error.

The plants of ancient history. Aside from neolithic or even preneolithic picture writing, preserved on bone and on the walls of caves, some of it certainly 10,000 or more years old, the first records of any economic plants are probably those of ancient Egypt and of China. Charles Pickering lists for as early as 4271 B.C. the recorded occurrence in the Middle East of many economic plants, including wormwood, palm, fig, gum, and a number of other species. Certainly flax is known, from the tombs of ancient Egypt, indicating a textile industry of at least 4,000 years antiquity. In China elaborate papers were being made while Europe was yet a barbaric province, and many food plants, including forms as strange to us as seaweed, were recorded as worthy of offering at shrines and altars.

Later, some few centuries before the birth of Christ, the Greek civilization observed and recorded a few facts about the plants of interest to that era. Theophrastus, in his *History of Plants*, about 340 B.C., and the renowned Aristotle made notes on useful plants of the world so far as it was then known. Subsequently the Roman naturalist Pliny the Elder, and his contemporary, the Greek Dioscorides, recorded for posterity a few accurate observations amidst a multitude of old wives' tales—Pliny's *Natural History* has been called a "repository for all the errors of antiquity." It was concerned chiefly with superstitions of the day and the reputed medicinal values of a number of plants. In fact, the majority of early writings stressed the medicinal or supernatural powers of strange "drug" plants, while disregarding as unworthy of comment—a matter for the attention of slaves only—agriculture and cultivated plants.

The Middle Ages saw but slight advance in

the study of economic plants, though populations expanded and agriculture was a necessary art. Interest in the medicinal and supposed supernatural virtues of plants continued, and in the late Middle Ages we find the herbalists expounding the doctrine of signatures. This curious doctrine held that plants were created expressly for man's use, and that the Creator, albeit sometimes cryptically, so designed a plant that its appearance reflected the use to which it could advantageously be put. Thus the walnut, resembling the convolutions of the brain, was thought to be helpful against afflictions of the brain; the bloodroot, with a reddish juice, a certain conditioner for the blood; the toothwort or pomegranate, endowed with a resemblance to teeth, an effective oral aid. In the elaborately written and beautifully illustrated herbals, botanical science had its utilitarian beginning. The start was made on what was eventually to be an extensive body of writings; and communication across national borders was begun. The way was now open for rapid expansion in plant exploration, collection, and classification.

Some landmarks in the science of economic botany. Classic in the development of economic botany are the works of Alphonse de Candolle and N. I. Vavilov. The Swiss de Candolle, in his *Geographical Botany* of 1855, and more particularly in his *Origin of Cultivated Plants* of 1883, was the first successfully to attempt integration of various types of evidence to determine, with a high degree of probability, the ancestral form, region of domestication, and history of most of the important cultivated plants. Dependent in his research upon the rather fragmentary knowledge of his time, de Candolle nevertheless made accurate deductions, for the most part confirmed later with but slight alteration. De Candolle's breadth of interest enabled him to draw upon discoveries in botany (particularly plant geog-

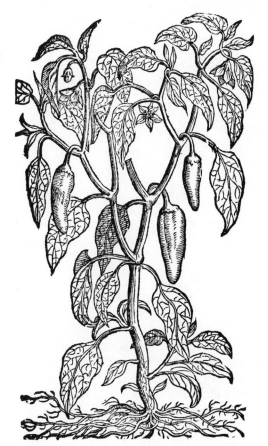

Chili pepper, *Capsicum annuum.* From the Lobel Herbal, 1581. (Courtesy Chicago Natural History Museum.)

raphy), archaeology, written history, and philology.

Vavilov, the Russian, was in a sense a de Candolle of the twentieth century. He adopted a worldwide approach and assembled immense quantities of data from many sources. Utilizing ideas from newer fields of science, particularly those of genetics and heredity, and more complete data in the older fields known to de Candolle, Vavilov was able to indicate with exactness presumed centers of origin of cultivated plants. In the light of more recent research, some of his generalizations have merited modification,

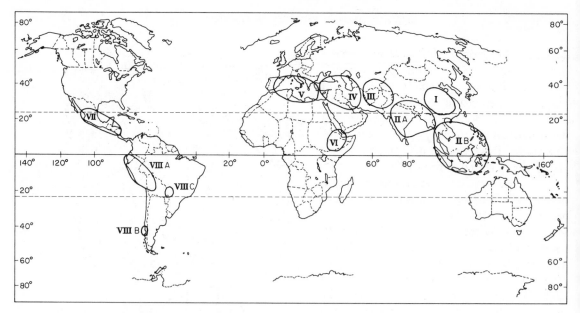

Figure 2-1 Chief world centers of origin of cultivated plants. (After Vavilov, *Origin, Variation, Immunity, and Breeding of Cultivated Plants,* Ronald Press, N. Y., 1951.)

but on the whole, economic botany is indebted to Vavilov for many basic principles.

Vavilov was convinced that most cultivated plants are quite dissimilar to their nearest wild relatives, and arose because of isolation caused by peculiar environmental or genetic conditions. The map of Fig. 2-1 depicts Vavilov's hypothesized centers of origin, and Table 2-1 charts the species presumed domesticated in each. Most centers are marked by a diversity of habitat, as noted for the Zagros foothills and Tehuacan Valley. Vavilov reasoned that under such circumstances considerable variation would occur in the wild ancestors of most cultivated plants, of which certain distinctive genotypes become successful and dominate as they spread from the center of origin. He believed that a successful cultivar* in new habitat would so dominate the plant population as to be something like a "pure line" kept free from cross-breeding. This would provide opportunity for recessive forms to become

manifest at the periphery of a cultivar's range, while dominant genes would tend to characterize the dispersal center. These conclusions were remarkably perceptive, especially considering the hearsay information about distant parts of the world in Vavilov's age of limited transportation.

The encyclopedias and old texts. Coordinate with the research of de Candolle and Vavilov in the advancement of economic botany was the assemblage of information in various encyclopedias and "practical" botanies. These are chiefly of historic interest today, but a number merit mention for their remarkable comprehensiveness in a day when the world was yet relatively unexplored by literate man. We shall mention here some of the encyclopedias perhaps outstanding even in modern times. In London, in 1908, Sir George Watt published his *Commercial Products of India*, following his earlier monumental *Dic-*

---

* The recently accepted term for a horticultural variety or selection (*"cultivated variety"*), as distinguished from the more basic botanical variety (a "subspecies").

Table 2-1

World Centers of Origin of Cultivated Plants

(after Vavilov, as translated, uncorrected for plant names)

I. **CHINESE CENTER:** The largest independent center which includes the mountainous regions of central and western China, and adjacent lowlands. A total of 136 endemic plants are listed, among which are a few known to us as important crops.

CEREALS AND LEGUMES	1. Broomcorn millet, *Panicum miliaceum* 2. Italian millet, *Panicum italicum* 3. Japanese barnyard millet, *Panicum frumentaceum* 4. Kaoliang, *Andropogon sorghum* 5. Buckwheat, *Fagopyrum esculentum* 6. Hull-less barley, *Hordeum hexastichum* 7. Soybean, *Glycine hispida* 8. Adzuki bean, *Phaseolus angularis* 9. Velvet bean, *Stizolobium hassjoo*
ROOTS, TUBERS, AND VEGETABLES	1. Chinese yam, *Dioscorea batatas* 2. Radish, *Raphanus sativus* 3. Chinese cabbage, *Brassica chinensis, B. pekinensis* 4. Onion, *Allium chinense, A. fistulosum, A. pekinense* 5. Cucumber, *Cucumis sativus*
FRUITS AND NUTS	1. Pear, *Pyrus serotina, P. ussuriensis* 2. Chinese apple, *Malus asiatica* 3. Peach, *Prunus persica* 4. Apricot, *Prunus armeniaca* 5. Cherry, *Prunus pseudocerasus* 6. Walnut, *Juglans sinensis* 7. Litchi, *Litchi chinensis*
SUGAR, DRUG, AND FIBER PLANTS	1. Sugar cane, *Saccharum sinense* 2. Opium poppy, *Papaver somniferum* 3. Ginseng, *Panax ginseng* 4. Camphor, *Cinnamomum camphora* 5. Hemp, *Cannabis sativa*

II. **INDIAN CENTER:**

A. **Main Center:** Includes Assam and Burma, but not Northwest India, Punjab, nor Northwest Frontier Provinces. In this area, 117 plants were considered to be endemic.

CEREALS AND LEGUMES	1. Rice, *Oryza sativa* 2. Chickpea or gram, *Cicer arietinum* 3. Pigeon pea, *Cajanus indicus* 4. Urd bean, *Phaseolus mungo* 5. Mung bean, *Phaseolus aureus* 6. Rice bean, *Phaseolus calcaratus* 7. Cowpea, *Vigna sinensis*

Table 2-1   (cont.)

VEGETABLES AND TUBERS	1. Eggplant, *Solanum melongena* 2. Cucumber, *Cucumis sativus* 3. Radish, *Raphanus caudatus* (pods eaten) 4. Taro, *Colocasia antiquorum* 5. Yam, *Dioscorea alata*
FRUITS	1. Mango, *Mangifera indica* 2. Orange, *Citrus sinensis* 3. Tangerine, *Citrus nobilis* 4. Citron, *Citrus medica* 5. Tamarind, *Tamarindus indica*
SUGAR, OIL, AND FIBER PLANTS	1. Sugar cane, *Saccharum officinarum* 2. Cocoanut palm, *Cocos nucifera* 3. Sesame, *Sesamum indicum* 4. Safflower, *Carthamus tinctorius* 5. Tree cotton, *Gossypium arboreum* 6. Oriental cotton, *Gossypium nanking* 7. Jute, *Corchorus capsularis* 8. Crotalaria, *Crotalaria juncea* 9. Kenaf, *Hibiscus cannabinus*
SPICES, STIMULANTS, DYES, AND MISCELLANEOUS	1. Hemp, *Cannabis indica* 2. Black pepper, *Piper nigrum* 3. Gum arabic, *Acacia arabica* 4. Sandalwood, *Santalum album* 5. Indigo, *Indigofera tinctoria* 6. Cinnamon tree, *Cinnamomum zeylanticum* 7. Croton, *Croton tiglium* 8. Bamboo, *Bambusa tulda*

B. **Indo–Malayan Center:** Includes the Malay Archipelago. Fifty-five plants were listed, including the following:

CEREALS AND LEGUMES	1. Job's tears, *Coix lacryma* 2. Velvet bean, *Mucuna utilis*
FRUITS	1. Pummelo, *Citrus grandis* 2. Banana, *Musa cavendishii, M. paradisiaca, H. sapientum* 3. Breadfruit, *Artocarpus communis* 4. Mangosteen, *Garcinia mangostana*
OIL, SUGAR, SPICE, AND FIBER PLANTS	1. Candlenut, *Aleurites moluccana* 2. Cocoanut palm, *Cocos nucifera* 3. Sugar cane, *Saccharum officinarum* 4. Clove, *Caryophyllus aromaticus* 5. Nutmeg, *Myristica fragrans* 6. Black pepper, *Piper nigrum* 7. Manila hemp or abaca, *Musa textilis*

Table 2-1   (cont.)

III.   CENTRAL ASIATIC CENTER: Includes Northwest India (Punjab, Northwest Frontier Provinces, and Kashmir), Afghanistan, Tadjikistan, Uzbekistan, and western Tian-Shan. Forty-three plants are listed for this center, including many wheats.

GRAINS AND LEGUMES	1. Common wheat, *Triticum vulgare* 2. Club wheat, *Triticum compactum* 3. Shot wheat, *Triticum sphaerocoecum* 4. Pea, *Pisum sativum* 5. Lentil, *Lens esculenta* 6. Horse bean, *Vicia faba* 7. Chickpea, *Cicer arietinum* 8. Mung bean, *Phaseolus aureus* 9. Mustard, *Brassica juncea* 10. Flax, *Linum usitatissimum* (one of the centers) 11. Sesame, *Sesamum indicum*
FIBER PLANTS	1. Hemp, *Cannabis indica* 2. Cotton, *Gossypium herbaceum*
VEGETABLES	1. Onion, *Allium cepa* 2. Garlic, *Allium sativum* 3. Spinach, *Spinacia oleracea* 4. Carrot, *Daucus carota*
FRUITS	1. Pistacia, *Pistacia vera* 2. Pear, *Pyrus communis* 3. Almond, *Amygdalus communis* 4. Grape, *Vitis vinifera* 5. Apple, *Malus pumila*

IV.   NEAR EASTERN CENTER: Includes interior of Asia Minor, all of Transcaucasia, Iran, and the highlands of Turkmenistan. Eighty-three species were located in this region.

GRAINS AND LEGUMES	1. Einkorn wheat, *Triticum monococcum* (14 chromosomes) 2. Durum wheat, *Triticum durum* (28 chromosomes) 3. Poulard wheat, *Triticum turgidum* (28 chromosomes) 4. Common wheat, *Triticum vulgare* (42 chromosomes) 5. Oriental wheat, *Triticum orientale* 6. Persian wheat, *Triticum persicum* (28 chromosomes) 7. *Triticum timopheevi* (28 chromosomes) 8. *Triticum macha* (42 chromosomes) 9. *Triticum vavilovianum*, branched (42 chromosomes) 10. Two-row barleys, *Hordeum distichum*, *H. nutans* 11. Rye, *Secale cereale* 12. Mediterranean oats, *Avena byzantina* 13. Common oats, *Avena sativa* 14. Lentil, *Lens esculenta* 15. Lupine, *Lupinus pilosus*, *L. albus*

Table 2-1   (cont.)

FORAGE PLANTS	1. Alfalfa, *Medicago sativa* 2. Persian clover, *Trifolium resupinatum* 3. Fenugreek, *Trigonella foenum graecum* 4. Vetch, *Vicia sativa* 5. Hairy vetch, *Vicia villosa*
FRUITS	1. Fig, *Ficus carica* 2. Pomegranate, *Punica granatum* 3. Apple, *Malus pumilo* (one of the centers) 4. Pear, *Pyrus communis* and others 5. Quince, *Cydonia oblonga* 6. Cherry, *Prunus cerasus* 7. Hawthorn, *Crataegus azarolus*

V. MEDITERRANEAN CENTER: Includes the borders of the Mediterranean Sea. Eighty-four plants are listed for this region.

CEREALS AND LEGUMES	1. Durum wheat, *Triticum durum expansum* 2. Emmer, *Triticum dicoccum* (one of the centers) 3. Polish wheat, *Triticum polonicum* 4. Spelt, *Triticum spelta* 5. Mediterranean oats, *Avena byzantina* 6. Sand oats, *Avena brevis* 7. Canarygrass, *Phalaris canariensis* 8. Grass pea, *Lathyrus sativus* 9. Pea, *Pisum sativum* (large seeded varieties) 10. Lupine, *Lupinus albus*, and others
FORAGE PLANTS	1. Egyptian clover, *Trifolium alexandrinum* 2. White Clover, *Trifolium repens* 3. Crimson clover, *Trifolium incarnatum* 4. Serradella, *Ornithopus sativus*
OIL AND FIBER PLANTS	1. Flax, *Linum usitatissimum*, and wild *L. angustifolium* 2. Rape, *Brassica napus* 3. Black mustard, *Brassica nigra* 4. Olive, *Olea europaea*
VEGETABLES	1. Garden beet, *Beta vulgaris* 2. Cabbage, *Brassica oleracea* 3. Turnip, *Brassica campestris, B. napus* 4. Lettuce, *Lactuca sativa* 5. Asparagus, *Asparagus officinalis* 6. Celery, *Apium graveolens* 7. Chicory, *Cichorium intybus* 8. Parsnip, *Pastinaca sativa* 9. Rhubarb, *Rheum officinale*

Table 2-1 (cont.)

ETHEREAL OIL AND SPICE PLANTS	1. Caraway, *Carum carvi* 2. Anise, *Pimpinella anisum* 3. Thyme, *Thymus vulgaris* 4. Peppermint, *Mentha piperita* 5. Sage, *Salvia officinalis* 6. Hop, *Humulus lupulus*

VI. ABYSSINIAN CENTER: Includes Abyssinia, Eritrea, and part of Somaliland. In this center were listed 38 species.

GRAINS AND LEGUMES	1. Abyssinian hard wheat, *Triticum durum abyssinicum* 2. Poulard wheat, *Triticum turgidum abyssinicum* 3. Emmer, *Triticum dicoccum abyssinicum* 4. Polish wheat, *Triticum polonicum abyssinicum* 5. Barley, *Hordeum sativum* (great diversity of forms) 6. Grain sorghum, *Andropogon sorghum* 7. Pearl millet, *Pennisetum spicatum* 8. African millet, *Eleusine coracana* 9. Cowpea, *Vigna sinensis* 10. Flax, *Linum usitatissimum*
MISCELLANEOUS	1. Sesame, *Sesamum indicum* (basic center) 2. Castor bean, *Ricinus communis* (a center) 3. Garden cress, *Lepidium sativum* 4. Coffee, *Coffea arabica* 5. Okra, *Hibiscus esculentus* 6. Myrrh, *Commiphora abyssinica* 7. Indigo, *Indigofera argente*

VII. SOUTH MEXICAN AND CENTRAL AMERICAN CENTER: Includes southern sections of Mexico, Guatemala, Honduras, and Costa Rica.

GRAINS AND LEGUMES	1. Maize, *Zea mays* 2. Common bean, *Phaseolus vulgaris* 3. Lima bean, *Phaseolus lunatus* 4. Tepary bean, *Phaseolus acutifolius* 5. Jack bean, *Canavalia ensiformis* 6. Grain amaranth, *Amaranthus paniculatus leucocarpus*
MELON PLANTS	1. Malabar gourd, *Cucurbita ficifolia* 2. Winter pumpkin, *Cucurbita moshata* 3. Chayote, *Sechium edule*
FIBER PLANTS	1. Upland cotton, *Gossypium hirsutum* 2. Bourbon cotton, *Gossypium purpurascens* 3. Henequen or sisal, *Agave sisalana*

Table 2-1   (cont.)

	MISCELLANEOUS
1.	Sweetpotato, *Ipomea batatas*
2.	Arrowroot, *Maranta arundinacea*
3.	Pepper, *Capsicum annuum, C. frutescens*
4.	Papaya, *Carica papaya*
5.	Guava, *Psidium guayava*
6.	Cashew, *Anacardium occidentale*
7.	Wild black cherry, *Prunus serotina*
8.	Cochenial, *Nopalea coccinellifera*
9.	Cherry tomato, *Lycopersicum cerasiforme*
10.	Cacao, *Theobroma cacao*
11.	*Nicotiana rustica*

VIII.   SOUTH AMERICAN CENTER: (62 plants listed)

   A.   Peruvian, Ecuadorean, Bolivian Center: Comprised mainly of the high mountainous areas, formerly the center of the Megalithic or Pre-Inca civilization.

      a.   Endemic plants of the Puna and Sierra high elevation districts included:

ROOT TUBERS

1. Andean potato, *Solanum andigenum* (96 chromosomes)
2. Other endemic cultivated potato species. Fourteen or more species with chromosome numbers varying from 24 to 60.
3. Edible nasturtium, *Tropaeolum tuberosum*

      b.   Coastal regions of Peru and nonirrigated subtropical and tropical regions of Ecuador, Peru, and Bolivia included:

GRAINS AND LEGUMES

1. Starchy maize, *Zea mays amylacea*
2. Lima bean, *Phaseolus lunatus* (secondary center)
3. Common bean, *Phaseolus vulgaris* (secondary center)

ROOT TUBERS

1. Edible canna, *Canna edulis*
2. Potato, *Solanum phureja* (24 chromosomes)

VEGETABLE CROPS

1. Pepino, *Solanum muricatum*
2. Tomato, *Lycopersicum esculentum*
3. Ground cherry, *Physalis peruviana*
4. Pumpkin, *Cucurbita maxima*
5. Pepper, *Capsicum frutescens*

FIBER PLANTS

1. Egyptian cotton, *Gossypium barbadense*

FRUIT AND MISCELLANEOUS

1. Passion flower, *Passiflora ligularis*
2. Guava, *Psidium guajava*
3. Heilborn, *Carica candamarcensis*
4. Quinine tree, *Cinchona calisaya*
5. Tobacco, *Nicotiana tabacum*

   B.   Chiloe Center (Island near the coast of southern Chile)

1. Common potato, *Solanum tuberosum* (48 chromosomes)
2. Wild strawberry, *Fragaria chiloensis*

Table 2-1   (cont.)

C. Brazilian–Paraguayan Center

1. Manioc, *Manihot utilissima*
2. Peanut, *Arachis hypogaea*
3. Rubber tree, *Hevea brasiliensis*
4. Pineapple, *Ananas comosa*
5. Brazil nut, *Bertholletia excelsa*
6. Cashew, *Anacardium occidentale*
7. Purple granadilla, *Passiflora edulis*

From N. I. Vavilov, *The Origin, Variation, Immunity and Breeding of Cultivated Plants*, trans. by K. Starr Chester, The Ronald Press Co., N. Y., 1951, with permission. (Arrangement after Carroll P. Wilsie, *Crop Adaptation and Distribution*, W. H. Freeman and Co., San Francisco, 1962.)

tionary of the Economic Products of India of 1885–94. In this alphabetic listing of economic plants, Sir George gives data on common names, habitat, varieties, cultivation, production, costs, markets, qualities, and uses. *The Chronological History of Plants: Man's Record of His Own Existence Illustrated Through Their Names, Uses and Companionship*, by Charles Pickering, published at Boston in 1879, was sixteen years in the composition. Pickering discusses in chronological order various plants as they are recorded in man's writings. Three indices, comprising about 150 pages, make this large volume useful as a reference. A third encyclopedia, of somewhat more recent times, is *Sturtevant's Notes on Edible Plants*, edited by U. P. Hedrick, published in Albany, N. Y., in 1919. This book lists various food plants alphabetically, with interesting comments on their palatability and use taken from the notes of Lewis Sturtevant, first Director of the New York Agricultural Experiment Station until his retirement in 1887.

The old texts are usually quite laborious, and are "practical" only in a rather remote sense. Often any discussion of economic plants is preceded by a review of botanical knowledge of the day—a commentary on stems, leaves, cells, and the like. Among

these texts are T. C. Archer's illustrated *Popular Economic Botany* (London, 1853) and his *Profitable Plants* (London, 1865). John Lindley's tedious *The Vegetable Kingdom*, also published in London, had gone through three editions by 1853; it is as much a taxonomic treatise as an economic one. W. Rhind's voluminous *A History of the Vegetable Kingdom* (London, 1857) lists economic plants according to use and classification. Warburg and Brand published in Leipzig about 1908 the beautifully illustrated *Kulturpflanzen der Weltwirtschaft*. Considered collectively, these early texts, despite their voluminous nature, did little to further the study of economic botany, other than to indicate its immensity and the diversity of its ramifications.

Trends in Economic Botany. With population pressure mounting, the well-being of nations rests upon ever-greater practical use of plants. In technologically advanced parts of the world a constant flow of new better-yielding and disease-resistant cultivars makes agriculture increasingly efficient. Within a few decades after introduction, the oriental soybean became a leading cash crop in the United States, a major source of farm income. Gaines wheat, bred with Japanese germplasm for sturdy stalks that do not lodge under high fertility, sparked unheard-of yields in the Palouse wheatlands of the Pacific Northwest. Even esthetics are not overlooked; expeditions to the Far East have brought back new rhododendrons and other ornamentals; a sophisticated seed industry annually acclaims its all-American selections; and orchids are routinely flown by jet to mainland markets from Hawaii. Just as the meat packer uses "everything but the squeal," so the forest industries now use entire trees in various species (even the bark as a mulch and soil conditioner), whereas formerly only the boles of select kinds were milled.

The imminent strain on world resources which the population explosion poses puts economic botany right in the mainstream of human concern. Regardless of what levels population may reach, food production is already insufficient (or at least insufficiently well distributed) for two-thirds of the world. And in spite of increasing acceptance of birth control, world population continues to expand at an annual rate of well over 1 per cent, and is estimated to reach 6 billion or more by the turn of the century. A crash program for increased food is needed merely to forestall mass starvation until social measures, hopefully, have time to bring population under control. So, man must intelligently muster the millions of species, some of them still untested or even unidentified, to adequately feed, clothe, and house an over-populated world. Applied sciences are confronted not only with problems of quantity, but with inadequate quality in the diet in many lands (especially protein intake). Can economic botany come up with new, economical, and more abundant sources of protein, acceptable to peoples most needing it? Are our economics sufficiently flexible worldwide to balance production with consumption, one means for reducing pressures from world population imbalances and resource disparities?

Economic influences. With few of its people any longer self-sufficient, the modern world is much motivated by economics. Production, marketing, and product usage all require a series of economic decisions. Output and price are determined by input costs, and relate to consumer decisions which depend upon factors such as income, preference, and availability. The price obtainable in the marketplace is an informational feedback which regulates not only the quantity of a commodity that will be produced, but details such as how much fertilizer can profitably be used to produce it. In the end economics decides such pervading matters as whether a particular society may enjoy the luxury of derived food such as meat, or whether there must be more efficient direct consumption of vegetable matter (which yields more energy per acre if not subjected to an extra cycle of animal feeding).

Thus, more and more, economic botany becomes involved with business decisions. In technically advanced economies, where labor is costly, agriculture is fast becoming a rural industry, which calls for expert management from financing through research and production to marketing. In fact the independent farmer is gradually being superceded by organizations, often a corporation, dealing in photosynthesis. Teams of specialists are employed to provide exact data for decision making beyond the capacity of any one individual. Field and laboratory findings may even be programmed into a computer for system analysis. Thus soil moisture, fertility, weather predictions, cost of materials, crop variety, seasonal disease incidence, market demand, and so on can be integrated for decisions on which crop to plant, when, and even at what spacing in the field!

Economic patterns are also changing in the less-developed parts of the world. Underdeveloped regions were once largely self-sufficient in food production, but have become net importers of food while stressing specialities such as beverage plants and agricultural raw materials for export. These agricultural exports often constitute the chief source of foreign exchange. During the first half of the twentieth century the trend was toward efficient plantation production, at least in the Tropics. Trained managers, generally from technologically advanced countries, supervised operations and introduced improved cultivars and cropping practices. Rising nationalism and resentment of absentee ownership reversed the trend in many parts of the world, and often curtailed some of the technical progress. In some cases familiar plantation crops such as sugar, tropical fruit, rubber, and natural fibers have been substituted for at least partly by other sources or by synthetics, as a result of market shifts, political developments, or other factors unrelated to plantation cropping itself.

In the long haul plant products seem certain to become increasingly more important. Plants offer the only renewable source of energy, as fossil fuels (coal, petroleum) and minerals become relatively exhausted. Eventually there should be a resurgence of interest in plant species many of which are not now utilized. New cultivars adapted to mechanization might even free less-developed lands from subsistence living. Tropical climate offering maximum entrapment of solar energy through photosynthesis may prove more valuable than underground assets, with obvious implications for expanded interest in economic botany.

SUGGESTED SUPPLEMENTARY REFERENCES

Books

Ames, Oakes, *Economic Annuals and Human Cultures*, Botanical Museum, Harvard University, Cambridge, Mass., 1939.

Anderson, E., *Plants, Man and Life*, Little, Brown Co., Boston, 1952.

Callen, E. O., *Diet as Revealed by Coprolites*, in Brothwell, D., and E. Higgs (eds.), *Science in Archaeology*, Thames and Hudson, London, 1963.

Danhof, C. H., *Change in Agriculture: The Northern United States 1820–1870*, Harvard Univ. Press, Cambridge, Mass., 1969.

De Candolle, Alphonse, *L'Origine des Plantes Cultivées*, Paris, 1883 (*Origin of Cultivated Plants*, London, 1886, reprinted Hafner Pub. Co., N. Y., 1959).

Dimbleby, G. W., *Plants and Archeology*, Humanities Press, N. Y., 1967.

Faegri, K., and J. Iversen, *Textbook of Pollen Analysis*, 2nd ed., Oxford Univ. Press, London, 1964.

Hutchinson, Sir J. (ed.), *Essays on Crop Plant Evolution*, Cambridge Univ. Press, N. Y., 1965.

Leighton, A., *Early American Gardens: For Meat and Medicine*, Houghton Mifflin, Boston, 1970.

Ochse, J. J., M. J. Soule, M. J. Dijkman, and C. Wehlberg, *Tropical and Subtropical Agriculture*, Macmillan, N. Y., 2 vols., 1961.

Purseglove, J. W., *Tropical Crops*, Vols. 1 and 2, *Dicotyledons*, Wiley, N. Y., 1968.

Sauer, C. O., *Agricultural Origins and Dispersals*, American Geographic Society, N. Y., 1952.

Ucko, P. J., and G. W. Dimbleby (eds.), *The Domestication and Exploitation of Plants and Animals*, Aldine Pub. Co., Chicago, 1969.

Vavilov, N. E.,* *The Origin, Variation, Immunity and the Breeding of Cultivated Plants*, Chronica Botanica, Waltham, Mass., 1950.

Yarnell, R. A., *Aboriginal Relationships Between Culture and Plant Life in the Upper Great Lakes Region*, Univ. of Michigan Press, Ann Arbor, 1964.

Encyclopedias

Bailey, L. H., *Standard Cyclopedia of Horticulture*, Macmillan, N. Y., 3 vols., 1935.

Bailey, L. H., *Manual of Cultivated Plants*, rev. ed. Macmillan, N. Y., 1949.

Uphof, J. C. T., *Dictionary of Economic Plants*, 2nd ed., Cramer, Wurzburg (Stechert-Hafner, N. Y.) 1968.

Periodicals

Agricultural publications such as the professional journals of the American Society of Agronomy (*Agronomy Journal, Crop Science*); popular periodicals such as *Farm Quarterly* (Cincinnati) and *Crops and Soils* (Madison, Wisc.).

Economic Botany, New York Botanical Garden, N. Y.
World Crops, London.

---

* Vavilov's active years were from about 1916 to 1936, marked by numerous publications (in Russia) for which it is difficult to obtain accurate citations. This is a collection of some of his more important writings. *Studies on the Origin of Cultivated Plants* was published in Leningrad in 1926, by the "Inst. Bot. Appl."

PART **II**

Products from the Plant
Cell Wall

THE NON-LIVING SUPPORTIVE WALLS OF PLANT CELLS HAVE BEEN USEFUL TO man from the beginning of his history. They were the main source of fuel, shelter, weaponry, tools, and fiber in early cultures, and to a great extent have remained so into the modern day. Civilization could hardly have arisen without the structural contribution from woody plants, at a time when metallurgy was in its infancy. In many parts of the world people still depend upon the forest for fuel, housing, and income, growing a few fiber plants from which cloth is woven and cordage string made.

CHAPTER 3

Forests Available to Man

THE MOST ABUNDANT SOURCE OF CELL WALL substance is the tree, and the most abundant source of trees is the forest. Forests are generally a rather dense assemblage of trees with a continuous or nearly continuous canopy of foliage well above ground level, sustained by sturdy single stems (the woody trunk of the tree). Herbaceous plants grow on the forest floor, especially seasonally in forests which shed their leaves for a part of the year. Shrubs, with numerous smaller stems, are also often a component of the forest, even approaching major significance in "brush" and "scrub" lands where for a variety of reasons the environment fails to support a good stand of trees. Thus definition of the forest is not exact, and even the heart of a forest belt may have openings occupied by herbaceous plants. Nor are forest boundaries clearly delineated. Dense forest usually gives way gradually to open forest (savanna), and eventually to grassland or desert with trees growing only in specialized environmental pockets, often dwarfed and slow-growing there.

In the battle for occupancy of the land, forests were remarkably successful—until man aided the herbs by felling the trees and starting fires. Trees tend to dominate the vegetation of tropical, temperate, and even boreal climates, wherever enough rainfall is received (usually 25 inches or more) and is adequately distributed. One time or another during geologic history almost all of the earth supported trees, even what is now ice-covered Antarctica. In recent millennia forest has been the natural cover for much of North America (the Great Plains excepted), South America (except limited southern deserts), Eurasia (except for deserts extending between the Arabian peninsula and western China, and some northern tundra), Africa (both along the Mediterranean coast and south of the Sahara), Australia (limited mainly to eastern and southern portions), and most of the islands of Oceania. During historical times much forest has been eliminated by man; even where forest would be the climax cover, land is today kept cleared for agricultural operations. At that, an inventory by the Food and Agriculture Organization of the United Nations indicates that forests occupy almost 40 per cent more of the earth's surface than does agricultural land. About one-third of this, over 4 billion hectares (10 billion acres), is still relatively isolated and inaccessible, the forests too costly to reach and exploit up to this time. But year by year more virgin forest is reached, and the time is not far distant when almost all forest with an abundance of large stock will be brought under exploitation.

Of the forest that is accessible, little more than half is being systematically handled.

29

Especially in South America and equatorial Africa there are large spreads of forests haphazardly used, frequently subject to slash-and-burn agriculture, not yet completely surveyed. Even in the technologically advanced parts of the world, such as in North America, Europe, and Japan, only about half of the forested land is subject to planned forestry. About two-thirds of the timber harvest in these technically advanced parts of the world is considered intelligently done, compared to less than 15 per cent in the tropical regions. In almost all regions there is room for considerable improvement in forestry practices, to make timber yields more remunerative. Table 3-1 shows that a forest can be equally as productive in trapping energy as is an agricultural crop. But much more so than with agriculture, silviculture seems not to be realizing its full potential.

Indeed, only within recent decades have forests come to be looked upon as a type of crop, where certain inputs are required to guarantee output for the future. So abundant were the forests that they have been long regarded as a capital resource freely available for the taking—if, indeed, not an "enemy" to be disposed of in favor of agriculture.

Table 3-1

Productivity of Coniferous Forest Compared to Agricultural Crops Blacksburg, Virginia

Type of Vegetation	Calculated Biomass Yield in Tons per Hectare per Year
Pinus virginiana (17-year-old planting)	14.3
Alfalfa—orchardgrass, forage sod	12.6
Corn, cultivated for grain	14.6

Adapted from H. A. I. Madgwick, *Ecology* **49**, 151 (1968).

Today, with availability of virgin forests running out, forestry must regard trees as a crop to be grown and managed, albeit with a lengthy cropping cycle. There must be attendant input costs the same as agriculture sustains. Not all forest will become a cellulose factory; recreation and conservation increasingly demand some "inefficiency" from the production standpoint for the higher purpose of esthetic enjoyment.

Knowledge about forest trees and their influence is still meager, compared to that concerning many intensively investigated farm crops. Worldwide it is still not clear what felling of the forest may entail, in terms of soil erosion, watershed stability, changes in the soil, and even in maintenance of a balanced atmosphere that could affect weather and climate. It seems evident that in most tropical climates the land is better adapted to permanent forest cover than to cultivation of the soil. Under the intense biological activity of the tropics most of the land's wealth is stored aboveground in the vegetation. Litter dropping to the forest floor quickly decays, and its nutrients are recycled into living plants, or else leached away by the abundant rainfall. Slash-and-burn agricultural openings in the tropical forest generally must be abandoned within a few years, and the lateritic soils turn brick-hard and unproductive if handled in the same fashion as is customary in temperate climates.

It is difficult to know the extent to which trees regulate the hydrology of a region. It is clear that forests have an important influence on water runoff within a watershed, and that they contribute considerable vapor to the atmosphere. They are also atmospheric filters, and a major supplier of oxygen. One may wonder if severe decimation of forests east of the Mediterannean and along the north coasts of South America and Africa, where wood has been the prime fuel since time immemorial, may not have been instrumental

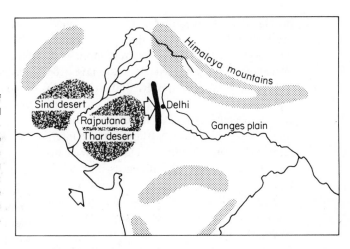

Figure 3-1 India: the eastward spread of the Rajputana deserts. When the Aryans entered India (*c.* 2000–1500 BC) these areas were woodlands; today, as a result of man's destructive activities over the centuries, they are barren desert. Further erosion caused by bad farming has encouraged the advance of the desert eastwards, and it is to check this menace that the 400 miles long by 2 miles wide belt of algaroba forest is being established. (After J. S. Douglas, *World Crops* **19**, 23 (1967).)

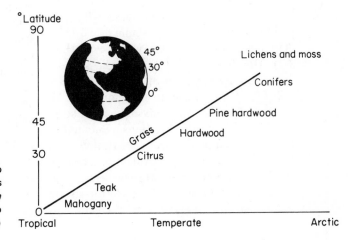

Figure 3-2 General relationship of forest to latitude. In drier habitat grass or desert prevails in all latitudes. (After W. C. Young, *Twenty-sixth Short Course on Roadside Development,* Ohio Dept. of Highways and Ohio State Univ., 1967.)

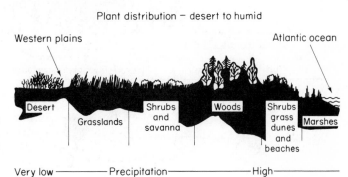

Figure 3-3 Generalized relationship of forest to rainfall as found in eastern North America. (After W. C. Young, *Twenty-sixth Short Course on Roadside Development,* Ohio Dept. of Highways and Ohio State Univ., 1967.)

in climatic change there. Certainly once-productive lands in these marginal climates have been becoming progressively more arid, unremunerative and sterile, as have the Rajputana deserts of western India (Fig. 3-1) and those of South Africa. Is the change in climate because of loss of forest, or is loss of forest inevitable (even without man's inter-ference) because of climatic change? We can-not be sure, and both influences may be at work.

The distribution of forest and its com-position are governed by many interacting ecological factors. In a general way tem-perature and rainfall are especially important. Temperature is basically determined by lati-tude and altitude, becoming cooler as these increase (Fig. 3-2). Rainfall varies with geo-graphical situation, and its abundance usually determines whether forest will prevail (Fig. 3-3). However, rainfall that might support forest in cool environment, or on clay-rich soils (which hold moisture well), might not be sufficient for trees on sandy or shallow soils, and in hot, windy locations. But unless natural conditions are drastically and permanently changed, equivalent to a change in climate, in time forest will regenerate (Fig. 3-4). As the last of the old-growth forests are felled, efficient regeneration becomes im-

perative to make tree crops sufficiently avail-able. More attention will have to be paid forest ecology in the future than has been accorded it in the past, in order to make the needed quantity of timber available, efficiently and economically.

The Kinds of Forests

Lumbermen and foresters alike recognize two convenient general categories of forests, *softwoods* and *hardwoods*, based upon the predominant type of trees contained. In the literal sense, the terms softwood and hard-wood are almost meaningless, for many softwoods are harder than many hardwoods, though on the whole the hardwoods as a group would undoubtedly average "harder" (more accurately, "denser"—greater density being usually a reflection of strength, ac-counted for by the sturdier cell walls of the individual cells). Botanically, a softwood is any member of the subphylum *Gymno-spermae*, while hardwoods belong to the subphylum *Angiospermae*.

The Gymnospermous or softwood forests. The Gymnosperms include several orders of plants, some of them extinct and known only from the fossil record. Of the orders extant, only the *Coniferales* (conifers),

Typical plant succession

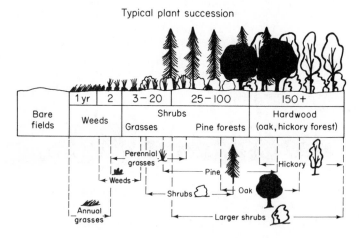

Bare fields	1 yr	2	3 – 20	25 – 100	150 +
	Weeds		Shrubs		Hardwood
			Grasses	Pine forests	(oak, hickory forest)

Figure 3-4 Trees are usually the climax vegetation where rainfall and soil are ade-quate. Abandoned, cleared land becomes forest in time. (After W. C. Young, *Twenty-sixth Short Course on Roadside Development,* Ohio Dept. of Highways and Ohio State Univ., 1967.)

including such valuable trees as the pines, spruces, and Douglas fir, is of considerable economic importance.

The conifers are for the most part readily distinguishable, even to the uninitiated, in that the leaves are usually in the form of needles or scales, and the reproductive structures are borne in cones. The pines or spruces or any of the common "evergreens" are familiar to everyone as being quite distinct from the common hardwoods, such as the oak, elm, maple, and similar species. Only in a few obscure genera might confusion result. Botanically the Gymnosperms are defined as bearing a naked ovule (potential seed) exposed upon the upper surface of a cone scale. This ovule, unenclosed in an ovary, is reached directly by the pollen. Concomitant with the type of ovule are other distinguishing features such as the frequent presence of resin and resin canals, rather uniform rays, and the lack of certain cell types in the wood.

Softwood forests, although seemingly on the decline in competition with the encroaching hardwoods, are yet of considerable extent. The Food and Agriculture Organization of the United Nations estimates that approximately 36 per cent (some 2.5 billion acres or 1 billion hectares) of the forest area of the world is predominantly coniferous. In the northern temperate latitudes, softwood forests on the whole cover considerably more acreage than do hardwoods. In the United States we find that approximately 62 per cent of commercial forest land is in softwoods, and about four-fifths of the timber crop comes from softwoods.

Gymnosperms may be regarded as occupying chiefly temperate to subarctic areas with moderate rainfall, although relatively minor stands are found in tropical climates, especially at higher altitudes. Thus in Central America, at higher elevations in Guatemala, Honduras, and Costa Rica; in central New Guinea; and more extensively in southern Brazil and northern India, significant areas may be dominated by coniferous vegetation. Western and northern North America (plus a smaller area dominated by yellow pines in the southeastern United States) and northern Eurasia provide the extensive, often nearly pure, softwood forests. In the U.S. colonization progressed westward to the Great Lakes region primarily through the exploitation of white pine. Today a final large-scale dividend from the original forest heritage is the Douglas fir of the Pacific Northwest.

Gymnosperms as a rule become increasingly commoner as the climate becomes progressively colder. Generally, in regions where both softwoods and hardwoods occur, the conifers show an increasing predominance from lower elevations to higher ones until the timberline is reached. Often these stands, especially at high altitudes, are found on poor, shallow, rocky soils—soils presumably not well enough developed to sustain Angiospermous growth. Such generalizations can apply, of course, only roughly to softwoods as a whole. Often local influences will outweigh more general factors to determine the type of tree dominant in a given locality: for example, in the Missouri Ozarks, typically a temperate hardwood region, shortleaf pine may dominate certain cherty, acid-soil ridge-tops although completely surrounded by the typical oak–hickory hardwood climax. Similarly, nuclei of hemlock may be found in the Appalachians and scattered here and there as far west as Indiana, usually on sandstone soils.

The Angiospermous or hardwood forests. In contrast to the comparatively few kinds of Gymnosperms, the Angiosperms are exceedingly diverse. A common lack of "pure stands" among hardwoods is one factor favoring exploitation of softwood forests first and by preference, since immense quanti-

The seeds of broad-leaved trees are encased in a protective body, examples of which are a pome, like the apple, or a nut, like the acorn. Broad-leaved trees belong to a large group of plants called "angiosperms" (meaning "vessel seeds") which also includes most species of garden and wild flowers.

The growth pattern of broad-leaved trees is complex. The trunk may grow straight for a considerable distance, or split two or more ways. Branches are essentially a further splitting up of the trunk. There is no terminal bud. Many branches grow upward at the same time.

The broad leaves of trees like the shagbark hickory shown here in both its summer and winter guise, require a great deal of water. So during winter, when moisture is sealed in snow or the frozen ground, they shed their leaves and seal themselves in. The name "deciduous," applied to such trees, comes from the Latin word meaning "to fall." In warm climates, the leaves of some species stay on the tree all year, but are replaced annually.

The wood of broad-leaved trees is more complex than that of the needle-leaved because the wood has special vessels or tubes to carry sap. All broad-leaved trees are known as "hardwoods" to lumbermen, although some, like basswood and aspen, have relatively soft wood.

Contrast between hardwoods and softwoods. (Courtesy St. Regis Paper Co.)

The seeds of narrow-leaved trees are carried in cones (whence the name "coniferous," meaning "cone bearers"). The seeds are at the bases of the scales of the cones, and fall out when the cones open. Since they are not encased in a protective body, the trees which bear them are called "gymnosperms," which means "naked seeds."

The growth pattern of narrow-leaved trees is simple and generally symmetrical. The trunk forms a relatively straight central stem with branches coming out more or less at right angles to the trunk. At the top of the trunk a single "terminal bud" does all the upward growing, adding to the height of the tree each year. Many narrow-leaved trees assume a pyramidal shape.

The needle-like leaves of such trees as the longleaf pine shown here are tough, leathery, and covered with weatherproof wax. Because such leaves can live through the winter, trees that bear them are frequently called evergreens.

The wood of narrow-leaved trees generally has simple cell structure. There are no specialized cells for carrying sap. In many cases, the wood is softer and lighter than in broad-leaved trees, so the lumber trade, for convenience, calls all narrow-leaved trees "softwoods." But some "softwoods," like the longleaf pine, have hard wood.

ST REGIS

Diorama of an Angiospermous temperate hardwood forest of the eastern U. S., as it appears in the early spring. (Courtesy Chicago Natural History Museum.)

ties of a standard, acceptable timber can thus be produced and marketed with a minimum of expense and inconvenience. The Angiosperms as a rule are easily distinguished from the Gymnosperms in having broad, deciduous leaves and in bearing flowers, not cones, as the reproductive structure. Only under environmental extremes—for example, in desert habitats, where some Angiosperms bear highly modified leaves or lack them entirely—is there likely to be any confusion between softwoods and hardwoods. Angiosperms have the ovule enclosed within an ovary, rather than exposed. Think of any common hardwood fruit—for example, the peach, walnut, locust pod, or acorn of an oak—and it is apparent that the seed (matured ovule) is contained within outer enclosing tissues that develop from the flower. Secondary characteristics, such as the presence of cell types in the wood not found in Gymnosperms, the absence of resin canals, and the diversity of rays, will be discussed in greater detail later on.

Hardwood forests are usually separated into two categories, the *tropical* and the *temperate hardwood forests*. The demarcation between these categories is chiefly geographical, and commonly, as in Mexico, there is a gradual transition from temperate to tropical types. On the whole, tropical hardwood forests contain a greater diversity of species (these seldom even approximating pure stands) than do their temperate counterparts. The trees are for the most part "evergreen," in the sense that ordinarily there is no winter dormant period. The leaves are normally shed annually, however, but are replaced by new growth before shedding or almost immediately thereafter. In some tropical forests where a prolonged dry season exists, as in northeastern Brazil, trees may remain without foliage for a few months, undergoing a cycle of foliation much as do the deciduous trees of temperate forests.

Temperate hardwood forests are apt to have fewer species represented, and these are usually more frequent and better known than those of tropical regions. Frequently interspersed are several species of pine, hemlock, and (in sunnier habitats) red cedar or juniper. Typically the hardwood trees shed their leaves during the colder portion of the year, remaining dormant until the following growing season. Temperate hardwood forests have lain in the path of man's civilization, and have been rather thoroughly exploited. Whereas many tropical forests are yet to be explored and utilized by man, the problem of temperate hardwood forests is now chiefly one of conservation and intelligent management. There has been much indiscriminate clearing of deciduous forest for agricultural purposes, sometimes where the land is suitable for nothing but forest.

Almost one-half of the forest area in the world, over 3.5 billion acres, is of the tropical hardwood type. On the whole, tropical forests are little investigated and poorly managed. But the tropical climate and year-round season encourages rapid growth, making tropical hardwood forests very productive, at least in terms of energy trapped. However, many practical difficulties stand in the way of efficient exploitation of this rich source of energy. Perhaps we should be thankful that these have insulated many tropical forests from such ruthless decimation as temperate forests have endured, thus conserving a valuable resource!

The main tropical forests, the "jungles,"

lie within the Amazon valley of South America and in equatorial Africa, lowlands with abundant rainfall. There are still essentially virgin forests in the East Indies, especially in Borneo and New Guinea, and much partly exploited forest in southeastern Asia. Relatively primitive peoples live in these regions, existing by subsistence slash-and-burn agriculture. Trees are ringed or felled, burned in the dry season, and the land handplanted to food crops which profit for a time from the fertility of the tree ashes. After a few years, fertility is exhausted and weeds become overwhelming, and the land is abandoned to an ecological succession that eventually becomes forest again. This seems an inefficient way of managing land, but until research provides more answers and social progress makes

Forest in British Guiana, typical of the tropical hardwood type. (Courtesy Chicago Natural History Museum.)

gains in these areas such cropping practices will prevail.

Temperate hardwoods comprise only about 16 per cent of the world forests, a little over 1 billion acres. Most temperate hardwood forests are found in east-central North America and middle latitudes of Eurasia, usually at lower elevations (softwoods dominate in the mountains). Generally, their original composition has been quite altered. Much temperate hardwood forest was felled to create farmland, and most of the rest opened to livestock. As one species, then another, became especially valuable, these were selectively harvested. Such high-grading leaves the forest to less valuable species and cull trees. In the United States, for example, prime white oak has been selectively removed for barrel staves, black walnut for gunstocks and furniture, hard maple for furniture and other household items. Moreover, what with modern man's peregrinations, afflictions have been introduced that cause wholesale elimination of certain species. Introduction of chestnut blight and Dutch elm diseases from Europe into North America wiped out the

American chestnut, and the magnificent American elm is progressively disappearing.

By and large the temperate hardwood forests are no longer a prime source of commercial timber, as valuable as they may be esthetically in a belt of dense human habitation. What remains of temperate hardwood forest is magnificent in its autumn color, and it is much used for recreational purposes. These forests have yielded many species now domesticated and conspicuous in the urban landscape. Scattered woodlots as well as regenerating forest lands serve as valuable wildlife habitat. Temperate hardwoods still constitute the main source of certain cabinet woods, furniture veneers, barrel staves, and specialty products such as baseball bats.

Forest Resources by Continent (Table 3-2).

AFRICA. Some of the world's most luxuriant tropical rainforest extends across the midsection of the "Dark Continent" to approximately 20° north and south of the equator centering on the Congo basin. Nevertheless, Africa is not heavily forested as a whole, less than 10 per cent of the land being densely wooded ("closed high forest") and only about 25 per cent (less than 2 billion acres) forested at all. Forests are almost entirely of the tropical hardwood type, subject to high-grading and the inroads of shifting agriculture. Their ecology is indeed fragile, especially where balance is delicate in savanna regions. Almost all timber exported comes from the western rainforest, but does not meet in total value the wood products imported into other sections of Africa.

A sparse forest belt botanically more like Europe than Africa (pine, cedar, oak) occurs in northern Africa, especially coastward from the Atlas Mountains (Fig. 3-5). The cutting of wood for charcoal and the grazing of goats there have been very destructive of what were once fine forests, especially in northern Morocco. Their denudation has seriously increased soil erosion, and perhaps alter-

Table 3-2

Estimate of World Forest Land by Continent and by Type

Region	Total	Conifers	Non-conifers
	Millions	of	hectares
North America	700	440	260
Central America	71	35	36
South America	810	10	800
Africa	680	4	676
Europe	137	80	57
U.S.S.R.	728	553	175
Asia	490	90	400
Pacific Area	88	4	84
World	3,704	1,216	2,488
More industrial	1,653	1,077	576
Less industrial	2,051	139	1,912

Adapted from FAO World Forest Inventory, 1963.

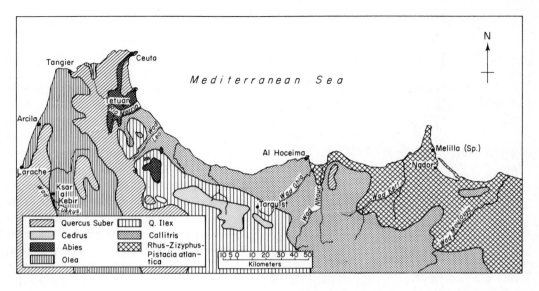

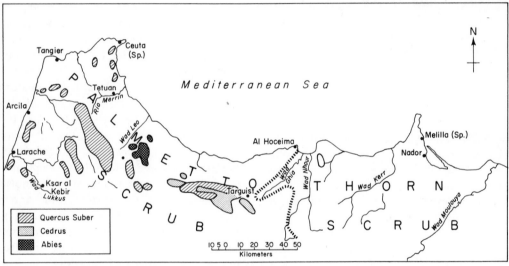

Figure 3-5 Contrast the primordial forests in North Africa (above) with their present extent (below). Deforestation and overgrazing have reduced much original forest to scrub. (After M. W. Mikesell, *Science* **132**, 443 (1960).)

ed climate for the worse. Isolated clusters of Aleppo pine (*Pinus halepensis*), Moroccan fir (*Abies pinsapo*), and Atlas cedar (*Cedrus atlantica*) occur in the mountains, and attest to the former grandeur. Cork oak (*Quercus suber*) is still fairly abundant since this tree is protected for its bark. Forest ceases abruptly south of the Atlas Mountains, where the awesome Sahara desert (constituting a fourth of Africa) supports scant vegetation of any type.

About 90 per cent of Africa's dense high forest occurs in equatorial western Africa. Already about half of the original forest there has been lost to shifting cultivation and settlement. From this belt come the renowned

Log extraction road in the equatorial forest of the Gold Coast, Africa. (Courtesy British Information Service. From *Unasylva*.)

not-inconsiderable reserve of over 200 million acres of dense high forest.

In the higher elevations of eastern equatorial Africa there is greater agricultural diversification and less demand upon the forest. Some intermixture of conifers (*Podocarpus gracilior, Juniperus procera*) occurs there, especially in southern Ethiopia, but almost all the rest of Africa's 800 million acres of forest is of the tropical hardwood type. Notable as cabinet woods are ebony (*Diospyros*), black stinkwood (*Ocotea*), black ironwood (*Olea*), African walnut (*Lovoa*), and rosewood (*Pterocarpus*). Veneer is made by peeling kararo (*Aningera*), of which a good supply exists in southwestern Ethiopia. Madagascar forests are rich floristically, with *Afzelia, Symphonia, Canarium*, and *Dalbergia* among the useful timber genera. Rhodesian teak, *Baikiaea plurijuga*, is important in parts of Zambia, Rhodesia, Botswana, and Angola.

Southward and northward the equatorial forest thins into a transition belt of drier savanna characterized by scrubby thorn trees such as *Commiphora* and *Acacia*, various legumes (*Parkia, Copaifera*), and the curious baobab (*Adansonia*) with huge elephantine trunks. *Isoberlinia* is abundant in the moister savanna north of the equator, while *Brachystegia* and *Julbernardia* dominate much savanna south of the equator. Savanna forests suffer a good deal from clearing and from burning (for "improved" grazing).

Industrialized southern Africa is a manufacturing center for forest products. South Africa is mostly savanna and desert, and supports little large-tree growth of its own except to a limited extent in mountains to the east. Such scattered forest as was indigenous has been badly decimated, and today the paper and paperboard industries depend much upon introduced trees grown on plantations (pines, such as *Pinus patula, P. elliotii, P. radiata; Eucalyptus*, such as *E. saligna, E. grandis, E. robusta; Cupressus*

export woods such as African mahogany (*Khaya*), obeche (*Triplochiton*), okoume (*Aucoumea*), sipo (*Entandrophragma*), limba (*Terminalia*), etc., some now becoming scarce. Especially *Khaya*—but also *Pseudocedrela* and *Entandrophragma* of the Meliaceae, *Aucoumea* of the Burseraceae, and *Afzelia* of the Leguminosae—may enter the trade as "mahogany." In addition to the mahoganies, other species of the Leguminosae, Combretaceae, Simarubaceae, and Euphorbiaceae are often exploited. *Didelotia idae*, Leguminosae, is the most frequently harvested species in parts of Sierra Leone and Liberia. The oil palm (*Elaeis*) is abundant along the west coast, and in tidal areas the mangrove (*Rhizophora*) is economically important as a source of tannin. In spite of limited internal demand for timber, accessible forest is fast being exhausted; a number of sawmilling operations occur near the coast and along the rivers, drawing upon nearby timber. The Congo, however, still retains the

lusitanica; Acacia; Grevillea). Planted pine forests are said to mature timber sufficiently large for lumber on a 40-year cycle.

Thus Africa, on the whole not heavily forested, does possess relatively unexploited equatorial forests, a reservoir of timber and forest products for at least a decade or two. Most of the forest land has not been settled, nor has it passed into private ownership, and except at the extremes (north and south) of the continent there is little forest regulation or conservational planning. Because of denser population, greatest use of wood in Africa is in Nigeria and Ethiopia, almost entirely for fuelwood and charcoal. The Ivory Coast and Gabon are the chief exporting nations.

Until recent times there has been comparatively little demand for wood other than for fuel (for riverboats, domestic firewood). About 90 per cent of all wood harvested is consumed locally for fuel, a use in which per capita consumption is about twice the world average. Construction and furniture are the next most important uses, destined to increase as the continent modernizes and transportation reaches the equatorial forests more readily. Foreign interests are chiefly responsible for commerce in the prized cabinet woods which go mostly to European markets (especially Germany and France).

ASIA. The forests of so vast a continent as Asia cannot be described in general terms, for immense variations in climate and topography occur. In Asia (including Ceylon, Japan, and the East Indies) 43 per cent of the forests are of the coniferous type (these almost wholly in Siberia and adjacent China and Japan); 27 per cent are of the temperate hardwood type (these more scattered, but chiefly in Siberia, China, Japan, and Asia Minor); and about 30 per cent are of the tropical hardwood type (these chiefly in India, the Malay area, and the East Indies). The coniferous forests of Siberia, comprising about 1 billion acres, are relatively unexploited. This forbidding and sparsely populated region affords little inducement to colonization, and doubtless scarcity of labor, supplies, and transportation will long delay complete exploitation of this largest world reservoir of softwood timber.

Although historically there was considerable forest in southwestern Asia, this has now been severely depleted and degraded. As in northwestern Africa, thorn scrub and eroded land replaces forest overgrazed by goats and livestock. What forest remains is confined mostly to the mountainous sections of Turkey and Iran. The steppe lands of the Jordan valley show remnants of what must have been dense stands of evergreen oak (*Quercus coccifera*), deciduous oak (*Q. aegilops*), and Aleppo pine (*Pinus halepensis*). *Pinus brutia*, similar to Aleppo pine, occurs ịn southern and western Turkey. Except for such natural deserts as the Arabian peninsula, it is clear that as late as biblical times the Near East was far more a garden spot than it is today. Probably not much more than 10 per cent of the land is now forested, this in mixed hardwoods and softwoods. Incidentally, a number of important fruit-tree domestications have come from Near Eastern forests.

North of the Black Sea occur some of the most productive forests in all Eurasia. Although this part of the Soviet Union may be regarded as "European Russia," it is convenient to include it with the other Soviet holdings which constitute the overwhelming mass of temperate forests in Asia. All told the U.S.S.R. has over 1 billion hectares ($2\frac{1}{2}$ billion acres) of forest, of which only about 15 per cent are managed, however. Seventy-eight per cent of the Soviet forests are coniferous, nearly half of the world's reserve. Some two-thirds are of mature timber, which, however, is not of such large dimension as is old growth in North America. On the whole the Soviet Union is considered about 60 per cent forested.

The relatively mild climate and ample rainfall west of the Ural Mountains makes forests there the most productive in the Soviet. However, this is a heavily populated region, with great demand for farmland as well as forest products. Thus it is not self-sufficient so far as timber is concerned, the deficit being made up through import from forests of lesser quality to the east. But where forest does remain, such as in the Caucasus and Carpathian highlands north from the Black Sea, wood reserves per unit area of land are as great as anywhere within the Soviet, on the order of 200 cubic meters per hectare. Mixed hardwoods and softwoods occur in this region.

Eastward, north of the Caspian Sea, arid climate checks forest luxuriance, and wood reserves per hectare are only about one-fifth of those farther west. Northward in the Ural Mountains the forests are moderately luxuriant, approaching but not quite reaching the high wood reserves of the southwestern forests. The Siberian and Mongolian forests are abundant but not of great quality, with wood reserves at most half as great as those of the southwest. Much of the northern forest is scrubby taiga, the season short, and the soil frozen most of the year. The annual increment for Siberian forests as a whole is no more than 1 billion cubic meters, less than one-fourth the yield of the productive forests of the southwest (annual increment there of 10 cubic meters per hectare per year).

The forest belts of the Soviet are typically stratified east–west according to latitude. The northern taiga contains pine, spruce, fir, larch, and "cedar," with birch the only hardwood of much consequence. The forest is interspersed with peat bogs and open areas of permafrost. The chief individual species are *Picea excelsa* and *P. obovota; Pinus sylvestris* and *P. siberica; Larix sukaczevi;* and a few species of *Abies.*

South of the taiga belt, mixed softwood–hardwood forests occur. The same coniferous species are found, but mixed among them are oaks (*Quercus*), maples (*Acer*), *Tilia,* and *Corylus avellana.* An especially rich assortment of hardwoods is found in the extreme eastern Soviet, from the Amur valley south to the Korean peninsula. Additional species of *Acer, Phellodendron, Juglans, Fraxinus, Carpinus,* and *Fagus* occur. The southern portion of this eastern forest has a mild climate, and supports subtropical broadleaf evergreens such as *Buxus, Laurus, Magnolia, Ilex,* and *Laurocerasus.* Other relatively tender species, of importance for their fruit as well as for their wood, are the walnut, *Juglans regia;* the chestnut, *Castanea sativa;* and several types of beech, *Fagus.*

On the whole, the pines, especially *Pinus sylvestris,* are the most important timber species in northern Asia. Spruces no doubt rank second. Larches (*Larix*) constitute a greater volume of standing timber than either pines or spruces, constituting about 40 per cent of the forest cover (about 28 billion cubic meters of bulk), but the wood is less useful. The birches (*Betula*) are the most abundant hardwood, constituting about 13 per cent of the forest cover. After larches, pines, spruces, and birches come cedars, firs, aspens, oaks, beeches, lindens, alders, and poplars—about in that order of ranking.

China, once luxuriantly forested, has been so heavily populated for millennia that most of the land is now denuded. There has been some effort to protect remaining snatches of forest, and reforestation has taken place in certain areas. It is estimated that 50 million acres of land have been planted to trees and bamboo, and that about 165 million acres of accessible forest existed in 1965. But small trees are often taken from new plantings almost before they can begin growth, so desperate is the need for firewood. China is not likely to become self-sufficient in forest products, even though per capita figures for timber usage in China are the lowest in the world. In the less populated interior highlands of China, massive shelterbelt plantings are being undertaken, meant to ring the deserts with a "great green wall."

Teak, among the most important of Asiatic timbers, at Madras in southern India. (Courtesy Chicago Natural History Museum.)

Southward near and in the Himalayan mountains occur conifers such as *Pinus armandii*, *P. griffithii*, and *P. yunnanensis*, and at lower elevation mixed hardwoods. Southeastward toward Burma and Vietnam occur diversified hardwoods such as oak, ash, elm, and chestnut, often intermixed with pines, firs, and larch. *Pinus massoniana* is prevalent in southeastern China, and is much planted there. Many well-known subtropical species exotic to the western world—camphor, *Cunninghamia*, and teak—also grow in this region.

India, like China, would be generally forested were it not for a concentrated population which requires most of the land for food production. Three-fourths of the original natural cover is now gone, and the remaining forests subjected to heavy pressure. In northern India–Pakistan, along the Himalayas, occur a few pines (e.g., *Pinus roxburghii*, at elevations generally below 7,000 ft; *P. armandii*, a white pine of moderate and high elevations). In middle and southern India scattered stands of tropical hardwoods occur, including such famed species as teak (*Tectona*) and many of the fine timber genera now abundant only in the East Indies (Leguminosae, Dipterocarpaceae). What the forests of India must have been like originally is best visualized from the relatively unmolested forests of sparsely populated Borneo.

From Burma southeastward to Singapore the tropical forests have not been so severely decimated as in India (but more so than in Borneo). Parts are protected by inaccessibility in the mountains that run from southern China the length of Vietnam. There are some 30 million acres of forest in South Vietnam, of which only one-fourth is primary forest with large trees. Such forestry as is practiced is crude, undertaken by hand with primitive tools and elephant power. When Indochina was under French rule forest reserves were organized and protected, to supply fuelwood and small poles on a sustained-yield basis. The coastal mangrove and *Melaleuca* forests were reasonably managed. After colonial rule ended, forest management lagged, and war rather than forestry occupied the region's attention. Unfortunately, Vietnam forests suffered considerable so-called "defoliation" (chemical killing) by plane during the Vietnam war. Considering the mainland of Southeast Asia as a whole, perhaps one-third of the land remains forested. The favorable growing conditions which prevail cause rapid regeneration of trees, however, wherever pressure on the land can be relieved, even for a few years.

Indonesia, the Philippines, and New Guinea support magnificent tropical hardwood forests except where these have been cleared in populated areas. In Borneo and New Guinea much of the forest is still unexplored, the timber trees only partly identified. Commercially the most important timber comes from the Dipterocarpaceae family, several genera of which provide excellent logs for export, usually called "Philippine mahogany." Perhaps 150 different species are involved. Typical of tropical forests, few pure stands occur, and a tremendous diversity of species is represented. Lack of transportation and manpower in the jungle keeps these forests "underutilized." Selective and careless logging is commonplace. Many useful trees are undoubtedly ignored simply because they are not known in foreign trade, and thus not requested by purchasers. Local facilities for handling and processing are crude. One custom is to girdle and kill large trees left after selective logging, in the belief that they will be overmature by the next cycle of rotational harvesting. But little is known of the prevailing ecology—which species will regenerate, and what are their shade requirements. In Malaya and the Philippines, where forestry is somewhat more advanced than in Borneo, it has been found that smaller trees of the Lauraceae, Myristicaceae, Rosaceae, Leguminosae, and Fagaceae are useful for providing cover that allows a new stand of the valued dipterocarps to emerge. And typical with tropical rainforest soils, it has been found ecologically disadvantageous to clear-cut the forest (as for plantations). Soil structure seems to worsen, and the water table lowers; the planted crop suffers slow debilitation.

Teak (*Tectona grandis*, Verbenaceae) is the best known timber tree of Southeast Asia. A good deal of teak has been planted to plantations. Most other timber trees are less familiar to the Western world. In addition to the Dipterocarpaceae family, many species of the Leguminosae and Sapotacae are used. *Shorea, Parashorea, Dryobalanops, Dipterocarpus, Koompassia, Cratoxylon,* and

Anthocephalus supply much of the export timber.

Japan and the more northerly islands of the western Pacific run the typical north–south gamut of forest types. At high elevations, and in northerly portions, conifers such as fir, spruce, larch, and pine predominate. *Pinus densiflora* occurs throughout Japan and Korea, while *P. thunbergiana* grows in coastal locations in both countries. In southern Japan hardwoods intermix with the conifers, with beech and oak becoming important. Taiwan and other islands farther south have tropical and subtropical hardwood forests, including plantations of camphor, fig, and other especially valuable species.

Asia, in spite of being over-populated in the south and east, is on the whole a tremendous reservoir of timber. Although the continent is only about one-quarter forested (with much desert and cleared land), still several billion acres of forest land are involved, an amount more than for any other continent. It is unlikely that the forests of Southeast Asia will be managed for sustained yield, but a good deal of progress is being made with Soviet forests to the north. Siberia alone could supply sufficient wood at present levels of consumption for years to come. Soviet forests constitute the second greatest timber reserve in the world today (after South America), and afford by far the greatest balance of export over import timber found anywhere. However, export costs are high because of transportation difficulties, and quality less than with old-growth timber from North America. Indonesia, including Borneo and western New Guinea, holds the greatest reserve of tropical hardwood timber in Asia, with Burma (and adjacent parts of India-China) not far behind. Greatest exporters of tropical hardwoods are the Philippines and Malaysia.

AUSTRALIA. Australian forests (Fig. 3-6) are predominately of the tropical and subtropical hardwood type, totaling nearly

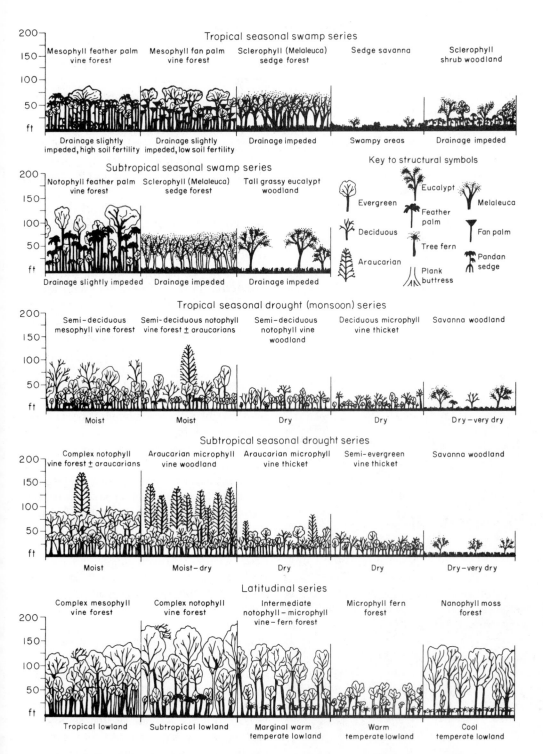

Figure 3-6 Forest profile sketches to illustrate structural changes along different gradients. (After L. J. Webb, "Environmental Relationships of the Structural Types of Australian Rain Forest Vegetation," *Ecology* **49**, 308 (1968).)

200 million acres but limited to a coastal fringe on the continent in the east and southwest. Most of the interior contains desert or scrub forest called malee. *Eucalyptus* is far and away the most important genus, several hundred species being indigenous to Australia. *E. regnans* is prominent in the uplands of Victoria, karri (*E. diversicolor*) and jarrah (*E. marginata*) prominent in southwestern Australia. Savanna that includes scattered stands of *Acacia*, with *Casuarina* along the watercourses, constitutes the grazing lands of central and eastern interior Australia. Tasmania, to the south of New South Wales, was originally in fine *Eucalyptus* forest similar to that of southeastern coastal

Kauri pine, *Agathis australis,* in New Zealand. Probably the best known tree of the island, but today largely cut. (Courtesy Chicago Natural History Museum.)

Australia. *Eucalyptus* did not reach so far south as New Zealand, which was almost solidly in forest of *Podocarpus*, *Dacrydium*, and *Nothofagus*, and with 1 million acres of the famous Kauri (*Agathis australis*) now largely eliminated by careless logging, fires, and gum extraction. A goodly portion of Australian timber needs are met by imports, primarily coniferous structural timbers from New Zealand (*Podocarpus*), North America, and northern Europe. Commendable forest programs have been initiated in Australia and New Zealand, including establishment of forest plantations, especially the introduced Monterrey pine, *Pinus radiata*. With limited availability of suitable climate for forests, however, Australia would seem destined to be a deficit producer as she becomes more populated and as demands increase.

EUROPE. Primeval Europe was almost totally forest; today the continent is less than one-third in forest, this for the most part heavily utilized and intensively managed. About one-third of a million acres of forest exist, over half in conifers. Europe has had exceptional influence upon world forestry, not only because she has been the modern center for learning in the science, but because European nations long controlled much of the forest resources of the world as colonial powers. Western civilization itself reflects development in forested habitat, a benevolent environment providing self-sufficiency that gives rise to political individualism. Forests have been especially important to the economy of Scandinavia, and some of the outstanding research on forest management has been undertaken there. Except in Scandinavia and the central and southern mountains, little remains of Europe's indigenous forests. The British Isles, for example, were heavily forested as late as the seventeenth century; today they are less than 4 per cent so, the remaining forests mainly in the Scottish highlands. Even there the fits and snatches constitute altogether less than 2 million acres of woodland, scarcely 9 per cent of the land area. The lowland oaks and highland pines that once underwrote the ships that ruled the world are no longer of commercial moment. In addition to native Scotch pine (*P. sylvestris*) and English oak (*Quercus robur*), there have been extensive plantings through the years of ash, birch, larch, Norway spruce, and more recently lodgepole pine, Sitka spruce, and Douglas fir from western North America. Much of the timber raised is used for pulp, with structural timbers largely imported.

Pinus sylvestris is the most important coniferous tree through northern Europe, accompanied by spruce, fir, and larch in Scandinavia. Only in Scandinavia are there extensive natural forests still to be drawn upon, a resource that is gradually diminishing in spite of reforestation and management second to none. Because of its abundance in Scandinavia, coniferous acreage exceeds that of hardwoods in Europe as a whole. The lowland areas from the Baltic to the Mediterranean have as their principal trees beeches (*Fagus*), oaks (*Quercus*), and lesser quantities of maple (*Acer*), ash (*Fraxinus*), linden (*Tilia*), and elm (*Ulmus*). Birch (*Betula*) mingles with the conifers in the subalpine areas of southern Germany, France, Switzerland, Austria, and Yugoslavia. The mild climate of the Mediterranean coast permits evergreen hardwoods there, in addition to the famous cork oak, *Pistacia*, and olive (*Olea*). *Pinus pinaster* and *P. halepensis* are of some importance in Portugal, Spain, coastal Italy, and scatteringly in Greece. The Austrian pine, *P. nigra*, ranges widely through southern Europe from Spain into Turkey. Weak hardwoods such as *Populus* and willows (*Salix*) are not overlooked in Europe. But on the whole Europe is underforested, and except in Scandinavia, an importer of timber. Where forest exists it is generally well managed and provides sustained yields, notably in Germany. Compared with other continents, Europe has the highest percentage of private forest holdings, with state ownership often devolving to lands needing reforestation. Greatest reserves occur in Sweden and Finland, the countries also leading in production (followed by France, Germany, Rumania, and Poland).

MEXICO, CENTRAL AMERICA, AND THE WEST INDIES. In general the flora is much like that of tropical South America, with many of the same genera, if not species, predominating. Mexico, however, the largest land mass of the region, possesses in its mountains and central plateau many genera and species common in the forest belts of the western United States. Throughout the region tropical hardwood forests dominate at lower elevations, usually grading through a narrow deciduous hardwood belt at moderate elevations to excellent, if limited, coniferous forests at high elevations. A large number of pine species are found at higher elevations in central Mexico, and *P. oocarpa* in nearly pure stands of excellent timber in the highlands of Honduras. *P. caribaea*, once thought to embrace the slash pine of the southern United States, is a lowland species from the gulf coast of Nicaragua and Honduras through the Bahamas.

The dense tropical hardwood forests extend more or less uniformly from coastal areas of middle Mexico through Tehuantepec, northern Guatemala, and British Honduras to Panama, except at the high elevations in central Guatemala, Honduras, and Costa Rica. Over most of the West Indies this same general forest type likewise prevails. The stands are typically luxuriant, with many diverse genera and species. Mangrove swamps, chiefly of *Rhizophora*, are very common in tidal zones throughout the area. The more important timber trees include: mahogany, *Swietenia*, and "cedar" (*Cedrela*) of the Meliaceae; logwood, *Haematoxylon*, rosewood, *Dalbergia*, and *Enterolobium*, *Pithecolobium*, *Caesalpinia*, *Andira*, *Gliricidia*, *Hymenaea*, and others,

Table 3-3

A Few of the Characteristic Species of a Central American Rainforest
(such forests are exceedingly heterogeneous)

Region or Type	Selected Examples
Coastal beaches and swamps	Palms such as *Cocos* and *Bactris*, *Citharexylum*, *Coccoloba*, *Conocarpus*, *Ficus*, *Hibiscus*, *Hippomane*, *Pterocarpus*, *Rhizophora*, *Symphonia*, *Ximenia*
Evergreen lowlands	*Anacardium excelsum*, *Brosimum spp.*, *Cedrela mexicana*, *Ceiba pentandra*, *Chrysophyllum mexicanum*, *Enterolobium cyclocarpum*, *Ficus spp.*, *Pterocarpus officinalis*, *Spondias mombin*, *Sterculia apetala*, *Swartzia panamensis*, *Tabebuia spp.*, *Terminalia lucida*, *Virola spp.*; in the understory *Andira*, *Capparis*, *Carica*, *Casearia*, *Castilla*, *Cecropia*, *Ficus*, *Inga*, *Lonchocarpus*, *Ochroma*, *Ocotea*, *Randia*, *Triplaris*, *Ximenia*
Lower montane (below 2,500 ft elevation)	*Schizolobium* and *Vochysia* in colonies, an understory of palms, *Anacardium*, *Andira*, *Licania*, *Hura*, *Ormosia*, *Pouteria*, *Simaruba*, *Sloanea*, *Sweetia*, *Tabebuia*, and many genera also in the lowlands
Upper montane (above 2,500 ft elevation)	Increasingly dominated by *Quercus* and *Cedrela*

Adapted from Paul H. Allen, *The Rainforests of Golfo Dulce*, Univ. of Florida Press, Gainesville, 1956.

A mahogany tree in British Honduras, one of the world's premier timber trees. (Courtesy Chicago Natural History Museum.)

of the Leguminosae; fustic, *Chlorophora*, of the Moraceae; lignum-vitae, *Guaiacum*, of the Zygophyllaceae; *Calycophyllum*, of the Rubiaceae; aceitillo, *Zanthoxylum*, of the Rutaceae; and a great many other genera and families. Paul Allen, discussing the rainforests of the Golfo Dulce region of Costa Rica, lists characteristic species according to habitat (Table 3-3).

Mexico as a whole is comparatively sparsely forested. Immense areas of northern, northeastern, and central Mexico are desert or semiarid, lacking forest, or with only a sparse scrub forest (particularly mesquite, *Prosopis*, of the Leguminosae). Perhaps a bare 15 per cent of the total land area is timbered, and a fair part of this is not of commercial value. In the western mountains and the southern "meseta central," the very important pine forests are found. In the coastal and eastern lowlands, typical tropical hardwood forest prevails. In spite of abundance of forests in many localities, other sections of Mexico are deficient in timber, and the paradoxical practice of exporting large quantities of timber while importing much finished lumber, primarily from the United States, prevails. Official Mexico has in recent years pursued a commendable course of forest regulation, although this is somewhat a case of locking the barn after the horse has been stolen, since for several centuries cutting (for firewood, at least) and clearing have gone on under primitive conditions and by wasteful methods.

Central America, except locally in the high central mountains and in the narrow savanna belt of the Pacific slope, is heavily forested with the tropical hardwood forest previously mentioned (better than 60 per cent of the total land area). At higher elevations pines and other conifers are dominant and important, but not extensive and accessible enough to offer much commercial interest. Some localities, such as southeastern Panama (Darien) are relatively unexplored, while others are densely populated and mostly cut-over. Here, as everywhere, there is roughly an inverse correlation between density of population and persistence of the forest. Once-forested Salvador, central Costa Rica, and, in the West Indies, Jamaica, Puerto Rico, and most of Cuba and Haiti are today practically devoid of forest. Nor is there, in most cases, any effective attempt at conservation of remaining forest areas or at reforestation. All countries utilize their forests chiefly for fuelwood. Guatemala and Nicaragua rank after Mexico as sources of industrial wood, and Nicaragua after Mexico in forest reserves.

The West Indies have much the same basic forest type as does Central America, with the same genera and often the same species predominating. As in Central America, coniferous stands, mostly pine, are comparatively rare, and pure stands are small in extent, being confined chiefly to mountainous areas of Cuba, Haiti–Dominica, and the larger Bahama Islands. The denudation of once-rich forest land is particularly serious in much of the West Indies, where dense populations have existed for centuries. Cuba contains the greatest forest reserves, about 6 million acres.

SOUTH AMERICA. South America is particularly rich in tropical hardwoods; it is only rather locally forested with softwoods and temperate hardwoods. Total area in forest exceeds that of any other continent, being over 2 billion acres and accounting for about one-fourth of world forests.

The Amazon valley forest (Fig. 3-7) forms the largest unbroken forest belt in the world. Within this vast area of well over 1 billion acres of "jungle" land are found about 5,000 species of trees, including many still unknown. A number of these are commercially valuable, and as many others are yet to be profitably utilized. Typical of tropical forests, this belt is wanting in pure stands, and exploitation has been chiefly of selected specialty species. Particularly useful are certain members of the Leguminosae (*Vouacapoua, Andira, Piptadenia, Copaifera, Apuleia, Peltogyne, Dalbergia, Machaerium, Hymenaea, Platymiscium, Enterolobium*); Meliaceae (*Carapa, Cedrela*); Lauraceae (*Nectandra, Ocotea*); Bignoniaceae (*Tabebuia = Tecoma?*); and Euphorbiaceae (*Hevea, Hura, Heironyma*). In addition, useful or potentially useful species can be found representing practically any family of woody Angiosperms.

Yet little has been done to utilize the forests of Amazonia, aside from the well-known rubber exploitation and local specialized timber production for export. The reason for this is apparent, for nowhere in the world is there a comparable region so little explored and so poorly populated. Except for precarious and expensive river travel, no transportation of consequence is known to the Amazon country. Some of the land is still in the hands of hostile Indian tribes.

Similar to the forests of Amazonia are those of the narrow coastal belt of eastern Brazil, which extend inland for as much as

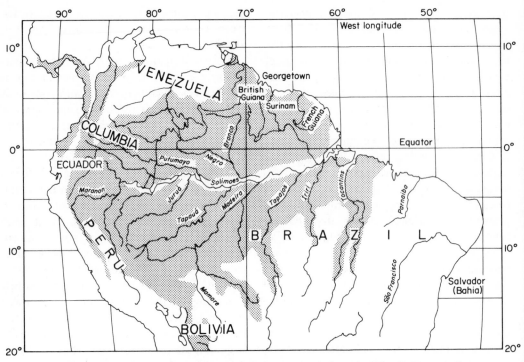

Figure 3-7 The range of the Amazon rainforest extends from the Atlantic Ocean to the eastern slopes of the Andes, covering an area more than 3,000 miles long and up to 1,000 miles in width. (After Llewelyn Williams, *Economic Botany* **15,** 224 (1961).)

Hauling logs with oxen in less developed rain forest area. Pará, Brazil. (Courtesy Chicago Natural History Museum.)

200 miles and southward from the northeastern tip of Brazil, around Brazil's southern bulge. In some sections, such as in southern Baía and Espírito Santo, the forests have been scarcely touched. Elsewhere, as along rail lines in northern and eastern Brazil and throughout the state of São Paulo, the land has been utterly denuded of trees. In this portion of Brazil, considerable attention has been given the flora, and much more is known concerning vegetational types and their ecology than is the case in Amazonia. Principal genera include *Aspidosperma*, of the Apocynaceae; *Couratari* and *Lecythis*, of the Lecythidaceae; *Astronium*, of the Anacardiaceae; and most of the genera previously mentioned as important in the Amazon area, especially those of the Leguminosae. Adjacent to and inland from the tropical hardwood coastal belt is found semiarid "sertão" country, especially in northeastern Brazil. In these semiarid regions, palms frequently dominate, and they often constitute a main source of income for the population. Important genera of the Palmae are *Orbignya*, the "babassu"; *Copernicia*, the "carnauba"; *Cocos*, the "ouricuri"; and *Attalea*, the "piassava." Interspersed with the abundant palms and forming a transition to the interior plains of Goyaz and Mato Grosso are scattered scrub forests, the "caatinga," in which various Euphorbiaceae and Leguminosae usually predominate.

In southern Brazil, extending from São Paulo to Paraguay and south to the state of Rio Grande do Sul, is found the only important Gymnospermous forest of South America. Here the distinctive Paraná pine, or "monkey-puzzle" tree, *Araucaria*, dominated until recently (with stands now becoming exhausted). To the west of the Paraná pine stands is found further tropical hardwood forest, including among its species the famed Paraguay tea, "yerba maté" (*Ilex*, of the Aquifoliaceae), and west of the Paraguay River the open "quebracho" (*Schinop-*

sis and related genera of the Anacardiaceae) forests of the chaco region.

Forests of the northern Andes differ comparatively little from those of Amazonia, but at higher elevations become rather open. Quinine, *Cinchona*, of the Rubiaceae, is one of the important trees of this area, as is also the balsa, *Ochroma*, of the Bombacaceae. Tidal areas, particularly in Ecuador, Colombia, and the north coast, possess abundant stands of mangrove (*Rhizophora* and other genera). Southward, in the southern Andes, the northwestern states of Argentina harbor rather poor, open, semiarid forests similar to those of northeastern Brazil. Prominent genera of the more moist localities include *Cedrela*, of the Meliaceae; *Calycophyllum*, of the Rubiaceae; *Ocotea* and *Nectandra*, of the Lauraceae; *Machaerium*, *Piptadenia*, and *Enterolobium*, of the Leguminosae; *Tabebuia*, of the Bignoniaceae; in the drier belts occur many Leguminosae (*Prosopis*, *Acacia*, *Caesalpinia*); *Celtis*, of the Ulmaceae; and *Zizyphus*, of the Rhamnaceae. In the extreme southern Andes occur rather poor, scrubby, slowgrowing forests of the rough mountainous areas; there are found the Gymnospermous genera *Araucaria*, *Fitzroya*, *Libocedrus*, and *Podocarpus*. Most prominent among the hardwoods are the "beech," *Nothofagus*, of the Fagaceae; and in Tierra del Fuego and Patagonia, the "magnolia," *Drimys*, of the Magnoliaceae.

Thus South America as a whole is extensively forested, and is a potential export source of large quantities of forest products even though currently an importer of timber on balance. Only portions of the pampas of Argentina and Uruguay, the Pacific coast of Chile, Peru and the highlands of Bolivia, and northeastern Brazil are inconvenienced by lack of sufficient local timber. South America is notably poor in coal resources, so that wood has been the primary fuel utilized during 400 years of colonization. As a result, many of the forests, particularly the scrubby, marginal, semi-arid forests of northeastern Brazil and those of São Paulo, have been severely depleted to provide fuel (charcoal and wood) for cooking, for running the few rail lines, for powering steamboats, and even for providing electric power to the large coastal cities. Perhaps even the climate has been altered for the worse because of extensive cutting of scrub forest protecting the watershed in regions where water is always the limiting factor to life.

Most of South America's forest land is inaccessible for profitable exploitation. There logging methods are primitive, involving hand felling, sawing, and hewing, and transport by oxen to wood-burning railroad spurs. Very little of the forest land is privately owned, Argentina being the exception in this respect, and some of it, although nominally state property, is even held by hostile Indian tribes. In contrast to the abundant hardwoods, softwoods occur in merchantable quantities only in the Paraná pine belt of southern Brazil and to a minor extent in the Chilean and Argentinian south Andes. Considerable reforestation has been practiced in south Brazil (São Paulo) and Argentina, usually with exotic species (mostly *Eucalyptus*). Brazil has by far the greatest forest reserves in South America, and is the dominant producer of industrial wood.

UNITED STATES AND CANADA. Forests in the Unites States and Canada are of considerably more interest and importance to the American student than are those of other continents. A more detailed discussion of them will be given in Chapter 4. Forests of North America constitute slightly less than

Western coniferous forest of the United States. Chester, California. (Courtesy Caterpillar Tractor Co.)

20 per cent of the world's forest acreage. Of the six forest belts of the United States, four are coniferous; the most important forests of the United States today are those west of the Great Plains, and these are all coniferous. Of the eastern forest regions, a broad belt from the Midwest, extending east and northward to New England, is in hardwood cover. All told, the United States is about 26 per cent forested; this percentage is decreasing daily, since annual cut usually exceeds annual increment. Canada is about 25 per cent timbered, almost exclusively with softwood types.

The United States and Canada are the premier lumber-producing countries of the world today, not only because they are endowed with rich forest land (which today, however, is approaching exhaustion), but because modern mechanized means of logging and milling can produce timber products in unprecedented quantity. Some mills of the Pacific Northwest have sawed more than 1 million board feet of lumber in a single day. When this performance is compared with that of hand logging and even hand sawing in many unmechanized, chiefly tropical, regions, it is easy to see why North American lumber has been able to invade foreign markets.

In contrast to the situation in most other continents, the forest lands of North America are to a great extent privately held. Only within the last few decades has the federal government assumed the responsibility of establishing extensive national forests. A great deal of forest research has been carried on toward making forest holdings continuously productive, and efforts to reseed and replant have been initiated. Funds have also been provided for fire control and other forest services. The United States has for decades been the world's greatest consumer of lumber, as well as the foremost producer; the per capita consumption of North America is about five times that of any other continent.

SUGGESTED SUPPLEMENTARY REFERENCES

Books

Allen, Paul H., *The Rainforests of Golfo Dulce*, University of Florida Press, 1956.

FAO, *Timber Trends and Prospects in Africa*, Rome, 1967.

——————, *World Forest Inventory*, Rome, 1963.

Haden-Guest, Stephan, John K. Wright, and Eileen M. Teclaff (eds.), *A World Georgraphy of Forest Resources*, Ronald Press, N. Y., 1956.

Heske, I. F., and E. Otremba (eds.), *Weltforstatlas*, Stechert-Hafner, N. Y., 1951.

Kaul, R. N., (ed.), *Afforestation in Arid Zones*, W. Junk N. V., The Hague, Netherlands, 1970.

Mirov, N. T., *The Genus Pinus*, Ronald Press, N. Y., 1967.

Penfold, A. R., and J. L. Willis, *The Eucalypts*, Leonard Hill, London, 1961.

Richardson, S. D., *Forestry in Communist Russia*, Johns Hopkins Press, Baltimore, 1966.

Zon and Sparhawk, *Forest Resources of the World*, McGraw-Hill, N. Y., 1923.

Periodicals

Various publications of the Division of Forestry, FAO (Food and Agriculture Organization of the United Nations), Rome. *Unasylva* was issued bimonthly from 1947 to 1949, and quarterly from 1950 on. *The Yearbook of Forest Products Statistics* is issued annually.

CHAPTER **4**

The Forest Belts
of North America

THE FORESTED LAND OF NORTH AMERICA IS exceeded in extent by that of Asia and South America, but North America surpasses all other continents in industrial timber harvest. (If local usage of wood for fuel were considered, the wood utilization from tropical and temperate forests of Asia combined would result in appreciably greater "produc- tion" there than for North America.) There are slightly more than 700 million hectares of forest land in North America north of Mexico, the timber of which is approximately two-thirds softwood. Nearly 400 million cubic meters of timber are harvested annually, of which five-sixths is softwood. Approxi- mately one-third of this is for making pulp,

Figure 4-1 Principal forest types of the United States. (Courtesy USDA Forest Service.)

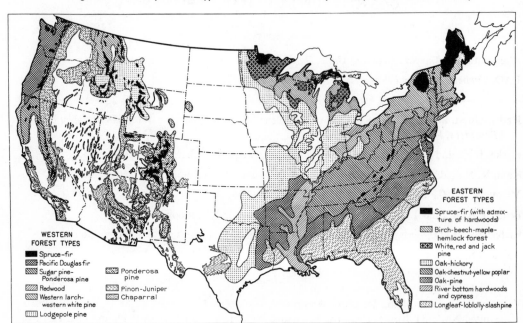

WESTERN
FOREST TYPES

- ■ Spruce–fir
- Pacific Douglas fir
- Sugar pine–
  Ponderosa pine
- Redwood
- Western larch–
  western white pine
- Lodgepole pine

- Ponderosa pine
- Pinon–Juniper
- Chaparral

EASTERN
FOREST TYPES

- ■ Spruce–fir (with admix-
  ture of hardwoods)
- Birch-beech-maple-
  hemlock forest
- White, red and jack
  pine
- Oak-hickory
- Oak-chestnut-yellow poplar
- Oak-pine
- River bottom hardwoods
  and cypress
- Longleaf-loblolly-slashpine

Table 4-1

Approximate timber volume in billions of board feet, of principal timber species in United States forests as of the 1960 decade.

Douglas fir	600	Principally in the Pacific Northwest
Western hemlocks	270	Pacific Northwest
Ponderosa (western yellow) pine	240	All western forests
True firs	235	Various western forests
Southeastern yellow pine (longleaf, slash, shortleaf, and loblolly)	200	Southeastern states
Red oaks, all types	90	Mainly central and upper South
Engelmann and related spruces	80	Rocky Mountain belt
White oaks, all types	73	Mainly central and upper South
Sitka spruce	76	Pacific Northwest
Lodgepole pine	53	Rocky Mountain belt
Western white and sugarpine	53	Northern Rockies and Pacific Southwest
Redwood	31	California
Hickory	28	Central and upper South
Sweetgum	26	Central and upper South
Tupelogum and blackgums	26	South-central
Hard maple	26	Deciduous hardwood belt
Eastern white and red pine	21	Northern coniferous belt

Adapted from *USDA Agricultural Statistics*, 1967, Wash., D.C.

two-thirds for wood products. About three-fourths of the timber harvest comes from the United States, one-fourth from Canada.

Figure 4-1 maps the principal forest types in the United States. Figure 4-2 is an adaptation of the St. Regis Paper Company's depiction of the twelve most important forest trees in the United States, considering their roles in abetting settlement of the nation. These species are not all of great commercial importance any longer, although the Douglas fir and ponderosa pine are leading timber species of the United States. Table 4-1 lists the volume of sawtimber on commercial forest land in the United States according to principal tree types. Douglas fir constitutes by far the greatest timber reserve in the nation. It is evident that western species occur in greater volume than eastern ones, on the whole reflecting the old-growth and virgin stands which still remain in the West. Volume is in part a reflection of the characteristics of a species. St. Regis' comparison of height and shape of ten familiar forest trees is given as Fig. 4-3. California redwoods hold the record for height (over 370 feet) and for greatest volume of wood per acre (up to a million board feet).

Colonization of North America was greatly aided by the luxuriant forests, and the early economy of the struggling thirteen states was quite dependent upon forests and forest products. The virgin forests of the east coast provided for European settlers a source of building material, shelter for game, and a resource of some value for export. At the same time the forests were a nuisance, requiring felling and burning in order to make way for the planting of familiar food crops. Colonization spread westward through such familiar habitat, and not until the colonists came upon the more arid Great Plains was it necessary for them to learn new ways of settlement not in the tradition of the Old World. For some time there was delay in settlement of the grasslands, until technology and a cooperative mode of existence advanced sufficiently to permit mastering of this rich but less benevolent environment.

■ **EASTERN WHITE PINE** — An important tree in Colonial America, for homes and for the masts and spars of ships. It is still an important timber tree, and one source of knotty pine decorative paneling.

■■ **DOUGLAS FIR** — Averaging 150 to 200 feet in height, the Douglas fir is considered the most important timber tree in the world. It is used principally for construction lumber and plywood, and for telephone poles.

■ **BALSAM FIR**—The most popular of all Christmas trees; used extensively for pulp and for wooden boxes and crates. It is also the source of a very clear oil called Canada Balsam, used in high power microscopy.

■■ **SHAGBARK HICKORY**—Tough hickory lumber supplied spokes for the wheels that crossed the continent and the ax handles that cleared the wilderness. Hickory smoke is used for curing ham.

■■ **PONDEROSA PINE** — A tall, straight tree found in many parts of western U. S.; second only to Douglas fir as a timber tree. Its wood is used for lumber, railroad ties, telephone poles, posts, and mine timbers.

Figure 4-2 Major forest species abetting settlement of the United States. (Courtesy St. Regis Paper Co.)

56

■■ **SWEET GUM** — The sweet part of the sweet gum is a sticky substance that oozes from cuts in the trunks and branches. The strong, light wood is widely used in furniture. Also an important pulping tree.

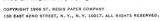

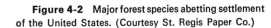

■WESTERN HEMLOCK—A tall, graceful tree, the western hemlock is important for timber and is also a prime source of pulpwood; and tannin, used in tanning leather, can be extracted from its bright red inner bark.

■WHITE SPRUCE—Spruces are the major pulping trees of eastern U.S., and Canada. White spruce has strong, resilient wood widely used for lumber and, because of its resonant qualities, for many musical instruments.

■■YELLOW POPLAR—The yellow poplar, the tallest hardwood in the east, is a favorite ornamental tree, and an important source of wood for musical instruments, television cabinets, and other furniture.

■■WHITE OAK—The oaks have always been our most important hardwood timber trees. Early American mansions, public buildings, and ships' hulls, were of oak. White oak was the traditional wood for whiskey barrels.

■LONGLEAF PINE—Found throughout southeastern U.S., this tree is a major source of turpentine, pitch, and resin; its lumber is also widely used in building construction, and in railroad cars.

■SUGAR MAPLE—This is the tree that maple syrup and maple sugar come from. But it is also an important source of fine hardwood, prized for furniture. It grows in the eastern half of the U.S., and Canada.

Figure 4-2 Continued.

57

272'

221'

162'
160'

122'
116'

102'
96'

78'

Giant Sequoia
Sequoia Natl. Park, Cal.

Douglas Fir
Olympic Natl. Park, Wash.

Ponderosa Pine
Lapine, Ore.

White Birch
Lake Leelanau, Mich.

American Elm
Trigonia, Tenn.

Figure 4-3 Comparative size and shape of prominent North American forest species. (Courtesy St. Regis Paper Co.)

How big is a tree? Each of these ten is believed to be the biggest of its kind now growing in the United States.

The "bigness" of a tree is defined as the sum of three dimensions: height, the circumference of the trunk 4½ feet from the ground, and one quarter of the crown spread.

A tree keeps growing as long as it lives, but not necessarily in all three dimensions. Some species stop growing upward after a time, but spread their crowns ever wider and add layers of wood to their trunks. Then too, each species has its own rate of growth and its own normal life span. These factors explain why even a mature, healthy oak, for example, can never be as big as a similarly favored sequoia.

The trees shown on these pages have been recorded by the American Forestry Association as the biggest of their species now standing.

Such records are as interesting to St. Regis as they are to you, because of our deep concern for the forests of America. These forests are our basic resource. They provide us with the wood from which we make printing papers, kraft papers and boards, fine papers, packaging products, building materials, and products for consumers.

To a large extent, then, the life of the forest is St. Regis' life. That is why we—together with the other members of the forest products industry—are vitally concerned with maintaining the beauty and usefulness of America's forests for the generations to come.

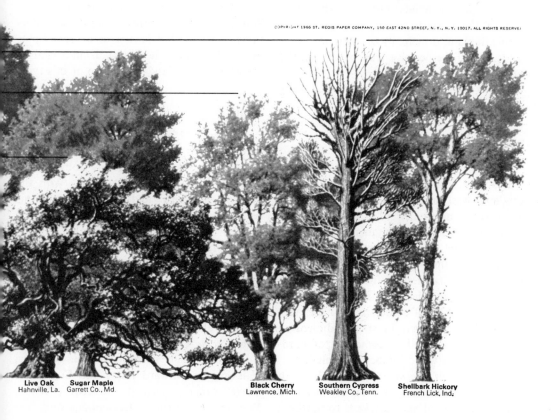

Live Oak
Hahnville, La.   **Sugar Maple**
Garrett Co., Md.   **Black Cherry**
Lawrence, Mich.   **Southern Cypress**
Weakley Co., Tenn.   **Shellbark Hickory**
French Lick, Ind.

These Great Plains have had not only historical influence in presenting a barrier that prevented easy access to the West (and exploitation of the western forests), but they form a natural separation demarking eastern from western forest belts. The eastern forests, of much historic importance, are today largely exhausted of large trees. A good deal of pulp, small dimension stock, and specialty hardwood is harvested in the East, but large timber comes mainly from the western forests, shipped cross-continent to eastern markets. Even though the eastern forests have been cut-over, considerable second growth is now available and much of the forest land is being given improved and increasingly intensive forestry management. In the southern pine belt alone over 1 billion cubic feet of timber are being harvested annually and the industry increasing; yet increment exceeds the cut. The eastern forests are overwhelmingly in private hands, but somewhat over half of the western forests are still in the public domain, principally on national forest lands. This is especially true of the Rocky Mountain belt.

North American forests fall naturally into six belts:

East

1. *Northern Coniferous Belt.* In the United States this forest type extends from New England westward to the Great Lakes in Minnesota, and southward at high elevations of the Appalachian Mountains to Tennessee. Its main extent, however, is in Canada, where in the northernmost part it is transcontinental, merging with the western forests in northern Alberta on into Alaska. It includes perhaps 200 million acres of commercial forest land, much of it, however, in smaller-dimension trees. About 100 billion cubic feet of hardwood and softwood timber is included.

2. *Deciduous Hardwood Belt.* This belt extends from New England (mostly as mixed stands) southward and westward to northern Georgia, northeastern Texas, and southern Minnesota, the higher elevations of the Appalachians excepted. It also occurs along watercourses in the plains region, and in Canada to the northeast of the Great Lakes (southern Ontario and the St. Lawrence Valley). This deciduous belt embraces around 100 million acres of commercial forests, containing less than 100 billion cubic feet of timber almost entirely of the hardwood type.

3. *Southeastern Coniferous Belt.* This important forest belt extends from Virginia southward to Florida east of the Appalachians, and westward through southern Georgia, Alabama, Mississippi, and Louisiana to Texas–Arkansas. This belt contains about 200 million acres of commercial forests, with more than 100 billion cubic feet of growing stock about evenly divided between softwoods and hardwoods.

4. *Subtropical Evergreen Hardwood Belt.* This belt is of minor extent in the United States, including southern Florida and portions of the Gulf Coast west to Texas plus Puerto Rico and the Virgin Islands. All of the Mexican lowlands would be of this type.

West

5. *Rocky Mountain Coniferous Belt.* A heterogeneous mixture of forests extends through the Rocky Mountains from Mexico north to Canada where it merges with the moister northern and Pacific coniferous belts. From the Great Plains westward through the Great Basin (especially Nevada) isolated forest occurs on numerous mountain ranges in spite of the region being generally quite arid. Nearly 100 million acres of commercial forest land is involved, containing about 100 billion cubic feet of growing timber, almost exclusively softwoods.

6. *Pacific Coniferous Belt.* This is a luxurious forest belt extending from southern California northward along the Coast and Sierra ranges, through western Oregon, Washington, British Columbia, and the

Yukon to Alaska. It includes over 100 million acres of commercial forest land, containing over 250 billion cubic feet of growing stock, almost wholly softwoods.

Each of these forest belts will be considered individually to note the important species, principal wood products, and economic usefulness.

1. Northern coniferous forests. Scattered among the predominating softwoods can be found a number of hardwood genera, principally aspen, *Populus tremuloides* (the most widespread species, ranging coast to coast), birch, *Betula*; beech, *Fagus*; and maple, *Acer.*

The principal Gymnosperms are: pines, *Pinus* (with four important species: *P. strobus, P. resinosa, P. banksiana, P. rigida*); larch or tamarack, *Larix* (with a single eastern species: *L. laricina*); spruces, *Picea* (with three important native species: *P. rubens, P. glauca, P. mariana*); hemlock, *Tsuga* (with a single important species; *T. canadensis*); fir, *Abies* (with a single important species: *A. balsamea*); arborvitae, *Thuja* (also with a single important species: *T. occidentalis*).

THE PINES. The genus *Pinus* is divided into two categories, the red or hard pine group and the white or soft pine group.

Reproduction of cover and title page of *An Act for the Preservation of White and other Pine Trees growing in Her Majesties Colonies.* 1710. (Courtesy *Chronica Botanica.*)

(387)

Anno Regni

A N N Æ

REGINÆ

Magna Britannia, Francia, & Hibernia,

N O N O.

At the Parliament Begun and Holden at *Weſt-minſter*, the Twenty fifth Day of *November*, *Anno Dom.* 1710. In the Ninth Year of the Reign of our Sovereign Lady *A N N E*, by the Grace of God, of *Great Britain, France,* and *Ireland,* Queen, Defender of the Faith, *&c.* being the Firſt Seſſion of this preſent Parliament.

L O N D O N,
Printed by the Aſſigns of *Thomas Newcomb,* and *Henry Hills,* deceas'd ; Printers to the Queens moſt Excellent Majeſty. 1711.

Anno Nono

Annæ Reginæ.

An Act for the Preſervation of White and other Pine-Trees growing in Her Majeſties Colonies of *New-Hampſhire,* the *Maſſachuſets-Bay,* and Province of *Main, Rhode-Iſland,* and *Providence-Plantation,* the *Narraganſet* Country, or *Kings-Province,* *Conneᶜticut* in *New-England,* and *New-York,* and *New-Jerſey,* in *America,* for the Maſting Her Majeſties Navy.

Whereas there are great Numbers of White or other Sort of Pine-Trees, fit for Maſts, growing in Her Majeſties Colonies of New-Hampſhire the Maſſachuſets Bay, and Province of Main, Rhode-Iſland, and Providence-Plantation, the Narraganſet Country, or Kings-Province, and Conneᶜticut in New-England, and New-York, and New-Jerſey, fit for the Maſting Her Majeſties Royal Navy : And whereas the ſame growing near the Sea, and on Navigable Rivers, may commodiouſly be brought into this Kingdom for the Service aforeſaid : Wherefore, for the better Preſervation thereof, Be it Enacted by the Queens moſt Excellent Majeſty, by and with the Advice and Conſent of the Lords Spiritual and Temporal, and Commons, in this preſent Parliament Aſſembled, and by the Authority of the ſame, That from and after the Twenty fourth Day of September, which ſhall be in the Year of our Lord, One thouſand ſeven hundred and eleven, no Perſon or Perſons within the ſaid Colonies of New-Hampſhire, the Maſſachuſets-Bay, and Province of Main, Rhode Iſland , and Providence-Plantation, the Narraganſet Country, or Kings-Province, Conneᶜticut in New-England, and New-York, and New-Jerſey, or any of them, do or ſhall preſume to Cut, Fell, or Deſtroy any White or other ſort of Pine-Tree
Eeeee 2 fit

Mature white pines in Connecticut, typical of the forest that greeted early colonists of the northern coniferous belt. (Courtesy USDA.)

These groupings can be readily separated in the northern coniferous forests in that the only soft pine native there has five (rarely four) needles per fascicle, whereas the hard pines have two (or in one eastern species only, three).

Pinus strobus, the white pine. Perhaps the most famous pine in the history of the United States, the eastern white pine originally occurred in extensive, often nearly pure, stands from the New England seaboard west in the northeastern United States and southern Canada to Minnesota and the Great Lakes states, and southward at high elevations in the Appalachians as far as northern Georgia. Limited natural stands are also found in southern Mexico, and the species is widely planted as an ornamental. It has been said of the virgin stands of this pine that a squirrel could run a squirrel's life exclusively through white pine and never have to come to ground. White pine was highly prized for masts of the sailing ships of the Royal Navy while northeastern America was still a colony. In fact, penalties for the cutting of this towering pine on the royal preserves was one of the minor irritations that cumulatively caused the rebellion of the colonies and the birth of a new nation. After this successful rebellion, it was chiefly exploitation of the white pine that sent colonization westward as far as the Wisconsin Territory. For more than a century thereafter white pine remained the most cut, most used, and most prized timber in the New World. Only the twentieth century has seen this once-great pine fall into a position of secondary importance, with final cutting-over of a great forest, and with the introduction from Europe of the deadly white pine blister-rust. But in spite of man and disease, white pine has reseeded itself over parts of its original range, and second-growth stands offer some respite to a fading economic species now less than one-fiftieth as abundant as in pre-Columbian America.

The versatile white pine lumber has been used for every purpose, from ship's mast to matchstick. Ease of working and lack of resin have made it especially suitable for window sashes, interior finish, patterns, boxes, and crates. White pine lumber on the market today comes chiefly from *Pinus strobus*' western relative, *P. monticola*. Some reforestation has been attempted with *Pinus strobus*, both in America and Europe, but the omnipresent white pine blister-rust makes such planting a risk. *P. strobus* demands a comparatively good soil environment. Hence, repeatedly logged or burnt-over white pine land is apt to be invaded by the less desirable jack pine (and ultimately by red pine) until the soil is once again built up.

Pinus resinosa, the red pine or Norway pine. This hard pine has essentially the same range as the more-sought-after white pine, except that it is not found in the Appalachians south of Pennsylvania. Never as common as white pine, it has nevertheless been logged with *P. strobus* and has suffered much the same sort of depletion. Densest stands probably occurred and still occur in the Great Lakes states. It is a species becoming more used in reforestation. Red pine wood is heavier than white pine, and may be used for general construction, mill work, crates, ties, pulp, and so on. In tolerance and in its soil demands, the species is intermediate between the jack and white pines.

Pinus banksiana, the jack pine. The jack pine is the most northern of all American pines, and its east–west range through Canada is practically transcontinental. In the United States it is common in the northern Great Lakes states; in Canada it is frequent from the Atlantic Ocean and Hudson Bay northward to the Arctic Circle and westward about 2,500 miles to within a few hundred miles of the Pacific Ocean. In usefulness the jack pine is a decidedly inferior species. A colonizer of open, sandy barrens, and frequently stunted toward the northern limits of its range, the jack pine usually has twisting, gnarled trunks with many persistent low branches, unsuitable for any quantity of clear lumber. The species is finding increasing use, however, as a pulpwood, locally for rough lumber and fuel, and for crates. Jack pine thrives on poor sandy soils incapable of sustaining the more desirable pines. It is thus important as a colonizer, building up run-down soils until these permit invasion by the more valuable species. It is intolerant of shading, and is seldom found in prominent association with other pine species.

Pinus rigida, the pitch pine. Pitch pine is a species exclusively of the East, being found along the Atlantic coast of the United States from Georgia north through New England, but not west of Ohio or of eastern Kentucky and Tennessee. The species is very resinous; in prerevolutionary times it was a source of pitch (tar and turpentine). The wood makes a hot fire, and at one time was frequently used to fire steam engines. The wood is said to be durable, and was used in colonial times to make water wheels for the grist mills. A small, scraggly tree, the pitch pine is today comparatively unimportant economically, but can be used for rough construction, props, fence posts, crating, and pulp. Table-mountain pine, *P. pungens*, has a similar range in the Appalachians from Pennsylvania to Tennessee, and Virginia pine, *P. virginiana*, from Pennsylvania to Alabama; both are of little economic importance.

The northern coniferous forest belt as it appears in Vermont. (Courtesy USDA.)

LARCH OR TAMARACK. A single species of tamarack, *Larix laricina*, is native to the northeastern coniferous forest. Its extensive range is almost identical with that of the jack pine, extending from all of eastern Canada northwestward to the Arctic Circle and Alaska, and southward in the United States through New England and the Great Lakes states as far south as West Virginia. It is the only northeastern conifer with annually deciduous leaves. Characteristically, it grows in swampy areas where competition is less severe, for the light requirements of tamarack are unusually high. The wood is durable in contact with soil, and consequently it has been much used for posts, ties, and sills. In the days of wooden ships, "knees" formed by the abrupt horizontal growth of tamarack roots on shallow soils were much sought for joining ribs to deck timbers. Tamarack bark has been used as a source of tannin.

THE SPRUCES. The spruce genus, *Picea*, contains about forty species, most of them Asiatic. Three species are native to the northeastern coniferous forest, all of moderate importance.

Picea rubens, the red spruce. This spruce is restricted to the East, being found in southeastern Canada and New England and southward along the Appalachians as far as North Carolina. The species was the favorite pulpwood throughout its range for many years. It is also important as a timber tree, being stiff and strong while relatively light in weight. It is much used to make ladder rails, paddles and oars, and sounding boards for musical instruments. Small trees are often cut for Christmas trees.

Picea glauca, the white spruce. White spruce is found much more extensively throughout the northern coniferous forest belt than is red spruce, occurring from Labrador, Newfoundland, and New England westward and northward through the Great Lakes states of the United States and most of forested Canada to Alaska. Thus, with jack pine and tamarack, white spruce is a characteristic species throughout the range of this forest belt. Perhaps its chief use is for pulpwood, much of which is imported into the United States from Canada; but, like red spruce, it is also utilized for lumber and other wood products, and for Christmas trees.

Picea mariana, black spruce. Black spruce has essentially the same range as white spruce, but occupies poorer sites, usually swampy, where it may be a companion of the tamarack. It is not commercially so important as the previous two species, being a smaller, slower-growing tree. Its chief commercial use is for paper pulp, but like the other spruces it is also used for lumber, diverse wood products, and Christmas trees.

HEMLOCK. The only important hemlock of eastern North America is *Tsuga canadensis*; its range is similar to that of white pine, a species with which it is often associated. Hemlock is particularly characteristic of cool, moist, shady slopes from New England to Minnesota, and through the Appalachians southward to Tennessee. The species is quite tolerant, and indeed the seedlings cannot endure strong light. Hemlock wood has been little utilized, but the bark (comprising some 15 per cent or more of the volume of the tree) for years was an important source of tannin. Formerly many hemlock stands were stripped of bark, and the wood left to decay in the forest. The species is still fairly common in the northeast, though not nearly so abundant as formerly. Often it is found in small stands scattered through the northern portion of the hardwood belt. Hemlock is an exceedingly graceful tree, able to stand shade and trimming, and thus lends itself readily to ornamental planting.

THE FIRS. The only important fir of the northern coniferous forest is the balsam fir, *Abies balsamea*. The range of balsam fir is that of the tamarack and the white, red, and black spruce, all of which are characteristically found throughout eastern Canada to or almost to Alaska, and from the northeastern United States almost to the Arctic Circle. The balsam fir is smaller than most spruce species, and less important economically. Its major use is for paper pulp, although lumber for interior trim, crates, and boxes is also obtained from it. It is outstanding as a Christmas tree, holding its needles much longer than do the spruces. Oil of fir may be distilled from its bark or needles; and the aromatic needles are frequently made into balsam pillows. Blisters on the bark yield Canada balsam, once much used in microscopy because it has the proper index of refraction for a mounting medium.

THE ARBORVITAES. Two similar genera, *Thuja* and *Chamaecyparis*, are usually con-

A balsam fir understory beneath 95 year old red pine on the Cutfoot in Minnesota. (Courtesy USDA.)

Mixed hardwood forest in North Carolina. (Courtesy USDA.)

sidered arborvitae (or, incorrectly, "cedar"). Of these, only one species is frequent in the northern coniferous forest.

Thuja occidentalis, the northern white cedar or arborvitae. This is the only species of the genus in northeastern America. The tree is comparatively small and slow-growing, and the lightweight wood splits rather readily; hence it has restricted use for lumber. But it is very durable in contact with soil, is easily worked and fashioned, and has little tendency to shrink or warp; thus it is useful for making shingles, ties, poles, posts, cooperage, and boats. It is highly desired for ornamental planting, and many cultivars have been introduced into horticulture.

2. Deciduous hardwood forests. The deciduous hardwood forest is exceedingly diverse as regards genera and species. Seldom do the high frequencies of a few species that are typical of softwood forests prevail. Thus selective cutting (high-grading) rather than clear-cutting has usually been practiced. Much of the original area of the eastern hardwood forest is now given over permanently to agriculture, with isolated woodlots representing formerly widespread species. Virgin stands are insignificant, and cut-over or partially timbered land left to be re-generated naturally has usually been worsened in composition because of the invasion by less valuable, unmarketable species. Second-growth timber is reclaiming exhausted land in New England, where second-growth mixed stands of softwoods and hardwoods once again grace a countryside denuded a century ago. Table 4-1 shows the deciduous hardwoods to be generally less voluminous than the major softwood species. Essentially in the order of their greatest abundance come the red oaks, the white oaks, hickory, sweetgum, tupelo gum and blackgum, hard maple, yellow-poplar or tuliptree, beech, soft maple, ash, yellow birch, poplars, basswood, and black walnut. Volume does not necessarily reflect value, however, and the last ranking species in this listing, black walnut, brings very high prices as a cabinet and veneer wood, while species of beech, soft maple, and poplar have little or no commercial value. The more important genera of this belt are *Salix, Populus, Juglans, Carya, Betula, Fagus, Castanea, Quercus, Ulmus, Liriodenderon, Liquidambar, Platanus, Prunus, Robinia, Acer, Tilia, Nyssa,* and *Fraxinus.* Equally typical of the belt are many other hardwood trees such as the hop hornbeam (*Ostrya*), the hackberry (*Celtis*), the osage orange (*Maclura*), *Magnolia, Sassafras,* honey locust (*Gleditsia*), holly (*Ilex*), buckeye (*Aesculus*), dogwood (*Cornus*), persimmon (*Diospyros*), *Catalpa*, and so on.

It scarcely seems possible that there is hardly an unaltered remnant left of this great forest belt. As the forest was invaded and cleared for homesteading it scarcely occurred to authorities even to record what the virgin forest was like. For much of the deciduous hardwood forest belt we have only fits and snatches of information on tree frequency, size, and other ecological considerations of the virgin forest, these mostly retrieved from land surveys that used trees for marking corners. An attempt to reconstruct the original forest composition in central Ohio (Delaware county) indicates that the dominant trees (as evidenced by percentage of basal area) were beech (35.7%), white oak (16.0%), sugar maple (15.7%), and red oak (14.0%). These dominants were followed by white ash, hickory, red oak, American elm, basswood, soft maple, ash, and buckeye. Table 4-2 shows the average stem diameter of the predominating trees in Delaware county, Ohio, as recorded in the 1832 survey.

THE WILLOWS, *Salix*. Many species of willow occur in North America, some of them extending as far north as trees will grow. Species frequently hybridize and are difficult to distinguish. Willow wood is too soft to have much commercial value.

Table 4-2

Average Stem Diameter in Inches, and Average Distance Between Pairs of Trees in Feet, for Townships in Delaware County, Ohio, According to the 1832 Survey

Average stem diameter

Township	Average dist. between pairs	Beech	Sugar maple	White oak	White ash	Hickory	Red oak	Black oak	American elm	Lynn (Basswood)	Soft maple	Blue ash	Red elm	Buckeye	Black ash
Harlem	33.2	13.0	—	14.5	—	—	30.0	—	10.0	—	—	—	—	—	—
Trenton	26.7	10.8	10.0	18.0	—	—	—	—	—	—	—	—	—	—	—
Porter	28.9	11.1	—	—	14.0	—	—	—	6.0	—	14.0	—	—	—	—
Genoa	32.6	14.0	12.0	—	—	—	—	—	—	—	—	—	—	—	—
Berkshire	27.7	11.4	8.0	—	—	12.0	18.0	—	—	—	—	—	—	—	—
Kingston	26.1	11.4	—	24.0	12.6	—	—	—	—	—	—	—	—	—	—
Orange	28.9	13.9	—	—	12.0	—	—	—	10.0	—	—	—	—	—	—
Berlin	27.3	13.2	15.0	—	11.5	—	16.5	—	—	—	—	—	10.0	—	—
Brown	26.0	14.0	11.3	30.0	13.5	12.0	—	—	—	15.0	—	—	—	—	—
Oxford	21.8	11.2	14.0	—	12.0	—	—	12.0	12.0	—	—	14.0	—	—	8.0
Liberty	23.8	14.1	16.0	—	14.8	12.6	—	—	12.0	—	—	—	—	—	—
Delaware	20.7	—	11.6	20.0	14.0	—	24.0	—	—	—	—	—	—	—	—
Troy	19.7	15.3	13.6	12.0	—	—	—	8.0	—	13.5	—	—	—	—	—
Marlboro	20.0	12.7	6.0	6.0	—	—	—	10.0	4.0	—	—	—	—	4.0	—
Concord	(14.9)	—	—	—	9.0	14.0	—	—	—	(Partial survey)		—	—	—	
Scioto	(28.4)	—	11.0	—	—	—	—	—	—	(Partial survey)		—	—	—	
Radnor	23.9	16.4	14.3	—	14.0	8.0	—	—	—	—	12.0	—	—	—	—
Thompson	—	—	—	—	—	—	—	—	—	(No survey record)		—	—	—	

Adapted from J. Gordon Ogden, III, "Early Forests of Delaware County, Ohio," *Ohio Journal of Science* **65**, 34 (1965).

THE POPLARS, *Populus*. Several species of poplar occur in the hardwood forests, and also as minority members of the coniferous belts. Of especial interest is the quaking aspen or popple, *P. tremuloides*, more common to open portions of the northern coniferous and Rocky Mountain forests than the hardwood belt. Its soft, weak wood is little used for lumber, but it is satisfactory for pulp and is in demand by paper mills. The ubiquity of the species and its quickness to regenerate mark it as potentially important. The cottonwood, *P. deltoides*, is found over most of the eastern half of the United States, where it is common along watercourses and is often planted as a windbreak in the plains. It is a very rapidly growing species, but the wood is soft and weak and warps badly.

THE WALNUTS, *Juglans*. Black walnut, *J. nigra*, and the butternut, *J. cinerea*, are the two species of the genus found in the eastern hardwood forests. Of the two, black walnut is much the more important. This species is the foremost cabinet wood of North America, having been used for fine furniture and interior paneling since colonial times. It had been used for gunstocks and airplane propellors early in the century. The wood is easily worked, withstands shock and strain, and is unusually durable. The edible nuts may be harvested for confections. The walnut is not uncommon, but it has become exceedingly scattered.

THE HICKORIES, *Carya*. The pecan hickory, *C. illinoensis*, is native to and extensively planted in the southern Midwest, where it is a source of pecan nuts. Other species are noted for the hard, tough, shock-resistant wood they produce, which has been extensively used for spokes and axe handles. Of these, *C. ovata*, the shagbark hickory, is perhaps the most common and important. It is found throughout the extent of the hardwood belt, and in the Midwest is a codominant in the climax vegetation.

THE BIRCHES, *Betula*. Several important

birch species are native to eastern North America, two of them (paper birch, *B. papyrifera*, and yellow birch, *B. lutea*) being as common to the northern forests of northeastern United States and southeastern Canada as to the hardwood belt. The economically important *B. lutea* is used for furniture, boxes, molds, flooring, veneer, interior finish, and so on; but the extant supply of first-class yellow birch is quite limited. *B. papyrifera*, famous for supplying the birch bark of Indian canoes, is utilized somewhat for making spools, clothespins, toothpicks, novelties, and as pulp. Other species, more southern in range, are less common and less utilized, but may be useful locally.

THE BEECH, *Fagus*. A single species, *F. grandifolia*, is found here and there in moister

Shagbark hickory, *Carya ovata*, at Corrick's Ford near Parsons, W. Va. (Courtesy USDA Forest Service.)

localities throughout the eastern United States, usually in association with sugar maple. The species is extremely tolerant, and seedlings are able to develop with but one-eightieth of normal day illumination; hence, beech forests become very dense and dark. The wood was long neglected because of difficulty in seasoning and a high percentage of defectives. It may, however, be utilized for furniture, flooring, novelties, boxes, and barrels. The tree is slow-growing and its frequency decreasing.

THE CHESTNUT, *Castanea*. The only large native species, *C. dentata*, was once extremely abundant through much of the hardwood belt, particularly in the Appalachians. However, introduction of the deadly chestnut blight from Asia early in this century has practically eliminated the chestnut as a commercial tree. During its period of commercial importance, the durable wood of the chestnut was used for structural purposes, interior trim, posts, ties, furniture, and the like. Wood from blight-killed trees is still gathered, and until recent years was an important source of tannin.

THE OAKS, *Quercus*. The oaks are the most economically important of all the hardwood genera. A number of species of *Quercus* can be found in all parts of the hardwood belt, and some as minority members of the southeastern and northern coniferous belts. In the eastern United States, more than twenty species occur, with hybridization among them common. Two general categories of oaks are recognized: the white oaks (with rounded leaf lobes, acorns on new wood, glabrous shells, lighter bark), and the red or black oaks (with pointed leaf lobes, acorns on second-year wood, hairy shells, darker bark). Of the two, the white oaks are the more important, although represented by a smaller number of species. White oak species include the chinquapin, chestnut, swamp white, post, bur, and white oaks: red oak species include the willow, shingle, blackjack,

scrub, scarlet, black, pin, red, Spanish, Hill's, live, and other oaks. In the first category, *Q. alba*, the white oak, is the most important species. The wood is about twice as heavy as white pine, durable, and attractive. It finds a multitude of uses: for cabinet work, trim, and flooring; tight cooperage (whiskey barrels); veneers; ties, poles, and pilings; boats and general construction. Because of its especial value and the heavy demand for it, large trees of this once very abundant oak are fast disappearing. Of the red or black oaks, *Q. rubra*, the northern red oak, is the largest, most widely distributed, and commercially most important species. Lighter in weight and weaker than white oak, the wood is nonetheless used for many of the same purposes, including general construction, flooring, furniture, interior finish, ties, posts, and fuel.

THE ELMS, *Ulmus*. Perhaps half a dozen species of elm are native to the eastern United States; of these, American elm, *U. americana*, slippery or red elm, *U. rubra*, and rock or cork elm, *U. thomasi* are the most important, all being fairly general in the eastern hardwood belt. Elm wood is so cross-grained as to be nearly unsplittable. It has been used for hoops and staves, sporting goods, crates, posts, and ties. Because of its graceful, fountainlike shape, *U. americana* has been highly prized as a street and shade tree, but it is now threatened with the same fate as the chestnut by Dutch elm disease.

TULIPTREE, *Liriodendron*. The tuliptree, *L. tulipifera*, is one of the largest and most valuable trees of the eastern states. Its range is from middle New York through Georgia, and from Louisiana to northern Illinois. In many southern areas where it was formerly abundant the practice of high-grading has nearly eliminated the tuliptree. The soft wood, called yellow-poplar, is used for many kinds of construction, interior finish, crates, woodenware, excelsior, veneer (especially as a core), and even for pulp. A rapid-growing tree with attractive foliage, it is utilized considerably as a shade tree.

SWEETGUM OR REDGUM, *Liquidambar*. The only native species, *L. styraciflua*, flourishes from southern Connecticut to Florida and west to Texas, but attains its greatest size and frequency in the bottomlands of the southern Mississippi valley. In volume of sawtimber cut annually in this belt, sweetgum ranks about third. The wood is strong and stiff, easily worked, and with an unusually attractive grain, but warps and twists readily. It is much used for furniture and interior trim, ties, flooring, veneer (specially stained to imitate more expensive woods), barrels, and pulp. The star-shaped leaf and burrlike fruit of the species make it popular as an ornamental or shade tree.

SYCAMORE, *Platanus*. The native sycamore, *P. occidentalis*, is one of the largest and most distinctive of the hardwoods. It is found through most of the eastern half of the United States, particularly along watercourses. The wood contains interwoven fibers and is thus tough and difficult to split, a characteristic favoring its use for butcher's blocks.

Quarter-sawed, it is attractive for furniture, veneer, and interior trim.

WILD CHERRY, *Prunus*. The only large species is *P. serotina*, which is widely scattered throughout the eastern half of the United States and southern Canada. This tree furnishes wood of the highest quality for furniture, interior finish, veneers, and handles, ranking close to walnut in this respect. The species is readily seeded by birds; this fact, combined with its usually scattered distribution, permits it to hold its own as a component of the hardwood forest fairly well.

LOCUST, *Robinia*. The largest and best-known species is the black locust, *R. pseudoacacia*, native to the southern Appalachians and the Midwest, but now extensively planted and naturalized in almost every state of this country and in Europe. The heavy wood scarcely shrinks or swells with moisture changes, and is used for such specialty items as dowels and pins to support glass insulators. The species is much utilized in soil reclamation.

THE MAPLES, *Acer*. Several species of maple are native to the hardwood belt,

Virgin sweetgums in the "Big Thicket" section of northeast Texas. (Courtesy National Park Service.)

Mixed second growth hardwoods along old Natchez Trace running from Tennessee to Mississippi. (Courtesy National Park Service.)

including silver, red, sugar, and boxelder maples. Of these the sugar or hard maple, *A. saccharum*, is the most important, being found in every state east of the Great Plains, and in southern Canada. Over much of the northern hardwood belt it is dominant or codominant in the climax forest. This species is responsible for most of the lumber marketed as maple, and usually ranks second only to white oak in quantity cut. No distinction is usually made in the trade between sugar and red maple. The wood, although lighter in weight than oak, is even stronger. It is used for flooring, furniture, musical instruments, veneers, and a wide variety of products for which a strong, close-grained wood capable of taking a polish is desired. Sugar maple also produces the maple syrup and sugar of northeastern "sugar bushes," and is a highly desirable shade and ornamental species.

BASSWOOD, *Tilia*. Some confusion exists regarding the species of this genus, but the name *T. americana* probably embraces most of the large trees of the hardwood forest known as basswood. Basswood is particularly abundant in the northern portion of the hardwood belt, where it is sometimes codominant in the climax forest. It is utilized for containers, pulp, and small articles for which a soft, light, easily worked wood is desired.

THE GUMS, *Nyssa*. This genus is not to be confused with sweetgum, the wood of which is marketed as "redgum." Two species, black tupelo or sour gum, *N. sylvatica*, and the cotton gum, tupelo, or water tupelo, *N. aquatica*, are moderately frequent in the moister southern and eastern portions of the hardwood belt and scattered through the southeastern coniferous forest belt in swampy localities. The wood is extremely tough; its greatest use has been for cheap veneers and veneer cores, flooring, wood rollers, and gunstocks.

THE ASHES, *Fraxinus*. Several species of ash are native to the hardwood belt, including the blue, black, white, red, and green ashes. The white ash, *F. americana*, is the most valued ash of this forest belt. Its range is from Nova Scotia to southern Minnesota, south to eastern Texas and Florida. The wood is hard and strong, and is utilized for tool handles, oars, sporting goods, furniture, and interior trim.

3. The southern coniferous forest. This forest occupies for the most part the sandy coastal plain of the southern Atlantic states and the deep South. It is by no means a solid, continuous stand of Gymnosperms. Cut-over areas of the southern coniferous belt usually become invaded by the more aggressive, less valuable hardwoods. A familiar forestry practice is to eliminate competitive scrub hardwoods with herbicides, effecting release of the pines. Four species of pine are especially important, known collectively as "southern yellow pine." *Taxodium*, the bald cypress, is important in swampy coastal areas and the Mississippi delta. The various hardwoods which extend into these southern coniferous belts were mostly reviewed in the previous section.

THE PINES. The southern yellow pines are of great commercial importance in the southeastern coniferous belt, and the basis for a revived, modernized forest industry in the Southeast. Many large timber firms have secured holdings in this region, where timber reserves are being built for sustained yield production, especially of pulpwood. Southern pines are still important timber trees, but with development of pulping technology that permitted their use for kraft paper and paperboard, southern pine has become the leading pulpwood cut in the United States, accounting for nearly half of the domestic supply. The principal pine species have overlapping ranges and hybridize with one another. Longleaf and slash pine occur mainly on the coastal plain, where they intermingle and cross. Longleaf also crosses with loblolly pine. The range of the loblolly and shortleaf pines extends farther north, the shortleaf reaching Missouri and southern Ohio and straggling into southern Connecticut. The loblolly pine is believed to hybridize naturally with shortleaf, pitch, and pond pines, as well as with longleaf. The shortleaf pine also crosses naturally with the pitch pine.

Pinus palustris (*P. australis*), *the longleaf pine.* Longleaf pine was originally the dominant and most abundant tree throughout the southern coniferous forest. As white pine declined in the North, longleaf pine was "discovered" in the South, and at its peak was probably the foremost lumber-producing species of the country. Because of large size and great strength, longleaf pine was favored for construction purposes. It is also utilized for ties, interior finish, pulp, and naval stores (turpentine and rosin) production. Because of a slow-growing seedling or "grass" stage, when it is palatable to hogs, longleaf suffers in competition with other species in areas where "razorbacks" are allowed to roam. Nor do seedlings survive transplanting well. These factors, combined with heavy cutting, poor turpentining, fires, and lack of conservation planning in the past, have depleted the stands of longleaf pine.

Pinus elliottii, the slash pine. Slash pine, originally with a more restricted southeastern range than longleaf, has become increasingly important northward. It is an unusually aggressive, fast-growing species, and the equal of longleaf in the production of naval stores. The wood is the heaviest of all the southeastern yellow pines, but on the market is not distinguished from longleaf or other southeastern species. It is utilized for the same purposes as longleaf, and, the smaller trees particularly, for ties and pulp. Today more naval stores probably come from slash pine than from longleaf. Total stands seem on the increase, because of the aggressiveness of the species in colonizing deforested lands and because of widespread planting for reforestation. At one time *P. elliottii* was considered synonymous with *P. caribaea*, the Caribbean pine, which grows in the Bahamas, West Indies, and Central America.

Pinus echinata, the shortleaf pine. Shortleaf pine is the widest-ranging of the southeastern yellow pines, found in the drier, hillier areas, where other species are not so tolerant of the colder winter temperatures. The wood is slightly less strong than that of longleaf pine, but is seldom distinguished from it in marketing. Shortleaf pine is used primarily for construction purposes, but also for furniture, mine props, agricultural machinery, crates, and pulp.

Pinus taeda, the loblolly pine. Loblolly pine has a range of approximately that of longleaf, but favors moister situations. It is the most abundant species today, in

Mature grove of loblolly pine in South Carolina. (Courtesy National Park Service.)

Louisiana bald cypress, *Taxodium distichum.* (Courtesy USDA Forest Service.)

growing stock equaling the volume of all other southern pines together. The wood is similar to the preceding species, and is utilized for the same purposes as other yellow pines. A good seeder, this species may quickly reforest abandoned agricultural land.

BALD CYPRESS. The only important species north of Mexico is *Taxodium distichum*, frequent in the swamps and bayous from coastal Virginia through Florida and west to central Texas, with a northern extension of range in the Mississippi valley as far as southern Illinois. "Gum" (*Nyssa*) often occupies the same habitat, and is harvested along with bald cypress. Bald cypress wood is very resistant to decay: fallen logs lying in the swamps are still merchantable after a thousand years. The species has been utilized mostly for structural purposes, water tanks, ships, ties, shingles, coffins, and other uses for which resistance to decay is of importance. Because of its tendency to favor the less competitive swamp environment, bald cypress has been somewhat protected from cutting; but today, modern methods of logging have seriously depleted the stands.

RED CEDAR. A single arborescent species, *Juniperus virginiana*, is native throughout

Forested hummock in Everglades region of south Florida. (Courtesy National Park Service.)

most of the eastern United States. Its light requirement is high, and the species is characteristic of open, seasonally dry, limestone glades and cut-over hillsides. The wood is very durable, and in demand for posts and poles; it also finds considerable use for specialty items such as cedar chests and closet linings, lead pencils, woodenware, and novelties. Red cedar is aggressive in colonizing suitable open areas. This and the fact that the tree is too scattered, small, and gnarled to be of great commercial interest have enabled the species to hold its own.

4. Subtropical forests of the United States. Most genera and families listed in the previous chapter for Mexico and Central America will grow in southern Florida, along the Gulf Coast, and in southern Arizona and California. Except for recreational purposes, the native forest (e.g., the Everglades and Florida Keys) has never been significant in the North American economy. In a few cases, introduced exotics suitable only to subtropical climates have become important: for example, *Citrus*, of the Rutaceae (oranges, lemons, grapefruit); *Aleurites*, of the Euphorbiaceae (tung oil); *Cinnamomum* (camphor) and *Persea* (avocado), of the Lauraceae; *Phoenix*, of the Palmae (dates); and several others, including many tropical ornamentals.

Some 500 species of tropical trees are native to Puerto Rico and the Virgin Islands, including some that are typical giants of the rainforest, and others shrubby types of dry habitat and mountain summits. Puerto Rico was originally almost completely forested, but today is mostly cleared. Perhaps 10 per cent remains forested, with only about

A kapok tree in the Virgin Islands. (Courtesy National Park Service.)

1 per cent of these forests in relatively virgin condition (protected by steep slopes or difficult location). Even those lands presently in forest are mostly second-growth, no doubt with composition appreciably altered from the virgin forest. Few trees as much as 1 foot in diameter exist. Harvest is mainly for fuel, but small trees are used for posts in rural areas. Good timber species, once abundant especially in northern Puerto Rico, include maria (*Calophyllum brasiliense*), ausubo (*Manilkara bidentata*), roble (*Tabebuia heterophylla*), and tortugo amarillo (*Sideroxylon foetidissimum*). Mahogany (*Swietenia mahagoni*) was not indigenous to Puerto Rico, but has been introduced and naturalized. Hawaiian forests are also of the tropical type, but are of limited extent and much cut-over.

5. The Rocky Mountain Forests. The broad Rocky Mountain forest "belt" merges northward in Canada with the northern coniferous forest (characterized there especially by white spruce, *Picea glauca*, black spruce, *P. mariana*, and jack pine, *Pinus banksiana*), and with the Pacific forest belt. It is perhaps the least exploited of all North American forest regions, mostly because of its relative inaccessibility and remoteness from markets. A few species have attained nationwide economic significance, and undoubtedly these forests will come to be utilized more as the finer Pacific forests are depleted. In general there is stratification of species according to altitude (Fig. 4-4). Low elevations have "nut" pines, juniper, and scrub hardwoods; western yellow pine is characteristic of slightly higher altitudes; still higher, lodgepole pine and Douglas fir dominate; and at the higher elevations to timberline, fir and spruce dominate.

Inasmuch as they are confined to the mountains, the distribution of these forests is extremely unequal: for example, only a small portion of the states of Nevada, Wyoming, and Utah, but almost half of the state of Colorado, is forested. In contrast to

Figure 4-4 Coniferous trees in the area centered on eastern Washington and northern Idaho, arranged vertically to show the usual order in which the species are encountered with increasing altitude. The horizontal bars designate upper and lower limits of the species relative to the climatic gradient. That portion of a species' altitudinal range in which it can maintain a self-reproducing population in the face of intense competition is indicated by the heavy lines. (After R. Daubenmire, *Science* **151**, 295 (1966).)

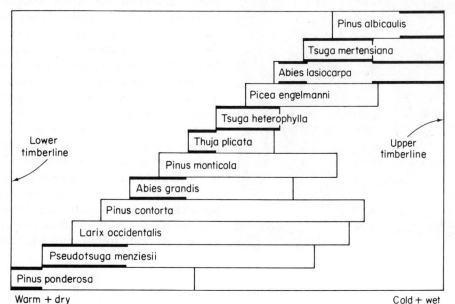

the eastern forests, the great majority of which are privately held, Rocky Mountain timber is largely in national forests.

THE PINES. The important western white pine, *Pinus monticola*, does occur in the northern Rocky Mountain forests, but it is chiefly a Pacific belt species and will be discussed under that heading. The pinyon

A virgin stand of ponderosa pine and sugar pine. (Courtesy USDA Forest Service. From *Unasylva*.)

Mature ponderosa pine. (Courtesy St. Regis Paper Co.)

pines include several species mainly found in Mexico; *P. monophylla* is prominent in California–Nevada–western Utah, and *P. edulis* important in Utah–Arizona–New Mexico–Colorado. *P. edulis* provides most of the edible pinyon nuts that come to market. These were once a staple of the southwestern Indian diet. Pinyons grow abundantly on the dry foothills where other trees have difficulty surviving, but they are gradually giving ground as death from natural causes, use for fuel and nuts, and competition from grazing and urbanization all take their toll. The trees are small, and of importance locally for posts and occasional construction as well as fuel. Of the yellow pines, two species, ponderosa and lodgepole, are important.

Pinus ponderosa, the ponderosa or western yellow pine. This species currently furnishes more lumber than any other pine. It is found in commercial quantities in every state west of the Great Plains. It is reasonably tolerant of dry conditions, but not as much so as pinyon pine; in northern Arizona, for example, pinyon pine-*Juniperus* usually dominate below about 6,000 feet, ponderosa pine at higher elevations. The trees are quite large; some have been reported with a diameter of as much as 8 feet. The wood is fairly light in weight, but is hard and strong. It is utilized chiefly for general construction, interior finish, boxes, and crates. The tree is slow-growing, but because of its wide range and often inaccessible location, has not been so seriously depleted as many other species.

Pinus contorta, the lodgepole pine. The lodgepole pine is only slightly less widespread than ponderosa, being found in the northern Rockies and scattered through most of the Pacific belt. The tree is small and frequently gnarled, and were it not for its extreme abundance would be of little commercial interest. The wood is used for ties, construction lumber, posts and poles, and fuel. The species is a prolific seeder and very aggressive; hence, like jack pine in the Northeast, it is of importance in colonizing poor or burned-over areas for watershed protection.

WESTERN LARCH. The largest and most important of all larches of the New World is the western larch, *Larix occidentalis*, found abundantly in the northern Rocky Mountains. The wood is among the heavier of the softwoods, works well, and is extremely durable. It is used for posts and poles, ties, mine timbers, interior finish, boats, boxes, and furniture. The light requirement of western larch is high: the species is usually favored where less resistant trees have been burned out, or where felling has cleared the forest, but for some reason the aggressive lodgepole pine has not taken over.

THE SPRUCES. Two well-known spruce species of the Rocky Mountain forest belt are the Engelmann spruce and the blue spruce.

Picea engelmannii, the Engelmann spruce. This species furnishes most of the annual cut of spruce in the United States. It is a feature of the high Rockies from Arizona to the Yukon. The wood is quite light in weight, lighter than white pine, but is strong for its weight. Uses include poles, doors, sash and interior trim. Engelmann spruce is tolerant, and often becomes a dominant or codominant in the climax forest.

Picea pungens, the blue spruce. Blue spruce is much more limited in range than Engelmann spruce, occurring chiefly in Colorado and to a lesser extent in adjacent Rocky Mountain states. The tree is scattered, usually at moderate elevations. The wood is of little importance except locally, but the species is deserving of mention because of its wide use for ornamental planting.

DOUGLAS FIR. The smaller, Rocky Mountain form of the Douglas fir, *Pseudotsuga taxifolia*, is one of the more important Rocky Mountain timber trees. The species will be discussed under "The Pacific forests."

THE FIRS. Alpine fir and white fir are primarily of the Rocky Mountain forest belt, although they do occur also in the Pacific belt. The grand or lowland white fir, red fir, and noble fir, although found to some extent in the northern Rocky Mountains, are more characteristic of the Pacific belt and will be discussed under that heading.

Abies lasiocarpa, the alpine fir. Alpine fir is found at high elevations from Mexico and New Mexico northwestward to Alaska and the Yukon, where it is frequently the companion of Engelmann spruce. The species has little economic importance except as watershed cover at high elevations. The wood is utilized locally for rough lumber.

Abies concolor, the white or Colorado fir. This species is perhaps the most important of all western firs. It ranges from Mexico northward through Colorado and Utah to Idaho, and through the Sierras of California and Oregon, where it attains greater size. The wood is not durable, but does possess moderate strength. White fir timber is marketed indiscriminately with other species, and is utilized chiefly for small construction, boxes, crates, interior trim, and pulp. The species is moderately tolerant, and is frequently a codominant in western forest areas. It is also planted ornamentally in the East.

6. The Pacific forests. These are the forests of the Sierras, Cascades, and Coast Ranges of California, Oregon, and Washington, extending northward through British Columbia to southern coastal Alaska. The densest and most spectacular temperate forests of the world are included within this belt, now the most heavily lumbered area of North America. Even though over half of the forests have been cut, a sufficient reserve remains to maintain the West Coast as a leading lumber center. Except locally in the drier parts of California and the coast, coniferous species dominate. The most prominent trees of the Pacific belt are: Douglas fir (the most important); redwood; western white pine; sugarpine; sitka spruce; western hemlock; lowland white, noble, and red firs; incense cedar; western red cedar; Port Orford cedar; and Alaskan cedar.

DOUGLAS FIR. Douglas fir, *Pseudotsuga menziesii*, is the premier wood of North America, constituting about half of the standing timber of the western forests and furnishing about one-fifth the annual cut. It is found throughout the Rocky Mountain belt, but attains its greatest magnificence and importance in the Pacific belt. Douglas firs, growing in a favorable Pacific environment, are among the tallest trees known (heights above 350 ft have been reported) and are sometimes as much as 15 ft in diame-

ter. Dense stands of such trees are rivaled only by the redwoods in board feet of timber per acre (cruised at more than 400,000 ft B.M.). The immense sizes and heights of Douglas fir permit production of knot-free boards of large dimensions in great quantity. Moreover, the wood is among the strongest of all American softwoods in comparison to its weight, works well, can be attractively stained, and is noted for its durability. The species is in heavy demand by the lumber and building trades, and is used for all kinds of construction (especially where long, sturdy beams are needed), for ties, poles, plywood (Douglas fir is the most-used plywood species), and interior trim. Douglas fir is not overly tolerant, but it is a prolific seeder and fast grower. Second-growth stands, protected from fire, can yield 30,000 ft B.M. in fifty years. At present the supply of large, virgin timber is rapidly approaching exhaustion, and Douglas fir can be expected to go the way of eastern white pine and southern yellow pine. Already a deficiency in the supply of first-class veneer logs has been felt. There is no reserve of comparable timber remaining once the Douglas fir is cut as there was in the case of the white and southern yellow pines. The "inexhaustible" forest reserves of America have now been worked from coast to coast.

THE REDWOODS. The redwood genus, *Sequoia*, includes the largest and perhaps oldest of living things. Most notable among them are the "Big Trees" of the California Sierras, *S. gigantea*, as much as 4,500 years old and weighing as much as 6,000 tons. These spectacular trees are, however, of no commercial importance other than as a tourist attraction. More important commercially is a second species, *S. sempervirens*, the coastal redwood, found along a narrow coastal fog belt of northern California and southern Oregon. These too are among the largest of trees, attaining heights over 370 ft and a diameter of as much as 27 ft. Stands

Old growth redwoods in the coastal forest of California. (Courtesy USDA.)

may be up to 80 per cent pure and cruise better than 400,000 ft B.M. to the acre. The wood of these gigantic trees is brittle, and in logging about 40 per cent wastage can be expected owing to breakage. The wood is soft, light, fairly strong, and easily worked. It is resistant to decay. As is the case with its eastern relative, bald cypress, fallen logs have endured for many centuries. Redwood is used for many purposes, including ties, large timbers, tanks, flumes, silos, posts, shingles, doors, general millwork, caskets, and furniture. A goodly portion of the original redwood forest has been cut, but stands of these forest giants sufficient for a few more years of logging remain. Fortunately, the species is able to regenerate from stump sprouts, and second-growth stands become marketable in as little as 50 years. Because the tree is so spectacular, much popular attention has been focused on it, with the result that efforts at conservation have met with greater than customary success and protected redwood parks established.

There has been a great deal of concern about exhaustion and replacement of the redwood forests, the ecology of which is not

Mixed coniferous forest in the Cascade mountains of Oregon being reseeded after a burn. (Courtesy Bell Helicopter Co.)

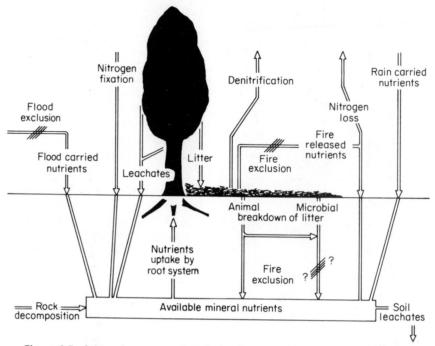

Figure 4-5 Schematic representation of mineral cycling in a stand of mature alluvial-flat redwoods and the blockage (denoted by cross-hatching) that occurs after the exclusion of flood and fire. (After Stone and Vasey, *Science* **159,** 160 (Jan. 12, 1968).)

completely understood and the exact needs for survival of mature trees not entirely clear. Competing species are the Douglas fir, tan oak (*Lithocarpus densiflorus*), grand fir (*Abies grandis*), and bay (*Umbellularia californica*). Control of fire, floods with occasional siltation, and other factors affecting survival and regeneration are matters for some disagreement. A diagramatic representation of factors affecting redwood growth is given in Fig. 4-5.

THE PINES. Two white pines, *Pinus monticola* and *P. lambertiana*, are of especial importance to the Pacific belt, in addition to the widespread yellow pine, *P. ponderosa*, already discussed as the chief pine of the Rocky Mountain belt.

Pinus monticola, the western white pine, is most common in the mountains of southern British Columbia, Washington, and eastward into northern Idaho and extreme western Montana, but also southward in the Sierras to southern California. The species is similar to eastern white pine, but its wood is slightly stronger and possesses a high commercial value. The wood is widely utilized for structural purposes, frames, moldings, and miscellaneous other products. Like its eastern relative, the western white pine is subject to attack by blister-rust.

Pinus lambertiana, the sugarpine. The sugarpine is the largest of all the pines, having been recorded up to 245 ft in height and 18 ft in diameter. It is confined to the Sierras and Coast Ranges of California and Oregon, where it grows in association with ponderosa pine, Douglas fir, and the various "cedars" and true firs. The wood is similar to that of the eastern white pine, although not quite so strong or stiff, and is used for similar purposes. The large boards it is possible to obtain from so large a tree make it especially

Coastal coniferous forest near Juneau, Alaska. (Courtesy USDA.)

useful for general construction and interior trim. The species grows rapidly in a favorable environment and is moderately tolerant, but, as are other five-needle pines, is subject to attack by blister-rust.

SPRUCE. The only important spruce of the Pacific belt is Sitka spruce, *Picea sitchensis*, occupying a narrow coastal strip from northern California to Alaska. It is one of the major timber and pulpwood species of the Northwest. The wood is utilized for boxes and crates, and for all kinds of millwork. The trees are the largest of North American spruces (maximum measurement, 300 ft tall, 16 ft in diameter) and, contrary to the usual habit of the genus, is a low-elevation (tidewater) species. In certain northern sections the species forms nearly pure stands.

HEMLOCK. A single species of hemlock is important in the Pacific belt, the western hemlock, *Tsuga heterophylla*, found principally in southeastern Alaska, coastal British Columbia, and western Washington. It is one of the four chief timber- and pulp-producing species of the Northwest. It is the favorite source of pulp for making rayon, and is much used for light construction, mill products, boxes, paper pulp, and (treated) ties. As seems common among Pacific belt

species, *T. heterophylla* is the largest North American representative of its genus, having been recorded as attaining 259 ft in height and 9 ft in diameter. The species prefers moist, cool locations, where it may occur in nearly pure stands. It is tolerant, and where abundant moisture is available seedlings may usurp much of the forest floor.

THE FIRS. Eight species of fir are found in the forests of the West; two of these have been discussed as members of the Rocky Mountain belt. An additional three species are worthy of mention as fairly important trees of the Pacific belt: the lowland white, noble, and red firs.

Abies grandis, the lowland white or grand fir. This species is found from British Columbia to middle coastal California, often at low as well as at moderate elevations. Grand fir is smaller than the two subsequent species, and is sometimes confused with white fir, *A. concolor.* The wood is of little commercial importance, but is used to some extent for pulp, slack cooperage, boxes, and rough construction.

Abies procera, the noble fir. Noble fir, a rival of red fir in size, is relatively restricted in range, occurring in the mountains of Washington and Oregon and extreme northern California. The species produces perhaps the best timber of any of the firs, and can be used for flooring, interior finish, millwork, boats, and boxes. It is prized as a Christmas tree, for which it is often cultivated. The species is naturally a tree of the mountains, but its requirements are not too exacting.

Abies magnifica, the red fir. This species is usually marketed indiscriminately with the more northern and more abundant Pacific silver fir, *A. amabilis.* Its range is confined to the high mountain slopes of Oregon and California. It is probably the largest of the American firs, with a maximum measurement of 230 ft tall by 10 ft in diameter. The wood is soft, light, and (as is true of most firs) inferior to spruce. Red fir is used largely for rough lumber, boxes, and mine timbers.

INCENSE CEDAR. A single species of incense cedar, *Libocedrus decurrens,* occurs in North America. It is confined to the mountains of Oregon and California and to lower California, where it is often a companion of sugarpine and the Big Tree (*Sequoia gigantea*). The wood is extremely durable when seasoned and is widely used for posts, ties, shingles, "cedar" chests, pencil slats, doors, and frames. Incense cedar is quite tolerant of shade, but prefers moist habitats.

ARBORVITAE. The only species of the genus *Thuja* native to the western forests is *T. plicata,* the giant arborvitae, more usually called western red cedar. It is frequent from coastal California to coastal British Columbia, and eastward through Washington to Montana and Alberta. The species is much larger than its eastern cousin, *T. occidentalis,* and is considerably more important economically. The wood is unusually durable, but soft and brittle. It is used primarily for making shingles. Other uses are for lumber, poles, pilings, boats, and exterior trim. Western red cedar is seldom found in pure stands; more commonly it is a minority species in Douglas fir or western hemlock forest. It prefers moist habitats.

OTHER "CEDARS." The Pacific forests contain two fairly large species of *Chamaecyparis,* of some commercial importance: *C. lawsoniana,* the Port Orford white cedar or Lawson cypress, and *C. nootkatensis,* the Alaska yellow cedar or yellow cypress. Both are coastal species, the former being confined to southern Oregon and northern California, the latter extending from Oregon to Alaska. Because of its resistance to acids, the durable Port Orford cedar wood has been utilized for separators in storage batteries. Millwork, boat construction, plywood, ties, timbers, and handles are more conventional uses.

SUGGESTED SUPPLEMENTARY REFERENCES

Collingwood, G. H., and Warren D. Brush, *Knowing Your Trees,* rev. Butcher, American Forestry Assoc., Washington, D. C., 1964.

Fowells, H. A. (ed.), *Silvics of Forest Trees of the United States,* Agriculture Handbook 271, Forest Service, USDA, Washington, D. C., 1965.

Harlow, W. M., and G. S. Harrar, *Textbook of Dendrology,* 5th ed., McGraw-Hill, N. Y., 1967.

Little, E. L., and F. H. Wadsworth, *Common Trees of Puerto Rico and the Virgin Islands,* Handbook 249, Forest Service, USDA, Washington, D. C., 1964.

Randall, C. E. (ed.), *Enjoying Our Trees,* American Forestry Assoc., Washington, D. C., 1969.

USDA Yearbook, *Trees,* Washington, D. C., 1950.

Univ. of Montana Center for Natural Resources, *Proceedings of Symposium: Coniferous Forests of the Northern Rocky Mountains,* Missoula, Montana, 1969.

CHAPTER 5

Wood and Its Uses

Wood, what it is. WOOD IS SO MUCH A part of man's everyday life, from toothpicks to massive furnishings and the very structures in which he dwells, that he seldom gives thought to just what constitutes this extremely common and useful substance. Everyone realizes that wood comes from trees or other woody plants and was therefore at one time a living substance. But he may not be aware of the detailed structure of wood, or of how and where it came to be formed by the plant.

With certain possible minor exceptions (e.g., virus), all living stuff is composed of cells, the stems of woody plants offering no exception. Not that all the cells of a tree trunk are "alive"—i.e., contain living protoplasm within the cell; only a minor percentage are in this state at any given instant. But at one time every cell within the trunk of the largest of trees is filled with active protoplasm, protoplasm which later disappears after having built an outer, rigid, encasing shell, the cell wall. This cell wall may be thin and weak, as it almost always is in the parent cells from which typical wood cells are derived, or it may, in age, become extremely thick and hard, making an almost solid strand of the cell. The thickness, composition, and position of the cell wall determines the qualities of the wood, for wood is the sum total of multitudes of these individual cells, each of distinctive size, shape, and strength.

As a young twig grows, there are found at its tip certain undifferentiated cells capable of repeatedly dividing. The train of cells produced or "left behind" by this meristematic (proliferating) group of cells assume particular shapes, sizes, and functions. Most of these cells soon become incapable of further dividing themselves. They are then recognizable as formed, mature cells, components of primary tissues "deposited" by the growing tip. From the outside inward toward the center of the twig, the following tissues are generally recognizable: epidermis, cortex, pericycle, phloem, cambium, xylem, and pith. As the twig continues to grow, the epidermis, cortex, pericycle, and phloem become part of the bark. The cambium, remaining meristematic—i.e., undifferentiated and capable of repeated cell division— continues alive throughout the life of the plant, producing additional (secondary) phloem toward the outside and additional (secondary) xylem in great quantities toward the inside. The primary xylem and pith remain throughout the life of the tree as an inconspicuous central core to the twig or log, the main mass of which is secondary xylem.

It is the xylem tissue, then (mostly, in quantity, secondary xylem), that constitutes wood. In the tree it consists of dead cells, except for a shallow zone toward the outside, next to the cambium. The centermost cells,

How does a tree build a trunk that can live for centuries – and hold up a weight of many tons?

All of a tree trunk's growing is done in a thin layer of living cells that surrounds the wood. This layer creates new wood on one side of itself, and new bark on the other. It thus, in effect moves outward, pushing the bark before it, leaving wood behind.

The marvelous chemistry of life tells this living layer just how many wood cells will be needed to support the leafy crown, and how much bark to build in order to protect the wood beneath it.

This complex process, infinitely repeated, has given the world its forests. To St. Regis it is of the greatest interest because trees are our basic resource. From them we derive the wood that gives us our products. We make printing papers, kraft papers and boards, fine papers, packaging products, building materials, and products for consumers.

Essentially, the life of the forest is St. Regis life. That is why—together with the other members of the forest products industry—we are vitally concerned with maintaining the beauty and usefulness of America's forests for the generations to come.

Parts of a tree trunk. Heartwood and sapwood are traditionally the valuable commercial substances. (Courtesy St. Regis Paper Co.)

The outer bark is the tree's protection from the outside world. Continually renewed from within, it helps keep out moisture in the rain, and prevents the tree from losing moisture when the air is dry. It insulates against cold and heat, and wards off insect enemies.

The inner bark, or "phloem" is the pipeline through which the food is passed to the rest of the tree. It lives for only a short time, then dies and turns to cork, to become part of the protective outer bark.

The cambium cell layer is the growing part of the trunk. It annually produces new bark and new wood, in response to hormones that pass down through the phloem with the food from the leaves. These hormones, called "auxins," have the power to stimulate growth in cells. Auxins are produced by leaf buds at the ends of branches as soon as they start growing in spring.

Sapwood is the tree's pipeline for water moving up to the leaves. Sapwood is new wood; as newer rings of sapwood are laid down on top of it, its inner cells lose their vitality and turn to heartwood.

Heartwood is the central, supporting pillar of the tree. Although dead, it will not decay or lose strength while the outer layers are intact. A composite of hollow, needlelike cellulose fibers bound together by a chemical glue called lignin, it is in many ways as strong as steel. A piece 12" long and 1" by 2" in cross section, set vertically, can support a weight of twenty tons.

which are usually darker in color and have thick cell walls often lined with tannins, gums, dyes, crystals, or the like, make up the heartwood. The sole function of heartwood is support, although at one time, of course, before the cambium laid on newer outer layers of xylem, the cells were alive or physiologically active. In a zone just outside the heartwood, usually narrower, the lighter-colored sapwood is found. Sapwood cells are "wet," for, although mostly without living protoplasm and thus not strictly "alive," they do conduct sap up the stem. Their function is both support and conduction, and as newer sapwood is formed on the outer side by the cambium a compensatory amount on the inside loses its physiological activity and becomes additional heartwood. As lumber, heartwood is usually more desirable than sapwood: for one thing it is usually drier, and hence less apt to shrink, warp, or check. In addition, it is often more resistant to decay or has been otherwise benefited by the deposition of tannins or various other chemical substances in or on the cell wall.

Xylem tissue contains several cell types. The frequency and kinds of these cells are distinctive for various species of wood, and a wood sample can usually be identified as to species by means of proper keys.* The kinds of cells contained in xylem are: tracheids, vessels, fibers, ray tracheids, and xylem parenchyma. The first three cell types run longitudinally in the stem; the rays run horizontally; and xylem parenchyma may occur either as an occasional longitudinal cell or as the chief or sole component of the rays. Longitudinal tracheids are typically linear-fusiform cells of modest diameter (perhaps 0.04 mm wide and 1 to 4 mm long),

with bordered pits usually predominating (simple pits connecting with ray parenchyma are the case in softwood); vessels are much wider, stockier, open-end cells, typically almost barrel-shape (as wide as 0.3 mm and not infrequently about 1 mm long), with many simple but sometimes bordered pits.

Fibers are much like narrow tracheids (being perhaps 0.01 mm wide and 1 to 2 mm long), with exceptionally thick walls and simple obscure pits. (The intermediate fiber-tracheids, however, possess bordered pits.) Ray tracheids of softwoods are typically elongate-rectangular and shorter than the longitudinal tracheids, being perhaps 0.03 mm wide and 1 mm long, and are usually thin-walled with bordered pits. Xylem parenchyma is typically slightly smaller than the tracheids, being much like the cambium from which it is derived, with a particularly thin cell wall. Tracheids serve both as supportive and conductive elements; vessels, chiefly for conduction; fibers, chiefly for support; ray cells, for horizontal transmission; and xylem parenchyma, for storage and, to a certain extent, transmission.

In all plant-cell types, the cell wall consists principally of two chemical substances: cellulose (including alpha, beta, and gamma forms, and hemicellulose; these are sometimes conveniently termed *holocellulose*) and lignin (a collective term including a number of similar compounds). Of these, holocellulose predominates (forming typically about 70 per cent of the dry cell wall) and is the more important economically. Lignin (constituting 20–35 per cent of the cell wall) seems to fill the spaces within the cellulose network, particularly in the middle lamella (pectinaceous core on which primary cell wall is laid down, the secondary cell wall

---

* The student who is interested in this phase of the subject may wish to examine a book containing keys for wood identification such as *Textbook of Wood Technology* by Panshin, DeZeeuw, and Brown (N.Y.: McGraw-Hill, 1964); or the *Forestry Handbook*, edited for the Society of American Foresters by Forbes and Meyer (N.Y.: Ronald Press, 1956).

Figure 5-1 Linkage of glucose units in a short section of the cellulose molecule. Dark circles represent carbon atoms, light ones oxygen. (After R. D. Preston, *Scientific American* **218**, 106 (June, 1968).)

being laid down on the primary cell wall) and the first-deposited portion of the secondary wall. Lignin may also accompany other metabolic "wastes" deposited in old xylem cell walls, the inactive heartwood.

Cellulose (Fig. 5-1) is a polymerized carbohydrate, a particular linkage of "glucose residues" $(C_6H_{10}O_5)_n$ of great molecular length. In this sense it is nature's "plastic," similar to such polymers as yield synthetic fibers like nylon, acrylics, and polyesters. Because of its abundance in nature, and consequent inexpensiveness, there has been no incentive for man to synthesize cellulose other than as an academic exercise. Through the use of electron microscopy and other techniques, much is known about cellulose, although the exact dimensions and physical chemistry of the polymers of which collectively cellulose is comprised are not always clear. They are monosaccharides, and those which are particularly important have molecular weights substantially exceeding 10,000. The main macromolecule contains several hundred beta-D-glucose units, and may contain small quantities of monomers such as mannose, galactose, xylose, and arabinose. The strands of glucose residues (macromolecules) assemble in aggregates of a higher order called micells by early researchers. Micells are in turn organized into units of a still larger order, the fibrils, which are oriented in differing patterns to form the layers of the cellulosic cell wall. Even today there is not complete agreement on the valence bonds or forces which determine the qualities of cellulose, such as its unusual mechnical strength, insolubility, and so forth. A

thorough discussion of the chemistry of cellulose can be found in such special texts as *Chemistry of Cellulose* by Ott and Spurlin, 1954.

Lignin (Fig. 5-2), the second ranking component of wood, almost half as abundant as cellulose, has long been both a puzzle and stepchild. It is generally agreed that lignins are polymers of phenylpropanoid units. They assume final form late in the process of cell wall formation, infiltrating the cellulose lattice, where they are chemically bonded to the cellulose and serve for stiffening of the cell wall. Lignin decomposes less readily than cellulose, and has traditionally been dismissed as the humuslike residue left after wood decays. There are several types of lignin, and the exact chemical structure of all of them is far from certain. Considering the amount of wood used for such purposes as paper pulp (where lignin is a leftover waste), it is apparent that great rewards will await anyone discovering a worthwhile use for lignin. There have been some attempts to alter it chemically to yield chemical intermediates similar to vanillin (certain intermediates such as cinnamic acid are important constituents of essential oils), and it has been recovered from pulping operations as an inexpensive binder for soil (especially in road construction). It would have some value as fertilizer except that the cost of processing and handling makes it quite noncompetitive on a plant nutrient basis. As a residue from pulping operations it is gummy and foul-smelling, and constitutes more of a problem in disposal than an asset.

Wood of softwoods is typically simpler

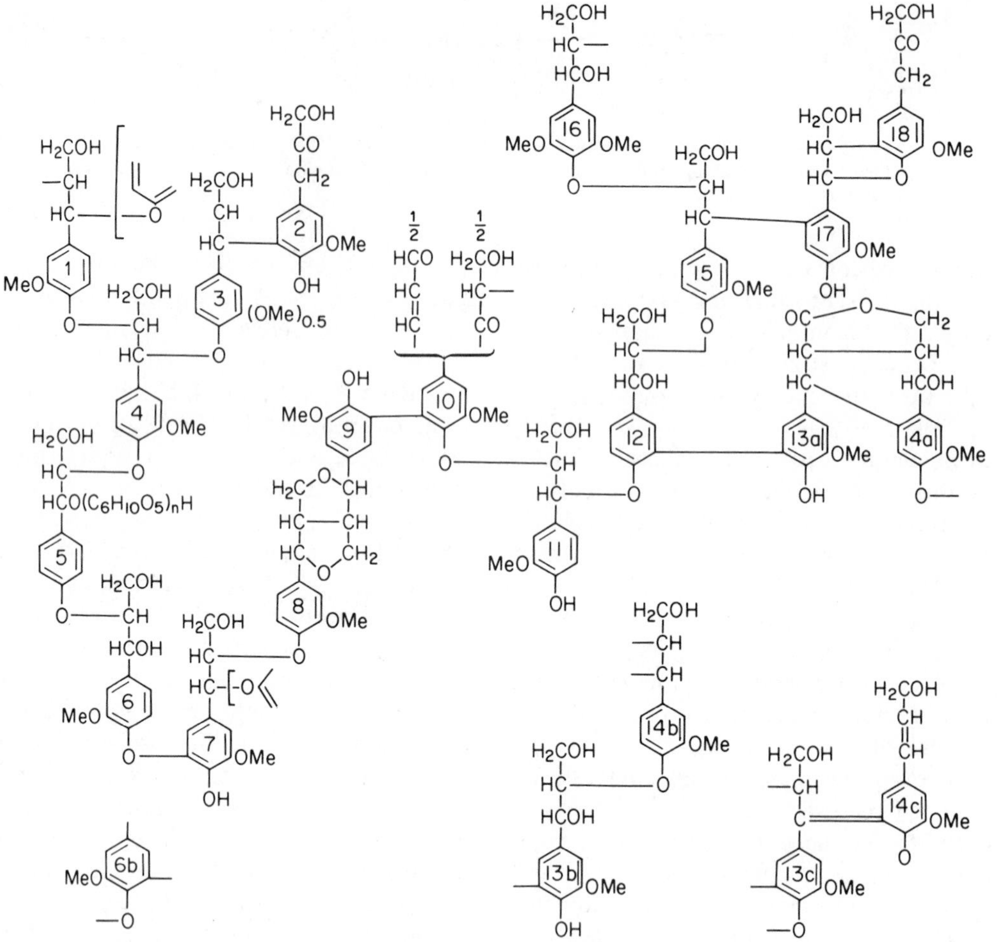

Figure 5-2 Suggested model for chemical structure of a spruce lignin molecule. (After Freudenberg, *Science* **148**, 599 (April 30, 1965).)

in structure than that of hardwoods (see Figs. 5-3 and 5-4). Most (about 90 per cent) of its volume consists of longitudinal tracheids, these being aligned in nearly straight rows with a minimum of distortion and diversity. Rays are rather thin (for the most part uniseriate—i.e., only one row of cells wide) and uniform. Resin canals may or may not be present, but in no case are true fibers or vessels ever found. Lack of the large, conspicuous vessels causes softwoods to be termed "nonporous." On the other hand, "porous" Angiospermous wood is charac-

terized by presence of vessels (these comprising some 55 per cent of the wood volume in sweetgum, for example), and fibers (fiber-tracheids comprising some 26 per cent of the wood volume in sweetgum). In hardwoods companion cells occur in the phloem (new bark); they do not so occur in softwoods. Hardwoods also usually possess wider multiseriate rays (comprising from 5 to 33 per cent of the wood volume in sweetgum); show distortion of the radial alignment of the cells, owing mostly to vessel development; and characteristically lack resin canals. So

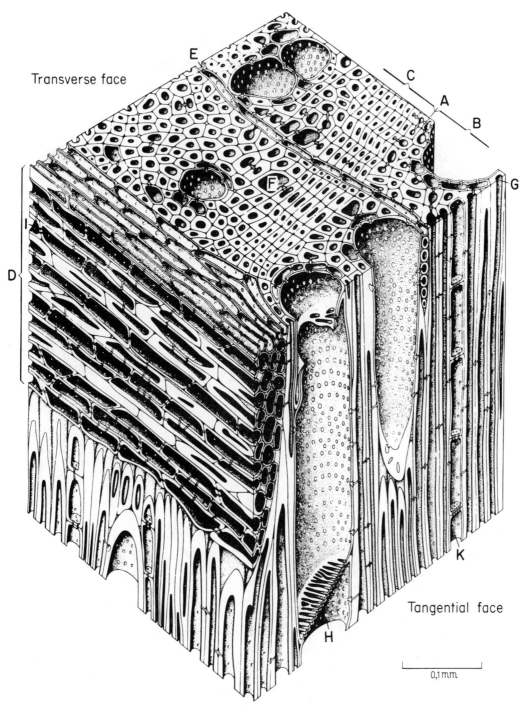

Figure 5-3 Diagram of a block of beechwood, a hardwood. A, boundary between annual rings; B, springwood; C, summerwood; D, ray (uniseriate); E, ray (multiseriate); F, tracheid; G, fiber; H, vessel; I, ray parenchyma cell; K, longitudinal parenchyma cell.

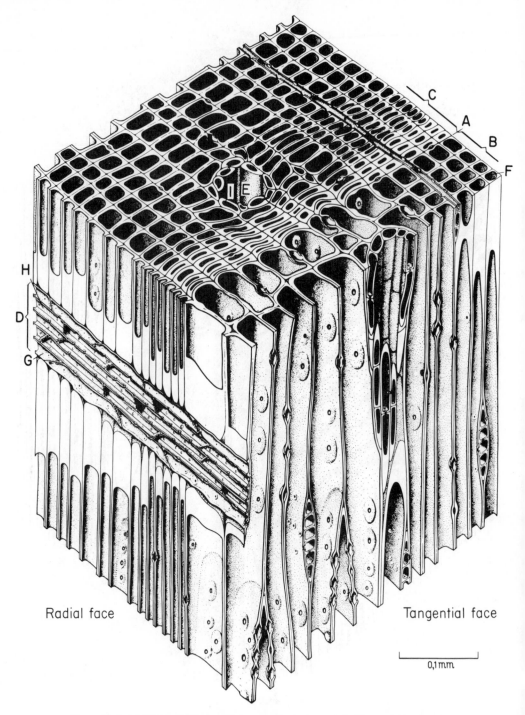

Radial face   Tangential face

0,1 mm

Figure 5-4 Diagram of a block of sprucewood, a softwood. A, boundary between annual rings; B, springwood; C, summerwood; D, ray; E, resin canal; F, longitudinal tracheid; G, transverse tracheid (or ray tracheid cell); H, ray parenchyma cell; I, epithelial cell.

different is the typical cellular arrangement of hardwood timber from that of softwood that one can readily learn to distinguish them with only brief experience.

As any tree grows under conditions of seasonal climate, a difference in cell size, cell type, and thickness of cell wall is observable in the wood formed during the year. In north temperate regions, active tree growth is during spring and summer, with a dormant or semidormant period during winter. In spring, with the physiological need for rapid conduction through the xylem, large, comparatively thin-walled cells are developed. Gradually, as the season progresses, the most recently formed cells become smaller and denser, until the growing season terminates with a layer of xylem of comparatively small and dense cells, which contrasts sharply with the next spring's large, thin-walled cells. The whole year's (growing season's) addition of xylem is called an annual ring. The uniformity of annual rings, their width, and the proportion of spring and summer wood they contain are important factors affecting strength and figure in lumber. In softwoods very rapid growth tends to increase the lighter springwood proportionately more than the heavier summerwood, and may affect the quality of the timber (strength being at least partly dependent upon the thickness of cell walls). In hardwoods, particularly the ring-porous types, the reverse is true. Favorable growth alters the amount of springwood laid down very little, but proportionally increases the amount of summerwood, thus producing stronger, denser timber. Hence hardwoods growing under favorable conditions usually produce better-quality lumber than those less ideally situated.

Hardwoods may be either ring-porous or diffuse-porous; occasionally they are intermediate in this respect. "Ring-porous" refers

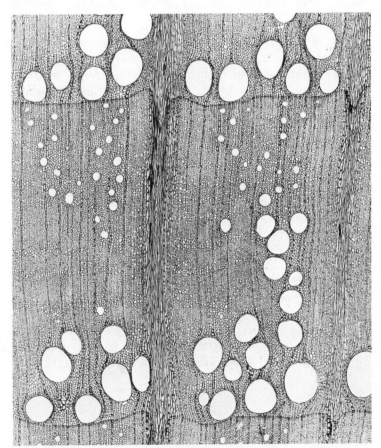

Figure 5-5 Transverse section of red oak, *Quercus rubra,* a ringporous wood. (Courtesy USDA Forest Service.)

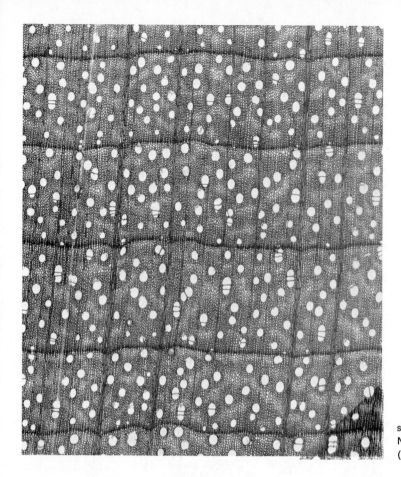

to wood in which large vessels are conspicuous in one portion (springwood) but absent in the rest of an annual ring (see Fig. 5-5). Most oaks, hickories, walnuts, elms, and ashes are of this type. Diffuse-porous woods, on the contrary, have generally smaller and nearly uniform vessels distributed throughout the annual ring (see Fig. 5-6). Maple, basswood, tuliptree, birch, and many cherries fall in this category. The appearance of finished lumber reflects, to a considerable extent, the type and arrangement of the vessels in the wood.

The appearance of finished lumber is also a question of the way in which a log has been cut. Three types of cut or surface are possible. Most obvious is the transverse or cross-section: such a cut is made in the felling of a tree or the sawing into sections of a large timber. In a transverse surface (Figs. 5-3, 5-4, 5-5, and 5-6) all longitudinal cells of the log (i.e., most tracheids, all vessels and fibers) are cut crosswise or transversely, whereas the rays are cut longitudinally. Annual rings are conspicuous as concentric circles. A second type of cut is that made longitudinally through the center (pith) of the log (Figs. 5-3 and 5-4), with the plane of the cut following a radius as seen in tranverse section. Such a surface would be obtained if one split or sawed a log directly through the center; it is termed a radial cut or radial section. In radial section, all cells, both longitudinal and ray, are cut longitudinally; none are cut transversely. Annual rings appear as parallel lines crossed by the wood rays. The third and most common type of surface is the tangential surface (Figs. 5-3 and 5-4). It is formed when a log is cut longitudinally but *not* through the center— i.e., not along a radius. All longitudinal cells are cut longitudinally, but ray cells are

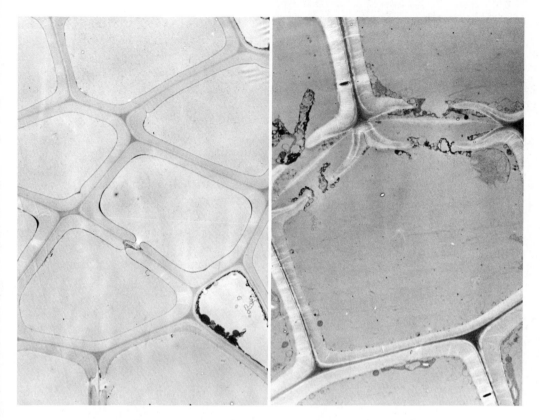

Left: Cross section of white pine tracheids (*Pinus strobus,* magnification 3060×).
Right: Cross section of vessel cells (*Tilia americana,* magnification 17,000×).
Note that the cells are joined by an intercellular substance or middle lamella; on the inner
side of the middle lamella is the primary wall, followed by the secondary wall next to the cell
cavity. (Courtesy Forest Products Laboratory, USDA Forest Service.)

cut transversely. Annual rings appear as subparallel but often wavy bands, interspersed with compact groups of ray cells which are seen in end view. The significance of the type of surface exposed will become apparent when veneers and ornamental woods are discussed.

The qualities of wood. Availability and inexpensiveness cannot alone account for the popularity of wood. It is frequently procured from the far corners of the earth by the richest of men and cut with precision tools costing thousands of dollars. For, in addition to being abundant, usable, and cheap, wood has an aesthetic attraction, its "warmth" and beauty being equaled by few decorative or constructive materials. No metal furnishings or stuccoed walls can compare with the walnut suite or oakpaneled room.

Then too, wood has a purely practical appeal. It is a relatively poor conductor of heat and can be useful as an insulator, holding heat in a house while keeping out the cold much more satisfactorily than a metal surface would. Nor does wood conduct an electric current. Moreover, it can be used over and over again, being easily fastened and unfastened, sawed and reshaped. Properly

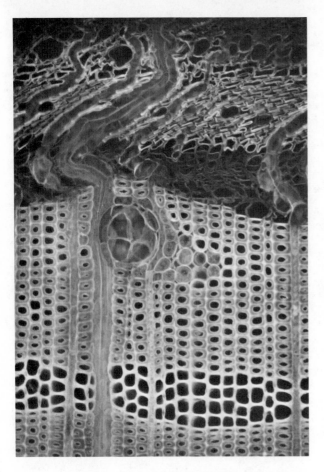

Cross section of southern yellow pine wood showing one full ring and some bark. Distorted section at top is the bark. Photographed at 220 diameters and further magnified by photographic enlargement. (Courtesy Forest Products Laboratory, USDA Forest Service.)

DENSITY. Density of wood refers to its comparative weight—technically its weight per unit volume. This weight is usually compared to that of an equal volume of water, and is stated as a fraction of unity (water weight), the *specific gravity* of the wood. Most woods have a specific gravity of less than unity, 1.0, owing to the considerable air space contained within the cells. Such woods are consequently lighter than water and able to float. The range of specific gravities of known woods varies from 0.04 to 1.40. Woods with specific gravities of less than 0.50 may be considered light; specific gravities of 0.50 to 0.70 indicate moderately light to moderately heavy wood; those above 0.70 indicate a heavy wood. The lightest important commercial wood is balsa, *Ochroma*, with a specific gravity of about 0.12. The heaviest are perhaps South American ironwood, (*Krugiodendron ferreum*), specific gravity 1.3, and lignum-vitae, *Guaiacum*, and quebracho, *Schinopsis*, with a specific gravity of about 1.25. Table 5-1 shows the comparative densities of the commoner or representative North American woods.

Actual wood substance—i.e., the material constituting the cell walls—is denser than water. Its specific gravity is about 1.53. If a block of wood were so completely crushed as to contain no air whatsoever, it would immediately sink. Everyone is familiar with the tendency of waterlogged wood—wood into which water has infiltrated to fill some or all of the pore space, replacing the more buoyant air—to sink or float low.

Density depends mostly upon the weight and quantity of cell wall material, since nearly all pore space is occupied by "weightless" air and the amount of "held" water (water "absorbed" in the cell wall) is low. We have seen that all cell walls are made up of essentially the same materials, and hence have similar weights. Consequently, it is the quantity of the cell wall that is chiefly re-

protected, it is enduring; it will not rust or crystallize. Wood is comparatively strong, seldom becoming shattered or scarred beyond repair by normal household use. It can be smoothed and sanded, and repainted many times over; scars and dents can be readily removed.

Wood of certain species is preferred to that of others for certain uses because the cellular composition is particularly suited to those uses. Wood samples can be readily tested, and comparative ratings have been formulated on a number of scores. Let us examine briefly in what ways woods qualify in certain important physical categories.

Table 5-1

Density of North American Woods

Name	Weight per Cubic Foot Green	Air Dry	Specific Gravity
SOFTWOODS:			
Cedar, incense (*Libocedrus decurrens*)	45	26	0.35
Cedar, Port Orford (*Chamaecyparis lawsoniana*)	36	29	0.40
Cedar, eastern red (*Juniperus virginiana*)	37	33	0.44
Cedar, western red (*Thuja plicata*)	27	23	0.31
Cedar, northern white (*Thuja occidentalis*)	28	22	0.29
Cypress, bald (*Taxodium distichum*)	50	32	0.42
Douglas fir (*Pseudotsuga menziesii*)	38	34	0.45
Fir, balsam (*Abies balsamea*)	45	26	0.34
Fir, lowland white (*Abies grandis*)	44	28	0.37
Hemlock, eastern (*Tsuga canadensis*)	50	28	0.38
Hemlock, western (*Tsuga heterophylla*)	41	29	0.38
Larch, western (*Larix occidentalis*)	48	36	0.48
Pine, jack (*Pinus banksiana*)	50	30	0.39
Pine, lodgepole (*Pinus contorta*)	39	29	0.38
Pine, longleaf (*Pinus palustris*)	50	41	0.55
Pine, northern white (*Pinus strobus*)	36	25	0.34
Pine, shortleaf (*Pinus echinata*)	51	38	0.49
Pine, slash (*Pinus elliottii*)	56	48	0.64
Pine, sugar (*Pinus lambertiana*)	51	25	0.35
Pine, western white (*Pinus monticola*)	35	27	0.36
Pine, western yellow (*Pinus ponderosa*)	45	28	0.38
Redwood (*Sequoia sempervirens*)	55	30	0.41
Spruce, black (*Picea mariana*)	32	28	0.38
Spruce, Engelmann (*Picea engelmanni*)	39	23	0.31
Spruce, white (*Picea glauca*)	35	28	0.37
Tamarack (*Larix laricina*)	47	37	0.49
HARDWOODS:			
Ash, white (*Fraxinus americana*)	48	42	0.55
Aspen (*Populus tremuloides*)	43	27	0.35
Basswood (*Tilia glabra*)	41	26	0.32
Beech (*Fagus grandifolia*)	54	45	0.56
Birch, paper (*Betula papyrifera*)	50	39	0.48
Birch, yellow (*Betula lutea*)	57	43	0.55
Butternut (*Juglans cinerea*)	46	27	0.36
Catalpa (*Catalpa speciosa*)	41	29	0.38
Cherry, black (*Prunus serotina*)	46	35	0.47
Chestnut (*Castanea dentata*)	55	30	0.40
Corkwood (*Leitneria floridana*)	..	..	0.21
Cottonwood (*Populus deltoides*)	49	28	0.37
Elm, American (*Ulmus americana*)	54	36	0.46
Gum, black (*Nyssa sylvatica*)	45	35	0.46
Gum, red (*Liquidambar styraciflua*)	50	34	0.44
Hackberry (*Celtis occidentalis*)	50	37	0.49
Hickory, bitternut (*Carya cordiformis*)	63	46	0.60
Hickory, shagbark (*Carya ovata*)	64	51	0.64
Ironwood, black (*Rhamnidium ferreum*)	86	80	1.04
Locust, black (*Robinia pseudacacia*)	58	48	0.66

Table 5-1   cont'd

Density of North American Woods

Name	Weight per Cubic Foot		Specific Gravity
	Green	Air Dry	
Locust, honey (*Gleditsia triacanthos*)	61	44	0.60
Maple, silver (*Acer saccharinum*)	45	33	0.44
Maple, sugar (*Acer saccharum*)	56	44	0.57
Oak, red (*Quercus rubra*)	63	44	0.56
Oak, white (*Quercus alba*)	62	48	0.60
Osage orange (*Maclura pomifera*)	62	52	0.76
Persimmon (*Diospyros virginiana*)	63	52	0.64
Sassafras (*Sassafras albidum*)	44	32	0.42
Sycamore (*Platanus occidentalis*)	52	35	0.46
Tuliptree (*Liriodendron tulipifera*)	38	28	0.38
Walnut, black (*Juglans nigra*)	58	39	0.51
Willow, black (*Salix nigra*)	50	26	0.34

Figures are taken largely from *Technical Note No.* **218** of the Forest Products Laboratory, Madison, Wisconsin.

sponsible for density. Wood with a preponderance of thick-walled cells, or cells in which there is little pore space, will be denser than wood with many large vessels or with much parenchyma; locust or oak is considerably denser than fir or basswood.

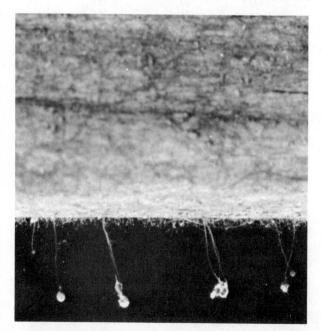

Wood-destroying fungi, *Poria incrassata* and *Merulius*, showing fruiting bodies on southern white pine. The narrow, irregular lines and spider-web-like growth, usually dark brown or black, accompany the activities of decay-producing fungi. (Courtesy Dow Chemical Co.)

Since density thus reflects to a large extent the collective thickness of the cell walls, it is often an indicator of strength in wood, since strength is likewise usually a reflection of stoutness in the cell walls.

DURABILITY. Natural durability is a reflection of wood's ability to withstand the attacks of decay organisms and, to a lesser extent, certain kinds of insects. Decay organisms (almost exclusively various fungi) derive their sustenance by dissolving wood substance through secretion of hydrolyzing enzymes. Typically they can survive well only in the presence of relatively abundant moisture, and drying wood to below 15 per cent moisture content is ordinarily sufficient to inhibit decay. On the other hand, completely wet (immersed, waterlogged) wood is not readily subject to decay, since most decay organisms also need air. Decay is more rapid in warm climates or seasons than in cold. Ideal conditions for decay are present in situations in which wood is in contact with moist soil or is subject to frequent showers; hence the problem is particularly serious for posts, ties, mine timbers, and tropical construction. The

Table 5-2

Comparative Natural Durability of Some North American Woods

SOFTWOODS

Cedar, eastern red	150–200	Pine, Norway	45–60
Cedar, southern white	80–100	Pine, pitch and sugar	45–55
Cedar, other species	125–175	Pine, western white	65–80
Cypress, bald	125–175	Pine, white	70–90
Douglas fir, dense	75–100	Pine, western yellow, pond, loblolly,	
Douglas fir, average	75–85	lodgepole	35–50
Firs	25–35	Redwood	125–175
Hemlock	35–55	Spruces, various	35–50
Larch, western	75–85	Tamarack	75–85
Pine, jack	35–45	Yew, Pacific (western)	170
Pine, southern yellow	60–100		

HARDWOODS

Ash	40–55	Hickory	40–55
Aspen	25–35	Locust, black	150–250
Basswood	30–40	Locust, honey	80–100
Beech	40–50	Magnolia, evergreen	40–50
Birch	35–50	Maple	40–50
Butternut	50–70	Mulberry, red	150–200
Catalpa	125–175	Oaks, red oak group	40–55
Chestnut	100–120	Oaks, white oak group	100
Cottonwood	30–40	Oak, chestnut	70–90
Elder	25–35	Osage orange	200–300
Elm, cork and slippery	65–75	Poplar, yellow (tuliptree)	40–50
Elm, white	50–70	Sycamore	35–45
Gum, black and tupelo	30–50	Walnut, black	100–120
Gum, red (sweetgum)	65–75	Willow	30–40

White oak = 100. Scores above 100 indicate greater durability than white oak, those below 100 lesser durability. Table adapted from *Engineering News-Record*, September 28, 1922.

Remarkable durability of wooden lintels of the sapote, *Achras sapota*, in the ruins of Uxmal, Yucatan. These Maya ruins are undated but the estimated age is about 800–1000 years. (Courtesy Paul H. Allen.)

comparative natural durability of some common North American timbers is given in Table 5-2. Woods that may be considered durable are: catalpa, almost all cedars, chestnut, bald cypress, juniper, black locust, red mulberry, Osage orange, redwood, sassafras, and black walnut. Woods notably low in resistance to decay are: aspen and cottonwood, basswood, fir, and willow. Intermediate are such species as: Douglas fir, eastern white pine, sweetgum, western larch, many oaks, southern yellow pine, and tamarack. Of high-intermediate status are: honey locust and white oak; of low-intermediate status are: ash, beech, birch, hemlock, maple, red oaks, and spruce.

The degree of "palatability" of the xylem cell wall to decay organisms will largely determine the degree of natural durability of wood. Other factors, such as the imperviousness of the wood (for example, when cell cavities are blocked by tyloses), may have a lesser influence. If the cell wall contains certain chemical compounds toxic or depressive to decay organisms, that wood will be comparatively resistant to decay. One of the most frequent as well as effective natural preservatives in wood is tannin. Logs with an abundant tannin content— say, 15 per cent—may lie on the forest floor for prolonged periods and yet remain sound: witness blight-killed chestnut logs decades old. Typically, heartwood is more durable than sapwood, both because heartwood is usually less moist and because various waste materials of the plant have been deposited in and on the cell walls. Depositions are frequently toxic to decay organisms or unpalatable to insect forms and thus are quite effective as natural preservatives. Since the nature of the cell wall determines durability, it is obvious that season of cutting, method of seasoning, utilization of dead or living trees, or even utilization of green or seasoned timbers has little or no effect upon the durability of the wood.

GRAIN AND FIGURE. To the layman, grain and figure in wood are more or less synonymous. Precisely, however, *grain* refers to the position or arrangement of the cells, whereas *figure* refers only to the design or pattern presented by the grain. The word *grain* is also used, with some distortion, to refer to porosity: for example, when a painter terms wood coarse-grained, meaning that it has large cell cavities which absorb a good deal of paint. Obviously, the grain of wood depends first of all upon the species of tree from which the wood was taken and secondly upon the manner in which the individual tree grew. Figure is, of course, primarily dependent upon grain, but may be highly influenced by unusual growth increments (e.g., burls), accidents or injury, and uneven infiltration of coloring materials in the secondary cell walls. We have also seen how plane of sawing drastically influences the figure, *plain-sawing* cutting across the rays (tangential surface), and *quarter-sawing* cutting all cells longitudinally (radial surface). Variations from plain- to quarter-sawn figure occur, even in a single board, and some veneers are cut to approach a transverse surface (cone cut).

In some hardwoods (often tropical species), unusual formations of grain are occasionally found. *Interlocked grain* (causing *ribbon* or *stripe figure* in quarter-sawn boards) occurs where, for some unaccountable reason, the spiral of the longitudinal cells reverses direction at repeated intervals in the newer annual rings. Even where basic slope of grain is maintained in all xylem, secondary undulations in cell arrangement within the annual ring can cause the *wavy* and more pronounced *curly* or *fiddleback* figure, particularly common in maple. If similar undulations occur radially, causing corrugation of the annual ring, *blister* or *quilted figure* results. Conical indentations in the growth increment, of unexplained origin, particularly in maple but also in birch and ash, produce *bird's-eye figure* when the log is cut tangentially. Unusual twisting of xylem cells at crotches, burls, buttresses, and stump swells may produce locally highly ornamental but unpredictable figures termed *crotch, swirl,* and so on. Similarly, the presence of knots (bases of old branches imbedded in newer xylem) influences figure (knotty pine); knots, however, are commonly considered defects. Most other deviations from normal in the grain not listed above, such as compression wood (abnormal xylem formed on lower side of branches and leaning trunks), brashness (weak cell walls, due to thinness, poor orientation of fibrils, and the like), shakes and frost cracks (rupture of cells in standing tree), pitch and bark pockets or streaks in wood, and various discolorations or stains are likewise considered defects.

HARDNESS. Hardness is again a reflection in part of the sturdiness of the cell walls, owing both to qualitative and quantitative factors, and consequently shows a marked positive correlation with density. Hardness tests with precise mechanical instruments under controlled conditions are possible, but approximate hardness, as tested with thumbnail or penknife, is usually sufficient for practical purposes. Light woods like balsa, basswood, buckeye, poplar, willow, fir, and white pine can be readily dented, whereas hickory, locust, Osage orange, persimmon, and oak cannot. Table 5-3 shows the comparative ranking of common North American timbers under five categories of hardness.

LUSTER, COLOR, AND POLISH. Luster, color, and polish are of comparatively minor importance in wood, except in logs that are to become cabinet wood, furniture, or interior trim. Luster depends upon the ability of the cell wall faces to reflect light, which in turn depends upon cell wall structure and composition (particularly infiltration materials—oily or waxy coatings usually inhibit luster), plane of cut (radial surface is usually more lustrous because ray flecks reflect light), and angle of incidence of light. Most woods

Table 5-3

Comparative Hardness of Some North American Woods

Soft	Moderately Soft	Moderately Hard	Hard	Very Hard
SOFTWOODS	SOFTWOODS	SOFTWOODS	SOFTWOODS	SOFTWOODS
fir, balsam	cedar, incense	bald cypress	southern yellow	none
spruce, various	cedar, Port Orford	Douglas fir	pine,	
	cedar, western red	hemlock	some Juniper	HARDWOODS
	firs, western	southern yellow pine		hickory
HARDWOODS	pines, lodgepole,	tamarack and larch	HARDWOODS	locust, black
basswood	ponderosa, red,		ash	locust, honey
buckeye	sugar, white		beech	Osage orange
"poplars"	redwood	HARDWOODS	holly	persimmon
	spruce, several	birch	hard maple	
		cherry	mulberry	
		chestnut	red oaks	
	HARDWOODS	elm	white oaks	
	butternut	blackgum	black walnut	
	catalpa	hackberry		
	tuliptree	magnolia		
	willow	soft maple		
		sassafras		
		sweetgum		
		sycamore		

are of intermediate luster, but in comparison certain kinds have definitely more sheen than others. For example, eastern spruce, white ash, catalpa, hard maple, and sweetgum are conspicuously more lustrous than bald cypress, white pine, elm, sassafras, and tuliptree. Color in wood is an extremely variable factor, even within a species. Notable differences in color are usually found between heartwood and sapwood, heartwood ordinarily becoming darker as a result of the deposit of infiltration products in and on the secondary cell wall. Most distinctive perhaps are the chocolate color of black walnut heartwood, the purplish or red heartwood of red cedar (*Juniperus*), the yellow of Osage orange and mulberry, and, among the exotics, the black of ebony or the wine red of mahogany. The ability of wood to take a polish—i.e., permit the imparting of a smooth, glossy surface to the faces of exposed cell walls—frequently determines its usefulness as a cabinet wood. Some North American woods known to polish well are, among the softwoods, incense cedar, Port Orford cedar, red cedar (*Juniperus*), white pine; and, among the hardwoods, beech, yellow birch, cherry, hackberry, holly, some oaks, maple, persimmon, sweetgum, walnut.

MOISTURE CONTENT. The moisture content of the green heartwood of temperate North American trees varies considerably, ranging from as much as 70 per cent to as little as 7 per cent of green weight. Of course, sapwood is typically more moist than heartwood (especially in softwoods), and wood from the base of a tree may be somewhat more moist than that higher up. Woods noted for a comparatively high moisture content in green wood are, among the softwoods, sugarpine, redwood, eastern hemlock, balsam fir, incense cedar, jack pine, bald cypress; and, among the hardwoods, willow, chestnut, buckeye, poplars, butternut, basswood, elm, catalpa.

Generally, other factors being equal,

woods of low moisture content would be preferred to those of high moisture content, in that on the whole less work would be required in seasoning them. But moisture content of green wood is perhaps no more significant than is the tendency of the wood to hold or reabsorb moisture (white oak and gum, for example, are notably difficult to dry out). Water contained in the cell cavities (pores) is first to evaporate. Woods with thin-walled cells and much pore space ordinarily have a higher free-water content in comparison to wood substance than heavier, harder woods, and are more apt to become "soaked" if left outdoors unprotected—as firewood often is. The imbibed or bound water of the cell wall may be later evaporated; under normal outdoor atmospheric conditions it becomes stabilized at about 12 per cent (of dry weight). Wood in which the moisture content of the cell wall is low is stronger than wood in which the moisture content is high.

POROSITY. Porosity, the quantity of pore space of wood, is a reflection of the size and abundance of the cell cavities (lumen) in the wood. We have already seen how the word *porous* has been used in a restricted sense, with reference to vessel cells only, to define nonporous, ring-porous, and diffuse-porous woods. Ordinarily, wood with thin cell walls and large cells is notably porous. There is thus, usually, an inverse correlation between heaviness and porousness. In some woods, such as white oak, the large cells, though in fair quantity, are blocked or obstructed by balloonlike growths, the tyloses. Such woods, though apparently porous in texture, are in actuality nearly impermeable, and are prized for barrels to hold liquids. All woods are more porous in transverse than in radial or tangential surface, in that a greater quantity of cells are cut across the lumen, thus permitting comparatively ready passage to gases or liquids. Porosity is of particular concern in the

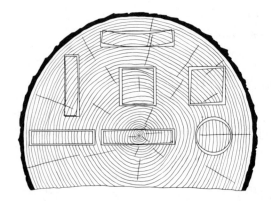

Diagram of different cuts of same log showing shrinkage or distortion as affected by direction of annual rings. (Courtesy USDA Forest Service.)

painting and preservative treatment of wood.

SHRINKAGE AND SEASONING. Wood as it comes from the forest contains appreciable moisture, often as much as the dry weight of the wood itself. Much of this water is loosely held in the xylem cell cavities and is considered "free" water in that it is held by capillary and similar forces which do not require much energy to be overcome. In addition the cell wall usually contains 25–30 per cent of "bound" water which must be reduced more difficultly by drying if the wood is to be practically useful.

Shrinkage in wood occurs as the moisture content of the cell wall decreases. There is no shrinkage so long as only the free capillary water—that of the cell cavities or pore space —evaporates. The point at which no free water remains, but the imbibed water in the cell wall is at its maximum, is called the fiber saturation point. Loss of water beyond the fiber saturation point, up to the oven-dry condition, may result in shrinkage of as much as 20 per cent (of log volume). This shrinkage is chiefly across the grain (at right angles to the long axis of the majority of cells), and about twice as marked tangentially as radially (primarily because of the strengthening rays in a radial plane.)

Most lumber, especially of the more valuable hardwoods, is dried or seasoned to reduce the moisture content found in newly felled trees. This saves shipping (weight) costs, prevents later shrinkage during use, increases strength, inhibits stain fungi and early decay, and favors application of glues, paints, or preservatives. Such seasoning may be accomplished in the open air, in drying sheds (usually with air circulation by fan), or in kilns. Controlled air humidity, rapid circulation of the air, and high temperatures favor rapid evaporation of water from the wood. The proper regulation of this evaporation is critical in helping to prevent checking, warping, twisting, case-hardening, or honeycombing. If evaporation from one surface is favored (a transverse face dries most rapidly, and should be shaded if possible in open-air seasoning), or if outer layers season rapidly while inner layers retain most of their moisture, stresses and strains will occur that may produce defects.

In open-air seasoning, distortion can be largely prevented by proper piling (both to hold the planks firm and to encourage good circulation of air) and by assuring comparable exposure of (and hence comparable evaporation from) all surfaces. In kiln drying, the same factors apply, but humidity, circulation, and temperature can be rigidly controlled. A drying kiln properly controlled can safely and quickly season wood to optimum moisture content. A moisture content of about 7 per cent for hardwood cabinet woods, 12–15 per cent for softwood construction lumber, is desirable. There is no reason to reduce moisture content below such a figure, for wood will soon reabsorb from the air sufficient moisture to reach this equilibrium, expanding somewhat in the process. Most defects in seasoning occur because of too rapid drying. Most kilns operate in a temperature range of 110–180°F, but in special cases go as high as 230°F. After kiln drying wood may be conditioned

WOOD AND ITS USES

Table 5-4

Shrinkage of Some North American Woods

Species	Volumetric	Radial	Tangential	Ratio—Tangential to Radial
		Percentage Shrinkage		
SOFTWOODS:				
Cedar, western red (*Thuja plicata*)	7.7	2.4	5.0	2.08
Cypress, bald (*Taxodium distichum*)	10.5	3.8	6.2	1.63
Douglas fir (*Pseudotsuga taxifolia*)	11.8	5.0	7.8	1.56
Pine, longleaf (*Pinus palustris*)	12.3	5.3	7.5	1.42
Pine, northern white (*P. strobus*)	8.2	2.3	6.0	2.61
Pine, shortleaf (*P. echinata*)	12.6	5.1	8.2	1.61
Pine, western white (*P. monticola*)	11.8	4.1	7.4	1.81
Redwood (*Sequoia sempervirens*)	6.8	2.6	4.4	1.56
Spruce, Sitka (*Picea sitchensis*)	11.5	4.3	7.5	1.74
HARDWOODS:				
Ash, white (*Fraxinus americana*)	13.3	4.9	7.9	1.61
Basswood (*Tilia glabra*)	15.8	6.6	9.3	1.41
Beech (*Fagus grandifolia*)	16.3	5.1	11.0	2.16
Birch, yellow (*Betula lutea*)	16.7	7.2	9.2	1.28
Elm, American (*Ulmus americana*)	14.6	4.2	9.5	2.26
Gum, sweet (*Liquidambar styraciflua*)	15.0	5.2	9.9	1.90
Hickory, bigleaf shagbark (*Carya laciniosa*)	19.2	7.6	12.6	1.66
Maple, sugar (*Acer saccharum*)	14.9	4.9	9.5	1.94
Oak, red (*Quercus rubra*)	13.5	4.0	8.2	2.05
Oak, white (*Quercus alba*)	15.8	5.3	9.0	1.70
Tuliptree (*Liriodendron tulipifera*)	12.3	4.0	7.1	1.78
Walnut, black (*Juglans nigra*)	11.3	5.2	7.1	1.37

From J. S. Mathewson, *The Seasoning of Wood*, Forest Products Laboratory, Forest Service, U. S. Department of Agriculture, Madison, Wisc. Table III.

in heated rooms to assure uniform moisture content throughout the piece, and increasingly wood is packaged in waterproofed paper for shipment to market on open flat cars. Table 5-4 lists shrinkage values, from green to oven-dried condition, for a number of common North American woods.

STIFFNESS. Stiffness refers to the ability of wood cells collectively to withstand bending or distortion. It is thus a reflection of both the quality and quantity of xylem cell wall substance, and of the binding surface between cells. Woods classed as very stiff include black locust and Osage orange; those classed as stiff include, among the softwoods, Port Orford cedar, northern white cedar

(brittle), hemlock (brittle), southern yellow and red pines, and tamarack (brittle), and, among the hardwoods, ash, hard maple, red oak, and black walnut; classed as moderately stiff are the softwoods bald cypress, Douglas fir, redwood, and some spruces, and the hardwoods magnolia, soft maple, sycamore, and tuliptree.

STRENGTH. As applied to wood, strength is a very general term. For precise comparison, the word must be qualified to indicate the kind of strength in question. Thus we have *crushing strength* (resistance to endwise compression or crushing of the cells); *tensile strength* (ability to withstand tension or pulling apart: force is applied in the opposite direction to that applied in testing crushing strength; the tensile strength of wood is greater than its crushing strength, and when wood breaks under load the upper cells fail or crush before the lower cells tear apart); *shearing strength* (ability of cells to withstand forces tending to slide them by one another: dependent upon adhesion of cell walls); *crossbreaking strength* (ability to sustain load at right angles to grain, hence a reflection of several types of strength mentioned); and so on. Of these several types, crushing strength is perhaps most critical. As indicated, it is less than tensile strength, and is first to give way in cross-breakage: it will thus largely determine crossbreaking strength or failure under load for most uses of wood. In Table 5-5 some common North American woods are classified as to their crushing strength. Bending strength parallels crushing strength.

TEXTURE. Texture is, of course, largely a reflection of grain—i.e., it depends upon the size and quality of the cells of the wood.

Often a primary consideration in ornamental woods, texture is of little or no importance in construction timbers. Most woods will wear or saw smooth, but others present the frayed or roughened ends of cell walls on any cut surface. Notable for *harsh texture* in this sense are incense cedar, hemlock, elm and poplars; in contrast, red cedar (*Juniperus*), Port Orford cedar, beech, and hard maple are *smooth-textured*. Texture may also refer to the size of cell lumen (and width of rays), in which case wood may be spoken of as *fine-textured* or *coarse-textured*. Red cedar (*Juniperus*) and other cedars, for example, are fine-textured in comparison with the coarse-textured bald cypress, redwood, or sugarpine; chestnut or oak is coarse-textured in comparison with basswood or gum.

TOUGHNESS. TOUGHNESS is a general term with several shades of meaning. Ordinarily, a tough wood is one which will split or tear apart only with difficulty. This characteristic is a reflection of grain (intertwining of fibers)

Table 5-5

Crushing Strength of Some Common North American Woods

Very High	High	Moderately High	Moderately Low	Low
HARDWOODS	SOFTWOODS	SOFTWOODS	SOFTWOODS	SOFTWOODS
hickory	Port Orford cedar	incense cedar	western red cedar	northern white cedar
black locust	bald cypress	red cedar	balsam fir	spruce
honey locust	Douglas fir	(*Juniperus*)	pines, lodgepole,	
persimmon	pines, southern	firs, western	ponderosa, sugar,	
	yellow and red	hemlocks	white	
	redwood	western white pine		
	tamarack and larch			
	HARDWOODS	HARDWOODS	HARDWOODS	HARDWOODS
	ash	birch	elm	basswood
	beech	chestnut	hackberry	buckeye
	cherry	gum	holly	catalpa
	hard maple	soft maple	tuliptree	poplars
	red oak	sweetgum		willow
	white oak	sycamore		
	Osage orange			
	walnut			

as well as of inherent tenacity of the cell walls. More precisely, toughness is resistance to shock—the ability of wood to "absorb" sudden blows or stresses without failing.

MISCELLANEOUS CHARACTERISTICS. Among the miscellaneous characteristics of wood might be listed odor and taste. Odor is dependent upon volatile compounds, such as the ethereal oils of cedar, which are not in themselves part of wood substance. They are usually more pronounced in green than in seasoned wood. Woods with distinctive odors include: incense cedar, Port Orford cedar, western red cedar, red cedar, (*Juniperus*), northern white cedar, bald cypress, Douglas fir, pines (lodgepole, ponderosa, red, southern yellow, sugar, white, and western white), basswood, buckeye, catalpa, sassafras, and black walnut. Distinctive taste in wood is less usual. It is dependent upon soluble infiltration products or depositions on the cell wall. Other miscellaneous characteristics often of importance are: cleavability (ease of splitting: tendency of wood cells to part, particularly along rays or annual rings), ease in working or machining, ability to take glues and hold paints, nail-holding capacity, and so on.

Uses of Wood

Worldwide about as much wood is used for fuel as for industrial purposes, slightly over 1 billion cubic meters annually. This is largely in undeveloped parts of the world, and is not so serious a timber drain as it might first seem since most of the fuelwood consists of small sticks and cull trees which would be of little commercial value otherwise. Wood used industrially worldwide also exceeds 1 billion cubic meters, about 65 per cent of it for lumber and related construction materials such as plywood, about 25 per cent for pulp. Utilization is chiefly in the technically advanced countries. According to the United Nations' FAO, utilization of

wood as a raw material (early 1960's) ran about 69 per cent for sawlogs, 24 per cent for paper and paperboard, 5 per cent for plywood, and 2 per cent for fiberboard or particleboard. Timber use in the United States is itemized by commodity in Table 5-6. In such an affluent society per capita consumption averages nearly 80 cubic feet annually, roughly 1,000 board feet weighing over 1 ton. It is evident that a technically advanced economy encourages more sophisticated use, with less timber serving for fuel and more for formed and processed wood products. There is encouragingly greater use of wood wastes, too, such as for particleboard or chemical conversion.

Table 5-6

Approximate Annual Drain of Timber in the United States in the Late 1960's by Major Use

Total Timber Used	12 Billion Cubic Feet
Sawlogs for lumber	49%
Pulpwood	27
Fuelwood	9
Veneer (including plywood)	9
Other (ties, posts and poles, mine timbers, cooperage, shingles, excelsior, charcoal, etc.)	4
Exports	2

Unprocessed timber—that not chemically changed, although it may be glued or treated with preservatives—will be discussed first. Most unprocessed timber becomes fuelwood or lumber. Furniture constitutes refined and high-value application of lumber, even though total timber drain for furniture is not great. Veneers and plywood are increasingly important timber manufactures, as are particleboards and laminated structures. Timber for posts, mine timbers, poles and pilings, railroad ties, cooperage, and miscellaneous minor usages (shingles, excelsior, sporting goods, etc.) continue as steady if minor markets.

Wood conversion products, primarily pulp and paper, is a second category of timber usage. Some wood is still processed for rayon, although synthetic fibers derived mostly from petrochemicals now dominate the fiber industry. Some wood is used in plastics and linoleum, and a little (mostly residues) for distillation, hydrolysis, and fermentation.

Finally, the forest yields products of secondary commercial interest, like cork and Christmas trees. The many extractives possible from trees will be discussed in appropriate later chapters dealing with essential oils, resins, dyes, tannins, fruits, honey, pharmaceuticals, and so on.

1. Fuelwood. As important as is wood for fuel worldwide, such use needs little discussion here, being mostly a simple local operation of felling and gathering. In North America only about 10 per cent of the timber harvest goes for fuel, but about 30 per cent in Europe and nearly 90 per cent in Latin America and Africa. Asia, exclusive of the U.S.S.R., probably consumes for fuel about three-fourths of its timber harvest.

In colonial times the United States relied almost exclusively on wood as a source of energy. Gradually coal, oil, gas, and electric power superceded wood. What little wood is used for heating today occurs mostly in remote forested areas of the South and the West. Cordwood, however, is popular for the suburban fireplace, and wood products make possible both the vacation campfire and the home cookout. A tidy charcoal industry caters to the home cookout, developing as America moved to the suburbs. Over a quarter-million tons of charcoal are used annually in the United States, most often as briquets (which can be prepared from wood wastes and often contain coal). Charcoal is made by smoldering wood under confinement so that sufficient air is present only for partial combustion (which drives off gaseous components). Charcoal is often the primary cooking fuel in less developed countries.

Parts of the world lacking in coal and petroleum resources may depend upon wood for power. The utilities of Brazil's major cities, for example, are energized by burning faggots brought in from scrub forest that is becoming increasingly depleted near the urban centers. Indeed, wood is frequently the only readily available source of energy for much of the tropics. There charcoal is crudely produced by firing piles of timber (usually sticks and cull logs of little commercial importance) under a mound of earth.

One of the major uses of wood is fuel, especially in the Tropics. This wood will be used for cooking in El Salvador. (Courtesy USDA.)

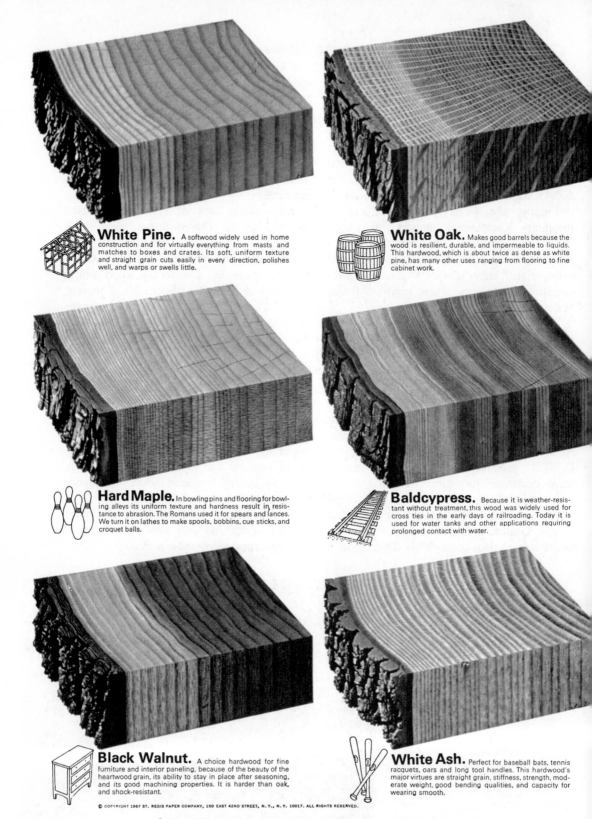

White Pine. A softwood widely used in home construction and for virtually everything from masts and matches to boxes and crates. Its soft, uniform texture and straight grain cuts easily in every direction, polishes well, and warps or swells little.

White Oak. Makes good barrels because the wood is resilient, durable, and impermeable to liquids. This hardwood, which is about twice as dense as white pine, has many other uses ranging from flooring to fine cabinet work.

Hard Maple. In bowling pins and flooring for bowling alleys its uniform texture and hardness result in resistance to abrasion. The Romans used it for spears and lances. We turn it on lathes to make spools, bobbins, cue sticks, and croquet balls.

Baldcypress. Because it is weather-resistant without treatment, this wood was widely used for cross ties in the early days of railroading. Today it is used for water tanks and other applications requiring prolonged contact with water.

Black Walnut. A choice hardwood for fine furniture and interior paneling, because of the beauty of the heartwood grain, its ability to stay in place after seasoning, and its good machining properties. It is harder than oak, and shock-resistant.

White Ash. Perfect for baseball bats, tennis racquets, oars and long tool handles. This hardwood's major virtues are straight grain, stiffness, strength, moderate weight, good bending qualities, and capacity for wearing smooth.

Some well-known woods and their uses. (Courtesy St. Regis Paper Co.)

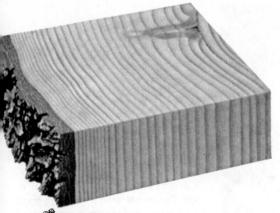

Red Spruce. A favorite for violin sounding boards because of its high resonant qualities. A softwood, it is easy to work, and is light in relation to its strength and stiffness. These qualities also make it eminently suitable for ladder rails, canoe paddles, and oars.

Hemlock. This relatively soft, light, straight-grained, resin-free wood, with its uniformly long fibers, is becoming one of the most important species for paper pulp. It is also used for structural lumber and plywood, and for boxes, barrels, and concrete forms.

Hickory. A hardwood unsurpassed for the handles of impact tools like axes and hammers, and for skis, because of its hardness, strength, toughness, and resiliency. In horse-and-buggy days it was widely used for wheel spokes and rims, singletrees, and buggy shafts.

Wood for barrels, books, or baseball bats: but which wood is best for what—and why?

Many qualities determine the choice of a wood for a specific job: weight, density, moisture content, stiffness, toughness, or the presence of knots and resin. A resinous wood, for example, does not paint well. A dense wood holds nails better.

"Hardwoods," those from broadleafed trees like oak and maple, are generally used in furniture and implements. "Softwoods," those from trees with needles, are used mostly for construction or for paper products. Softwoods have a simple cellular structure with most cells lined up and down the tree. Hardwoods tend to have complex cell structure, and more solid cell wall material.

For St. Regis, all the qualities of a tree are significant. Trees are our basic resource. From them we derive the wood that gives us our products. We make printing papers, kraft papers and boards, fine papers, packaging products, building materials, and products for consumers.

Essentially, the life of the forest is St. Regis' life. That is why we—together with the other members of the forest products industry—are vitally concerned with maintaining the beauty and the usefulness of America's forests for the generations to come.

Making charcoal in a remote forest in India. The natural forest is felled and made into charcoal, and the site is then planted with more valuable species. (Courtesy Forest Research Institute, Dehra Dun, India. From *Unasylva*.)

In industrially advanced temperate lands wood may be used, too, to smoke meats, flavor food with an attractive "outdoor" odor, and to cure tobacco.

Wood for fuel has the least exacting requirements of any timber use. All species may be utilized, their efficiency as a fuel depending upon their density and dryness, and on the efficiency of the stove or furnace in which they are burned. Other things being equal, the heavier the log, the more wood substance available per unit volume. We have already seen how all wood substance is essentially similar and hence yields, pound for pound, nearly equivalent quantities of heat (about 4 calories per gram burned). There are three stages in burning. First, any moisture contained in the wood is evaporated. This evaporation consumes a proportional amount of heat, so that the drier the wood —i.e., the more effectively it has been seasoned and stored—the better are its fuel qualities. Second, volatile materials are vaporized, causing the flame of the fire as they burn; these are essentially equivalent in fuel value in all species, but a poorly laid fire or inefficient stove may smoke—i.e., fail to effect complete combustion so as to utilize all potential heat. Furthermore, "resinous" vapors from many woods suffering incomplete combustion may accumulate on chimneys, creating a fire hazard there, or may blacken utensils used over an open fire. Third, the basic cell wall material (mostly cellulose) burns or glows without flame. This is the stage of maximum heat production, and sufficient draft must be maintained to effect complete combustion. It is at this stage that denser woods (those with more cell wall substance) demonstrate their superiority in quantity of heat produced per volume used. Woods like Osage orange, black locust, honey locust, and hickory are, for heating purposes, superior to such kinds as poplars, willows, basswood, and catalpa.

A discussion of fuelwood should not end without mentioning coal. In a sense, coal *is* wood—a residuum including fallen logs principally of the late Paleozoic era of some few hundred million years ago. Natural forces have compressed and altered the wood substance of the cell walls of many strange and ancient plants until nearly pure carbon remains. Indeed, coal, lignite, peat, and perhaps even portions of petroleum are a resource bequeathed to modern man by extensive forests of past ages—forests quite dissimilar, of course, to those of today. What energy those forests of long ago accumulated during countless ages man is now releasing in a few brief centuries.

2. Sawlogs and lumber. Well over a half-billion cubic meters of timber are harvested worldwide annually for sawlogs. The U.S.S.R. and the United States lead in production, with Europe as a whole not too far behind. There is relatively little sawlog production from Latin America, Africa, or most of Asia. In less developed parts of the world, timber cutting and marketing is still primitively done, with even large timbers being hand trimmed and sawed. At the other extreme, transportation and mechanization are highly perfected in North America and Europe. Machines are available in the United States capable of opening roads even in rough terrain, to say nothing of felling, bucking, and loading trimmed logs onto huge transports for quick hauling to the mill. The hardy lumberjack of the Paul Bunyan tradition, lost in the wilderness for the winter, has yielded to the machine operator who goes home at the end of the working day like anyone else.

Likewise, sawmills have become increasingly efficient, and are often mechanized to the extent of computers directing the operations. More and more, before ever a log is handled, systems analysis routes it to the economically soundest use. Logs of suitable quality can be directed to highest-grade usage, such as for veneer, taking into consideration market conditions (there may be a glut of certain species or specifications, so the mill can turn attention to other items in greater demand). The need for such efficiency has caused consolidation of the traditional small mills into better capitalized units which are often operated by large corporations. Yet, in south Asia, over half of the lumber is still hand sawn, and capital is insufficient both there and in most of Africa and Latin America for the building of modern sawmills. In underdeveloped parts of the world rough timber is often sent to mills in more technologically advanced places for sawing. Even in technologically advanced countries unprocessed logs may be shipped great distances for sawing: in the late 1960's Japan imported from North America much timber as logs rather than lumber, which could be more economically sawed in Japan than in the United States with its higher labor costs.

An abundant supply of good quality (preferably large-size) timber is important to the lumber industry, since bringing the log to the mill represents from 50 to 70 per cent of the cost of production. Thus milling operations tend to arise close to the forest. Timber suffers some loss from shattering when felled, as much as 40 per cent with gigantic, brittle redwoods. Even after delivery to the mill output of lumber is seldom more than 60–70 per cent of the timber input. The remainder is accounted for in slabs, edgings, trimmings, shavings, and sawdust. But wood hand sawed in Burma is said to retrieve only 35–45 per cent yield, varying, of course, with quality of log and species. In the United States trimmings and slab that used to be burned as waste are increasingly used for pulp and particleboard. Even the bark is being marketed (mostly as a mulch and soil conditioner). In modern

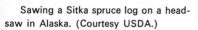
Sawing a Sitka spruce log on a headsaw in Alaska. (Courtesy USDA.)

Will America run out of forests?
No! One reason is that we are making
every log work harder.

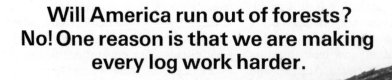

America's demand for lumber and paper grows by about 4 per cent each year. Can the forests meet this challenge? Yes. Good forest management helps us grow a continuous supply of the forest crop. And we get more from every log we use.

Sawmills have long known how to derive the maximum amount of lumber from a log. Nevertheless, at one time up to 50 per cent of some logs went unused. Today these "waste" pieces are routinely converted into chips for pulp and paper mills. Many pulp mills depend upon these chips for a good part of their supply. With millions of tons of chips so used each year, hundreds of thousands of acres of forest land need not be touched.

To St. Regis, conservation of the forest is a matter of prime concern. Its trees give us the wood for our products. We make printing papers, kraft papers and boards, fine papers, packaging products, building materials, and products for consumers.

Essentially, then, the life of the forest is St. Regis' life. That is why we—together with the other members of the forest products industry—are vitally concerned with maintaining the beauty and usefulness of America's forests for the generations to come.

ST REGIS

How lumber is sawed from a log. (Courtesy St. Regis Paper Co.)

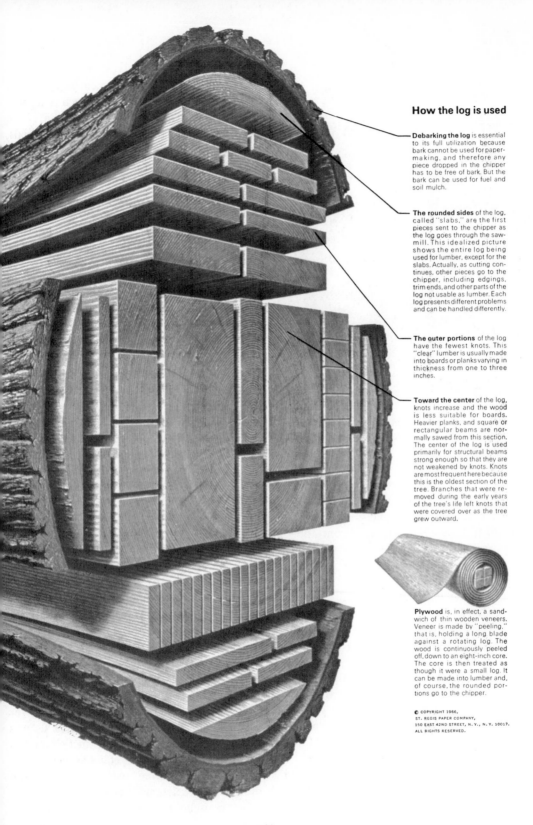

How the log is used

Debarking the log is essential to its full utilization because bark cannot be used for paper-making, and therefore any piece dropped in the chipper has to be free of bark. But the bark can be used for fuel and soil mulch.

The rounded sides of the log, called "slabs," are the first pieces sent to the chipper as the log goes through the saw-mill. This idealized picture shows the entire log being used for lumber, except for the slabs. Actually, as cutting continues, other pieces go to the chipper, including edgings, trim ends, and other parts of the log not usable as lumber. Each log presents different problems and can be handled differently.

The outer portions of the log have the fewest knots. This "clear" lumber is usually made into boards or planks varying in thickness from one to three inches.

Toward the center of the log, knots increase and the wood is less suitable for boards. Heavier planks, and square or rectangular beams are normally sawed from this section. The center of the log is used primarily for structural beams strong enough so that they are not weakened by knots. Knots are most frequent here because this is the oldest section of the tree. Branches that were removed during the early years of the tree's life left knots that were covered over as the tree grew outward.

Plywood is, in effect, a sandwich of thin wooden veneers. Veneer is made by "peeling," that is, holding a long blade against a rotating log. The wood is continuously peeled off, down to an eight-inch core. The core is then treated as though it were a small log. It can be made into lumber and, of course, the rounded portions go to the chipper.

111

Douglas-fir sawlogs being loaded for hauling to the mill in the Pacific Northwest. (Courtesy St. Regis Paper Co.)

milling it is important not to mar or mark the logs during handling, lest their usefulness for veneer and other high-quality uses be diminished—again a far cry from the old lumber drives down the rivers at time of spring thaw and use of peaveys in the mill-pond.

In the United States lumber production in recent years has been at somewhat less than 40 billion board feet annually, of which at least three-fourths is softwood. Sawlog consumption was approximately 6.2 billion cubic feet in 1967, of which about 0.6 billion cubic feet were imported. About three-fifths of the harvest in the United States is from the western states, most of the remainder from the southeastern pine belt. Most lumber goes for construction, maintenance, and repair, activities closely linked to the state of the economy and the demand for housing. Since World War II housing starts have

exceeded 1 million units per year, and in times of housing boom approach 2 million annually. In spite of competition from metals, masonry, plastics, and other materials, wood has remained the favorite construction material for homes, and is widely used in the interiors even of houses with nonwood exteriors. Wood is a remarkably useful and adaptable material. Frame houses built in colonial times, such as the Fairbanks house of Dedham, Massachusetts, still endure after several hundred years of exposure to the elements. It is not uncommon to find, in old dwellings and barns that are razed, sound timbers decades old of dimension not available today, and which might be reused except for their unstandardized dimensions and intractability from long drying out.

Today's lumber can be made fire-resistant, decay-resistant, and termite-proof by chemical treatment, increasing its usefulness. There are also improved gluing and joining techniques that make suitable smaller structural elements than formerly thought necessary, and which enable formation of arches and trusses that span distances greater than the length of a single board. A conventional modest-sized frame house is said to contain about 10,000 board feet of lumber. By using modular materials and prefabricated components a house of similar proportions might be built with as little as half as much lumber. Laminated timbers stronger than solid wood and rigid box panels for walls are made possible by modern glues. These will, no doubt, play a greater part in future construction.

A considerable amount of lumber was formerly used for crating. During World War II, for example, crating accounted for more than a third as much lumber as was used for houses. Except for such things as heavy machinery, the trend has been away from wooden crating, with paperboard and fiberboard replacing lumber. For produce, small household items, chemicals, and so on,

Wooden trusses of perhaps the world's largest bridge, supported by glued-laminated arches, soar over U.S. 16 near Mount Rushmore, S.D., at Keystone "Y." (Courtesy Monsanto.)

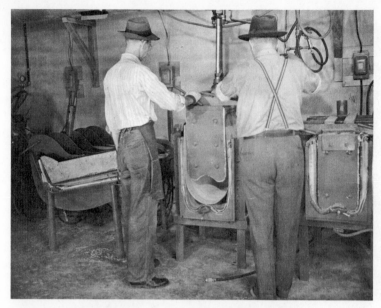

Much modern furniture is formed from molded plywood. A double "U" and a right-angle mold are shown here. (Courtesy Hardwood Plywood Manufacturers Assoc.)

Matched hardwood veneers for table tops. (Courtesy Hardwood Plywood Manufacturers Assoc.)

paper cartons and plastics have almost entirely replaced wooden containers, being more economical, easier to handle, and lighter to ship.

A steady demand exists for lumber to make manufactured articles. No substitute has ever been found able to match the warmth and beauty of wood for furniture. Along with furniture, handles, boats, musical instruments, caskets, pencils, woodenware, sporting goods, signs, and similar uses account for some 3 billion board feet of lumber consumption annually. A few com-

ments are in order concerning the use of woods for furniture.

Both beauty and reasonableness of price in furniture depend upon the utilization of veneers. The greatest craftsmen in furniture design have always been masters of veneering. Perhaps the finest desk ever made, that commissioned of Jean Oeben by Louis XV of France, took two master craftsmen nine years to produce its unrivaled marquetry; today it is in the Louvre, valued beyond price. There has often been popular prejudice against veneers, solid wood being thought superior. Yet no solid wood can show the striking figure and matched design produced by the veneer knife cutting logs of selected grain at exactly the most favorable angle. Nor is solid furniture so resistant to change in dimension and cracking or splitting as is well-made veneer. On the other hand, solid-wood furniture does give its owner the satisfaction of knowing that the wood throughout the piece is the same as that on the surface. If furniture is to be carved, solid wood is almost invariably preferred. Those parts subject to abrasion and wear had best consist of solid wood. Pianos not only take advantage of the beauty of wood, but utilize its acoustic properties as well, its resonance enhancing the pleasing tone of the music.

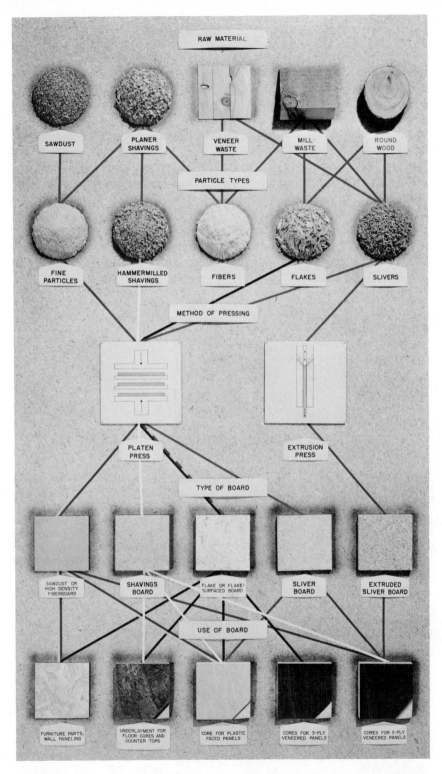

Wood residues are increasingly being utilized, as in the production of particleboard. (Courtesy Forest Products Laboratory, USDA Forest Service.)

3. Veneers, plywood, and panels. No forest products industries have grown more rapidly than plywood and particleboard manufacture. Plywood is now accepted throughout the world as a substitute for lumber, preferable especially where large surfaces are needed. Plywood success had to await development of modern veneering techniques, and invention of glues capable of bonding the wood or wood particles permanently under all conditions of use. Well over 20 million cubic meters of plywood were being produced annually in the mid-1960's, half in North America. This was three times as much as a decade earlier, and a tenfold expansion is predicted by the end of the century. Particleboard production shows an almost equally amazing growth record. Particleboard is made by compressing wood chips treated with a bonding agent into panels, and usually surfacing them with an attractive veneer. Several million cubic meters of particleboard are produced annually, chiefly in Europe where particleboard is as much used as is plywood. Similar in principle is fiberboard, made by compressing ground wood, a process that will be more fully discussed under pulp. Several billion cubic

Gigantic mauls of ring barker gnaw bark off peeler blocks in first step of plywood manufacture. Eight-to-ten-foot logs are bared to solid wood after leaving mill pond and then are taken to lathe for conversion to veneer. (Courtesy American Plywood Assoc.)

Barked logs are moved to hoist which readies them for chucking into lathe at left. Chucks are inserted either at geometric center or at heart center of peeler block. Block is turned slowly against veneer knife until perfect cylinder is formed. (Courtesy American Plywood Assoc.)

meters of fiberboard are used annually in North America and Europe, especially for containers and for insulation board.

Veneer is a thin slice of a log cut in sheet form. It may be cut in any of several fashions. The most common and important veneer is *rotary-cut*, so called because the bolt (log), after a softening by boiling or steaming for as long as 60 hours, is gripped by chucks on powered shafts and rotated against a large, stationary knife which cuts a thin sheet of desired thickness from the outer surface. The knife is set to move continuously inward as the veneer "unpeels" (much like the

unwinding of a roll of paper), so that there need be no break in the sheet until only a small unmanageable core of the log remains. There is very little log waste when this method of slicing veneers is used. It is used for all types, but particularly the cheaper veneers. The figure, of course, will be that of a tangential surface.

Another way of slicing veneer (producing what is called *sliced* veneer) is to draw the flitch (log to be veneered), fastened on a movable carriage, obliquely against the knife, so that a slice of veneer is cut off on each downstroke. On the upstroke, the flitch

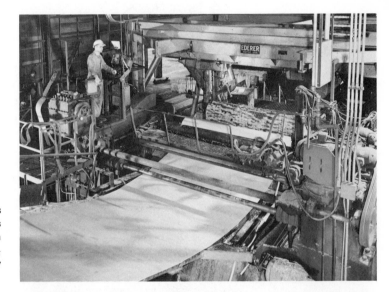

Eight-foot-wide wooden ribbon flows into plywood works as peeler block rotates against blade in lathe. Second operation in fabrication of plywood removes veneer $\frac{1}{10}$ to $\frac{3}{16}$-inch thick from blocks. (Courtesy American Plywood Assoc.)

Core veneers with grain at 90 degrees to that of face material are fed between rolls of this glue spreader. Sheet layers at right will carry face veneer over core layer's head. Plywood panels are built up of odd numbers of plies, usually three, five, or seven, and derive strength from cross-banding of grain. (Courtesy American Plywood Assoc.)

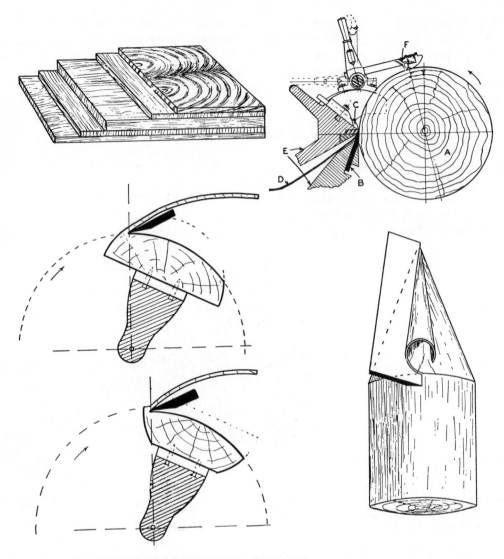

Diagrams of plywood and of veneer cutting. Upper left shows 5-ply construction; upper right, rotary veneer lathe; middle left, stay-log cutting; lower left, back cutting; lower right, cone-cutting. (Courtesy *Economic Botany*.)

recedes and then automatically advances the set distance to produce the next veneer sheet. As with sawed veneers, long, narrow strips are produced, but in all sliced veneers the inner side next to the knife shows minute tears or checks, so that for face stock perfect for matched paneling (in which the reverse side of alternate sheets must face up) sliced

veneer is inferior to sawed veneer. Slicing is much more economical than sawing (there is no sawdust waste; extremely thin sheets are possible), and is used for making veneer of many decorative cabinet woods such as mahogany, walnut, oak, and others.

The first modern method developed for making veneer was by sawing, but it is the most wasteful one. Sawing necessitates the reduction of about half the bolt to sawdust. To produce sawed veneers, thin sheets of the log are sawed just as is lumber. After development of the slicing methods sawed veneers continued to be produced only where highly figured, expensive and special-use woods were to be cut on a particular plane without fear of tearing the fibers. In modern practice, a stationary, power-driven, disklike saw blade, thicker in the center than at the circumference, cuts slices of an automatically regulated thickness from the flitch mounted to the steel "stay log" on a movable carriage that approaches and recedes from the saw. A bevel block throws the newly cut veneer sheet to the side, where it is automatically piled. The process is most used to produce quarter-sawn (radial face) paneling veneer, in which either side may face outward to produce matched paneling.

Commercial veneer varies in thickness, ranging from as little as $\frac{1}{110}$ inch in thickness to as much as $\frac{3}{8}$ inch, but it is commonly cut about $\frac{1}{30}$ inch thick (sliced veneer) or about $\frac{1}{4}$ to $\frac{1}{8}$ inch (rotary veneer). For most purposes newly cut veneer must be dried, which is usually accomplished by heating the veneer while it is pressed flat in any one of a number of contrivances for the purpose (rollers, hot plates, endless belts). Cheap veneers may be simply air-dried. After drying, the veneer is usually run through a clipping machine, which removes imperfections and glues on patches stamped from the same die.

Veneers have been known since almost 1,500 years before the birth of Christ, and are found among the royal possessions in the tombs of the Pharaohs. How these ancient veneers were made is not definitely known, but presumably they were laboriously formed by the abrasion of split logs. Throughout history, master craftsmen of fine furniture have preferred veneers for the remarkable figure thus obtainable.

Plywood is simply three or more (always an odd number) sheets of veneer glued or bonded together, with alternate sheets having the grain at right angles to adjacent sheets. Since wood proves to be strongest along the grain, this alternation of ply produces a sheet or plank more uniformly strong than is a comparable size of solid wood. Each sheet of veneer reinforces the adjacent sheet. The center ply is made twice the thickness of its adjacent plies so that quantitatively as much grain runs in one direction as another. In most cases the center ply or plies consist of cheaper "core" wood such as tuliptree or gum. The more expensive, figured "face" veneers are glued to the outside. The very fine cabinet woods, such as mahogany and walnut, are used for very thin veneers, which are simply glued to a strong, solid core of cheaper materials.

Until comparatively recent times, efficient glues and binding materials for plywood were not always to be had, with the result that plies frequently failed or peeled, particularly in moist or in alternately moist and dry situations. Within the last two decades, however, remarkable advances have been made in glues and glueing techniques. Thus plywoods capable of withstanding outdoor exposure, marine use, and various indoor stresses, formerly fatal, have been developed.

Synthetic-resin glues (phenol–formaldehyde, urea–formaldehyde, resorcinol–formaldehyde, and melamine–formaldehyde as thermosetting types, and polyvinyl acetate and polyvinyl butyral as thermoplastic types) permit the production of remarkably serviceable plywoods and laminated structures. Thermosetting types set irreversibly at curing temperatures up to 320°F.; thermoplastic types, however, soften at high temperatures and are thus of limited usefulness. All types are comparatively complex, being more or less specific in their requirements as to hardeners (catalysts to aid setting), pH, fillers, extenders, solvents, curing temperatures, storage life, and working life. Plywood made with thermosetting synthetic-resin glues is usually produced by the hot-press method: the sheets (or coatings) of resin glue are placed between the plies, which are then cured under heavy pressure (about 200 lb per sq in.) at the proper temperature between the platens of the hot-press machine. Curing occurs in a matter of minutes, yielding a water-resistant, high-quality, all-purpose plywood.

Special resin glues have also been developed for bonding metal to wood. These are usually of the thermosetting type. "Sandwich panels" (thin surface sheets of aluminum, veneer, or other material, bonded rigidly to a light core, usually of resin-impregnated, corrugated paper) give promise of economy in mass-produced, cheap housing. These panels are very light in weight, easily handled and assembled, a natural insulator in themselves, and stiff enough to need little or no framing in construction. Similar in principle also is "stressed-skin" construction, in which faces of plywood are rigidly glued to a frame of small-size lumber (1×1, for example, instead of 2×4).

With modern glues plywood is said to be stronger than steel, pound for pound. In the United States about 15 billion square feet of plywood ($\frac{3}{8}$ inch basis) is manufactured annually, about 85 per cent of it from softwoods. Between 2 and 3 billion square feet of hardwood plywood is also imported. Nearly 6 billion board feet of softwood timber is used for veneer production, and slightly less than 1 billion board feet of hardwood. Douglas fir plywood made in the Pacific Northwest is the most used type, especially for construction (roof and wall sheeting, subflooring, shelves, and cabinets), boxcar linings, pallets, boxes, signs, etc. Quality veneer logs of ornamental woods such as walnut, maple, and cherry have become increasingly scarce and expensive, selling in the Midwest for as much as $1,000 per 1,000

board feet and more. Inexpensive veneer used in single thickness for fruit boxes, or bonded into cheap plywood, is often made from sweetgum, tupelo gum, and southern yellow pine.

Because of the rapid expansion of the plywood industry some overcapacity results when construction demand lessens. A number of plywood plants in the Douglas fir area of the Pacific Northwest have had to close down or consolidate. Those remaining are lean and efficient, having undergone vertical integration to include chip handling facilities and profitable disposal of by-products. Technological developments make possible the use of smaller peeler logs, thus reducing dependence on the butt section of old-growth timber (which can be reserved for top-quality lumber). With modern steaming methods even sapwood makes acceptable veneer, necessitating careful handling of the logs to prevent their scarring. It is now possible to form plywood into graceful curves much like molded plastic. And using outdoor glues with plastic overlays on the surface, plywood can make attractive exteriors as well as interiors.

4. Posts, mine timbers, poles, and pilings. Timber used for posts, mine timbers, poles, and pilings in the United States amounts annually to over a half-billion cubic feet. About 300 million posts are needed annually, principally for new fencing and replacement on farms. Much of the supply comes from farm woodlots and is used on the same farm where it is produced. Often cull trees or inferior lumber species are utilized. Traditionally, durable woods such as cedar, cypress, Osage orange, and black locust have been preferred for posts, but nearly all species are utilized depending upon local availability. Preservative treatment is usually given nondurable woods, materially lengthening the life of the post.

Mine timbers account for annual consumption of about 80 million cubic feet of timber in the United States. Mine timbers are used for props to support overhead burden in tunnels as ore is removed, as lagging placed between props and walls to prevent infall of debris, and for rail ties for carts and small trains within the mine. As with posts, culls and inferior species can be utilized. Preservative treatment is much practiced where mine timbers are utilized in permanent or semipermanent passages, considerably lengthening their notoriously short life (mines are notably moist and conducive to decay).

Almost 50 million miles of telephone and telegraph wire, and millions of miles more of electric utility wire, now span the United States aboveground, held by nonconductive, noncorroding wooden poles. Over 6 million new poles go to market annually. In addition to being good insulators, wooden poles are also strong and elastic under storm stress, and, properly treated, are quite durable. Practically all poles utilized today are treated with preservatives. Some are treated in their entirety, others given butt treatment only.

Softwoods have traditionally furnished species most suitable for poles, but, depending upon local availability, almost any species of sufficient dimension and strength can serve. Most used are perhaps Douglas fir, southern yellow pine, and lodgepole pine.

Over 1.5 million pilings are used annually in the United States. Since pilings must be large and sound, practically 100 per cent of the supply must be obtained from sawtimber, mostly softwoods. Pilings are used below building foundations on unstable soils, for bridges and trestles, and for wharves and docks. Where sunk in soil below water table, no preservative treatment is needed, since complete exclusion from air is as effective a preservative means as is exclusion from moisture. In such situations the pilings outlast the structures they support. Gound-level pilings and those subject to marine-borer attack (chiefly that of mollusks) should be given preservative treatment.

PRESERVATIVES AND PRESERVATION METHODS. Reduction of the moisture content of wood to below about 15 per cent, and keeping it there, is an effective method of preserva-

Mechanized loading of tree-length pine sawlogs in Texas. Quality timber such as this is used for poles and pilings. (Courtesy Omark-Prentice Hydraulics.)

Utility poles being loaded into a hot-cold immersion tank. Poles are first soaked in a hot bath of preservative and later transferred to a cold bath. The process is said to accelerate the penetration of the wood preservative by causing a partial vacuum through the change in temperature. Notice the removal of the sapwood and incising at the ground-line area to insure deep penetration of the preservative at this vulnerable spot. Pentachlorophenol in oil is being used. (Courtesy Dow Chemical Co.)

mediate contact with moist soil, as are posts, ties, or poles.

More effective is the introduction into the wood of chemicals toxic to decay organisms. These are of diverse types, but can be conveniently grouped in three categories: (1) creosotes and creosote mixtures, (2) other toxic organic oils, and (3) water-soluble inorganic chemicals. Of these, the first type, comprising various forms of creosote, is at present the most important for posts, poles, pilings, mine timbers, and ties. No other preservative has yet proved more satisfactory than creosote, where the odor, color, and "bleeding" of the latter is not objectionable. Of the several organic compounds in the second category, toxic organic oils, the most successful have been the chlorinated phenols, particularly pentachlorophenol. Chlorinated phenols are insoluble in water; hence wood once adequately impregnated with petroleum solution of a chlorinated phenol loses little of the preservative through leaching, and may remain in exposed situations with relative immunity to stain, decay, and termite attack. It will not have the objectionable odor and paintproof characteristics of wood that has been creosoted. Of the water-soluble inorganic chemicals comprising the third category, several have found some use, including mercuric and zinc chlorides, various arsenicals, chromates, copper compounds, and so on. Such preservatives ordinarily have no objectionable odor or color, and do not interfere with the application of paint. They are easily introduced in aqueous solution, but unless precipitated in some fashion within the wood are just as easily leached away under exposure.

Preservatives may be introduced into the wood in a number of ways. Obviously, the

tion. The reduction may be accomplished by air seasoning, kiln drying, boiling in oil (which quickly vaporizes and expels the contained moisture), or even by a treatment using superheated steam, dielectric current, vacuum drying, and vapor drying.* Thereafter the wood can be kept more or less dry in moist situations by application of various coatings. These are in no case 100 per cent effective: primer paints and linseed oil have little effect; ordinary paints and varnishes, moderate effect; and aluminum or asphalt and pitch paints (when applied in several coats), considerable effect. All in all, application of coatings (paints) is of little practical use in preserving wood that is in im-

---

* This process subjects logs, particularly those used for ties, to vapors of boiling organic chemicals, which mix with vaporized moisture from the log and are then removed, usually under vacuum, to be separated from the water and reused; the process reduces seasoning time of refractive woods like gum and oak from months to hours. Much the same principle is involved in "solvent seasoning," in which liquids such as acetone are used to extract moisture (and rosin) from the wood.

method that provides better penetration will insure better protection. Penetration is most easily effected along the grain, since the cell cavities opening on a transverse surface are proportionally more abundant and more uniformly distributed than are those opening on a tangential or radial face.

Some preservative treatments depend upon the natural diffusion of water-soluble preservatives in green or even in standing timber. One method involves standing the newly cut log in a bucket or barrel of preservative: evaporation at the top draws upward first the column of sap and then the preservative solution. Preservative can be introduced by borings or appropriate channeling into the sapwood of a growing tree, to be conducted upward throughout the sapwood. In this method the comparatively impervious heartwood is little affected. Other methods involve application of preservative powders or pastes to green lumber. These are dissolved by the wet surface and diffuse into the wood in the free water of the cell lumen. Sometimes force of gravity is utilized to introduce preservative solutions into wood: commonly a rubber inner tube or some other leakproof container of preservative is fastened to the higher end of the log and the solution is allowed to filter downward for a period of days or weeks.

More commonly, preservatives are simply applied to the surface of seasoned wood, either by brushing or spraying or by dipping or cold-soaking the stock. Cold-soaking involves immersion of the wood beneath the solution of preservative for some length of time, and, of course, effects greater penetration than does mere dipping.

The most efficacious way of introducing preservatives into wood is under pressure or vacuum. The hot-and-cold-bath process involves heating and cooling of the wood to produce a natural vacuum within the wood cells. The charge is brought to a high temperature in a vat of boiling preservative and is then immediately transferred into a second comparatively cold preservative solution. The expansion and expulsion of contained gas during the heating creates a vacuum in the cells upon cooling, drawing the preservative solution into the lumen. Typically, the charge is enclosed in a sealed chamber to which controlled pressure or vacuum can be alternately applied, and into which the preservative can be introduced or withdrawn. In one process a preservative containing copper, chromium, and arsenic is forced into the cells at pressures up to 150 p.s.i. forming insoluble compounds "locked in" the cell wall that will not leach and which provide lasting protection against termites and decay even to completely exposed wood. Certain woods that are comparatively impervious, such as gum, white oak, beech heartwood, and some others, may be treated more rapidly if lateral incisions are first made by passing the wood through rollers equipped with incising teeth, thus permitting more ready lateral entrance of the solution.

Akin to preservative treatments is the fireproofing of wood. Fire-retardant chemicals may be applied either as a coating or in an impregnating solution. Neither will prevent charring of wood at very high temperatures, but both are capable of preventing wood from self-kindling. Among the coatings utilized for indoor use are linseed oil paints in which borax has been incorporated, sodium silicate solutions, alginate gels, methyl cellulose, magnesium oxychloride, casein plus borax mixtures, synthetic resins, and fire-retardant chemicals similar to those used in impregnation; for outdoor use, zinc borate, chlorinated rubber, and chlorinated paraffin have found some use. Chemicals effective as impregnating solutions include aluminum sulphate; various ammonium, barium, cadmium, calcium, cobalt, magnesium, and zinc salts, particularly their phosphates or borates; boric acid and borates; and various phosphates.

5. Other uses of raw wood. Although railroad trackage is declining rather than increasing in the United States, there is still a steady demand for replacement of wooden railroad ties. No metal or synthetic has yet proved a satisfactory substitute for the wood tie, which is resistant to deterioration, economical, and allows easy fastening of the rail. To save shipping costs ties are sawed from local timber (where trees of sufficient size are available), and are almost invariably treated with preservatives to extend their life. A life of 30 or more years is not uncommon for a tie. But even then, what with nearly 130 ties to each mile of track, replacement calls for about 30 million ties per year, perhaps 1 billion board feet. Railroads also consume appreciable wood for boxcar linings (especially plywood) and for car repairs.

A moderate-sized cooperage (barrel-making) industry persists in the United States, largely because the law requires that whiskey be aged in new white oak barrels. Metal drums, plastic containers, and multi-wall paper bags have replaced wooden cooperage for most products formerly shipped in barrels, such as cement, flour, produce, and chemicals. Barrels suitable for containing liquids are termed tight cooperage, those that need not be watertight, slack cooperage. The premier wood for tight cooperage is white oak, in which tyloses block the cavities of the heartwood cells and make the wood nearly impervious to the passage of liquid. Furthermore, white oak does not impart a disagreeable odor or taste to the contents. Although glass-lined steel tanks are being substituted for barrels and large casks or vats in the breweries and wineries, the demand for whiskey barrels is rapidly eating up large specimens of white oak in the midwestern hardwood belt.

In barrel making, the headings are generally cut in one location, and the staves (on special drum-shaped saws to give proper curvature) in another. These are sent to an assembling point where "raising" occurs. The staves are heated or steamed (to make the wood flexible), and forced together (a process termed windlassing), after which charring and final touching-up are completed. Such barrels, profiting from the mechanical efficiency of the double arch, are exceedingly resistant to shock and external impact. The basic principle of barrel making has not been improved upon in over 2,000 years, since the Babylonians and Egyptians made

Constructing barrels from white oak staves. This white oak stock was cut from the Ozark National Forest, Arkansas. (Courtesy USDA Forest Service.)

Bourbon barrels in a char cooker. These barrels were made from white oak cut from the Ozark National Forest, Arkansas. (Courtesy USDA Forest Service.)

what are the most ancient barrels known. While whiskey barrels cannot be reused, vats and tanks made of redwood in California have been in continuous service for holding wines for as long as a century.

Wood shingles are still produced, for their artistic effect. Softwoods such as poplar and basswood may be shredded into excelsior on scoring and shaving machines; excelsior is proving useful as a mulch for turf seeding (as along the roadside). There is demand for small bolts of wood for turnery, and in woodworking industries that produce dowels, spools, bobbins, sporting goods, and so on. Hardwoods are of especial attractiveness for ornamental bowls. Baseball bats are mostly made from ash.

Sawdust and shavings, once mostly used for boiler fuel, are finding higher-value employment. They work well as an absorbent for liquids (including manure on the farm), as a mild abrasive (in hand soap, metal polishes, etc.), as a cushioning and packaging material, and as an insulating filler or ex-

tender. Large chips may be included in pulping charges. Fine particles (wood flour) serve as a filler for pressed board, felts, low-grade pulp, and linoleum. Wood flour used in explosives contributes to the explosive power of dynamite.

WOOD RESIDUES. Efforts to gain the last iota of profit from the log, and new concern about pollution from traditional means of residue disposal (chiefly by burning), have initiated utilization of wood residues that were once considered waste. Percentage of residue increases as old-growth timber is replaced by smaller stock; it is apparent that there is more bark, for example, per unit of usable wood in a small log than in a large one. On the whole there are about 20 cubic feet of bark per 1,000 board feet of lumber. Additionally, in sawing a log, about 10 per cent of the volume is lost to sawdust. With smaller logs the percentage of edging and slabs increases as wood is cut to the typical rectangular shape of lumber. Thus with the smaller-dimension, second-growth stock, perhaps only 50 per cent of the raw material is recovered as lumber.

Although residues are still used for fueling boilers, much is now being employed for the manufacture of pressed hardboard. As noted earlier, particleboard is one of the fastest-growing categories of wood product, and a profitable outlet for mill residues. A great deal of attention is being given bark disposal. Bark makes an excellent mulch and soil additive, but because of its bulk it can seldom be shipped more than a few hundred miles and still be profitably sold. But the trend is unmistakably toward greater utilization of all parts of the tree. One day the lumbermill may make a claim similar to that of the slaughterhouse, where "everything but the squeal" is utilized.

6. Pulp and paper from wood conversion. Pulp and paper account for about 95 per cent of all converted timber. About 80 million tons of woodpulp are produced annually worldwide, of which slightly more than one-fourth is the mechanical type and slightly less than three-fourths chemical pulp. Nearly half of the pulp goes to make paper other than newsprint, about a third to make paperboard (primarily for containers), about one-sixth for newsprint, and only about 6 per cent for fiberboard. Usage of pulp products ties in pretty well with standard-of-living, those countries having a high degree of literacy and technological advancement utilizing most of the production. By continent North America is the leading producer and consumer of pulp, followed by Europe and then Asia. The United States alone produces more than one-third of the world's pulp products, around 40 million tons, nearly half of which goes into paperboard cartons.

Since time immemorial the cellulose cell wall has supplied the stuff of which paper and pulp are made. But it is only within the last few centuries that this material has been derived from wood. Even today, where timber is scarce, esparto grass, kenaf, bamboo, or similar plants may supply the necessary cell wall substance for the manufacture of paper. Basically, paper consists of many separate, fine, overlapping fibers (plant cells or groups of plant cells) that have been matted or felted into thin sheets and then dried. Pulp is an unprocessed mixture of the loose fibers before felting. Pulp can be made from a diversity of fibrous materials: true fibers such as flax or hemp (directly, or indirectly in the form of rags or rope waste); entire stems of grass; the residue from sugar cane (bagasse); wheat straw; fibrous coatings of stems, leaves, and needles; etc. The abundance and availability of trees make wood pulp the most used type in timbered lands, however.

Pulp products have had phenomenal growth, production about doubling each 15 years in recent decades. In the future, rate of growth can be expected to tail off in the developed countries, but speed up in less developed parts of the world where literacy is likely to increase. In the United States nearly 60 million cords of pulpwood were consumed in 1968, of which over one-third came from the southeastern pine belt. Only a few years ago southern pine was considered unsuitable for paper making, but now the southeastern yellow pines are the most important pulp species, used especially for kraft papers. Western softwoods (including spruce, firs, hemlocks, and Douglas fir) provide about one-fifth of the pulp in the United States, and of this three-fifths is derived from chips that are often a by-product of the lumber mills. Hardwoods from the eastern forests contribute nearly one-fifth of the pulp, weak-wood species such as aspen, cottonwood, yellow poplar, and the gums being most cut. There is increased use of chipped residues here, too. About 7 million tons of pulp is also imported from Canada.

Wood may be converted into pulp simply by grinding the wood (mechanical pulp), or chemical dissolution of the binding materials (mostly lignin) may free the fibers which are then largely stable cellulose. The dissolved lignin has little economic value,

Mechanical loading and hauling of aspen pulpwood. (Courtesy Omark-Prentice Hydraulics.)

Caterpillar diesel tractor pulls 24 sleighs, each carrying four cords of pulpwood logs, from bush cutting in Ontario, Canada. (Courtesy Caterpillar Tractor Co.)

and indeed is usually a liability in that it presents a problem of disposal. Chemical conversion for the production of rayon and plastics will be discussed in a subsequent section, but it is pertinent to review here the several recognized processes for converting wood to pulp for the making of paper, paperboard, and fiberboard.

One of the earliest efforts to make pulp from wood was by Keller, in Germany, with a wood-grinding machine for which a patent was obtained in 1840. In 1841 some groundwood paper was made in Nova Scotia, from spruce. But not until after the Civil War was there production of groundwood pulp in the United States. Chemical processing of wood seems to have been discovered in England in 1852, where Hugh Burgess used caustic soda to separate the fibers. Two years later Burgess brought his soda pulp process to the United States and secured a patent, although not until more than a decade later was there commercial production of soda pulp (in Philadelphia). About this time, also in Philadelphia, G. B. Tilghman invented the first sulphite pulping process. This was not perfected until after 1870, however, in Sweden. The sulphate process for making pulp, used chiefly for kraft paper, was not invented until about 1879, in Germany, by

Dahl. Although common in Europe, sulphate mills were not established in the United States until 1908. Today the greatest tonnage of pulp in the United States is produced by the sulphate process, utilizing chiefly the southeastern yellow pines. A still newer idea is "holopulping," in which wood chips are subjected to mild oxidation such as with chlorine dioxide, and the lignin extracted with a solvent such as sodium hydroxide; almost none of the cellulose is lost.

The groundwood process; mechanical. Logs are sawed into short sections, debarked, and cleaned. The cleaned bolts are then fed into the grinding mill, where a large millstone, cooled and lubricated by a continuous flow of water, shreds the wood into a fibrous condition. The loose fibers are floated to a sliver screen, riffler, centrifugal pulp screen, and pulp thickener for concentration. The pulp is fed into the beaters or paper making machine described for sulfite pulping, below. The mechanical process gives the highest yield (90 per cent or more by weight) and is the cheapest method, but the pulp is weak and paper made from it yellows readily. Usually mechanical pulp is mixed with a certain percentage of chemical pulp. Newsprint, cheap magazines and catalogues, paper toweling, building boards, and the

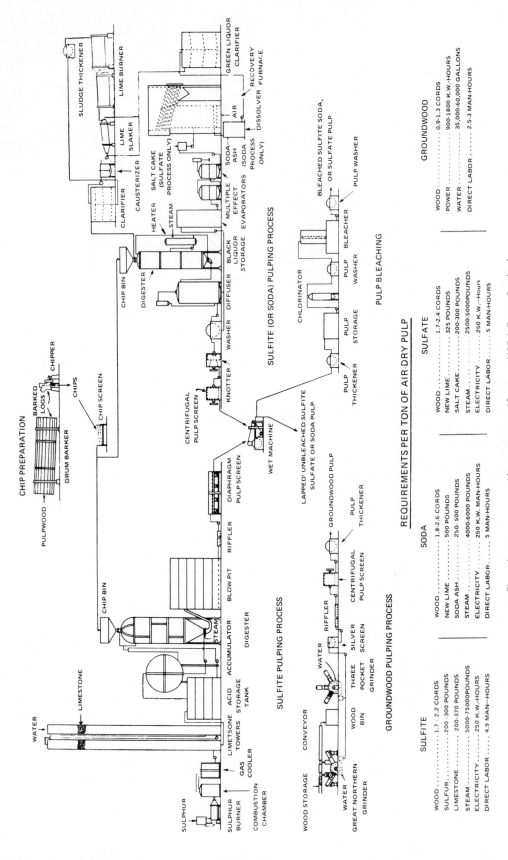

Flow sheet or paper-making processes. (Courtesy USDA Forest Service.)

like are largely composed of groundwood pulp. The principal woods used to produce it are the long-fibered, light-colored softwoods, including spruce, fir, and some pine and hemlock; little hardwood is used because of the shortness of fiber.

The soda process; chemical. Cleaned and debarked logs are chipped in special machines to small shaving-size fragments. These fragments are then introduced into a digesting tank and cooked with a sodium hydroxide solution at a temperature of about 240°F. Lignin and other noncellulose constituents are dissolved by the alkali solution. The remaining cellulose fibers are concentrated, dried, and stored for papermaking. The yield from the soda process is only 40 to 48 per cent. The pulp may be used alone for making bulky papers, such as blotters, for which strength requirements are not high, or it may be mixed with sulfite pulp to produce book and envelope papers. Woods utilized are chiefly hardwoods, including aspen and cottonwood, basswood, beech, birch, maple, and gum.

The sulfite process; chemical. Bolts are sectioned, debarked, cleaned, and chipped as described under the soda process. The chips are then discharged into the top of huge steel digestors lined with acid-resisting brick. Digestion is accomplished by an acid solution consisting of bisulfites and sulfurous acid pumped in from below. After thorough cooking under pressure, the digested chips are piped into a blow pit, where the acid solution containing the dissolved lignin is drained off. The remaining cellulose fibers, now nearly pure cellulose, pass through a riffler and pulp screen, and are washed, dried, and either stored as lapped pulp or sent immediately to the beaters and papermaking machines. Some high-quality pulps may be bleached with chlorine and calcium hypochlorite before being concentrated on the pulp screen (wire-cloth cylinder rotating in a vat) of the "wet machine."

In the sulfite process a yield of less than 50 per cent is attained, but the pulp formed is strong and durable. Woods used in this process are softwoods low in resin and a few hardwoods, common kinds being spruce, fir, hemlock, tamarack, some pines, aspen, and birch. Sulfite pulp may be used for a variety of purposes: for book, bond, wrapping, and tissue papers; for rayon; and, in combination with cheaper pulps, in newsprint and many other papers.

The sulfate process; chemical. This process is similar in principle to the sulfite process, but differs from it in that the chipped wood is digested by an alkaline solution of sodium hydroxide and sodium sulfide. This digesting solution is capable of dissolving resins, waxes, and fats from the wood, and hence the process can make use of a great variety of species, including the pines, that cannot be pulped by other methods. Formerly, sulfate pulp was used unbleached, but recently bleaching methods that do not weaken the pulp have been devised. Consequently, in addition to its original use for bag and wrapping (kraft) papers and boxboard, sulfate pulp is also used for book, magazine, writing, bond, and specialty papers. As in the sulfite process, the yield is not high, being less than 50 per cent. The chief advantage of the process is that by this method many otherwise useless pulp species, including the abundant southern yellow pines and all hardwoods, may be readily pulped. A byproduct of such pulping is tall oil, a mixture of rosins, fatty acids, and other wastes, useful for many chemurgic purposes.

The neutral sulfite process; chemical. This process, basically a modification of the soda process, is also known as the monosulfite process. It utilizes as a digesting solution sodium sulfite in an alkaline or neutral medium. The addition of the sulfite to what might otherwise correspond to the soda digesting solution not only speeds the pulping but produces a stronger, more durable fiber.

Yield is poor, but the pulp produced is of high quality. A variety of both softwoods and hardwoods may be utilized, and the pulp is amenable to bleaching.

The semichemical process; mechanical and chemical. In the semichemical processes wood chips are first softened or weakened by steam or chemicals (e.g., neutral sodium sulfite) and are then reduced to pulp mechanically. As might be expected, yields are intermediate between those derived from strictly mechanical and strictly chemical means, ranging as a rule from 70 per cent to 80 per cent of original pulpwood weight. Semichemical pulping makes use of hardwoods, for the most part, and is utilized chiefly for corrugated board stock, low-grade wrappings, roofing felt, and insulating board.

PAPER MAKING. The sequence of steps in making paper from pulp varies according to the mill and the type of product produced. In general, pulp is first sent to the beater, a rectangular vat equipped with a "paddle wheel" which thoroughly separates the pulp fibers and distributes them uniformly in the proper amount of water. At this stage, certain fillers (to add weight and body to the paper) and sizes (to impart a smooth, ink-impervious surface to the paper) may be mixed with the pulp, and other pulp, such

A 70-ton load of wood chips from Montana is delivered to a kraft paper mill in Tacoma, Wash. (Courtesy St. Regis Paper Co.)

as groundwood, may be blended in. Common fillers are clay, talc, and alum; sizings may consist of starch, gluelike compounds, or various resins. Dyes may also be added at this point, to produce colored papers.

After thorough mixing in the beater, the pulp is sent to the papermaking machine. This machine consists of an endless belt of wire cloth, the Fourdrinier screen, agitated in two directions (to insure crossing and intermeshing of the fibers) and moving continuously below the flow box, from which a regulated flow of pulp is discharged upon it. The fibers of the pulp are "strained out" by the Fourdrinier screen, the water filtering through to a collection trough. To effect still further drainage, the Fourdrinier screen is passed over a series of suction boxes. At this stage the matted or felted pulp is incipient paper. Still moist, it is picked up at the end of the Fourdrinier screen by another endless belt and subjected to a series of pressures and high temperatures between rollers that further dry and consolidate the fibers. Additional external sizing may or may not be applied at this time. When sufficiently dry, the paper sheet is guided into the huge calendering machine, where it is subjected to a final regulated pressure to control the finish. Finally, it is wound in large rolls preparatory to being cut to desired size for specific uses.

The art of paper making has had a long and notable history. In pre-Christian times, paper of a sort was made in Egypt by beating to paper-thinness laminated stems of the papyrus plant. Apart from this, and from parchment made of animal skins, no important writing substance was known until the invention of paper in China, probably by Ts'ai Lun, at the beginning of the second century A.D. Traditionally, the early Chinese scholars had laboriously written with a stylus on strips of wood, or, later, upon woven cloth, particularly silk. Perhaps it was the wasted trimmings from silk cloth that suggested the principle of paper making:

Top: "Wet end" of the Seminole Chief at St. Regis Paper Company's Jacksonville, Florida kraft mill. The pulp slurry enters this gigantic paper-making machine here. Bottom: "Dry end" of the Seminole Chief, where finished paper issues in tremendous rolls. (Courtesy St. Regis Paper Co.)

why tediously weave these fibers together by hand if they could just as well be floated on to a suitable surface, pressed flat, and dried? Ts'ai Lun may well have noted that frayed silk fibers floating on the water's crest could be picked from the surface in a matted, tangled condition and spread as one mass in the sun to dry. It was only one more step to devise a suitable sievelike tool for lifting the fibers from the water. The first such sieve—the original "Fourdrinier screen" of the paper industry—seems to have consisted of a bamboo frame with a cloth mesh; the principle persists to this day, although the technological improvements based upon it are legion. Even the characteristic imprint on the fiber mass made by the Chinese mold (sieve), or specific designs incorporated in the mold or screen by later craftsmen, are retained to this day in the form of the watermark of modern papers.

During the Middle Ages the art of paper making was introduced into Europe, perhaps as early as the eighth century, but knowledge of the art was not widespread until much later. In some quarters, although known, it was rejected as "un-Christian." The advantages of paper over parchment, however, soon dictated its acceptance once it became widely known. But the European quill pen required a harder surface than the Chinese character brush, and so necessitated the production of harder, coated (sized) papers. For making this type of paper, linen rags (linen had been recently accepted in Europe for undergarments, in place of wool) served admirably as the fiber source. The demand, however, soon outdistanced the supply of linen rags, and a more abundant source of a cheap fiber was sought. Perhaps the papermaking wasp (which builds its nest from macerated wood) first supplied the hint as to how paper might be made from wood, but in any event trees soon became the primary source of fiber for paper. With the advent of the printing press and the subse-

quent heavy demand for paper, only wood could supply the quantities of fiber that permitted the civilization of the eighteenth, nineteenth, and twentieth centuries to take its characteristic form—that made possible the stored knowledge of our libraries and the packaged goods of our supermarkets.

Moreover, additional uses for paper are still being discovered, some of them spectacular. An entire house has been constructed of impregnated paper; treated paper is used for water-resistant bags that hold feeds, flour, cements, and other materials damaged by moisture; flameproof paper is used for drapes and other home furnishings—even for ash trays; "wet-strength" paper is replacing burlap for farm produce and to serve as container for such diverse articles as laundry, lobsters, and orchids; laminated papers made with plastic glues have shown, pound for pound, twice the tensile strength of steel; even paper clothing has created a flurry of interest.

Of course not all pulped fiber is used to make paper; a minor portion goes for insulating board and fiberboard. It is reported that the Masonite Corporation, a leading producer of fiberboard, has been able to produce at a relatively low cost a "molasses" high in pentose and hexose sugar as a by-product of hardboard production. The process involves treatment of pulpwood chips with high-pressure steam, and recovery of the carbohydrate from solution by vacuum filtration. Research shows this molasses to be as useful for cattle feed as is the conventional molasses from sugar.

Woodpulp substitutes. With the increasing demand for pulp, and the diminishing availability of old-growth trees, attention has been given to other possible sources of fiber. Some authorities believe that it will prove economically advantageous eventually to derive fiber from annual crops or faster-growing perennials than are trees. Bamboo has been suggested as a perennial which would yield a frequent, abundant harvest from persistent rootstocks without replanting. Agriculturists suggest that an annual crop amenable to mechanized-farming techniques would be more logical. Interest has centered especially upon two fiber plants that are grown to a limited extent for textile fibers in some parts of the world, kenaf and sunn.

Kenaf is *Hibiscus cannabinus* of the Malvaceae, and sunn is *Crotalaria juncea* of the Leguminosae. Kenaf is the more promising as a source of pulp fiber, growing well under agricultural handling and being competitive against annual weeds. The bast fiber has been shown to be quite suitable for paper making. Under good growing conditions 7 to 10 tons of dry matter per acre can be obtained from kenaf. The entire stalk is used as the pulping material. In the United States the crop seems best suited to the southeastern states, income-per-acre generally not matching corn in the cornbelt without irrigation.

7. Rayon and other cellulose derivatives. Rayon, the first of the commercial synthetic fibers, still finds considerable use for textiles and tire cord in spite of increasing competition from petrochemical synthetics (these will be mentioned further under textile fibers). Over 4 million tons of woodpulp are used annually in the production of rayon, the primary markets being Europe and North America. This involves as little as 4 per cent of the North American woodpulp but as much as 10 per cent of the Japanese; included are cellophane and other dissolved cellulose materials as well as rayon specifically. While the future may seem bleak for rayon in technically advanced economies, because of the competition from other synthetic fibers, the ready availability of cellulose and the relatively simple technology involved in rayon production should favor its continuing importance in the developing nations.

Rayon was first manufactured about 1910, as a substitute for silk, and was referred to as "artificial silk." It is made by dissolving suitable grades of woodpulp, the solution being extruded and hardened. Other products from dissolved cellulose include cellophane, nitrocellulose, acetate plastics, photographic film, and various molded "plastics" such as tool handles and spectacle frames. Most are being replaced by plastics from other sources. About 1 million tons of rayon and other cellulose derivatives are produced in the United States annually.

There are four common chemical methods of making plant cellulose into rayon, differing from one another primarily in the solvent used to dissolve the cellulose of the plant cell walls. In all four processes the subsequent procedure is the same: the viscous solution of cellulose is drawn through small apertures or dies, after which it is hardened by plunging it into a hardening bath, or by evaporating the solvent from it. A number (from 13 to 270) of the fine filaments so produced are twisted together as they leave the hardening chamber to form rayon thread or yarn. If the filaments are cut into short sections (1 to 6 inches in length) and then spun as are cotton or other natural textile fibers, spun rayon with a woolly feel and pleasant softness is produced.

The principle behind the production of rayon is centuries old; that cellulose might be so transformed was predicted in the seventeenth century. It was not until 1855, however, that the Swiss chemist Audemars patented a process for transforming dissolved nitrocellulose into fine threads. Chardonnet, a pupil of Pasteur, instituted commercial production in 1891. The first commercial production in the United States was begun in 1911. It has made exceedingly rapid strides in a very short time. By 1923 world production of rayon had equaled that of silk, with more than 50,000 tons produced annually.

The viscose and the cellulose-acetate

SHEETS OF CELLULOSE—The form in which cellulose is received at the rayon plant.

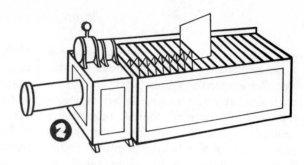

STEEPING PRESS—The sheets of cellulose are steeped in a solution of sodium hydroxide.

AGING THE CRUMBS—The crumbs are stored in cabinet-like containers for aging under constant temperature. During the aging the molecular structure of the cellulose undergoes a change.

XANTHATION—Carbon bisulphide is blended with the crumbs in this rotating churn. During the process the crumbs turn orange. The orange colored crumbs are known as cellulose xanthate.

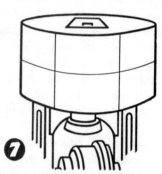

PREPARING THE SPINNING SOLUTION—The cellulose xanthate is mixed with a dilute solution of sodium hydroxide to form the spinning solution.

(Inset) Close-up of Spinneret (Drawing is about the size of the actual spinneret). Each hole in the spinneret forms an individual filament of yarn.

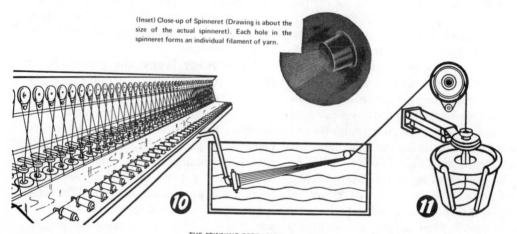

SPINNING MACHINE—The actual formation of rayon yarn takes place in this machine. (An enlargement of a single unit is shown in the next illustration.)

THE SPINNING OPERATION—The spinning solution is pumped through a spinneret immersed in a hardening bath. As the solution leaves the spinneret it hardens in the form of tiny filaments.

FORMING THE RAYON CAKE—The filaments are guided up over a rotating glass wheel and down into the whirling collectinb bucket. At intervals the yarn is removed from the bucket in the form of a circular cake.

Flow sheet for rayon manufacture. (Courtesy E. I. DuPont de Nemours & Co.)

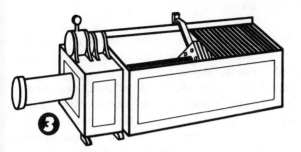

3

REMOVING EXCESS LIQUID—After the cellulose sheets have softened and soaked for a predetermined length of time, a hydraulic ram squeezes out some of the liquid. At this stage the sheets are called alkali cellulose.

4

SHREDDING THE ALKALI CELLULOSE—This shredding machine reduces the moist sheets to fine crumbs.

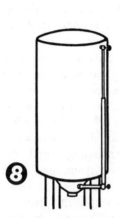

8

RIPENING THE SPINNING SOLUTION—The spinning solution is allowed to mature or ripen in huge tanks.

9

FILTERING THE SPINNING SOLUTION—The spinning solution is passed through several filters to remove all foreign matter.

12

WINDING INTO CONES—The cake of rayon yarn is then washed and dried, after which it is wound into cone form.

In addition to cones, the finished rayon yarn is also shipped in the form of skeins and tubes as desired by the weaving or knitting mills.

CONTINUOUS FILAMENT RAYON YARN—Rayon yarn in the form of a continuous strand.

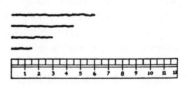

RAYON STAPLE FIBER—Short lengths of rayon, usually from 1½ to 6 inches. The staple is formed into spun rayon yarn by any one of several spinning systems.

methods of making rayon account for approximately 95 per cent of all rayon yarn production in the United States. The original nitrocellulose method was soon discarded, and the more recent cuprammonium process (in which cellulose is dissolved in copper sulphate and ammonium hydroxide) has been used but little.

The viscose process. Cellulose high in alpha-cellulose (formerly cotton linters or more recently selected chemical woodpulp especially from spruce and western hemlock) is first cleansed in alkali. The clean material is steeped in caustic soda (sodium hydroxide), after which it is pressed and shredded, and then aged under controlled temperatures. After aging, it is treated with carbon disulphide to form cellulose xanthate, and the xanthate is dissolved in dilute caustic soda to form the viscose spinning solution. At this point oils or pigments may be added to reduce luster. The solution is filtered, then forced through a spinneret (cap of precious metal perforated with tiny holes scarcely visible to the naked eye). The filament so produced is coagulated or hardened in a bath of sulphuric acid, sodium sulphate, zinc sulphate, and (sometimes) glucose. A rotating glass wheel then leads the filaments into a whirling bucket, which twists them into yarn. If necessary, the yarn is further washed and bleached.

The cellulose-acetate process. In the preliminary stages the procedure is the same as in the viscose process. Thereafter the linters (only recently has pulp been used in this process) are treated with acetic acid and allowed to age. After aging, the material is mixed with acetic anhydride in the presence of a catalyst to form cellulose acetate. The cellulose-acetate flakes are precipitated by running the mixture into water. They are then washed, dried, and dissolved in acetone to form the spinning solution. After passage of the spinning solution through the spinneret, the filaments are coagulated by evaporation of the acetone in warm, humid air, and then twisted into yarn in the same manner as viscose rayon.

Plastics. Dissolved cellulose also ushered in the modern age of plastics. The industry came into being in 1868, when cellulose nitrate from cotton was combined with natural camphor to yield the substance known as celluloid, a product used for detachable collars, billiard balls, toys, handles, and many other articles. Not long after, it was found that nitrated cellulose combined with castor oil and pigments could be formed into artificial leathers. Modern industry has since learned to combine cellulose in a variety of forms with many other substances to yield a number of different plastics. A second line of development uses wood (in the form of wood flour) in combination with various binders to yield plasticlike materials. Fiberboard, of course, is made by exploding wood chips heated with pressurized steam and then molding the resultant fine fibers under heat and pressure. Presumably the steam softens the lignin bond and allows separation of the cellulose chains; on subsequent molding under pressure the cellulose chains are again welded together through a lignin bond.

Similar in principle to plastics are various "modified woods." These are usually formed by compression, often after impregnation with resins of various kinds. Heat and pressure cause the wood to assume a dark color and become extremely hard, heavy, and durable. A product known as "impreg" has phenolic resin impregnated throughout the wood, so that after treatment all of parts are chemically alike. Impreg is remarkably resistant against swelling and shrinking, and to attack from decay or termites.

Other chemical wood conversions. Wood fiber contains up to 20 per cent of hemicellulose. Hardwood hemicellulose is about two-thirds pentose and one-third hexose sugars, but softwood hemicellulose is less

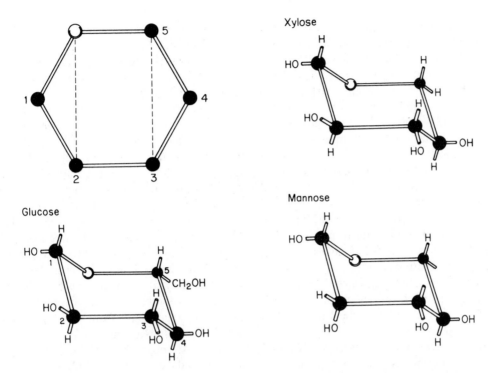

Xylose

Mannose

Glucose

Figure 5-7 Molecular configuration of pentose sugars. Xylose and mannose are typical subunits of polysaccharides other than cellulose (viz. xylan, mannan), part of the cementing matrix of the cellulose cell wall. (After R. D. Preston, *Scientific American* **218**, 105 (June, 1968).)

than one-half pentose sugar. Mention has already been made of the usefulness of these sugars in wood molasses, a by-product of hardboard production. Acid hydrolysis of cellulose breaks it down to 80 per cent or more of hexose and some pentose sugar. These chemical derivatives can be the source of a variety of products if economical acid conversion can be achieved. Pentose sugar (Fig. 5-7), for example, might be converted to furfural, a new material for plastics such as nylon and synthetic resins. Possible derivatives of hexose sugars are sorbitol, propylene glycol, ethylene glycol, and glycerine.

Carbohydrates derived from wood waste can be used to grow *Torula* and other yeasts that are excellent sources of proteins, vitamins, and even fats. So far the processing is marginally profitable, but may become more important as food shortages occur in an increasingly overpopulated world. Acid hydrolysis of wood yields carbohydrate suitable for fermentation by yeast to yield ethyl alcohol, and other solvents and acids procurable through fermentation. This is an "ace in the hole" for the fermentation industries any time other raw materials become too expensive. As this is written there is not yet large-scale profitable use for the tremendous amounts of dissolved lignin produced in wood conversion. Some has been used as a dispersing agent in concrete, as a soil stabilizer, in storage-battery plates, as a linoleum adhesive, and so on. Also, chemical breakdown of lignin yields vanillin-like substances and other chemical intermediates. Unfortunately, demand for such purposes is very small compared to the large supply of lignin available from pulping liquors.

Historically wood distillation was a fairly important forest products industry, to yield chemicals such as wood alcohol, acetic acid, and various "tars." Today most such chemicals are more economically derived from other sources, and seldom are wood distillates collected even as a by-product of charcoal making. Old pine stumps are still being distilled to some extent in the South as a source of turpentine and rosin (naval stores, further discussed in Chapters 11 and 9). Hardwood heartwood has been preferred as the source for pyroligneous acid, the name given the liquid condensate from destructive distillation. It is collected from vapors that result from heating of wood in a closed chamber, typically large steel ovens into which several steel cars covered with cordwood are run on rails; as the charge heats, gases escape at the top and are conducted through water-cooled condensers. Gas that does not condense is used to help heat the oven, or evaporate the pyroligneous acid. Yields per cord of hardwood run about 50 bushels of charcoal, 180 pounds of calcium acetate (after neutralization of the pyroligneous acid with lime), and about 10 gallons of methanol.

8. Secondary forest products. Perhaps the greatest secondary use of the forest today is for recreation, and for holding a few portions of this globe somewhat wild and free from the pollution that urbanization brings. This will be touched upon in the following chapter. Tangible substances from the forest include maple syrup and maple sugar, naval stores, medicinals, resins and tannins, foods such as nuts and fruits, game, and various ornamental materials. Often the use or gathering of these is of a recreational nature not economically competitive with domesticated crops grown agriculturally (primitive peoples, however, especially in the tropics, may rely greatly upon the bounty of the forest). Most such usages will be touched

upon in the appropriate chapters devoted to extractives and foods. There are two small industries worth mentioning here, however, as examples of secondary forest products— cork and Christmas trees.

COMMERCIAL CORK. Cork is the bark of the cork oak tree, *Quercus suber* of the Fagaceae. The cork oak grows naturally in the western Mediterranean region, abundantly enough to be of commercial importance in parts of Portugal, Spain, and northwestern Africa. A little cork is harvested in southern France, Italy, Corsica, and other Mediterranean islands, and the tree has been introduced into California. At suitable times of the year the bark is pried away from the tree in a way such as to not damage the cambium, a process called stripping. New bark will regenerate from the cambium, permitting a subsequent stripping in about 10 years. Although the cork oak forests are thinner and less extensive than formerly, cork continues to be produced in locations where it has been harvested for nearly 2,000 years. About 5 million acres of cork oak forest is involved, most of it with native rather than planted trees. Only in recent times has the cork oak forest received much management, including reforestation of decimated areas and protection from fire. In most countries the government now regulates stripping to assure the health of the forest and perpetuation of the industry.

Cork finds a steady world market, although in industrialized parts of the world its use for seals, gaskets, insulation, and so on is being contested by plastics and rubber. But much of the world has no synthetic substitutes for cork, which is uniquely resilient, light of weight, easy to grasp when wet, and imparting no odor or flavor to food and drink. Nor does cork catch fire readily. It lasts almost indefinitely, and is an excellent insulator and nonconductor of electricity. These properties eventuate from the structure of cork, a mass of nearly isodiametric cells

Bales of cork from Europe being received in the United States. (Courtesy USDA.)

that are like so many little air-filled "balloons" each coated with impervious wax. The ancient Greek and Roman civilizations utilized cork for floats aiding soldiers to cross streams, among other things. It has been used as a seal for jars since early times, but not until the glass bottle was invented in the fifteenth century did it become widely known.

Quercus suber is a subtropical evergreen, its thick bark adapting it well to the semiarid climate and the desiccating winds which sweep the western Mediterranean lands. Trees grow about 60 feet tall with a spreading, branching habit, the trunks reaching 10 feet or more in circumference. Each year the tree lays down new installments of cork bark, somewhat like annual rings in the wood. The outer part of the bark becomes rough and deeply fissured, and is scraped away to reveal the smoother inner bark that becomes the cork used in bottle stoppers. The bark is stripped in early summer when tree sap is flowing. The cork cells themselves are dead and contain no sap, but they separate from the trunk more easily at this season. Circular cuts are made several feet apart on the trunk or larger branches, connected by a vertical slit in the length of the section. Workmen pry the cylinder of bark away from the trunk with special hatchets. Sleeves of bark are removed from the larger branches

by climbing into the tree. Only bark an inch or more thick is worthy of stripping. It is said that large trees peel as much as a ton of bark. A workman can strip up to 200 pounds of bark a day, which is stacked in the vicinity and left to season for several weeks. Then it is boiled in huge tanks to soften it (leaching tannins and extraneous materials), and the rough outer surface is scraped away by hand. Then comes drying, trimming, grading, and baling. Bark ready for market consists of flat sheets of several square feet.

Cork is widely used for bottle closures, gaskets, liners in bottlecaps, floats (such as life preservers), nonslip walkways, handles, and so on. Ground cork is used in some linoleums. Pulverized and compressed cork makes corkboard, used for flooring and insulation. Often ground cork is treated with adhesive and pressed in large, heated molds. Cork tiles for flooring can be made by baking fragmented cork under pressure without an adhesive; the natural resin in the cell walls serves to bind them when baked for about 10 hours at temperatures between 450 and 600°F.

World cork production amounts to about one-third million tons annually, about half of this from Portugal. Spain and Morocco account for most of the remainder. The United States usually imports 100,000 tons or so of cork in a year. Although the cork oak has been introduced into California, where it grows well, cost of producing cork there cannot compete with the industry in the native cork forests.

CHRISTMAS TREES. The harvesting and sale of Christmas trees has become a sizable seasonal business in the United States, with nearly 50 million trees sold annually for decorative purposes. Over half are grown and harvested in the United States, the remainder imported from Canada. About five-sixths come from natural woodlands (although trimming and shaping may be involved), while about one-sixth are specially grown on plantations. Many farmers are planting Christmas trees as a seasonal source of income when other farm activities are limited.

The harvest of Christmas trees from natural forest lands is not the drain on timber that might first seem to be the case. Often this is a thinning operation where the remaining trees are benefited by reduced competition. Natural seeding is apt to be overly dense and irregular. Young trees are generally pruned back at least once and shaped for symmetry, in response to increasing sophistication of the market. The crop usually requires 4 to 8 years before reaching harvest-

able size. The cut tree will generally regenerate new stems from buds in the axils of the lower branches, which when reduced to a single leader provides a second crop within a few years.

The species of Christmas trees in demand varies from region to region, depending upon local availability and past custom. Canadian balsam fir, *Abies balsamea*, is quite popular in the midwestern United States, having a delightful fragrance and holding its needles well. Spruces, *Picea*, appearing similar to balsam fir to the uninitiated, do not hold their needles so well. Spruce can be distinguished from fir by the rough needle stubs on the older branches, fir bark being quite smooth. White, red, and Scotch pines are often used as Christmas trees in the East, Douglas fir in the West, red cedar (*Juniperus virginiana*) in the upper South. One of the premium Christmas trees grown in farm plantings in the West where the elevation is sufficiently high is the noble fir, *Abies procera*. In forest country the growing of Christmas trees provides excellent use for land that is often too rocky, steep, or otherwise inappropriate for other cropping purposes. The Christmas tree custom is relatively recent, the idea seeming to have gained acceptance in Germany about the middle of

A stand of 7 year old Douglas fir Christmas trees being pruned in Oregon. (Courtesy USDA.)

the nineteenth century, where a family project at Christmastime was to select and cut a tree from the forest.

Decorative foliage other than Christmas trees also comes to market at Christmas time from forest and farm. Evergreen boughs, holly branches, ferns, magnolia, mistletoe, Mahonia, and other ornamentals are all used. The planting of holly (*Ilex*) orchards in Oregon solely for the production of Christmas greenery has become a rural enterprise of some significance.

SUGGESTED SUPPLEMENTARY REFERENCES

Brauns, F. E., and D. A. Brauns, *The Chemistry of Lignin, Supplement Volume*, Academic Press, N. Y., 1966.

Browning, B. L. (ed.), *The Chemistry of Wood*, Interscience, N. Y. and London, 1963.

Calkin, J. B. (ed.), *Modern Pulp and Papermaking*, Reinhold, N. Y., 1957.

Cooke, G. B., *Cork and the Cork Tree*, Pergamon Press, London and N. Y., 1961.

FAO, Rome: various studies, such as no. 16, *Wood: World Trends and Prospects*, 1967.

Hunt, G. M., and G. A. Garratt, *Wood Preservation*, McGraw-Hill, N. Y., 1967.

Hunter, D., *Paper making, the History and Technique of an Ancient Craft*, 2nd ed., Alfred Knopf Co., N. Y., 1947.

Kazlowski, T. T., *Growth and Development of Trees*, vols. 1 and 2, Academic Press, N. Y. 1971.

Marchessault, R. H. (ed.), *Proceedings of the Sixth Cellulose Conference*, Wiley, N. Y., 1969.

Panshin, DeZeeuw, and Brown, *Textbook of Wood Technology*, McGraw-Hill, N. Y., 1964.

——————, Harrar, and Bethel, *Forest Products—Their Sources, Production and Utilization*, 2nd ed., McGraw-Hill, N. Y., 1962.

Stamm, A. J., *Wood and Cellulose Science*, Ronald Press, N. Y., 1964.

Stephenson, J. N. (ed.), *Pulp and Paper Manufacture*, McGraw-Hill, N. Y., Vol. I, 1950, Vol. II, 1951.

Tsoumis, George, *Wood as a Raw Material: Source, Structure, Chemical Composition, Growth, Degradation and Identification*, Pergammon Press, Elmsford, N. Y., 1968.

USDA Forest Service, Forest Products Laboratory, Madison, Wisc.: various publications, such as no. 106, *The Demand and Price Situation for Forest Products*, 1967, and Miscellaneous Publication 861, *Products of American Forests*, 1961.

Zimmermann, M. H. (ed.), *The Formation of Wood in Forest Trees*, Harvard Univ. Press, Cambridge, Mass., 1964.

Zivnuska, J. A., *U. S. Timber Resources in a World Economy*, published for Resources for the Future, Johns Hopkins Press, Baltimore, 1967.

CHAPTER 6

Forests, Man, and the Future

IN THE PRECEDING CHAPTERS SOMETHING of the material importance of world forests, especially those of the United States, has been pointed out. There also exist, however, many less tangible factors that cannot be measured either by number of trees or by dollar value. The kinship of man and forest historically, the recreational values of the forest, the stabilizing influence of the forest upon the environment, and the multiple uses of the forest that the future is certain to bring are all of interest and concern. The forest has sponsored civilization, as a source of materials and an environmental haven less harsh than the prairie, desert, and tundra. At the same time it has been an impediment to developing agriculture, requiring much energy for its clearing and release of its soils for conventional food crops. There can be little doubt that virgin forests of any great extent will not much longer withstand mankind's expansion. More and more, forests of the world will come to be managed for their "tree crop" just like other lands. Man's problem is to strike a proper balance between his material needs for forest products, and the recreational, esthetic, wilderness, and environmental protection wants which the forest has traditionally supplied.

Compared to the natural condition in which half the world's land surface was in old-growth trees, and within recent centuries when fine forest clothed about half of North and South America, a third of Eurasia and a fourth of Africa—nearly 4 billion hectares all told—rather little virgin forest is left in the world (see Table 6-1). The once-abundant forests of North America were rapidly and wastefully exploited progressively westward from the east coast, and only in the western belts are there relatively meager stands of large virgin timber remaining (see Table 6-2). Paul B. Sears philosophizes ("What Worth Wilderness," *Bull. to the Schools*, State Univ. of N.Y., Mar. 1953).

"The State of Ohio, containing about 40,000 square miles, was once a magnificent hardwood forest. . . . But somehow it never occurred to anyone to set aside a square mile, much less a township, of primeval vegetation for our future generations to see and enjoy. This could have been done for less than the cost of a single pile of stone of dubious artistic and cultural merit. . . . We ought to remember, too, that in a large measure our power and leadership are based upon the lavishness of nature, building undisturbed through milleniums. . . . Is it unworthy in our enlightened day to commemorate, by generous preservation, the natural wealth which has been the life blood of our economy? . . . The biologist and others who would design intelligent land-use, must have their norms or standards of measurement. And these norms, to a large degree, are to be found in

Table 6-1

Status of World Forests about 1965.

	Growing Stock (in billions of cubic meters)	Production (in millions of cubic meters)
North America	44, about 80% softwoods	370, about 85% softwoods
Latin America	79, nearly all hardwoods	35, about 60% hardwoods
Europe	12, about 75% softwoods	233, about 75% softwoods
Africa	4, nearly all hardwoods	36, about 85% hardwoods
U.S.S.R.	79, about 80% softwoods	277, about 90% softwoods
Asia (non-Soviet)	17, about 60% hardwoods	128, about 55% hardwoods
Australia and Oceania	4, nearly all hardwoods	18, about 65% hardwoods

Data from FAO *Yearbook of Forest Products Statistics*, 1966. Softwoods are preferentially harvested, and the timber drain (compared to growing stock) greatest in industrialized parts of the world.

Table 6-2

Timber Reserves in the United States about 1963.

Region	Growing Stock* in millions of cubic feet			Sawtimber† in millions of board feet		
	All Species	Softwoods	Hardwoods	All Species	Softwoods	Hardwoods
New England	30,695	16,003	14,692	54,428	30,368	24,060
Middle Atlantic	46,611	5,664	40,947	103,704	13,744	89,960
Lake	31,896	8,350	23,546	59,705	18,213	41,492
Central	27,267	1,289	25,978	91,937	4,302	87,635
North	136,469	31,306	105,163	309,774	66,627	243,147
South Atlantic	40,516	17,053	23,463	117,554	54,697	62,857
East Gulf	24,663	14,854	9,809	69,890	44,661	25,229
Central Gulf	30,359	13,379	16,980	93,619	49,746	43,873
West Gulf	38,548	17,402	21,146	131,007	75,653	55,354
South	134,086	62,688	71,398	412,070	224,757	187,313
Pacific Northwest	203,085	191,867	11,218	1,091,929	1,055,470	36,459
Pacific Southwest	55,518	54,861	657	304,634	302,298	2,336
Northern Rocky Mts.	61,738	60,855	883	271,192	269,645	1,547
Southern Rocky Mts.	36,986	32,505	4,481	147,200	139,225	7,975
West	357,327	340,088	17,239	1,814,955	1,766,638	48,317
All regions	627,882	434,082	193,800	2,536,799	2,058,022	478,777

* Net volume in cubic feet of live sawtimber and poletimber trees from stump to a minimum 4-inch top (of central stem) outside bark or to the point where the central stem breaks into limbs.

†Net volume of the sawlog portion of live sawtimber trees in board feet log scale, International $\frac{1}{4}$-inch rule. Live trees of commercial species containing at least one sawlog. Softwoods must be at least 9.0 inches in diameter breast height, except in California, Oregon, Washington, and coastal Alaska where the minimum diameter is 11.0 inches. Hardwoods must be at least 11.0 inches in diameter in all states.

From *USDA Agricultural Statistics*, 1967. Although growing stock is not much greater in the Pacific Northwest than in other regions, the Pacific Northwest accounts for by far the greatest reserve of softwood sawtimber.

the complex pattern of inter-relationship represented by the undisturbed natural community."

This is practical understanding as well as sincere appreciation of the disappearing virgin forest! The forests of India are three-fourths gone, and even forests as remote as those of the Amazon valley and central Africa are being degraded by selective removal and gradual elimination of the finer, more important species (a process called high-grading). The loss is not only of the great trees, but of all associated fauna and flora dependent upon the natural habitat. Thus not only will the virgin logs so in demand for rotary veneer and large timbers become depleted, but many other forest species (the usefulness of which may not even yet have been discovered) suffer decimation and possible extinction.

What influence forest decimation has upon climate, hydrology, and watershed management is not entirely clear. Doubtless it varies greatly in different parts of the globe. In an age of increasing air pollution, the forest may come to be regarded more highly as a purifier of the atmosphere than solely as a source of wood products. The air over densely populated, urbanized parts of the world, such as Japan, already shows an oxygen deficit; more oxygen is consumed and more carbon dioxide emitted than the vegetation can restore to balance by photosynthesis. Only when the air mass moves on to extensive forest or agricultural lands is the oxygen replenished, the pollutants dissipated.

Some of the early accepted relationships between forest and stream flow are open to question. The absorption and release of rain from a watershed is due more to soil quality than to the vegetational cover, although of course the vegetation does influence soil structure. There is no question but that the forest prevents soil loss by erosion, but the forest may or may not retard the melting of snow, increase precipitation,

retard evaporation, or aid water infiltration (by breaking impact of the rain, slowing its access to the soil, and channeling it to litter-covered or permeable spots). Little is known of the relationships of albedo and of energy exchange for various forest dominants, differing stands and conditions; this may have influence upon water absorption and transpiration. Certainly the forest canopy, especially important in the tropical forests, is deserving of more research. And how much rainfall is to be left for the forest, how much to be drained off with the greatest possible efficiency for domestic purposes in the valley below, is a crucial question in urbanizing areas.

Heretofore man has little reckoned what might be the more subtle consequences of felling the forest. Tree litter contributes about 3,400 kilograms per hectare annually in Wisconsin, and as much as 40 per cent of the nitrogen thus cycled is temporarily held in the herbaceous understory. Clear-cutting of a forest upsets these balances and causes many side-effects. Nitrogen and various cations are quickly lost, entirely aside from any soil erosion which may also occur (see Table 6-3). This can lead to marked change in the quality of the streams draining the area, even to eutrophication. With tropical lateritic soils, removal of the forest in favor of cultivation causes them to "set up" brick-hard

Table 6-3

Nutrient Loss Associated with Clear-Cutting a Forest in New Hampshire (forest cut in winter of 1965–66)

Forest Area Tested	Nitrate–Nitrogen Loss or Gain (kg of N/ha)
1. Test forest year before cutting	1.5 gain
2. Test forest year after cutting	53.6 loss
3. Adjacent uncut forest same year as 2	3.0 gain

Data from F. H. Bormann, G. E. Likens, D. W. Fisher, and R. S. Pierce, "Nutrient Loss Accelerated by Clear-Cutting of a Forest Ecosystem," *Science* **159**, 882 (Feb. 1968).

in a relatively few years. In most cases such soils are suited only to uncultivated tree crops.

Although the forest is beneficent to neighboring urban areas, the reverse is scarcely true. The lethal effect of dusts and gases released by ore processing (such as copper, sulphur, and fluoride compounds) are well known, affecting forest "downwind" for miles. More recently it has been realized that exhaust gases and ozone from the city (e.g., Los Angeles smog) are at least partly responsible for debilitating nearby forest (ponderosa pine stands in the mountains east of Los Angeles). Needle blight and tipburn of white pine in the Northeast became evident only as urbanization proceeded in that region. "Decline" of many native species (maple, ash, oak) is noted in the forests of the eastern United States, and is no doubt due to ecological upset as well as from introduction of virulent pests. Pressure upon the land will continue to increase; a problem for forestry is to decide which species are best adapted to the forests of the future, sufficiently productive in the changing environment. Whittaker finds that volume increment (measured in cubic centimeters per square meter per year) varies as greatly as from 258 for white oak in oak–hickory forests of eastern Tennessee to as little as 21 for hard maple and shortleaf pine in the same environment. Yet at the mountain tops in the Great Smokies nearby, *Abies fraseri* and *Picea rubens* together yield above 500! Such local considerations will have to be taken into account for efficient cropping of the forest.

The economic importance of forests is obviously great. Worldwide forest products were estimated to be worth more than forty billion dollars for 1965. Net export value (in millions of dollars) ran about 630 for North America, 430 for the U.S.S.R., and 120 for Africa; net import costs were 950 for Europe, 370 for Asia (exclusive of the U.S.S.R.), and 210 for Latin America. In the United States

nearly as many people have been employed in some phase of timber production and fabrication as in farming, with the timber industries accounting for some 3 or 4 per cent of the national income. Forestry is especially important to the economy of such states and regions as Oregon (with 30 million acres of forest) and the southeastern pine belt (collectively about 150 million acres in several states). The Pacific Northwest and the Southeast offer typical contrast of younger forest lands and those longer settled.

The "big timber" country of the Pacific Northwest is awesome for its abundance of wood. Much is still public land, and questions relating to management arise that include grazing, mineral, water, and recreation rights. Even in this land of abundant forest, some still virgin, the accent is now upon good management. A thinning of second-growth forests once clear-cut is possible in as little as 50 years, and 80-year Douglas fir offers

A young stand of Sitka spruce in the ecological succession of an Alaskan forest. (Courtesy USDA.)

prime yields. Logging roads are increasingly made all-weather, and specialized equipment is used for cutting, skidding, and loading, thus helping to protect the land. Permanent custodianship aimed toward continuing productivity is gradually replacing a "cut out and get out" philosophy. The mills are of tremendous capacity, heavily capitalized, and "tree farming" a way of life.

The forests of the Southeast are thoroughly cut-over, and today mostly managed for small timber, especially pulpwood. Yet the value to the economy of the region approaches one billion dollars annually. New paper mills of great capacity are being built throughout the region, as the technology mentioned in previous chapters permits economical use of local species for pulpwood and plywood. Most timber reserves are privately held, and managed today on a sustained-yield basis. In some areas trees are being harvested faster than they are being replaced, and there is a dearth of quality sawtimber stock. In Georgia, for example, 27 per cent of the softwood sawtimber volume was in large trees in 1931, but only 14 per cent by 1967. There is a similar decrease in sawtimber-size hardwoods, too. This contributes to a relative decline in lumber production vis-à-vis pulp and plywood. Obviously, the state of the forest has quite a bearing on how this natural resource will be employed.

Silviculture. Forestry finds itself in transition away from "finding" timber to growing it—silviculture. The transition is relatively advanced in the more industrialized parts of the world such as Europe and North America, still quite retarded for most tropical rainforests. In advanced areas the forest trees are now well known, although subspecific variation and genetic qualities are still little investigated. Their life history, physical qualities, range of adaptation, and general growing requirements are all fairly well understood. Site qualifications are receiving more

attention, as a guide to tree cropping. "Cellulose forestry" concerned mainly with total growth is giving ground to more specific aims and sophisticated utilization.

The need for more efficient use of forest resources is especially recognized by the large corporations coming to dominate the forest products industry. With time running out for utilization of the capital accumulation that virgin timber represents, sustained-yield forestry is imperative. Most of the big operators have accumulated immense holdings to which they apply modern practices that grow timber efficiently. While they are building these reserves of growing stock they buy timber from the national forests, small landholders, and farm woodlots to tide them over. Thus large commercial holdings are generally being conserved, small holdings exploited. The latter are typically less well managed, and their eventual exhaustion as a significant source of timber seemingly inevitable. Whether the larger, better-managed holdings will be sufficient to supply the timber needs of the future depends upon many factors, not the least of which will be the cost of end products made from wood vis-à-vis competing substances such as plastics and metals.

Many facets of science are brought to bear these days on the growing of forests. Productivity has been increased, as we have seen, by better utilization of all components of the forest and all parts of the tree. No doubt productivity will be further improved as studies of energy entrapment by different forest ecosystems progresses. Then forest efficiency can be better compared to crop efficiency, with alternatives weighed for use of any given parcel of land. In the meanwhile trees will contine to be grown to a harvestable age as depicted in Fig. 6-1, by such appropriate techniques as are pictured in Fig. 6-2.

Regeneration of tree stands has received much attention. While nature will eventually deliver a climax forest, natural regeneration can be slow and erratic. And in many cases the forester wants a subclimax stand, as with the southeastern yellow pines. Techniques for achieving such ends will vary with the site and the species, but with conifers especially initial establishment relates primarily to mechanical qualities of the seedbed. Whether a site is best reseeded naturally from seed-bearing trees left uncut, or seeded by man (perhaps with select stock, perhaps distributed from the air), or by planting seedlings grown in a nursery will be determined largely by economics and accessibility. Logged-over regions of New Zealand, for example, have been efficiently reseeded from helicopter with *Pinus radiata*, the seed coated with a fungicide and bird repellent.

Generally, direct seeding represents the most economical means for regenerating a forest. This is particularly appropriate for large, inaccessible areas seeded from the air, and with species such as longleaf pine which transplant poorly. Sometimes the site must receive preliminary preparation on the ground, such as discing, to make it receptive to the seed. Almost inevitably there is some loss of seed to wildlife, so that tree seeds are generally treated with repellents and poisons. Even wild pigs can be deterred from eating acorns treated with a substance such as cresol.

Only within recent years has the production of tree seed become much of an industry. Tree seed is largely collected from felled timber, and in some areas (such as in Douglas fir forest) from squirrel caches. Where the trees are not so large, as with slash pine in the southeastern belt, mechanical tree shakers have been developed which cause the cones to drop without injury to the tree. It is said that such a trunk shaker can remove in a few seconds as many cones as a man might pick in an hour. In tests in Mississippi trees up to 20 inches in diameter have yielded nearly a bushel of cones in less than 10 seconds by shaking. Most tree seed is still collected from wild trees, but select genetic sources are gradually being provided, especially in Europe. The difficulty in accumulating proven lines is obvious, considering the length of a tree's life cycle, the paucity of genetic information about most species, and the fact that specimens which yield seed best are not necessarily the fastest growing or best in timber quality. Nevertheless one can expect the forests of the future to be increasingly planted to semi-cultivated stock, grown from seed that is carefully stratified and stored for optimum germination.

Nursery seedlings are often used for replanting, especially of small woodlots restocked

Tree shaker used to harvest cones for seed. (Courtesy Gould Bros. Inc.)

A tree is born.
It grows. It dies. But a forest
can go on forever.

In a commercial forest, each tree should be harvested while still healthy, and converted into useful products for man before it deteriorates and decays. This process, is in fact, the essence of modern forest management. Some trees are cut at maturity; some earlier, to give their neighbors more food, water or light. And to keep a forest growing perpetually, sturdy seedlings continually start their new growth cycle.

Replacing the trees we cut with healthy new stock is important to St. Regis, because wood is our basic resource. From it we make printing papers, kraft paper and board, fine papers, packaging products, building materials—even paper plates and school supplies.

Thus, the life of the forest is St. Regis' life. That is why we, together with other members of the forest products industry, are vitally concerned with maintaining the beauty and utility of America's forests for the years to come.

ST REGIS

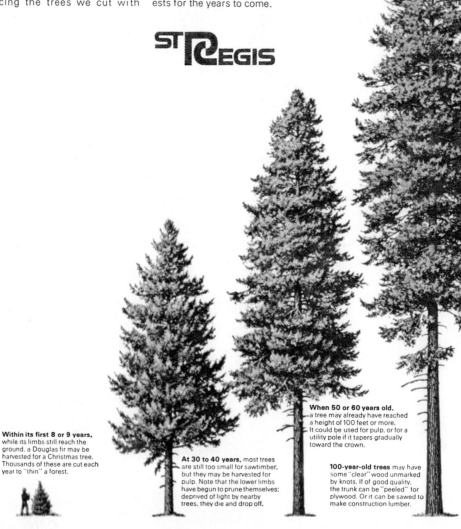

Within its first 8 or 9 years, while its limbs still reach the ground, a Douglas fir may be harvested for a Christmas tree. Thousands of these are cut each year to "thin" a forest.

At 30 to 40 years, most trees are still too small for sawtimber, but they may be harvested for pulp. Note that the lower limbs have begun to prune themselves: deprived of light by nearby trees, they die and drop off.

When 50 or 60 years old, a tree may already have reached a height of 100 feet or more. It could be used for pulp, or for a utility pole if it tapers gradually toward the crown.

100-year-old trees may have some "clear" wood unmarked by knots. If of good quality, the trunk can be "peeled" for plywood. Or it can be sawed to make construction lumber.

Figure 6-1 Maturation and decline of a forest tree. (Courtesy St. Regis Paper Co.)

Trees continue to grow after their first centennial, but at a slower rate. In virgin forests, some Douglas firs survive for hundreds of years and reach heights of over 300 feet.

A declining tree, whether decades or centuries old, is easy to detect. Its branches are dying, its bark scaling off. Rot has begun to spread throughout the trunk.

Upper branches fall away, leaving a bare spike. At last the tree is weak enough to be blown down in a high wind. Or its own weight may be enough to topple the rotted trunk.

149

It is easy to destroy a forest. But treat it well, and it will yield its treasures gracefully—forever.

A forest responds to care. In many cases we do not even need to plant new trees. A forest will reseed itself perpetually—if we know when to cut and how to cut. When and how depend on many things. The kinds of trees you are growing. The way their seeds are spread. Their best growing conditions.

Knowing these things—and a thousand more—is vital if the life of the forest is your concern. And the life of the forest is St. Regis' life. To its treasures we add the resources of modern technology to create noteworthy printing papers, kraft papers and boards, fine papers, packaging products, building materials, and products for consumers.

By planning their operations with intelligence and restraint, St. Regis and other members of the forest products industry are helping to nurture the usefulness of America's forests, and to enhance the enjoyment and inspiration to be derived from them for generations to come.

Figure 6-2 Methods of timber regeneration. (Courtesy St. Regis Paper Co.)

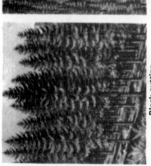

Block cutting

Trees like the Douglas fir need full sunlight for proper growth. Also, their seeds are carried a good distance by the wind.

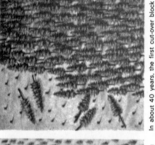

For these reasons, the accepted way of harvesting Douglas fir is to cut all marketable trees in a block of 100 acres or less. Ample standing timber is left between blocks.

In time, seeds blow in from the surrounding forest and new trees appear. The new stand is thinned periodically to provide growing space and to remove undesirable trees.

In about 40 years, the first cut-over block is mature enough to provide seed for the neighboring block. After a few more decades the first block can be harvested again.

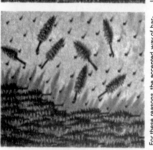

Seed-tree cutting

Southern pines also need sunlight, and their seeds are carried by the wind. But isolated trees are relatively windfirm.

It is therefore possible to remove all trees from a large area except for 4 or 5 per acre, which act as seed trees. These remain until the new seedlings are established

When the seedlings have grown enough so that they are reasonably safe from fire—5 to 10 years—the seed trees can be removed in their turn.

After about 20 years the new stand of pine should be thinned to prevent crowding. Under favorable conditions, the pines can be harvested 30 years after original cutting.

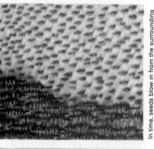

Selective cutting

Young red spruce will grow in the shade of larger trees. Thus trees of all ages and sizes are found growing side by side.

COPYRIGHT 1968 ST. REGIS PAPER COMPANY, 150 EAST 42ND STREET, N.Y., N.Y. 10017. ALL RIGHTS RESERVED.

In harvesting red spruce, the most mature and marketable trees are selected for cutting. (In this picture, they are the trees shown in a slightly darker color.)

After they are removed, their neighbors provide seed for new trees to replace them. Meanwhile, other trees are maturing, and, after 10 to 20 years....

...these trees are ready to be harvested in their turn. A forest of red spruce, if carefully harvested, can therefore provide a continuous supply of marketable trees.

A pine seedling nursery for replanting the forest. (Courtesy St. Regis Paper Co.)

through conservation agencies. Accessible burned-over lands, such as parts of the famed Tillamook burn in Oregon, are often restocked with nursery seedlings. The nurseries employ modern horticultural methods, such as soil fumigation to reduce root-rot. Douglas fir and Scotch pine in Washington and Oregon have shown nearly double the absorptive root area when grown in fumigated seedbeds as compared to unfumigated.

The emphasis today is upon sustained-yield timber production. Forest holdings are parceled out such that only so much timber is cut in any given year as is replaced by new growth held in reserve. If the climate, site, and kind of tree are such that a cycle of 100 years is required to replenish the stand with adequate timber, then ideally only one one-hundredth of the acreage would be harvested in any one year. Thus the forest would renew itself in perpetual cycle. Whether the logging be "selective"—cutting only mature trees and letting other stock grow—or whether it be clearcutting in blocks will depend upon type of tree and its requirements for regeneration. The seedlings of species such as Douglas fir have a high light requirement; the stand regenerates poorly if unfelled trees are left to shade the ground, so clear-cutting is practiced. On the other hand, redwood regenerates well from the stumps of old trees even under relatively shaded conditions, as do maple, hemlock, and beech from seed.

Douglas fir trees become merchantable for lumber when they reach about 100 feet in height, achievable on good sites in as little as 50 years. More often Douglas fir, hemlock, redwood, cedars, and other western conifers require 60 to 100 years to reach sawlog size. Early growth is typically slow, with later growth more rapid. Pulpwood can be obtained in as little as 20 years in the Southeast, and in a tropical climate in as little as 5 or 6 years. Thinnings are usually made from heavily stocked land each several years for pulpwood and poles, "releasing" the remaining timber for more rapid growth. Trees make better lumber where densely grown, the lower limbs being shaded out early to give clear (i.e., branchless) boles free of knots.

Logging the forest. Cutting the timber and getting it to market has been the main expense in forest exploitation. In thinly settled lands the virgin forests are free for the taking or nearly so; the main costs are labor and transportation. Even under modern sustained-yield forestry, these are the most important cost factors. Primitive logging was done by the settler himself, using axe and saw to fell the trees that became shelter and fuel. As civilization advanced a hardy breed of specialist arose in the United States, the lumberjack—as much a legend in the once-great lumbering states of Wisconsin, Michi-

Elephants dragging timber in India. The elephant pulls the log by a rope held in his mouth so he can see what he is doing. Elephants also perform this work in harness. (Courtesy Forest Research Institute, Dehra Dun, India. From *Unasylva*.)

gan, and Minnesota as has been the cowboy in the western plains. The lumberjack is worth a moment's reflection. Although he ruthlessly annihilated the forest, his toughness and tenacity advanced settlement westward. Tales about the lumberjack are as tall as the trees he cut,* and his endurance is legendary.

The lumberjack. The lumberjack was a distinctive product of natural selection in the rough environment of the wilderness. He was of necessity proved in the rough-and-tumble school of physical prowess. To have lasted even one year in the profession he must have been durable enough to have survived incredibly severe working conditions and to lead a life that was both primitive and

brutal. The working day was from dawn until dusk, with the lumberjack like as not wet to the waist from standing in sleet or slush. Concoctions passing for liquor in the camps, the wild fights, and the omnipresent venereal disease seemed never to faze a real lumberjack. Precarious loading on ice-covered trails in a day of few safety devices and the grinding logs of the mad spring drive on swollen rivers were always a hazard. These men spent some 8 or 10 months in the wilderness, sleeping on bough beds in a single fire-lighted bunkhouse, bathing and shaving rarely if at all, merely conversing, playing cards, or perhaps fighting for amusement—but all the while accumulating a jackpot of wages. Is it any wonder, then, that after the spring drive was in, the accumulated wages were invested in one ephemeral but magnificent spree? The lumberjack had lived months in the wilderness awaiting the moment when he might "blow her in." It was not uncommon that after a first night in Bangor, Saginaw, Muskegon, or Seattle, or in any of a number of other towns whose names are famed in logging history, a paid-off lumberjack might awaken, if he could, with pockets as empty as when he left for the woods several months before. Thus did the logger first cut his way through Maine, then west to the Great Lakes states, thence haphazardly into the southeastern pine forest but in force into the Pacific Northwest where the last of his kind made a final but futile stand before the inexorable invasion of "civilization."

Today mechanization has invaded the forest. For the most part the loggers are quite ordinary in appearance and behavior, often commuting daily from their homes to the place of felling over logging roads traversable by modern automobiles. The "wild bull of the timber" has become a family man; and so has passed a distinctive American type,

---

* These loggers were so tough that when they needed a shave they'd hammer the whiskers inward and bite them off!

once numbering between one and two hundred thousand souls.

Modern Logging Methods. Logging today is not of the type that Paul Bunyan would recognize. In place of the hand axe there is a powered saw, and huge tractorlike machines manipulate giant tree shears without the operator ever setting foot on forest soil. Another machine lifts the felled tree and snakes it through a tunnel of saws and blades that lop off the limbs, cuts the trunk into segments, and even shucks the bark. There is other equipment that can feed the logs into a chipper reducing it right in the woods to a pile of chips ready for the pulp mill. Where once the "high climber" precariously ascended one of the tallest and strongest trees, to cut away its top for the fastening of pulleys

The modern portable spar of metal for fast high-lead logging. (Courtesy St. Regis Paper Co.)

geared to snake felled logs to a central point, today there are extensible metal spars on huge rubber-tired vehicles. Where once the bullwhacker pulled fallen logs from the forest with oxen, and later with a stationary "donkey engine" operating cables to the spar tree, today machines both big and small scurry about the forest cutting and trimming trees, moving them to loading points. Where once the lumberjack skidded logs to frozen streams, for washing downstream in spring torrents, today massive vehicles lift the logs aboard huge semitrucks for delivery to the mill. Although logs still may be stored in a mill pond, and moved economically down waterways as huge rafts, it is trucks that bring them out of the forest.

It has paid the timbermen to mechanize; where a lumberjack might harvest 1 cord a day, a man with today's machines harvests as much as 7. There is saving of time, too; where it took several days to top and rig a spar tree, the new portable units with steel masts are ready for high-lead logging within 2 or 3 hours of arriving at the site. The spending of tens of millions of dollars annually for special logging equipment is not extravagance but sound business. One device, over 60 feet high, costing nearly $100,000, delimbs trees, severs them at the base, and groups them for skidding; it can fell a tree a minute even in rough terrain. Mobile chippers capable of consuming 120 feet of log per minute may cost $250,000. One mammoth tree-chewing machine weighing 40 tons, toted into the timberlands on two semitrailers, strips the bark off a felled tree in 30 seconds with steel-studded rollers, then passes the bare log to a chipping unit that reduces the log to a blizzard of chips almost instantaneously. The cost of such equipment precludes its use by the small operator, and points to increasing dominance of the industry by large, well-capitalized corporations. Still, with half of the available timber in the United States on woodlots of 20 acres or less

One-cord grapple unloading 100″ pulpwood at Northern Wisconsin paper mill. (Courtesy Omark Industries.)

(as of 1968), it will be some time before a man with a power saw is rendered completely obsolete!

Farm forestry. By and large forestry practices on the farm are seldom the best. The farm woodlot is usually of secondary concern, primary attention being given agricultural crops. Oftentimes the woodlot is degraded by turning livestock into it, compacting the soil and destroying seedlings (the cattle gain little graze from the shaded understory). Many farmers are unaware of good timbering practices, even of the value of trees they have. Black walnut of prime size may be worth as much as $2,000 per 1,000 board feet, cherry, maple and white oak suitable for veneer not too much less. Not infrequently trees of such species in sawtimber size are chopped into cordwood or fence posts. Usually the farm woodland is not sufficiently dense nor properly thinned for maximum erect growth; spreading, low-limbed wolf trees usurp much of the space while producing little bole. However, most states do employ foresters analogous to the Extension Service county agent, available for consultation by the small timber owner.

Corporation forestry. The future of forestry seems to lie more with the well-capitalized corporation than with the individual landholder, at least for forests in technically advanced countries. Only thus can specialists be teamed with expensive equipment, and modern business management provided. Operations in the forest become better organized, affording full utilization of the tree, and the company integrates vertically from growing the timber all the way to manufacturing the finished product. A glance at the annual report of any major timber company indicates the magnitude of operations. Typically there are many divisions, with offices throughout the world. One company, with sales approaching $1 billion annually, has several manufacturing plants in each of these categories: *pulp, paper, and paperboard* (kraft paper and pulp, boxboard, printing papers, fine papers, specialty papers); *packages, packaging systems, and packaging materials* (corrugated shipping containers,

wirebound wooden boxes, bags, folding cartons, plastic packaging, grocery bags, laminated and reinforced paper); *converted paper
products* (laminated, coated, and reinforced
papers for use in construction, notebooks,
paper plates, and trays); *lumber, plywood, and
other wood products* (boards and finished
lumber, sheeting, decorative plywood, overlaid panels, poles, posts, pilings, ties);
industrial chemicals (tall oil and sulphate
turpentine, arabinogalactan gum, *Torula*
yeast, lignosulphonates); *packaging machinery and food-processing systems* (pas-
teurizers, freezers, heat exchangers, systems-
engineering for dairies and food processors,
bulk tanks, etc.); *international* (with branches
or affiliates in 20 other countries to include
worldwide licensing and technical agreements). This is indeed a far cry from the
individual landholder and lumberjack,
responsible for the timber reaching market
only a half-century ago. A corporation of
this type will spend hundreds of millions of
dollars annually just on equipment and
improvements. It will employ tens of thousands of people, control millions of acres of
timberland, produce hundreds of millions
of board feet of lumber and of plywood, and
millions of tons of pulp and paper.

Enterprises of such dimension inevitably
fall heir to sideline obligations. With population pressure upon the land for recreation
and other purposes heavy, it is poor public
relations to hold immense acreages "unused"
while growing trees on a better-than-50-year
cycle. Most timberholding corporations allow
hunting on and hiking through their forest
lands, and may even provide picnic or
camping facilities. Some have branched into
the recreation business, operating ski lodges
and lake resorts.

With such massive activity, modern business methods and tight financial control
become necessary. Whereas formerly the
logger did not count the cost of growing his
timber, exploiting as he did a capital asset

provided by nature, operations today must
consider all of the costs of doing business
from the amortization of land and the
cutting and drying of wood to the manufacture and sale of consumer products. Systems
analysis is in order, and computer aid for
intelligent programming. Computerized
information about the current market makes
possible instant production-line adjustment,
such as calling for more cutting of hemlock,
less of fir, reserving more veneer peeler cores
for poles instead of sending them to the
chipper, diverting more wastes into kraft
pulping because of a surge in demand, and
so on. There is even electronic lumber
grading and guidance of milling to yield
proper proportions of differing lumber
dimensions. The computer may decide what
type of wood fits a particular need; plywood
sheathing, for example, need not be "pretty,"
only strong, and can be made from other than
high-grade logs. Perhaps the computer will
point to suitability of spruce for interior ply
in place of fir, with a quicker curing cycle
as an added economic advantage. Flow of
operations is directed to assure maximum use
of equipment, avoiding bottlenecks such as
a low-capacity glue spreader at the plywood
plant.

Overall profit depends upon maximum
economic use of each log. It is not likely that
the modern lumbermill will use steak to make
hamburger, so to speak, routing sound
sawtimber or veneer stock into pulp. Where
an item is in excess supply, research attempts
to find profitable uses for log wastes, cull
stock, and species once considered unsuitable.
For most efficient use of the timberlands it is
necessary to be engaged in all facets of wood
use—again something possible for the
corporation but not for the farmer or small
landholder.

Tropical forestry. Except on scattered
plantations tropical forestry is quite undeveloped. There is the added commercial

Hand sawing of deodar cedar railway sleepers (ties) in a hill forest of India. The forests are remote so the sawing must be done on the site. The sleepers are carried to the nearest stream and floated to the depots. (Courtesy Forest Research Institute, Dehra Dun, India. From *Unasylva*.)

A tow of mahogany logs on the Escondido river, Nicaragua. They will be loaded on an ocean vessel at a seaport. (Courtesy USDA.)

handicap of a tremendous diversity of species yielding few merchantable trees of standard type per acre. Half of the world's timber is felled for fuel and in the tropics prime logs are often utilized for this purpose. Many more great trees are sacrificed to slash-and-burn preparation of the land for crops. In less advanced lands there may be wastage of prime timber simply because there are not tools and transportation able to cope with large timbers! Smaller branches may be brought manually to market, the fine butt logs left in the forest to rot. The difficulties are due to multiple causes owing largely to educational, population, and material deficiencies. In tropical America, for example, selective logging (high-grading)—since the Spanish conquest—has impoverished the stands. No thought is given to means for the continued reproduction of valuable species. As a result, sought-after species such as mahogany, balsa, chiclé, and the like have been largely eliminated in accessible areas. The quality of the forest stand is consequently diminished, and additional specimens of the desirable species must be sought far afield at drastically increased expense.

Climatic factors, too, make it difficult to pursue economical forest practices in the tropics. Mechanical equipment quickly deteriorates what with lack of roads, trained mechanics, and repair facilities. Owing to floods, disease, and the impassable condition of roads during the rainy season, logging operations must often be seasonal. Labor is apt to be transient, listless, unhealthy, and often extremely scarce; and supervision of what labor can be secured is nearly impossible in the extensive jungle. The necessities of life are scarce, so that much of the laborer's effort is expended in winning a livelihood from the jungle rather than in doing his job. Transportation is undependable; often supplies can be moved only on human shoulders, or tediously up-river by canoe. Railroads and roads passable to a modern automobile usually penetrate little farther than the outskirts of the inhabited coastal cities. As a result, shipping costs are high and shipping itself is risky.

Long-range problems confront tropical forestry. With selected old-growth stock rapidly being decimated, where is adequate seed stock for future replenishment of these species to come from? In what way can selective cutting be altered to encourage valuable tree species having a high light requirement, which cannot adequately reproduce their kind in the density of the tropical forest? Some progress is being made in many localities, but the way is long and difficult where improved practices are contrary to the established way of life of tropical peoples.

The future of forestry. Changes are apt to be even more drastic in forestry than in agriculture. Under expected population pressure, much of the better forest lands will no doubt be devoted to food production, relegating "tree cropping" chiefly to steep, inaccessible, rocky, and other secondary sites. As the virgin stands are exhausted, an era of small-tree utilization is imminent. Perhaps small stock will be converted chemically and reformed, or perhaps laminated into larger pieces. It is estimated that in the United States over 100 million acres of present forest land will be retired to other use, leaving somewhat less than a half-billion acres available for timber; the maximum timber estimated producible from this would be about 40 billion cubic feet annually. Practical capacity would be more like 20 billion cubic feet, 10 per cent under estimated needs by the year 2000. Almost certainly wood products will become more costly, since there is lower yield and increased wastage from small stock compared to big trees. Also, the cost of holding forest land can be expected to increase.

More prefinished wood products mass-produced in the factory can be expected. Technology should yield wood for interior decoration that is essentially free of maintenance, with surface coatings that need only to be wiped as would a painted wall. There should be increasing use of residues, especially for hardboard and particleboard. One would expect greater demand for modular lumber, precut to specified size in response to high labor costs at the construction site. Prefabrication can be expected to gain. There should be greater use of panelization, such as to make movable room dividers that are not only attractive but capable of bearing a heavy load. There should be more derived products from wood, such as newer packaging materials and various chemical treatments to fortify lumber against insects, decay, fire, shrinkage, and so on. Impregnated woodpulp will doubtless find a new use. Laminated lumber is now accepted, and "nailing" with glue gaining favor.

Tree breeding. Genetic understanding of forest species has been slow to accumulate because of the prolonged life cycle for trees; 25 or more years are needed for even preliminary performance indications. A few economically important species from the forest, such as rubber and certain fruit or nut crops, have had improved cultivars selected. There has been little crossing and analysis of genotypes, however. In fact, the traditional scheme of harvesting the best trees and leaving the culls has probably worsened the genetic condition of the native forests.

Tree breeding stock increasingly is being accumulated, especially in Europe, the United States, and Japan. It will be some while before there are practical results, even until hybridization, back-crossing, inbreeding, heterosis, ploidy, and similar features are understood. Some seed is collected from superior parent trees for propagation in nurseries, where additional rogueing for characteristics detectable in the seedling stages can be practiced. Although most trees are open-pollinated, it should be possible to group advantageous parent types to yield seed of superior average quality en masse.

Populus is one of the more intensively investigated genera, especially in Europe. The poplars grow rapidly, and should be useful for pulp, small dimension stock, boxes, excelsior, and other uses for which a weak wood suffices. The species are mostly native to North America and eastern Asia (there are only four European species). Aspen has become a dominant in the cut-over pine lands of the Great Lakes area in the United States, and is being increasingly planted under intensive management in northern Europe where most of the cultivar selection has taken place. Natural hybridization occurs and the species are not clearly demarked. Diploid-chromosome count is 38, but there are some natural triploids and tetraploids within species. *P. tremuloides*, *P. tacamahaca*, and *P. aegeiros* have shown promise for interspecific crossing. Some controlled hybridization has been accomplished by means of twigs cut just prior to flowering, and by graft transplants.

Since forest trees are highly hetrozygous, there would seem to be less possibility for utilizing heterosis than with agricultural crops. Accumulation of inbreds is difficult because of the lengthy life cycle. However, it has been possible to select, from natural populations, trees that have resisted canker (*Populus*), blister-rust (*Pinus*), chestnut blight (*Castanea*), Dutch elm disease (*Ulmus*), and various oak afflictions (*Quercus*). Quercus would seem to be an especially fruitful genus for investigation, being circumboreal and for centuries an important timber tree. There are some 200 to 300 species, all diploid (24 chromosomes). In the United States much natural introgression occurs among oaks; directed crosses have been made in Russia. In northern Europe a number of "Dutch oak" cultivars have been selected out of *Quercus robur*.

Forest protection. Fire protection for the forest is now widely practiced. In the United States firewatch towers and organized groups of firefighters are accepted public services. Although the chance of a forest fire becoming started increases as forested areas are more used, the techniques for controlling fires are also improving. Modern transportation means for reaching the fire quickly, often by airlift, help ground crews to contain fires before they get out of hand. Fire-retarding chemicals as well as water may be applied from the air in appropriate instances. Aircraft is increasingly used, too, for pest control, as well as for reseeding. A helicopter can carry 1,200 pounds of pesticide, remain aloft for as much as 4 hours, with a range of over 300 miles. Extensive aerial sprayings have been made against spruce budworm, but not without controversy (insecticide can have undesirable effects upon the general ecology, even though the pest in question may be controlled).

Spraying the forest by helicopter. (Courtesy Bell Helicopter Co.)

Fighting forest fire with aircraft in Arkansas. A fire retarding slurry is dropped. (Courtesy USDA Forest Service.)

While one or another disease attacks almost all forest trees, it has not been economically practical to spray fungicides generally in the forest. More often disease control is attempted by biological programs, such as attempted eradication of wild gooseberry, the alternate host for white pine blister rust. Nearly sixty million dollars have been spent for ground crew efforts. There is some selective killing of hardwoods in the pine belt of the Southeast with herbicide, both from the ground and by air. Care must be taken that the spray not reach valuable farm acreage. Herbicidal control of brush and trees along powerline rights-of-way is routine, and in some instances chemical soil sterilization is practiced (such as around fuel storage tanks and along railroad roadbeds).

Forest fertilization. The mineral requirements of forest trees are less well known than those of agricultural crops. It is evident, however, that fertilization can stimulate growth, and in certain instances may be economically justified, especially in that forests of the future will more often be relegated to poorer planting sites. However, other management practices such as release of conifers from the competition of hardwoods by herbicidal spraying can be more valuable in increasing yield than fertilization; fertilizer is advantageous to the competing species as well as to the timber crop. On the whole, tree fertilization

Fighting forest fire on the ground. Payette National Forest, Idaho. (Courtesy USDA.)

has been confined pretty much to nurseries and orchards in the United States, where high-value tree crops justify the expense. Newer forms of light weight fertilizer which can be applied from the air may make forest fertilization more practical in the future, a practice that is growing in Scandinavia. In Sweden about 200,000 acres were fertilized from the air in 1966, mostly with pelleted urea applied about 200 pounds to the acre. Plans were formulated to fertilize over one million acres in Finland about this same time. Phosphate provides good growth response on the peaty soils of Finland, nitrogen (urea) on the drier land. It is suggested that forest be fertilized about once every 5 years, thereby increasing yield as much as 20 per cent. Whether this increased volume of timber is of sufficient value to justify the cost of the fertilizations will vary with differing forest belts and economic conditions.

Afforestation. Planting forests anew is a matter of concern in many parts of the world. Trees have long been used for shelterbelts, to break the force of the winds, and to ameliorate the microclimate. Many tree windbreaks have been planted in the mid-continental plains of the United States. Similar efforts have been made to ameliorate harsh desert climates in Israel and other parts of the Near East. Perhaps the most dramatic effort, however, is the attempt to reclaim the Rajputana desert in the state of Rajasthan, India. This "Great Indian Desert" was created from land once forested, by man's decimation of protecting vegetation. Rainfall is limited, and with cover gone the sandy soil has blown into raw dunes. These are first planted to grass strips, to stabilize the sand. Next, hardy trees such as *Prosopis spicigera* are

A new stand of Douglas fir nearly 40 years after replanting the Cispus burn in Washington. (Courtesy USDA)

Esthetic enjoyment is an increasingly important forest use, as in the Redwoods National Park, California. (Courtesy U.S. Dept. of Interior.)

planted. So long as overgrazing and tree cutting can be controlled, the windbreaks, shelterbelts, and grass plantings reinforce one another; eventually some of the trees once native can be expected to reestablish. *Acacia senegal*, and shrubby species of *Zizyphus*, *Grewia*, and *Commiphora* are being planted; *Eucalyptus* (numerous species) are being readied in plantations for later planting.

Esthetic and recreational considerations. In the United States the demand for recreational facilities is expected to double by the year 2000 over what was available in 1970. It has been estimated that some 60 million acres will be needed. Forests are going to have to bear a large share of this load, if for no other reason than that they are almost the only space left where people can "get away" from civilization. Thus forests must be regarded as having special environmental value, entirely distinct from their usefulness as a source of product. It can be safely predicted that in an overcrowded world a locale for "spiritual fulfillment" and renewed appreciation of nature will become ever more highly valued. Thus the forest, as a community of wild fauna and flora, should be

increasingly cherished. It might be anticipated, too, that the custodians of timberland would become more interested in the ecology of the forest. Studies such as those undertaken in northern Arizona, where the pinyon pine and ponderosa pine belts overlap, reveal the physiological responses that explain why here one, there the other, species dominates. Understanding critical responses of this sort is important not only for establishment of seedlings, but also for proper management of the watershed to prevent deterioration in the carrying capacity of the land.

The forest is peculiarly a record of the climatic history of a region. Note in Fig. 6-3 how the history of an individual tree can be read from its growth rings. Rainfall especially affects the width of the annual ring, the cell diameter, and the thickness of the cell wall. Studies of the growth-ring patterns in a series of living trees (consideration given to stresses due to site) compared to timbers found in archaeological ruins enable the expert to date habitation quite accurately. It is not possible to discuss details of dendrochronology here, but examination of Fig. 6-4 should make the principal clear. Figure 6-5 diagrams the relating influences which might produce

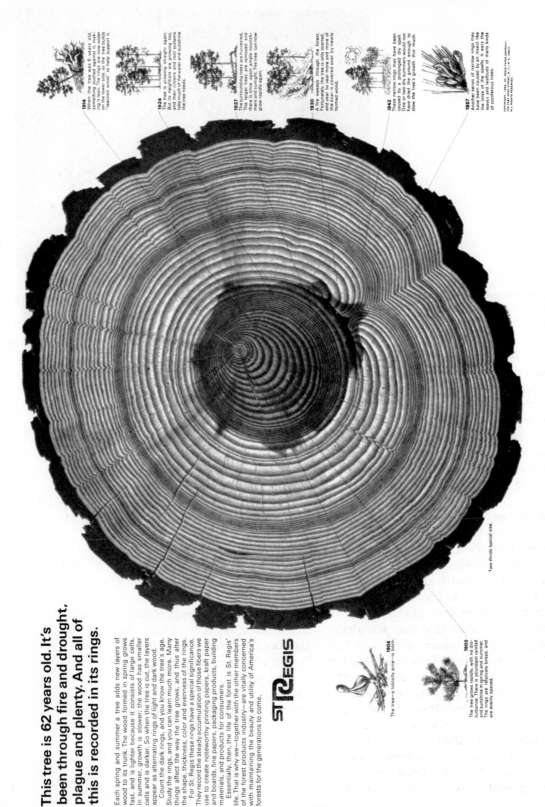

This tree is 62 years old. It's been through fire and drought, plague and plenty. And all of this is recorded in its rings.

Each spring and summer a tree adds new layers of wood to its trunk. The wood formed in spring grows fast, and is lighter because it consists of large cells. In summer, growth is slower; the wood has smaller cells and is darker. So when the tree is cut, the layers appear as alternating rings of light and dark wood.

Count the dark rings, and you know the tree's age. Study the rings, and you can learn much more. Many things affect the way the tree grows, and thus alter the shape, thickness, color and evenness of the rings.

For St. Regis these rings have a special significance. They record the steady accumulation of those fibers we use to create noteworthy printing papers, kraft paper and boards, fine papers, packaging products, building materials, and products for consumers.

Essentially, then, the life of the forest is St. Regis' life. That is why we—together with the other members of the forest products industry—are vitally concerned with maintaining the beauty and utility of America's forests for the generations to come.

ST REGIS

1904
The tree—a loblolly pine—is born.

1909
The tree grows rapidly, with no disturbance. There is abundant rainfall and sunshine in spring and summer. The rings are relatively broad, and are evenly spaced.

1914
When the tree was 6 years old, something pushed against it, making it lean. The rings are now wider on the lower side as the tree builds "reaction wood" to help support it.

1924
The tree is growing straight again. But its neighbors are growing too, and their crowns and root systems take much of the water and sunshine the tree needs.

1927
The surrounding trees are harvested. The larger trees are removed, and there is once again ample nourishment and sunlight. The tree can now grow rapidly again.

1930
A fire sweeps through the forest. Fortunately, the tree is only scarred, and year by year more and more of the scar is covered over by newly formed wood.

1942
These narrow rings may have been caused by a prolonged dry spell. One or two summers would not have dried the ground enough to slow the tree's growth this much.

1957
Another series of narrow rings may have been caused by an insect like the larva of the sawfly. It eats the leaves and leafbuds of many kinds of coniferous trees.

Two-thirds typical size.

162

Figure 6-3 The climatic history of an area can be read in the growth rings of its trees. (Courtesy St. Regis Paper Co.)

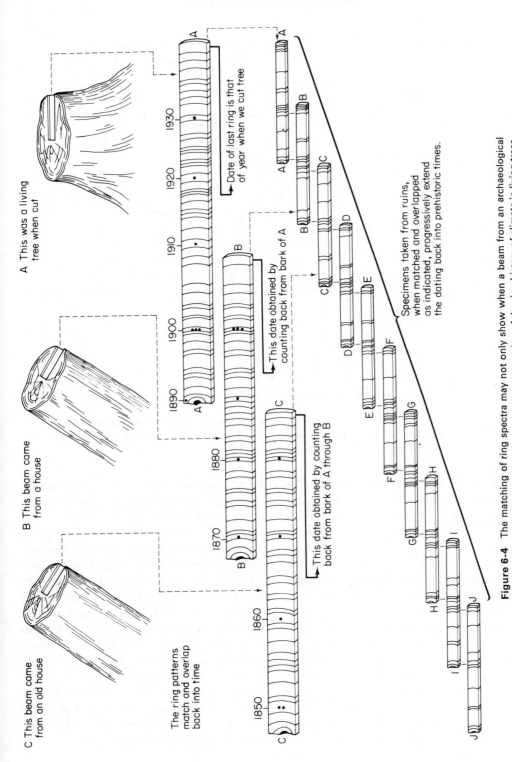

A This was a living tree when cut

B This beam came from a house

C This beam came from an old house

The ring patterns match and overlap back into time

Date of last ring is that of year when we cut tree

This date obtained by counting back from bark of A

This date obtained by counting back from bark of B

Specimens taken from ruins, when matched and overlapped as indicated, progressively extend the dating back into prehistoric times.

Figure 6-4 The matching of ring spectra may not only show when a beam from an archaeological site was felled but may also lead to a long backward extension of the ring history of climate in living trees. (After Stallings, *Dating Prehistoric Ruins*.)

163

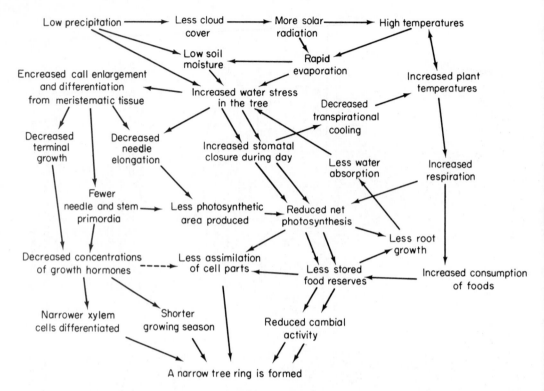

Figure 6-5 Schematic diagram of the model hypothesized for the relationship between low precipitation and high temperature and the production of a narrow tree ring. (After H. C. Pritts, *Science* **154**, 976 (Nov., 1966).)

narrow annual rings. Especially exciting has been the discovery in recent years of ancient bristlecone pines (*Pinus aristata*) in eastern California and western Nevada, some of them over 4,000 years old. The species provides a 7,100-year tree-ring chronology, utilizing both living and long-dead specimens.

The forest is source of many indirect benefits. Of course it is a wildlife refuge, protecting not only game but the watershed. Streams flowing out of forest are usually clear and cool, habitat for prized fish such as trout. For both esthetic reasons and streambank erosion control it would seem logical that waterways be forested, and that their headwaters be protected by coves of natural vegetation. The forest is a source of wild foods: mushrooms, berries, nuts, and tree

fruits. It is unfortunate that so many youths no longer experience the joys of collecting walnuts, butternuts, pecans, hickory nuts, beechnuts, chestnuts, hazelnuts, pawpaws, plums, persimmons, crabapples, and similar delicacies from the wild; as food their value is negligible but experiences connected with their gathering is beyond price. Most species which yield wild food have not been brought into cultivation and others have yielded cultivars only slightly different from those occurring in the forest. In tropical forests collection of wild fruits and nuts may contribute appreciably to sustenance of the people, and at least in the case of products such as coconut, Brazilnut, and cashew may be a significant item of commerce.

SUGGESTED SUPPLEMENTARY REFERENCES

Allen, and Sharpe, *An Introduction to American Forestry*, 3rd ed., McGraw-Hill, N. Y., 1960.

American Forests, periodical of the American Forestry Association, Washington, D. C.

Barrett, John W. (ed.), *Regional Silviculture of the United States*, Ronald Press, N. Y., 1962.

Davis, K. P., *Forest Management*, 2nd ed., McGraw-Hill, N. Y., 1966.

Duerr, W. A., *Fundamentals of Forest Economics*, McGraw-Hill, N. Y., 1966.

Fowells, H. A. (ed.), *Silvics of Forest Trees of the United States*, Agriculture Handbook 271, Forest Service, USDA, Washington, D. C., 1965.

Gerhold, H. D., R. E. McDermott, E. J. Schreiner, and J. A. Winieski, *The Breeding of Pest-Resistant Trees*, Pergammon Press, Elmsford, N. Y., 1966.

Hanson, Herbert C., and E. D. Churchill, *The Plant Community*, Reinhold, N. Y., 1961.

Hidy, R. W., et al., *Timber and Men*, Macmillan, N. Y., 1963.

Osmaston, F. C., *The Management of Forests*, Stechert-Hafner, N. Y., 1968.

Pentoney, Richard E., and W. L. Webb (eds.), *The Challenges of Forestry*, a compilation of addresses and papers from Fiftieth Anniversary Celebration, College of Forestry, Syracuse Univ. Press, Syracuse, N. Y., 1961.

Stoddard, Charles H., *Essentials of Forestry Practice*, 2nd ed., Ronald Press, N. Y., 1968.

Storer, John H., *The Web of Life*, Devin-Adair, N. Y., 1953.

Tennessee Valley Authority, *Forest Fertilization, Theory and Practice*, Symposium on Forest Fertilization April 1967 at Gainesville, Fla., Muscle Shoals, Ala. 1968.

USDA Yearbooks, pertinent titles such as *Trees*, 1949, and *Outdoors USA*, 1967, especially; but also *Water*, 1955; *Soil*, 1957; *Land*, 1958; *A Place to Live*, 1963; *Farmer's World*, 1964; *Consumers All*, 1965.

Wackerman, A. E., et al., *Harvesting Timber Crops*, McGraw-Hill, N. Y., 1966.

CHAPTER 7

Fibers

"FOOD AND FIBER" ARE CUSTOMARILY LINKED in agricultural citations, suggesting importance for fiber on the order of food. Many millions of tons of fiber enter commerce each year and more are of incidental local use. Without fiber there would be neither cloth nor cord, neither tires nor brooms. Any raw

Hanks of well-cleaned stem fiber. (Courtesy USDA.)

material out of which yarns and woven materials can be made, or ropes, brushes, baskets, mattress stuffing, and so on manufactured, is a potential source of commercial fiber. Hundreds of substances are involved. A previous chapter discussed wood fiber, and the synthetic fiber rayon, derived from chemical manipulation of plant cellulose. This is the heyday of the synthetic fiber, although most are now derived from petrochemicals rather than plant cell wall. Animal and mineral fibers have been important since time immemorial, especially wool and silk, and asbestos. Our concern here, however, is with natural vegetable fibers. Along with wool they have been the mainstay for textile making since spinning and weaving were first invented.

With usable fiber to be found in hundreds of genera, in scores of plant families, it is no wonder that now one, now another, has gained ascendency during the course of history. To a great extent fortuitous circumstances rather than cell wall qualities determine which fibers become commercially successful—such factors as the cost of production, the simplicity of freeing the fiber from unwanted tissues, the adaptability of mechanical devices to processing the species, and suitability of the plant to prevailing agricultural procedures. So nearly "perfect" a fiber as ramie has never become a significant factor in world

166

trade mainly because of the difficulty in separating the fiber from the stem! Within the last century cotton has become the most important natural fiber, being superlative for the weaving of soft, absorbent, durable cloth that is washable.

Crude fiber, if nothing more than strips of bark, must have been used when man took his first faltering steps toward civilization. Along with animal sinew such materials could be used to bind sticks or supporting elements in structures, aid in carrying and forming containers, and find practical applications in snares and weapons. Archeological evidence is somewhat scanty, since fibers do not preserve so well as do pottery and bone. Ancient findings are mostly preserved in arid caves, such as in the drier areas of the western United States. In the Tehuacan valley of Mexico cotton and *Agave* fibers nearly 8,000 years old have been discovered (Fig. 7-1), and vegetable fibers of other sorts nearly as old in locations

Figure 7-1 Plants recovered from the excavations at Tehuacán, Mexico, came into cultivation at different times over a 7,000 year time period. The fiber plants *Agave* and *Gossypium* (right side) were nearly as early as the food plants. (After C. Earle Smith, Jr., *Economic Botany* **19,** 79 (Jan.–Mar., 1965).)

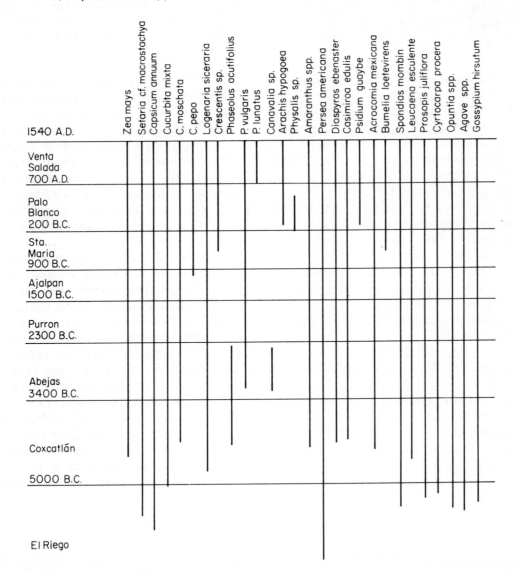

Even a thatched roof of palm is a use of fiber. Mixed plantings about a house in Panamanian back country. (Courtesy USDA.)

so diverse as Peru, Utah, and the Near East. Fragments of a woven palm leaf fabric, perhaps 12,000 years old, have been uncovered in Mexico. *Yucca* and *Typha* mats almost as old have been found in a Nevada cave, in a region where certainly the early Indian tribes wove baskets. Swiss lake dwellers of 8000 B.C. grew and wove flax. Ancient Egyptians 70 centuries before the birth of Christ used flax as well as papyrus. Records show that the Chinese of 2000 B.C. cultivated hemp; and cotton cloth is recorded to have been used in India 3,000 years ago. According to Herodotus (445 B.C.), the clothing of Xerxes' army was of cotton.

Fiber production has a history of gradually improved technology. *Agave* fragments nearly 9,000 years old from Coxcatlán cave in Mexico seem to show chewing away of soft tissue to extract the fiber. Other fibers were mechanically scraped by hand much as they still are today in remote areas, and some even submerged in water to ret (a process which will be discussed shortly). A net of *Apocynum* fibers complete with drawstring, from Danger Cave, Utah, has been dated about 5000 B.C.

These are bast fibers from the stem which not only had to be carefully separated, but skillfully knotted to create the net. Was not the discovery of joining fibers by knot, and later the spinning of yarn to make fabric weaving possible, on a par with the discovery of fire and the invention of the wheel for furthering development of civilization! More will be said about ginning, spinning, and weaving, as individual fibers (especially cotton) are discussed.

The best textile fibers, like the most useful wood fibers, are nearly pure cellulose. So adequately and inexpensively do plants provide cellulose fiber, that man has had little incentive to synthesize it. Rather, cellulose from the plant cell wall has been transformed into various types of rayon, an industry experiencing remarkable growth just before mid-century. Rayon manufacture is a reasonably simple means of utilizing a raw material available in the less technically advanced parts of the world; it will no doubt hold its own, although in recent decades other synthetic fibers have surpassed it in importance. By 1969 use of synthetic fibers

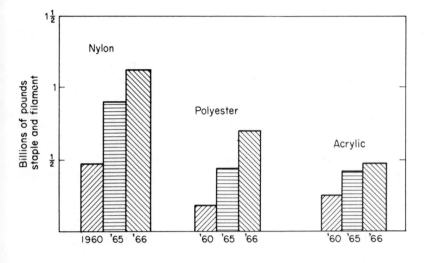

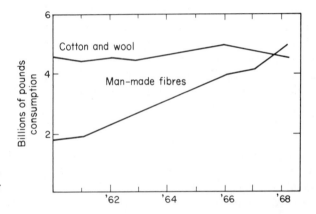

Figure 7-2 Growth of leading synthetic fibers in early 1960's. (After *Business Week*.)

more than equaled that of natural fibers. Figure 7-2 shows the growth in production of three familiar classes of synthetic fibers from 1960 through 1966. Synthetic fibers accounted for over 4 billion pounds of production in the United States alone before the end of the decade.

Price has considerable influence on the widespread acceptance of a fiber. Cotton commands a great international market because by and large it is quite inexpensive (in the United States, however, various subsidies and supports have often kept the domestic price above the world price). Most of the synthetic fibers have found acceptance at a premium over cotton, although severe competition among them frequently drives prices down to the general range of natural fibers. Special characteristics, such as the great strength of nylon, and the permanent-press feature of polyesters (e.g., Dacron), make them preferred for certain uses instead of cotton, or perhaps in combination with it. While we are not here primarily concerned with synthetic fibers, it must be realized that their competition drastically influences natural fiber production. Synthetic fibers with special attributes will no doubt continue to make inroads in the near future; but for the long haul, what with increasing population bringing higher prices and gradual exhaustion of nonrenewable resources, fiber from plants should eventually gain most of the market.

It may be well to review more fully just what is meant by the word *fiber*. There are a great many shades of meaning and degrees of inclusiveness, depending upon the particular sense in which it is used.

THE FIBER—BOTANICALLY. Previous chapters have described certain plant cells called fibers. Fibers are important constituents of wood, giving to it many of the mechanical and physical properties discussed in Chapter 5. In every case, however, it was the mass of wood cells, not the fibers alone, that was important. In fact, the word *fiber* is often used very loosely to describe any and all cells; for example, even paper pulp derived from softwoods, technically without fiber cells, is spoken of as consisting of fibers. Fiber cells, as considered distinct from vessel, tracheid, and parenchyma cells in wood, are morphologically distinguished by their slender, tapering form, very thick walls, and usually simple pits (perforations) in the secondary wall. As identified by these criteria, fibers are also found in tissues other than wood (xylem), notably in pericycle, phloem, or even cortex. No matter from what tissue derived, fibers are typically 1 to 250 mm. long, only about 1/100 to 6/100 mm. wide, with walls frequently 1/100 mm. thick. Almost invariably they are dead cells in maturity, with an exceedingly narrow lumen (cavity where living protoplast had been). They usually occur in groups or bundles and are closely cemented to one another, particularly at the fusiform tips which dovetail obliquely with the next (longitudinally) adjacent fiber.

THE FIBER—CHEMICALLY. Chemically, the cell wall of fiber cells consists, as do most cell walls, chiefly of cellulose, with percentages of lignins, hemicelluloses, and occasional other substances. Usually, however, the wall is more nearly pure cellulose, uncontaminated by dyes, tannins, resins, and other substances,

and with lesser percentages of lignin than is common for most plant cells. Fibers are thus formed chiefly of carbon, hydrogen, and oxygen. Flax, hemp, and ramie have the cellulose largely associated with pectic materials, while jute and the so-called "hard" fibers are more lignified. The high percentage of cellulose in fibers correlates positively with such desirable characteristics as strength and durability. Likewise, low moisture content of the cell wall is usually indicative of physical superiority.

THE FIBER—COMMERCIALLY. In commerce, "fibers" include practically all small, thin, slender fragments of many substances. There are fibers of mineral origin (asbestos, spun glass, and so on) and of animal origin (wool and other animal hair, silk, feathers, and so on), as well as the more important plant fibers. In the latter category, anything from multicellular twigs and roots to unicellular seed or epidermal hairs may be called fibers in the trade. Dried Spanish moss, the husk of a coconut, the flowering stalk of broomcorn, the outer bark of many palms, all are marketed as fibers. Fibers of commerce may have exceedingly diverse origins, cellular composition, and microscopic morphology.

Classification of fibers. Three important groupings of fibers, exclusive of wood fibers, may be made, based upon the type of fiber and from what part and kind of plant it is derived.

1. SOFT, STEM, OR BAST FIBERS. Fibers found grouped outside the xylem, in the cortex, phloem, and pericycle (bark), are termed stem or bast fibers, or, in the United States, soft fibers. None of these labels has much to recommend it, and all have on occasion fallen into disrepute. Fibers of this category are developed by many dicotyledonous plants. The cells themselves are true botanical fibers. They are typically found grouped into clusters of several or of many cells, and the whole cluster may in some cases serve as the "fiber" in spinning. Each fiber cell

is tenaciously cemented to the adjacent fibers by the pectic middle lamella. The strands so formed are very strong and durable. For man's use, these fibers and bundles of fibers are ordinarily freed from other stem tissues by retting, a process utilizing the action of microorganisms in a suitably moist environment to rot the weaker cells. Retting may be accomplished under water, as, for example, in India where jute is immersed in ponds and rivers. In this case both aerobic and anaerobic organisms co-operate to dissolve away the cementing materials and free the fibers. Or dew retting may be practiced, as is done for flax, where the cut stems of the plant are subjected to aerobic conditions by merely being left in the field where it was harvested until microbial activity, under the influence of moisture from nocturnal dews and fogs and rains, frees the fibers. Many of the most important fibers (flax, ramie, hemp, jute, and the like) are of the soft (stem, bast) type. They are typically very durable, able to withstand bleaching or other harsh treatment.

2. HARD, LEAF, OR STRUCTURAL FIBERS. Fibers termed hard, leaf, or structural are in reality strands of small, short cells found in monocotyledonous plants. They constitute the supportive and conductive strands, principally in the leaf (few monocotyledons have conspicuous "woody" stems), and as such are botanically termed fibrovascular bundles. The fibrovascular bundles consist of both xylem and phloem and various ensheathing cells, and are found more or less scattered through a weak pithy matrix of the leaf or stem. The cells are lignified to a greater or lesser degree, and hence are "hard" in comparison with the dicotyledonous "soft" fibers in which the cellulose is largely associated with pectic materials.

The whole fibrovascular bundle serves as a unit fiber. Such fibers are usually separated, not by retting, which would dissolve the cementing substances, freeing the then useless small cells, but by being mechanically scraped free of the pithy matrix through which they are scattered. Hard fibers cannot ordinarily be bleached or chemically treated, and on the whole are less durable than soft fibers. Because the mechanical scraping to free the fibrovascular bundles must in large part be done tediously by hand, hard fibers are grown almost exclusively in regions where cheap labor is available. Most of the important cordage fibers (viz. Manila hemp, sisal, New Zealand hemp, Mauritius hemp, istle, and the like) fall into this category.

3. SURFACE FIBERS. The only other important category of fibers besides hard and soft fibers (and, of course, wood fibers, previously mentioned) is that of surface fibers (those borne on the surface of stems, leaves, fruits, seeds, and so on). The most important surface fibers are seed hairs, consisting of single-cell growths from the seed surface which are diurnally thickened by deposition of nearly pure increments of cellulose. Quality of seed-hair fiber is high. Separation of the fiber is by ginning,* or sometimes by hand pulling. The most important plant fiber in the world today, cotton, is a surface fiber. No other surface fiber is of first-rank importance, although kapok has for years been a standard stuffing fiber. A shortage of kapok during World War II led to the development of milkweed seed fibers as a substitute. Surface fibers may be derived from any of a great diversity of plants.

Fibers are frequently classified according to the use to which they are put. Aside from paper making, a topic discussed in a previous

---

* The principle of modern ginning is identical with that devised by Whitney in 1793, though with innumerable refinements, of course. It involves the catching of the surface fibers on a battery of toothed disks or combs and concomitant pulling through slots too narrow for passage of the seed. The fibers are in this manner torn from the seed, and are then brushed loose or blown free from the disk or comb.

chapter, a general grouping into four categories is commonly made:

A. TEXTILE OR APPAREL FIBERS. This category includes all fibers used for the manufacture of fabrics. The quality of fiber needed for textile fabrics is usually high, with strength, pliability, pleasant feel, durability, and the like all being of importance in determining its suitability for the manufacture of wearing apparel, lace, canvas, and other commercial fabrics. The important fibers so used are principally cotton, with lesser quantities of flax, ramie, and hemp. For coarse fabrics of inferior quality, such as burlap, bagging, sacking, and so on, jute is the principal fiber utilized, with smaller quantities of cotton, flax, hemp, and a few of the hard fibers also finding some use.

B. CORDAGE FIBERS. Cordage fibers are those utilized for making tying twines or rope and binder twines. For the first, the softer, more flexible fibers are most desirable; consequently, soft fibers, especially jute, cotton, and hemp, but also lesser quantities of flax and several hard fibers, are used. For rope and binder twine, hard fibers, principally abacá, sisal, New Zealand, and Mauritius hemps, but also lesser quantities

Examples of native use of braiding fibers from wild plants. These baskets were woven from *Ischnosiphon* leaves (Marantaceae) in the isolated Vaupes valley of Colombia. (Courtesy Paul H. Allen.)

of cotton and several soft fibers, are utilized.

C. BRUSH OR BRAIDING FIBERS. A great variety of miscellaneous twigs, leaves, barks, splints, and true fibers may be utilized to make brushes and brooms and various braided articles such as hats, mats, baskets, rugs, and the like. For brushes and brooms, istle and sisal (hard fibers), piassava (surface fiber of palm leaf and stem), broomcorn (plant inflorescence), or a number of kinds of twigs and other fibers may be used. For braided materials, all the hard fibers and various grasses, leaves, sectioned stems, and so on commonly serve as weaving material.

D. FILLING FIBERS. Filling fibers are those used for stuffing (upholstery, mattresses, life preservers), caulking (seams between planks, barrels, plumbing), and reinforcing (wallboards, compressed pulp, plastics). Chief stuffing fibers are kapok, cotton, several hard fibers, jute, Spanish moss, and locally innumerable grasses and other materials. Hemp and jute are the principal caulks. Sisal, istle, hemp, and jute, commonly in tow form (fibers too short for spinning), are perhaps the chief reinforcing fibers.

Spinning and weaving. Although the first coverings that man used were doubtless animal skins, he early turned to the lighter, cooler and more flexible clothing that textiles provide. In order to have cloth and cordage, a weaving technology had to develop.

The weaving of fabrics involves first the spinning of yarn. Originally, spinning must have been done entirely by hand (as it still is today in out-of-the-way places), by straightening and thinning the masses of fibers between the fingers and then twisting them into yarn between thumb and fingers. Yarn is a continuous thread of twisted, interlocking fiber strands. Soon mechanical means of spinning yarn were developed. First there must have come the simple distaff and spindle. The distaff consisted of a short

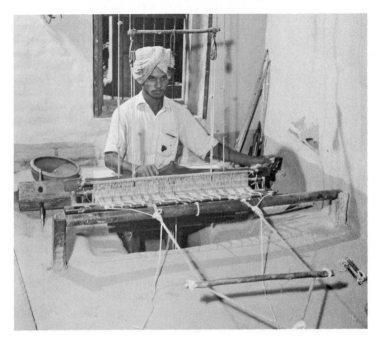

Traditional handlooming in northern India, still an important cottage industry. (Courtesy USDA.)

stick, one end of which bore loosely wrapped fiber, the other end of which was tucked beneath the arm to free the hands. One hand of the spinner drew and thinned the fiber from the distaff; the other revolved the spindle, to which the thinned yarn was drawn, twisted, and wrapped. This spindle, simply formed from a round stick of wood about a foot long, was revolved (to twist the yarn) by pushing by hand against the leg and thigh of the operator. So fundamental is this principle of spinning that modern power spinning machines still follow it.

An improvement over the distaff-and-spindle system was the invention of the spinning wheel. In the spinning wheel, a band or belt from a hand wheel serves to rotate a spindle mounted horizontally in the frame, eliminating the difficult rotation of the spindle between palm and thigh. The spinning wheel remained the standard means of spinning until the eighteenth century. Then, with the heavy demand for threads, laces, bags, sails, ropes, and twines, several inventions were made which revolutionized the textile industry. Drawing rolls and pins were devised for straightening and thinning the fiber

A modern high-speed spinning frame, typical of the mechanization that has changed textile making from a cottage industry. (Courtesy USDA.)

173

previously attenuated by hand; multiple spindles were introduced; and the whole spinning–weaving process came to be performed by elaborate machines, mechanically powered. The only fibers generally utilized in the early spinning and weaving machines were flax and hemp, grown principally in Europe. Later, slave labor in America and invention of the cotton gin brought cotton to the fore; enlargement of Indian commerce made cheap jute generally available; and in the early 1800's abacá from the Philippines showed its superiority over hemp and flax for cordage.

The mechanization of textile making has not been without some disadvantages. Production processes, such as ginning of cotton, may reduce average staple length of the fiber; personalized selection of uniform fiber is not possible as with hand spinning. It is interesting to note that in modern-day textile fabrication in Ethiopia, where this is still a cottage industry in the home, spinners usually prefer hand picked fiber to that mechanically harvested. There is likewise waste in mechanized handling, not only through some loss of fiber, but because of inclusion of detritus which later must be cleaned out and disposed of. One of the disadvantages a natural fiber such as cotton suffers in competition with synthetic fibers is a few cents per pound cost for properly cleaning up the fiber.

Several of the more important natural fibers will be discussed in the following pages. It is evident that not all useful species can be

Table 7-1

Comparative Ranking of Five Leading Hard Fibers and Four Leading Soft Fibers

(Surface fiber, cotton not included)

Characteristic	Rank				
	1	2	3	4	5
CORDAGE OR HARD FIBERS					
Durability	Abacá	Sisal	Istle	New Zealand	Mauritius
Tensile strength of strands ..	Abacá	Sisal	New Zealand	Istle	Mauritius
Length of strands	Abacá	Mauritius	New Zealand	Sisal	Istle
Fineness	New Zealand	Mauritius	Abacá	Sisal	Istle
Uniformity...............	New Zealand	Mauritius	Abacá	Sisal	Istle
Pliability.................	Mauritius	New Zealand	Abacá	Sisal	Istle
All characteristics	Abacá	Sisal	New Zealand	Mauritius	Istle
TEXTILE OR SOFT FIBERS					
Durability	Ramie	Flax	Hemp	Jute	
Tensile strength	Ramie	Hemp	Flax	Jute	
Length of fiber cell	Ramie	Flax	Hemp	Jute	
Cohesiveness	Flax	Hemp	Jute	Ramie	
Fineness	Ramie	Flax	Hemp	Jute	
Uniformity	Flax	Ramie	Hemp	Jute	
Pliability	Flax	Ramie	Jute	Hemp	
Color	Ramie	Flax	Hemp	Jute	
All characteristics	Flax	Ramie	Hemp	Jute	

From L. Weindling, *Long Vegetable Fibers*, New York: Columbia Univ. Press, 1947.

mentioned. Over 300 fiber plants have been recorded in use in East Africa alone! Those attaining commercial prominence are valued for various attributes. One comparison of prominent fibers according to selected attributes was made by Weindling (Table 7-1). Should any of these fibers fail, for reasons of cost, availability, or other cause, not only are the synthetics ready to supplant them, but hosts of additional natural fibers as well.

Cotton, *Gossypium spp.*, Malvaceae. The most important vegetable fiber in the world today is cotton, derived from the seed hair of several species native to both the Old and New Worlds. Cotton has been used for weaving fabrics for many centuries, and it must have been in cultivation for several millenia. It has been suggested that one of the earliest uses for cotton may have been consumption of the small green bolls, which are rich in sugar. Seed fiber was woven into fabric as early as 2500 B.C. (Huaca Prieta

The cotton plant, showing a branch, flower, and mature but unopened boll. (Courtesy USDA.)

excavations in Peru), and boll fragments even older are known from the Tehuacan diggings in Mexico. Long before the Christian Era cotton and cotton cloth were known in India and the Near East. Interestingly, one of the most serious pests threatening modern cotton, the boll weevil, also infested pre-Colombian cotton in Mexico, as proved by preserved remains from an Oaxacan cave dated about A.D. 900. Today nearly 12 million tons of cotton fiber (nearly 50 million bales) are produced annually throughout the world (Fig. 7-3), with the United States the leading producer, followed by the U.S.S.R., India–Pakistan, Brazil, Mexico, and the United Arab Republic.

The ancestry of cultivated cotton is quite obscure. Genetic evidence indicates that the most used species, the American tetraploids *G. hirsutum* and *G. barbadense* (upland, and Sea Island or Egyptian-type, cottons respectively), contain one diploid set of New World chromosomes and one genome from the Old World. Apparently the crossing that resulted in the tetraploid species took place so long ago (fragments in the excavations mentioned above seem to be of the tetraploid type) that transport across the Atlantic or Pacific by human beings is unlikely. Perhaps in those ancient days the present Old World species were also represented in the New World, but have since become extinct, leaving only their tetraploid descendents (from crossings with New World species) as a reminder. In historical times there has been additional crossing of different species, both natural and directed by man, and, of course, various kinds of cotton have been widely distributed around the world. Species delimitation is often uncertain, although it seems agreed that cottons fall into five diploid genome groups ($2n = 26$) and one tetraploid combination ($2n = 52$, a combination of the A and D diploid groups). In addition to the important New World tetraploids named above, *G. arboreum* and *G. herbaceum* are

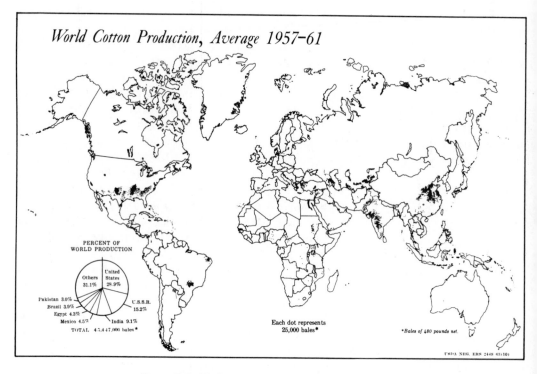

Figure 7-3 World cotton production. (Courtesy USDA.)

much grown in the Old World. There has been some speculation that *G. herbaceum* var. *acerifolium* may have been the earliest cultivated cotton along the trade routes from the Indus valley into northern Africa. This and many other entities still occur in the back country of Ethiopia, where certainly cotton was grown as early as A.D. 350. A close wild relative there is *G. anomalum*. It appears that cotton was not introduced into Egypt until the Arab Conquest of A.D. 640, and that the modern long-staple cotton for which Egypt is famed is a still more recent introduction of *G. barbadense* from South America in post-Columbian times. Recently hexaploid cotton from crossing *G. hirsutum* with the wild Australian diploid *G. sturtianum*, and doubling the triploid with colchicine, has been obtained, said to have very high yarn strength. A map indicating the distribution of Gossypium taxa, after Phillips, appears as Fig. 7-4.

Cotton's great importance is relatively recent. The Greek and Roman civilizations largely ignored it, relying mostly upon flax, wool, and silk. Of course the tropical nature of cotton made it inappropriate for much of the Roman Empire, although it certainly was grown in northern Africa and in the Near East. When the Moors invaded Spain in the tenth century they introduced a highly developed cotton technology, and Barcelona became a center for textile manufacturing, especially of sailcloth for the Spanish fleet.

Thereafter cotton's star rose rapidly. As early as the fourteenth century much raw cotton was being shipped from the Mediterranean area to the Low Countries of northern Europe for spinning and weaving. Shortly thereafter trade was established with the Far

East, and much cotton cloth manufactured in India was brought to Europe. With the first stirrings of the Industrial Revolution in England textile weaving became important, although during the 1600's weaving was still largely a cottage industry. Fabrics were usually of mixed fiber, linen for the warp and cotton the woof. Upon the invention of a series of spinning and looming devices by the mid-nineteenth century, machine weaving sup-

planted the cottage industry, and demand for cotton fiber rose tremendously. The long-staple cotton from the United States became especially in demand, leading to an economic system there which eventually spawned the Civil War. The widespread and intensive growing of cotton in the southern states during these years led to ruination of land much of which should never have been subjected to clean cultivation.

Figure 7-4 Distribution of diploid taxa of *Gossypium*. (After Lyle L. Phillips, *Amer. J. Bot.* **5**, No. 4, 328–35 (1966).)

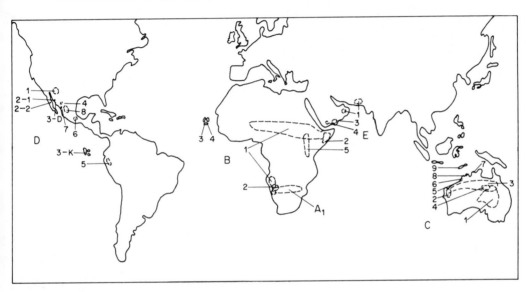

Distribution of the diploid taxa of *Gossypium*

A_1	*G. herbaceum* L.	D_2-1	*G. armourianum* Kearney
B_1	*G. anomalum* Wawra and Peyr.	D_2-2	*G. harknessii* Brandegee
B_2	*G. triphyllum* Hochreutiner	D_3-K	*G. klotzschianum* Andersson
B_3	*G. barbosanum* Phill. and Clem.	D_3-D	*G. davidsonii* Kellogg
*B_4	*G. capitis-viridis* Mauer[a]	D_4	*G. aridum* (Rose and Stand.) Skov.
C_1	*G. sturtianum* Willis	D_5	*G. raimondii* Ulbrich
C_2	*G. robinsonii* Mueller	D_6	*G. gossypioides* (Ulb.) Stand.
C_3	*G. australe* Mueller	D_7	*G. lobatum* Gentry
*C_4	*G. bickii* Prokhanov	*D_8	*G. trilobum* (DC) Kearney
*C_5	*G. custulatum* Todaro	E_1	*G. stocksii* Masters
*C_6	*G. populifolium* (Benth.) Muell.	E_2	*G. somalense* (Gurke) J. Hutch.
*C_7	*G. cunninghamii* Todaro	E_3	*G. areysianum* Deflers
*C_8	*G. pulchellum* (Gardn.) Fryxell	*E_4	*G. incanum* (Schwartz) D. Hillc
*C_9	*G. timorense* Prokhanov	*E_5	*G. longicalyx* Hutch. and Lee
D_1	*G. thurberi* Todaro		

[a] (Genomic affinity of taxa with designation (*) not established cytologically; sub-genome designation used for convenience)

Perennial cotton in Brazil. (Courtesy USDA.)

Cultivating cotton in Egypt where hand labor substitutes for mechanization. (Courtesy USDA.)

A cotton packet boat of the 1800's. (Courtesy Chicago Natural History Museum.)

An individual cotton fiber magnified approximately 2,000×. (Courtesy USDA.)

When cotton supplies were cut off by the American Civil War, the British introduced long-staple New World strains into their colonies in various parts of the world. Most of the cotton crop derives from these species, perhaps 80 per cent from upland cotton, *G. hirsutum*. Most cotton is grown as an annual, although not infrequently in North Africa and Brazil the plantings are maintained several years as perennials. In much of the world cotton is still hand picked, but only in remote areas is spinning and weaving still a cottage industry. The cotton plant may find uses other than for fiber, too, since reputedly all parts of the plant have "medicinal value" (whether consumed or applied externally). In northern Africa a fermented cake contains cottonseed, perhaps a foretaste of greater use of cottonseed oil and protein in foods from industrialized parts of the world. It has been found that cottonseed can yield about 18 per cent of an edible flour that is 65 per cent protein by a "liquid-cyclone" processing that first involves removal of the oil. As much as 25 per cent can be used successfully to enrich wheat bread.

For every 500-pound bale of cotton, 900 pounds of seed become available to furnish oil and fodder cake and some 70 to 95 pounds of linters for cellulose and rayon. But in spite of the utility of cottonseed and linters, *Gossypium* has always been cultivated primarily for the seed hairs or lint. The delicate seed hairs each consist of a single cell 1,000 to 6,000 times as long as it is wide, with a narrow lumen the length of it. The seed hairs first attain maximum length, then increase daily in cell wall thickness. Wall thickness and hence fiber quality is partially influenced during this period by weather or growing conditions. Final structural composition of the wall is indicated in Fig. 7-5. In 1 pound of cotton there are some 90 million seed hairs. As the seed ripens, the drying hair collapses into a twisted filament of nearly pure cellulose. Because of the marked cohesiveness

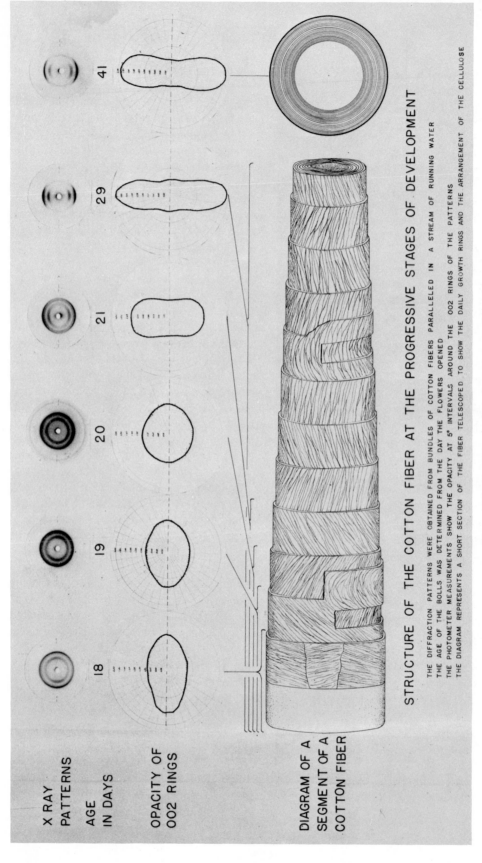

STRUCTURE OF THE COTTON FIBER AT THE PROGRESSIVE STAGES OF DEVELOPMENT

THE DIFFRACTION PATTERNS WERE OBTAINED FROM BUNDLES OF COTTON FIBERS PARALLELED IN A STREAM OF RUNNING WATER
THE AGE OF THE BOLLS WAS DETERMINED FROM THE DAY THE FLOWERS OPENED
THE PHOTOMETER MEASUREMENTS SHOW THE OPACITY AT 5° INTERVALS AROUND THE OO2 RINGS OF THE PATTERNS
THE DIAGRAM REPRESENTS A SHORT SECTION OF THE FIBER TELESCOPED TO SHOW THE DAILY GROWTH RINGS AND THE ARRANGEMENT OF THE CELLULOSE

X RAY PATTERNS

AGE IN DAYS

18   19   20   21   29   41

OPACITY OF OO2 RINGS

DIAGRAM OF A SEGMENT OF A COTTON FIBER

Figure 7-5 Drawing of cotton fiber. (Courtesy USDA.)

created by such twisting, plus surprising durability, strength, and pliability, these seed hairs lend themselves extremely well to spinning—so well, in fact, that cotton provides more clothes and cloth for man than all other textiles combined.

Since the early days of the Republic, the United States has been the world's leading producer of cotton, grown entirely in the southern third of the country. Soil exhaustion in the Deep South, together with the passing of slavery, caused production to shift westward, until today most cotton comes from lands west of the Mississippi, often under irrigation. Remunerative cotton growing calls for a long frost-free season, moderately abundant moisture, and either extensive mechanization or ample cheap labor. The vast stretches of India, the leading cotton country for 3,000 years, and of Russia and Brazil, include much suitable environment.

In North America *G. hirsutum* is grown as an annual, being planted from March 1 to June 1 and harvested in early autumn. The plant grows about 2 feet tall, and develops many side branches which progressively bear from the base outward large yellow flowers that turn orange the second day. Large seed capsules, the bolls, mature from the flowers, splitting in maturity to reveal many seeds, each with its mass of cellulosic seed hairs 16 to 21 microns wide and up to 5 cm long in Sea Island types (staple length in typical upland cottons averages only little more then 2.5 cm). Approximate percentages by weight of all the plant parts are as follows: roots, 9 per cent; stems, 23 per cent; leaves, 20 per cent; capsule husks, 14 per cent; seed, 23 per cent; lint, 11 per cent.

Quite in contrast to the early growing of cotton in the United States, and in parts of the world today where labor is inexpensive, is the businesslike production of cotton in the southwestern United States, extensively mechanized and even automated. Yields under irrigation in such locations as the Imperial Valley of California run as high as 6 bales to the acre, double or triple top production in the Southeast. Even higher yields are recorded with skip-row planting patterns, utilized under acreage restriction by the government to control production. Skip-row patterns vary from planting two rows and skipping one (often the practice in the Southeast) to planting four and skipping four (frequently the practice in the Southwest). Yield may be increased as much as 50

Foreign cottons showing different lengths of staples. (Courtesy USDA.)

per cent so long as only the rows planted are counted as acreage. However, solid planting is generally more economical, considering capitalization of the land and better opportunity to utilize equipment fully without loss of time for moving from field to field.

Cotton planting begins about the first week of March, and may be staggered to avoid seasonal glut at harvest. De-linted seed of a modern cultivar such as Deltapine, treated with fungicide, is planted about 50 pounds to the acre with machines capable of sowing many rows at a time. The rows are mechanically thinned as the cotton sprouts to assure only one plant each few inches. Ample phosphatic fertilizer and nitrogen (usually anhydrous ammonia) are incorporated into the seedbed before planting, with supplementary fertilization a few weeks later and additional fertilizer applied in the irrigation water about every two weeks. As much as 600 pounds of nitrogen to the acre may be used to mature a high-yielding crop. Pre-emergence herbicides are often used to control weeds, and directed post-emergent spray or flaming later if need be. It is not uncommon to spray insecticides by airplane, at costs running so high as $50 per acre annually, often done at

night because the insecticide is more effective then. Large cultivators loosen the soil within an inch of the row, many rows at a time, until layby. Harvest, too, is completely mechanized, with a cotton picker (the most familiar type has rotating spindles onto which the cotton fibers stick) generally going twice over the field, followed by a "scrapper" machine which can salvage an additional fraction of a bale of leftover fiber that has escaped the picker. In most cases a chemical defoliant precedes picking to free the plants of interfering foliage. All of this is mentioned to illustrate what a far cry cotton production in a technically advanced farm area is, compared to the laborious hand growing of yesteryear and in many parts of the world today. Precise management is required, and even computerized programming to assure maximum efficiency of time and machinery.

Heterosis (hybrid vigor) is manifest in cotton, with increased yields double or triple that of the inbred parent strains. Unfortunately cotton is not reliably cross-pollinating; self-sterile lines have been found but little developed. There are indications that radiated cotton seed often grows into plants that show some pollen sterility, and

Self-contained cotton harvester in action. (Courtesy Hesston Co.)

Mechanized picking of cotton in Texas. (Courtesy USDA.)

Cotton bales ready for shipment in Texas. (Courtesy USDA.)

resistant and yield long-staple fiber will be needed.

After harvest, cotton is brought to the gin. In ginning, rotating serrated disks catch the lint, pulling it through narrow slits through which the seeds cannot pass. A series of brushes, rotating in the same direction as the disks, but at a much faster speed, serve to sweep the fibers from the teeth of the disk. These are then blown to a convenient storage place and are eventually baled under heavy compression. The bales are the basic market item of cotton commerce, upon which the trading of the large cotton exchanges depends. Samples from bales are examined by graders who pass judgment on its quality (cleanliness, staple length, color, and so on). A great many named grades have been established.

that some regulated hybridization could be achieved through radiation or chemical treatment. Many environmental factors, however, overshadow genetic considerations, and a hybrid that responds well one year may show little advantage another. For cotton to maintain its position in the world of fibers, certainly improved cultivars that are disease-

Cotton purchased for making fabrics next goes to the mill, where a series of beaters and "pickers" fluffs the fiber and frees it of foreign particles. The fiber must then be carded, combed, and drawn, a sequence of operations accomplished by a series of barbed cylinders into which the fiber is fed. The process basically involves having each cylinder rotate at a slightly higher speed than the preceding one, thus drawing out and thinning the cotton. It has been calculated that in the milling process 1 inch of original

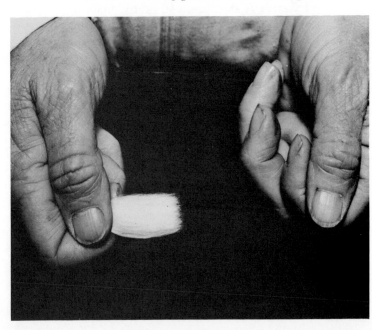

Classifying cotton. (Courtesy USDA.)

Typical large-scale cotton operation in the southwestern United States. In the foreground are bales of cotton linters, the short fibers adhering to cotton-seed after ginning. The seeds from which they have been removed are in the background to be used for oil. The linters are used in the production of photographic and x-ray films, plastics, gunpowder, felts, yarns, mattresses, writing paper, etc. (Courtesy USDA.)

cotton is drawn out to a length of 183,272 miles. A final combing assures parallel orientation of the fibers and nearly uniform staple length. Slivers of the prepared fiber are then twisted (spun) into yarn and thread, as has been described.

Cotton has had a profound influence upon the lives of all people. In the United States it brought about a one-crop economy for the South that led to social and economic disaster. Cotton for the mills of England, perhaps more than respect for a new nation, caused Britain to recognize the Monroe Doctrine. Certainly the sale of cotton goods to the Indian and African markets dictated Britain's policy of dominating the sea lanes during the nineteenth and early twentieth centuries.

Brazil, Egypt, and India have used cotton exports to offset unfavorable trade balances.

Cotton came suddenly into world prominence, and soon superseded flax and wool because of several lucky coincidences—particularly the development of the cotton gin and the mechanization of textile production during the Industrial Revolution. Because it was adaptable to mechanization, cotton could be produced more cheaply and abundantly than competing fibers. Thus an intrinsically rather useless seed hair became abruptly valuable due to manufacturing innovation!

Flax, *Linum usitatissimum*, Linaceae. The fifty or so species of *Linum* are found in both the New and Old Worlds. They are often difficult to distinguish. The section containing *L. usitatissimum*, the only species of much economic interest, has a diploid chromosome complement of 30, although the majority of species exhibit a haploid genome of 9. *L. usitatissimum* is said to differ from *L. angustifolium*, *L. africanum*, and *L. decumbens* by only one chromosome translocation. The cultivars of *L. usitatissimum* chosen for fiber are different than those for seed oil (linseed oil). Long-stemmed, little-branched selections are in demand for flax fiber, and the plants are densely sown to encourage slender stems.

Figure 7-6 shows world flax acreage just after World War II, and is still roughly indicative of where the species is grown today if oil as well as fiber varieties are included. Fiber comes almost exclusively from Europe (Russia included). World production generally runs over $\frac{2}{3}$ billion ton, the U.S.S.R. accounting for more than half. Other leading producing countries are France, Poland, Belgium, and the Netherlands. Ireland, like Belgium famous for its linen fabrics, once grew considerable flax but today imports almost all its fiber.

Flax is generally conceded to have the highest quality of any bast fiber, with a tradition of being used for the finest fabrics. Such designations as "bed linen" and "table linen" are derived from the use of flax for household purposes, and are retained even though today most home "linens" are of cotton or synthetics. Flax was used by the prehistoric lake dwellers of Switzerland, and to weave soft, lustrous linens for the Pharaohs of Egypt as much as four millennia ago. Linen preceded cotton cloth in India. During the Middle Ages flax was the leading fiber in Europe, where its weaving centered in the Low Countries. It was brought to North America by early colonists. Flax fiber is adapted to many uses because of its length, strength, and beauty; sheer handkerchiefs, durable suits and dresses, beddings, up-

holstery, towels, decorative articles, sewing threads, fishnets, and many other familiar products are made from the fiber.

Flax grows best in cool climates having moist summers. It is grown as an annual, with planting and care much as for small grain. It lends itself well to mechanization except for extraction of the fiber, for which considerable hand labor is needed. Even today harvest is usually by hand pulling rather than cutting, in order to conserve full length of stem. Pulled plants are stacked in the field to dry prior to freeing of the fiber by retting. In some cases the flax is simply left in the field for dew retting, but more often it is taken to streams or ponds, or to tanks having controlled water temperature. The river Lys in Belgium has been renowned for over 1,000 years as an excellent retting stream, supposedly because of the favorable characteristics of the water. Tank retting under controlled temperature is said to produce superior fiber extractable after only a few days (as compared to days longer with stream retting, and several weeks with dew retting).

The retted stem is dried and then run through breaking and scutching machines which crush the woody core and remove the non-fibrous parts. A combing (called hackling) follows. Broken fiber is kept separate for making coarse fabrics, rope, upholstery stuffing, and so on. The long fiber is bundled and baled for shipment to weaving centers. Flax fiber is stronger wet than dry, one reason for its preference in fabrics subject to weather and for towels. Weaving is as with other textiles, involving cardings and combings to draw out the fiber before it is spun into yarn. Poorly retted fiber yields thread of inferior quality, often dark in color. Spinning can be either by a "wet" or "dry" process, in the former method the fiber being drawn through warm water to soften it. Yarn may be bleached or dyed before weaving. Today most linen is woven on elaborate power looms, although there is still some traditional craft weaving

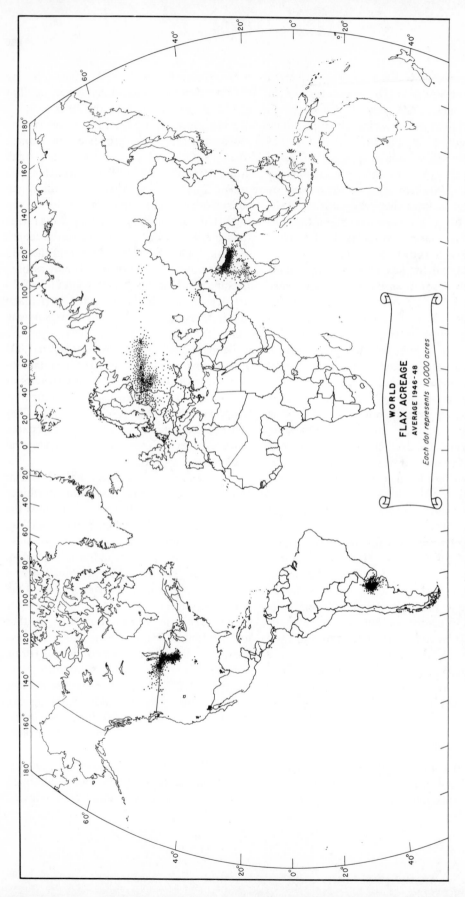

Figure 7-6 World flax acreage. More than 80% of flax acreage is in four areas—the United States, the Soviet Union, Argentina, and India. Of these centers, the Soviet Union is important in commercial production of flax fibers. (Courtesy USDA.)

WORLD
FLAX ACREAGE
AVERAGE 1946-48
Each dot represents 10,000 acres

of fine embroidery and lace. Most cigarette paper is made from flax fiber (stems of oil seed flax can be used) because the nearly pure cellulose burns evenly and without objectionable taste or odor.

Cytogenetic studies have been conducted on many species of *Linum.* Cytoplasmic male sterility is known in *L. usitatissimum;* if economics warranted the expense, certainly hybrid flax seed could be produced even though the species is normally self-pollinating.

Hemp, *Cannabis sativa*, Moraceae. Hemp, apparently native to the Far East, has been grown in China for nearly 5,000 years. The story goes that an emperor of China, 28 centuries B.C., sponsored its first cultivation. It was utilized for weaving in Greece centuries before Christ, and was grown in western Europe during the Middle Ages (more for its seeds, utilized for food, than for its fiber). The same hemp plant which yields both an important bast fiber and an edible or oil seed (commonly used in the United States in birdseed mixtures) is also the source of marijuana (the resinous foliage or inflorescence). In order to control marijuana commerce, the United States government requires a special license to buy seed and to grow and market hemp.

Of all vegetable fibers, hemp is most like flax, and to a degree has been substituted for it. Today it is grown mainly in Europe (including Russia). World production is probably not much over 300,000 tons annually with the U.S.S.R. the leading producing nation followed by Yugoslavia, Hungary, and several other southern European countries as well as India and China.

The hemp plant is dioecious—i.e., has separate male and female plants. Stems are generally about 8 feet long; they contain a conspicuous pith surrounded by a thin xylem cylinder and then the commercially valuable bark with its bast fiber bundles. The fiber bundles are generally 40 to 80 inches long, lustrous, pliable, but less fine than flax. Stem growth responds to gibberellin spray by developing more and longer fibers. The individual fiber cells are typically less than 5 centimeters long and about $\frac{1}{40}$ millimeter wide. The thick secondary wall contains more lignin and pectic materials than are present with flax. Hemp fiber has great strength and durability but because of the lignin is less flexible and elastic than is flax. Thus flax has been favored for textiles while hemp became the first important cordage fiber. It has since largely been displaced by abacá and sisal, and by cotton for fabrics. However a great deal

Hemp planted for fiber in China.
(Courtesy USDA.)

of hemp is still woven into work clothing by rural peoples.

Hemp aided its own downfall in the New World. The colonists brought hemp with them to America, and it was extensively grown for homespun cloth and for the ropes and caulk (oakum) that kept Yankee bottoms plying the seas. But these same clipper ships soon brought to America the cheaper fibers of the Far East, in competition with which hemp gradually diminished in importance. Today little hemp is used in America, although some is imported into northern Europe (especially from Italy, which has the reputation for producing the best-quality fiber).

Hemp requires a 4-month growing period to yield fiber, several weeks more to produce seed. In the Orient hemp is frequently planted to paddy land ahead of the rice crop as part of a double-cropping system. A small part of the crop is left in a corner to produce seed for the next year. Seeding is broadcast or in rows, the latter said to give somewhat higher yields. A recommended spacing is 30 centimeters between rows, 4 cm between plants in the row. When fertilization is possible yield is increased up to 200 per cent. Weeding is by hand hoeing. On more advanced farms there may be insecticidal spraying to control a stem borer. Because of the temptation to get the subsequent foodcrop started, hemp may be cut before it is mature enough for maximum yield. Harvest is by hand, with a sickle, about the time that pollen is shed. Leaves remaining on the upper stalk are flailed off and utilized as green manure.

Hemp fiber may be freed either by pond or steam retting. The latter is quicker but more expensive and generally produces poorer fiber. In Korea cut stems are tied in bundles and placed in a closed chamber where water is poured over hot stones to produce the steam. Two or three hours is sufficient to loosen the bark, which is then stripped from the stem tissues by hand. Pond retting, on the other hand, requires one week or more, and is satisfactory only when the water temperature ranges between 11° and 28°C. Following pond retting the stems are dried in the open, then crushed to permit shaking the stem core loose from the fibrous bark. In rural areas the fiber may be twisted into yarn and woven on hand looms. In much of the East hemp cloaks are traditional for mourners. Also, hemp cloth is often used to wrap the corpse. Because of customs such as these a good deal of hemp continues to be grown. There is also some industrial use for canvases, twines, and sewing threads, and as a packing material.

In Europe yields generally run 30 or 40 per cent higher than in the Far East. Mechanical harvesting with a reaper has now become commonplace. Pond rather than steam retting is practiced, and scutching is usually by hand. For homespun cloth the freed fiber is utilized without further treatment, but that sent to mills is combed to produce finer yarn (which may be mixed with other fibers in weaving).

Ramie, Boehmeria nivea, Urticaceae. Ramie, "China grass," or rhea (var. *tenacissima*) has long been one of the most promising of fiber plants—yet has never lived up to its promise. The chief difficulty has been inability to extract the fiber easily from the gummy ensheathing cells. Yet the fiber itself is in most ways superior to other natural fibers, being eight times stronger than cotton and four times stronger than flax.

Ramie is a nettlelike, shrubby plant, with slender stems and cordate leaves that are often whitish below. The stems become 6 to 8 feet long and 12 to 20 millimeters thick at the base. The plant, presumably native to China, has been cultivated in Asia for centuries, and was probably also known in ancient Egypt. It was tried out in Europe several times without success; in 1855 it was introduced into the United States and grown in Florida, Louisiana, Texas, and California.

Ramie fiber is a typical bast fiber. Properly prepared, it is extremely durable, strong, lustrous, fine-textured, resistant to water and mold, and of excellent color. The imported fiber, normally kept for long periods without degumming, proved more brittle than was desirable, but apparently fibers degummed when fresh meet flexibility requirements as exacting as those for tire cord. The cementing gums of the fiber bundles are unfortunately not generally amenable to destruction by ordinary bacterial activity, so that the fiber can seldom be cheaply separated by ordinary retting. Instead, it must be separated by pounding and scraping, and even then contains 12 to 30 per cent "gum" or "resin" (pectic and other materials), which must eventually be dissolved away chemically. The fiber cells are the longest among natural commercial fibers, being 6 to 30 cm long and about $\frac{1}{30}$ mm in diameter. They occur most frequently in the lower portions of the stem. Their walls are very smooth—so smooth that spinning by machinery developed for other fibers becomes difficult and so discourages wider use of ramie.

Ideally, the ramie plant is best suited to warm, moist climates and rich, loamy soils. It is perennial, one planting sufficing for several years' harvests. Usually, propagation is by division of the clumps, although planting from seed or from cuttings is possible. With repeated cutting (as many as six harvests annually), heavy fertilization is usually necessary to maintain the vigor of the stand. Harvest is by hand in Asia, often one stem at a time, as the proper degree of maturity is attained.

Bark is usually stripped by hand and cleaned by repeated pulling over a knife-edge. Yields of as much as 1,400 pounds of cleaned fiber per acre may result from well-established fields of ramie. Almost half of this dry weight, however, is lost in further degumming. Many degumming methods have been devised, necessarily harsh because of the persistence of the pectic adhesions, some chemical (e.g., boiling in lye, bleaching and treating with acid), others bacteriological (retting with proper bacterial species). Degummed ramie is washed and dried, then typically treated with a softening agent. In industrialized regions, ramie spinning is commonly done on machinery developed for silk, cotton, or wool, often with less than satisfactory results. Principal uses for ramie have been for fishing nets, firehose, hats, various fabrics (reportedly mothproof), upholstery, thread, gas mantles, and sometimes paper.

World ramie production is difficult to estimate, since much of it is locally used in China and the Far East, and even that entering commerce often being mixed with other fiber. It has not gained favor in the West because of extraction difficulties, and in the East, being perennial, it is not so remunerative as are annual crops with a shorter growing season that permit a second planting.

Jute, *Corchorus capsularis and C. olitorius,* Tiliaceae. Jute is the final of the "big four" bast fibers (flax, hemp, ramie, and jute); of them, it is first in quantity and last in quality. The fiber is rough, weak, and of relatively low cellulosic purity. Jute has been grown in India since 800 B.C., but it first gained prominence as sackcloth during the sixteenth century. Yarns were generally spun by peasants and woven on hand looms at home. As British commerce with India increased, great demand arose for gunny sacks, and huge jute mills were built in Calcutta, still the center for the industry. During the 1800's much raw jute was also exported to Dundee, Scotland, where it was spun and woven on flax machinery. Whale oil was used to soften the fiber. The first American jute mill, at Ludlow, Massachusetts, is still in operation. Jute fabric, mainly as burlap and gunny sacks, still brings much of the world's produce to market—even bales of rival fiber cotton. It is also used for furniture webbings, twines, backing for carpets and linoleum, and for upholstery. An advantage of burlap, aside from its inexpensiveness, is that it is not torn by sharp hooks in handling. World production of jute fiber is more than 3 million tons, two-thirds of which comes from India–Pakistan, and some from Amazonian Brazil.

There are about 40 species of *Corchorus*, having a diploid chromosome number of 14. *C. olitorius* is believed native to northern Africa and India, and *C. capsularis* probably to southern China. Fiber from the former is generally finer and softer but more yellowish than that from the latter, although there is much variation among the many cultivars of both species. *C. capsularis* is earlier maturing and most grown. Jute is an annual, the stems (up to 5 meters tall) maturing in from 3 to 5 months. Fibers occur in long wedgelike bundles in the bark, soft and pliable when first extracted but deteriorating upon aging. The individual fiber cells are only a few millimeters long, the walls thick but not uniform and considerably lignified. Jute accepts dyes readily, but cannot be bleached, and in time the fiber becomes progressively harsh and darker.

Jute grows best on deep alluvial soils where moisture is ample. It is planted as an annual during the rainy season, usually broadcast. Stands are thinned by hand to populations of less than 200,000 per acre, and weeds are hand hoed at the same time. After harvest the fields are generally planted to a second and even third crop. Cutting of the stems is by sickle when the plant is mature enough to have seed pods forming. Sometimes the fields become flooded, which entails cutting under water. The stems are dried to eliminate foliage, then tied into bundles for submergence in pools or ditches for retting. It takes 1 to 4 weeks to disintegrate the soft tissues binding the fibers, although in India a number of the bacteria responsible have been isolated and proven able to hydrolize the pectins in as little as 5 days with innoculation.

Hand separation of the fiber from the stalks follows retting, usually accomplished while the operator stands in the pond and "floats" the fiber on the surface, swishing it to free it from fragments of bark. The resulting fiber is about 6 per cent of the green weight of the stem, and chemically is about 75 per cent cellulose, 11 per cent lignin, and 12 per cent xylan. It is stronger dry than when wet. In India, after drying it is usually sold to traveling merchants and eventually to the Calcutta market for weaving or export. Mills test fiber strength at time of purchase, which can be poor if retting has been careless.

Other bast fibers. Although not commercially of great moment, a number of other species yielding bast fibers merit mention. All are potentially important, and may find greater use as market demands change (such as possibly for annual sources of fiber for paper pulp).

HIBISCUS, Malvaceae. Kenaf, *H. cannabinus*, and roselle, *H. sabdariffa*, are promising substitutes for jute in bags and rough fabrics. World production is on the order of 1 million tons annually, chiefly in India, Thailand, and China. Other species of *Hibiscus* such as rama, *H. lunarifolius*, and common okra, *H. esculentus*, also contain stem fiber suitable for rough cloth and crude cordage. Kenaf (also known as mesta, bimili jute, and Ambari hemp) is well adapted to a variety of growing conditions, and yields a crop quickly (90–120 days) with reasonably little care. Kenaf was apparently domesticated in western Africa where, after cotton, it is the most widely grown fiber plant. It is a diploid with 36 chromosomes. It has been suggested in the United States as a possible annual crop source of paper pulp fiber, but so far has not proven so remunerative generally as other crops competing for the same land. Roselle (also known as Java jute) is a tetraploid native to central Africa; it is somewhat slower growing than kenaf (requiring about 180 days), and a bit more difficulty is experienced in separating the woody stem core from the bast. It is not only a source of fiber, but its seeds are often roasted and ground into meal, its leaves and shoots cooked as a vegetable, the flowers made into jellies and confections, and a fermented drink derived from the juice. Most *Hibiscus* fiber is utilized domestically, and numerous species besides those cited show promise at least of making genetic contributions (see Table 7-2).

SUNN HEMP. Sunn hemp, *Crotalaria juncea*, Leguminosae, is also known as san hemp or sun. It is second only to jute in India as a source of bast fiber. The fiber is used mainly for cordage; being resistant to deterioration in water, it is suitable for fishnets. Production is nearly 100,000 tons annually, almost entirely from India. Small amounts of fiber are imported into the United Kingdom for making paper, and it is used somewhat in the manufacture of cigarette paper in the United States. The species is planted for fodder and green manure as well as for fiber.

Crop handling is about the same as for jute.

URENA. Urena fiber, from *Urena lobata*, Malvaceae, is much like kenaf and a substitute for jute. The species is native to China, but it has naturalized in many tropical locations where today it grows wild. Fiber production is primarily from western Africa and tropical America, where it is known by such trade or common names as Congo jute, paka, cadillo, malva blanca, guaxima vermelha, carrapicho, and (in Florida) as Caesar weed. About 15 million tons are produced in Brazil, mostly used for coffee bags. About equally as much is exported from the Congo, mostly to northern Europe. *Urena* culture and fiber removal (by retting) are about the same as for jute, and the fiber serves the same commercial purposes.

Other genera of the Malvaceae also yield commercial quality bast fiber, although they are little grown at present. Among them are

Table 7-2

Fiber Yield and Fiber Quality Properties in 13 Species of Hibiscus

(Means Ranked According to Fiber Yield)

Species	No. of Lines	Fiber Yield, g/stem sample	Fiber Quality Properties* texture	Fiber Quality Properties* color	Amount of Bark
H. sabdariffa (edible)	3	25.3 a	2.78 c	2.77 b	2.88 b
H. cannabinus "Everglades 71"	1	22.4 a–b	2.67 bc	2.00 ab	1.33 a
H. cannabinus (wild)	3	21.2 a–c	2.78 c	2.89 b	2.88 b
H. diversifolius	2	18.0 a–d	2.33 bc	3.00 b	3.00 b
H. maculatus	1	16.7 a–d	2.67 bc	2.67 b	3.00 b
H. furcellatus	3	16.5 a–e	2.39 bc	2.44 ab	2.88 b
H. sabdariffa (fiber)	1	15.3 a–e	2.33 bc	2.33 ab	2.33 ab
H. acetosella	5	14.1 a–f	2.33 bc	2.26 ab	2.80 b
H. radiatus	9	13.7 b–f	2.57 bc	2.42 ab	2.61 b
H. dongolensis	1	11.0 c–f	2.00 a–c	2.00 ab	1.50 a
H. bifurcatus	2	10.2 c–f	2.00 a–c	2.16 ab	2.67 b
H. meeusei	1	9.8 d–f	1.33 ab	1.33 a	2.33 ab
H. rostellatus	2	9.4 d–f	2.16 a–c	2.33 ab	2.67 b
H. surattensis	1	5.4 e–f	1.00 a	2.00 ab	2.00 ab
H. sudanensis	1	3.1 f	3.00 c	3.00 b	3.00 b

* Fiber quality properties: 1 = superior to check (*H. cannabinus* "Everglades 41"); 2 = same as check; 3 = inferior to check.

From F. D. Wilson, Economic Botany **21**, 134 (April–June 1967).

Sida (*S. rhombifolia*, "Cuba jute"; also *S. acute* of Mexico and *S. tiliaefolia* of China), *Abutilon* (*A. avicennae*, "China jute" or "Indian mallow"), *Pseudabutilon* (*P. spicatum*, the "paco-paco" of Brazil), *Cephalonema* (*C. polyandrum*, the "punga" of Africa), and others.

Urtica dioica, and *U. urena*, Urticaceae, have been used in western Europe for textiles. Historically other members of the nettle family have been utilized in northern Europe, Italy, France, and Hawaii, but have been supplanted by cotton in recent centuries. The fibers are obtained by retting, or by decorticating with machines, much as with ramie.

Both the milkweeds, *Asclepias*, Asclepiadaceae, and the dogbanes, *Apocynum*, Apocynaceae, are noted for their bast fibers. *Apocynum* has been cultivated in the U.S.S.R. as a substitute for flax and to make "Kendyr paper." *A. cannabinum*, Indian hemp, was widely utilized by the American Indians. The seed hairs from milkweeds have served as a substitute for kapok under emergency conditions, too.

A number of trees yield fiber from the inner bark. Species of basswood, *Tilia*, Tiliaceae, provided fiber for the American Indian; the fiber was derived by peeling the bark from young trees, boiling it with wood ashes, and then drawing it across a sharp bone before drying and twisting it into yarn (which was generally woven into rough fabrics or baskets). Perhaps even more famous is the paper mulberry, *Broussonetia papyrifera*, of the Moraceae. South Sea Islanders make a fabric from the bark without spinning, known as tapa or masi. The inner bark is freed, cleaned, then laid evenly in several layers while still damp and allowed to dry. The webbing so formed is beaten with a wooden mallet on a smooth plank until it spreads and mats together into a clothlike fabric. In Japan this same fiber is cut into strips, twisted into yarn, and woven with other fibers to make typical cloth. Said to be used in much

this same fashion are inner barks of *Antiaris*, Moraceae, and *Lagetta*, the "lace bark," Thymeliaceae.

Hard, leaf, or structural fibers. Fibers of this group are also legion, and come from several families, all of them Monocots. They are chiefly utilized for cordage, twines and ropes. Abacá and the *Agave* species serve well to characterize the group. The fibers, being entire fibrovascular bundles, tend to be long, coarse, and harsh; they are usually quite strong, although, in spite of the group designation, really no "harder" than the bast fibers just reviewed. Approximately 1 million tons of hard fiber are produced annually, worldwide, of which more than three-fourths comes from *Agave* (principally sisal and henequen), and nearly one-fifth from abacá.

The preliminary process for making cordage is essentially the same as for spinning soft fiber yarn for textiles. The hard fibers are first combed, then attenuated and twisted on complicated machines to interlock the fibers into yarn. The yarn may be used as such (as tying twine), or a few yarns may in turn be twisted together (with an opposite twist to that of the yarn) to yield cord (wrapping twine). If a few cords are similarly twisted together for greater strength, rope is formed; and in turn a few ropes may be twisted together to yield cable. Obviously, the number of fibers to the yarn and number of yarns to the cord, and so on, will largely determine the strength of cordage. Regulation of the twist in making and combining yarns is very important. Under strain the fibers tend to untwist, but this tendency in the yarn should be counterbalanced by the opposite twist in the cord, and so on up to cable-size cordage. Each successive combination is, of course, always twisted in the opposite direction to the last. During manufacture of larger cordage, an oil is commonly applied to the fiber. Oiled fibers are softer and less dusty, and oiled rope is easier to handle. Various waterproofing or preservative substances are sometimes added with the oil where special qualities in the rope are desired.

For binder twine, only hard fibers have been found both resistant enough and cheap enough to satisfy requirements. Abacá is used chiefly for the finer grades, sisal (sometimes mixed with abacá) and henequen for coarser grades. Wrapping twines of two, three, or four strands (yarns) also commonly use abacá for finer grades and other hard fibers for coarser grades, or soft fibers and cotton for "soft" strings for small (indoor) packaging. Rope can be made from almost any fiber, but before synthetics was usually made from hard fibers; three-fourths of all hard fiber used for this purpose is abacá. Ropes are commonly three-, four-, or six-strand, and are oiled or otherwise impregnated.

Abacá or Manila hemp, Musa textilis, Musaceae. Abacá is closely related to the banana, which it resembles. Like the banana it is strictly a tropical plant, and grows best

Cut "stems" of abacá on conveyor belt leading to the crushing hopper. Honduras. (Courtesy Paul H. Allen.)

where rainfall is heavy and mean annual temperature around 80°F. Abacá is native to the Philippine Islands, from which 95 per cent of the more than 100,000 tons of annual world production comes. The fiber is sometimes used for weaving textiles there, but mostly it is exported for the manufacture of high-quality cordage.

Abacá became commercially important relatively recently. It was not even known to the Western world until 1686, and was not an item of commerce much before a century ago. Although synthetics such as nylon contest abacá today, it has gained quite a following since the turn of the century. The fibers are long (from 2 to 4 meters), quite strong, light in weight, and resistant to deterioration in salt water (therefore of special demand for marine rope and ship caulking). Early production was from scattered native habitations utilizing wild plants. Later, however, more efficient, large abacá plantations were developed cultivating abacá as a crop.

Unlike the banana, abacá yields viable seed. However, propagation is quicker from separation of the stem clusters, or by cutting off and planting side shoots ("suckers"), or by sectioning rhizomes. The starts are planted into individually dug holes made among the "brush" from newly felled forest. Food crops may be interplanted until the abacá becomes large. As is typical with shifting agriculture in the tropics, fertility tends to become exhausted after a few years. Experience has shown potassium additions to be especially important for maintaining abacá.

The abacá plant is technically an herb, although it may have as many as two dozen "stems" (clusters of leaves) rising 8 meters or more. These grow from a perennial rootstock or rhizome. About a third of the stems reach flowering age each year. The stems (petioles) are cut for fiber at about the stage when the inflorescence emerges. Care is taken not to damage young shoots toward the center of the clump, or developing side shoots. The inner

Clean abacá fiber being sorted before drying. Honduras. (Courtesy Paul H. Allen.)

Of the *Agave* fibers, SISAL, from *A. sisalana*, is the most important, traditionally accounting for over half of the world's production of hard fiber. Sisal is grown chiefly in Africa (Tanzania, Kenya, Angola, Mozambique, Madagascar) and in Brazil. HENEQUEN, from *A. fourcroydes*, is grown chiefly in Cuba and Mexico, and usually accounts for no more than one-fourth of annual *Agave* fiber production. There are several other somewhat less planted species, especially CANTALA from *A. cantala*, MEXICAN HENEQUEN from *A. lurida*, and LETONA from *A. letona*. All of the commercially important *Agaves* seem to be polyploids and are sterile. Their base chromosome number is 30, although investigations of sisal and henequen polyploids show chromosome counts ranging up to 149. No doubt the modern cultivars, now vegetatively propagated, represent a great deal of ancestral intercrossing.

petiole sheaths yield a higher-quality fiber than do the outer ones. The petioles are cut with machetes into about 5-foot lengths, and taken to local centers for extraction of the fiber. Sometimes the sheaths are simply cut longitudinally into strips, and the strips scraped between a block of wood and a knife to get rid of most of the pulp. On plantations, however, powered decorticators draw the strands between various scraping devices.

Agave fibers, Amaryllidaceae. The genus *Agave* and several related genera have long been the primary source of hard fibers, accounting for about four-fifths of their production. Commerce has centered chiefly in Central America and eastern Africa. The development of synthetic fibers during the 1960's, especially polypropylenes, established what amounted to a price ceiling for hard fibers that prevented recognition of rising costs on the extensive sisal plantations of Africa, Mexico, and Brazil. Less efficient operations went out of business, and the dominance of the cordage industry by natural fibers lessened.

Harvesting Henequen in Central America. (Courtesy USDA.)

Agaves are well adapted to arid environments, and require tropical warmth. Ample rain or irrigation, however, does increase production on well-drained soils where *Agave* grows best. Most sisal is produced on plantations, with propagation from the bulbils that form on the inflorescence, or from separation of side clumps ("suckers") that form on older plants. Sisal is native to Central America, even though mostly grown in Africa today. Sisal cultivation was tried in Florida in 1836, but the industry did not thrive. Shortly thereafter the species was successfully introduced in the West Indies and a modest industry established. Only with advent of plantation growing in Africa in the twentieth century did sisal come to dominate the hard fiber industry, however.

An *Agave* plant produces about 300 leaves during its lifetime, after which it flowers and dies. Under favorable growing conditions this life cycle may be completed in just a few years, or where growth is not rapid it may take half a century. An *Agave* leaf is ready for harvest when it is mature enough to have attained an essentially horizontal position. On sisal plantations in Africa first cutting is generally in the third year. With henequen in Yucatan first cutting is generally 5–8 years after planting. The leaves are cut by hand with a machete. If spiny (as is the case with several species), the spines are cut off in the field. The cut leaves are brought to a decorticating center where the pulpy tissues are scraped away mechanically. On plantations machines first crush the leaves, making the hand scraping somewhat simpler. The crudely separated fiber is then washed, dried, and cleaned more thoroughly by shaking and brushing (often with the aid of powered machinery). Fiber yield is 4 or 5 per cent of the green leaf weight.

During the rough treatment that *Agave* fiber sustains in its cleaning, the internal xylem and phloem of the fibrovascular bundle are generally destroyed, so that the fiber consists chiefly of the sheathing cells. Considerable difference occurs in the size of the strand and its degree of lignification, depending upon the age of the leaf, the bundle position in the leaf, and of course upon species and growing conditions. Short-fiber and off-grade types are separated for making into paper or fiberboard, while the better-quality fiber is reserved for cordage twine. On the African plantations the waste pulp from processing is returned to the field as a fertilizer. Incidentally, sisal sap contains hecogenin, a source material for synthesis of cortisone.

Sisal is named for an old shipping port on the Yucatan peninsula, from which much fiber came when sisal first excited attention about 1839. Use of the fiber goes back millennia, however, as proven by archeological findings. Sisal was much utilized by the Indians of Yucatan when the Spaniards first arrived in 1509. Henequen has also been grown in Mexico since prehistoric times. In the early records there is considerable confusion between sisal and henequen, and other of the *Agave* fibers. Today 90 per cent of the world's henequen is still produced in southern Mexico, on the dry, limy soils of Yucatan. Around the middle of the twentieth century the Mexican government took steps to restrict large henequen plantations, and to encourage its growing on small, individual farms. The cordage factories of Yucatan were also consolidated into one corporation in 1961, making the henequen industry pretty much a ward of the Mexican government. Most henequen is exported to the United States.

LECHUGUILLA, *A. lophantha var. poselgaeri* (*A. lecheguilla*), known also as Tula istle or Tampico fiber, resembles a small sisal plant. It is native to the arid mesas of northern Mexico at elevations from 3,000 to 6,000 feet. Lechuguilla istle fiber is coarse and stiff, and is often used in the manufacture of scrubbing brushes as well as cordage, upholstery and carpet pads, and locally for rugs, saddle blankets, and the like. Up to 15,000 tons of lechuguilla istle are produced annually in Mexico, under government control. Harvest is mainly from wild plants by rural workers when no more remunerative work is at hand. The central stalk (containing about a dozen budding leaves) is pulled off by means of an iron ring attached to a stick, the fiber later freed by scraping the fleshy leaves between a heavy knife and a block of wood at some central location.

Another *Agave* fiber, JUAMAVE ISTLE, *A. heterocantha* (*A. funkiana*), is also produced in relatively small quantity in the mountains of Tamaulipas, Mexico. The fiber is of finer quality than with most species of the genus.

Other hard fibers. MAURITIUS FIBER OR MAURITIUS HEMP, *Furcraea gigantea* (*F. foetida*), Amaryllidaceae. This is a species native to eastern Brazil where it is known as piteria; it derived its common name upon introduction into Mauritius about 1790, where it first became commercially important. The fiber is used for sugar bags in Mauritius, and also domestically for cordage. Less than 2,000 tons of fiber are produced annually. FIQUÉ, from *F. macrophylla*, is similar, with small production centering in Colombia.

FORMIO OR NEW ZEALAND FLAX, *Phormium tenax*, Liliaceae. This is grown in Argentina, New Zealand, Chile, and some of the oceanic islands near Africa. For a hard fiber, it produces a rather soft and pliable fiber. When Captain Cook visited New Zealand in the late 1700's he found this fiber being used to weave baskets, clothing, and cordage. The plant has since been spread to various parts of the world, and production of no more than 15,000 tons annually comes chiefly from Argentina and New Zealand, much of it consumed domestically. Formio fiber deteriorates rather quickly when damp, and is often mixed with sisal or abacá when made into rope.

AFRICAN OR GUINEA BOWSTRING, *San-*

Yucca brevifolia of the southwestern deserts of the United States and Mexico, a leaf fiber. (Courtesy U.S. Dept. of Interior, National Park Service.)

sevieria metalaea (*S. guineensis*), Liliaceae. This is a relative of the familiar sansevieria house plant. It is grown for fiber principally in tropical Africa, where a small amount of strong, white, but brittle fiber is extracted. It is used for cordage, fishnets, bowstrings, mats, and coarse cloth, mostly locally. Another fiber plant of the Liliaceae is *Samuela carnerosana*, a relative of *Yucca* (which it is often called), of which there is some production in Mexico.

CAROÁ, *Neoglaziovia variegata*, Bromeliaceae. This has long been used in Brazil where it grows wild throughout the extensive desertlike northeastern section of the country. The thorny leaves are cut by hand and brought by burro to a processing station where the long, silky fiber somewhat finer than sisal is extracted. It is manufactured into cordage, nets, fishlines, bags, or (mixed with other fiber) cloth.

As much as 12,000 tons have been produced when jute supplies were cut off, although usual production is only a few thousand tons, mainly used for weaving coffee bags. The PINEAPPLE, *Ananas comosus*, also of the Bromeliaceae, yields a leaf fiber that is sometimes consumed locally.

BANANA, *Musa*, Musaceae. Species of the banana can provide weak petiole fibers not nearly so satisfactory as those from its relative abacá. However, ENSETE or ABYSSINIAN BANANA, *Ensete ventricosum* (*Musa ensete*), is widely cultivated in the Ethiopian highlands for fiber and food (cooking young shoots or rhizomes, or fermenting pulp). It is propagated by vegetative "suckers" (which form when the flowering core is cut out of a rhizome) rather than by seed, the new plants usually started in nurseries for transplanting to the field. Women extract the fiber by hand scraping of cut petioles with a bamboo "knife." It is used for weaving sacking, and for cordage.

Brush, braiding, filling, and miscellaneous other fibers. Rather few fibers other than those already discussed have attained much commercial importance. Fairly well known are kapok, coir, and in limited areas various other stuffing materials broadly classed as "fiber." The Panama hat "palm" has gained considerable recognition as a specialty, and various piassavas and broomcorn are important for brooms, scrubbing brushes, and so on. Almost any strong but pliable material can be used for purposes mentioned, including twigs or splints cut from wood and bamboo; the distinction between "fiber" and other structural material becomes vague. Only a few of the more recognized specialty fibers can be reviewed in this section, and such familiar substances as cereal straws, reeds, and cane stems (bagasse) are little more than mentioned.

KAPOK OR SILK-COTTON TREE, *Ceiba pentandra*, Bombacaceae. Kapok is native to tropical America, but now widely spread throughout the tropics of both the Old and New Worlds. World commerce in kapok has run about 30 million tons, but in industrialized parts of the world it is becoming replaced by synthetic materials such as foam rubber and plastic foams. Kapok is a large rainforest tree with striking basal buttresses (see the photograph on p. 74), the bark covered with spines when young. The leaves are palmately divided, borne on a spreading crown which after flowering bears numerous capsular fruits 4–10 inches long that stand out when the foliage is shed. Each capsule splits into five valves, revealing many small seeds surrounded by a dense mass of silky hairs, the kapok

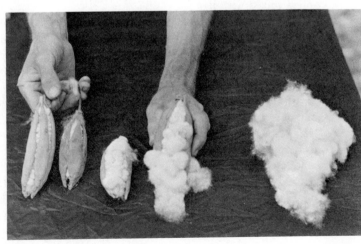

Kapok seed pods at various stages of maturity. At right, the silky fiber from one pod. This kapok, grown in Ecuador, was introduced from Java. (Courtesy USDA.)

fiber. Like cotton the fiber is nearly pure cellulose and of excellent intrinsic quality; but
it is quite smooth and lends itself poorly to
spinning. Thus it has been mostly used as a
stuffing material, for mattresses, life preservers, insulation, and so on. Its resiliency
and imperviousness to water result in its
being much used for outdoor and marine
equipment, including sleeping bags. Most
commercial kapok is produced in the Far
East, where the tree is now extensively planted. The fruits are gathered by hand, often
being cut from the tree with long pole-knives,
and the fiber separated by hand from the
capsule and seed. Kapok was widely planted
on the east coast of Africa about the time of
World War II, when considerable demand
arose for the fiber in life preservers. Demand
fell off after the war and the plantings were
neglected, until a new local demand arose
about 2 decades later as the peoples of newly
emancipated countries, no longer content to
sleep on thatch mats, demanded mattresses
(locally made ones stuffed with kapok were
economically within reach). Java supplies
most of the true kapok fiber, but a few thousand tons of an Indian kapok come from a
similar species, *Salmalia malabarica*, in
India. Another seed floss called akund is
obtained from *Calotropis procera* and *C.
gigantea* which grow wild in India, the fiber
often mixed with kapok.

COIR, *Cocos nucifera*, Palmae. The outer
husk of the coconut fruit provides a harsh
fiber once much used for stuffing. World
commerce often runs 100,000 tons annually
or more, with the coastal belts of India accounting for about half. Coir production is a
cottage industry, the nut generally retted in
salt water for a lengthy time, after which it
is beaten to separate the fiber, further washed,
and hackled to separate the stiff fibers. They
are used chiefly for inexpensive stuffing and
mat making. Coconut palms are strand plants
growing nearly 100 feet tall without branches.
The nuts are sometimes gathered by agile

climbers ascending the inclined trunk with the
aid of a rope looped around the ankles. Nut
clusters are cut one at a time and let fall to
the ground, where the husk is typically removed by thrusting the coconut against a
stationary metal spike. It is said that a worker
can husk as many as 2,000 nuts a day.

OTHER PALMS. "Crin vegetal" is a fiber
from the base of a small fan-palm, *Cha-
maerops humilis*, abundant in northwestern
Africa. It is often a substitute for horsehair.
Piassava fiber is a coarse, tough, water-
resistant fiber used for brushes, brooms, mats,
and various types of cordage, obtained from
the leaf sheaths of several palms, especially
Attalea funifera and *Leopoldinia piassaba*
of Brazil. As much as 20,000 tons of piassava
fiber have been produced annually. A west
African piassava is derived from *Raphia
vinifera*. A Madagascar piassava is obtained
from *Dictyosperma fibrosa;* this fiber is much
used for sweeping brooms in Europe.
Palmyra fiber comes from *Borassus flabel-
lifera*, a fan-palm native to Ceylon. Kitul is a
relatively soft and pliable palm fiber, from
Caryota urens, found in southeastern Asia.
Palmetto fiber is obtained from *Sabal palmetto*, which grows wild in the southeastern
United States and the Caribbean area.
The fiber is much used for brushes, but also
in ropes and mats. Immature leaves of
Astrocaryum chambira are twisted into string
or "chambira fiber" in Colombian Amazonia,
mainly used to weave hammocks and carrying bags.

A relative of the palms is *Carludovica
palmata*, of the Cyclanthaceae, a shrubby,
palmlike plant of tropical America. It yields
PANAMA HAT FIBER, also called TOQUILLA and
JIPIJAPA. "Panama" hats are actually made in
Ecuador, but derived their name from
trans-shipment through Panama to California
during the goldrush of 1849. *Carludovica*
fiber is obtained from young leaves just beginning to fan out, from which the coarse veins
are stripped and the leaves then boiled. The

Weaving Panama hats in the American Tropics. The fibers, commercially speaking, are strands from leaves of the Panama hat palm, *Carludovica palmata*. (Courtesy Chicago Natural History Museum.)

leaves are shred by hand into strips, which are dried, rolled into cylindrical strands, bleached in sulphur smoke, and further cleaned before weaving into hats. Hat making is a traditional cottage industry in Ecuador, with a specific design woven into the center of the hat crown signifying the locality or the maker. The "fiber" is kept damp while being worked. Several hundreds of thousands of hats are exported from Ecuador annually.

SPANISH MOSS, *Tillandsia usneoides*, Bromeliaceae. This is an epiphyte in the Gulf Coast area of the southeastern United States. Since colonial times it has been collected, fermented in piles, then ginned free of scale and debris. It was much used for furniture stuffing, but is becoming replaced by synthetic paddings.

BROOMCORN, *Sorghum bicolor*, Gramineae. Broomcorn is the traditional Temperate

Zone source of fiber for brushes and brooms. Special cultivars yield the stiff panicles, which are harvested when about 15 inches long and still green (to prevent brittleness and coarseness). Other members of the grass family also yield brush fibers such as the broomroot, *Epicampes macroura*, of which the tough, crinkly roots are utilized after treating to remove the outer coatings.

Luffa aegyptiaca (*L. cylindrica*), Cucurbitaceae. This plant yields a spongy core from the gourdlike fruit, which after retting and cleaning is used as a "vegetable sponge" and for oil filters or strainers. Most production comes from Japan and Central America, and usage is suffering from competition with synthetic sponges.

Stems of various grasses and sedges yield fibrous material, often utilized in paper. Indeed, the original paper, PAPYRUS, was derived by beating *Cyperus papyrus* stems in ancient Egypt. Cereal straw and sugar cane residues (bagasse) are used for making fiberboard and paper in various parts of the world, as well as are ESPARTO GRASS (*Lygeum spartum*, *Stipa tenacissima*), REEDS (*Phragmites communis*, etc.) and several other grasses. Table 7-3 compares fiber characteristics of several non-woody sources of fiber with woodpulp.

Various BAMBOOS, also of the grass family, are an excellent source of cellulosic plant materials because of the persistent growth and high yields. Bamboo has long been important for various structural uses in the Orient, but has been considered as a source of pulp and plastic raw material in the United States as well. Mechanized means for handling the crop, however, have so far limited its usefulness. In Taiwan fresh bamboo shoots are sectioned and split, followed by mechanical crushing and shredding on spike-studded drums to yield a "fiber" much like excelsior. This is used for brushes and brooms, as a filling material for mattresses, as an abrasive for cleaning floors, and in making paper. Chinese ceremonial "joss paper" has some

Table 7-3

Some Dimensional and Compositional Characteristics of Selected Pulpwood and Non-woody Plants Processed for Pulps and Paper

Species	Fiber Length, mm	Fiber Width, μ	Lumen Width, μ	Cell Wall Thickness μ	Crude Cellulose Content, %	Alpha-Cellulose Content, %
Coniferous woods:	2.0–5.7	32–44			60	39–45
Englemann spruce	3.5					
Longleaf pine	4.0				58	
Douglas fir	4.5				61	
Deciduous woods:	0.6–1.9	38–50			45–62	38–49
Aspen poplar	0.4–1.9	10–40			63	
Blackgum	1.7–1.8					
White maple	1.7				61	
Paper birch	0.8–2.7	20–30				
Non-woody plants:						
Sugar cane	0.3–3.4	9–45	16	5	48–58	30–37
Depithed	—	—			56–63	37–41
Cereal grain straws	1.1–1.5	9–13			43–54	29–39
Rice straw	0.5–2.5	4–15	3	3.5	43–47	26–36
Bamboos	1.1–3.8	12–22	2–7	4–9	44–62	30–43
Reeds	1.0–1.8	8–20			46–54	36–43
Flax	4 –8.5	10–37				
Esparto grasses	0.5–3.5	7–18	5.6–6.7	2.7	49–50	33–38
Mulberry, bast	2.3–9.9	17		5–8		

From P. F. Clark, *Economic Botany* **19**, 396 (Oct.–Dec. 1965).

Bamboo is an important source of fiber and structural material in the Orient, and has been suggested for paper pulp in the United States. (Courtesy USDA.)

chemical rice straw pulp added to the mechanical bamboo fiber. A number of bamboos are used in the Orient, including species of *Phyllostachys*, *Bambusa*, and *Dendrocalamus*.

RICE PAPER is a soft velvety substance much used for making realistic artificial flowers. It is not properly named, for it is derived from the spongy pith of *Tetrapanax papyriferum*, Araliaceae, and not from rice. The species is native to eastern China, but has been introduced into most tropical and subtropical parts of the world where it is often planted as an ornamental. Production of rice paper is chiefly from Taiwan and China, where propagation is usually by the transplanting of side shoots from old plantings. The stems are cut when two or three years old, the leaves and small twigs removed, the stems then soaked in running water to loosen the pith (which is said sometimes to be forced out of the stem cylinder with a blunt stick, and in other cases removed by splitting the stem). The pith is promptly dried and tied in bundles for bringing to the rice paper factories. There it is cut into ribbons with a sharp knife much as veneer is sliced from a log. Since this is a hand operation, much dexterity is required. Only in the Orient is labor inexpensive enough for production of rice paper. Rice paper has been used in China as a medium for painting, but most of it is utilized for the manufacture of imitation flowers, often in Japan or Hong Kong, or for shipment to the United States for sale through craft shops.

Vegetable fibers and the future. There is no reason to suppose vegetable fibers will diminish in importance in underdeveloped parts of the world, where synthetic fibers would be too costly if available at all. And, long-term advantages would seem to lie generally with renewable resources from crop plants rather than with increasingly scarce mineral raw materials. However, there is not likely to be any letup in pressure from synthetics in the near future, and improved technology may challenge the very usage of fibers in the traditional way. Fiber production, processing, and manufacture into a finished product (spinning and weaving) is laborious and thus costly. Already the market is experiencing a technological shortcut in nonwoven fabrics derived by techniques similar to paper making or plastic extrusion, for items such as disposable garments. Certainly plastic bags and covers are supplanting woven ones for commodities and fertilizers, and are being made tough enough to withstand the rigors of loading and unloading in worldwide shipment. Perhaps the future belongs to chemical derivatives from a raw material pool that well might be plant cellulose, obtained from any of numerous economical sources.

So far as traditional fiber usage is concerned, cost as much as quality determines acceptance. Any reasonably adequate fiber that can be produced more cheaply than others is almost assured of a market. Consequently, the important fiber-producing areas have tended to coincide with cheap labor regions, or where ample mechanization reduces cost. Often fibers come to be produced exclusively in a limited portion of the globe as a specialty of the region.

It is curious how small a part the cost of the fiber plays in the price of the ultimate manufactured article. Commercially important fiber is seldom utilized in the locality where grown, and may pass through the hands of a series of middlemen before being manufactured in some distant country—a country which in turn may use but little of the manufactured article, and ship it again for sale halfway around the world. The multitudinous handling, manufacturing, and shipping charges come to obscure the comparatively negligible price originally paid for the fiber. Various tariffs, duties, and cartel arrangements further complicate the matter.

Growing fiber plants is usually less costly than is extraction of the fiber, which is typically imbedded among other tissues difficult to remove. Thus the development of labor-saving machinery to accomplish this task can have a profound influence upon the fate of a fiber. The familiar retting and scraping processes have long been recognized as inefficient, but so far no economic substitute for them has been found. Various chemical extractions and controlled retting processes have been tried, even "explosion removal" under sudden release of a high pressure, and ultrasonic treatments with high-energy sources. Much cleaning and combing is still done by hand, although machinery of differing degrees of efficiency has been developed to aid in this process. The trend can be expected to continue in industrialized nations, where high labor costs must be compensated for through increased efficiency made possible only through employing capital (for elaborate machinery) in place of man-hours.

SUGGESTED SUPPLEMENTARY REFERENCES

Brown, H. B., and J. O. Ware, *Cotton*, 3rd ed., McGraw-Hill, N. Y., 1958.

Elliot, Fred C., M. Hoover, and W. K. Porter, Jr. (eds.), *Advances in Production and Utilization of Cotton*, Iowa State Univ. Press, Ames, 1968.

Hess, K. P., *Textile Fibers and their Use*, 5th ed., J. B. Lippincott Co., Philadelphia, 1954.

Heyn, A. N. J., *Fiber Microscopy*, Interscience, N. Y., 1954.

Kirby, R. H., *Vegetable Fibres*, Leonard Hill, London, and Interscience, N. Y., 1963.

Lock, G. W., *Sisal*, Wiley, N. Y., 1962.

Matthews, J. M., *Textile Fibers, Their Physical, Microscopical and Chemical Properties*, 5th ed., Wiley, N. Y., 1947.

Mauersberger, H. R. (ed.), *Matthew's Textile Fibers*, 6th ed., Wiley, N. Y., 1954.

USDA Yearbook, Washington, D. C., *Farmer's World*, 1964; see especially pp. 218–260.

Weindling, L., *Long Vegetable Fibers*, Columbia Univ. Press, N. Y., 1947.

PART III

Cell Exudates And Extractions

Most plant cells, sometime during their existence, secrete or excrete complex metabolic products other than the cell wall. These are very often viscous or liquid, "exuding" from the cells or tissue, and hence are termed *exudates*. Other depositions within or about the cell are crystalline or solid: e.g., alkaloids, oxalates, and starch. In most cases the exact function of the exudates in the plant is uncertain; in some instances it seems to involve waste materials. Exudations may take the form of resins, gum, and latex; or, in the broader sense of the term, even suberin, waxes, oils, tannins, dyes, pigments, alkaloids, and the like. But the latter must usually be extracted by man from the plant tissues, and hence are more appropriately termed *extractives*. Exudations and particularly extractions may be found interspersed through the intermicellar spaces of the cell wall (as with certain tannins and dyes from wood), or they may be channeled completely into special intercellular conducting tubes or through the lumen of dead cells (as with naval stores and rubber).

In some cases, as with vegetable oils and sugar, or fruit juices and fermentations, the distinction between extractive and food or beverage (Part IV) is far from clear-cut. Included in Part III are most instances of massive processing of plant material (and usually considerable commerce in the extractive), even though it may end up largely as food or as an ingredient in foods. With the foods of Part IV such "processing" is largely accomplished internally by the animal body!

On a dry-weight basis less than one-third of plant tissue is ordinarily protoplast, seat for extractive formation. However, exudates and extractives usually become concentrated in intercellular spaces, or may be dispersed through the cell wall lattice itself. On the whole exudates and extractives represent a rather small proportion of the total plant bulk, and much plant material must be expended to retrieve relatively little of the valued extract. It is no wonder, then, that exudates and extractives for the most part have been highly esteemed and expensive items of commerce. Indeed, the essential oils (spices and perfumes), dyes, gums, resins of the ancients, and wild rubber in the twentieth century, were prizes for which empires were risked. Even such substances as oil and carbohydrate extractives, commonplace today, depend upon a highly perfected technology (or in days of old, slave labor, as with the early sugar plantations) to make them inexpensive enough for mass market.

Exudates and extractives suffer the same competitive pressures noted for cell wall substances. Other species or synthetics can substitute for almost all sources of plant derivatives to be reviewed in the next seven chapters. Indeed, many of the items are largely of historical interest, and others at least partially displaced in the modern market. Synthetic rubber, synthetic resins, chemical tannin substitutes, analine dyes, chemically derived flavors, man-made medicinals, insecticides and herbicides—all of these synthetics dominate modern commerce in their respective classes. On the other hand, probably half of the world's medicinals are still derived from plants, and industrial fermentations are of importance in specialized areas; there is as yet no substitute for tobacco; and although vegetable oils, sugars, and starches some day may be gained from plants other than those now used, substitution by synthetics from outside the plant kingdom is unlikely. Indeed, of all nonfood plant materials, industrial usage of vegetable oils and carbohydrates would seem most promising. Produced as they are agriculturally, their energy without cost from sunlight, they well may become basic raw materials for future plastics and synthetic fibers that will shape our material civilization.

CHAPTER 8

Latex Products

A GREAT MANY PLANT EXUDATES TAKE THE form of latex, which is a milky liquid, a complex colloidal mixture of water, salts, hydrocarbons, and various other organic compounds (alkaloids, resins, oils, proteins, sugars, starch, and like materials). The usefulness of latex to the plant is uncertain, but possibly it aids in closing and healing injured parts and acts as a storage medium for nutrients. Lactiferous plants are found in at least twenty plant families, notably the Euphorbiaceae, Sapotaceae, Moraceae, Compositae, Apocynaceae, and Asclepiadaceae. Lactiferous plants embrace small herbs and large trees, native to temperate and subarctic regions as well as to equatorial forest belts, growing in varying climates from arid desert to humid rainforest. The latex is produced by living cells in the protoplasm, and apparently accumulates in the vacuoles. The production may take place in leaf, stem, or perhaps even root; the latex so formed escapes from the vacuoles and is carried through the plant in latex vessels—laticifers—either dead cells or cavities formed between cells.

There are two general categories of lactiferous products: *rubber*, which contains a high percentage of cis-polyisoprene, an elastic polymer; and *balata*, in which the principle hydrocarbon is trans-polyisoprene (with rather abundant resin included), an

inelastic polymer. Both result from coagulation of the latex, in which the solid coagulum is precipitated and separated from the useless liquid serum. Because of its use in familiar consumer products, rubber is the better known, although chiclé balata, used in chewing gum, is not unfamiliar.

Although latex may be found in various parts of diverse species, most often it is extracted from the bark of tropical trees. Tapping varies with the source and local custom: with the Pará rubber tree, *Hevea*, an efficient tapping scheme has been worked out, reviewed under discussion of that species. With *Castilla* and *Manilkara* the whole tree is typically felled, the trunk girdled at intervals, and the latex allowed to drip into containers placed below the gashes. *Achras*, and most secondary rubber sources, are commonly left standing but gashed from top to bottom; although the bark is badly disfigured, tapping is usually not lethal. *Palaquium* balata is increasingly extracted from the foliage with solvents. *Parthenium* is macerated to free the rubber polymer physically.

Coagulation generally occurs naturally if latex is allowed to stand for several days, permitting acidification through fermentation and enzymatic change. Foul-smelling serums are avoided and a cleaner coagulum obtained quickly, however, if coagulation is induced artificially. In some cases the latex is heated, in others salts or acids added (in primitive localities these may be sea water or fruit juice). The repelling forces between colloidal particles ends under the influence of the electrolyte, and a soft, whitish coagulum precipitates from which the serum is poured off. The coagulum is further treated in various ways, to eliminate excess liquid and better preserve the product. Rubber is typically smoked to become a tough, dark, elastic substance, while balatas turn hard and brittle naturally (and in some cases are stored under water to prevent further oxidation).

Rubber.

A number of lactiferous plants produce rubber, but only *Hevea brasiliensis* has had a vital influence on the world's economy. *Landolphia* and other Apocynaceous vines of equatorial Africa were the first exploited rubber plants. The vines were cut to drain the latex and of course supplies became depleted within a few years, spelling doom for the African rubber boom. Not only were the vines wantonly exploited, but the native tribes were "taxed" increasing quotas of rubber by colonial powers. When these were not met, atrocities that are today unthinkable were inflicted upon the Africans. As rubber supplies from Africa diminished, the boom spread to tropical America, where *Castilla* in Central America and northern South America, *Hevea* in Brazil, and African introductions (such as *Funtumia*) in the West Indies were all highly regarded potential sources. For various causes *Hevea* eventually came to dominate the natural rubber industry. Of all rubber plants, few yield quality rubber in sufficient quantity to make exploitation profitable. Of these none has rivaled *Hevea* rubber in adaptability to cultivation and ability to withstand repeated tappings. The story of *Hevea brasiliensis* is essentially the story of rubber.

Pará rubber, *Hevea brasiliensis*, Euphorbiaceae. Other species such as *H. benthamiana* and *H. guianensis* might be included, too, since they contribute to modern rubber cultivars. Probably there has been hybridization in the wild; perhaps even the seeds that Wickham sent to Kew, the basis for the cultivated rubber of the Far East, were of hybrid stock.

Rubber was first heard of in Europe as an unnamed plaything of the Mexican Aztecs, and as the legendary "caoutchouc" of the Amazon country—a strange substance, soft yet resilient, of which the tropical American Indians fashioned balls, torches, and watertight implements. By 1600 natives and Spanish settlers alike were waterproofing shoes, clothing, and hats with latex. La Condamine, in 1734, coming down the Amazon from what is now Ecuador, gave the name "cauchuc" (*caoutchouc* in French) to rubber utilized by Guiana Indians, and reported "heve" or "jeve" to be the native name for rubber in the Andean region. From this, Aublet later coined the generic name *Hevea*; strangely enough the "heve" reported by La Condamine must have been not *Hevea* (cauchuc) but *Castilla*, of the Moraceae. La Condamine sent samples of rubber to Paris, where it remained largely a curiosity for a century. Priestly then utilized it for rubbing out pencil marks, and dubbed the strange substance rubber, by which name it has been known ever since in English. In 1823 MacIntosh discovered that rubber was soluble in naphtha, and initiated the first experiments in waterproofing with rubber in regions remote from the rubber tree, whose uncoagulated latex had heretofore been necessary for impregnating fabrics. Sixteen years later Goodyear's discovery of vulcanization, the addition of sulfur to rubber at proper temperatures to yield a generally useful product, finally opened the way to widespread commercial applications. Until some way had been discovered to keep it

A jebong half-spiral panel correctly tapped with the jebong knife.

faster than supply and prices consequently skyrocketing, it was perhaps inevitable that the famed Pará rubber tree would be introduced into colonial empires in other parts of the world. Sir Henry Wickham in 1875 sent 7,000 seeds from the lower Tapajos to England, and later Robert Cross (who had been sent to Central America for seeds and plants of *Castilla*) shipped young trees. In England, at the Kew Gardens, young *Hevea brasiliensis* was started, and eventually the seedlings carefully shipped to Ceylon. From these few seedlings, grown with little enthusiasm in Ceylon, eventually sprang the gigantic Far Eastern rubber industry, spawning thousands of plantations, principally in Ceylon, the Malay area, Java, and Sumatra. With cheap but good Far Eastern labor, efficient European management, economical plantation systems, intelligent breeding programs, and ideal climates and soils, it was not long before production centered in the Far East. Production from wild sources in the Amazon languished, except briefly during World War II when vital rubber supplies were cut off from the Far East. After World War II production centered once again in the Far East.

from becoming sticky when warm and brittle when cold, rubber could have little importance. Development of the pneumatic tire was the final impetus needed to make rubber collection and commerce an important world industry.

At first rubber was collected entirely from wild plants. With demand increasing ever

Hevea is native to tropical South America, occurring in greatest frequency in the Amazon valley. Wild species are found from Bolivia,

Hevea rubber sheets hung to dry in shed, in Peru. (Courtesy USDA.)

eastern Peru, and Colombia eastward to the Guianas and the states of Maranhão and Mato Grosso in Brazil. The plants vary in size from small to very large forest trees as much as 150 feet tall—typically of the lowlands and often standing in water for many weeks at flood season. The bark is smooth except where injured or tapped, and when cut yields a white or yellow latex. In some species the latex flows well; in others, scarcely at all. The leaves are palmately compound, with a long petiole and three elliptic, petiolulate leaflets. Trees are monoecious, having separate male and female flowers on the same inflorescence. The flowers are small and inconspicuous. Bloom is ordinarily during the dry season. The fruit is a conspicuous three-lobed capsule containing large beetle-like seeds similar to castor beans.

Hevea brasiliensis has been introduced into various parts of the world, including Central America and the West Indies, eastern Brazil (Baía), Africa (particularly Liberia), Ceylon, the Malay archipelago, Sumatra, Java, and other East Indian localities. In most areas where introduced it is grown as a commercial crop under plantation conditions. In South America and parts of Central America and the West Indies, *Dothidella* disease, the South American leaf blight, has made growing of the high-yielding strains developed in the Far East impractical. Strains resistant to disease but still possessing the high-yield characteristics of Far Eastern stock have since been bred.

Latex in *Hevea brasiliensis* contains on the average about 30 per cent dry rubber; the rest, the serum, is mostly water. The latex is found in numerous latex vessels in the bark. Since the inner bark is comparatively richer in this respect than the outer bark, to obtain maximum yields in tapping the bark must be cut as close to the wood as possible without injuring the delicate growth layer, the cambium. The latex vessels in the bark do not run exactly vertically, but spiral to the right at about a 30° angle from the perpendicular. Thus to obtain maximum yield in tapping, by cutting across as many latex channels as possible per unit of incision, a tapping cut should run from the upper left to the lower right, at a 30° inclination. Only in this fashion will the latex vessels be cut transversely and their severed ends occur in maximum frequency. The latex flows readily from the vessels severed in tapping, from distances up or down the trunk as far as 1 meter from the cut. Flow is longer and faster (hence yields somewhat greater) in the cool of the morning than at midday or after. Consequently, plantation tapping starts at daybreak and ends by noon. Best yields are not obtained until the tree has been tapped several times. The inherited qualities of the tree, its vigor and physiological condition, and the season

Hevea latex from wild trees is emptied into receptacle before being made into smoked ball rubber like that at the left.

of the year all have an effect on latex yields, as does the system of tapping. Under experimental conditions, and with pedigree trees, yields per tree of as much as 1 ton a year have been obtained.

Tapping systems vary greatly in various parts of the world, reflecting chiefly the degree of supervision that can be given tappers. On the scientifically managed plantations, particularly in the East Indies and Malaya, "jebong" tapping is exclusively practiced. A special jebong knife of high-quality steel is used. This knife has a V-shape head sharpened on both edges, and the neck is so curved that with practice an easy, quick stroke from left to right will remove the proper thickness (about 1 millimeter) of bark. A typical opening cut at 30° inclination is made halfway around the tree (half-spiral panel) or completely around it (full-spiral panel). Depending upon the vigor of the tree and the number and extent of panels opened each day, every alternate day, or every third or fourth day, the tapper progressively cuts another sliver (1 mm) of bark from the lower edge of the cut or panel, severing the sealed ends of the latex vessels and renewing the flow of latex. During the course of a year about 15 cm of bark will be thus consumed. One side of a tree may be tapped for 10 or 12 years before ground level is reached (assuming the panel was started as high as a tapper could reach, 6 or more feet above the ground). By that time, if the tapping was carefully done and the cambium not injured, a new bark will have formed over the tapped panel. This new bark is then ready for a second series of tappings. Usually two half-spiral panels, one high on one side, the other low on the other, are opened on a tree and tapped every other day. Low panels give better yields than high panels. During months of foliage drop the tree is given a rest from tapping.

Jebong tapping is seldom practiced where uneducated and unsupervised tappers scour the jungles for wild trees (in Brazil and other South American countries). In the Amazon valley, wounds were originally made in the bark with small hatchets (the "machadino" method), and the dripping latex collected by winding a vine about the trunk to conduct it to a container. Sometimes the latex was collected simply by letting it run down the bark into leaves, cavities, or whatever containers might be available at the base of the tree. By the time of the great rubber boom, small tapping knives with a U-shaped head (the "Amazonas" method) had been introduced. In Amazonas tapping, a series of separate, parallel incisions are made, one each time the tree is tapped. Compared to jebong tapping, such a system is extremely wasteful of bark, but it is still practiced along the backwoods "estradas" of the Amazon valley.

On rubber plantations latex is carefully kept free of bark or dirt. Before renewing the tapping cut (jebong system), the tapper pulls away all fragments of coagulated latex from the previous tapping, to be kept separate as scrap. A clean, specially designed latex cup is hung below the lower end of the panel, from a support designed not to injure the bark. A small metal spout is inserted in the outer bark to conduct the latex to the cup. After tapping his "task" (designated number of trees), the tapper retraces his steps and carefully empties each cup of latex into a covered pail. Latex cups are cleaned and inverted on the holder to avoid the accumulation of rain during the usual afternoon showers. In marked contrast, there is little organized care in wild-rubber regions. The latex, accumulated in any kind of receptacle, often merely cupped leaves, is carried to camp in any available container or liquid-proof sack.

Treatment of accumulated latex varies greatly, depending upon the locality, custom, and use to which the rubber is to be put. On plantations, filtered latex was typi-

Freshly poured latex coagulating on surface of ball as it is held in the smoke. Amazon valley.

cally coagulated with acid under controlled conditions and the coagulum pressed thin and free of water through a series of roller presses, the last one of which is corrugated. The sheet of rubber so produced is dried in a "smoke house" to become "smoked sheet." The vapors from the selected woods burned in the smoke house (pyroligneous acid) and heat (about 45°C) darken and preserve the rubber. Quality smoked sheet is translucent, free from impurities, and remarkably durable. It is baled with waste or low-grade sheet wrapping for shipping to the manufacturer. About 1965 a granular rather than sheet rubber, "heveacrumb," was introduced to the trade. This can be produced exactly to the specifications of the purchaser, and shipped in conveniently handled polyethylene-wrapped blocks. There has also been increasing tendency to centrifuge the latex at the plantation to an extremely concentrated

form, treat this with an anticoagulant (ammonium hydroxide), and ship the resultant concentrated latex cheaply to consuming countries in large tankers.

By contrast, in the Amazon valley latex from wild trees traditionally has been coagulated over a smoky fire. After his morning rounds, the tapper or a member of his family sits in a thatched hut beside the smoldering fire and tediously pours latex onto a paddle (flattened stick) held in the smoke. Each thin film of latex is coagulated by the smoke, and gradually there is built up a gigantic ball of rubber weighing up to 200 pounds. Such a yield may take several weeks and involve the tapping of many widely spaced trees. Smoked ball rubber is of excellent quality if the latex has been kept clean, in spite of the crude way in which it is produced, but it is questionable whether any quality in rubber is worth the human discomfort (especially smoke irritation to eyes) attendant upon smoked ball production.

The price of smoked sheet and smoked ball rubbers varies with the grade. Graders base their decision largely upon moisture content of the rubber and frequency of impurities, but also consider uniformity, locality where produced, care in smoking and handling, and so on.

Wild rubber exhibits great variations in yield, and it is not uncommon to see *Hevea* plants left untapped after an experimental cut or two has shown unsatisfactory yield. In the Amazon valley, however, it is unusual to find *Hevea* trees entirely unscarred from previous tapping during one or another of the rubber booms (the most recent during World War II).

On the plantations of the Far East, *Hevea* was early subjected to intensive selection for yield and other characteristics, and various breeding programs were undertaken. Assessments on rubber shipments provided funds adequate to maintain a research staff

Rafts of smoked ball crude rubber on a tributary of the Amazon.

of experts devoting their entire energies to improvement of rubber. In perhaps a shorter time than that in which any other wild plant has ever been domesticated, the plantation rubber trees of the Far East were made far superior to the wild trees of the Amazon. Average yields were increased far beyond the highest ever known in the wild, and techniques in growing, maintenance, tapping, and processing were standardized along the most efficient lines. Production became so efficient that it was possible to land rubber in New York at actual cost of less than 4 cents a pound in the 1930's. Obviously wild rubber slowly and tediously produced in the Amazon valley could not compete, and it is understandable how 90 per cent of the world supply of rubber came to be produced on Far Eastern plantations.

In propagating plantation rubber only seedlings selected for vigorous growth are utilized for stock. On these are then budded trunks of pedigree clones selected for high yield. By this method each bud of the high-yielding tree can quickly become another tree with characteristics identical to those of the parent. In budding, a slit or slits are made in the bark of the stock and a lateral bud cut from the desired clone is inserted under the flap of bark. Careful binding holds this

in place and prevents desiccation. After the bud has grown to the stock, the top of all branches of the stock seedling are cut off and the selected bud is permitted to effect terminal growth. The trunk thus produced from the bud will have all the characteristics of the high-yield tree from which it came. Even more elaborate is the production of double-budded trees. The same procedure described above is followed and the resulting high-yielding but not disease-resistant sapling is protected by spraying for a few years. Then a bud from a disease-resistant clone is budded to the top. There results a tree with vigorous rootstock from one parent, high-yielding trunk from another, and disease-resistant top from still another.

The rubber industry. The uses for rubber are legion, including stoppers, washers and gaskets, erasers, soles and heels, balls, bands, rollers, elastics, gloves, waterproofings, cushioning, wire insulation, conveyor belts, and so on. The greatest use, however, is for penumatic tires. A brief sketch of the processes involved in tire manufacture (Fig. 8-1) may not be out of place here. The steps in making a tire involve a preliminary cleaning and maceration of solid rubber. This is then mixed with sulfur, carbon black, pigments, and certain other materials and thoroughly pulverized in huge "Banbury mixers." The compounded rubber is eventually applied to tire cord, a tough fabric usually woven of rayon or other synthetic fiber. The cord is dipped in liquid rubber and thereafter sent through complex "calender" machines which press additional compounded rubber onto both surfaces. The rubberized cord is then cut into oblique sections termed *plies*. Several plies (four in ordinary passenger car tires) are applied, one on top of the other, on a circular building drum, with each consecutive ply oriented 90° from the preceding one, just as the grain of veneers is oriented in making plywood. To the top ply is added a thick coating of wear-resistant rubber, which becomes the tire tread. The plies, now formed into a barrel-shaped cylinder on the building drum, are placed in a forming machine to be pressed into doughnut shape; then, still soft, in a mold resembling a huge waffle iron. It is there vulcanized or "cured" by steam, assuming the outline, tread, and form dictated by the mold. All the parts become fused into a single unit, the tire.

In the short time since its domestication some three-quarters of a century ago, rubber has had a tumultuous history. At the turn of the century, when the demand for rubber began to grow, only wild rubber was to be had. So rapidly did the demand increase that at one point, during 1910, rubber for a moment reached the all-time high of $3.06 per lb. Those were wild times up the Amazon. Thousands of miles inland from the sea the sweltering rubber city of Manaos was built on the Rio Negro near the Amazon, in the midst of the greatest uncharted forest belt in the world. Cost was of little concern; a great opera house was erected there for which materials and talent alike were imported from far-off Italy. This remains today, gradually deteriorating in a city surrounded on all sides by thousands of miles of jungle, as a gaudy reminder of an era of ephemeral

Amazonian glory. Still told today in Brazil are many stories, true or otherwise, of this first rubber boom. One relates that strategic up-river spots (from which Indians in canoes laden with rubber could conveniently be shot in passing and their booty seized) sold for fabulous sums. But the decade of boom quickly ended before 1920, as the Far East developed the plantation system and by it produced more rubber than the world could consume.

Low prices prevailed for rubber for several years during the early 1920's. From a high of over $3.00 per lb, rubber fell to as little as 14 cents. As a consequence, British producers of the Far East developed the so-called Stevenson plan, an agreement among themselves to restrict rubber production. Once again rubber prices rose, and a boom was on. Prices reached a peak of $1.23 per lb, and an annual average of more than 70 cents per lb in 1925. Then American purchasers began to search for their own lands for growing rubber, and the Dutch, not restricted by the Stevenson plan, began to match British production. By 1928 Britain was forced to repeal the restriction act, as their control of world production vanished. Rubber prices dropped at once, and by 1932 hit an all-time low of 3 cents per lb. Then an international agreement among the British, Dutch, and French once again established restriction quotas, but of a milder sort than under the Stevenson plan. Rubber prices were stabilized at approximately 20 cents per lb, where they remained until World War II.

With the outbreak of World War II, Far Eastern plantations quickly fell to the Japanese. Necessary rubber for a mechanically fought war was cut off. With only Ceylon, a portion of India, and Africa still producing plantation rubber for the Allies, every supplementary source of rubber was sought in desperation. Even though the development of synthetic rubber was destined to supply most of the war demand, a second tropical American rubber boom was on. The Amazon valley again felt the flow of wealth, with the base price of rubber pegged at 60 cents per lb by inter-American agreement. Trees bearing the scars of the first rubber-boom tapping again felt the tapper's knife. But this second, milder, boom was as certain to break as did the first. As soon as Far Eastern plantations were once again in Allied hands, world rubber prices dropped to the 15-to-30-cent-per-lb level.

Discussion of the rubber industry cannot be concluded without a few words concerning synthetic rubber (Fig. 8-1). At the beginning of World War II, synthetic rubber was largely a curiosity. In the United States less than 8,000 tons had been made by 1941; by 1944, 1 million tons could be produced, sufficient to supply more than the normal peacetime demand for all rubber in the country. Chemists had long realized that rubber was a concatenation of hydrocarbon molecules, but they did not know just how the molecules are put together. However, small molecules such as isoprene could be strung together (polymerized) to yield rubberlike substances (elastomers). Butadiene and styrene, similar to isoprene, copolymerize to form an elastomer not identical with natural rubber but very similar to it. Both butadiene and styrene are readily made from coal, petroleum, or alcohol. Today there are special synthetics for almost any conceivable use.

World production of natural rubber is nearly 3 million tons annually, mostly from Malaysia and Indonesia, consumed especially in Europe. Synthetic rubber production is equally as great, especially used in North America; its cost effectively puts a price ceiling on natural rubber, which today can be economically produced only in regions of inexpensive labor. A percentage of natural rubber is blended with synthetic for many types of tires and other rubber products. Brazil, Pará rubber's homeland, sends only about 20,000 tons to market nowadays, and all of Africa (especially Liberia and Nigeria) about 150,000 tons.

Rubber-Yielding Species of Lesser Importance. PANAMA RUBBER, *Castilla.* Among the species of *Castilla,* of the Moraceae, were probably the first rubber plants noted by La Condamine as the "heve" of the Ecuadorian Andes. *Castilla* is also the probable genus of rubber plants used by the Central Americans and Mexican Indians, and until the middle of the nineteenth century remained of greater importance than *Hevea.* Some ten species occur from Mexico to Bolivia and Amazonian Brazil. Most important in the Amazon valley is *C. ulei*; most frequent in Central America is *C. elastica,* Panama rubber. The trees characteristically grow in drier habitats than *Hevea,* and become equally large, with pronouncedly buttressed trunks.

A felled and girdled *Castilla* rubber tree of the upper Amazon valley.

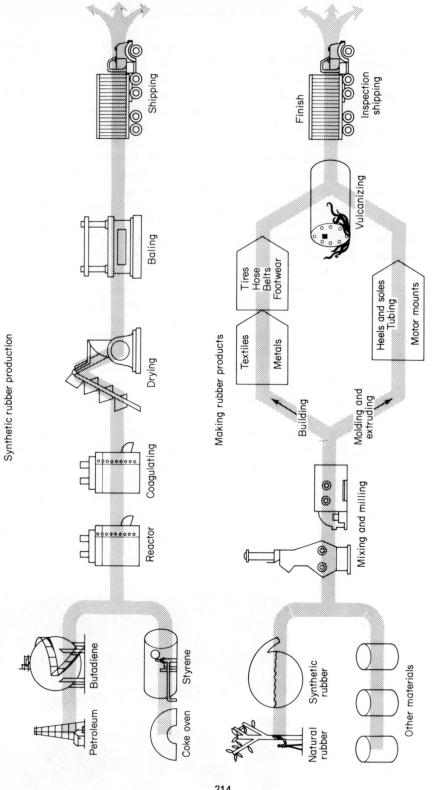

Figure 8-1 Flow sheet of rubber manufacture. (Courtesy Rubber Manufacturers Assoc.)

Synthetic rubber production

Petroleum
Butadiene
Coke oven
Styrene

Reactor
Coagulating
Drying
Baling
Shipping

Making rubber products

Natural rubber
Synthetic rubber
Other materials

Mixing and milling

Building
Molding and extruding

Textiles
Metals

Tires
Hose
Belts
Footwear

Heels and soles
Tubing
Motor mounts

Vulcanizing

Finish

Inspection shipping

Leaves are oblong, simple, alternate, with encircling stipule scars. Flowers are individually small, in catkinlike clusters. The plant is monoecious. Latex vessels occur in the inner bark, and evidently do not anastomose. Unlike *Hevea*, once the latex canals in *Castilla* are severed, they continue draining until no more latex is left in the tree. The tree then dies, or takes many months to recuperate sufficiently to allow a second tapping. Thus in practice trees are often tapped but once, and may be felled for this purpose. In the wild, *Castilla* has been largely eliminated in accessible areas owing to felling by rubber gatherers.

After locating the wild trees, tappers typically fell these forest giants and girdle them at intervals of a few feet in order to drain them completely of latex. One tree may yield several gallons of latex forming as much as 70 pounds of rubber. Or, if not felled, the tree may be ascended by means of a climbing rope and a series of slanting gashes cut in the bark to drain the latex; a U-headed knife or a machete are the usual tapping instruments. The draining latex is collected in a suitable container, or, in backwoods areas, simply in a hole in the ground as it flows to the base of the tree. The quality of the rubber is regarded as lower than that of *Hevea*, owing more, perhaps, to generally poor collection practices that allow dirt to reach the latex than to inherent inferiority of the rubber.

During the first rubber boom high expectations were held for *Castilla*. Botanists of the Royal Botanic Garden, testing the genus in Trinidad, felt that its rapid growth and tolerance to varied habitat offered greater promise than with *Hevea*. There was an interval of great demand, even black-market distribution of seedlings, through much of the West Indies. But inexperience with growing, tapping, and handling perhaps more than inherent deficiencies of the trees led to abandonment of West Indian rubber plantations. Scattered naturalized plants are still found in the islands today.

Maniçoba, *Manihot*. The genus *Manihot*, of the Euphorbiaceae, includes not only the invaluable cassava or manioc, staple food of many tropical peoples, but also a number of lactiferous species known as maniçoba. Principal species tapped for rubber are *M. glaziovii*, the Ceará maniçoba of northeastern Brazil; *M. heptaphyla* and *M. piauhyensis*, Piaui maniçoba, also of northeastern Brazil; and *M. dichotoma*, Jequié maniçoba of eastern Brazil (southern Bahía). At least the first has been introduced into Africa and the Far East, but because of smaller size and lesser yields has never seriously competed with *Hevea*. All these species are native to Brazil's semiarid "sertao," a region unsuited to more useful rubber plants. The maniçobas are small to medium trees with alternate, palmately lobed leaves, inconspicuous monoecious flowers, and fruit capsules much like those of *Hevea*.

Maniçoba latex is often more viscous and thus slower-flowing than that of *Hevea*. It is found in latex canals in the inner bark as in *Hevea*, and responds to similar tapping schedules. The horny outer bark is so hard, however, that jebong tapping cannot be practiced. With Ceará and Jequié maniçoba, V-tapping, open-wounding, or a modification of the herringbone system of tapping is practiced on the trunk, with latex collection in latex cups. With Piaui maniçoba, a hole is dug at the base of the plant and the root tapped, the latex draining onto a chalk-like powder used to line the hole to prevent undue adherence of sand. The quality of the rubber is good except for unavoidable impurities, but production is small and uncertain. When rubber prices were at their peak, maniçoba tapping was profitable enough to cause extensive plantation planting in Brazil's semiarid hinterland, but competition of *Hevea* from the Far East soon forced abandonment of these plantations.

GUAYULE, *Parthenium. Parthenium argentatum*, guayule, of the Compositae, is the only plant native to the United States that has been commercially utilized for rubber. The species occurs in north-central Mexico and Texas, where it is found in semiarid locations. It was first described botanically as recently as 1859. The plant is a slow-growing perennial about 2 feet tall and lives many years. It has simple, silvery gray leaves and small heads of inconspicuous flowers. Rubber is found not in the latex canals but contained within the cells in all tissues. It is particularly abundant in the phloem and xylem rays of the stem and root. Extraction of the rubber is by cutting the plant, macerating the tissues, and separating the rubber thus freed by flotation in a liquid bath of proper density. By selection, the usual average yield of 7 per cent of dry weight in wild plants can be increased to over 20 per cent or more.

Guayule was first exploited for rubber

Guayule plant, *Parthenium argentatum.* (Courtesy Chicago Natural History Museum.)

during the first rubber boom at the turn of the century. German interests built several extraction plants in north-central Mexico at that time. During the World War II emergency, the United States government undertook an elaborate investigation of guayule and its cultivation, but later abandoned the program as synthetic rubber became available. For most purposes guayule rubber is inferior to that from other plants, owing to a high resin content. Production from wild sources in Mexico has never run more than 5,000 tons annually.

MANGABEIRA, *Hancornia. Hancornia speciosa*, of the Apocynaceae, is native to the sandy, flat-top coastal elevations or "taboleiros" of northern, eastern, and southern Brazil, extending to Paraguay. In peacetime, the latex of mangabeira is sought domestically to some extent, for making rubber cement and, locally, for waterproofing cloth. With proper coagulation the latex yields a rubber, which has, however, poor keeping properties and is definitely inferior to *Hevea*, *Castilla*, and *Manihot*. Mangabeira is a small, slow-growing, gnarled, and twisted tree, with simple, entire leaves and a soft bark. The flowers are attractive, white, fragrant; the edible fruit resembles a small persimmon and is sought as a delicacy when ripe. The plant has never been adapted to cultivation, however. The several hundred tons of mangabeira rubber it was possible to procure during World War II came entirely from wild sources.

The latex system of *Hancornia* bark must resemble that of *Castilla*, for one tapping serves to drain the tree thoroughly, and it cannot then be successfully tapped again for a few months. Normally, mangabeira is not felled in tapping as is *Castilla*, but is subjected to a series of V-shaped or inclined gouges or spiral channelings, followed by knife-edge cuts to the wood. The latex flows for several minutes. It is typically collected in latex cups stuck in the bark just below the tapping cut. The latex is then carried to a central depository, where it may be coagulated by boiling; by combination with alum, salt, or other chemicals; or by fermentation. The coagulum is, preferably, pressed free of excess water and marketed at once, to prevent deterioration (tackiness). Production is for the most part by completely uneducated nomadic tappers who move from place to place in the desolate taboleiros, since no one locality can sustain tapping for long.

LANDOLPHIA AND OTHER APOCYNACEOUS VINES. These vines of the Apocynaceae native to equatorial Africa were, for a few years during the first rubber boom, a modest source of rubber. *L. kirkii, L. heudelotii,* and *L. owariensis* are reported to be the most utilized species. The plants are large jungle vines, which must be cut at ground level and the draining latex collected, or pulled down and segmented. Either way the plants are killed. The latex is crudely coagulated by heat, fermentation, or plant juices (acids).

Extraction of *Landolphia* latex involves one of the more unsavory episodes of rubber's history. It is said that during the height of Africa's wild-rubber boom, Leopold's government gave procurement concessions to ruthless groups who "taxed" peaceful tribes quotas of rubber. If quotas were not met women were taken and chain-gangs roamed the jungles. Purportedly after the rubber boom, the Belgian Congo had many thousands of mutilated natives, part price for a few thousand tons of rubber.

PALAY RUBBER, *Cryptostegia*. Two species of *Cryptostegia*, of the Asclepiadaceae, have been introduced into the New World from Madagascar: *C. grandiflora* and *C. madagascariensis*. These, and hybrids of the two, have seemed potentially useful rubber sources. The plants are rapid-growing vines with large, showy flowers. Latex is usually gathered by cutting off the tips of the growing stems and collecting the small amounts which then exude from the latex tubes of the bark. Chemical or mechanical extraction from all parts of the plant is also possible. During World War II the Board of Economic Warfare initiated an elaborate *Cryptostegia* plantation project in Haiti for emergency production, against the advice of experienced rubber growers and with disappointing results.

RUSSIAN DANDELION, *Taraxacum*, Compositae. *Taraxacum kok-saghyz.* This is the Russian dandelion or chew-root, a potential source of rubber, discovered in Russia in 1931 and introduced into the United States in 1942. It much resembles the common dandelion (which also contains a latex). Selected strains of *Taraxacum* can be made to yield several times the average latex quantities of ordinary stock. The plants are dug up at maturity and the latex extracted from the tuberous root by crushing and maceration.

OTHER RUBBER-PRODUCING PLANTS. Various goldenrods, *Solidago* (Compositae), have been experimented with as rubber plants, originally by Thomas Edison. It has also been shown that a number of Temperate Zone milkweeds, *Asclepias* (Asclepiadaceae), are capable of producing rubber. In the tropics, *Ficus* (Moraceae), *Alstonia, Forsteronia, Funtumia, Mascarenhasia, Odontadenia,* and *Tabernaemontana* (Apocynaceae), and *Euphorbia* (Euphorbiaceae) have at times held promise and have given commercial production of minor importance for short periods.

Balatas

The term *balata* may be broadly applied to several latex products which are neither elastic nor resilient upon coagulation. Such products as chiclé (chewing gum), guttapercha, juletong, and various other tropical "balatas" fall into this category. None is of large-scale importance.

Ascending tree with aid of a climbing rope, this Colombian tapper taps *Sapium tolimense*, the caucho rosado. (Courtesy Paul H. Allen.)

CHICLÉ, *Achras zapota*. Chiclé, nispero, sapodilla, and chewing gum are among the more than twenty vernacular names used for the species in Central America. Originally the peculiar balata utilized for chewing gum, hard and brittle until masticated, came exclusively from several strains of *Achras zapota* (and other doubtfully distinct species of the genus), of the Sapotaceae. *A. zapota* occurs scattered through Central America, the West Indies, and northern South America, in greatest frequency in the Yucatan peninsula of Mexico and Guatemala. It has also been widely planted throughout the tropics for its edible fruit (sapodilla). Exhaustion of the accessible stands in the Yucatan area, a result of excessive tapping, has made necessary the procurement of latex from other sources to supply chewing gum demand and supplement the diminishing production of true chiclé. Several other latices behave much like that of *Achras*, including *Couma, Stemmadenia, Tabernae-*

montana, Thevetia, Plumeria, and *Cameraria*, of the Apocynaceae; certain species of *Calocarpum, Sideroxylon, Dipholis, Bumelia, Lucuma, Chrysophyllum, Manilkara, Mimusops*, and *Sapium*, of the Sapotaceae; *Castilla, Brosimum, Pseudolmedia*, and *Ficus*, of the Moraceae; and *Jatropha*, of the Euphorbiaceae—all indigenous to the New World.

Achras zapota in the wild is a large, smooth-boled tree of the rainforests up to 4 feet in diameter. It possesses alternate, elliptic leaves and inconspicuous, perfect flowers. The pearlike fruit is esteemed by tropical peoples. Latex of *A. zapota* occurs in a series of canals or vessels in the inner bark just as do latices of rubber. It was first extensively used as a masticant after attempts to vulcanize it during the first rubber boom failed but revealed the ease with which it could be compounded with adulterants, sugars, and flavors. The superiority of chiclé to Temperate Zone spruce and cherry gums was quickly recognized, and within half a century raw gum in annual quantities of 7,000 tons was being imported into the United States to supply the chewing gum industry.

Wild chiclé is tapped by "chicleros" who invade the tropical forest during the rainy season (July to February in the Yucatan area). These tappers must be skilled in the use of the machete and the climbing rope, for the trees are tapped the entire length of the trunk by a series of zigzag gashes, made with a sharp machete, half encircling the tree. On older trees the harder outer bark may be first scraped away, to facilitate the more exacting work of channeling to the wood. If the tapping is carefully done, without too much injury to the bark, the scars heal and the tree becomes available for future tapping some years later. Careless and over-severe tapping, however, kills a goodly number of trees annually. During tapping the latex spills through the series of guiding gashes in the bark and is channeled by a leaf inserted in a flap in the bark at the base of the tree into a convenient vessel, usually a skin or canvas bag. Yields may be as high as 60 pounds of latex from one tree; the average, however, is 2 to 10 pounds. Just as with *Hevea*, tapping must be done in the early morning, roughly from daybreak until noon, for not enough latex to make collection worthwhile is obtained after noon, as temperatures become warmer. The afternoon is usually spent in locating trees for the next day's tapping. In locating new trees, a sample cut is made to determine whether or not profitable yield can be expected.

Latex is assembled at a nearby camp, usually located near a source of water. Water is needed not only for personal use by the tappers, but also in the molding of the latex. Ordinarily the latex is brought in the same day the tapping takes place, but collection sacks may be left on the tree until the next day if no danger of afternoon rain exists. After sufficient latex is secured, it is cooked in large cauldrons over an open fire, usually once a week. This boiling down ordinarily takes about 2 hours, during which time the latex is stirred continuously. When it has become thoroughly viscous it is removed from the fire and further worked until nearly firm. It may then be molded by hand into blocks, soap and water being used freely to prevent it from sticking to the hands or to utensils. The blocks of coagulum contain 40 to 50 per cent moisture at this stage, but further drying reduces moisture content to 25 to 35 per cent by the time they are ready for shipping.

Demand for chiclé and related balatas comes chiefly from the United States. The gum is purified by crushing and washing, then melting and centrifuging. To this base of pure chiclé are then added varying amounts (up to 95 per cent) of natural or synthetic adulterants. Powdered sugar, corn syrup, and flavor are added and blended in huge mixers. The mixture is extruded from a "kneader" as a thin sheet, which is in turn automatically cut and packaged to yield chewing gum, a social nuisance particularly peculiar to the United States.

CHILTÉ, *Cnidoscolus.* Species of *Cnidoscolus,* Euphorbiaceae, grow in desertlike environments of northern Mexico and northeastern Brazil. The red chilté, *C. elasticus,* and the lowland or white chilté, *C. tepiquensis,* have both been minor sources of latex in southern Mexico since preColombian times, used for molding small articles. Production is entirely from wild trees (see Table 8-1). Red chilté occurs in the mountains of west-central Mexico, lowland chilté in coastal Jalisco. The latex is gathered by gouging the bark herringbone fashion with a series of interconnecting channels. It is typically left to coagulate naturally by

Red chilte tree (*Cnidoscolus elasticus*) five years old, tapped to determine latex yield. (Courtesy USDA.)

Table 8-1

Yields of Balata from Cnidoscolus tepiquensis

Material Measured & Test No.	Tap	Latex Yield			Moisture Content	Dry Solids Content per	
		Trees Tapped	Total	Average per Tree Tap		Tap	Tree/Season*
	Sequence	No. of	cc	cc	percentage	grams	grams
Latex:							
1	First	47	25,004	532	62	202.3	2,426
2	Second	36	18,612	517	62	196.5	2,358
Total		83	43,616				
Average per tree				525	62	199.5	2,394
Coagulum:			grams	grams			
3	Second	52	19,000	365	54.6	165.6	1,987
4	Second	43	20,430	475	54.6	215.6	2,587
Total		95	39,430				
Average per tree				415	54.6	188.4	2,261
Combined ave. latex and coagulum						193.9	2,327

* Estimated on basis of 12 tappings per season.
From L. Williams, *Economic Botany* **16**, 68 (1962).

fermentation. Pressed into rectangular blocks for export, small amounts are used as an extender for more important balatas.

GUTTA-GUM, *Couma*. *Couma* (Fig. 8-2), of the Apocynaceae, is ubiquitous in tropical America (Fig. 8-3). *C. rigida* is the mucugé of Brazil; *C. macrocarpa*, the leche-caspi or sorva, has been a familiar adulterant of rubber latex during rubber booms. It is believed that *Couma* could be tapped as is *Hevea*, but more often the trees are felled and drained by girdling as with *Castilla*. The latex is typically coagulated by boiling, the coagulum immersed in water to prevent oxidation. The coagulum has a high resin content, and is locally used as a caulk. Export is chiefly to the United States for compounding in chewing gum. The fresh latex is said to be potable, and may be used as is "cream" in coffee.

GUTTA-PERCHA, *Palaquium*. Gutta-percha (Fig. 8-4) comes principally from *P. gutta*, Sapotaceae. It grows naturally from India to the central Pacific. Gutta-percha has been

The "vacahosca" or "arbol de vaca" (*Couma macrocarpa*) is frequent in the forests of Venezuela. It is extremely laticiferous, and the coagulum is used locally for caulking canoes. (Courtesy USDA.)

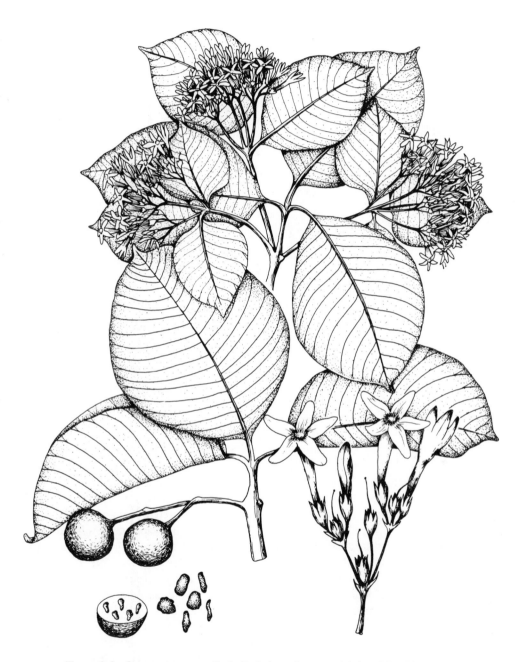

Figure 8-2 *Couma macrocarpa* Barb.-Rodr. Large leaves and fruit × 2/3 (U.S. Nat. Herb. 2,195,172) ; flowering branch and seeds × 2/3 (U.S. Nat. Herb. 1,441,683) ; flower detail × 2. (After L. Williams, *Economic Botany* **16**, 252 (Oct.–Dec., 1962).)

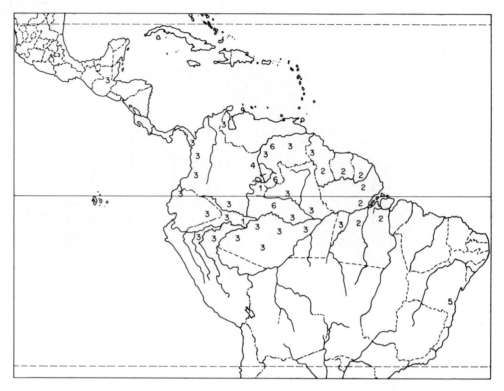

Figure 8-3 Distribution of *Couma* species: 1, *C. catingae*; 2, *C. guianensis*; 3, *C. macro-carpa*; 4, *C. multinervis*; 5, *C. rigida*; 6, *C. utilis.* (After L. Williams, *Economic Botany* **16,** 253 (Oct.–Dec., 1962).)

well known for years. It is pliable when heated, but turns hard and dark when exposed to air. Because it is an excellent nonconductor and impervious to water, it has found widespread use as insulation for transoceanic cables. Miscellaneous other uses are for golf ball centers, acid-resistant receptacles, and temporary dental fillings. For most purposes gutta-percha has been supplanted by synthetic plastics. Although gutta-percha had long been used by the native peoples of southeastern Asia, its botanical identity was not recognized until a little over a century ago. Production origin-ally came chiefly from Singapore, but as supplies became scarce *Palaquium* trees were sought out in the remoter sections of Borneo, the Philippines, and other southwestern

Pacific islands. There has been some planta-tion growing and tapping schemes developed to protect the tree (rather than felling, the original method). In recent decades balata has been extracted from the foliage. Leafy twigs are pulverized and washed in cold water; small threads of balata float to the surface and are skimmed off for processing. Blocks of gutta-percha are preserved under water to avoid surface oxidation. Early in the century nearly 1 million pounds of gutta-percha were produced annually, but today competition from synthetic materials and substitution by gutta-gum from South Ameri-ca has reduced the industry to only a fraction of this.

JULETONG, *Dyera.* A few species of *Dyera*, principally *D. costulata*, of the Apocynaceae,

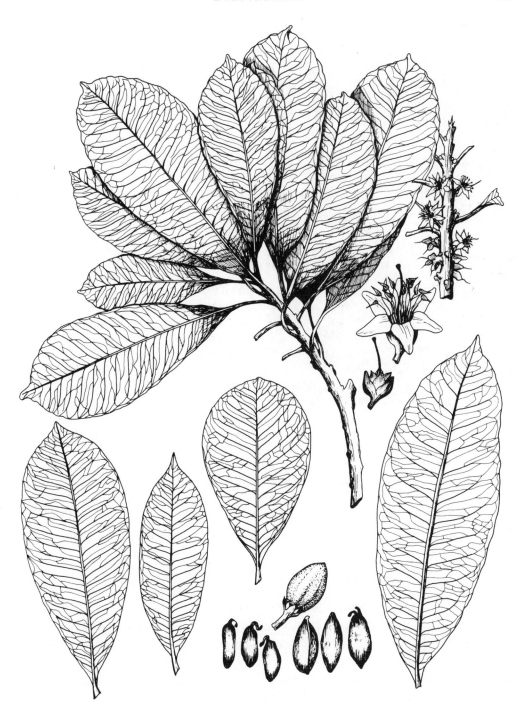

Figure 8-4 Gutta-percha (*P. Gutta* [Hook. f.] Baill.). All elements × 2/3, except flower detail, × 2. (Adapted from original which was drawn by Regina O. Hughes, USDA.) (After L. Williams, *Economic Botany* **18,** 9 (Jan.–Mar., 1964).)

Juletong tree (*Dyera costulata*) in Borneo. (Courtesy USDA.)

yield the balata known as juletong. Probably also involved are species of *Alstonia* and *Rauwolfia*. All these species are Far Eastern. *D. costulata*, indigenous to Malaya, is reported to be one of the best-yielding of all lactiferous plants. Shortly after the turn of the century, juletong was utilized to produce a very poor grade of rubber (75 to 80 per cent resin; 19 to 24 per cent rubber). Since that time it has been imported into the United States as an adulterant and substitute for chiclé. The balata is similar to gutta-percha, and is similarly handled.

OTHER BALATAS. Many other lactiferous plants are potential source of balata. Indeed their tapping may be of local importance, and if market conditions warrant they become a temporary item of commerce. Genera of the Sapotaceae are especially important, including *Bumelia*, *Calocarpum*, *Chrysophyllum*, *Dipholis*, *Lucuma*, *Manilkara*, *Mimusops*, and *Sideroxylon*. In the Apocynaceae *Cameraria*, *Plumeria*, *Stemmadenia*, *Tabernaemontana*, *Thevetia*, and other genera are of interest. *Brosimum*, *Ficus*, and *Pseudolmedia* are balata-yielding genera of the Moraceae.

SUGGESTED SUPPLEMENTARY REFERENCES

Baum, Vicki., *Weeping Wood*, Sun Dial Press, N. Y., 1945.

Collier, Richard, *The River that God Forgot, the Story of the Amazon Rubber Boom*, Collins, London, 1968.

Dijkman, M. J., *Hevea—Thirty Years of Research in the Far East*, Univ. Miami Press, Coral Gables, Florida, 1951.

Polhamus, L. G., *Rubber*, Leonard Hill, London, 1962.

Rubber Research Institute of Malaya, Kuala Lumpur, Malaya: various reports, such as that of 1966.

USDA, ARS, Production Research Report No. 10, *Rubber Content of Miscellaneous Plants*, Washington, D. C., 1957.

Williams, Llewelyn, series in *Economic Botany*, especially **15**, 3 (1961); **16**, 1, 2, and 4 (1962); **17**, 2 (1963); and **18**, 1 (1964).

CHAPTER **9**

Pectins, Gums, Resins, Oleoresins, and Similar Exudates

A HODGEPODGE OF TRUE GUMS (POLYSAC-charides), related pectic materials, and various resinous products (see, for example, Fig. 9-2) are the focus of this chapter. Few are of great commercial importance, and very few derived from widely cultivated plants. Consequently, there is apt to be unpredictable supply, inexact quality specifications, and intermittent production. Pectic and gumlike exudates or extractions may be secured from commercial fruit wastes or from the wild. Because pectins are so common, many potential sources exist, economics dictating which will be more used; there is no outstandingly important source plant, and the trend is toward widespread chemurgic utilization of several vegetable wastes. Gums are chiefly employed in adhesives and sizings. Most natural resins come from wild plants, although the important pine resin is increasingly a by-product from paper pulp manufacture (tall oil). About 10 per cent of the families of higher plants contain species which produce resin in appreciable quantity. Resins are typically dissolved in suitable solvents to form coatings (varnish), compounded into plastics and rubber, altered chemically (as to make soap or

similar esters), used as fillers or sizings, and utilized in small quantity for medicinal purposes.

Gums and Pectins

Gums and pectins occur naturally as cementing substances between cells (middle lamella), or as decomposition products of cellulose, usually in tissues undergoing disintegration. They are simpler chemically than are resins, and more easily substitutable. Although widely used, they have generally commanded neither the price nor the human interest extended resins. Gums consist mainly of polymerized carbohydrate subunits, and, unlike resins, are hydrophyllic (soluble or easily dispersed in water). They form viscous colloidal gels useful as sizings, thickeners, and stabilizers. Those containing little resin, generally polysaccharides of arabinose, galactose, and other sugars, are quite mucilaginous.

Gums are usually insoluble in common reagents such as alcohol, ether, and various oils. They will char but not burn freely as do resins. In many plants, gums exude naturally

A new water-soluble gum made from corn sugar using a fermentation process. Potential uses are in oil well operations, forest fire fighting solutions, and in food, pharmaceutical, and other products. (Courtesy USDA.)

from stems, roots, or leaves, and in other cases they are produced in quantity only when the plant is bruised or wounded, exuding from the lesions. Gums may serve the plant as a seal over wounds, as a mild antiseptic, or in chemical and physical retention of water. Exploited gums come largely from tropical and subtropical species of the Leguminosae. Pectins have been commercially extracted for the most part from citrus rinds and from the apple pomace of cider mills.

Agar is a long-known gum, an alginate from seaweed produced mainly in Japan. Attempts have been made to discover economical substitutes for agar among crops adaptable to agriculture. The Leguminosae,

Plantaginaceae, Cruciferae, and Convolvulaceae contain many promising genera. The seed endosperm of many species in these families (over 30 species in the Leguminosae alone) contains 20 per cent or more of gum, frequently galactomannans. Extraction is relatively uninvolved, requiring little more than the grinding of the endosperm to the fineness of flour. Galactomannans are used as paper additives, for waterproofing explosives, as a textile sizing, for flocculation of sewage, and in cosmetics and drugs.

The finer grades of gum are utilized in clarification of liqueurs, "finishing" of silk, and preparation of quality watercolors. Intermediate grades are used in confections, pharmaceuticals, and printing inks, as a size, and in certain dyeing processes. Least costly grades go into commercial emulsions, inks, matches, and similar products. In cosmetics and pharmaceuticals gums serve to protect and soothe membranes and skin (demulcents, emollients), or to emulsify or bind mixtures in creams, lotions, ointments, and medicines. Their property of leaving a "plastic coating" over objects upon evaporation of the solvent makes them useful in hair lotions, wave-setting preparations, and the like. In the paint industry, hydrophilic colloids serve as thickeners, emulsifiers, stabilizers, or even as film-formers or binders, particularly in water ("casein") paints. Preparation of cotton yarn for the looms often involves use of gums to stiffen, toughen, and bind the fibers. A great many textiles of all kinds are "finished" with gums or dyed from pastes formulated with gum. Many gums provide body and bulk to foodstuffs—commercial ice cream is a notable example. Gums may also serve as paper size, in formulation of adhesives, as "plastic" coatings, and in many other ways.

Commercial pectins. The most common and by far the most familiar use for high methoxyl pectins is in the making of jellies and marmalades, to insure jelling. Less well known is the similar use of pectins in pharmaceuticals, cosmetics, medicines, textile sizing, adhesives, certain rubber latices, steel hardening, fibers and films, and so on. Candy gumdrops, formerly utilizing tough gum arabic, are frequently made of softer, clearer pectin. Ice creams and sherbets utilize considerable quantities of pectin as stabilizers, and soda-fountain syrups or crushed fruits have pectin added to them as a thickener. In any juice or frozen food, pectin not only thickens but also helps prevent granulation, crystallization, or separation of pulp. In tomato juice and catsup, pectin adds viscosity and improves appearance. In salad dressing, pectin assists in stabilization and emulsification. It is similarly useful as an emulsificant in dairy products, such as cheese and milk. In bakery goods, pectin not only improves texture and yield but retains moisture as well, preventing quick staling. In pharmaceuticals and cosmetics, pectin chiefly serves as a jellifier, emulsifier, or binder. It has been so employed in dentifrices, salves and pastes, cosmetic creams and lotions, and as a binder in pills. As a medicinal, pectin has been utilized to alleviate both diarrhea and constipation. It has also been used in treatment of burns or soft-tissue wounds, and to check bleeding. Bacteriologically it has served as a culture medium. As an adhesive, pectin products have been found to take effective hold on glass and metal as well as on wood or paper. Cigars are commonly tip-treated with pectin. For adhesives, a preservative is usually compounded with the pectin. In the "creaming" of rubber latex, pectin is useful in coating the rubber particles and causing their separation from the nonrubber constituents. In quench-hardening of steel, 0.5 to 4 per cent pectin solutions impart a range of hardness from brittle to tough.

Pectins for this diversity of industrial uses come chiefly from citrus and apple fruit wastes. Potential tropical sources include the guava (*Psidium:* Myrtaceae), the pink shower pod (*Cassia:* Leguminosae), and possibly other fruits, such as mountain apple and surinam cherry (*Eugenia:* Myrtaceae), soursop (*Annona:* Annonaceae), carambola (*Averrhoa:* Oxalidaceae), tamarind (*Tamarindus:* Leguminosae), natal plum (*Carissa:* Apocynaceae), and pomelo (*Citrus:* Rutaceae). Of temperate and semitemperate sources, most important are apple pomace (*Malus:* Rosaceae), which contains up to 2.5 per cent pectin; and lemon and orange pulp and grapefruit waste (*Citrus:* Rutaceae), which contain up to 5.5 per cent. A great many other fruits contain considerable pectin, as evidenced by the natural jelling of various fruit jellies.

Commercial pectins are generally extracted by mild acid hydrolysis under controlled temperatures and pH, which converts water-insoluble forms to the water-soluble state. The resulting liquor is clarified by enzymatic hydrolysis of any starch present, decolorization with carbon, and filtering with filter-aids. The pectin is then precipitated with alcohol or colloidal aluminum hydroxide. Drying, grinding, and standardization follow, with the final product being marketed in either powder or liquid form. Several thousand tons of pectin are produced annually in

Peels, cores, stems, apple chunks, and cut apples move on conveyor for processing and pectin extraction. (Courtesy USDA.)

the United States, offering considerable competition to the imported gums.

Gums. GUM ARABIC. Gum arabic is derived from a few wild species of *Acacia* (Leguminosae) native to and often the most abundant plants of the dry, sandy, thorn scrub areas of North Africa, Arabia, and India. Most important is *A. senegal* (*A. verek, A. arabica?*), a small, gnarled, thorny tree growing to heights of about 20 feet in rather heavy stands in the semiarid portions of Africa north of the Senegal River. The tree is often also cultivated over much of North Africa. *A. senegal* possesses a gray outer bark and a fibrous inner bark; hooked thorns on the branches; "feathery" bipinnate leaves 3 to 4 centimeters long; small, yellow, spikate flowers; and a prominent, flat legume containing significant amounts of tannin. The trees are tapped at the close of the rainy season (winter to early summer) when leafless, by incising the outer bark with a small hatchet and peeling strips of this bark away. A viscous exudation forms as tearlike blobs upon the exposed underbark, probably a result of fungoidal infection. These exudations, after they have hardened on the outside, are collected by hand, about once every ten days. On the average, trees yield about 150 grams of gum per annum. The fragments are cleaned of adhering debris, and usually bleached in the sun. Gum from the Sudan is typically brought to buying centers by camel train and sold at government auction.

Gum arabic has an empirical chemical formula similar to that of cane sugar, and upon hydrolysis yields arabinose, which is largely insoluble after precipitation and drying. The odorless and tasteless gum is widely used as an adhesive, in confections, polishes, inks, sizes, and medicinals. No estimate is available as to how much gum arabic is utilized for various purposes, but probably no more than 25,000 tons are exported annually from the Sudan, mostly to Europe. Various adulterants, such as the gum of *Prosopis*, are frequently mixed with true gum arabic. Gum arabic has been utilized by man since before the birth of Christ. Apparently it served ancient Egyptians as a glaze in painting, and it was later mentioned in the works of Theophrastus, Celsus, Dioscorides, and others. The name "gum arabic" is reported to have been coined by Dioscorides.

Other African gums derived from various species of *Acacia*, often including the same species as yield "true" gum arabic (but marketed under different local names), include Sudan or Kordofan gums, Senegal or Berbera gum, Sunt gum, Suakim or Tahl gum, and many other local types, all extracted in the manner described for gum arabic. Cape gum is obtained from *A. karroo* (*A. horrida*) and *A. giraffae*, of South Africa. *A. arabica, A. modesta*, and other species yield Indian gum arabic. An Australian gum derived from *A. decurrens* and several other species is marketed as Wattle gum. All in all, much confusion exists as to the botanical identity of the various *Acacia* gums entering commerce, and grading and use reflect more closely handling methods and shipping source than inherent differences between the gums of different species.

GUM TRAGACANTH. Tragacanth gum comes chiefly from the arid highlands of Iran, derived from several species of *Astragalus*, Leguminosae. Production has reached as much as 2,000 tons annually, although declining of late because wild stands, the only gum source, are becoming exhausted (in spite of a conservation law designed to protect them, they are often appropriated for fuel). Species grow to the size of a shrub, scattered over the inhospitable arid lands north of the Zagros Mountains. Gum collection is seasonal, by nomadic gatherers. The most valued species, such as *A. gossypinus* and *A. echidenaeformis*, develop a mass of gum in the center of the stem where pith tissue would

A camel caravan passing date palms in Iran. Resins and gums of older times were frequently carried in this fashion. (Courtesy Chicago Natural History Museum.)

normally be expected. In summer this cylinder of gum is under pressure, and will exude if the outer tissues are perforated. At least twenty-three species are known to behave in this fashion. Some exudation occurs if the bark is scraped, but gatherers have learned that incisions to the central gum cylinder provide a better yield. Typically the earth is dug away from the taproot, and one or two incisions made into the upper part of the root sufficiently deep to reach the central gum mass. The ribbons of gum which exude are collected every few days for a period of about 2 weeks, after which another tapping is usually made. Gum collection starts in June and tapers off with the onset of autumn rain. It is said that a large *Astragalus* plant may yield about 10

grams of gum on its first tapping, though an average of 3 grams is likely.

Gum tragacanth has been collected since time immemorial, having been employed by the ancients as a bitter medicinal, recommended by Greek physicians as early as the seventh century B.C. In times of stress the gum may also have served as an emergency food. The better quality "ribbon gum" is used in pharmaceuticals, liqueurs, cosmetics, dental creams, and confections. "Bitter gum" of lesser quality may be used for dyes, sizings, waterproofing, and similar industrial purposes.

KARAYA GUM. Gum karaya (gum Kadaya, katilo, kullo, kuteera, bassora, Indian tragacanth, India gum) is the dried exudate from *Sterculia urens* (Sterculiaceae), of India. It is similar to tragacanth but cheaper, and has been much used in textile, cosmetic, and food industries within the last half century. *Sterculia urens* is a tree about 30 feet high, found chiefly in the central provinces of India. The trees are tapped during the dry season, particularly March to June. Several incisions or blazes about 2 feet long are made, often to the heartwood, from which the gum oozes. This is collected in tear form as it hardens, often as an odd job or when other work is scarce. Marketing is through Bombay. Certain African and other Asiatic species of *Sterculia* yield similar gums of limited importance.

GHATTI GUM. Gum ghatti (ghati, Indian gum) is the exudate from the stem of *Anogeissus latifolia*, of the Combretaceae, native to India and Ceylon. The tears of gum are collected in late spring. Uses of ghatti gum are the same as for gum arabic. *A. schimperi* of Africa is also suggested as a source of gum.

LOCUST BEAN GUM. Locust bean gum (carob gum, St. John's bread, swine's bread, gum hevo, gum gatto, jandegum, lakoe gum, lupogum, luposol, rubigum, tragon, tragosol) is derived from the fruit of the carob tree, *Ceratonia siliqua*, of the Leguminosae, widely grown in the Mediterranean area. The legumes are rich in sugars and protein, and have been used since ancient times for cattle and human food and fermented beverages. The gum is extracted in Europe from the seeds, following a process of sun-drying the pods. About 35 per cent of the seed is gum, and about 10 per cent of the legume is seed. Extraction typically involves dehulling, perhaps partial crushing, roasting (at about 150°C.), boiling for a few hours in water, and evaporation of the resultant solution to yield the brittle gum, consisting largely of mannose with lesser concentration of galactose. Locust bean gum is used in paper making, and as a stabilizer in food products.

Similar to locust bean gum are certain other hemicellulose gums obtained by much the same basic method of extraction, from flaxseed (*Linum usitatissimum:* Linaceae), psyllium seed (*Plantago psyllium:* Plantaginaceae), quince seed (species of *Pyrus* or *Cydonia:* Rosaceae), and iceland "moss" (*Cetraria islandica*, a lichen).

GUAR. Guar is obtained from *Cyamopsis tetragonoloba* of the Leguminosae, a plant resembling its relative the soybean. It is an ancient crop in India where the seed is used for food. A colloidal gum is derived from the seed, the endosperm yielding a polymer of galactose and mannose. Guar gum shows high viscosity in relatively weak concentration, swells in cold water, is useful over a broad pH range, and has excellent film-forming and stabilizing qualities. Chief uses are as a flocculent and filter-aid in the mining industry, and in the manufacture of cosmetics, drugs, papers, and certain foods.

Guar is one of the few gum sources lending

Guar plants ready for harvest. (Courtesy USDA.)

itself well to modern agriculture. It is a promising crop plant for the arid southwestern United States because of its drought resistance. Indeed, it has been grown there for years as a minor forage and green manure. Guar gained impetus as a gum source during World War II when foreign supplies were cut off. Selections were made from old fields in Arizona and New Mexico, and from these have been released several gum-yielding varieties. In parts of Texas facilities have been established for processing and handling guar, in anticipation that interest will spread as it did with the soybean. Conventional farm equipment handles guar, and the crop fits well into rotation with cotton. Estimates indicate that guar gum demand in the United States could exceed 50,000 tons annually.

AGAR AND AGAR-LIKE GUMS. Agar or agar-agar (Japanese isinglass, Japanese gelatin, kanten, and so forth) is well known as a culture medium in the bacteriological laboratory (an inert jellylike substance solidifying after melting at about 40°C.). It also finds use in laxatives and other pharmaceuticals; as a stabilizer and thickener in ice creams, sauces, and other foods, and to hold moisture and improve texture in pastries; and as a size. World production of agar in 1967 was about 2,000 tons. Agar is extracted from several species of seaweeds, the principal genera being *Gracilaria* (*G. confervoides*, etc.), *Gelidium* (*G. cartilagineum*), and *Suhria* of the red algae, the Rhodophyceae. Chemically agar consists mostly of the carbohydrate gelose, a galactan ester, and is largely nitrogen-free.

Agar as an extract was reportedly first discovered accidentally by a Japanese who had thrown out certain seaweed jellies. Natural freezing and thawing destroyed the gel structure and yielded the characteristic translucent flakes of dry agar. As the industry developed, the dried algae were carried to the mountains, where the proper combination of freezing nights and sunny days yields agar flakes from the gel formed by boiling the algae. In the limited production in the United States, mechanical freezing and thawing is substituted, and chlorination is used in place of sun-bleaching. In California divers sometimes harvest algae on the sea bottom (about 1 to 1½ tons of seaweed per day per man), or where possible it is hand raked, after which it is dried, baled, and shipped to the factory. There, and in Europe, it is washed and pressure-cooked in a series of extractions. The cooking liquor, after filtering, is allowed to gel in open tubs. The gel is macerated and then frozen for two days at about 14°F. Upon thawing, the agar flakes may be filtered from excess water and dissolved impurities. The flakes, still 90 per cent water, then undergo further hot-air drying and bleaching in a hypochlorite solution. After a final washing and drying, the product is ready for market, at an ultimate moisture content of about 20 per cent.

Similar to agar in production and in service are the hydrophilic colloids algin (the extract from various brown algae, such as the kelps *Laminaria* and *Macrocystis*, Fig. 9-1) and carrageenin (the extract from Irish "moss," *Chondrus crispus* and other red algae). About 10,000 tons of alginates were produced worldwide in 1967, and half as much carrageenin. Algin and alginates find greatest service as ice cream stabilizers, but also serve generally as emulsifiers, and agents for jellying, thickening, suspending and adding "body." They have long been used in dentifrices, shaving creams, hand lotions, and various medicinals. An unusual property amplifying their usefulness is ability to change viscosity in direct proportion to the concentration of calcium ions. Thus viscosity of mucilages can be easily controlled by addition of simple salts such as calcium citrate. Carrageenin finds its largest use as an emulsifier in nonsettling chocolate drinks. Irish "moss" is usually hand raked along the New England and European coasts

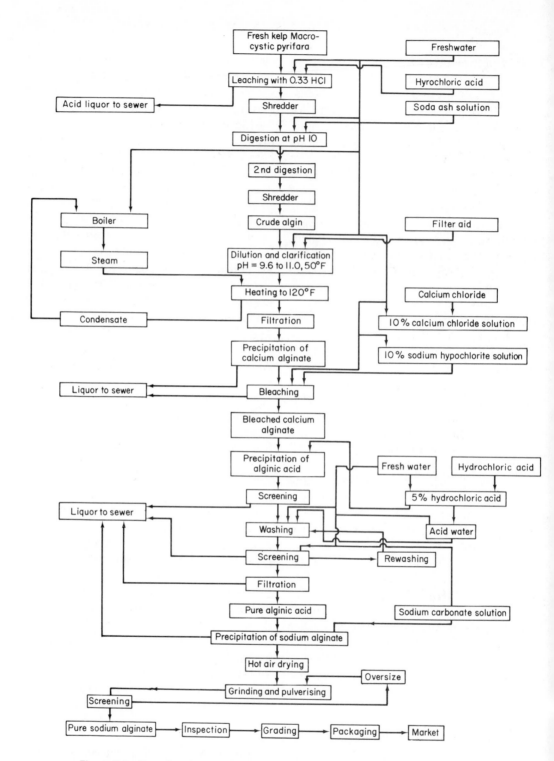

Figure 9-1 Flow sheet for production of sodium alginate from the giant kelp, *Macrocystis pyrifera,* in California. (Courtesy *Economic Botany.*)

during the summer months. Kelps for producing algin are usually gathered by grapples from power boats, or by mechanical cutters, in both the Atlantic and Pacific.

MISCELLANEOUS GUMS. The following plants are sometimes utilized as sources of gum: species of *Butea*, Leguminosae (Dhak, Palas, or Bengal kino), of India; *Caesalpinia*, Leguminosae (Brea), of Argentina; *Chukrassia*, Meliaceae (Chittagong), of India; *Cochlospermum*, Cochlospermaceae (Kutira, India gum, Bassorin), of India; *Cycas*, Cycadacae (Cycas gum), of Asia; *Feronia*, Rutaceae (Feronia gum), of India and the Far East; *Melia*, Meliaceae (Azadirachta) of the East Indies; *Moringa*, Moringaceae (Moringa gum), of the Eastern tropics; *Picea*, Pinaceae (spruce gum), of temperate regions; *Prosopis*, Leguminosae (mesquite gum), of Mexico and South America; *Prunus*, Rosaceae (cherry, shiraz, and cerasin gums) of temperate regions; and various named or unnamed gums from *Aegle*, Rutaceae; *Albizzia*, Leguminosae; *Ailanthus*, Simaroubaceae; *Anacardium*, Anacardiaceae; *Bauhinia*, Leguminosae; *Bombax*, Bombacaceae; *Buchanania*, Anacardiaceae; *Cedrela*, Meliaceae; *Enterolobium*, Leguminosae; *Madhuca* (*Bassia*), Sapotaceae; *Piptadenia*, Leguminosae; *Pterocarpus*, Leguminosae; *Sapindus*, Sapindaceae; *Stereospermum*, Bignoniaceae; *Symphonia*, Guttiferae; *Terminalia*, Combretaceae; and many more.

Resins and Related Exudates

Resins have found use since time immemorial, particularly among Malayan and Pacific peoples. Their combustible properties made resins useful as all-weather torches; and their waterproofing properties, for service as boat caulk. In the Near East resins have been used since antiquity as medicinals, incense, and in crude varnishes. The Egyptians used them for coating mummy cases, and there is evidence that the ancient South American Incas utilized resins in their embalming mixtures. Early painters commonly employed resins in their favorite paint formulae.

Resinous secretions are usually found in special ducts and canals in a wide variety of plants and plant parts. As a constituent of latex, resin appreciably lowers its value for rubber. The mode of resin formation, or the utility of resins to the plant, is little understood. Presumably they originate through reduction and polymerization of carbohydrates such as starch. Many are oxidative products of essential oils, terpenoids consisting of three to six isoprene units (Fig. 9-2). Most resins do not seem to enter actively into the metabolism of the plant, and may be a secondary product of "main-line" metabolic activity. As surmised of gums, they may be of some service in preventing desiccation and in aiding the healing of wounds. It has been suggested that they are evolutionarily closely linked to the attraction and repulsion of insects.

Resins are typically derived from stems and roots following chipping, incising, or bruising. They are brittle, amorphous, more or less transparent, insoluble in water but more or less soluble in ordinary reagents such as alcohol, ether, and turpentine. They possess luster, are ordinarily fusible, and burn with a smoky flame when ignited. In contrast to the carbohydrate nature of gums, resins consist largely of resin acid, anhydride, and ester mixtures. So diverse are the types and so many the plants from which resins are derived that there has never been a satisfactory classification and grouping of resins or a convenient pigeonholing of resin-yielding plants. The more important commercial resins come especially from the Pinaceae, Leguminosae, and Dipterocarpaceae. Far Eastern resins from the Dipterocarpaceae are usually considered "damars," whereas the turpentines (Pinaceae) and related substances

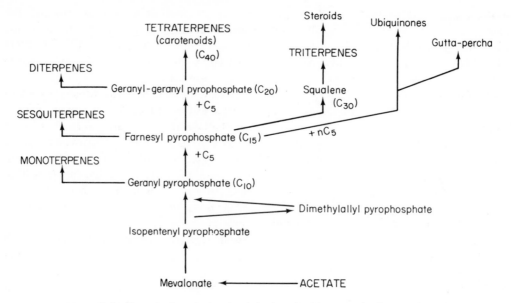

Figure 9-2 Biosynthetic pathways for derivation of mono-, sesqui-, di-, and triterpenoids which comprise resins. (After Langenheim, *Science* **163**, 1158 (1969).)

containing essential oil are spoken of as "oleoresins." "Copal" refers to several resins primarily from the Caesalpinoideae subfamily of the Leguminosae, but also from unrelated sources, some fossilized. Highly aromatic oleoresins (often from the Leguminosae, but also prominently from the Burseraceae), technically containing benzoic or cinnamic acids, are termed "balsams"—a term, however, loosely applied to a number of resins. There are also "gum resins," containing percentages of gums, and "kinos," which are soluble in water and astringent. Resins exposed to slow chemical change for a few centuries are known as "recent fossil resins," whereas those formed thousands of years ago are termed "fossil resins." Fossil resins are for the most part polymerization products of terpenes, of which amber is the best-known example.

Great advances have been made in the compounding of synthetic resins, so that it is likely that natural sources will be of decreasing importance. Synthetic resins are used in all-weather plywoods, for plastics, in tin can linings, paints and varnishes, safety glass, "silent" gears, waterproofing cloth, binders for abrasives, etc. Synthetic resins are largely synthesized from phenol and formaldehyde, glycerin, urea, vinyl materials, and esters of various other organic compounds. Natural resins find similar uses, and in addition are much utilized as sizing; in making soaps, rubber, linoleum, floor coatings, pharmaceuticals, fireworks, plastics, incense; in lithographic work; and so on.

Resins may be compounded with oils (oil varnishes—utilizing chiefly linseed, tung, oiticica, or perilla oils) or volatile solvents (spirit varnishes—utilizing various alcohols or other organic liquids). As a group, oil varnishes are superior in resistance to spirit varnishes. The latter are essentially resin solutions that dry upon evaporation of the solvent. Hard resins are especially favored for spirit varnishes, giving a higher gloss than do soft resins. Spirit varnishes have found greatest use as paper size, furniture and leather finishes, sanding sealers, and shellac substitutes. The term *lacquer* originally indicated natural plant exudations used directly as a coating, but has come to signify spirit varnishes in which a cellulose derivative is the major film-forming substance. World trade in natural resins is estimated to approach 1 million tons annually.

Turpentines. Turpentines consist of a mixture of resin in essential oil, found particularly in the resin canals of the inner bark and sapwood of various coniferous trees. The crude product is often spoken of as pitch. Pitch or crude turpentine is commonly obtained by tapping or chipping the tree, and is distilled to yield its component essential oil and resin (colloquially called "rosin," Table 9-1). Destructive distilla-

tion of coniferous woods also yields these products. Pitch was known to the ancients, and was then and later much used in caulking or treatment of wooden ships. Turpentine production and processing have consequently received the appellation "naval stores" industry. Although spirits of turpentine and rosin are produced in many parts of the world, the naval stores industry has attained its greatest importance in the southeastern United Sates, where it is one of the oldest of North American enterprises.

Turpentining in the United States was the basic source of income of the Carolinas in colonial times. In fact Carolinians became known colloquially as "Tarheels" because, in the early practice of extracting pine resin by smoldering logs in a pit covered with soil, the resin, which accumulated amidst dirt and ashes, stuck to the workers' feet as they removed it. Gradually "gum" extraction spread southward and westward through the southeastern coniferous belt, later centering in Florida and Georgia. Crude and inefficient means of extraction were followed, and remained little improved until quite recently.

Crude turpentine is largely derived from several species of pine (*Pinus*). In the United States the species most used are the longleaf

Table 9-1

Composition of Pine Rosin

Component	Percentage
Neutral components	10
Abietic-type acids	53
Dihydroabietic acid	11
Tetrahydroabietic acid	18
Dehydroabietic acid	3
Oxidized acids	5

Naval stores operation in Louisiana covers 10,000 acres of longleaf pine. The center tree has been faced 4 years, and soon will be cut for timber. (Courtesy USDA.)

pine (*P. palustris*) and the slash pine (*P. elliottii*). Ponderosa pine (*P. ponderosa*) has been experimentally tried as a turpentine source in the western United States, with uneconomical results. In Mexico several species are tapped. *P. maritima* is the European pine most utilized for turpentine; it is quite widespread in southern France and Spain. Other species of that region utilized to greater or lesser extent are *P. halepensis*, *P. nigra*, and *P. pinea*. In eastern Europe a low-grade turpentine has been obtained from the Scotch pine, *P. sylvestris*. Turpentine from pines, especially *P. roxburghii*, is also reported obtained in India and southeast Asia.

Crude turpentine occurs in interconnecting ducts or resin canals, particularly in the sapwood. Major canals run vertically; smaller ducts connecting with these open to the bark. Severance of any canal permits the crude turpentine to exude. The old system of tapping was to "box" the lower trunk of the tree—i.e., to chop a cavity a few inches wide and several inches deep at the base of the tree to collect the turpentine —and then to "chip" the bark and outer sapwood above this box, thus opening the resin ducts so that pitch would exude and drip down into the box. Chipping by this method typically follows a V-form, with the apex of the V situated directly above the box. When the pitch dries and oxidizes, the resin ducts become sealed, necessitating renewed chipping at weekly intervals. The turpentining wounds never heal, since the cambium is removed in chipping. As soon as the bark has been stripped from most of the lower tree trunk by a sequence of boxings and chippings on all sides, ordinarily over a period of 10 years, the tree is abandoned. Since open wounds permit ready entrance of fungi, the turpentining system promotes wood decay. All this was of little importance in early America, when thoroughly chipped trees were abandoned

as having no further use. As lumber gradually became more valuable, trees chipped by the old methods were used but degraded, because of the deep chipping streaks on the finest part of the log.

Recommended practices in turpentining involve several modifications of the traditional system. Boxing is discouraged. Instead, tin gutters and spouts are tacked to the first streak (tapped area), which guide "gum" to a suitable pail. No severance of resin canals connecting root with trunk results, and additional yield of 27 per cent the first year and 22 per cent the second year over similar trees chipped by the old boxing method is possible.

Acid or 2,4-D treatment of the streak has been shown to increase yield and prolong flow. In traditional practice, chipping must be done once a week. If, however, the newly exposed surface is sprayed with sulfuric acid, flow is stimulated for as long as three weeks. With acid treatment, bark chipping

Close up of acid spraying of pine chipped for turpentining in Florida. (Courtesy USDA.)

Pine gum processing plant in Florida. Raw gum is poured into canted charge tank. Beneath it is the blow tank from which gum is carried by steam pressure to melters (two white tanks on lower level). Gum is washed with water prior to distillation. (Courtesy USDA.)

is to be recommended over wood chipping. The small lateral gum canals extending to the bark then effectively drain the gum in as great quantities as do the larger canals severed in wood chipping. Chipping becomes easier and faster, and the wood is not degraded because of mutilation.

Within the last few decades, tree or gum turpentining has declined. A high point of almost 2 million drums (520 pounds each) of rosin was obtained in 1909; by 1967 production had declined to 210,000 drums. Meanwhile turpentining by wood distillation increased from practically nothing in 1910 to about 940,000 drums of rosin by 1967. This yield has come largely from the processing of stumps of pine left after generations of logging in the southeastern forest belt. The stumps are freed by dynamiting or are pulled free by massive cranes or other machinery. Much of the stump supply, how-

ever, has now become exhausted, and a decline in wood turpentine production is to be expected. Most future production can be expected from the processing of tall oil, mainly a by-product of kraft paper pulping; tall oil yielded 720,000 drums of rosin in 1967.

Crude turpentine must be distilled. Old-time stills were simple kettles placed directly over an open fire and charged with fresh "gum" plus any chips, bark, and pine needles at hand. Scorched, low-grade rosin was not infrequently had. Centrally located stills utilizing steam distillation later became feasible, the crude turpentine brought to the still from the surrounding area. The raw "gum" is first diluted and filtered. It is then chemically treated (usually with acid) to produce lighter, more expensive grades of rosin, and the treated "gum" jetted into water to remove all traces of acid prior to actual

237

distillation. Steam distillation for 2 or 3 hours follows, in which spirits or oil of turpentine fractionates off above the water and the rosin remains as a heavy residue. While still hot, the rosin is screened, cooled, and hardened. Spirits of turpentine finds its greatest use in the chemical industry; rosin, as a sizing and in the manufacture of varnishes, inks, waxes, seals, lubricants, and the like.

Tall oil turpentine is condensed from volatilizations during pulp cooking; up to $4\frac{1}{2}$ gallons of foul-smelling crude turpentine (due to pulping liquor) are derived per ton of pulp. About 4 or 5 per cent of resin is also procured, saponified by the sulfate cooking and residual in the black liquor. These resin salts can be purified by washing and precipitation, after which acidification regenerates resin comparable to that found in pine "gum" (also fatty acids, primarily oleic and linoleic).

Naval stores in the paint and varnish industries. Drying oils, resins, and thinners are the basic ingredients of varnishes. Oil paints, both flat and enamel, consist of at least two-thirds varnish plus certain selected pigments and other minor products. Drying oils may come from a variety of sources to be discussed later. Resins, the "solid" portion of varnish at ordinary temperatures, include various synthetics, pine rosin, and to a lesser extent types discussed on subsequent pages. Thinners, in addition to spirits of turpentine, include petroleum distillates and organic solvents such as alcohols, acetates, and the like. One formula for compounding a varnish will serve as an example. Several hundred pounds of pine rosin are esterified with glycerin (heated at 500°F until neutral, in stainless steel kettles) to form "ester gum." Seventy pounds of this "gum" are cooked with 300 pounds of tung oil at 560°F until thickened. Thereupon 30 pounds additional "ester gum" are quickly added, and the mixture vigorously stirred to prevent gelation. Four hundred pounds of turpentine are then slowly added. Slight amounts of metallic salts may be included to hasten drying. The hot varnish is finally filtered to insure clarity.

Other important or interesting resins. AMBER. One of the most interesting of resins is the fossil resin, amber. Classically it has been regarded as formed from exudations of now-extinct species of pine, especially *Pinus succinifer*, of the Pinaceae. Spectroscopically, however, it is closer to modern *Agathis* resin, and may even have accumulated under rather tropical conditions. This classical amber is found in subterranean deposits, particularly in northern Europe. In ancient times amber was a decidedly mysterious substance, being variously believed to be generated by the sun's rays, solidified lynx urine, mineralized honey, petrified whale sperm, and so on. It was, however, correctly designated a tree resin by Aristotle, Pliny, and others. Ancient peoples esteemed amber for ornaments and charms, or as a static electricity plaything, and at times even utilized it as a medium of exchange. In recent times it has been used for jewelry, and somewhat in electrical apparatus.

The resin is a hard, brittle, amorphous compound, having an empirical formula of $C_{40}H_{64}O_4$, a specific gravity slightly heavier than water, good insulating properties, and easy workability in carving or machining. "True amber" must contain at least 3 per cent succinic acid, and often as much as 8 per cent. It varies in color from yellow to black, the lighter types darkening upon continued exposure. Amber is capable of taking a high polish, one reason for its popularity for jewelry. It is mined chiefly in Germany, and is also found washed up on the beaches of northern Europe and elsewhere as underseas deposits weather. A few hundred tons have been collected in favorable years, but as

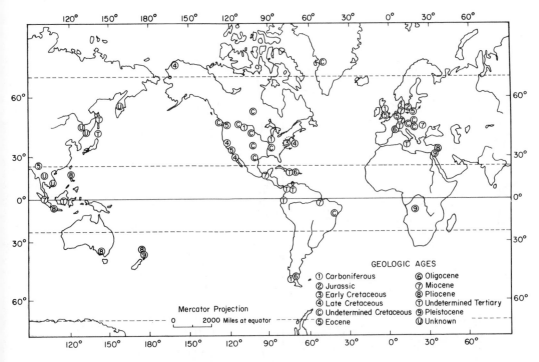

Figure 9-3 Geographic distribution of ambers according to geologic age. Note how widespread accumulation of fossil resins has been. The earliest are from Coniferales species. (After Langenheim, *Science* **163**, 1159 (1969).)

deposits have been worked out production has declined. That amber was once liquid is attested to by the fact that now-extinct insects are sometimes found trapped in it. The largest reported piece of amber weighed 18 pounds and was valued at $30 million (before World War II).

Hymenaea, of the Leguminosae, is probably the source of a Miocene amber found in southern Mexico, Colombia, and Brazil. Spectroscopically it resembles resin from the tropical forest tree, *H. courbaril*. Spectroscopic comparisons indicate probable sources for other ambers, too, some of which date back to the Carboniferous (Figs. 9-3, 9-4).

BALSAM-OF-PERU. Balsam-of-Peru is derived from *Myroxylon balsamum*, of the Leguminosae, and its varieties (var. *pereirae* of Central America is the chief source). The

balsam serves primarily as a medicinal for the healing of persistent sores and ulcers, skin diseases, and bronchitis or coughs. It has also been used as a fixative in perfumes, for consecrated oils in the Roman Catholic Church, in soaps, as a component of arrow poison, and as a vanilla substitute. The aromatic resin is collected mostly from wild trees, these ranging from Mexico into northern South America. The tree has also been introduced into the Far East. The chief producing area is El Salvador, from which supplies of balsam-of-Peru have been exported since colonial times. First use of balsam-of-Peru was by the Indians of the New World, but the Spanish "conquistadores" soon discovered its efficacy and it became a highly prized export from the colonies to Europe, where at times it was

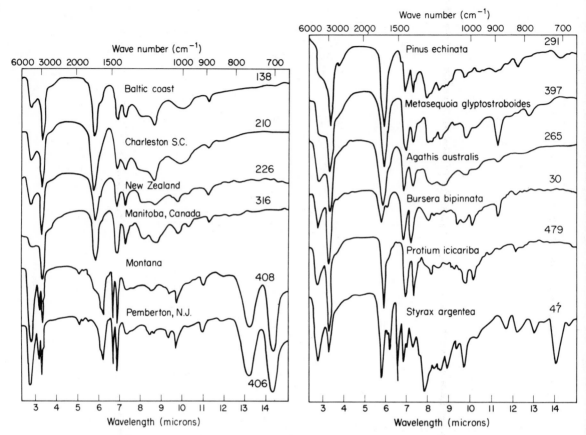

Figure 9-4   (*Left*) Infrared spectra of representative ambers from six sites. (*Right*) Infrared spectra of the nonvolatile fraction of resins from extant populations of six species. Similarity is in some cases striking. (After Langenheim, *Science* **163**, 1162 (1969).)

worth its weight in gold. Production from Central America was transshipped to Spain via Callão, Peru, and thus the product came to be known as balsam-of-Peru, even though little has ever been produced in Peru proper.

Myroxylon balsamum (var. *pereirae*) is a beautiful forest tree particularly common on the Pacific slopes of Central America. It may exceed 100 feet in height, and has a high, flat crown and a slender, smooth trunk. The small, shiny leaflets are gland-dotted; the flowers small and scarcely attractive; the fruit a winged, one-seeded legume resembling a gigantic "maple squirt." The "balsameros" or balsam gatherers cut a

small panel about 4 inches wide in the lower trunk of the tree, removing the bark to the wood. The bark immediately above this panel is then beaten or scorched, after which the resin exudes in a wound response. It is generally believed that heat from balsam wood itself produces the best yields, so that burning fagots split from old balsam stumps are usually applied for the purpose of wounding the bark. Experience is necessary in order to know just how long to apply the flame to obtain best yield and yet not char the bark. Absorbent rags are fixed in the upper panel to absorb exuding resin during the subsequent 1 to 3 weeks. At the time these rags are collected, a new panel is also made in the previously injured bark above the old panel; and the whole process is then repeated. Ultimately it becomes necessary to ascend the tree to open new panels. This may be accomplished by means of climbing ropes or by "rungs" of balsam wood driven into the trunk in crude borings. Needless to say, the collection system puts a severe strain on the trees, but they may continue to live for years as long as sufficient strips of live bark are left up and down the trunk.

After collection the redolent rags soaked full of balsam are boiled in water at centralized locations. The resin becomes fluid, and small quantities of it sink to the bottom of the cauldron to be retrieved later. The rest is extracted by expression, usually in locally made devices that cause twisting and tightening of a coarsely woven sack charged with the hot rags. The filtered product thus extracted is then marketed directly. Annual yield per tree amounts at best to a few pounds of crude balsam.

COPAIBA BALSAM. Another of the oleoresins that have attained modest importance is copaiba (or copaiva), a balsam derived from several species of *Copaifera* (Leguminosae), of South and Central America. Perhaps *C. reticulata* yields most of Brazil's copaiba, and *C. officinalis* most of that from northern South America and Central America. This resin has been utilized as a fixative in scenting soaps and perfumes; occasionally in varnishes; in photography; and as an antiseptic medicinal. It is collected from wild trees, fairly abundant through the Amazon valley to southern Central America. The trees are of medium size, with gland-dotted, asymmetrical leaflets and a large, one-seeded legume.

Collection is typically by isolated tappers, financed by local "barraconistas," much as in collection of wild rubber. These collectors may remain in the jungle for weeks or even months on end. Yields are reported to be best during the dry season, but tapping may be undertaken at any time of the year. Tapping consists of drilling several holes or hacking incisions well into the trunk. The wounding serves as stimulus for the accumulation of large quantities of resin in and about the cavities. Yields of as much as 11 gallons of resin have been reported attained from a single tree. The resin is aromatic and with a bitter taste; and is thin, of a clear, pale color, soon becoming more viscous and yellowish.

Tappers bring the resin to smaller centers in payment of debts or to barter it for various necessities. It is then shipped to larger exporting cities such as Manaos (Brazil) or Maracaibo (Venezuela), where exporters usually further remove bark fragments and other debris. At ultimate markets, the copaiba is usually distilled to yield a transparent oil (35 to 85 per cent of the copaiba by volume) utilized for purposes previously mentioned. Substitutes for copaiba are reported obtainable from several Far Eastern members of the Dipterocarpaceae and from African species of *Daniella* (Leguminosae).

COPAL RESINS. The most important copals are those of considerable hardness and comparatively high melting point. Their greatest use is in varnishes and as a plastic binder. A great many copals formerly entered commerce, but today only a few are exploited.

Congo Copal, probably mostly from *Copaifera demeussei*, Leguminosae, is commerically the most important of the copals. Production from central Africa runs close to 20,000 tons annually. The resin is a fossil or semifossil type; collectors gather it from the forest floor after locating it by prodding with iron-tipped sticks. It is usually shipped to Europe for cleaning by air-blast methods. The species from which it comes is still an important forest tree, and sometimes fresh resin is collected by wounding the tree. Considering the diversity of the flora and the uncertainty as to what long-decayed tree may have been responsible for a given piece of resin, it is no wonder that lack of uniformity exists in the commercial product. West African Copal, from other species of *Copaifera*, was once of considerable commercial importance. It occurs both in fossil form and as a product of the wasteful tapping of living trees.

East African or Zanzibar Copal, *Trachylobium verrucosum*, Leguminosae: Natives collect both the fossil and fresh resin of this bifoliolate forest tree, especially in Tanzania. The small, warty legume is reported to contain 20 per cent resin.

South American Copals, chiefly *Hymenaea courbaril*, Leguminosae: Both fresh and fossil types are locally collected; the fossil types are of superior value and hardness.

East Indian or Manila Copal, *Agathis alba*, Pinaceae: This species of the East Indies and Malaysian area yields fresh, semifossil, and fossil resins; the latter types, however, have become increasingly uncommon.

DAMAR RESINS. These Dipterocarpaceous resins were long used in the East for torches, boat caulk, and incense. In contrast to the copals, they are freely soluble in turpentine, although less hard, and enter especially into manufacture of "spirit varnishes" for wallpaper and decorative work. The origin

Damar resin being extracted from a Dipterocarp in Thailand. (Courtesy USDA.)

of many damars is quite obscure. Apparently a number of species of *Hopea*, *Shorea*, *Balanocarpus*, *Vateria*, and other genera are involved. *Balanocarpus heimii*, of Malaya, is tapped on the upper trunk by natives who ascend the tree on rattan ladders supported by pegs driven into the sapwood. *Shorea robusta*, of northern India, yields fresh resin with tapping, but also is the source of fossil and semifossil types, the latter in stalactitic form from natural exudation.

ELEMI. There occur numerous oleoresins from diverse or unknown sources termed *elemi*. Most of these probably come from the family Burseraceae. Manila elemi is derived by gashing the trunk of *Canarium luzonicum*, a large Philippine tree. This soft, adhesive resin has been much used for native torches and boat caulk, and finds some value in the Western economy for printer's ink and for reducing brittleness in varnishes. Species of *Amyris* (Rutaceae), of southern Mexico, yield an elemi that may be substituted for Canadian balsam. Brazilian elemi apparently comes from three or four species of *Bursera* and *Protium*, *P. heptaphyllum*, ranging from Colombia to Paraguay, constituting the source of "brea branca" or "tacamahaca." *Calophyllum* and *Symphonia* (Guttiferae) of the same region are said to yield the bastard or yellow elemi. West Indian elemi comes from *Dacryodes hexandra*, and African elemi from *Canarium schweinfurthii*. Several additional elemi-producing species of *Canarium* are known from Mauritius and the East Indies.

FRANKINCENSE. Frankincense or gum olibanum comes from various species of *Boswellia*, Burseraceae, especially *B. carteri* and *B. frereana*. This oleo-gum-resin is of little importance nowadays, but because of its frequent mention in the scriptures and by classical writers, it merits brief discussion. Frankincense was one of the important trade articles of the Phoenicians, and of the

Gathering sticklac in Thailand. (Courtesy USDA.)

Near East with the Far East. It was used by the Egyptians in fumigation, embalming, and medicine. So esteemed was it in Biblical times that frankincense constituted one of the memorable gifts of the Magi to the Christ child. The species are small trees or shrubs of dry habitats in northeastern Africa and the Arabian peninsula adjacent to the Gulf of Aden. Collections are apparently from wild trees only. Collectors incise or scrape the stems during the dry season, and collect the tearlike exudations as they harden. What little use frankincense finds today is mostly as ceremonial incense for the church.

KAURI RESIN. Kauri resin, from *Agathis australis*, Pinaceae, is one of the most valuable of natural hard resins, and was the original resin employed in making linoleum. It is obtained in the fossil or semifossil form from accumulations in present or former Kauri forests. Lumps weighing as much as 100 pounds have been collected. Kauri resin has been exported from New Zealand for a century, mostly to the United Kingdom and North America. The peak year was 1899 (some 11,000 tons), with production since diminishing as supplies approach exhaustion. The Kauri is the monarch of the New Zealand forests, where it ranked as one of the world's foremost timber trees.

LAC. A resin secreted by a scale insect supplies lac. This insect, *Laccifer* (*Tachardia*) *lacca*, abounds in India and East Asia, feeding on a number of plants, including *Butea*, *Cajanus*, and *Acacia*, of the Leguminosae; *Schleichera*, of the Sapindaceae; *Zizyphus*, of the Rhamnaceae; and *Ficus*, of the Moraceae. *Butea* was cultivated for lac-growing at least as early as the second century A.D. The reddish-orange, translucent, resinous secretion encasing the insect during its life (except for brief intervals when males emerge to seek the females) is affixed to the twigs of the parasitized plant. The young emerge in swarms as minute, sluggish, red-colored insects about 1 millimeter long. They seek succulent twigs and suck sap through a beak thrust into the soft plant tissues. After the insects swarm, twigs containing the old lac encasements are removed and the resinous encrustation (stick lac) scraped away, ground, soaked, and washed. After this treatment it is known as "seed lac," and when cooked with other resins and minerals becomes "shellac." India is the chief producer of shellac. Shellac is used in high-polish interior "spirit" varnishes, certain waxes,

243

inks, and lacquers; as a size, stiffener, and glaze; and for ornaments.

LACQUER. A few genera and species of the Anacardiaceae are responsible for the resinous natural lacquers of China, Japan, and other Eastern countries. *Rhus verniciflua* is a handsome Chinese tree now widely cultivated. It is propagated by seed or root cuttings. The bark is tapped in midsummer, and the milky sap, a watery mixture of resin, gum, and albuminoids, collected. This lacquer must be applied in a damp atmosphere, necessary for the hardening of the resin. The Japanese and Chinese have carried the art of lacquering to a high degree of perfection. The finishing of a single article may require hundreds of coats of lacquer, over a period of years, each applied and then dried and polished. Later coats may contain coloring matter, metallic dusts, or pulverized mother-of-pearl. A Burmese lacquer similar to that of China and Japan is obtained from *Melanorrhoea usitata*. In southeast Asia and Taiwan lacquer is said to be obtained from *Rhus succedanea, R. ambigua,* and *Semecarpus vernicifera.*

MASTIC. "Gum" mastic is an oleoresin taken from the shrub *Pistacia lentiscus* (var. *chia*), of the Anacardiaceae, native to the eastern Mediterranean. It is small, dioecious, compound-leafed evergreen growing to about 10 feet. Ordinarily the male trees alone are propagated, it being said that they yield superior quantities of resin. Propagation is almost entirely by cuttings, and the plant grows so slowly that 50 to 60 years are required for it to attain full maturity. Cultivation and fertilization are helpful, but are not always supplied.

Tapping is restricted (by law in the chief producing region, the island of Chios) from July 15 to October 15, during which two intensive tapping periods are observed. Before any collection is made, the soil about the shrub is cleared, leveled, brushed, beaten hard, and covered with a white soil that will not discolor fallen mastic. Ten to twenty horizontal wounds about 2 centimeters long are opened in the bark twice weekly for 5 to 6 weeks, an awl-like tool being used. Resin oozes from the cuts, falling to the prepared earth or dripping over the bark to be collected at the end of tapping. The gathered resin is sorted and cleaned before grading and marketing. Exceptional trees may yield as much as 1 kilogram of resin a year. Mastic is used in high-grade varnishes designed for protecting pictures, and in perfumes, pharmaceuticals, and even chewing gum. Another Eastern species. *P. cabulica*, yields Bombay mastic.

MYRRH. Myrrh, from *Commiphora* (*Balsamodendron*), Burseraceae, a large genus of chiefly African species, is as famous in the writings of antiquity as is frankincense, and similarly is little used today. What collection there is comes largely from Somali and southern Arabia at the Gulf of Aden, said to be from *C. myrrha*. Schizogenous canals yield the resin upon wounding of the bark. The resin is collected as tearlike particles when it is dry. Myrrh is an oleo-gum-resin that in ancient times found use in incense, perfumes, salves, medicines, disinfectants, and embalming mixtures.

SANDARAC. African sandarac is obtained from *Tetraclinus articulata*, Pinaceae, and is used primarily for special varnishes such as those used for pictures, leather, and metal. The small coniferous tree grows in the mountains of Algeria and Morocco. Australian sandarac, from species of *Callitris*, Pinaceae, is similar to the African type, but finds little use outside of Australia. Both kinds are collected from natural exudations on the trunk.

STORAX. The original storax (sometimes written *styrax*) came from Asiatic species of *Liquidambar* (*L. orientalis*), of the Hamamelidaceae. There has also been intermittent production from the common sweetgum or redgum (*Liquidambar styraciflua*), which grows wild from New England to Guatemala.

In Asia storax is reportedly obtained by bruising the bark of the trees, then boiling the resin-soaked inner bark in sea water to free the storax. In the southern United States the bark is peeled for about 1 foot along half or more of the circumference of the lower trunk. The oleoresin exudes from the sapwood, and at 2- to 3-week intervals is scraped free and collected.

Collected storax is melted, strained, then allowed to solidify. It is a chocolate-colored, fragrant, balsamic oleoresin, somewhat resembling beeswax in consistency. It is utilized in medicinal and pharmaceutical preparations, for incense (particularly in China), in perfuming powders and soaps, as a fixative in "heavy" perfumes, and in flavoring tobacco.

A solid resin from *Styrax officinalis*, Styracaceae, has at times been marketed in the Mediterranean area as "storax."

Miscellaneous minor resins

Abies, Pinaceae—*A. balsamea*, Canada balsam; northern North America; optical work.

Ailanthus, Simaroubaceae—*A. malabarica*, mattippal; India; medicinal for dysentery.

Anacardium, Anacardiaceae—*A. occidentale*, the cashew, yields a by-product resin from the shells discarded by the cashew nut industry; this serves as a phenolic compound in making synthetic resins.

Anisoptera, Dipterocarpaceae—*A. thurifera*, palosapis; Philippines.

Auracaria, Pinaceae—Brazilian and Australian species.

Bursera, Burseraceae—*B. simaruba*, incense tree; West Indies; the fragrant "gum" has been used for church incense.

Cistus, Cistaceae—labdanum (or ladanum) comes from several species; Mediterranean area; perfumery.

Convolvulus, Convolvulaceae—*C. scammonia*, scammony; Mediterranean area; drug from root

now largely replaced by similar ones from *Ipomoea spp.*

Demonorops, Palmae—This jungle palm genus of Malaya and the East Indies yields a red resin from the surface of immature fruits, known as dragon's blood. In ancient times a "dragon's blood" was collected from stem incisions on *Dracaena spp.*, Liliaceae. It found use in the varnishes of the great Italian violin makers of the eighteenth century.

Dipterocarpus, Dipterocarpaceae—many Eastern species; torches, caulk, drugs.

Dorema, Umbelliferae—*D. ammoniacum*, ammoniacum; western Asia; stimulant, and used in perfumery.

Elaeagia, Rubiaceae—*E. utilis*, pasto lacquer; Colombia; resin from buds.

Euphorbia, Euphorbiaceae—*E. resinifera*, gum euphorbium; Morocco; vesicant.

Ferula, Umbelliferae—asafoetida from root of *F. foetida* and *F. rubicaulis*, Persia; carminative and sedative; galbanum and sagepenum from other species.

Garcinia, Guttiferae—several tropical species yield gamboge, a resin–gum mixture, formerly used as a cathartic but most important as a water-paint pigment and in "spirit" varnishes for metal.

Gardenia, Rubiaceae—few species of Southeast Asia and the Pacific area yield bud resins used in native medicine.

Guiacum, Zygophyllaceae—Caribbean lignumvitae species yield the wood resins used as stimulants.

Ipomoea, Convolvulaceae—a few Mexican species, such as jalap, *I. purga*, yield purgative resins from the roots.

Laretia, Umbelliferae—*L. acaulis*, Chile; gum resin similar to *Ferula*.

Larix, Pinaceae—*L. decidua*, Venice turpentine, a gum resin obtained by boring to heartwood; northern Italy.

Picea, Pinaceae—*P. abies*, Burgundy pitch; Europe; varnish and medicinal plasters.

Podophyllum, Berberidaceae—North American and Indian species; extracted from rhizome with alcohol; purgative.

Schinus, Anacardiaceae—*S. molle;* subtropics; oleoresin.

Sindora, Leguminosae—Philippine species yield "supa" oils from wasteful hacking of cavities in the wood and frequent firing of these cavities.

Spermolepis, Myrtaceae—*S. gummifera;* New Caledonia; a high-tannin resin.

Styrax, Styracaceae—tropical Asiatic species yield benzoin on tapping; incense, balm, carminative, expectorant, and diuretic.

Thapsia, Umbelliferae—*T. garganica;* Mediterranean area.

Xanthorrhoea, Liliaceae—several species of this Australian genus commonly referred to as "grass trees" yield acaroid or accroides resins from the stem about the leaf bases.

The future for gums and resins. Modern chemistry makes it feasible to synthesize polymers which substitute for the natural products, utilizing any of a great number of inexpensive raw materials. It can be anticipated that many of the sources mentioned in this chapter will gradually become obsolete, replaced perhaps by plants grown agriculturally. In the search for new crop plants there are already being considered: species of *Crepis*, Compositae, source for crepecynic acid, useful in resins for industrial coatings; vernolic and other epoxy fatty acids from *Vernonia*, Compositae, used in coatings and plasticizers; seed gums and resins from *Euphorbia lagascae. Cassia occidentalis* is said to yield 1 ton of seed per acre, a potential source of gum similar to but darker than guar. The seed is toxic, but roasting detoxifies it.

The ready availability and abundance of such sources certainly puts a ceiling upon resin and gum prices. It will become progressively less feasible to gather wild gums and resins as supplies grow scarce and labor becomes more costly in various parts of the world. Synthesis will no doubt provide "improved" polymers that can withstand greater extremes of temperature and outdoor exposure than the natural substances (perhaps making possible strong adhesives for outdoor use that will substitute for nails, screws, and rivets!).

SUGGESTED SUPPLEMENTARY REFERENCES

Anderson, A. B., "Recovery and Utilization of Tree Extractives," *Economic Botany* **9**, 108–140 (1955); includes a bibliography of 58 titles.

Barry, T. H., *Natural Varnish Resins*, Ernest Benn Ltd., London, 1932.

Glicksman, M., *Gum Technology in The Food Industry*, Academic Press, N. Y., 1969.

Howes, F. N., *Vegetable Gums and Resins*, Chronica Botanica, Waltham, Mass., 1949.

Mantell, C. L., *The Water-Soluble Gums*, Reinhold, N. Y., 1947.

Percival, E., and R. H. McDowell, *Chemistry and Enzymology of Marine Algal Polysaccharides*, Academic Press, N. Y. and London, 1967.

Whistler, R. L., *Industrial Gums—Polysaccharides and Their Derivatives*, Academic Press, N. Y., 1959.

CHAPTER **10**

Vegetable Tannins
and Dyes

TANNINS AND DYES ARE SECRETION PRODUCTS found almost universally in plant tissues. In spite of their universality, they are not sufficiently concentrated in most plants to permit economical extraction; in consequence relatively few plants have been of commercial importance. Tannins and dyes are not overly complex chemicals, usually being compounds of carbon, hydrogen, and oxygen (plus some nitrogen in dyes) which in many cases man has learned to synthesize more cheaply than they can be obtained from plants. This is particularly true of dyes, vegetable dyes being today largely of historical interest since the development of coal-tar derivatives. Only a fraction of the dyes used today are of natural origin. Vegetable tannins, on the other hand, have held their own to a greater extent. Synthetic tannins cannot be made cheaply in competition with natural sources, although chromium salts and other tanning chemicals have become increasingly important as substitutes for vegetable tannins. Also, newer leather substitutes such as the porometrics have gained favor. Natural tannins are used principally for hard sole leathers, the leathers so tanned exhibiting less shrinkage on drying than the soft, thin chrome leathers commonly used for uppers. Even sole leather is

receiving increasing competition from rubber and composition materials.

Tannins

The term *tannin* is applied to a variety of astringent substances, usually glucosidal. The empirical formula of one tannin (Chinese nutgall) is $C_{76}H_{52}O_{46}$. Natural tannins have been characterized as either catechol or pyrogallol types, the distinction rather obscure and seemingly based upon origin rather than exact chemical structure. Another classification depends upon whether the tannin is hydrolized by acids and enzymes or not. Tannins are water-soluble, and those utilized for transforming hides to leather are obtained by water extraction from a variety of barks, woods, leaves, roots, and fruits. The usefulness of tannins to the plant is uncertain: they may constitute discarded wastes or metabolic by-products, or they may be effective as "antiseptics" or preservatives. Tannins utilized commercially may be purchased in solid or liquid form, and are made by evaporating the original water extraction to the desired concentration in evaporators and vacuum pans. Their chief usefulness stems from their ability to combine with the proteins of hides (animal skin) to yield stable leather, resistant to air, moisture, temperature change, and bacterial attack. Vegetable tannins also react with iron salts to form dark-bluish or dark-blackish compounds, the basis of ink.* Sometimes tannins find use in medicinals, for their astringent qualities. In the United States a significant portion of tannin consumption is as a dispersant to control the viscosity of mud in oil-well drilling.

In tanning animal hides, the entire thickness of hide must receive the tannin. Plunging the hide in a strong tanning solution would cause the leather to become relatively impervious on the surface, thus preventing ready access of the tannin to the inner parts. Consequently, hides are ordinarily first soaked in spent tanning liquors, the acidity of which causes swelling and permits entrance of gradually increased concentrations of tannins. Tanning may take anywhere from a few hours to several months, depending upon conditions and upon the thickness of the hide. After tanning, slivers of the leather are cut to ascertain the degree of tannin penetration. The leather is then treated with oil or grease to give it softness, pliability, and waterproof characteristics; dyed to the desired color; and perhaps "finished" with a coating of gum, resin, or wax.

The tanning industry is centuries old. A thousand years before Christ the Chinese were already making leather, and the tanning of skins was practiced extensively in the Roman Empire. American Indians utilized a variety of plants to tan buffalo hides. The first tannery in the United States antedated the Revolution by a century and a half. The leather industry in North America soon centered in New England, utilizing abundant and cheap hemlock bark available locally. Boston became the nucleus of the leather trade, until depletion of hemlock stands shifted the tanning industry westward to Pennsylvania. There was subsequently further decentralization southward and westward, but by and large the tanning center of the United States remains in Pennsylvania, where tanneries, having exhausted local tannin sources (chestnut, oak), rely upon imports of tannin concentrates and upon substitute chemicals.

Development of the tanning industry. The story of the tanning industry, from perhaps an original chance softening of wild animal pelts in streams charged with twigs and bark that provided some preservative effect, to an imposing industry turning billions of pounds of green-salted hides into hundreds of types of leather, is one of persistent though unspectacular growth. Egyptian vegetable-tanned sandals more than 3,300 years old are on display in one museum. Greeks and Romans anticipated present-day methods of preparing hides for tanning by soaking the skins in limewater to loosen the hair (a fermentation process) and then dehairing with a scraping knife and beam. Various unprocessed barks and berries, gallnuts, and sumac leaves were applied as the tanning agents—an uncertain method continued through the Middle Ages. Gradually it was found that leaching the tannins into circulating hot water before tanning gave more uniform results and shortened the tanning process. The door was then open for development of tannin concentrates and stan-

---

* Ink has been known in Egypt and China for almost 5,000 years. Tannin inks, resulting from the combination of iron salts with tannin and usually stabilized with gums, were known from at least the eleventh century. Logwood inks utilize both the tannin and the dye components of logwood extract, and function partly as a tannin ink and partly as a color ink. Colored inks today are made almost entirely from synthetic dyes. India ink is an insoluble combination of carbon black with a glue. Printing inks are made from finely pulverized varnish–pigment mixtures.

dardized dry extracts. Inasmuch as these are easily transportable, the leather industry was enabled to move out of the forest.

The other ingredient of leather besides tannin is the hide. Since Roman times, animal skins from slaughterhouses and other sources, preserved with salts, have been prepared by soaking in a lime solution to loosen the hair. Special machines now dehair the hide. The prepared hide is then thoroughly soaked in the tanning solutions. The tanned hide is washed, bleached, and oiled before being hung for drying. Final treatment varies with the type of leather to be produced, and may involve scouring of the grain, application of coatings, or compacting of the fibers under heavy brass rollers to give a polished and lustrous finish.

Natural tannin sources. In the 1968 edition of his *Dictionary of Economic Plants*, Uphof lists nearly 100 genera as worthwhile sources of tanning materials, certainly representing several hundred species (*Acacia* and *Quercus* alone provide numerous species of at least local importance in various parts of the world). Production figures for even the more important sources are not readily available, so diffuse is tannin production, and so frequently from uncultivated plants that are only haphazardly harvested. In toto, hundreds of thousands of tons of tannin must be utilized worldwide. Much enters international commerce as a hidden constituent of such things as fruits and beverages (tannin content has an important influence upon tea quality, for example). Some sources of tannin are listed for their historical interest (e.g., chestnut wood, hemlock bark), and others because of their potential as a cultivated crop (e.g., canaigre, sumacs). Of present-day commercial importance wattle and mangrove barks, and chipped quebracho wood, stand out. These, along with a few additional established trade items, serve as examples of tannin plants. Obviously if

economic conditions warrant, any source could be supplanted by another from among the hundreds of potential plants that contain tannin in quantity.

The traditional reliance upon wild plants for tannin is fast changing. Even so ubiquitous (and relatively difficult-to-harvest) a source as mangrove, growing in swampy tidal mudflats throughout the tropics, is apt to decline in importance as labor and shipping costs rise. Stands of quebracho trees will be exhausted shortly, and this tannin of declining importance. Only wattle, of which nearly 1 million acres have been planted in southern Africa alone, seems to have prospects for competing successfully with the newer chemical tanning methods and leather substitutes being introduced.

CANAIGRE, *Rumex hymenosepalus*, Polygonaceae. Canaigre is a dock native to the western United States. The fleshy taproot resembles a slender carrot or parsnip, and is reported to contain up to 35 per cent tannin on a dry-weight basis. This can be extracted from the pulverized root with water. Fibrous residues are removed by centrifugation, and the soluble sugars from the solution by fermentation. Canaigre plants are propagated by crown separations planted in autumn. Were the crop to become commercially important, no doubt seed production and mechanized planting could be initiated. The crop can be handled as an annual one, the roots dug the summer following an autumn planting. Canaigre yields are said to run as high as 10 tons of roots to the acre under favorable conditions.

CHESTNUT, *Castanea dentata*, Fagaceae. During the 1930's more than 100,000 tons of chestnut tannin extract was produced annually from fallen trees killed by chestnut blight in the eastern North American forests. Largely because of the high tannin content the chestnut logs lay well preserved on the forest floor for years. The wood was chipped, the tannin leached in hot water and concentrated by evaporation under vacuum. Commercial quantities of fallen wood no longer exist, and modern-day costs of gathering would price chestnut tannin above imports in any event.

HEMLOCK, *Tsuga canadensis* AND *T. heterophylla*, Pinaceae. The eastern hemlock, *T. canadensis*, was the most important colonial source of tannin in the United States, and remained so until chestnut wood supplanted it, which was only after 1900. Trees were wastefully felled, stripped of their bark in the woods, and the logs left to rot. Hemlock bark contains from 8 to 14 per cent tannin, easily extracted after pulverization. Hemlock tannin was used chiefly for heavy leathers such as sheepskins and soles.

The hemlock tanning industry was at its peak during the nineteenth century, during which most of the hemlock from Maine to the Catskills became exhausted. The Catskill Mountains in New York were a center of the industry from about 1820 until after the

Hemlock forests of New England were long the principle source of domestic tannin in the United States. (Courtesy USDA.)

Civil War, with hides from South America being distributed to many small tanneries there. About 1 cord of hemlock bark was required for every 10 hides, on a weight basis about 6 times as much bark as leather. Consequently hides were hauled to the rural areas for tanning rather than shipping bark to industrial centers. The industry was prominent from Maine into Pennsylvania; south of Pennsylvania hemlock peters out, although oak bark often substituted for it in southerly locations. This was an era when leather was important for harnesses and even breeches (widely worn by workmen in those days) as well as for shoes.

Hemlock bark was cut in late spring and summer when it would peel more readily. Felled trees were girdled at 4-foot intervals, sheets of bark stripped off with a tool called a "spud." Enough bark was cut during this short season to supply the local tannery throughout the year; supplies of bark were often left in the woods and hauled to the tannery on sleds in winter. Hemlock tanning reached its peak in Maine about 1880, when the value of leather products approached 10 million dollars; 50 years later the industry was less than a tenth this important. Hemlock tanning declined in the Catskills shortly after the Civil War, as the once-abundant hemlocks became thoroughly depleted and replaced by hardwoods. In the Pennsylvania mountains hemlock tanning continued well into the twentieth century, however.

Western hemlock, *T. heterophylla*, now felled for timber in appreciable quantities in the Pacific forest belt, would appear an acceptable substitute for the eastern species. There has been some bark accumulation at the pulping mills of the Pacific Northwest, but tannin extraction has seldom proved economically worthwhile considering the special handling costs required. Moreover, where logs are floated in a millpond or transported as rafts in brackish water, leaching of the soluble tannin occurs, and there

Mangrove, *Rhizophora mangle*, at low tide. The bark is an important tannin source. (Courtesy Chicago Natural History Museum.)

is some impregnation with objectionable salts. Removal of bark in the forest, and its separate handling, has seldom proven economically feasible, although its potentiality as a tannin source exists should the demand ever become sufficiently great.

MANGROVE, *Rhizophora mangle*, Rhizophoraceae. The red mangrove is an abundant colonizer of brackish coastal belts throughout the world tropics. At low tide the many, arching prop-roots of this small tree form an impassable tangle above the ooze of the mudflats. Workers usually penetrate the mangrove swamps in small boats, cut the trees, and strip the bark. Even the leaves, containing appreciable amounts of tannin, are locally utilized. The chief export, however—one of the world's least expensive tannins—is obtained from the bark. The bark is transported back to centralized depositories, and air-dried in sheds for several weeks. It is then pulverized and exported directly to tanneries, or more commonly is taken to local extracting factories, where it is boiled in copper extractors. The resultant liquor is evaporated to a dark-red, vitreous solid, the commercial extract, containing 55 to 58 per cent tannin. Produc-

tion comes mostly from the East Indies, East Africa, and Central America. Because of its omnipresence in the tropics, mangrove has been considered an "inexhaustible" source of tannin. The tannin is mainly used for sole and industrial leathers, to which it imparts a reddish tinge.

QUEBRACHO,* *Schinopsis lorentzii*, Anacardiaceae. Quebracho tannin is a comparative newcomer to the tanning industry, having been utilized commercially for less than a century. Production comes from the "Gran Chaco" of Argentina and Paraguay to Bolivia, where the depleted supply presages an inevitable future decline in production.

Quebracho is indigenous to the open forests west of the Paraguay and Paraná Rivers in northern Argentina and Paraguay. The climate is subtropical, the terrain flat and often swampy, the forest seldom continuous but rather interspersed with savanna. *Schinopsis* trees prefer the lowlands, seldom being found except where the ground is swampy part of the year and the ground water often impregnated with salt. Trees take about 100 years to reach maturity. The wood serves a number of uses, such as

---

* Literally "axe-breaker" (*quebrar*, to break; *hacha*, axe), in reference to the exceedingly hard wood, one of the hardest and heaviest known.

for posts, construction timber, cabinet wood, and the like, but is most valued as a tannin source. About 22 per cent of the tree substance is tannin.

Logging takes place in the Chaco wilderness, where transportation, labor, and supplies are all deficient. Thus only large, adequately capitalized companies can undertake quebracho exploitation. Logs felled by axemen (largely Indian) are usually dragged to transportation points on two-wheel carts drawn by oxen. At railheads or watercourses they are sent on their way, originally to foreign markets for tannin extraction, but more recently to local extraction plants along the Paraguay and Paraná. The extraction mills are usually portable, but demand ready water supply and adequate labor, both of which are uncertain in the Chaco wilderness.

Only the heartwood of quebracho is used as a tannin source. Bark and sapwood, containing only 3 to 8 per cent tannin, are usually trimmed away in the forest. Heartwood yields about 30 per cent tannin by weight. The wood is chipped on special machines designed to cope with its extreme hardness. The chips (or sawdust) are mechanically conveyed to extraction chambers, where steam at high pressure is played upon them to produce a heavy, dark liquor. This liquor is treated with bisulfide and evaporated to solid form. The cake so produced contains 62 to 68 per cent tannin. The tannin is of excellent quality and acts quickly in tanning to give strong, firm, tough leather. Quebracho tannin is usually blended with other tannages. Annual production has reached nearly a half-million tons at times, accounting for about half the world commerce.

SUMAC, *Rhus coriaria* AND OTHER SPECIES, Anacardiaceae. *R. coriaria* is a Sicilian species, the tannin from which has been used for soft, light-colored leathers. It has been of modest importance in world trade for many years, production running a few

hundred tons, mostly from Italy. The tannin is extracted from the leaves, which are reported to contain up to 35 per cent tannin.

The usefulness of this species in Europe has encouraged investigation of New World sumacs in the United States. *R. aromatica* leaves have been shown to contain 20 per cent tannin (dry weight), *R. copallina* up to 33 per cent, *R. glabra* and *R. typhina* about 25 per cent. Tannins from these species give a poorer leather color, however, than with *R. coriaria*, but are suitable for pale leathers such as moroccos and roans for bindings, gloves, and the like. Were tannin prices ever to justify the production costs, no doubt these and other species of *Rhus* could be adapted to cultivation as a perennial source of tannin from the foliage.

WATTLE, *Acacia spp.*, Leguminosae. Wattle, or acacia negra, is the name given to the bark from several species of *Acacia* (*A. decurrens* and its varieties, Australia; *A. albida*, North Africa; etc.), now planted in various parts of the world (Africa, Ceylon, South America). Propagation is usually by seed. Tannin from wattle bark has been popular for some time in British tanneries and is coming to be much more widely used in the New World as well. Bark stripping is by hand when the trees are from 5 to 15 years old. Wattle bark is said to contain as much as 50 per cent tannin, and may be used directly in dried-powder form or as an extract. Extraction is much the same as for other tannin barks, involving drying, pulverization in hammer mills, and hot-water leaching. A gum similar to gum arabic can further be extracted from the extract by precipitation with alcohol. Millions of wattle trees have been planted in southern Brazil and southern Africa, supplying hundreds of tanneries there and providing export of over 100,000 tons annually to Europe and the United States. Wattle is usually employed in the tanning of soft leathers.

Other recognized tannins. DIVI-DIVI. Throughout lowland tropical America occurs a small tree, the divi-divi, *Caesalpinia coriaria*, of the Leguminosae, from whose fruit is extracted the divi-divi tannin. The short, coiled pods, said to contain up to 50 per cent tannin, are gathered mostly from wild trees and subjected to hot-water extraction. The tannin imparts a yellowish tone to leather.

GAMBIER. Gambier or white cutch is a resinous extraction from the leaves of *Uncaria gambir* of the Rubiaceae, native to the Malay area and the East Indies. The shrubs are cropped several times yearly, and the "resins," containing 35–40 per cent tannin, extracted in boiling water. It finds use in dyestuffs, masticatories, and medicinals as well as a tanning agent for all kinds of leather.

MYROBALAN. Tannin is derived from the fruit of the myrobalan, embracing two or more species of *Terminalia*, of the Combretaceae, small trees native to India. The fruits are harvested unripe. The tannin finds use in blends, principally for calf, goat, and sheep skin tanning.

OAK. Various oaks, *Quercus spp.*, Fagaceae, contain moderate amounts of tannin in the bark. This has been used in many parts of the world where more economical sources of tannin are not available. *Q. velutina* was widely utilized in the eastern United States during the colonial period, and *Lithocarpus densiflora*, the tanbark oak, in California. Oak bark was the most general source of tannin in North America until about 1800 when improved transportation permitted concentration of tanning in the hemlock areas of the Northeast. The Bay Psalm Book of 1640 was, in fact, bound in Massachusetts oak-tanned leather. The colonial tanneries were simply holes dug in the ground and walled off with heavy planking. Hides were placed within, covered with oak bark, and allowed to remain 6 months or longer.

Galls of several oaks in the Mediterranean area, especially *Q. lusitanica*, *Q. ilex*, and *Q. tauricola*, have been a commercial if somewhat local source of tannin. The galls, caused by wasps depositing eggs on the developing leaves, have an abnormally high tannin content. Other oaks of Asia Minor, such as

Peeling oak bark for tannin in Costa Rica. (Courtesy USDA Forest Service. From *Unasylva*.)

the Turkish oak, *Q. macrolepis*, yield **valonia** tannin, from the acorn cup. After sun-drying the cup is said to contain about 45 per cent tannin, used in tanning blends for finer grades of leather.

Dyes

Dyes are often found associated with tannins, and extractions from such plant parts as oak galls and bark, gambier and sumac leaves, logwood and Brazilwood heartwood provide both brown dye and tannic acid. Unlike tannins, however, the importance of natural dyes remains largely with the past. When the Englishman Perkin made an indigo-like dye in 1856 from "useless" coal-tar ingredients, the death-knell sounded for botanicals, and synthetic dyes have increasingly come to dominate the industry. Following Perkin's lead, German chemists perfected aniline dyes from coaltar and quickly established the synthetic-dye industry. By the turn of the century most once-great plant-dye ventures had fallen into oblivion. In all but a few cases synthetics are superior and significantly less expensive than botanical dyes, even when the latter are derived from abundant wild plants for which the only costs are those of collection. But dye plants have not been without their influence upon history; along with spices the search for dye plants encouraged exploration of the far corners of the globe and the rise and fall of empires.

Egyptian tombs yield some information on the earliest known dyes of the Western world. Colorings on mummy cerements include saffron and perhaps indigo. The excavations of Pompeii revealed elaborate dyer's shops there that no doubt stocked weld, fustic, and cutch. Woad ornamented the skins of the inhabitants of primeval Britain, long before Caesar's legions crossed the Channel. Indeed, every land seems to have discovered indigenous dye plants. Many became important items of commerce and exchange, and are often referred to in the classical compendiums such as of Herodotus, Dioscorides, Pliny, and Ovid. For centuries camel caravans through Baghdad and Damascus brought exotic dyestuffs to Europe from the East. Relatively few native European and Mediterranean dye plants had been utilized in western Eurasia since before historical times. From the sixteenth to eighteenth centuries Portuguese, Dutch, and British explorers brought dramatic oriental dyes to the European market, including the famed but enigmatic "Turkey red" (basically madder, but often a blend of dyes that reputedly required a month in which to develop the color) and safflower.

Dyes were perplexing substances to the ancients. What manner of coloring materials occurred in roots, stems, fruits, and leaves, and were so exceedingly fugacious in delicately colored blossoms? Why did some leach readily, whereas others formed semipermanent combinations with certain fabrics, and still others developed color only belatedly upon exposure to air? The dye chemist now realizes that a dye must be applied to fabric in a soluble form, but must be made "fast" there (to prevent it from "running") by chemical combination with impregnating mordants or, much more rarely, by direct combination with the fibers themselves. Not only does the mordant, ordinarily a metal salt such as alum, cause the dye to "take," but it also helps develop the characteristic color. Several colors are obtainable from the same dye by use of differing mordants, a feat difficult to accomplish with certainty before the advent of pure chemicals in diverse kinds. Yet even primitive man had learned empirically which dyes were fast for cotton, which for flax, and which were too fleeting to be useful with any fiber. In addition to natural salts and minerals he had discovered the usefulness of cream of tartar, wood ashes, tannic acid (from sumac), and even manure and urine as mordants.

Plants capable of yielding dye are legion, and many hard-pressed peoples have had to utilize obscure local sources. On the frontiers of colonial America, for example, walnut and butternut, smartweed and osage orange, and many other familiar plants supplied the shades of yellow, red, and brown for home-woven fabrics. Still today there is at least hobbyist interest in natural dyes from native plants; Mary Colton discusses dye plants and recipes of the American Southwest in her *Hopi Dyes*, and a Plants and Gardens handbook from Brooklyn Botanic Garden reviews dye plants worldwide (see the References at the end of this chapter). Only a few have ever been of international exchange and world significance. Those of especial historical repute are reviewed on subsequent pages.

Many dyes, whether synthetic or natural, are not overly complex chemically. Maclurin, from fustic, *Chlorophora tinctoria*, for example, is a benzophenone with the empirical formula $C_{13}H_{10}O_6$. With an alum mordant it yields a yellow color; with iron, gray; with chromium, yellow–green. In other cases the dyes are a variety of glucosides, quinones, flavones, or similar compounds, and, like tannins, are apparently metabolic by-products whose use to the plant is obscure. Dye plants occur in all plant families; perhaps the Leguminosae family contains the greatest number of prominent examples.

The dye plants. INDIGO. Indigo, or anil, the brilliant blue "king of dyestuffs," is derived from leaves of *Indigofera tinctoria* and other species of *Indigofera*, of the Leguminosae, extensively grown in former times in India but known in the form of one or another similar species throughout most of the world. The dye possesses great natural fastness to both light and water and for 4,000 years was one of the most popular dyes. Indigo does not take on its characteristic color until it undergoes aerobic fermentation or oxidation. Consequently, the process of preparation involves steeping the freshly cut plants in water for half a day and aerating the drawn-off solution, usually by means of beaters, until the color develops. A blue precipitate settles out as beating is discontinued, and is heated to prevent further fermentation. The filtered sludge, after drying, is formed into the indigo cakes of commerce. The color factor is known as indican, a colorless, soluble glucoside until oxidized.

The true indigo plant is a tall herb with narrow, pinnate leaves, small, papilionaceous flowers, and slender pods. In ancient India, however, almost any blue dye prepared by a fermentation process was indiscriminately called indigo. Such dyes were used medicinally as well as for coloring. That the Indian commerce was of considerable importance is shown by the fact that in 1631 three Dutch ships brought to Holland on the same day some 333,000 pounds of indigo from India. Yet so obscure was the actual plant source of indigo that as late as 1705 a British patent was granted for "obtaining indigo from mines," in spite of the fact that the cultivated plant had been introduced into Europe as early as the fifteenth century and was already displacing woad in Italy.

Although indigo could be proved superior to woad, influential woad growers and distributors caused indigo to be prohibited over the entire European continent for more than a century. To quote from a promulgation of 1577 in England, concerning indigo: " . . . prohibited under the severest penalties is the newly invented, harmful, balefully devouring, pernicious, deceitful, eating and corrosive dye known as 'the devil's dye' . . . used instead of woad." A local law in Nuremberg required dyers to take an annual oath that they would not use indigo. Yet, because of its merits and in spite of all attempts to discourage its use, by the middle of the seventeenth century indigo had become the most popular dye in Europe. It was largely imported from the Orient and the New World colonies. In Carolina, Eliza Pinckney tried to cultivate indigo, and in spite of a temporary setback when a hired grower from the West Indies intentionally ruined her plantings for fear that they would undercut his homeland's export trade, she produced indigo of such distinction that an exhibit before Parliament in London in 1744 influenced it to establish a sixpence bounty per pound on American colonial indigo. Scientific growing of indigo in the New World was quickly copied by enterprising Indian producers, and once again in the post-Revolutionary period the center of production shifted to India as rice and cotton became more profitable crops than indigo in the United States. A successful indigo industry continued on the banks of the Ganges until the advent of aniline dyes.

WOAD. The woad plant, *Isatis tinctoria*, of the Cruciferae, contains in its leaves the same colorless glucoside, indican, as its unrelated rival indigo. Woad plants are slender biennial or perennial herbs growing 2 to 5 feet tall, with simple, sessile stem leaves and a large panicle of small, yellow flowers. The species appears to have been indigenous to southern Russia, but was in cultivation and escaped throughout Europe in very early times. Indeed it is speculated that early German names for woad (weedt, etc.) gave rise to the modern word "weed." Seed is either drilled or broadcast, with thinning of the young plants to about 10 inches. Hand picking of the leaves commences in early summer. When the woad industry was one of importance, the freshly picked leaves were taken to the rude woad mill and crushed by specially designed wooden rollers. The resulting pulpy mass was formed by hand into crude balls, which were dried in an open, roofed shelter. These balls were later pulverized at the convenience of the operators, wetted, and allowed to ferment for several weeks. During fermentation, extremely foul odors are emitted, and so offensive were these that an English decree forbade processing of woad within 5 miles of the country estates of Queen Elizabeth. The fermented woad was dried, packed in wooden barrels, and dispatched to the dyer.

The dyer provokes further fermentation of the processed woad (or indigo) to make it soluble. Insoluble indigotine is said to be converted to the colorless indigo white, which serves to treat the fabric. The characteristic blue color develops only upon oxidation of the indigo white, which occurs with exposure of the dyed fabric to air.

Obviously woad production can find little place in the modern economy because of the immense amount of hand labor it requires. Yet in Europe after the Middle Ages woad was a basic item of commerce, particularly in Central Europe. The Saxon town of Erfurt, Germany, for example, became wealthy enough through the woad trade to establish and maintain its own university in 1392. Large exports continued to be made to England throughout the fourteenth, fifteenth, and sixteenth centuries. Woad has been used in the paints of some of the world's most famous painters, and, among the ancients, as a medicinal for ulcers and many other sorts of ailments.

MADDER. The brilliant red of madder or "turkey red," obtained from the root of *Rubia tinctorum* and other cultivated species of the Rubiaceae, has been a symbol of courage and a favorite of man since earliest times in the Far and Near East. Madder is a weak, herbaceous shrub, with angular, armed stems; opposite, entire leaves; yellow flowers; and blackish, berrylike fruits. The dye is found principally in the cortex of the long, slender roots, in the form of glucosides that are readily broken down to yield alizarin (the most important dye material) and purpurin. Plants are ordinarily grown 1 to 2 years before the roots are dug. The roots are washed, dried, pulverized, and, formerly, were marketed in the powdered form in casks or linen sacks. An infusion from the powder gives the brilliant scarlet of "turkey red." Depending upon the choice of mordant, madder can be made to yield red, pink, orange, lilac, black, or brown shades, fast to both light and water.

Madder was well known as a dye plant in India before historical times, and later was an important item of commerce from India and the Near East (Baghdad) to western Europe. In the late Middle Ages the plant was introduced into Europe, which resulted in a diminishing market for oriental roots. From about 1500 almost until the advent of synthetic dyes, the foremost madder-growing country was Holland. The canny Hollanders took to vegetative propagation of the madder, to insure continuance of the best hereditary lines, and practiced the finest agricultural methods known to the world of that day. A very minor demand for madder still exists, by painters desiring natural pigments and by some physicians who still credit madder with therapeutic virtues.

LOGWOOD. Logwood dye, from *Haematoxylon campechianum*, of the Leguminosae, serves as an example of an important dye that is extracted from wood. The species is a small, gnarled, and thorny Caesalpinaceous tree, with pinnate leaves, small leaflets, and modest yellow flowers. Logwood grows naturally throughout Mexico, Central America, and the Caribbean area, and is cultivated in Mexico, Santo Domingo, and Cuba. It has also been introduced into other tropical regions. Propagation is by seed.

The red heartwood, after removal of bark and sapwood, yields the dark-purple haematoxylin dye and appreciable amounts of tannin. Extraction from chipped wood, either processed near its source or exported, may be accomplished by steam under pressure, by boiling under atmospheric pressure, or by simple diffusion in standing water.

Logwood was long an important European import, although not to the degree that indigo, woad, and madder were. The use of this dye fell off with the development of synthetic substitutes, but during World War I, when German aniline exports were cut off, logwood again began to appear in American ports in considerable quantity. It still finds some use as a histological stain, and as a special-purpose dye.

ANNATTO. Annatto of commerce comes from the orangish pulp surrounding the seeds of *Bixa orellana*, of the Bixaceae, said to be native to Brazil but today naturalized through most of the world tropics. The small, shrublike trees are quite ornamental, having heart-shaped leaves, spacious terminal panicles of large, mallowlike, pinkish flowers, and globose capsules armed with soft spines. The plants seem to thrive in

Annatto, *Bixa orellana*. (Courtesy Chicago Natural History Museum.)

moist locations; for example, they are quite abundant along drainage ditches in the eastern Chaco of Paraguay. Under cultivation the species is propagated by seed.

Reportedly, two pigments, bixin and orellin, occur in the seed pulp. In extraction the seeds and pulp are stirred in water (or other suitable solvent, such as chloroform) until the pulp is well dissolved. Debris is filtered away and the solvent evaporated to a pastelike consistency. This paste is used principally as a food coloring (for cheese, butter, and margarine); a condiment (tropical peoples often use annatto on rice—to the benefit of the average tropical diet, usually deficient in vitamin A, for the dye contains about 2 per cent of that vitamin); a yellow or red dye (for soap, pomades, various fabrics, chocolate, and the like; and, among primitive tribes, as a skin paint, perhaps to some degree also useful as an insect repellent); and locally as a remedy for skin diseases and burns.

SAFFRON. Saffron was the principal yellow dye of the ancients. It is derived from the elongate stigmas of the saffron crocus, *Crocus sativus*, of the Iridaceae, an autumn-flowering plant native to Asia Minor. About 4,000 stigmas are needed to yield 1 ounce of dye. It is no wonder that saffron soon relinquished much of its market when comparable yellow dyes from other, cheaper sources were developed. Saffron is used in medicinals, as a flavoring, and in perfumes, as well as a dye; and for these purposes it is still cultivated today in limited quantities. Spain is the chief producing country, but over 4,000 acres are planted to saffron in the Vale of Kashmir in India, where inexpensive labor (by women) permits the daily picking of the saffron flowers.

In the Middle Ages saffron enjoyed great popularity as a drug, although modern medicine finds the virtues attributed to it to be of little merit. The dye is used today chiefly as a food coloring. Many Hindus apply a spot of saffron to the forehead as a "good omen." Etymologically "saffron" is derived from "azafran," the Arabic name for the plant introduced into Spain along with the species early in the tenth century. Saffron is propagated from daughter corms that form at the base of the mother stem. It grows best in subtropical climates, on well-drained, loamy soil. Corms are planted about 6 inches deep, in rows. With adequate rainfall or irrigation, yields of 5 or 6 pounds of saffron can be expected per acre. In India the plantings are maintained as perennials, but in southern Europe replanting is every few years (or sometimes annually in Italy).

The main ingredient of saffron is crocin, a yellowish-red pigment, itself a combination of several glycosides. With potassium hydroxide saffron yields reddish crystals, and with sulfuric acid a bluish color. One part of commercial saffron is sufficient to color 10,000 parts of water. In addition to the several glycosides saffron contains a bit of essential oil, some starch and sugar, and several other hydrocarbons.

SAFFLOWER. The safflower, *Carthamus tinctorius*, of the Compositae, is a thistlelike plant, the first important substitute discovered for saffron. The dye is obtained from the yellow or orange flower heads. Individual flowers are hand picked when at full bloom, and either dried in the sun or kneaded in water and then dried. The red dye derived is weak, and serves chiefly to fix soluble yellow colors. The species is believed indigenous to southern Asia, and has long been cultivated there and in Egypt, where it has served as food and medicine, as well as dye. During the Middle Ages the plant was cultivated in Europe. In the days of Spanish hegemony it was introduced into the New World colonies. But owing to the tedious labor required in handling the flowers, and the impermanent nature of the dye, the plant was abandoned as an important dye source even before the development of synthetics.

TURMERIC. Turmeric, like saffron, finds more use today as a condiment and a food color than as a textile dye. The orangish, fleeting dye it yields is obtained by solvent extraction from the powdered rhizome of species of *Curcuma*, of the Zingiberaceae, long grown in China, India, and the East Indies. Once used to dye cotton, wool, and silk, turmeric serves today many of the same purposes as annatto—in the coloring of foods and in the making of curries, and as a chemical indicator.

Partial listing of other vegetable dyes

Agrimony—yellow dye from rhizome of *Agrimonia striata*, of the Rosaceae.

Alder—brownish dye from bark of *Alnus sp.*, of the Betulaceae.

Alkanna—red dye from roots of *Alkanna tinctoria*, of the Boraginaceae.

Archil, orchil, orseilles, or cudbear—important red dye from fermentation of the Lichen genus, *Roccella*.

Barwood or camwood—red to blue dye from wood of *Baphia nitida*, of the Leguminosae.

Bearberry—yellowish dye from leaves of *Arctostaphylos uva-ursi*, of the Ericaceae.

Bedstraw—yellow to red dye from roots of *Galium sp.*, of the Rubiaceae.

Bloodroot—red dye from root of *Sanguinaria canadensis*, of the Papavaraceae.

Blue ash—faint blue dye from leaves and twigs of *Fraxinus quadrangulata*, of the Oleaceae.

Blueberry—blue or gray dye from fruits of *Vaccinium sp.*, of the Ericaceae.

Brazilwood—reddish dye from the wood of species of *Caesalpinia*, of the Leguminosae, originally much sought-after following the discovery of Brazil and responsible for giving that country its present name.

Broomsedge—yellowish dye from leaves of *Andropogon virginicus*, of the Gramineae.

Buckthorn—greenish dye from fruit of *Rhamnus cathartica*, and *R. infectoria*, of the Rhamnaceae.

Buckwheat—blue dye from stems of *Fagopyrum esculentum*, of the Polygonaceae.

Buffaloberry—red dye from the fruit of *Shepherdia argentata*, of the Eleagnaceae.

Butternut—yellow to gray dye from bark or husks of *Juglans cinera*, of the *Juglandaceae*; black walnut, *J. nigra*, can be similarly used.

Cochineal—red dye from a cactus parasite, the insect *Coccus cacti*.

Cutch—brown or olive dye from stem exudation of species of *Acacia*, of the Leguminosae; also from *Uncaria gambir*, of the Rubiaceae.

Fustic—yellowish or olive dye from heartwood of *Chlorophora tinctoria*, of the Moraceae: young fustic is a substitute from the twigs of *Cotinus coggygria*, of the Anacardiaceae.

Gamboge—yellow resin–dye from species of *Garcinia*, of the Guttiferae.

Greenwood, dyer's or Kendall green—yellow to green dye from all parts of *Genista tinctoria*, of the Leguminosae.

Henna—orange dye from leaves and young shoots of *Lawsonia inermis*, of the Lythraceae.

Litmus—pink to blue dye obtained from the Lichen, *Roccella tinctoria;* partly synonymous with archil.

Lokao—green dye from wood of species of *Rhamnus*, of the Rhamnaceae.

Madrona—brown dye from bark of *Arbutus menziesii*, of the Ericaceae.

Maple—rose dye from bark of *Acer rubrum* and *A. platanoides*, of the Aceraceae.

Osage orange—yellow-orange dye from wood of *Maclura pomifera*, of the Moraceae: like fustic.

Persian berries—yellow to green dyes from unripe fruit of *Rhamnus infectoria*, of the Rhamnaceae. See *Buckthorn*.

Poke—red ink from fruits of species of *Phytolacca*, of the Phytolaccaceae.

Quercitron—yellow or buff dye from bark of species of *Quercus*, of the Fagaceae.

Sandalwood, red, or red sanderswood—red dye from the wood of *Pterocarpus santalinus*, of the Leguminosae.

Sappanwood—red dye from heartwood of *Caesalpinia sappan*, of the Leguminosae.

Weld or dyer's rocket—reddish to yellow dye from all parts of *Reseda luteola*, of the Resedaceae, a type of mignonette.

Summary of the vegetable dye industry. The heyday for natural dyes lies in the past, when vegetable dyes made fortunes for growers and traders, influenced political trends, and were instrumental in opening up parts of the world to colonization or in bringing other parts into the orbit of empire. Coloring materials have been in such demand throughout the history of mankind, and their use is so ingrained in man's customs that they can be thought of today as a necessity. The demand for dyes will never fail; but there is little likelihood that natural dyes will ever again offer appreciable competition to the synthetics. Of the hundreds of possible dye sources in plant roots, stems, leaves, flowers, and fruits, many scores have at some time or other been utilized and have been of at least local importance (see References). But all except a few of these have been duplicated synthetically, and of the great dyes of international trade from the Middle Ages until the nineteenth century —particulary woad, indigo, and madder— none have been able to withstand competition from their cheaply produced, standardized, aniline counterparts. A few artists and isolated advocates of natural dyes maintain that synthetic dyes are garish and harsh, but the inexorable factors of cost and facility of production dictate acceptance of synthetics in any modern economy.

SUGGESTED SUPPLEMENTARY REFERENCES

Adrosko, R. J., *Natural Dyes in the United States*, Natl. Mus. Bul. 281, Smithsonian Inst., Washington, D. C., 1968.

Colton, M. R. F., *Hopi Dyes*, Museum of Northern Arizona Bul. 41, Northland Press, Flagstaff, Ariz., 1965.

Gore, T. S., et al. (eds.), *Recent Progress in the Chemistry of Natural and Synthetic Colouring Matters and Related Fields*, Academic Press, N. Y., 1962.

Haslam, E., *Chemistry of Vegetable Tannins*, Academic Press, N. Y. and London, 1966.

Howes, F. N., *Vegetable Tanning Materials*, Butterworths, London, and Chronica Botanica, Waltham, Mass., 1953.

Hurry, J. B., *The Woad Plant and its Dye*, Oxford Univ. Press, London, 1930.

Kierstead, S. P., *Natural Dyes*, Bruce Humphries, Inc., Boston, 1950.

King, L. J., "The Distribution of Woad in North America," *Proceedings of the Northeastern Weed Control Conference* **21**, 589–93 (1967); contains a 38-item bibliography.

Leechman, D., *Vegetable Dyes From North America*, Webb Pub. Co., St. Paul, Minn., 1945.

Leggett, W. F., *Ancient and Medieval Dyes*, Chemical Pub. Co., N. Y., 1944.

Mairet, E. M., *Vegetable Dyes*, Chemical Pub. Co., N. Y., 1939.

Shetky, E. J. (ed.), *Dye Plants and Dyeing*, Plants and Gardens Handbook, Brooklyn Botanic Garden, N. Y., 1964.

CHAPTER 11

Essential Oils for Perfumes, Flavors, and Industrial Uses

MORE ESTHETIC THAN TANNINS, MORE ROMANtic than dyes, essential oils have been among the most coveted of botanical products, notably as perfumes and spices. Especially was the spice trade to influence the course of empires and stimulate colonial rule in the Far East, the repercussions of which the world is still experiencing. Not only do extravagant essences grace milady's presence, and not only do spices turn prosaic foodstuffs into artistic creations, but essential oils serve medicine and industry, too.

Unlike the workaday products of previous chapters, one generally substitutable for another according to economic dictate, essential oils are mostly specific, prized for individual attributes. The fragrance of the violet is not that of the rose, nor the flavor of peppermint that of spearmint. If santonin is a cure for worms, wormwood (in spite of the name) is not necessarily so. Since distinctive essential oils are found in multitudes of species throughout the world, here indeed is a rich mother lode for perusal. Almost as much as with medicinals (a topic of the following chapter), numerous essential oils have been utilized by even the meanest of cultures. Our chapter would extend unendingly were even the more familiar essential oils reviewed at length. A few of the major sorts are discussed as examples, others listed with but a thumbnail sketch. Persons inter-ested in more extensive coverage are urged to consult the Suggested References at the end of the chapter.

These days most essential oils are derived from cultivated plants. Those of temperate climates are usually grown by modern agricultural techniques. In the tropics both plantation cropping and cottage industries can be found. Even in the United States there is still some back-country production of essential oils from wild plants, such as cedarleaf oil from twigs of *Thuja occidentalis* in New England, sassafras oil from roots and stumps of *Sassafras albidum* in the upper South and border states, sweet birch oil from *Betula lenta* in the Appalachian Mountains, wintergreen from *Gaultheria procumbens* in Pennsylvania, and witch hazel from *Hamamelis virginiana* in Connecticut. But the meager production from wild plants scarcely compares with the tonnages of cultivated peppermint and spearmint, or with essential oils resulting as the by-product of other industries (such as citrus oil from citrus wastes, and fruit or seed oils from apricots and almonds).

In contrast, the essential oil industry of Australia still relies much upon native plants. Most prominent is *Eucalyptus*. But with a flora said to be 90 per cent in the aromatic Myrtaceae and Rutaceae families, other essential oil-yielding genera are to be

Sorting cloves. (Courtesy Chicago Natural History Museum.)

expected there also. For example, tea tree oil from foliage of *Melaleuca alternifolia*, sandalwood oil from stems of *Eucarya spicata*, and boronia flower oil from *Boronia megastigma* are all of commercial importance, mostly derived from wild stands. Oil extraction from wild plants is generally rather crude, accomplished with portable stills that are easily hauled and assembled. In the case of boronia flowers drums of solvent containing floral extract are shipped to central locations for final extraction.

Worldwide, exploitation of wild plants is gradually giving way to cultivation of essential oil crops. Essential oils "go a long way," and relatively limited quantities enter commerce compared to the huge volumes of food substance and vegetable oils. Contrary to the general trend, here is a facet of agriculture where the small producer in a circumspect market still has a future.

Essential oils are widely distributed through the plant kingdom, unusually abundant in several unrelated plant families such as the Labiatae, Rutaceae, Geraniaceae, Umbelliferae, Compositae, Lauraceae, Myrtaceae, Gramineae, and Leguminosae. Essential oils consist chemically of a variety of organic substances, including benzene derivatives, terpenes (C:H ratio usually 5:8 as $C_{10}H_{16}$; empirical formula of camphor, $C_{10}H_{16}O$), and various other hydrocarbons and straight-chain compounds.

Usually they occur as mixtures of more than one oil. In comparison with other oils, essential oils have smaller molecules, ordinarily less than 20 carbon atoms long. They are typically liquid and possess an aromatic fragrance, owing to their volatilization (evaporation) upon contact with air. They may be readily extracted from plant tissues by steam distillation, expression, absorption into fats, or solvent extraction; but they require mild separation techniques, since many of them are unstable and become altered under rigorous physical and chemical treatment. Upon crude separation the essence is commonly mixed with small quantities of various acids, bases, phenols, ketones, aldehydes, and the like, the presence of which may effect the delicate nuance of fragrance so desired in certain oils.

As with tannins, the utility of essential oils to the plant is obscure. They serve to some extent to attract insects (flower fragrance) or repel animal depredation (irritating or offensive secretions, like that of poison ivy). Possibly they also have an antiseptic or regulative value. In plant metabolism they appear useful as moderately reactive intermediates capable of being oxidized, reduced, esterified, or compounded in larger molecules. Frequently they are considered to be hydrolysis products of complex glucosides. By and large they can be thought of as by-products of carbohydrate and fat metabolism—rather readily accumulated, yet, once formed, not so ready to enter the mainstream of metabolic activity.

Essential oils are most frequently found as floral essences, but are not uncommon in leaves, bark, fruit, and seeds, or even in the wood of stems and roots. Frequently different parts of the same plant, such as leaves, bark, and root, yield different though similar essential oils. The oils are mostly secreted into special glands (the oil commonly forming in cells lining the gland or cavity, and passed by these through their cell walls), or between the cell wall and cuticle of epidermal hairs, where the slightest breakage of the cuticle permits volatilization and brings out the characteristic fragrance. Glands frequently appear as translucent dots in leaves or tissues (termed pellucid-punctate) when viewed against the light. Age, growth, and climatic conditions have all been shown to affect the frequency of glands and the quantity of essential oil contained in plant tissues.

Centuries before the birth of Christ the technical basis for essential oil production was well established in Persia, India, Egypt, and other parts of the Orient. The Greek historian Herodotus (484–425 B.C.) mentions the distilled oil of turpentine, and trade between Greece (and later Rome) and the Orient was carried on in essences that had been trapped in fatty oils (a primitive extraction or "enfleurage" process). By the late Middle Ages distillation of essential oils had become quite general in Europe, with the production of oils of juniper, rosemary, lavender, and turpentine being particularly prominent. The most important essential oil produced in North America has always been turpentine. Apart from turpentine, citrus, peppermint, and spearmint oils are the only other essential oils to have attained significant production in the United States. Several thousands of tons of essential oils are imported annually, however.

Essences of comparatively minor impor-

tance, derived from wild plants or small holdings tended by rural peoples, usually receive unsophisticated handling. Such family operations, with labor costs of little consequence and production haphazard, contrast with the technologically advanced perfume or citrus oil industries. There, skilled, mechanized production-line methods are used. Oils so produced are of uniformly high quality, and in constant supply at relatively stable prices.

Uses of essential oils are varied. Probably most notable is their use for perfuming soaps, deodorants, and toilet preparations, and for flavoring various food and beverage products and tobacco. Essential oils also sometimes serve as antiseptics and stimulants, as ingredients in medicinals, as a laboratory reagent, as solvents in the paint industry, as insecticides, and as a component of plastics, polishes, pastes, ink, glue, and the like.

Extraction or production methods. For mankind's multifarious needs, essential oils must often not only be separated from the solid portions of plant tissues, but must be further concentrated to comparatively pure form. Four basic systems for accomplishing this are in use: distillation, cold-fat (enfleurage) extraction, solvent extraction, and expression. Occasionally a combination of these methods best fits the need, as when expressed oils are purified by distillation. Distillation has always been the most widely practiced method of essential oil production; but enfleurage is called for to trap such delicate essences as are utilized in perfumery, which would be altered or destroyed by distillation. Solvent extraction is a more recent process. It is invading fields where distillation or enfleurage were formerly customary.

1. DISTILLATION. Two general distillation processes are practiced. Where the essential oil is immiscible (i.e., will not mix with

water), steam distillation or boiling may be practiced; the distillate separates out as a layer of oil separate from a layer of water. This type of oil extraction was mentioned under the discussion of turpentine. On the other hand, where water and various components of the essential oil mixture are miscible, it becomes necessary to use fractionation or rectification techniques. These depend upon different "boiling" or volatilization temperatures of the components.

In order to isolate an essential oil by the first method, hot water or live steam is injected into a still charged with an aromatic plant. Some stills simply have the plant material floating in boiling water; other types hold the plant material on a screen above the boiling water, in direct contact only with the water vapor (steam); yet others are equipped to introduce steam alone into the still, with no boiling water present. Under the influence of the boiling water or steam, essential oil will be freed from the oil glands in the plant tissue. Both water and essential oil vaporize, to be condensed by an adjacent condenser and drained into a proper receptacle, where the oil separates automatically, above or below the water, depending upon its density. Steam or water is continuously charged into the still until all essential oil has been vaporized and until the distillate forming in the condenser is essentially pure water. In most distillations, the plant material, unless very thin (flowers, leaves), is first macerated or fragmented (comminuted) to assure the exposure of all glandular tissue to steam or hot water. For example, seeds and fruits which contain the essences in deeper tissues are usually crushed before distillation in order to rupture as many cells as possible. Distillation must follow soon after crushing, however, for care must be taken in all stages from harvest to final distillation of any essential oil that undue exposure does not permit differential or excessive evaporation of ethereal elements.

In the second method, involving fractionation, plant materials may sometimes be directly distilled; but more commonly mixtures of essential oils produced by the foregoing method are isolated and purified. In its simplest form the process involves a gradual increase in temperature of the mixture, during which the more volatile oils distill first and the less volatile components progressively later. Several distillations may in some cases be necessary to attain the desired degree of purity in the distillate. The same result is readily achieved in industrial plants by use of a fractioning column in conjunction with a constant-pressure (vacuum) or constant-temperature retort.

Spent plant material from any distillation may be used for fuel, fertilizer, or in cattle feeds. Its value, of course, depends upon the material in question and the demand for it in the part of the world where it is produced.

2. ENFLEURAGE OR COLD-FAT EXTRACTION. Where distillation may have deleterious effects on an essential oil through hydrolysis, polymerization, or resinification, or where delicate oils become "lost" in large volumes of water or are no longer produced by flowers killed in boiling, enfleurage is usually practiced. The center of enfleurage extraction is Grasse, Provence, France. Cultivation of flowers for perfume is the chief agricultural pursuit in the vicinity of Grasse, with thousands of acres being devoted to nothing but the growing of perfume flowers. Basically, enfleurage consists of applying fresh flowers (or aromatic parts) to glass plates covered with pure tallow or lard, highly absorptive of floral essences. The plates are piled one above the other on square wooden frames or "chassis," so that all fragrance emanating from the flowers will be held in an essentially airtight compartment proximate to a layer of fat. Depending upon the species, the flowers continue to produce fragrance for as long as 2 days, after which they are removed and replaced by a fresh batch. Following absorption of the essence from several dozen

successive batches of flowers collected throughout the entire blooming season of the species, the saturated fat is subjected to alcoholic extraction, which dissolves the trapped essential oil but not the insoluble fat. The alcohol solution is then concentrated to yield the perfume oil. One pound of concentrated oil may be worth thousands of dollars.

Enfleurage is practiced in cool cellars. Special attention is given to the preparation of an odorless fat, of just the right degree of hardness: it must not engulf the flowers, yet it must be soft enough to come into intimate contact with the delicate floral parts. Usually a mixture of 1 part of tallow to 2 parts of lard is used. Purification and preparation of the absorptive fat, termed "corps," is carried on during the winter months in anticipation of the next flowering season. All flowers with which the corps is charged must be removed as soon as they begin to wilt, to avoid the introduction of objectionable odors. Few substitutes for hand labor have ever been devised for this process. Skilled women, tweezers in hand, remove flowers from the fat-covered glass plates and charge these anew with fresh flowers; this goes on daily for a period of as long as 2 or 3 months. There have been attempts to develop a suction apparatus for flower removal. From time to time the corps is scratched with metal combs in order to increase and change the surface. The saturated or nearly saturated corps is finally removed from the glass plate with a spatula, carefully melted, and stored in sealed containers. This final product is spoken of as "pomade," and in times past was utilized directly without further treatment. Alcoholic extracts are more commonly made, by stirring the pomade for several days in specially designed containers. This may be done at Grasse, or the pomade may be shipped to manufacturers in Paris, London, Berlin, and New York. Several washings with alcohol are performed, after which the fat is completely odorless. It cannot be reused for enfleurage, and is usually made into soap.

In olden times a hot-fat method of extraction was practiced. Certain flowers, such as orange, violet, and acacia, which hydrolize no further essence after they are picked, were macerated and immersed in hot fat (at a temperature of about 80°C.) Absorption of the essential oil took place the same way as in the enfleurage process, and the pomade was similarly handled. This method has been superseded by solvent extraction, a process discussed below.

3. SOLVENT EXTRACTION. Although solvent extraction is a "recent" process in essential oil production, it had been shown as early as 1835 that volatile solvents could be used to extract essences from flowers more conveniently than could hot fats. The solvent method, once developed, was quickly adopted for processing all types of perfume flowers that do not continue to produce fragrance once they are picked. The basic method of solvent extraction is very simple: fresh flowers are charged into specially constructed extractors at room temperature and treated with a carefully purified solvent. The solvent penetrates the flower tissues, dissolving out the essential oil along with traces of certain other "impurities." The solvent is then evaporated off, usually *in vacuo*, leaving a residue consisting largely of flower oil. At no stage is heat applied, so that the essence is not subject to heat alteration, as it might be during distillation. The chief disadvantages of the process are the need for comparatively elaborate and expensive equipment and for precision control by trained supervisors. Hence solvent extraction is not likely to replace distillation in rural areas or among backward populations. For low-price oils the process is uneconomical, in that a considerable amount of purified solvent is unavoidably lost.

Ideally the solvent used for solvent extraction should be cheap, nonabsorbent of water, chemically inert to the degree that it does not react with the essential oil while still being a complete and quick solvent for it, and sufficiently volatile to evaporate readily without heat but not so volatile as to result in undue waste through evaporation. No solvent yet discovered has completely satisfied all these requirements, the one most nearly fitting the need being highly purified ether. Ether does, however, have the disadvantages of flammability and high volatility. Less satisfactory than ether are benzol and alcohol.

Extractors for the process may be stationary or rotary. Three or four washings with solvent are usual, the later washings being advanced to earlier position with subsequent charges of flowers. About 5 hours are required for processing a batch of flowers, including final steam distillation of the spent material to recover solvent. The concentrated washings are filtered, then evaporated in a water-bath type of "still" in which the temperature never reaches 60°C. Final concentration is *in vacuo*.

4. EXPRESSION. Expression of essential oils is practiced in certain special cases, as in the production of citrus oils from the juices and waste rinds from citrus canning factories. In general, expression involves squeezing any plant material at great pressures in order to press out the oils or other liquids. The process is carried out by hand operated presses or crushers in isolated rural areas or by gigantic mechanical presses in industrial centers. In the production of citrus juices the oil is unavoidably expressed from the rind at the same time the juice is extracted from the fruit. The juice is then separated from the oil by centrifuging.

Storage of essential oils. Most essential oils deteriorate (oxidize and polymerize) upon prolonged exposure to air and light. Hence they are ordinarily kept in closed, completely filled containers, and perfumers in particular store their valuable essences in hermetically sealed bottles kept in dark, cool cellars. Without such precautions essences become less intense, grow darker and more viscous, develop a bleaching effect, and eventually change into a brown, odorless resin.

Perfumery Essences

The art of perfumery is as ancient as recorded history, and throughout the centuries has given rise to vogues and practices that at times reflected a practical need and at other times approached the ridiculous. In days when sanitary facilities and personal hygiene were little considered, essential oils not only masked offensive odors but doubtless served antiseptic purposes as well. At the other extreme, from times antedating the Christian era up to pre-revolutionary France, foppish use of essences was often carried to the point of absurdity. It is said that among the idle rich of the Greek civilization, manners dictated use of differing and special perfumes for various parts of the body; and at the French court of Versailles, the king himself supervised the proper blending of essences for the royal bath, for which a different formula was designed for every day of the year.

Essences seem to have been first used in the mysterious religious ceremonies of the pre-Christian world. In burial rites for Egyptian Pharaohs, and for the embalming practiced by many Eastern peoples, essential oils found special use. The ancient Chinese culture made use of incense in the temples, and the Chinese people are known to have perfumed their persons and flavored their foods with essences. Babylonian and Jew, Persian and Indian propitiated the gods with offerings of fragrant incense, which consisted at least in part of essential oils. In the Greek and Roman cultures, perfumery, once largely devoted to religious ceremony, became a treasure peculiar to the wealthy. This was the day of the lavish bath, of slave girls comforting guests with the scented fan, and of perfumed doves fluttering about the ornate banquet hall. Europe, the Near East, and the far corners of the Orient were searched for essences to satisfy the extravagances of the elite. Myrrh from Persia, spikenard from India, saffron from Spain, sandalwood from China, thyme from Algeria, and jasmine from Arabia were but a few of the essences imported into Rome. Nero's triumphal entrance into the city was over a road strewn with fragrance.

During the Middle Ages less attention was given to perfumery. But the Crusades of the twelfth century brought back to France from the exotic lands to the south strange perfumes and the secrets of perfumery. To supply the newly developed and widespread demand for essence in France, the perfume industry was created by royal charter, and to this day the center of the perfume industry remains in southern France. There master blenders, as skilled in combining many fragrances into an exquisite whole as the composer in creating a symphony from the sounds of many instruments, reign over an industry of high importance in social and economic life. Many favorite perfume recipes are highly secret,* but a sixteenth-century formula given in *Les secrets de Maistre Alexys le Piedmontois* calls for the following evil concoction to provide a magic lotion, the assured preservative of beauty: "Take a young raven from the nest, feed it on hard eggs for forty days, kill it, and distill it with myrtle leaves, talc, and almond oil."

In modern perfumes, pure, unadulterated essences are seldom used. Instead, these are fortified and extended with synthetic materials and, in the case of expensive essences, with cheaper oils. Fixatives are almost invariably used. Fixatives are oils, usually of animal origin (musk, ambergris, civet, castor), but sometimes consisting of fatty oils and balsams from plants. They retard volatilization of the essences and equalize volatilization of various components in the perfume so that none predominates at any given instant. Many of the final wash waters in distillation are utilized for their residual fragrance in the making of colognes and lotions. Fats remaining after enfleurage may similarly retain sufficient fragrance to merit their incorporation in pomades and creams.

France, of course, leads the world in perfume production, followed by England, India, Turkey, and the United States. France is estimated to manufacture annually hundreds of tons of scented oils from millions of tons of flowers, including especially orange, rose, and jasmine, but considerable violet, mimosa, tuberose, and many other kinds of flowers as well. In the perfume laboratories of France, it is said, workers never develop lung or bronchial disorders, a fact which perhaps bears out Pliny and Hippocrates, who classified perfumes among the medicinals.

Examples of perfume oils. ROSE OIL (OTTO OF ROSES, ATTAR OF ROSES). Rose oil is one of the most prized as well as one of the most expensive (nearly $1000 per lb in the late 1960's) of perfume oils. It is obtained from flowers of *Rosa damascena*, of the Rosaceae, and to a lesser extent from *R. alba* and *R. centifolia*, the latter two purportedly yielding a less fragrant oil than the first. These roses are small shrubs, grown chiefly in the Balkans (Bulgaria in particular), and to some extent in southern France, Asia Minor, and India. Blooming time, and hence the rose oil production season, extends from

---

* One of the simpler formulae listed by Poucher for a synthetic acacia perfume, "Acacia 1016," includes, in parts per thousand: anisic aldehyde, 340; paramethyl acetophenone, 50; bois de rose oil, 100; phenylacetic aldehyde, 50; citronellol, 50; benzyl acetate, 50; rose no. 1090, 45; jasmine no. 1053, 55; methyl anthranilate, 50; isobutyl benzoate, 80; musk ketone, 20; vanillin, 10; coumarin, 20; heliotropin, 70; undecalactone (10 per cent), 10.

April to July. Extraction is mainly by distillation (Bulgaria), but enfleurage and, more recently, solvent extraction are often practiced (France). In distillation, the otto floats to the top of the distillate and can be siphoned off, inasmuch as it is less dense than water and immiscible with it. Average yield is said to be less than 0.5 gram of oil from each 1,000 g of flowers. The principal constituent of rose oil is citronellol (40 to 65 per cent), with lesser quantities of geraniol, nerol, linalol, and other organic compounds making up the balance. One seldom finds pure rose oil (rose oil absolute) on the market, for it is more profitable to stretch the supply of this expensive oil with cheaper substances such as rhodinol, geraniol, citronellol, and so on from other sources. In the "Valley of Roses" east of Sofia, Bulgaria, over 200,000 people are employed in rose oil production. There the plants are grown in the light stony soils, often as a hedge about fields. As is typical of roses, bloom occurs on second-year wood. Canes are rooted and transplants made in the autumn or very early spring. Well-cared-for rose plants may remain productive for 20 to 30 years. The flowers are collected in early morning, usually by children and older women, in the late bud stage (at which time the fragrance is at its best) and are taken immediately to the distilleries. Several tons of rose oil are produced each season, most of which is exported to France for use in perfume.

JASMINE. An important perfume oil is obtained from several species of *Jasminum* of the Oleaceae, especially *J. grandiflorum*, native to southern Asia. This is the only species much cultivated in southern France, but in India *J. auriculatum* and *J. sambac* are also grown. Thousands of tons of jasmine flowers are harvested from July to October in southern France, with flower yields up to 2 tons per acre. In India yields are customarily only about a half-ton. Approximately 5,000 flowers are needed to make 1 pound. Propagation of jasmine is by cuttings, planted to well-fertilized open fields, up to 2,000 plants per acre. With favorable conditions and well-drained soils plantings may last 15 years or more. Maximum essence is said to be obtained from flowers picked at daybreak, but for practical reasons buds are often picked the evening before and allowed to open during the night.

Most jasmine oil is obtained by solvent extraction with ether or benzene. The "concrete" so obtained generally represents a yield of about 0.3% of the weight of the fresh flowers. The concretes are further concentrated to make "absolutes," distilled under precise temperatures and pressures to yield as much as 23 per cent volatile oil. About 15 per cent of jasmine oil is derived by enfleurage, and in India (for local products) an interesting adaptation of the method utilizes soaked *Sesamum* seeds interlayered with flowers to absorb the volatilizations. Synthetic jasmine similar to natural essence contains benzyl acetate as a base.

VIOLET. Violet essence is obtained mostly from the parma violet, *Viola odorata* var. *semperflorens*, of the Violaceae. The chief constituent of this essential oil is ionone, which is also obtainable from citral and affords a base for synthetic violet perfume. Violet growing centers are in the Grasse and Toulouse regions of southern France, and in the Taggia valley of Italy. Greatest fragrance comes from violets that are somewhat shaded; hence plantings are often protected by hedgerows of olive and orange trees. Violet plants are set in well-cultivated fields, the planting usually made in winter. The violets flower one year after planting. Flower picking for bouquets begins in autumn, for perfume mostly from January to April. Fragrance is greatest from blossoms picked at night or in the very early morning. They are hand gathered, typically by women, who can collect up to 15 pounds per day (nearly 2,000 flowers to the pound). Extraction is chiefly with solvents, but there is some enfleurage production. Over a half-ton of flowers is needed to yield 1 pound of essential oil concentrate.

PATCHOULI. Patchouli oil is obtained from species of *Pogostemon*, Labiatae, principally *P. cablin*. *P. cablin* is native to the Philippines and southeast Asia, and has been introduced into many tropical areas such as the Seychelles where there is a small but thriving patchouli industry. It is a good example of an essential oil produced by small tropical landholders, and as an intercrop on plantations while tree crops are maturing. Until the middle 1800's patchouli was a mysterious fragrance on silks and shawls imported by Europeans from India, serving in part as an insecticide. About 1850 the British East India Company brought patchouli leaves to England, where the essence was highly esteemed in court circles; it later fell into disrepute as being the hallmark of dissolute women. In modern perfumery it is considered one of the finest fixatives for heavy perfumes and soaps, blending well with other essential oils. Patchouli oil consists mostly of sesquiterpenes, with small amounts of other substances such as benzaldehyde, eugenol, and cinnamic aldehyde.

Patchouli is an herbaceous perennial, with glabrous, quadrangular stems and opposite, somewhat fleshy leaves that contain oil glands within the mesophyl as well as at the surface. Foliage contains from 2 to 6 per cent oil, said to be most concentrated in the three or four youngest leaves on a stem. Where labor is available for clipping just the terminal growth, oil yields are higher than with mass mowing usually practiced on plantations, a minor advantage for the small grower. In all cases the cut patchouli is dried ("fermented") for about 3 days before distilling, to effect rupture of the interior cells and better release interior oil.

Indonesia is the center for patchouli oil production. The plant grows best on virgin forest soils containing humus, but resents standing water. The species seldom if ever flowers, and is propagated entirely by cuttings. Cuttings planted about 1 yard apart make a solid stand within 4 to 6 months, during which period weeding should be practiced. New plantings are often shaded with coconut fronds as a protection against sun. Replanting to fresh ground every few years is recommended, thus escaping build-up of pests such as nematodes. Dried patchouli leaves are usually distilled locally by primitive means. Often the leaves are simply placed in or above boiling water, from which the steam carrying the aromatic oil is condensed. Ideally temperature and steam pressure would be varied according to the charge, with stemmy material requiring more prolonged distillation for adequate yield. Freshly distilled oil is golden yellow, but it darkens and becomes more viscous with age, developing more fully the odor for which it is esteemed.

ORANGE BLOSSOM OIL, OR OIL OF NEROLI. Oil from the flowers of certain species of orange should not be confused with "petitgrain" derived from distillation of leaves and twigs, or with orange oil derived from expression of fruit rinds. Orange blossom oil comes mostly from *Citrus aurantium*, of the Rutaceae, grown in Italy, Spain, and Portugal for the fruit, but in Provence, France, almost exclusively for the flowers. Propagation is by seed and by grafting. A mature tree in a good year may produce up to 15 kilograms of flowers, and the total orange flower crop for all of France may exceed 2,000 tons. Flowering normally occurs in May. About 80 per cent of the oil of neroli is produced by distillation, the remainder chiefly by means of volatile solvents. Chemically orange blossom oil consists largely of linalol and various terpenes (dipentene, pinene, and camphene), with some geraniol and other organic compounds. The oil serves chiefly as an extender and as an ingredient in synthetic perfumes and cologne. The oil itself can be synthesized cheaply and readily.

GERANIUM OIL. Geranium oil is one of the important "behind-the-scenes" perfume oils. It is extracted from the foliage of several species of the cultivated "geranium" (*Pelargonium*, of the Geraniaceae). *P. graveolens*, *P. odoratissimum*, *P. roseum*, *P. capitatum*, and *P. fragrans* are all said to be cultivated for perfumery purposes, primarily in North Africa and southern Europe. Plants are readily propagated by cuttings and survive for several years in frost-free climates. Harvest is by hand (Algeria) or by mowing machinery (France) in the spring, from April to June, and in the autumn, from October to November. The newly cut foliage is given a preliminary drying prior to steam distillation. Yields of about 1 gram of oil per 1,000 grams of foliage are reported. The principal constituent of geranium oil is geraniol, which finds its greatest use as an extender or substitute for rose oil and other more expensive essences, and as the source of fragrance for soaps and similar products.

LEMON GRASS OIL. Lemon grass oil is obtained primarily from *Cymbopogon citratus* and *C. flexuosus*, of the Gramineae. These are perennial tropical species probably native to India, cultivated from Paraguay to Florida in the New World and widely in the Far East. Plants are typically propagated by separation of the rhizomes, and are hand planted. Once the plantings become well established (second year,) three cuttings of foliage are possible annually. After a few years the planting deteriorates, and new plantings are started, preferably in new soil. Harvest is typically by hand, with bush knife or machete, but in more mechanized centers mowing machinery may be used. The freshly cut grass is subjected to steam distillation. In backwoods areas this may be of a crude sort, in wood-fired stills, but in industrialized localities it is performed with efficient equipment under thorough regulation. The oil separates off in the distillate as an immiscible layer above the water. Lemon grass oil consists chiefly of citral, much used in various flavorings and as a source of ionone, the base for synthetic "violet" perfume and "verbena" soaps. Most lemon grass oil goes into perfume, soaps, cosmetics, and mosquito repellents. Beta-ionone, derived from lemon grass oil, has been the source of commercial Vitamin A synthesis. A few score tons are produced annually in Guatemala and Honduras, with lesser amounts from elsewhere in the Americas, and several hundred tons from the East Indies and especially India. In parts of Malaya more lemon grass is grown for cookery than as a commercial oil plant. Oils similarly used and extracted from grasses by distillation include citronella, from *Cymbopogon nardus*; palmarosa or ginger grass, from *Cymbopogon martini*; and vetiver, from *Vetiveria zizanioides*.

Some other perfume oils.

Acacia—large tropical genus of the Leguminosae (discussed under "Gums" in Chapter 9; see also *Cassie*). The acacia perfume of commerce is based upon another genus, the fragrant *Robinia pseudoacacia*, common in the midwestern United States and blooming in May. The essence is easily synthesized.

Ambrette—obtained from seed of *Hibiscus abelmoschus*, of the Malvaceae, and used in small quantities as an exalting agent in fine perfumes.

Basil—obtained from foliage of *Ocimum basilicum*, of the Labiatae. The plant is grown principally in the Mediterranean area, and the oil extracted by distillation.

Bergamot—obtained from the rind of *Citrus aurantium bergamia*, of the Rutaceae, grown almost exclusively in southern Italy. The oil is obtained by expression. It is much used in colognes and will blend well with synthetics. It is also often substituted for petitgrain.

Bois de rose—see *Linaloe*.

Boronia—flower oil obtained from *Boronia*

megastigma, of the Rutaceae, which grows wild in western Australia, solvent extracted by dumping the day's blossom harvest into drums of ether which are subsequently transferred to factory centers for concentration in vacuo.

Broom or genet—obtained from flowers of the shrub *Cytissus scoparius*, of the Leguminosae, cultivated in Provence, France.

Calamus—obtained by distillation of the rhizome of *Acorus calamus*, of the Araceae, harvested along the banks of European rivers. The oil is reminiscent of patchouli.

Carnation—formerly obtained from several species of *Dianthus*, of the Caryophyllaceae, grown in southern France, but duplicatable by synthesis from the similar clove oil. The oil is quite valuable and finds use in only the most expensive perfumes.

Cassia or Chinese cinnamon—an essence unfortunately termed cassia, for it is not obtained from the genus *Cassia*, but from distillation of the bark of the Asiatic *Cinnamomum cassia*, of the Lauraceae. Nor should it be confused with Cassie, obtained from the genus *Acacia*.

Cassie—obtained from flowers of *Acacia farnesiana*, of the Leguminosae, a winter-flowering shrub found wild in the New World tropics and cultivated in the Mediterranean area. The oil has a delightful bouquet, and is much used in compounding "violet" odors.

Cedarleaf—obtained by distillation of the foliage of *Thuja occidentalis*, of the Cupressaceae. Processing cedarleaf oil is a minor industry in the northeastern United States. The oil is used to some extent in perfume blends and as a fixative. Some conifer oil is also obtained from the needles of black spruce, *Picea mariana;* hemlock, *Tsuga canadensis;* and balsam fir, *Abies balsamea*. Peak yields are obtained by distillation during January–March (about 1 per cent of needle weight).

Citronella—one of the famous grass oils, obtained by distillation from leaves of *Cymbopogon nardus*, of the Gramineae. The species is grown chiefly in Java, but also to a limited extent in the New World tropics and subtropics. The oil is widely used in compounding cheap perfumes and for making menthol.

Cyclamen—an important synthetic perfume, seldom obtained by solvent extraction from species of *Cyclamen*, of the Primulaceae.

Fern—obtained from foliage and rhizomes of various ferns (*Pteridophyta*), usually by solvent extraction.

Gardenia—obtained by solvent extraction from the flowers of several species of *Gardenia*, of the Rubiaceae. Commercial extraction from fresh flowers is chiefly in Reunion, but synthetic gardenia is made in a number of perfume centers.

Ginger—obtained by distillation of rhizomes of *Zingiber officinale*, of the Zingiberaceae, cultivated in various tropical localities. Traces of the oil are used in "oriental" perfumes.

Heliotrope—a common type of synthetic perfume, seldom obtained by actual extraction from flowers of *Heliotropum peruvianum*, of the Boraginaceae.

Hyacinth—based upon solvent extraction of flowers of *Hyacinthus orientalis*, of the Liliaceae, but synthetically compounded according to several formulae.

Labdanum—obtained from the resinous leaves of several species of *Cistus*, of the Cistaceae, usually by solvent extraction. The "oil," much employed as a perfume fixator, is actually an oleoresin.

Lavender—important oil obtained by distillation of flowers of the perennial *Lavandula officinalis* (*L. vera*), of the Labiatae, native to the southern Alps. Flowering is typically in August, and flower yield (mostly from wild sources in southern Europe) may amount to 4,000 tons in one season. Lavender oil is seldom used alone, but serves in many blends. Another species of *Lavandula*, *L. latifolia*, yields the spike lavender oil.

Lemon—obtained by expression of the peel of *Citrus limon*, of the Rutaceae. Chief constituent of lemon oil is citral. The oil blends well with other essences to give "verbena" perfume, and is much used in soaps and as a substitute for bergamot.

Lilac—lilac perfumes are mixtures of synthetics, not commercially extracted from the lilac, *Syringa vulgaris*, of the Oleaceae.

Linaloe—obtained by distillation of heart-wood of several members of the New World Burseraceae. The oil consists largely of linalol, an important constituent of many essences. Sometimes termed linaloe, but more commonly bois de rose, is the essence distilled from *Ocotea caudata*, of the Lauraceae, from South America, a source material in compounding several synthetic perfumes.

Mignonette—obtained by solvent extraction of the flowers of *Reseda odorata*, of the Resedaceae, native to the Mediterranean area.

Mimosa—not obtained from the genus *Mimosa*, but from species of *Acacia* (which see).

Narcissus—obtained from the flowers of *Narcissus jonquilla*, *N. poeticus*, and other species of the Amaryllidaceae, by solvent extraction or enfleurage. Commercial growing is chiefly in southern France.

Oakmoss—obtained by a solvent extraction from the lichen genus *Evernia*, of the Parmeliaceae, collected mostly in the Alps. It is much used in compounding face powders and is blended in fine perfumes.

Orange—expressed from the rind of the sour orange, *Citrus aurantium*, of the Rutaceae. Principal constituent is limonene, an extender for oil of neroli and much used in colognes and as a flavoring.

Orris—obtained from the rhizome of several species of *Iris*, including *I. germanica*, cultivated particularly in the Mediterranean area. Rhizomes are dug in summer, peeled by hand, and dried. Pulverized, these may be incorporated in "violet" and other perfumes.

Palmarosa—one of the grass oils, from *Cymbopogon martini*, much used in blending because of its cheapness. See discussion of lemon grass oil.

Petitgrain—obtained by distillation from foliage of *Citrus aurantium amara*, of the Rutaceae, grown chiefly in the Mediterranean area and in Paraguay. The oil finds much use in soaps, cologne, and for blending with other essences.

Rosemary—obtained by distillation of the inflorescence of *Rosemarinus officinalis*, of the Labiatae, grown in the Mediterranean area. It finds greatest use in cheap perfumes, hair washes, and soaps.

Saffron—obtained from distillation of the dried stigmas of *Crocus sativus*, of the Iridaceae (see discussion under dyes). Saffron oil is perhaps the most expensive of the essences, and only traces are used to give a characteristic bouquet of the "oriental" type.

Sage—obtained from foliage of *Salvia officinalis*, of the Labiatae, and used as an adulterant for rosemary and spike.

Sandalwood—obtained from distillation of heartwood and roots of *Santalum album*, of the Santalaceae, indigenous to India. The oil is used as a medicinal as well as in various perfume blends. Australian sandalwood is from *Eucarya spicata*, a diminutive tree of western Australia.

Spike—see *Lavender*.

Spikenard—obtained by distillation of the root of *Nardostachys fatamansi*, of the Valerianaceae, native to the Himalayan area. Oil from *Valeriana officinalis* is sometimes substituted for the true spikenard.

Sweet Pea—name for the synthetic essence: the natural sweet pea oil, coming from enfleurage of *Lathyrus odoratus*, of the Leguminosae, is not an item of commerce.

Thyme—obtained by distillation from foliage of various species of *Thymus*, of the Labiatae, grown particularly in Spain and France, and used in soap perfumery, mouthwashes and dentifrices, and as an anthelmintic. Thymol is today largely synthesized in the United States, from coal tar or eucalyptus oil.

Tuberose—obtained by enfleurage or solvent extraction from flowers of *Polianthes tuberosa*, of the Amaryllidaceae. The flowers are grown particularly in southern Europe and are generally harvested from August to October.

Valerian—obtained from distillation of rhizomes and roots of *Valeriana officinalis*, of the Valerianaceae, cultivated in Europe and Asia. See *Spikenard*.

Verbena—obtained from distillation of leaves of *Verbena triphylla*, of the Verbenaceae, cultivated in southern France and Algeria. It is used for blending with citrus oils, orris, rose, heliotrope.

Vetiver—essential oil obtained by distillation of the roots of *Vetiveria zizanioides*, Gramineae, native to Southeast Asia. Mats made of the roots give a pleasant odor in huts, and serve as something of a fumigant. Vetiver does not produce viable seed, and is propagated by division. The essential oil used in perfumery comes mainly from Reunion, Madagascar, and Haiti; it is useful as a fixative for "violet" odors.

Ylang-ylang—obtained by distillation of flowers of *Canangium odoratum*, of the Annonaceae, cultivated particularly in the Far Eastern tropics. The tree blossoms throughout the year, but the best flowers are generally picked in May and June. The oil is much used in expensive perfumery, and is blended particularly in "oriental" odors.

Spices and Flavoring Oils

Most spices owe their valued properties to essential oils. Thus added to the romance of perfumery is an intriguing story of exotic Eastern islands and swashbuckling buccaneers. The magic of spice-trade riches lured adventurers of the Western world to the Indian Ocean islands, and for centuries imperialistic governments squabbled for possession of a subcontinent and a corner of the globe where nature had willed most of the world's spices to grow (see Fig. 11-1 and Table 11-1). First Arabia, then Venice, and, by the 1500's, Portugal came to dominate the spice trade. Tiny Portugal brought to knee Ceylon, Malacca, and the splendid cities of the Malabar coast, and carved for herself a distant empire southeast of the China Sea, a half-year's voyage from the homeland. With conquered slaves to produce the cinnamon of Ceylon and Malay, the cloves of the Moluccas, the nutmeg, pepper, and opium of Java, the Portuguese East India Company poured such wealth into Lisbon as that venerable city had never known. The ancient "spices of Araby"—flavors, preservatives of meat, and luxuries of the rich—had never seen such widespread demand in Europe in the days of the Arabian caravan trade, or while Venice was a Mediterranean way station. And never was a monopoly more skillfully maintained than during Portugal's brief interval of world hegemony in the spice trade. But such factors as the union of Portugal with Spain, whose attention was directed to the New World, not the Old,

Figure 11-1   Spices from the Far East. (After Robert M. Necomb, *Economic Botany* **17,** 129–32 (1963).)

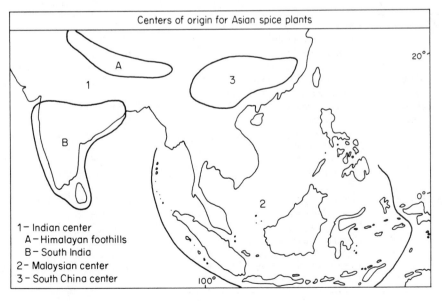

Table 11-1

Spices from the Far East

INDIAN CENTER

Family	Species	Common Name
	A) HIMALAYAN FOOTHILLS	
LAURACEAE	*Cinnamomum Tamala* (Buch.-Ham.) T. Nees & Eberm.	Indian Cassia
PIPERACEAE	*Piper longum* L.	Long pepper
RUTACEAE	*Murraya Koenigii* Spreng.	Curry leaf tree*
ZINGIBERACEAE	*Amomum aromaticum* Roxb.	Bengal cardamom
	Amomum subulatum Roxb.	Greater or Nepal cardamom
	Curcuma Zedoaria (Berg.) Rosc.	Zedoary*
	Zingiber officinale Rosc.	Ginger
	Zingiber Zerumbet Smith	
	B) SOUTH INDIA (MALABAR COAST AND CEYLON)	
LAURACEAE	*Cinnamomum Cassia* (Nees) Nees ex Blume	Cassia
	Cinnamomum zeylanicum Breyn.	True Cinnamon* (Ceylon)
MYRISTICACEAE	*Myristica fragrans* Houtt.	Nutmeg, Mace
	Myristica malabarica Lam.	
PALMACEAE	*Areca Catechu* L.	Betel-nut Palm
PIPERACEAE	*Piper Betle* L.	Betel pepper
	Piper Cubeba L.f.	Cubeb pepper
	Piper longum L.	Long pepper*
	Piper nigrum L.	Black pepper*
ZINGIBERACEAE	*Curcuma longa* L.	Turmeric
	Curcuma Zedoaria (Berg.) Rosc.	Zedoary
	Elettaria Cardamomum (L.) Maton.	True cardamom*
	Zingiber officinale Rosc.	Ginger
	Zingiber Zerumbet Smith	

MALAYSIAN CENTER

Family	Species	Common Name
LAURACEAE	*Cinnamomum Cassia* (Nees) Nees ex Blume	Cassia
	Cinnamomum Loureirii Nees	Saigon cinnamon
	Cinnamomum zeylanicum Breyn.	True cinnamon
MYRISTICACEAE	*Myristica fragrans* Houtt.	Nutmeg and Mace* (Moluccas)
MYRTACEAE	*Syzygium aromaticum* (L). Merr. & Perry (*Eugenia caryophyllata* Thunb.)	Clove* (Moluccas)
PALMACEAE	*Areca Catechu* L.	Betel-nut palm
PIPERACEAE	*Piper Betle* L.	Betel pepper*
	Piper Cubeba L.f.	Cubeb pepper*
	Piper longum L.	Long pepper
	Piper nigrum L.	Black pepper
	Piper retrofractum Vahl. (*Piper officinarum* C.D.C.)	Javanese long pepper*
ZINGIBERACEAE	*Amomum Kepulaga* Spr. & Brkil.	False cardamom*
	Amomum xanthioides Wall.	False cardamom
	Curcuma longa L.	Turmeric*
	Curcuma Zedoaria (Berg.)	Zedoary
	Elettaria Cardamomum (L). Maton	True cardamom
	Zingiber officinale Rosc.	Ginger*
	Zingiber Zerumbet Smith	

Table 11-1

Spices from the Far East

SOUTH CHINA CENTER

Family	Species	Common Name
LAURACEAE	*Cinnamomum Cassia* (Nees) Nees ex Blume	Cassia*
MAGNOLIACEAE	*Illicium anisatum* L.	Chinese anise* or Bastard star anise
	Illicium verum Hook. f.	Star anise*
ZINGIBERACEAE	*Amomum xanthioides* Wall.	False cardamom*
	Curcuma longa L.	Turmeric
	Curcuma Zedoaria (Berg.) Rosc.	Zedoary
	Curcuma Zerumbet Roxb.	Zedoary
	Zingiber officinale Rosc.	Ginger

* Most certain place of origin.
From Robert M. Newcomb, *Economic Botany* **17**, 127–132 (1963).

attacks on the spice fleet by the English and the Dutch, and haughty treatment of native populations by a graft-ridden ruling group brought about the decline of the fantastic Portuguese empire.

Destiny then dictated that by the close of the sixteenth century the Dutch would rule the Malabar coast, and that one by one the great Eastern strongholds of a decaying Portuguese empire would fall to privateers from Holland. Dutch imperialism and Dutch domination waxed in the East— for the Dutch were even more efficient and ruthless than the Portuguese. The Midas-touch of the spice monopoly poured wealth into Holland, and bestowed an unparalleled cultural bounty of painters and poets, musicians and scholars, upon an ascetic kingdom that had until then rarely known wealth and idleness; for where poet, like peasant, has to scurry for a living, cultural output is usually lean. The favors of the Orient sat well upon this second European empire in the East, and three centuries later the businesslike and no-less-determined Hollanders were still drawing heavy interest from that investment in tropical daring first made in the seventeenth century. But though the glories of an East Indian empire were ultimately more enduring for the Dutch than for the Portuguese, Dutch affluence was challenged and

many Dutch possessions conquered by a mighty Britain. British domination over the East Indies was more enlightened, but the importance of the spice trade was fast diminishing. Spices of all kinds had been smuggled away and planted the world around. No longer was there a scarcity or a monopoly. Ceylon's cinnamon groves were still important, and the growing of spices was still practiced in the East Indian islands (returned to Holland by Britain in accordance with a treaty of 1815), but their commercial importance has since been slight in comparison to that of the twentieth-century enterprises in rubber, quinine, and tropical staples.

Many essential oil flavors, such as vanilla and oil of lemon, are extracted by the methods discussed under perfumery. These are usually utilized as liquids or in solution. On the other hand, most spices consist of the plant parts themselves, from which the essential oil has not been extracted. These are ordinarily prepared by drying and grinding. Grinding on a commercial scale is done to a precise degree of fineness and with minimum loss from shrinkage by efficiently designed machines of a number of types (hammer mills, attrition mills, roller mills, limited mills, pulverizers). Packaging must be done properly to assure keeping qualities, for evaporation of the volatile essential oils would result in rapid deterioration. Most flavoring materials in the United States fall under the jurisdiction of federal Pure Food and Drug laws, so that today the adulteration practiced in the past is exceedingly rare. Indeed the entire spice industry, although it stems from a wild and romantic past, is today a sober and businesslike enterprise.

Examples of flavor oils. CINNAMON. Cinnamon is an ancient spice of the Orient that was in use long before the coming of Europeans. It is produced from the bark of *Cinnamomum zeylanicum*, of the Lauraceae, indigenous to Ceylon and India, with a diploid chromosome complement of 24. It was one of the aromatics used by Egyptian Queen Hatshepsut some 3,500 years ago. The cinnamon of Biblical narrative, however, may have been the bark of cassia ("cassia cinnamon," *C. cassia*), used somewhat interchangeably in the trade today. Cassia and cinnamon were ingredients of the holy anointing oils and perfumes used at the time of Moses, and familiar items of commerce in Babylon. When Nero's wife Poppaea died in Rome around A.D. 65, a year's supply of the most precious commodity in the imperial storehouse—cinnamon—was burned at the funeral. Some cinnamon may have been brought overland from India's Malabar Coast by caravan, but there was also probably commerce through the Persian Gulf and Red Sea, by both Arabian and Phoenician mariners.

The cinnamon tree is small and bushy, with thick, shiny, simple leaves and inconspicuous flowers. The bark contains up to 1 per cent of a volatile oil which is 55 to 75 per cent cinnamic aldehyde. Cinnamic aldehyde can also be derived from cassia bark and can be synthesized.

Cinnamon comes to market as the traditional bark, or as a distilled oil. The latter is mostly derived from foliage, although there is also some distillation of rough bark not suitable for other sale. Typically, young leaves are gathered from stump sprouts, cut, or plucked every 18 to 20 months. The leaves are stripped by hand, and sacked for transportation to the distillery while still fresh. There they are subjected to steam distillation, usually by passing pressurized steam through a cylinder full of leaves. The steam carries the volatile oil to a water-cooled condenser. The oil, heavier than water, is drawn off at the bottom. Cinnamon leaf oil is approximately 90 per cent eugenol, which can be converted into vanillin.

Bark stripping follows the monsoon season (the southwest monsoon brings heavy rains from May to August in southern Asia), and is preferably from second-year wood as a new flush of twigs develop. At this time the sap in the twigs makes for easy peeling. The cut twigs are taken to the peeling shed and the rough outer bark scraped away. The twigs are girdled about every 12 inches, and slit lengthwise on each side with a special knife. The bark is pried loose and dried, curling into tubelike sections called quills. Compound quills result when several bark sections are superimposed, making a nearly solid cylinder. Cull quills and trimmings are salvaged for powdered cinnamon. Yields average about $\frac{3}{4}$ ton of quills to the acre.

Cinnamon production comes chiefly from Ceylon and the Seychelles Islands of the western Indian Ocean. Cinnamon was first introduced into the Seychelles about 1767, and has since become widely naturalized. Much of the production there is from plantations. Seedlings are typically started in nurseries, planted two or three to a hole to give "bushiness" to the planting. Plantations in Ceylon are largely native-owned today, in contrast to the former European ownership; still earlier bark was obtained from wild trees as forced tribute to Singhalese princes. Although Ceylon and the Seychelles account for most cinnamon production, both cinnamon and cassia are widely grown in southern India, China, Malaya, Indonesia, and to a small extent in the New World tropics. Export is chiefly to northern Europe, the United States, and Japan. United States taste prefers cassia, of which the best quality was shipped out of Vietnam until the war there disrupted commerce in "Saigon cinnamon" during the late 1960's, and alternative supplies were sought in Sumatra. The spice is used for flavoring cakes and pastries, in beverages, candies, drugs, cosmetics, perfumery, and the oil sometimes in medicines.

CLOVES. Cloves are another exotic spice whose production was once confined by monopoly-minded Dutch to a single island of the Moluccas (Spice Islands) capable of close supervision, the clove tree having been eliminated from all other East Indian territory. The clove of commerce is the dried, highly aromatic, unopened flower bud and twig tip of *Eugenia caryophyllata*, (*Caryophyllus aromaticus*), of the Myrtaceae. Cloves were utilized in China before the time of Christ to perfume the breath of court officials before they addressed the sovereign. Today production comes chiefly from the Malagasy Republic and Indonesia.

Clove trees are rather small, attaining maximum heights of 40 to 50 feet, and have small, simple, opposite, punctate leaves and inconspicuous flowers. They flourish best on well-drained soils in maritime climates with 90–100 inches of annual rainfall. Propagation is from seed in beds given protective shading. An individual tree may yield up to 70 pounds of dry cloves, but production is variable because of a tendency toward alternate-bearing years. Clove trees begin to flower in the fourth to sixth year. Harvest is in late summer and again in winter. The crop of immature buds is picked by hand as they redden, by men and women who climb to the treetops. After picking, the buds are sun-dried and marketed as the familiar clove spice, or clove oil may be extracted by distillation. In Indonesia cloves are mixed with tobacco for smoking, this use accounting for nearly two-thirds of world clove production. Clove oil contains 80–95 per cent eugenol, often used for the synthesis of vanillin. In addition to its use as a flavor, the oil is also widely employed in disinfectants, toothpastes, and mouthwashes. In perfumery it serves as a "sweetener" and intensifier, blending well with various essences. Clove oil is also used as a clearing agent in histology.

NUTMEG AND MACE. Another of the tropical tree species is the nutmeg, *Myristica fragrans*, Myristicaceae. Its seed becomes the *nutmeg* of commerce, and a membrane called an aril that surrounds the seed becomes *mace*. The tree is native to the Molucca Islands, and is cultivated today principally in the West Indies (Grenada), Indonesia, and Ceylon.

A branch of the nutmeg tree, *Myristica fragrans*, with male flowers, mature fruit, and seed with aril. (Courtesy Paul H. Allen.)

Nutmeg is used for flavoring foods, sauces, and beverages. Its oil is used in pharmaceuticals, confections, condiments, perfumes, cosmetics, and soap. Mace is similar chemically, and is used for similar purposes. The nutmeg tree is dioecious, and the sexes cannot be distinguished until the trees flower a half-dozen or so years after planting. Most trees are started from seed. To insure sufficient fruit-bearing (female) trees, two seedlings are generally planted near each other and later male trees are removed except one for about every dozen female trees to insure pollination. Sometimes branches from male trees are grafted onto female trees. Nutmeg trees continue to produce for 50 years or more, attaining heights of 40 feet and top yields of several thousand fruits annually. The fruit resembles an apricot, the fleshy part splitting open to reveal the seed with its scarlet aril. Some seeds fall to the ground, but most are gathered with a long-pronged pole equipped with a basket. Over a half-ton of fresh nutmeg is generally harvested per acre. The nutmeg seeds are dried in the sun until the kernel rattles freely in the shell, after which the shell is broken and the nutmeg kernel ground.

VANILLA. The original vanilla extract was obtained from the pods of a tropical climbing orchid, *Vanilla planifolia*, Orchidaceae. The fully grown but unripened seed pod of this and related species is fermented to develop the characteristic vanilla aroma, and the essence (glycoside), mainly vanillin, is extracted in alcohol to yield the vanilla extract of commerce. Vanillin, 3-methoxy-4-hydroxy-benzaldehyde, can be synthesized from other materials such as eugenol. However, there is still considerable commerce in vanilla, principally from the islands off the southeastern coast of Africa (including the Malagasy Republic, Reunion, and the Seychelles) and Uganda. World trade amounts to several million pounds annually, mostly sent to the United States (which imports about 3 million pounds).

The vanilla orchid bearing young pods. (Courtesy Chicago Natural History Museum.)

Bernal Dias, an officer in the Cortez expedition, was perhaps the first European to become familiar with vanilla spice when he observed its use in Montezuma's court for compounding the chocolate drink "chocalatl." The vanilla beans were valued as tribute in the Aztec Empire, and after the conquest became a familiar export to Spain; even today there is moderate vanilla production in Mexico. Vanilla growing is rather complicated, requiring abundant, inexpensive labor. The plant needs a tropical climate, partial shade, ample rainfall, and rich soil. Generally land is cleared except for a few shade trees that also aid in support of the climbing vines. The soil is cultivated, cuttings of vanilla planted, suitable supports set up, and the vines trained to them. Vanilla roots are confined chiefly to the surface, so that a mulch at the base of the plant is recommended. A certain amount of pruning and pest control is necessary.

Vanilla flowers pollinate less than 1 per cent naturally, so that hand pollination becomes necessary. Experiment shows that treatment with herbicides such as 2,4-D and dicamba will cause fruit set, but the beans are generally not so large as when pollinated. Fruits may be thinned to avoid reduced size and weakening of the vine. When the tip of the pod begins to turn yellow the beans are picked for curing. Curing involves heating or freezing to kill the tissues, then traditionally a sweating (in which the beans are kept in the sun by day, rolled in woolen blankets by night) and a final conditioning or after-fermentation in metal-lined boxes at room temperature for several months until the characteristic fragrance develops. An oven-curing process has been originated in Uganda, however, that is said able to complete curing in 80 hours. One pound of vanilla pods suffices to make 1 gallon or more of extract; cut beans are immersed in 35 per cent (or stronger) alcohol and heated. Sugar and glycerin may be added to help preserve the flavor and aroma.

Vanilla is used for flavoring ice creams, chocolate, beverages, and various sweets, and in perfumery. Synthetic vanillin, which substitutes for many of these purposes, can be produced from pulping liquor, from coal tar, and from eugenol obtained from other spice plants. However, the additional components of natural vanilla extract impart a superior flavor, and the natural product is preferred to and more highly valued than synthetic vanillin.

PEPPER. Pepper, *Piper nigrum*, Piperaceae, is another example of the importance of tropical spice sources. Pepper is the most important of all spices in terms of world

trade, a taken-for-granted condiment on Temperate Zone dinner tables. Peppercorns are the dried fruit of this tropical climbing vine native to southeast Asia. Production comes chiefly from India and Indonesia, but to some extent from Brazil and the Malagasy Republic. It is one of the fabulous spices of the ancients, which lured adventurers to the Far East and enticed Columbus to his voyages. It was an important trade item of the Near East in Roman times.

The vine is perennial, with alternate shiny leaves and adventitious roots that enable it to cling to tree trunks up which it climbs as high as 30 feet. Many cultivars have been selected, bearing names sounding exotic to the Western ear. Pepper cultivation is largely a cottage industry, although sometimes it is intercropped on tea, coffee, or palm plantations, where it receives professional attention. Cuttings may be planted at permanent sites, or be started in nurseries for transplanting. Lateral branches bear insignificant flowers without petals, these sometimes perfect and sometimes only pistillate. Just before the

Cuttings from this pepper vine will provide 8 to 10 new plants. (Courtesy USDA.)

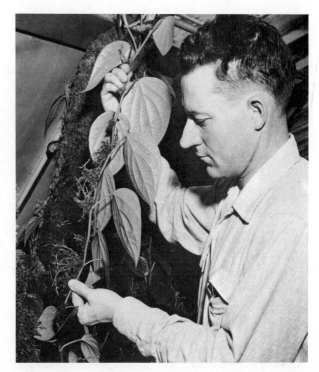

Field planting of young pepper on a farm near Mayaguez, Puerto Rico. (Courtesy USDA.)

Pepper plantation near Chanthaburi, Thailand. (Courtesy USDA.)

small green fruits ripen they are hand picked. Cut inflorescences are taken to the drying yard, heaped on a flat area, and trodden on with bare feet to remove the "stems." The loose fruits are sorted by hand, and left to dry with occasional stirring. Within 3 days the berries turn black, and a few days' additional drying makes the crop ready for market.

White pepper is produced by allowing the corns to ripen further, then removing the outer hull by boiling or fermentation. White pepper lacks the pungency of the black form. In addition to a volatile oil, pepper contains the alkaloid piperine as well as resin and other ingredients. The diploid chromosome number is said to be 128, indicating a polyploid condition. Plants begin bearing after 3 or 4 years, and yield up to 5 pounds of pepper annually per vine. World output of pepper probably averages about 100,000 tons annually, of which the United States utilizes about 40 million pounds.

GINGER. The sun-dried rhizomes of *Zingiber officinale*, Zingiberaceae, yield ginger spice, and upon distillation an essential oil which finds some use in perfumery. The species is native to tropical Asia, but is now cultivated throughout the world tropics. India and Taiwan supply about three-fourths of the world's supply. Ground ginger is used extensively for flavoring all sorts of food products, including soft drinks. Three types of ginger are recognized by the trade: West African (mainly cultivated in Sierra Leone), Indian (mainly from the Malabar Coast), and Jamaican. Peeled Jamaican ginger is generally considered the finest.

The ginger plant is a large, coarse, perennial herb. In Jamaica it is grown in the mountains at elevations from 1,600 to 3,000 feet as a cottage industry on small acreages. Planting from rhizome sections follows spring rains. A planting lasts for several years, then is usually rotated with other food crops before replanting. About 9

Whole plants of ginger, *Zingiber officinarum*. (Courtesy Chicago Natural History Museum.)

months after planting, ginger rhizomes are ready for harvest. They are dug with a hoe, broken into sections, and the larger pieces peeled for market, the smaller ones set aside for replanting. Yields run about 1 ton of dried ginger per acre. The peeled ginger is sun-dried and sold through exporters.

MUSTARD. Several types of mustard are derived from various species of *Brassica* and other genera of the Cruciferae, most of them native to Europe and western Asia. The condiment mustard much used for "hot dogs" and other meats is derived from the seeds of *B. juncea, B. alba* and *B. nigra.* The Greeks used mustard for a fish sauce, and the Romans believed it to have medicinal virtues. In the fourteenth century the English compounded crushed mustard seeds with vinegar, antecedent to modern mustard pastes. Mustard seed contains several pungent principles, including the glucocide sinigrin and the volatile allyl isothiocyanate (the "hot" ingredient). After harvest mustard seed is thoroughly dried at 115°F before milling to separate the bran from the flour (the crushed embryo and cotyledons). Mustard so produced is incorporated into various sauces, pickles, and so on, where in addition to contributing flavor it also acts as a preservative against mold. In medicine mustard has been used for poultices, and as an emetic. Mustard oils are extracted by expression from the seed of various species, used as are vegetable oils. Mustard plants grow rapidly and are frequently used as soil cover and for grazing. Most mustards are adapted to temperate climate, and are especially grown in Europe and China.

SAGE. Garden sage is *Salvia officinalis,* of the Labiatae. It is rather stiff, many-branched subshrub with slender, grayish leaves that are rough to the touch. The plant was probably cultivated in pre-Christian times, and is definitely described in the *Hortulus* of Charlemagne's botanist–physician. Sage tea was drunk in England before beverage tea became available, and sage was a familiar village garden plant in both England and North America. It is native to the Mediterranean area, one of over 500 species in this large genus. Most commercial production comes from Yugoslavia, around 1,000 tons of which are imported into the United States annually.

Sage is propagated either from seeds or by cuttings, the latter to maintain cultivar identity. Cuttings are field-planted about 2 feet apart, several thousand to the acre. The plant is perennial, the fields productive for up to 10 years. Sage foliage is harvested when the plants begin to bloom, with second and third cuttings possible up until frost. For economy sage is mowed and raked just like hay, although a higher-grade product would result from a gathering of the younger leaves only. In some areas sage is artifically dried.

The chief volatile oils of sage are cineol, two pinenes, salvene, a sesquiterpene, two borneols, camphor, ketones, and thujone. The dried leaf contains less than 2 per cent of essential oil. Sage leaf is used almost exclusively for flavoring foods, including meats such as pork sausage and poultry stuffing. Infusions of the leaf have been used medicinally as a tonic and gargle, and in treatment of many ills. Sage oil, obtained by distillation, is utilized in a number of pharmaceuticals.

DILL. Dill is a minor essential oil typical of several obtained from aromatic seeds in the Umbelliferae family. Dill is obtained from *Anethum graveolens,* a herbaceous annual widely used for flavoring pickles and kraut. Dill oil is produced by distillation of leaf, stem, and seed. The species is native to Europe, and is cultivated particularly there and in Asia. In the United States it is often grown in home gardens. Where cultivated agriculturally, dill is drilled in rows much as are beets in the early spring. Seed is harvested as with small grain. Plants cut for distillation are left to "cure" in the fields for a day or two before being brought to the still. Dill seed is much used for flavoring soups, salads, pickles, meats, and so on. The principal constituent of dill oil is carvone, also abundant in caraway seed.

HOPS. The hop plant is *Humulus lupulus* of the Moraceae. Commercial hops are the dried inflorescences of female plants. These contain substantial amounts of a complex, bitter essential oil in which fourteen or more constituents have been identified, including *lupulin*. Various parts of the hop plant, however, have been utilized in times past. The Romans used the species as a salad green, and still today young shoots are eaten in central Europe as a vegetable. Among the ancients, concoctions of hops were reputed to have broad medicinal virtues; indeed modern tests show hop extract to have antibacterial properties. The overwhelming use for hops today is for the flavoring of beer. Yet there is no evidence that beer produced before the eighth century contained hops. During the Middle Ages many monasteries became famous for their hopped beers, and

Dill. A drawing from the Lobel Herbal, 1581. (Courtesy Chicago Natural History Museum.)

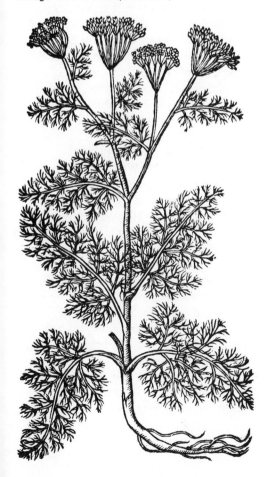

by the fourteenth century hopped beer was in general favor throughout northern Europe.

The hop plant is a perennial herbaceous vine. Pistillate plants bear leafy inflorescences about 2 inches long called "cones," the commercial hop. The essential oil is borne on glandular hairs from the epidermis of both perianth and bracts. If male plants can be eliminated, seedless cones result, especially esteemed for brewing. Hop quality is influenced by growing conditions, being best in equable climates with ample moisture. In the United States much of the hop growing centers in Oregon and Wisconsin, with European production in England and central Europe. Propagation is usually by rhizome cuttings, the vines trained up strings to arbors or trellises. Plantings are productive for about 7 years.

Hops are best picked when fully mature. In many parts of the world the cones are picked by hand, but in the United States mechanical pickers (machines with rotary drums) separate the cones from the vine. Fresh hops contain up to 80 per cent moisture, and so must be dried; careless heating, however, results in volatilization of the essential oil. Usually fresh cones are dried in kilns where a large volume of moderately warm air is blown through the loose hops for about 12 hours. Formerly sulfur dioxide was introduced, which bleached and helped preserve the hops; more recently preference has developed for green hops and sulphuring largely has been discontinued. Unless the baled hops are held cool and reasonably dry they undergo deterioration and lose value.

For brewing beer (discussed in a later chapter) hops are added to the wort as a flavoring, precipitant, and partial sterilant. The bitterness of beer largely depends upon the quality of hops used; in Europe a greater hop content is generally preferred than in America. The amount of essential oil in hops is relatively small, seldom as much as 0.5 per cent. Distilled hop oil, occasionally used for tobacco flavoring, is primarily humulene ($C_{15}H_{24}O_9$) and myrcene ($C_{10}H_{16}$). Hop concentrates are being increasingly prepared at the hop growing site, rather than sending baled hops to the breweries. Since 1 to 3 pounds of hops are used for each barrel of beer, the transportation savings can be appreciable. About 25,000 tons of hops are produced in the United States annually, and more than twice this much in Europe (especially Germany and the United Kingdom). World production is slightly less than 100,000 tons.

PEPPERMINT AND SPEARMINT. Peppermint, *Mentha piperita*, and spearmint, *M. spicata*, are the two most widely used mints. Spearmint produces seed, and many selections from progeny have been made for quality and yield. Peppermint seldom sets seed, and is thought to be a natural hybrid, possibly with *M. spicata*, *M. aquatica*, *M. silvestris*, and *M. rotundifolia*

Peppermint, showing characteristic leaves and flowers. (Courtesy USDA.)

any or all in its ancestry. Both species are familiar home garden plants, but their greatest importance is for flavoring and medicinal oils distilled from the foliage. There are records of peppermint distillation by the Egyptians only a few centuries after Christ, but little commercial demand for mint oils existed before the twentieth century. Menthol, widely used in lotions, antiseptics, dentrifices, cigarettes, and certain foods, is the most abundant constituent of peppermint oil. Commercial menthol, however, is derived mainly from the Japanese mint, *M. arvensis*, produced chiefly in Japan; domestic peppermint oil is used mostly for flavorings. Spearmint oil is largely carvone, which is usually more economically obtained from citrus wastes than by distillation of the mint.

Mint growing first became an important agricultural activity on the mucklands of northern Indiana and southern Michigan in the early 1900's. Verticillium wilt discouraged growing there, and production shifted chiefly to the Columbia river basin of Oregon and Washington. It is said that midwestern production is superior to that of the West, however, because the western oil must be fractionated to remove certain unfavorable flavors and aromas. Mints do well on moist, organic soils. In the West, where these are lacking, the crop is usually irrigated. Mint stolons are planted in rows, the first year called "row mint." Harvest is just before frost, after which the mint is plowed down to revive from underground parts the second year as a solid stand of "meadow mint." Such plowing keeps the crop suitably thinned and helps control disease. Mint is machine-mowed and allowed to dry in the fields until moisture is approximately 35 per cent. Then it is distilled on the premises, usually in simple stills that steam the charge for about 45 minutes. The spent mint is composted or spread directly back on the land.

Mint is very responsive to daylength,

spreading from stolons only under short days, and not producing optimum essential oil yields until daylength approaches 16 hours of fairly intense sunlight. Hence it is not likely crop yield will prove competitive in the South, where summer days are not this long. The mints are susceptible to diseases that are quite difficult to control, especially *Verticillium alboatrum* (spearmint less so than peppermint). World production of unrefined peppermint oil is probably about 1,500 tons annually, chiefly from the United States and Soviet Russia, and scatteringly elsewhere.

OREGANO. Oregano is a savory herb especially cherished by Mediterranean peoples, and introduced into America by immigrants from that area. *Origanum vulgare*, Labiatae, is "officially" the source of oregano, but commercially it appears that the spice is derived from other plants too, such as *Lippia graveolens* of the Verbenaceae. Mexican oregano seems to be from the latter species, European oregano from *Origanum* (although just which species is uncertain, since the *O. vulgare* introduced into the United States is not so flavorful as the import from Europe). Scores of species in over a dozen genera have been referred to as "oregano" in the literature. Whatever species "true" oregano may be, it seems to have had a southern European origin, and was known as early as the Greek civilization.

Some other flavor oils

Allspice—known also as pimento, Jamaica pimento, and Jamaica pepper. The spice is the dried fruit of *Pimenta officinalis*, of the Myrtaceae, a small tree native to the American tropics. Production is mostly from Jamaica. The spice finds use chiefly as a food flavor, eugenol being the principal oil component.

Almond—oil of almond is obtained by expression or distillation from seeds of the almond tree, *Prunus amygdalus amara*, of the Rosaceae, native to Persia and cultivated today in the Mediterranean area and the southern United States. It is used as a food flavor.

Angelica—angelica oil is distilled from roots or seeds of the biennial *Angelica archangelica* (*Archangelica officinalis*), of the Umbelliferae, indigenous to northern Europe and Asia. It is used in confections, preparation of liqueurs, perfumery, and medicinals.

Anise—an early known aromatic, the fruit (seed) from *Pimpinella anisum*, of the Umbelliferae, a herbaceous plant native to the Mediterranean area, and cultivated in many parts of the world. It is used as a food flavoring. The oil, consisting chiefly of anethol, is obtained by distillation of the fruits, and finds use in medicinals, dentifrices, perfumery, and beverages.

Balm—also known as melissa. The oil is obtained by distillation of leaves and tops of *Melissa officinalis*, a perennial herb of the Labiatae, native to the Mediterranean area and cultivated there and in the United States. The leaves find use in food flavoring, the oil in beverages.

Capers—the unexpanded flower buds of *Capparis spinosa*, of the Capparidaceae, a low, trailing shrub growing in the Mediterranean area. Capers are principally used for flavoring pickles, relishes, and sauces.

Caraway—another seed spice of the Umbelliferae, from the biennial *Carum carvi*, native to Europe and cultivated there and in North America. It is used as a bread, biscuit, cake, and sausage flavor, and in Europe for cheese, kraut, and kummel (alcoholic cordial). Carvone is the principal constituent of the oil, a ketone also found abundantly in dill and spearmint.

Cardamon—the highly aromatic dried fruit and seed of *Elettaria cardamomum*, of the Zingiberaceae, native to India and cultivated there, in Ceylon, and in Central America. Cardamon is an ingredient of curry powder, and a seasoning in many types of sausage. It also finds use in incenses, perfumes, masticatories, and medicinals. Cardamon plants are tall, coarse, large-leaved perennial herbs typical of the ginger family, and are often planted between tea or rubber on Indian plantations. Fruit is harvested in autumn.

Cassia—another essential oil with an unfortunate popular name, since it is not of the genus *Cassia* but rather from the bark of *Cinnamomum cassia*, of the Lauraceae, native to southeast Asia. Other species of *Cinnamomum*, the cinnamon genus, yield Saigon cassia and Batavia cassia. Oils from all types consist largely of cinnamic aldehyde; and the pulverized bark finds wide use in bakery confectionery, and other food industries. The distilled oil finds limited use in perfumery.

Celery—celery spice is the seed of the biennial *Apium graveolens*, of the Umbelliferae, native to Europe and cultivated there and in the United States. The seed is available whole or ground, as a kitchen flavor for soups, sauces, and the like, and for compounding with salt to produce celery salt.

Coriander—spice obtained from whole or pulverized seeds of *Coriandrum sativum*, of the Umbelliferae, native to Europe; similar in culture and use to caraway.

Coumarin—also known as tonka bean. An easily and today almost universally synthesized essence, once obtained from species of tropical American *Dipteryx*, of the Leguminosae, and also obtained for perfumery from *Liatris odoratissima*, of the Compositae, of southern North America. It has served as a tobacco flavor and, as a substitute for vanilla.

Cumin—another seed spice of the Umbelliferae, from *Cuminum cyminum*, native to the Mediterranean area, and used much as are the similar seeds of caraway, dill, celery, and coriander.

Eucalyptus—oil obtained by distillation of foliage of several species of *Eucalyptus*, of the Myrtaceae, native to Australia and today cultivated in many parts of the world. Used as a flavoring, for dentifrices, as a medicinal, and for synthesis to menthol or other essences.

Fennel—a seed spice from *Foeniculum vulgare*, of the Umbelliferae, native to Europe and serving culinary use similar to that of coriander.

Fenugreek—the seed of *Trigonella foenumgraecum*, of the Leguminosae, native to southern Europe and cultivated in both the Old and New Worlds. It is used for imitation maple extract,

Garlic. These bulbs are a favorite flavor for meats and sauces. (Courtesy USDA.)

certain pickle and chutney spices, in cattle feeds, and as a medicinal. The oil is extracted by distillation.

Garlic—a favorite flavor for meats and sauces, obtained from fresh or dried and pulverized bulbs of *Allium sativum*, of the Liliaceae. Similar to onion, *A. cepa*, and other species of *Allium* (leek, chives, shallots).

Horseradish—this condiment is obtained by grating the fleshy root of *Rorippa armoracia*, of the Cruciferae, native to southern Europe and much grown in both Europe and North America. Its pungent taste is due largely to the glucoside sinigrin.

Melegueta pepper—seeds of wild plants of *Aframomum melegueta*, Zingiberaceae, of West Africa, have since ancient times been used locally as a pungent food flavor, and as a substitute for true pepper in Europe in times of scarcity.

Mint, Japanese—this mint, *Mentha arvensis piperascens*, of the Labiatae, has already been considered (as grown in Japan) under the discussion of peppermint. It is the usual commercial source of menthol, the distilled oil being up to 80 per cent menthol. The menthol is extracted by crystallization under refrigeration. Japanese mint has been cultivated in Brazil and our Pacific states, as well as in Japan.

Parsley—a plant of the Umbelliferae, *Petroselinum (Apium) sativum*, of which, however, the vegetative parts, rather than the seeds, serve as a condiment. The species is native to the Mediterranean area, but is widely cultivated in the United States, particularly in Louisiana.

Pepper, capsicum—cayenne pepper, red pepper, paprika, Tabasco, and chili powder are all derived from the fruits of species of *Capsicum* (*C. frutescens* and *C. annuum*), of the Solanaceae, native to the American tropics and subtropics, and cultivated today in many parts of the world. The seasoning principle of capsicum pepper is not an essential oil, although this type of pepper is thought of more as a spice than as a food.

Poppy—poppy seed, frequent in bakery goods, is derived from the opium poppy, *Papaver somniferum*, of the Papaveraceae, indigenous to Asia and cultivated in many parts of the world. It is another "spice" by association, the oil being fixed, not essential.

Sarsaparilla—a flavor, formerly much used in beverages and medicinals, obtained from roots of wild species of *Smilax*, of the Liliaceae, native to the American tropics and subtropics.

Sassafras—flavor widely used in carbonated beverages and dentifrices, obtained by distillation of bark and chipped wood of *Sassafras albidum*, of the Lauraceae, native to the east-central United States. Almost 500 tons are marketed annually.

Savory—seasoning obtained from foliage of *Satureia hortensis*, of the Labiatae, native to southern Europe and cultivated there and in the United States.

Sesame—seed of the annual, herbaceous *Sesamum indicum*, of the Pedaliaceae, indigenous to Asia and cultivated there. Sesame is used as is poppy seed, and like the latter has a fixed but not essential oil.

Star anise—this spice is obtained from the aggregate of follicles (fruit) of *Illicium verum*, of the Magnoliaceae, a small tree native to southeastern Asia and cultivated in China. It is used by confectioners and in cough drops.

Tarragon—tarragon spice consists of the dried leaves and flowering tops of *Artemisia dracun-*

Turmeric (*Curcuma longa*), 14 months old. Finger-like rhizomes are used in condiments and for dyeing. El Recreo, Nicaragua. (Courtesy USDA.)

culus, of the Compositae, used particularly in pickling mixtures. The oil has minor value in perfumery.

Turmeric—turmeric is useful as a yellow dye, as well as a kitchen flavoring. It is obtained from the rhizomes of *Curcuma longa*, of the Zingiberaceae, native to the southeastern Orient and cultivated in India and China. The washed rhizomes are dried and then pulverized, a basic ingredient of curry powders.

Wintergreen or *Gaultheria*—natural wintergreen can be distilled from foliage of *Gaultheria procumbens*, of the Ericaceae, or from bark of the sweet birch, *Betula lenta*, of the Betulaceae, both found in eastern North America. The essence, largely methyl salicylate, is readily synthesized. It finds use in beverages, confections, and medicinals.

Zedoary—a spice consisting of the dried, pulverized rhizome of *Curcuma zedoaria*, of the

Zingiberaceae, native to and principally cultivated in India. It finds use in the manufacture of liqueurs, bitters, medicinals, and curry.

Essential Oils as Medicinals and in Industry

Although the chief uses of essential oils are for perfumery and flavoring, these by no means exhaust the ways in which essences serve man. A few additional uses, in medicinals and as an industrial raw material, are listed here.

Relatively few drugs consist entirely of essential oils. Medicinals containing aromatics were first regarded with awe and superstition, gaining an undeserved reputation. In a latter age of skepticism they were abandoned for newer drugs. In passing along the traditions of a remedy, our ancestors relied upon experienced fact as well as upon fancy. Medical science, once taught to scoff at the remedies of the ancients, now often finds traces of powerful and often unknown components in the drugs of the past. Perhaps, given time, man will decipher and utilize the secrets of little-known aromatics as he has those of antibiotics. Some volatile oils are powerful external and internal antiseptics, and others sedatives, stimulants, and stomachics. Some possess analgesic properties; others are haemolytic or antizymatic. Some essential oils, such as wormseed, are among the best known anthelminthics.

Essential oils have preservative value, too. Several essences are toxic to bacilli, and oils of oregano, cinnamon, clove, camphor, and garlic are germicidal. Thymol, carvacol, mustard oil, and clove oil are fungicidal to a degree, and some are synergistic (the fungicidal action of thyme is prolonged for years by the addition of terpineol, pine oil, and benzoin). It is not strange, then, that essential oils have played a part in embalming, extending back to Egyptian embalming over 5,000 years ago. Herodotus describes an "infusion of pounded aromatics" for preparing an eviscerated body, which was filled with crushed myrrh, cassia, frankincense, and so on; cedar oil was introduced into the bowel. Aromatic wood chips accompanied wrapping for mummification.

Nor have essences failed man on the industrial front. One of the first plastics, celluloid, was compounded of camphor. Many products with a disagreeable odor become fully acceptable and agreeable when treated with essential oils. Essences have been incorporated in synthetic rubbers, thus opening new and profitable manufacturing fields. Turpentine has been widely used as a solvent and raw material. In addition to these uses, essential oils are incorporated in small quantities in glues and other adhesives, animal feeds, automobile finishing supplies, insecticides and repellents, furniture polishes, janitor supplies, paints, paper and printing inks, petroleum and chemical products, textile processing materials, veterinary supplies, and many other accompaniments of modern living.

Moreover, one cannot disregard the economic significance of essential oil procurement, processing and marketing involving as it does a multitude of human beings, often in remote corners of the earth where products for foreign exchange are badly needed. The United States imports most of its essential oils, a stimulus to world trade with less industrialized countries. In all these ways essential oils influence man's daily life, to a degree out of proportion to the comparatively minor position essences hold in export–import volume.

Examples of medicinal and industrial oils. CAMPHOR. Camphor was one of the familiar essences of the ancients; it is recorded as having been in use in the Western world as early as the sixth century B.C., and it was doubtless utilized in the Orient long before that. The chief natural source is *Cinnamomum camphora*, of the Lauraceae, native to

Taiwan, China, and Japan. But the camphor molecule is not uncommon and has also been obtained from plants of the Dipterocarpaceae (*Dryabalanops*, from Borneo), the Labiatae (*Ocimum kilimandscharicum*, from Africa), the Compositae, and even from animal sources. As early as 1903 camphor had also been synthesized, and the synthetic form, usually prepared by oxidation of borneol, is today abundantly utilized. Camphor, unlike most essences, remains a white, crystalline solid at ordinary temperatures.

The common camphor tree is a large forest plant with alternate ovate leaves, small white flowers, and red drupaceous fruits. The old oriental method of extraction involved felling the trees, chipping the wood, and crudely distilling the chips. The wasteful destruction of trees once threatened the industry, a not uncommon happening in wild-plant exploitation. Reportedly as much as 3 tons of crude camphor have been derived from a single tree, the wood yielding as much as 5 per cent of the essence by weight. The crude product is further refined (typically by sublimation with quicklime and charcoal) at commercial or export centers. Taiwan was long the leading producer, at one time marketing thousands of tons annually. Camphor is still used in cold remedies and other medicinals, liniments, perfumery, insecticides, and formerly much in manufacture of nitrocellulose derivatives such as celluloid.

TURPENTINE. The naval stores industry has already received discussion in a previous chapter. This is merely to point out that an essential oil of no little importance, turpentine, as well as resin, is derived from distillation and processing of various species of *Pinus*, Pinaceae. In the United States only about 84,000 barrels of turpentine were produced by the traditional chipping of trees in 1967, about twice this amount by steam distillation of old wood and stumps, but about 427,000 barrels as a by-product from sulfate paper pulping. Almost all of this was used industrially, primarily for synthetic pine oil (used in insecticides) and polyterpene resins (much used for pressure-sensitive tapes).

CITRUS WASTES. Various industrial oils are obtained from citrus residues left after juice extraction from oranges, grapefruits, and other *Citrus* species of the Rutaceae. "Citrus stripper oil" is a mixture of essential oils, a by-product from processing the peels of oranges and grapefruits for molasses; it is 95 per cent D-limonene, a monocyclic terpene used as a raw material in making of fine organic chemicals. Synthetic spear-

Table 11-2

Composition of Citrus Fruit

Physical Component	Percentage	Chemical Component	Percentage
Juice	40–45	Water	86–92
Flavedo (outer peel)	8–10	Sugars	5–8
Albedo (inner peel)	15–30	Pectin	1–2
Rag and pulp	20–30	Glucoside	0.1–1.5
Seeds	0–4	Pentosans	0.8–1.2
		Acids (citric mostly)	0.7–1.5
		Fiber	0.6–0.9
		Protein	0.6–0.8
		Fat	0.2–0.5
		Essential oil	0.2–0.5
		Minerals (K, Mg, Ca, P, etc.)	0.5–0.9

From J. W. Kesterson and R. Hendrickson. *Economic Botany* **12**, 164 (1958).

mint oil, largely L-carvone, can be derived from stripper oil, which also finds use in paints, varnishes, plastics, and soap perfumes. "Citrus peel oil" is recovered from cannery refuse either by expression or distillation and is used for flavoring and scenting various products. The composition of citrus fruit is given in Table 11-2. While essential oil is a relatively modest component of the fruit, its total industrial potential becomes clear when it is realized that about 70 per cent of the citrus crop is processed, involving nearly 100 million boxes of fruit.

LONG PEPPER. Although the long pepper, *Piper longum*, Piperaceae, is of the same genus as condiment pepper, it is used for medicinal purposes rather than as a spice. Long pepper is of ancient usage in India by Hindu physicians. It was also recognized by the Greeks and Romans. It has been used for ills ranging from respiratory ailments to epilepsy, leprosy, and even snakebite. The long pepper fruits contain about 1 per cent volatile oil, as well as the alkaloid piperine and other ingredients.

The plant is a trailing, perennial herb with cylindrical seed spikes. It grows in the subtropical regions of India, both wild and cultivated. The fruit is gathered when mature but still green. Laborers can pick about 5 pounds in a working day. The fruits are sun-dried at collection centers before marketing. Roots and rhizomes of older plants are also collected for chipping into small sections sold as "piplamool."

WORMSEED. The anthelminthic volatile oil from American wormseed, *Chenopodium ambrosioides anthelminticum*, has been produced from this cultivated plant without interruption for more than a century in Maryland. The species is native to the American tropics and subtropics. Wormseed oil is distilled from the fruiting tops of the plant, the seed coats yielding the highest percentage of oil. Yields of up to 40 pounds of oil per acre are reported. The crop is handled much as is tobacco, being started in seedbeds and set in the field in rows. Harvest is by combine.

The oil has a single use—as a vermifuge. Its effectiveness in control of intestinal parasites seems to depend on its ascaridole content, which may vary from year to year or with variation in harvest time, rate of distillation, temperature of distillate, and other factors.

BAY RUM. Bay rum is obtained by distillation of the foliage of *Pimenta racemosa* (*P. acris*), of the Myrtaceae, native to the West Indies, where the bay rum industry centers. The bay rum lotions found in barber shops are soothing to the skin and presumably have some healing qualities, owing to bactericidal action by the phenols present. It is marketed as an alcoholic solution of about 1 per cent bay oil, and received the name bay "rum" from the former practice of distilling bay in rum and water.

The trees can be grown on rocky hillsides unsuited to agriculture, and harvest can await slack times in other agricultural pursuits. The trees retain their dark green, glossy foliage throughout the year. In the wild they may grow to 60 feet in height, but where utilized for bay oil, they are continuously cut back and seldom allowed to attain heights of more than 15 feet. Plants flower in March and April, and are usually propagated by seeds.

At harvest, leaves alone may be stripped from the tree by hand, or the smaller branches may be cut and removed with the leaves. The leaves are held in bins for about 3 days, then crushed and bruised in a chopper located above the distilling vat. Chopped leaves fall into the vat and are distilled for about 5 hours. Bay oil is drawn off from the condensation tank.

Some other essential oils for medicinal or industrial use

Buchu—the oil is distilled from foliage of several species of *Barosma*, of the Rutaceae, indigenous to South Africa. It is used as an antiseptic, stimulant, and tonic.

Cajeput—essence distilled from foliage of *Melaleuca leucodendron*, of the Myrtaceae, a small tree of Southeast Asia. It is widely used as a local remedy in the Orient, and is exported to some extent to Europe.

Cedarwood—the oil is obtained by distillation of wood of *Juniperus virginiana*, of the Cupressaceae, native to the eastern United States. The oil is cheap and finds extensive use in the soap industry, as a moth repellent, deodorant, component of polishes, and as the basis for several perfumes.

Chamomile—an old-time remedy obtained by infusion of the dried flower heads of *Matricaria chamomilla*, of the Compositae, native to Eurasia and grown there and in the United States. Reputedly a tonic and stimulant.

Cubeb—oil obtained by distillation of the dry, unripened fruits of *Piper cubeba*, of the Piperaceae, a clambering shrub of the East Indies, much used in catarrh remedies and in soaps. The dried fruits have also been employed as spices.

Hoarhound—a favorite domestic remedy in early America, prepared by infusing the dried tops of *Marrubium vulgare*, of the Labiatae, native to Eurasia but naturalized in North America. Hoarhound candy and lozenges are used for treatment of colds, dyspepsia, and other ailments.

Pennyroyal—obtained from the very aromatic *Hedeoma pulegioides*, of the Labiatae, native to the eastern United States. It finds use in medicinals, but because the plant is small and is collected with difficulty, it is a negligible item in commerce.

Santonin—santonin oil, from distillation of the flower heads of *Artemisia cina*, of the Compositae, native to western Asia, serves as an anthelminthic.

Tansy—tansy oil, from leaves and tops of *Tanacetum vulgare*, of the Compositae, is distilled to a limited extent in the north-central United States for use in medicinals.

Tea tree—tea tree oil, from distillation of foliage of *Melaleuca alternifolia*, of the Myrtaceae, growing in Australia, is said to have high germicidal properties. It is used in medicated soaps, dentifrices, and certain medicinals.

Witch hazel—the volatile oil from distillation of twigs of *Hamamelis virginiana*, of the Hamamelidaceae, common in the eastern United States, is used in external medicine.

Wormwood—wormwood oil, also known as absinthe, is obtained by distillation of leaves and tops of the perennial herb *Artemisia absynthium*, of the Compositae, native to the Mediterranean area and cutivated in the United States, particularly in Michigan and Oregon. Principal uses are for liniments and the liqueur absinthe.

Summary of the Essential Oil Industry

It is evident that there is no dearth of essential oils in the world; they can be found in many species and many plant parts. To some extent they can be chemically altered to substitute one for the other. Inexpensive sources of essential oil, such as by-products of other industries, will certainly find use of some kind in this technological age. However, the distinctiveness of most essential oils, the composition of which relates to a particular species, almost assures that the industry will not shift wholly to mass production of only a few essences. Rather, essential oils should remain a rich field for specialty products. For many the demand will be small, but the product so valued that higher costs easily can be sustained (e.g., essences, where minute traces suffice for flavor or aroma). It is pleasant to note a modern-day field in which there is still a place for small-scale, entrepreneurial activities, even sometimes involving wild plants. In the field as a whole, ample room still exists for discovery and investigation of essential oils, and the essential oil industry is not likely to be swamped with synthetic substances derived from fossil fuels and manufactured by massive technological operations.

SUGGESTED SUPPLEMENTARY REFERENCES

American Spice Trade Association, *A Treasury of Spices*, 5th ed., N. Y., 1956.

Arctander, S., *Perfumes and Flavor Materials of Natural Origin*, AVI Pub. Co., Westport, Conn., 1960 (two parts).

Brooklyn Botanic Gardens Plants and Gardens, *Handbook on Herbs*, Summer, 1958; *Japanese Herbs and Their Uses*, Summer, 1968.

Burgess, A. H., *Hops*, Leonard Hill, London, Interscience, N. Y., 1965.

Correll, D. S., "Vanilla—Its Botany, History, Cultivation and Economic Import," *Economic Botany* **7**, 291–358 (1953); contains a bibliography of 66 items plus 10 pages of "Additional Literature."

Guenther, Ernest, *The Essential Oils*, Van Nostrand, Princeton, N. J., Vols. 1–6, 1948–52.

Landing, J. E., *America Essence, A History of the Peppermint and Spearmint Industry in the United States*, Kalamazoo Pub. Mus., Kalamazoo, Mich., 1969.

Merory, Joseph, *Food Flavorings*, Chemical Pub. Co., N. Y., 1968.

Parry, J. W., *The Story of Spices*, Chemical Pub. Co., N. Y., 1968; Vol. II, *Their Morphology, Histology and Chemistry*, 1969.

——————, *Spices*, Chemical Pub. Co., N. Y., 1962.

Poucher, W. A., *Perfumes, Cosmetics and Soaps*, Van Nostrand, Princeton, N. J., 1942.

Rosengarten, F., *The Book of Spices*, Livingston Pub. Co., Wynnewood, Penn., 1969.

Verrill, A. H., *Perfumes and Spices*, L. C. Page and Co., Boston, 1940.

CHAPTER 12

Biodynamic Plants: Medicinals, Insecticides, Growth Regulants, Tobacco, Etc.

MEDICINALS, INSECTICIDES, AND HERBICIDES derived from plants are like organic dyes and essences in that, while they have repeatedly been of great historical importance, they are often today superseded by synthetic or newer counterparts. There is no dearth of drug plants, real or imaginary, as is revealed by even the small sampling of Fig. 12-1 and Tables 12-1 and 12-2. In fact there is scarcely any species which does not con-

tain minor components that might have physiological effects under certain conditions. But most of the vague and mystic remedies of the ancients and the herbalists were discarded as scientific experimentation developed. The drugs most frequently found in the medicine cabinet a few generations ago have been displaced to a great extent by synthetics such as the sulfa drugs first, and then the antibiotics derived from lower

Figure 12-1   Typical alkaloids. (After Leete, *Science* **147**, 1001 (1965).)

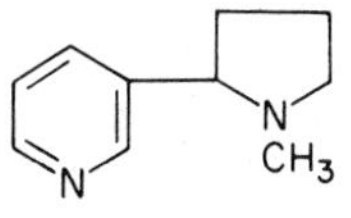

Nicotine (in Tobacco)	Strychnine (in Strychnos species)	Nocardamine (in Actinomyces buchanan)	Gliotoxin (in Trichoderma viride)
Codeine (in Opium poppies)	Quinine (in Cinchona bark)	Lysergic acid amide (in Rivea corymbosa)	Samandarine (in Salamander)

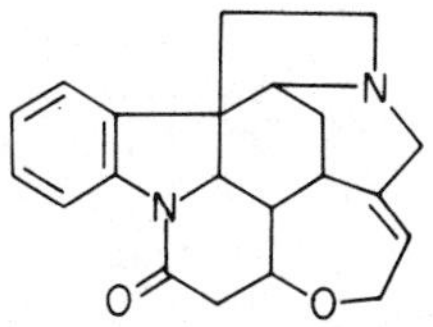

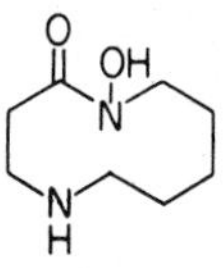

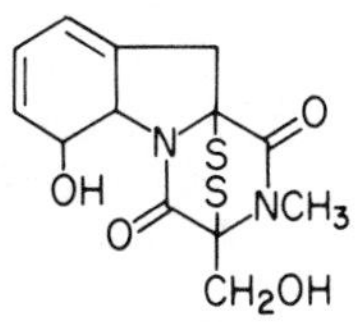

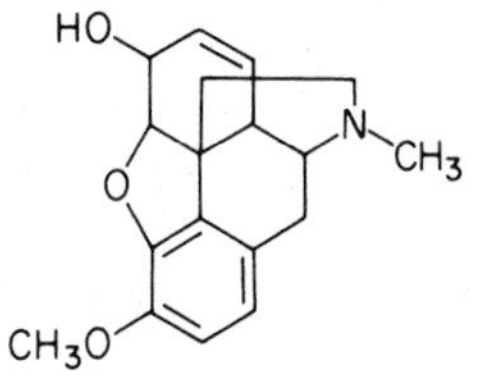

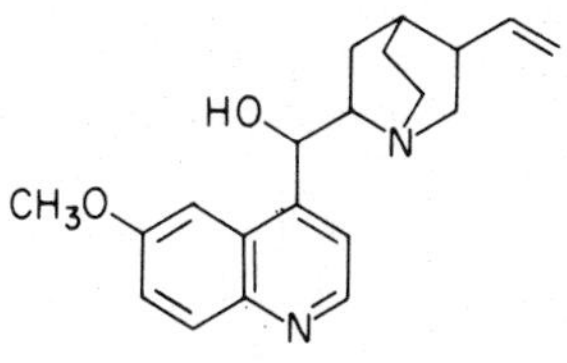

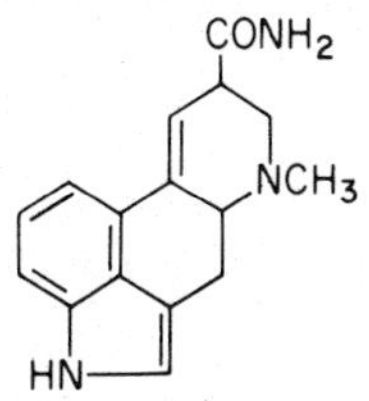

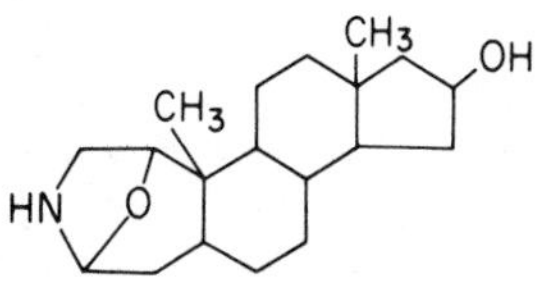

Table 12-1

Species of the Appalachian Mountains, U.S., Still Gathered from the Wild for Sale as Medicinals

1. *Acer spicatum* Lam.
2. *Achillea millefolium* L.
3. *Acorus calamus* L.
4. *Adiantum capillus-veneris* L.
5. *Adiantum pedatum* L.
6. *Aesculus hippocastanum* L.
7. *Aletris farinosa* L.
8. *Alnus serrulata* (Ait.) Willd.
9. *Amaranthus hybridus* L.
10. *Angelica atropurpurea* L.
11. *Aplectrum hyemale* (Muhl.) Torr.
12. *Apocynum androsaemifolium* L.
13. *Apocynum cannabinum* L.
14. *Aralia nudicaulis* L.
15. *Aralia racemosa* L.
16. *Arctium minus* (Hill) Bernh.
17. *Arctium* spp. L.
18. *Arisaema triphyllum* (L.) Schott.
19. *Aristolochia serpentaria* L
20. *Asarum canadense* L.
21. *Asclepias syriaca* L.
22. *Asclepias tuberosa* L.
23. *Baptisia tinctoria* (L.) R. Br.
24. *Berberis vulgaris* L.
25. *Betula lenta* L.
26. *Caulophyllum thalictroides* (L.) Michx.
27. *Ceanothus americanus* L.
28. *Chamaelirium luteum* (L.) Gray
29. *Chelone glabra* L.
30. *Chenopodium ambrosioides* L.
31. *Chimaphila umbellata* (L.) Bart.
32. *Chionanthus virginicus* L.
33. *Cimicifuga americana* Michx.
34. *Cimicifuga racemosa* (L.) Nutt.
35. *Cnicus benedictus* L.
36. *Collinsonia canadensis* L.
37. *Comptonia peregrina* (L.) Coult.
38. *Corallorhiza* spp.
39. *Cypripedium calceolus* L.
40. *Datura stramonium* L.
41. *Dioscorea villosa* L.
42. *Echinacea purpurea* (L.) Moench.
43. *Eryngium aquaticum* L.
44. *Euonymus atropurpureus* Jacq.
45. *Eupatorium perfoliatum* L.
46. *Eupatorium purpureum* L.
47. *Fragaria virginiana* Duch.
48. *Fraxinus americana* L.
49. *Galium aparine* L.
50. *Gaultheria procumbens* L.
51. *Gaylussacia frondosa* (L.) T & G
52. *Gelsemium sempervirens* (L.) Ait.
53. *Gentiana villosa* L.
54. *Geranium maculatum* L.
55. *Hamamelis virginiana* L.
56. *Hedeoma pulegioides* (L.) Pers.
57. *Hepatica acutiloba* DC.
58. *Hydrangea arborescens* L.
59. *Hydrastis canadensis* L.
60. *Jeffersonia diphylla* (L.) Pers.
61. *Juglans cinerea* L.
62. *Juglans nigra* L.
63. *Juniperus communis* L.
64. *Juniperus virginiana* L.
65. *Lactuca scariola* L.
66. *Leonurus cardiaca* L.
67. *Lindera benzoin* (L.) Blume
68. *Liquidambar styraciflua* L.
69. *Lobelia inflata* L.
70. *Lycopus virginicus* L.
71. *Marrubium vulgare* L.
72. *Menispermum canadense* L.
73. *Mentha piperita* L.
74. *Mentha spicata* L.
75. *Mitchella repens* L.
76. *Monarda didyma* L.
77. *Myrica cerifera* L.
78. *Nepeta cataria* L.
79. *Panax quinquefolium* L.
80. *Passiflora incarnata* L.
81. *Phytolacca americana* L.
82. *Pinus palustris*
83. *Pinus strobus* L.
84. *Plantago* spp.
85. *Podophyllum peltatum* L.
86. *Polygala senega* L.
87. *Polygonatum biflorum* (Walt.) Ell.
88. *Polygonum hydropiper* L.
89. *Populus tacamahacca* Mill.
90. *Prunella vulgaris* L.
91. *Prunus serotina* Ehrh.
92. *Pyrus americana* (Marsh) DC.
93. *Quercus alba* L.
94. *Radicula nasturitum-aquaticum* L.
95. *Rhus glabra* L.
96. *Rubus* spp. L.
97. *Rumex crispus* L.
98. *Sanguinaria canadensis* L.
99. *Salix alba* L.
100. *Salix nigra* Marsh.
101. *Salvia officinalis* L.
102. *Sassafras albidum* (Nutt.) Nees

Table 12-1 (cont'd)

Species of the Appalachian Mountains, U.S., Still Gathered from the Wild for Sale as Medicinals

103. *Scrophularia marilandica* L.	115. *Trillium erectum* L.
104. *Scutellaria lateriflora* L.	116. *Tsuga canadensis* (L.) Carr.
105. *Senecio aureus* L.	117. *Ulmus rubra* Muhl.
106. *Solanum carolinense* L.	118. *Veratrum viride* Ait.
107. *Spigelia marilandica* L.	119. *Verbascum thapsus* L.
108. *Stellaria media* (L.) Cyrill.	120. *Verbena hastata* L.
109. *Stillingia sylvatica* L.	121. *Veronicastrum virginicum* (L.) Farwell
110. *Tanacetum vulgare* L.	122. *Viburnum nudum* L.
111. *Tephrosia virginiana* (L.) Pers.	123. *Viburnum prunifolium* L.
112. *Tiarella cordifolia* L.	124. *Xanthorhiza simplissima* Marsh.
113. *Trifolium pratense* L.	125. *Xanthoxylum americanum* Mill.
114. *Trilisa odoratissima* (Walt) Cass	

From A. Krochmal, *Economic Botany* **22**, 333 (1968).

plants. The fame of aromatics, pyrethrum, and rotenone for insecticidal use was dimmed by DDT and subsequent chemical creations, although a revival of interest comes as DDT and other "hard" pesticides fall into disrepute for their toxification of the environment. Plant hormone research led to the renowned synthetic 2-4,D and its kin. Tobacco, first used as a medicinal, has become more popular for smoking. Medicinals are among the first plants to which man gave particular attention, and in medieval times their study constituted almost alone the science of botany. Insecticides received large-scale attention only after settlement of the world's lands and the perfecting of agriculture. Growth regulants did not become a tool for human use until almost the middle of the twentieth century, nor were hallucinogens much in the news until the hippie phenomenon of the 1960's.

Chemically, medicinals, tobacco, insecticides, and growth regulants from plants depend for their effects upon a variety of extractives. These are in some cases glucosides, in other cases alkaloids, and in many cases of varied compounds some still undeciphered. Many of the essential oils, dyes, latices, and even tannins and vegetable oils mentioned in other chapters are widely used as medicinals or occasionally as ingredients in insecticides. For example, the spice nutmeg contains hallucinogenic phenylpropenes including myristicin, and fiber-yielding *Agaves* steroidal sapogenins. Biodynamic substances are frequently products of living cells, al-

Searching for physiologically active plant products. Here divers are gathering algae for specific alginates from which to extract biologically active compounds. The search goes on in sea and on land. (Courtesy Okeanos, Inc.)

An alkaloid extracted from flowers (*left*) and developing fruit (*center* and *right*) of the Chinese tree, *Camptotheca acuminata,* Cornaceae, yields camptothecin, a material that has shown anti-tumor activity in tests on laboratory animals implanted with an experimental type of leukemia. (Courtesy USDA.)

though perhaps "wastes" or intermediate metabolites and not an integral part of the protoplasm. Many have no obvious utility to the plant.

By and large, botanicals for the drug and insecticide industries have been so diverse, so ephemeral, or so limited in use that widespread cultivation has seldom resulted. Even in the twentieth century many plants such as ipecac and ginseng come largely from wild sources. In World War II large supplies of quinine were again taken from the wild. Yet, as will become apparent in discussing the important medicinal and insecticidal plants, cultivation is coming more to be practiced. The day of the wild-herb hunter

has largely passed. And in the case of antibiotics, the entire culture of molds and extraction of the drug are elaborate, efficiently controlled laboratory procedures.

Plant sources of drugs and insecticides are sought in all corners of the globe, in all climates. They come from trees, shrubs, and herbs, and from primitive plants that are not even one of these. They are made from fruit, flower, leaf, stem, or root, into solids, liquids, infusions, or dusts. Most are gathered by hand and locally dried. Some must undergo preliminary treatment or curing at their source before being shipped to consuming centers. Seldom is any part of the world self-sufficient as regards drugs. The United

States, a country of high labor costs, is particularly dependent upon imports for the majority of botanicals used for medicine and for insecticides. About 7,000 tons of medicinal plant materials were imported in 1965, mostly from India. These are for the most part imported in crude form (milled drugs pay special duty) and are processed by skilled labor using machinery that is usually not available in the producing areas. Grading and testing for adulteration control quality. Practically all drugs require some milling before they are marketed—such as sifting, reconditioning, grinding, and blending. Extracts may be made by maceration, percolation, filtration, and evaporation, utilizing water, alcohol, or some other solvent. Concentrates may be made by precipitating specific components out of "a mixed solution." Several dissolutions and precipitations or crystallizations may be necessary to attain the desired degree of purification. Other extraction procedures (such as distillation, fractionation, and so on) are the same as for essential oils.

Medicinals

No one knows how man first came to use plant materials for his ailments thousands of years ago. Primitive man did not understand disease, viewing it as a malevolent influence of the gods, or due to supernatural spells of one sort or another. Disease was thus of a magico-religious nature. Perhaps some of his experimental foods had unexpected physiological effects that ameliorated the disease. Plant substance with a bitter taste that might cause vomiting or defecation, or cause hallucinations, could be interpreted as a means for expelling the demons responsible for the ill. Whatever the beginning, the interest and progression in curative plant products has been relentless, until today the manufacture and marketing of medicinals is a major industry. In the United States alone about $3 billion were being spent annually for pharmaceuticals in the late 1960's.

The history of drugs is largely the early history of botany. More than 4,000 years ago a few hundred drug plants were known to the Assyrians, and only slightly later Egyptian records indicate established preparation and usage of medicines. A few centuries before Christ, the Greeks gave a disproportionate amount of attention to medicinals. About 75 B.C. the famed Dioscorides discussed in detail several thousand botanicals in his outstanding *De Materia Medica*—a book destined to become the authoritative reference concerning medicinal plants for the next fifteen centuries. In the Middle Ages little new was learned, but the advent

Harvesting a giant tree lily from a forest in Ethiopia in the continuing search for biologically active extractives. (Courtesy USDA.)

of the herbalists in the fifteenth to seventeenth centuries heralded the beginning of modern botany. The herbals, magnificent tomes compounded of artistry, superstition, and the beginnings of observational science, dealt mostly with medicinal plants. The "Doctrine of Signatures" assumed that all plants were created for man's use and were endowed with certain forms and shapes that marked them for a specific use in treating similarly shaped organs in the human body. The walnut, for example, was a brain tonic; the bloodroot, a blood tonic. But with the advance of knowledge the superstition connected with most reputed medicinals was revealed, and gradually most of the remedies of the old apothecary shop fell into disrepute. Very few medicinal species have remained in favor for even so much as a century—only the giant drugs, such as quinine, opium, and belladonna. But strangely enough modern medicine now and then discovers in an ancient remedy unusual chemical compounds that exert unusual influence upon human physiology. Perhaps the lore of centuries past did not contain as much superstition as it has sometimes seemed. Table 12-1 lists 125 species still collected from the wild in Appalachia and offered through drug houses.

Botanical drugs may be classified according to the plants from which they are derived, the disease for which they are used, or their chemical nature. Perhaps the last has the most to commend it. In the following pages a few outstanding drugs of three or four chemical categories will be discussed and some of the world's commercial drug plants will be listed. Antibiotics, the newest and most glamorous of plant drug types, will be first discussed. The key ring structure of penicillin, the "type" antibiotic, contains a C—C—N—C—S linkage. All of the early antibiotics and most of those effective today are peptide derivatives.

Next to be discussed will be a few alkaloidal drugs, a category of high importance including quinine, opium, cocaine, strychnine, and belladonna. Great diversity exists among the alkaloids, more a categorization of convenience than of chemical relationships. Almost inevitably they taste bitter. The typical alkaloid contains an heterocyclic C—C—N—C—C ring, with nitrogen as an amino group. Theoretically alkaloids would seem derivable from certain amino acids. The ergot alkaloids from *Claviceps*, like many from the morning-glory family, appear to be derivatives of lysergic acid, and have hallucinogenic characteristics. The empirical formula for quinine is $C_{20}H_{24}N_2O_2$, for nicotine $C_{10}H_{14}N_2$, for strychnine $C_{21}H_{22}N_2O_2$. Figure 12-1 gives structural formulae for several alkaloids. The biological role of alkaloids is not at all clear. At least in some cases they appear to be in a dynamic state rather than the end product of metabolism. Around 3,000 have been identified in the plant kingdom, and it is estimated that tens of thousands more remain to be discovered. Certainly there would seem to be potential for additional important medicinals in this group. Morphine was the first alkaloid isolated in crystal form, over a century-and-a-half ago. Quinine was synthesized in 1944.

Among glycoside drugs, digitalis is perhaps most famous. In the glycoside (a type of glucoside) molecule there occurs a sugar (usually glucose) linkage. Other useful glycosides are obtained from *Dioscorea*, and rutin (for treating hemorrhagic disorders) from a variety of plants including buckwheat and *Eucalyptus*. Marijuana was once thought to be alkaloidal, but more recently has been found to derive its effects from several constituents, especially tetrahydrocannabinol. Drugs with the chemical nature of essentials oils have already been discussed in a previous chapter, and will not be elaborated upon here. Caffeine-producing plants, sometimes used as medicinals or stimulants, largely will be covered in a later chapter devoted to beverage plants.

Table 12-2

Major Hallucinogenic Plants and Their Active Principles

Plant	Family	Active Principle(s)
Cannabis sativa	Cannabinaceae	Tetrahydrocannabinol
Lophophora williamsii	Cactaceae	Mescaline
Piptadenia species	Leguminosae	Substituted tryptamines
Mimosa species	Leguminosae	Substituted tryptamines
Virola species	Myristicaceae	Substituted tryptamines
Banistereopsis species	Malpighiaceae	Harmaline, harmine
Peganum harmala	Zygophyllaceae	Harmaline, harmine
Tabernanthe iboga	Apocynaceae	Ibogaine
Ipomoea violacea	Convolvulaceae	d-Lysergic acid amide d-Isolysergic acid amide
Rivea corymbosa	Convolvulaceae	d-Lysergic acid amide d-Isolysergic acid amide
Datura species	Solanaceae	Scopolamine
Methysticodendron amesianum	Solanaceae	Scopolamine
Amanita muscaria	Agaricaceae	Pantherine, ibotenic acid
Psilocybe mexicana	Agaricaceae	Psilocybin

From N. R. Farnsworth, *Science* **162**, 1090 (1968).

Hallucinogens. The hallucinogenic or psychedelic ("mind-expanding") drugs have especially left their mark on the twentieth century, through widespread use by students and fringe groups beginning in the 1960's. Some are listed in Table 12-2. The active principles of such familiar types as marijuana have been known for some time. Mescaline and psilocybin have also been used by primitive societies for many decades, although not so intensively investigated. Only with synthesis and ready availability of LSD (lysergic acid diethylamide) did psychedelic use become widespread, with organized "trip" sessions fairly commonplace. If there are social problems arising from use of the hallucinogens, so also is there promise of medical advancement from their constituents, such as for the treatment of schizophrenia. Public concern in the United States about this group of drugs seems to have resulted in overreaction, especially with regard to marijuana. Hallucinogens have been used with little or no harm to the users in various societies since time immemorial, and there would seem no justification for the harsh penalties fixed by law against possession or sale of marijuana. Indeed, impartial research suggests that the hallucinogens are less habituating and less deleterious than is the smoking of tobacco, and less a social problem than is the drinking of alcohol. This is not to say that overindulgence in any drug may not prove harmful, but there is no evidence that the hallucinogens are addictive as is heroin or as habituating as nicotine. The difference between almost any drug being a cure (or a comfort) and being poisonous is quantitative.

There are many interesting cases of the use of hallucinogens by primitive people. The witch doctors of certain tribes in Siberia employed narcotics from *Amanita muscaria* to induce fits of "madness." Schultes comments ["Hallucinogens of Plant Origin," *Science* **163**, 246 (Jan. 17, 1969)]:

Many of these peoples have discovered that the intoxicating principles are excreted unaltered in the urine, almost as hallucinogenic as the original plant material. This discovery has given rise to the custom, sometimes ritually executed, of the inebriate's drinking his own or another's urine when he feels the intoxication waning, thus repeatedly effecting a continuation of narcosis.

In North America certain Indians of the Southwest traditionally use peyote to achieve ceremonial hallucinations; the Native American Church has maintained a continuing fight with government drug authorities for perpetuation of this custom in the United States on the grounds of religious freedom.

There are many additional cases throughout the world, but by way of example Michael Harner's experience with the Jivaro Indians in the Equadorian Amazon is illuminating ["The Sound of Rushing Water," *Natural History* (June–July, 1968)]:

> These shamans . . . take an hallucinogenic drink . . . in order to enter the supernatural world . . . prepared from segments of the vine *Banisteriopsis* . . . . The Jivaro boil it with the leaves of a similar vine . . . to produce a tea that contains the powerful alkaloids harmaline, harmine, tetrahydroharmine, and quite possibly dimethyeltryptamine . . . . These compounds have chemical structures and effects similar to LSD, mescaline, [etc.] . . . . I did not fully appreciate the psychological impact . . . upon the native view of reality, but in 1961 I had occasion to drink the hallucinogen with another Upper Amazon Basin tribe. For several hours after drinking the brew, I found myself, although awake, in a world literally beyond my wildest dreams.

Foliage of *Psychotria spp.* is simmered with *Banisteriopsis* sections to make uascá, an hallucinogenic beverage used for spirit-worship among the Auaris in Acré and other sections of western Amazonia, giving much the same response noted by Harner. The Sànama Indians of Amazonia find *Banisteriopsis* rare, and depend more upon *Virola* resin, which is taken as snuff. *Virola* bark is heated over fire until the resin oozes; it is scraped off with arrowheads, later pulverized, and usually administered into the nostrils by a companion through a small blowpipe. It is especially used by shamans, and generally at funeral ceremonies lasting several days.

Hallucinogens are found in a number of plant families, as is apparent from Table 12-2. These range from the primitive fungus family, the Agaricaceae, through the Moraceae, Aizoaceae, Myristicaceae, Leguminosae, Zygophyllaceae, Malpighiaceae, Cactaceae, Apocynaceae, Convolvulaceae, Labiatae, and Solanaceae, to the Compositae. Very few monocots yield hallucinogens, although there are stories of *Kaempferia galanga*, Zingiberaceae, being so used in New Guinea. How the hallucinogens in these diverse groups function is not entirely clear, but may involve substances that chemically transmit impulses across synapses between neurons. For a fuller review than is possible here of the newly prominent hallucinogen field, the Suggested Supplementary Readings at the end of this chapter may be consulted.

Antibiotics. Antibiosis, the inhibitory effect of the excretions of one organism on another, is commonplace in nature. Part of the survival system of desert plants, for example, is to prevent other plants or its own seedlings from growing and competing for water over the root system of an established plant. Inhibitory substances from root and foliage accomplish this. Likewise with many higher plants, there is often subtle repression of crop by weed, or of one component of a blended planting by another. It has been recognized for many years that juglone from the black walnut, for example, inhibits growth of species such as peach within the main root area of the walnut. But in pharmacology, "antibiotic" has come to refer almost exclusively to substances obtained from primitive plants of the Thallophyta, which are highly toxic to other microorganisms but seldom damaging (in proper dosage) to higher animals. As noted, most of the effective antibiotics so far isolated are peptides. In a chapter discussing medicinals, certainly these should be mentioned,

In these huge fermentation tanks antibiotics and other fermentation products are mass-produced. The tanks, filled with nutrient media, are inoculated with microorganisms grown first in smaller "seed" tanks. After fermentation is completed, the contents of the tanks will be filtered and processed to obtain the final product. (Courtesy Chas. Pfizer & Co., Inc.)

even though there will be further discussion of useful products from microorganisms in Chapter 21. The discovery of modern antibiotics is certainly a landmark in botanical science.

Knowledge of molds as healers is not entirely new, for the Chinese in 2000 B.C. are said to have applied green mold to relieve festering ulcers. In 1877, moreover, Pasteur and Jaubert observed that cultures of anthrax bacilli could be destroyed by contaminating organisms. But not until 1928, in a British hospital, when Sir Alexander Fleming noted that a green *Penicillium* mold contaminating his bacterial cultures caused lysis (dissolving and destruction) of the disease-causing bacteria, was one of the most spectacular fields of healing really opened up. The final chapters in the story of antibiotics remain to be written, for active research is today seeking ever more useful or specialized drugs from all sorts of molds the world over. Eventually harmless metabolic products from lower organisms may serve to curb many of man's most serious diseases.

Fleming's early work on penicillin was elaborated by Dr. Howard Florey and others, with Florey coming to the United States in 1941 to aid in setting up a program for quantitative production of the scarce and expensive drug, until then dissolved from tediously grown surface molds. At the U. S. Government Regional Laboratory in Peoria, Illinois, methods were developed for growing large quantities of the mold and extracting, purifying, and crystallizing the penicillin into stable form. Several private companies thereupon took up the production of penicillin, elaborating production techniques and turning out annually millions of dollars worth of the potent extractive from *Penicillium notatum* and *P. chrysogenum*, of the Ascomycetae, the most productive strain of which, it is reported, was not found in the distant parts of the world but prosaically upon a cantaloupe in the market at Peoria. Productivity of this strain has been further enhanced by development of X-ray and ultraviolet mutants.

Originally penicillin was simply the filtrate of nutrient broth upon which *Penicillium notatum* was grown. The mold grew only on the surface, for it is an aerobic fungus. Gradually techniques were developed for extracting the penicillin with solvents (ether, amyl acetate, and the like), from which in turn it was precipitated by appro-

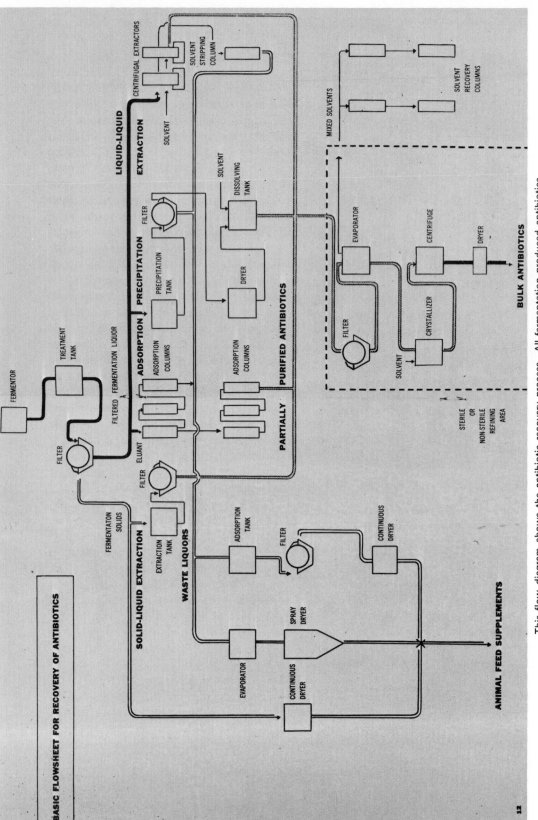

This flow diagram shows the antibiotic recovery process. All fermentation-produced antibiotics follow the general recovery scheme. (Courtesy Chas. Pfizer & Co., Inc.)

priate techniques. The purified penicillin proved a thousand times more potent than Fleming's original material and was far more stable. Most penicillin is prepared today as the stable crystalline salt, which may be kept for years unrefrigerated.

Although many antibiotics are now synthesized, the scheme for commercial production from live mold is essentially as follows. The highly selected strain of mold is cultured in small flasks of nutrient broth. These flasks are used to inoculate larger "seed tanks," which in their turn inoculate the huge (12,000-gallon) vats in which commercial production occurs. The food or medium supplied is typically of a sterile, standardized lactose-corn steep liquor base for *Penicillium* (or soybean meal for most actinomycetes). Sterile air is continuously bubbled through the tank (since the mold is aerobic). Eventually, after maximum growth of the mold, filtering, solvent extraction, and precipitation from the solvent are effected. Further crystallization or purification, and standardizing and packaging, are done quickly under aseptic conditions.

Antibiotics other than penicillin that have attained commercial success are streptomycin from *Streptomyces griseus*, which attacks many bacteria unaffected by penicillin (but is in turn ineffectual against certain types controlled by penicillin, and has other disadvantages); aureomycin, from *Streptomyces auriofaciens*; chloromycetin, from *Actinomyces*; terramycin, from *Streptomyces rimosus*; and several others, all of them from actinomycetes. Bacitracin and polymyxin are two of the relatively few antibiotics obtained from bacteria, but contain polypeptides that frequently cause allergic reactions. All told over 200 antibiotics have been described in the literature or patented. Most of these have been unable to compete with the more successful antibiotics now dominating the market, because of such factors as toxicity or side-reactions affecting the human organism, development of resistance among

certain pathogens, difficult or uneconomic production, and so on.

Neither penicillin nor any of the other antibiotics is a cure-all. Their action is first of all antibacterial and not effective against toxins already present. Moreover, the antibiotics are unpredictably lethal to certain bacterial types, harmless to others; and build up resistant strains in many cases. Until antibiotics are tried upon a whole array of diseases, their effectiveness against each one cannot be known. Penicillin, for example, can cure anthrax, pneumococcic pneumonia, various streptococcic infections, and the like, but fails against malaria, tuberculosis, poliomyelitis, and cancer. Likewise, antibiotics may cure diseases in certain body sites and not in others, or with certain means of administration and not with others. Much remains to be discovered about just how antibiotics work, but empirically used they have indeed been a breakthrough in man's age-old fight against disease, both for animal and plant.

The ledger strain of quinine, *Cinchona ledgeriana.* (Courtesy Chicago Natural History Museum.)

Quinine. One of the most remarkable drugs from plants is quinine, derived from species of *Cinchona*, of the Rubiaceae. Perhaps this antimalarial was used by the Andean Indians before European discovery of the New World; it is certain, however, that the *conquistadores* and particularly the Jesuits brought it to Europe, where centuries passed before its true merits were recognized. Doubtful legend has it that the wife of the Count of Cinchon, stricken with malaria in 1638, was cured by the native fever-bark, until then an obscure local remedy. It is said that the countess was so pleased that she had the bark of the genus that now bears her name sent to Europe, where it intermittently found favor, though not with a medical profession addicted to bleeding for all ills and condemnatory of anything new and unfamiliar—particularly if it cured a patient too quickly.

Cinchona is native to the Andean highlands of South America, from Bolivia to Colombia and scatteringly in Panama and Costa Rica. It is seldom found at elevations below 1,000 feet, preferring the coolness and abundant but distributed rains of well-drained mountain slopes. Many species have been described, although some botanists believe most to be but variants of *Cinchona pubescens* and *C. officinalis*. *C. lancifolia*, *C. oblongifolia* and *C. cordifolia* are familiar names. *C. ledgeriana* is the high-yielding strain for Far Eastern plantation stock (usually considered a strain of *C. calisaya*).

Cinchona habitat in Ecuador. Trees yielding quinine are growing at 5,000 feet or higher in these mountains. Rainfall here is over 100 inches. (Courtesy USDA.)

The related *Remijia pedunculata* has also given commercial yields of the quinine alkaloids.

Four main alkaloids are found in *Cinchona* bark, all useful as antimalarials (collectively used to make "totaquine"). These are cinchonine, cinchonidine, quinine, and quinidine. In unselected wild barks they occur (together) in amounts up to 7 per cent. By selection and breeding, from the high-yielding Bolivian plants secured by the servant (later killed for this effort) of one Mr. Ledger, British resident of Bolivia, and peddled to private British interests in India and to the Dutch government, strains yielding 17 per cent crystallizable alkaloid have been developed. Of this type, of course, are the commercial quinine plants of the twentieth century, against which the wild, unselected, and distant sources of South America cannot compete. During World War II atabrine, an antimalarial developed in Germany in 1932 from coal-tar sources, was used to supplement an inadequate supply of quinine from the wild trees of South America. Malaria seemed pretty well under control until the Korean and Vietnam Wars demonstrated that most antimalarials, especially the synthetic products, often only mask the symptoms rather than effect a complete cure.

The story of quinine thus strangely parallels that of rubber. It is a story of an important wild plant discovered in South America, harvested only from the wild for decades and never in the area of its origin given the benefits of domestication; of seed and growing stock taken to the Far East by European powers; of intense scientific attention to make the plantation plant superior to species growing in the wild; of monopoly restricting production and enforcing price maintenance until war and synthetics put an end to a lucrative trade. It is another striking example of the fact that wild plants cannot compete successfully with superior cultivars grown

Cinchona bark is removed from a seven year old tree. Guatemala. (Courtesy USDA.)

on centralized plantations abounding in supervised labor. For although in 1880 most production—some 3,000 or 4,000 tons of bark—came from South America, by the 1930's that continent produced less than 1,000 tons annually, whereas the East Indies produced about 12,000 tons. Before World War II about 90 per cent of world quinine came from Java.

In plantation practice vigorous rootstock is grown from seed, usually taken from the hardy "succirubra" strain. To this, at about 6 months, are grafted buds from selected clones of high-yielding "calisaya" types; the process is similar to that used for rubber. Considerable care must be taken to follow proper techniques of shading, disease control, budding, and the like. When about 5 feet tall the budded seedlings are planted in rows approximately 4 feet apart and are allowed to grow for some 6 years. Harvest is then begun as a thinning-out process. Each year more trees are utilized, until, in about 12 years, the entire plantation is consumed. Usually, in harvesting, the entire tree is uprooted, to provide bark from trunk, root, and branches. In some strains the roots yield more bark than the trunk. Usually the bark is first beaten with a mallet to loosen it and then is peeled by hand, with the machete or other knife. The peeled bark must be dried quickly to prevent loss of alkaloids—a phase of production that in the wild is rather haphazardly carried out in crude fire-heated shelters or in sunny clearings in the forest. On plantations properly regulated drying ovens are possible. The fully dried bark is then sent to the factories, usually located in the United States or Europe, for final solvent extraction of the alkaloids. In this extraction powdered bark is mixed with lime, and the alkaloids are removed with amyl alcohol or ether. These alkaloids are in turn extracted from the solvent by shaking in acidified water: they precipitate out when the water is made alkaline.

In the wild, quinine trees that have been cut down seem to come back from root sprouts, so that the "extermination" feared in many areas is a rather temporary condition. Many of the best quinine forests in South America, completely worked out in the 1880's, yielded again a great quantity of healing bark in the 1940's. During World War II these forests were systematically exploited by North American technical crews advising local technicians and labor. *Cascarilleros* (bark collectors) were taught to recognize the quinine tree and sent into the forests to fell *Cinchona* in order that Allied armies might fight a war that often centered in the tropics. Probably more was learned concerning wild quinine in this brief period than in all the preceding centuries since its discovery.

Although quinine was synthesized in 1944, synthetic production has not been commercially worthwhile. Substitutes such as atabrine, chloroquine, plasmoquine, paludrine, or emetine can now be readily and cheaply produced, putting something of a ceiling on quinine prices.

Opium. Another renowned botanical drug is opium from the opium poppy, *Papaver somniferum*, of the Papaveraceae. The species is believed to have been native to Asia Minor, where it is often encountered as a weed. It has spread throughout the civilized world in all climates similar to that of the Mediterranean area. The wild ancestral form is apparently not even known, so long has man cared for and cultivated the opium poppy.

Cinchona bark drying in the sun. Guatalon, Guatemala. (Courtesy USDA.)

A *Papaver somniferum* field in South Manchuria. (Courtesy USDA.)

Its legend is recorded in the hieroglyphics of Egypt, and Homer sings of it in his *Odyssey*. Hippocrates, Theophrastus, Pliny, Dioscorides, and Celsus all discuss it; and many a more recent author, not the least of whom is De Quincey (*Confessions of an English Opium-Eater*, 1822), has written with fascination of this great analgesic. The alkaloids derived from opium afford relief to millions of sufferers, at the same time that they breed addiction among other millions.

Crude opium reportedly contains the following components: gum, 25 per cent; rubber, 6 per cent; resin, 4 per cent; oils, 2 per cent; water, 10 per cent; pigments and other extraneous matter, 24 per cent; meconic acid, 1 per cent; morphine, 11 per cent; codeine, 1 per cent; and some thirty other alkaloids (narcotine, narceine, laudanine, meconine, and the like), 15 per cent. All the alkaloids and meconic acid have some physiological influence, but the basic attributes of opium are mostly due to the alkaloid morphine, $C_{17}H_{19}O_3N$. It was not long, of course, before modern chemistry learned to extract morphine from opium in nearly pure form; but it has not yet been able to synthesize this alkaloid (the similar strongly addictive heroin is an acetylated derivative of morphine). In extracting commercial morphine crude opium is treated with

calcium chloride to precipitate meconic acid and certain other impurities, while sodium acetate removes narcotine and papaverine. Finally crude morphine is precipitated out in alkaline solution and purified by repeated crystallization.

The opium poppy is cultivated for opium or as an ornamental in many parts of the world, especially India and China (where the opium is mostly locally consumed) and in the Near East and Mediterranean area (source of most of the world's legitimate supply of opium). It is a rigid, spiny-leaved annual, with large, four-parted, white to purple flowers and urn-shaped fruits (capsules). The very fine seeds are broadcast, often in two or three sowings, to assure a crop, for the seedling is notoriously susceptible to unfavorable weather upon germination. The poppy grows rapidly during spring and matures several capsules during June and July. Planting, cultivation, and harvest are almost universally done by hand, so that commercial growing of the poppy the world over must take place in cheap-labor areas.

The alkaloids are tediously produced by incising the maturing capsule, usually after sunset, by means of special tools designed not to cut too deeply (else the "milk" will be "lost" among the seeds), and scraping off

307

the next morning the coagulated drops of exudate that result. Many such scrapings are dried and then kneaded into dark yellow or brown, bitter balls of crude opium. These are carefully packed to avoid abrasion and shipped to buying centers. Yields of 25 to 40 pounds of opium per acre are possible under good growing conditions. Such production requires about 20,000 poppy plants of six capsules each. It is difficult to estimate the world production of opium, for much is surreptitiously collected and used. Imports to the United States amount to a few hundred tons annually, and world production has been estimated at about 10,000 tons, of which about 400 tons are estimated to be sufficient for medical needs, the rest being sold in illicit traffic.

The habit-forming characteristic of opium is well known. Small quantities, whether originally taken unwittingly or by intent, cause pleasant feelings of exhilaration and peace. Gradually, with habitual use, larger doses must be taken to produce the sensation, until such quantities as might have been lethal at the onset give no more than a mild effect. By this time so great is the user's desire for the drug that it will be obtained by violence if need be. Early efforts to control trade in the drug met with little success. In fact, contrary to popular belief, use of opium in the Orient (first by the Indians, who ate it; later by the Chinese, who smoked it) is relatively recent and was fostered by "enlightened" Europeans. Even with many bans upon the sale of nonmedical opium, opium dens flourish.

As a medicine, opiates are usually taken orally, rectally, or by means of a hypodermic, to render parts of the body insensitive to pain. Morphine is also used in a number of cough medicines, since it depresses the brain center involved in coughing. Many additional medical uses, as well, have been found for morphine and codeine, although their sedative and narcotic effects account for their chief importance.

Cocaine. Coca or cocaine, another famous alkaloidal sedative, is, like opium, capable of diminishing or suspending emotion and perception. It is derived from *Erythroxylon coca* and perhaps other species of the Erythroxylaceae. Besides yielding the local anesthetic cocaine, it is most famous as the peculiar stimulant of Andean porters and laborers, who, it is said, can work with superhuman endurance for days with little or no food if they chew coca. Actually, coca paralyzes the nerves that convey hunger pangs. Distances in the Andes are frequently reckoned in "cocadas," the span that can be traveled on one chew, instead of in miles. Many an Indian's cheeks have become permanently distended from his habit of continuously holding a coca cud or "aculli" in the mouth. With protracted and excessive use, coca may become habit-forming.

Native to the Andean region, coca has long been used there as a masticatory and as a cure for all the ills that flesh is heir to. It is typically chewed together with pow-

Coca or cocaine plant, *Erythroxylon coca.* (Courtesy Chicago Natural History Museum.)

dered lime or leaf ash, carried in the omni-present gourd or "poporo" and said to be necessary to get any effect from the coca. Significantly, commercial extraction of co-caine utilizes this principle of the addition of lime to extract cocaine alkaloids. Coca is sold in all the local markets, and is often one of the most profitable of cultivated crops. Many thousands of tons are consumed locally in South America each year. As with quinine, wild plants in South America proved insufficient to supply world demand when the utility of cocaine came to be known, with the result that the plant was introduced into the East Indies. Much of the supply of coca for international trade has come from plantations in Java. Imports to the United States amount to a few hundred tons annually, a good part of which, in "de-alkaloidized" form, goes into "cola" flavorings.

The coca plant is a small tree or shrub with alternate, ovate leaves, prominent stip-ules, and tiny white flowers. It is generally planted from seeds, the seedlings being kept shaded in the nursery for 8 to 10 months. Planted out, they are given greater or less attention, depending upon whether grown in a small local patch in the Andes or on a supervised plantation in Java. When the plant is about 6 feet tall, the leaves are collected by hand, especially in April and September, "sweated" (cured), and dried. Drying may be in the sun or in special drying sheds. The leaves are then powdered in a mortar for local consumption, or sold to extraction factories. Extraction of the cocaine alkaloid is accomplished by a solvent process similar to that described for opium. The chemical structure of cocaine, $C_{17}H_{21}O_4N$, has been fairly well deciphered, and although true cocaine is not synthesized, the similar novocaine and procaine have been artificially made, to become common standardized sub-stitutes available to dentists and doctors. As in the days of the Inca Empire, coca continues to be most used in its homeland—

Coca field being stripped of leaves. Area at left has been stripped. Fields are stripped five times each year, yielding about 3,500 pounds of dry leaves per hectare per year. Peru. (Courtesy USDA.)

as a narcotic, stimulant, and supposed panacea.

Belladonna. Another old but still impor-tant alkaloidal drug is belladonna, from the leaves of the deadly nightshade, *Atropa belladonna*, of the Solanaceae, indigenous to Europe. Related plants of the Solanaceae, notably *Datura meteloides*, *D. stramonium*, and *Hyoscyamus niger*, yield similar alka-loids often included in the "belladonna series." Belladonna has been extensively used in European medicine since earliest times. The deadly nightshade is repeatedly illustrated in the great herbals, and its toxic properties recognized in the magic potions of the alchemists. Until the nineteenth century collection was chiefly from wild sources, but since that time limited though significant cultivation has been practiced. Harvest of leaves is generally at flowering time, in late May or June, when alkaloidal concentration is greatest. Leaves (and roots,

Belladonna is a much-branched perennial herb 2 to 3 feet tall, the fruit a black, poisonous berry. (Courtesy USDA.)

leaves has been used in treatment of asthmatic paroxysms. Atropine or hyoscyamine are used alone to check secretion, stimulate circulation (in shock treatment, for example), counteract muscle spasms, dilate the eye, and relieve pain locally. Scopolamine does not stimulate the sympathetic centers but does markedly depress the parasympathetic, finding utility as a narcotic and anti-insomniac, and formerly (with morphine) as the obstetrical anesthesia, "twilight sleep."

Strychnine. Strychnine, from species of *Strychnos*, of the Loganiaceae, has attained greater fame as a poison and an arrow poison (curare) than as a medicinal. In proper dosage it can relieve paralysis and stimulate the central nervous system, although in medicine today it is superceded for these

if used) are dried for 2 to 15 weeks at cool temperatures and then pulverized. Solvent extraction follows, usually with ether or ethyl acetate, from which the alkaloids are crystallized out by appropriate techniques.

The deadly nightshade is a herbaceous perennial, with creeping rootstock; alternate, ovate, entire leaves; more or less hollow stem; purplish, campanulate corolla with reflexed lobes; and a brownish or blackish berry. All parts of the plant contain alkaloids in fractions of 1 per cent but are most concentrated in metabolically active cells. In 1809 the alkaloid atropine ($C_{17}H_{23}O_3N$) was isolated from belladonna, and later other alkaloids, including hyoscyamine (the isomer of atropine), apoatropine, belladonine, norhyoscyamine, noratropine, scopolamine, hyoscine, tropacocaine, and meteloidine. Atropine, hyoscyamine, and scopolamine are the three most used in medicine. They serve as stimulants to the sympathetic nervous system and hence as antidotes for opium alkaloids, as diuretics, in dilating the pupil of the eye, in treatment of palsy, and externally in ointment as an anodyne. Inhalation of burning belladonna

Branch and fruit of the strychnine plant, *Strychnos nux-vomica*. (Courtesy Chicago Natural History Museum.)

purposes by safer drugs. Commercial strychnine comes from the Oriental *Strychnos nux-vomica* and the Philippine *S. ignatii.* Some curares utilize species of *Strychnos* (as well as other plants), including *S. guianensis* and *S. toxifera.* The toxic and medicinal properties of *Strychnos* are predominantly due to two alkaloids, strychnine ($C_{21}H_{22}N_2O_2$) and the less poisonous brucine ($C_{23}H_{26}N_2O_4$). These are extracted from the seed, although other parts of the plant also contain them in lesser concentrations.

Seldom are *Strychnos* plants cultivated. The fruits are collected in off-hours or off-seasons by desultory labor, and the seeds removed by hand from the bitter pulp of the large, orangelike fruit. These seeds are washed, thoroughly dried, and then sold to local merchants. Eventually the seeds are shipped to extraction factories in industrialized countries. There the seeds are pulverized and subjected to extraction with boiling sulfuric acid and to expression. Strychnine, brucine, and other alkaloids are precipitated out from the solution and purified by crystallization in alcohol.

The *Strychnos* plants are usually woody vines or weak trees of tropical forest regions. The leaves are opposite, ovate, and usually hirsute; the flowers small; the fruit large and indehiscent. Strychnine is not an old drug: it was first introduced into Europe in the sixteenth century, but was little utilized until the nineteenth. Most production today comes from the Far East.

Perhaps the most mysterious use of the *Strychnos* alkaloids is in the compounding of curare poisons. These are the resinous extracts secretly brewed since antiquity in the jungles of South America for tipping poison blowgun darts and arrows. They are silent killers of the South American wilderness. In spite of intensive research, the exact compounding of curare has not been exactly revealed. Many plants usually go into the brew—*Strychnos* may be used, but certainly *Chondrodendron toxicoferum* in Amazonia, and various Menispermaceae, Leguminosae, and the like. Perhaps no two potions are exactly identical. They all have in common the ability to paralyze man or animal wounded by the treated dart. They

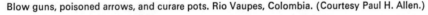

Blow guns, poisoned arrows, and curare pots. Rio Vaupes, Colombia. (Courtesy Paul H. Allen.)

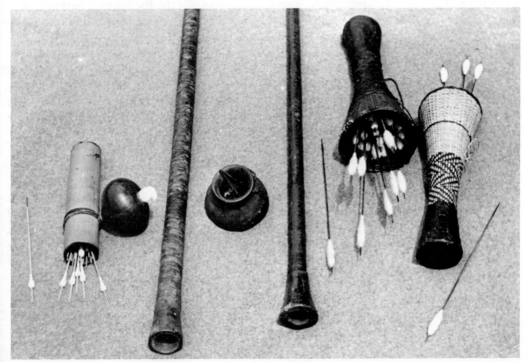

find medicinal use in regulated dosage for the relief of muscular spasms. Legend has it that in the secret production of curare among South American aborigines, certain old women skilled in the art were shut up in a house with the necessary ingredients to make the distillation. If, when the brew was ready after some two days' time, the women were not half-dead from noxious fumes, the poison was considered inadequate and the women punished.

Humboldt was perhaps the first reliable European eyewitness of curare preparation. He reports that the bark of the liana "bejuco de mavacure," probably *Strychnos*, was reduced to fine shreds and used to make an infusion that was in turn concentrated by boiling. This concentrate had a strongly bitter taste (curare is reportedly poisonous only if it enters the bloodstream, and not dangerous when taken orally). The brew was blended with "kiracaguera," unidentified botanically, to add body. With further boiling a black, tarlike substance resulted, easily applied to arrows and darts. It was said by Indians surviving battle wounds from poison darts that the curare caused "congestion of the head," "raving thirst," protracted nausea, and numbness about the wound. Of course many other plants besides *Strychnos* have been used in various parts of the world for arrow poisons, including lethal bacteria from certain swamplands (the first bacterial warfare), *Amorphophallus*, of the Araceae; *Antiaris*, of the Moraceae; many Leguminosae (*Abrus precatorius*, *Derris*, *Mucuna*, and the like); *Croton* and other genera of the Euphorbiaceae; and *Lophopetalum*, of the Celastraceae.

Digitalis. The foxglove, *Digitalis purpurea*, of the Scrophulariaceae, common European wildflower and introduced ornamental in many parts of the world, yields the glycosides digitoxin ($C_{34}H_{54}O_{11}$), digitalin ($C_{35}H_{36}O_{14}$), digitalein, and digiton, impor-

tant as heart stimulants and diuretics. The species is mentioned in the famous herbals, and is known to have been used medicinally as early as 1250, although precise knowledge concerning its action did not come until centuries later. William Withering, in England, gave special impetus to digitalis in publishing a treatise based upon experimentation and careful observation at the end of the eighteenth century. The plant is a beautiful, pubescent, herbaceous biennial or perennial, with condensed basal leaves and an erect leafy stem up to 5 feet tall, bearing apically a raceme of large, thimble-like, purplish flowers. The small seeds germinate with difficulty under adverse conditions; usually they are started under glass in February. Flowering does not take place until the second year, but leaves yielding the drug may be gathered at any time. These are dried in the shade. Yields of up to 600 lb per acre are reported. The dried leaves are pulverized and consumed directly, or the active glycosides are solvent-extracted, commonly with alcohol, to yield tincture of digitalis.

The most important effect of digitalis is its influence on circulation of the blood. In proper dosage the interval between heart contractions is lengthened and the contraction is made strong and regular, regularizing the pulse and increasing the blood pressure. Sufficient digitalis to meet medical needs can be grown on a few hundred acres, so there is little chance that the industry will become important from the quantitative standpoint.

Marijuana, Hashish, Bhang, Ganja, or Kif. Marijuana or marihuana is an ancient drug, described in a Chinese compendium on medicines as early as 2737 B.C. (herbal of Emperor Shen Nung). It comes from a resinous oil of *Cannabis sativa*, of the Moraceae, a plant previously discussed as source of hemp fiber, and also yielding a seed oil.

Hemp stalks, showing the leaf pattern around the female flowers. (Courtesy USDA.)

Tetrahydrocannabinol is the active principle, hallucinogenic under certain circumstances, bringing about cerebral excitation, illusions, and visions. Resinous hairs from the inflorescence of the female plants are the chief source of the hallucinogens. Tropically grown marijuana is "stronger" than that from hemp grown as a summer annual in temperate climates. Age also affects the potency of the active ingredients. Medicinally, marijuana may serve as an anodyne and in treatment of nervous disorders. As a narcotic-stimulant, marijuana is taken in several ways: the dried, pulverized product may be smoked, often in combination with tobacco; a beverage may be made from the powdered leaves; or the "resin" may be vaporized and inhaled, or eaten. Taken in any of these ways it can cause stupefaction and ecstasy. Strenuous efforts have been made to curb the use of marijuana in the United States, although its use has been common in Asia for 2,000 years. Tests seem to prove that marijuana is not addictive and less habituating than tobacco, being less injurious to health than either nicotine or alcohol, both accepted social "drugs;" the severe penalties for its possession and sale in the United States seem excessive and emotional. Hashish, the unadulterated resin from the female inflorescence (reputedly often collected by workers running through patches so that the fragments of oil cling to their clothing), is associated in folklore with a curious Mohammedan sect, the "hashishins" (whence "assassins"), whose avowed religious purpose was to murder "enemies of the faith" in order to enter paradise, and who presumably used hashish heavily as an intoxicant. The sect numbered tens of thousands in Syria and Persia until suppression by the Tartars in 1256. Bhang generally refers to the cut tops of the female inflorescence served as a decoction in water or milk, then drunk, or dried and smoked. Ganja is a carefully prepared bhang used in India. Kif is the general term in Morocco for all forms of marijuana, although smoking is the common form of use, mixed equally with tobacco. Kif is the main crop in sections of the Rif Mountains. In the United States the relatively mild leaf mix is known as pot or grass, and the cigarettes as reefers, joints, etc. Tetrahydrocannabinol, chief active ingredient in marijuana, has been synthesized.

Rauwolfia or Reserpine. *Rauwolfia* (also spelled *Rauvolfia*) *serpentina*, of the Apocynaceae, is the source of reserpine, chemically similar to serotonin, a substance in the brain related to LSD. Reserpine can terminate the schizophrenia-like symptoms from LSD or from mental illness. Reserpine has long been used in India for treatment of the mentally ill, and is widely employed today for hypertension and for the lowering of blood pressure. In cases of insanity it serves as a tranquilizer, and it has even been used for treatment of insect bites, fevers, and dysentery.

Rauwolfia is native to India and the East Indian forests. When demand for reserpine became strong shortly after mid-century, wild stands of the plant were largely exterminated. Most production now comes from cultivated plantings. The species is a small, perennial shrub tolerant of a broad range of climatic conditions. It may be propagated from seeds, cuttings, or plant separations. Seedlings are started in nurseries and transplanted to the field. Rauwolfia requires ample rainfall, or irrigation in drier areas. The roots, chief source of the drug, are ready for harvest in 2 or 3 years. Yields may run as high as 4 tons per acre, with an alkaloid yield of less than 2 per cent. More alkaloid is found in the bark than in the xylem. India is the center for production, with most of the crop exported to Europe and North America.

Some other drug plants.

Aconite, monkshood, or wolfbane—the herbaceous perennial, *Aconitum napellus*, of the Ranunculaceae, yields aconitine, a famous alkaloidal poison once used in the treatment of neuralgia and rheumatism. Native to the mountains of Europe and western Asia, it has been cultivated throughout the world as an ornamental and drug plant.

Alectra—*Alectra parasitica chitrakutensis*, Scrophulariaceae, is a parasite on *Vitex* roots in India, the rhizome used to make a paste eaten for treatment of leprosy and lesser ailments.

Aloe—fresh leaves of several Asiatic, Mediterranean, and Antillean species of *Aloe*, Liliaceae, including *A. barbadensis* and *A. ferox*, yield a glucosidal juice used to treat burns, particularly X-ray burns. When evaporated to a semisolid form, it finds use as a laxative. Both wild and cultivated plants are used: the fleshy leaves are cut and hung with the cut end downward to allow the juice to drip into suitable containers.

Amanita or fly-agaric—poisonous alkaloids from *Amanita muscaria*, a Basidiomycete fungus, serve as hallucinogenic narcotics and stimulants similar to marijuana.

Aconite, *Aconitum napellus.* (Courtesy Chicago Natural History Museum.)

Aloe vera, the juice of which has been used in the treatment of x-ray burns. (Courtesy Chicago Natural History Museum.)

Araroba—araroba powder, used for skin ointments, is obtained from cavities in the heartwood of *Andira araroba*, of the Leguminosae, growing wild in Baia, Brazil.

Arnica—all plant parts of species of *Arnica*, of the Compositae, particularly *A. montana*, a herbaceous perennial of the European Alps, find use for treating bruises and sprains.

Ashwagandah—an ancient drug of India from *Withania somnifera*, Solanaceae, used as a tonic and sedative; wild plants are uprooted and dried.

Balm of Gilead—buds of *Populus candicans*, Salicaceae, of the northeastern United States, are used as an ingredient in cough syrups.

Berenjene—A cultivatable source of steroids in the American tropics is *Solanum mammosum*, Solanaceae.

Betel—not strictly a drug; the world's most used masticatory, chewed by Eastern peoples of the Indian and Pacific Ocean islands and southern Asia since ancient times. Betel contains a harmless narcotic which may have some medicinal value (for counteracting overacidity, for example), and which produces mild stimulation and a feeling of well-being. The masticatory is compounded from the seed of a palm, *Areca catechu*, and the leaf of *Piper betle*, Piperaceae, often with various supplementary aromatics. Usually a fresh leaf of *Piper betle* is smeared with lime paste (often plus resin), together with sliced *Areca catechu* nut and possibly other flavoring (e.g., cloves, tamarind, and the like). The mass is worked in the mouth without swallowing; the process stimulates a copious flow of saliva, which is continuously expectorated. The saliva and eventually the teeth of the addict turn red or brown. To most Westerners betel chewing appears to be a disgusting habit, although it differs little from the tobacco chewing once widespread in the United States. Confirmed betel chewers have bad teeth, but it is difficult to pin the blame for this on betel chewing, for nutritional deficiencies may equally well be

Betel nuts from the betel palm, *Areca catechu*. (Courtesy Chicago Natural History Museum.)

Betel-nut palm growing in Thailand. (Courtesy USDA.)

the cause. The "makings" for betel chewing may be bought at any market in the East and are carried on one's person, as is tobacco in the Western world. Both the *Areca* palm and the betel leaf are widely cultivated.

Buchu—*Barosma betulina*, Rutaceae, is harvested from wild and cultivated plants in South Africa, for use as a diuretic.

Burdock—dried roots and seed of the common, troublesome weeds *Arctium lappa* and *A. minus*, Compositae, find medicinal use.

Caapi—the stem of *Banisteriopsis caapi*, Malpighiaceae, is boiled to yield a narcotic beverage, consumed by Indians of the upper Amazon for its hallucinatory effect. Often it is drunk alternately with chicha beer, vomiting and visions being induced after several draughts. It finds use in the flagellation ceremonies whereby youths are ushered into manhood, and doubtless helps the initiates withstand the pain. The alkaloids are harmaline, harmine, tetrahydroharmine, and dimethyltryptamine.

Cacao—the chocolate tree, *Theobromo cacao*, Sterculiaceae, will be discussed in a later chapter devoted to beverage plants. The alkaloid theobromine, however, is commercially extracted from chocolate tree seed wastes and finds some medicinal use as a stimulant.

Calabar bean—the fruit of the African *Physostigma venenosum*, Leguminosae, yields physostigmine, of value in ophthalmology for contracting the pupil. The plants are woody climbers found in the wild.

Camptotheca—*C. acuminata*, Cornaceae, a tree indigenous to mainland China, is the source of camptothecin, a promising remedy for cancer of the intestine. (Illustrated on page 296.)

Canafistula—a concentrated extract from the fruit pulp of the golden shower or canafistula tree, *Cassia fistula*, Leguminosae, finds use as a laxative.

Cascara, chittembark, bearwood, cascara sagrada, frangula, and Persian bark—extract of the bark of species of *Rhamnus*, Rhamnaceae, contains alkaloids, tannins, and other substances much used in laxatives. *R. purshiana* is well known in the wild in the western United States, and *R. frangula* is cultivated in southeastern Europe.

Chaulmoogra—the fruit of the Oriental *Taraktogenos kurzii* and *Hydnocarpus anthelmintica*, Flacourtiaceae, yields oils (organic acids) formerly used to treat leprosy.

Colocynth—dried pulp and seeds from fruits of *Citrullus colocynthis*, Cucurbitaceae, occurring both wild and cultivated in Asia and Africa, yield the well-known cathartic colocynth or "bitter apple."

Columba—the root of the perennial vine *Jateorrhiza columba*, Menispermaceae, is used to prepare tonics and bitters.

Cubebs—the dried fruit of *Piper cubeba*, Piperaceae, in addition to serving as a condiment, finds use as an antiasthmatic. It is cultivated in the East Indies.

Dioscorea—steroids from several species, primarily from Mexico and Central America, have been used to derive the sapogenin, diosgenin, utilized in manufacture of cortisone. Other species of the Dioscoreaceae are important food root crops in Africa.

Dragon tongue—dried foliage or a brew from it, of *Sauropus changiana*, Euphorbiaceae, of southern China, is used to treat respiratory ailments; an extract is antibiotic to certain bacteria.

Ephedrine—an alkaloid from stems of Asiatic species of *Ephedra*, Gnetaceae, was much used, until the synthetic form was developed, to make nasal and bronchial remedies for relieving congestion, and in treatment of low blood pressure. The plants are xerophytic, leafless members of an unusual Gymnosperm order. Ma-huang, a familiar medicine in China since time immemorial, is derived from *E. sinica*, *E. equisetina*, and *E. intermedia*. Plants are collected in autumn, sun-dried and sectioned. Before use the sections are boiled, often in a honey–water mixture, and usually further roasted until the water is evaporated. The bitter remedy resulting is prescribed for typhoid, colds, coughing, and various pains.

Ergot—long before penicillin and the antibiotics were known, another important drug from fungi was in use—ergot, from the sclerotium (spore-producing body) of *Claviceps purpurea*, a parasite on cereals, particularly rye. The sclerotium contains as the chief active ingredients ergotinine, ergotoxine, ergotamine, and histamine. These are useful in controlling hemorrhage after childbirth and in treating migraine. The ergot fungus is very frequent in the rye fields of western and southern Russia, the Balkans, and the Iberian peninsula. Before ergot was well known, many serious epidemics resulted from the consumption of the sclerotia along with rye grain. Gruesome tales are related of some epidemics of the 1700's. One relates that a woman suffering from ergotism in riding to a hospital brushed against nearby vegetation and caused her gangrenous limb to fall away at the knee. Fortunately, regulated dosages of ergot cause no ill effects and are extremely useful to medicine. A derivative of ergot is LSD, lysergic acid diethylamide, a favorite of the psychedelic cult.

Genista—the blossoms of *Genista canariensis*, Leguminosae, are smoked as a minor psychedelic.

Gentian, bitterroot, bitterwort—the root of *Gentiana lutea*, Gentianaceae, found in moun-

tainous regions of central and southern Europe, yields extracts much used in tonics and bitters.

Ginseng—one of the most interesting medicinal plants is ginseng, from *Panax quinquefolium*, of the Araliaceae, native to the shady hardwood forests of the northeastern United States. The fleshy roots of this palmately leaved herb are believed by the Chinese to possess extraordinary powers for curing all disease, although the species is little used in the United States. Its use in China traces back four millenia. The more closely the thick, spindle shaped root simulates the human form the more valuable the Chinese consider it. A glycoside, panaquilon, is contained, which pharmacologists credit with little medicinal value. Roots are usually collected from the wild in the United States and exported to China. Exports in 1966 were nearly 100 tons, worth more than $4 million. Numerous attempts at cultivation have been made but with indifferent economic success, since the plant is slow-growing (6 years from seed) and subject to disease. *Panax*

Ginseng, *Panax quinquefolium*, the root of which is believed by the Chinese to possess extraordinary curative powers. (Courtesy Chicago Natural History Museum.)

ginseng, of Korea and Manchuria, is apparently not distinguished from the New World species in the trade. Attempts have been made to cultivate this species in Asia.

Goldenseal—the roots and rhizome of *Hydrastis canadensis*, Ranunculaceae, have been collected in the woods of eastern North America to such an extent that the species has been almost eliminated. Seven or eight tons of roots annually come now from propagated plants started by seed, division of rootstock, or buds from larger roots. The alkaloids are tonic and used in treatment of inflamed mucous membranes.

Hellebore—this name denotes alkaloids from species of *Veratrum*, Liliaceae, sometimes used in treatment of neuralgia, or to glucosides from the roots of *Helleborus niger*, Ranuculaceae, used as a purgative.

Henbane—vegetative parts of various species of *Hyoscyamus*, particularly *H. muticus* or *H. niger*, of the Solanaceae, yield hyoscyamine and scopolamine (hyoscine), alkaloids similar to atropine from the belladonna plant. Henbane is also used as a poison; and it is sometimes smoked for its intoxicating effect.

Hermal—alkaloidal seeds of *Peganum harmala*, Zygophyllaceae, have been used in India as an anthelmintic and narcotic.

Ipecac or ipecacuana—the rhizome and roots of *Cephaelis ipecacuana* and *C. acuminata*, Rubiaceae, herbaceous perennials of Central and South America, yield the alkaloid emetine useful in amoebicides and emetics.

Ipomoea—hallucinating alkaloids are found in the seeds of both *Ipomoea* and *Convolvulus*, Convolvulaceae, of which *I. tricolor* (*I. violacea*) is perhaps most used, known in Mexico as badoh negro.

Jaborandi—leaves of *Pilocarpus jaborandi* and other species of the Rutaceae of tropical America contain alkaloids useful for inducing physiological actions opposite to those induced by atropine.

Jalapa—the root of the perennial herbaceous vine *Ipomoea purga* (*Exogonium jalapa*), Convolvulaceae, native to Mexico and cultivated in India and elsewhere, yields a laxative.

Jequerity—seeds of the jequerity bean, *Abrus precatorius*, Leguminosae, contain the poison abrin, which, like ricin from the castorbean and similar substances from other legume seeds serve as hemagglutinins. The bright red and black seeds are often used for ornamental jewelery in the American tropics, considered a hazard when brought into the United States (lest children consume the broken beans by chance).

Jimson weed, devil's trumpet, etc.—species of *Datura*, of the Solanaceae, yield from the seed, flowers, and foliage several alkaloids, including scopolamine (hyoscine), hyoscyamine, and atropine. The drug stramonium from *Datura*, is a source of "knockout drops." Medicinally it may be used to relieve asthma and treat bronchial complaints, or externally as an ointment. Scopolamine finds use as a preanesthetic in childbirth and surgery, and in ophthalmology. *D. stramonium* is recognized in the United States pharmacopoeia, but the Mexican *D. innoxia* is reported to be a still better source of scopolamine.

Joshua tree— *Yucca brevifolia*, Liliaceae, yields steroidal sapogenins suitable for cortisone and estrogenic hormones.

Kavakava—a beverage prepared from the masticated roots of *Piper methysticum*, Piperaceae, widely grown in the Pacific islands; it is an integral part of the religious and social life there, and has tranquilizing properties.

Khat, murmungu, or *mirra*—*Catha edulis*, Celastraceae, of eastern Africa and the Middle East is famous as a stimulant; the foliage of this tree is masticated or drunk as a tea. The principle alkaloid is cathine, $C_6H_5 : CHOH : CHNH_2CH_3$.

Krameria or rhatany—the dried root of *Krameria triandra* and other species, Leguminosae, from South America, finds medicinal use for its astringent properties.

Licorice—the dried rhizome and root of the Middle Eastern *Glycyrrhiza glabra*, Leguminosae, finds use as a demulcent and expectorant, although it is perhaps better known as a tobacco and confection flavor.

Lily of the valley—rhizomes and roots of *Convallaria majalis*, Liliaceae, yield a cardiac stimulant and diuretic similar to digitalis.

Lobelia—the dried foliage of the North American *Lobelia inflata*, Lobeliaceae, contains the alkaloid lobeline sulfate used as an antispasmodic, emetic, in treatment of respiratory disorders, and in antismoking pills. In cultivation the plants are mowed when in full flower.

Lupine—*Lupinus termis*, Leguminosae, seed contains over 2 per cent alkaloids usually leached out before consumption in the Middle East.

Mandrake—the root of *Mandragora officinarum*, of the Solanaceae, has been a famous remedy and pain-killer since Biblical times. Under the "Doctrine of Signatures" it was believed to resemble the human body; and when pulled from the ground, it was alleged to emit a horrible shriek sufficient to kill the collector. The plant contains several alkaloids, including podophyllin, mandragorin, and hyoscyamine.

May apple—roots and rhizomes of *Podophyllum peltatum*, of the Berberidaceae, yield podophyllin, experimentally used in recent years in treatment of paralysis, or as an emetic and cathartic. *P. emodi* of India has been used as a cathartic.

Ololiuqui—seeds and sometimes other parts of the plant of *Rivea corymbosa*, Convolvulaceae, of southern Mexico, contain intoxicating alkaloids and glucosides that were much used by the Aztecs as an euphoric and still find some use as a reputed pain-killer and local anesthetic. Included are lysergic acid amide and isolysergic acid amide.

Peyote, peyotl, or mescal buttons—*Lophophora williamsii*, of the Cactaceae, is a small, carrotlike, spineless cactus of the southwestern United States and Mexico that contains hallucinogenic alkaloids with effects similar to those of marijuana. The Indians chew the buttons until soft, then roll them into pellets to be swallowed for their narcotic effect. Peyote is believed by the Indians to be a panacea. An ethanol extract has broad antibiotic properties.

Pituri—vegetative parts of *Duboisia hopwoodii* and *D. myoporoides*, Solanaceae, are chewed and smoked by Australian natives for the hallucinating, narcotic effects. It also serves as a fish poison. The shrub is a commercial source of scopolamine.

Poison hemlock—the foliage of *Conium maculatum*, of the Umbelliferae, yields the famous alkaloidal poison drunk by Socrates. It causes paralysis, convulsions, and eventual death. It may be used medicinally as a sedative. The chief alkaloid is coniine, a medullary depressant.

Pomegranate—the bark from stem and root of *Punica granatum*, Punicaceae, cultivated particularly in Asia and the Mediterranean area, has served since time immemorial as a vermifuge.

Psilocybe—*Psilocybe mexicana*, Agaricaceae, yields psilocybin and psilocin; a widely used hallucinogen.

Psyllium—*Plantago ovata*, Plantaginaceae, yields a seed valued in India as a household remedy for treating intestinal troubles. About 75 per cent of the crop is exported to the United States and to Europe for use in laxatives.

Quassia—Jamaican quassia comes from *Picraena (Picrasma) excelsa*, and Surinam quassia from *Quassia amara*, both of the Simarubaceae. A bitter extract is made from the wood and used as a bitter tonic, pinworm remedy, and insecticide.

Saffron, meadow—the corm of *Colchicum autumnale*, Liliaceae, yields the alkaloid colchicine, once used to treat rheumatism and gout but today a research tool for doubling chromosome numbers.

Scammony, Mexican—dried roots of *Ipomoea orizabensis*, Convolvulaceae, yield a cathartic.

Senega, snakeroot, or milkwort—dried roots of *Polygala senega*, Polygalaceae, sometimes find use as an emetic and stimulant.

Senna—sun-dried leaves of *Cassia senna* and *C. angustifolia*, Leguminosae, find wide use as a cathartic. Production comes largely from cultivated trees grown in southern India.

Soapbark—the inner bark of *Quillaja saponaria*, Rosaceae, contains water-soluble glucosides (saponins) which give copious lather in water. In medicine it finds use as an emulsifier, expectorant, and cutaneous stimulant, but is toxic when taken in quantity, dissolving the blood cells. Other saponin-yielding plants are soapwort or bouncing Bet, *Saponaria officinalis*, Caryophyllaceae; soapberry, *Sapindus saponaria*, Sapindaceae; and soaproot, *Chlorogalum pomeridianum*, Liliaceae.

Squills or sea onion—Urginea maritima, a bulbous plant of the Liliaceae found in the Mediterranean area, finds use as a diuretic and heart stimulant (white variety), and perhaps more notably as a rodenticide (red variety). Mature bulbs take 5 to 6 years if grown from seed or 4 to 5 years from bulblets. Production comes almost entirely from wild sources. The bulbs are collected by hand and cut and dried before they are marketed.

Strophanthus—the seeds of the tropical African vines *Strophanthus hispidus* and *S. kombe*, Apocynaceae, yield alkaloids useful as a heart stimulant. In Africa they have also served as arrow poisons. *S. sarmentosus* is a source of cortisone.

Tagetes—the marigold genus, Compositae, contains a number of indigenous species from Argentina to the southwestern United States. The plants are very aromatic, copiously covered with glands containing pungent oils, esters, and phenols. Infusions and poultices have been used as analgesics, antiseptics, carminatives, stimulants, and vermifuges, although effectiveness in these respects is discounted by modern pharmacology.

Tamarind—the fruit pulp of the well-known tropical fruit tree, *Tamarindus indica*, Leguminosae, finds medicinal use as a laxative.

Tea and coffee—these beverage plants, to be discussed in a subsequent chapter, provide the alkaloid caffeine, a stimulant and diuretic.

Valerian or garden heliotrope—rhizomes and roots of the European *Valeriana officinalis*, Valerianaceae, yield a supposed nervine, carminative, and antispasmodic.

Velvetbean—seeds of *Mucuna spp.*, Leguminosae, yield a few per cent of L-dopa, useful in treatment of Parkinsons' disease.

Thousands of additional drug plants are recognized in the various pharmacopoeias of the world. Obviously it is impossible to list them here; and indeed many are continuously being dropped from succeeding editions. A number commonly collected in the northeastern United States, as apparently authentic medicinals, were listed in Table 12-1.

Insecticides

It is apparent that innumerable plant species yield products capable of exerting some physiological influence upon the mammalian organism. It is not unexpected that others affect the life processes of pests such as insects, even to the point of destroying them. Many plant extractives of this nature do exist, and serve the world as insecticides. The U. S. Department of Agriculture lists some 1,200 plant species as having insecticidal value. The great majority of these are of slight importance, and even the better among them tend to be overshadowed by synthetic insecticides such as DDT, diazinon, chlordane, carbaryl, parathion, malathion, and so on. Yet the synthetics are not the complete answer, for in many cases less toxic natural products are required, especially around the home and for foods, and as dusts and sprays for livestock or pets.

Peruvian field worker with *Lonchocarpus* root. From these roots comes rotenone, a valuable insecticide, used for centuries by South American Indians as a fish poison. (Courtesy USDA.)

Insecticides complement biological controls. Although, increasingly, predatory insect will be pitted against pest, sterilized males released to reduce the breeding populations, and hormonal substances applied to break the life cycle, when an insect attack does occur an insecticide is needed. Thus insecticides bid fair to continue in demand in a world destined to have its population outstrip the carrying capacity of the land. Crop monoculture, conducive to the spread of epidemics, intensifies the problem. Contact and stomach insecticides with adequate residual value, not poisonous to mammals, and against which insects build little or no immunity, such as the botanicals are, certainly have a place. Rotenone and pyrethrum are especially in the limelight now that the world has become increasingly concerned about environmental contamination and ecological degradation. Many additional botanical products are synergists, emulsifiers, adhesives, or stabilizers in insecticides.

The rotenone insecticides. Rotenone was first known as a mysterious and unidentified fish poison* or barbasco of the deep jungles of South America. There the natives collected roots of certain forest plants, and utilized them to "poison" small streams or pools, thereby causing the fish to float to the surface, where they were easily collected. The fish were in no way "poisoned" for human consumption by the barbasco, just as rotenone, a contact and stomach poison to insects, is in no way detrimental to human beings when dusted onto vegetables. For so long had the barbasco (also called "cube," in Peru; "timbo," in Brazil; "haiari," in British Guiana; "nekoe," in Surinam) been used by South American aborigines that the selected variety commonly planted today is reportedly unknown in the wild. This unusual plant is *Lonchocarpus nicou*, var. *utilis*,

The tiny white flower of the pyrethrum is helping to lift the economic status of people in the highlands of New Guinea. A woman picks pyrethrum in a field near Bukepena. (Courtesy Australian News & Information Bureau.)

of the Leguminosae. It is a tetraploid ($2n = 44$), which may partly explain its comparative sterility and absence in the wild. All other species so far tested have shown a $2n = 22$ chromosome number. Other species and varieties, especially *L. urucu*, grown mostly in Brazil, also contain rotenone, but seldom in as high proportion as *L. nicou utilis*. *Lonchocarpus* is a small, weak tree with alternate, pinnately compound leaves, racemes of modest papilionaceous flowers, and flat, indehiscent fruits.

---

* See S. S. Lamba, *Indian Piscicidal Plants, Economic Botany* **24**, 134–6 (1970), for a list of fish poisons embracing seventeen plant families, although the Leguminosae is by far the most important.

Derris elliptica growing in Ecuador. (Courtesy USDA.)

In the Far East the counterpart of *Lonchocarpus* is *Derris* ("tuba"), also of the Leguminosae. Taxonomists have been hard put to find any significant difference between these two genera, other than their geographical distribution, one being found in the New World and the other in the Old. Native populations in Malaya and the East Indies have for centuries extracted rotenone from *Derris elliptica*, and to a lesser extent from *D. malaccensis*, and were doing this commercially before South American peoples marketed *Lonchocarpus*. Good genetic stock of *Derris* now has been introduced into the West Indies and Central America. Before 1930 rotenone was a comparative unknown in American agriculture, but it has since been so widely accepted that imports of crude and pulverized roots into the United States alone amount to thousands of tons annually.

Derris and *Lonchocarpus* are propagated by cuttings, whether grown in local patches along the Amazon tributaries or in supervised plantations and experiment stations of more populated regions. The cultivation of *Lonchocarpus* has met with little success outside of its native environment, Amazonian

Lonchocarpus roots being harvested from a planting about 18 months old. (Courtesy USDA.)

Peru and Brazil. The care used in planting varies: at one extreme, stems of three to six nodes are rooted in nurseries; at the other, they are simply thrust at an angle into a stick hole in the ground and tapped in with the foot. Planting is typically on land newly cleared from the forest, and without aid of mechanical equipment. The "take" is evidently seldom less than 50 per cent and under good conditions comes to 80 per cent or more. The weak, vinelike shrubs or trees are allowed to grow 2 to 3 years before uprooting. During this time two or three weedings must be performed by hand. The plants then constitute a tangle about 8 feet high. The tops are cut away and the roots pried loose. The roots, averaging about 2 pounds each, are washed, cut into convenient lengths, and dried. During the first year there is interplanting of food crops in the barbascales. Negotiations for sale are often initiated before the roots are dug, and the grower may sell his crop months before harvest. Various contracts may be entered into for actual digging, and marketing may proceed through several intermediaries. Newly dug roots contain about 60 per cent moisture, but are reduced to about 20 per cent moisture content before export.

Rotenone has typically been processed in the consuming country, although high labor costs and the tropical nature of the plant would make preliminary extraction preferable in place of origin. Rotenone occurs as a mixture of six crystallizable rotenoids (complex ring compounds) of an oil-like composition, in association with other complex organic compounds, various saponins, starch, and tannins. It is largely insoluble in water but soluble in oil. The rotenone-bearing resin occurs most abundantly in the xylem and phloem rays and parenchyma, and in the pericycle. It ranges from 1 to 20 per cent (dry weight) of the root. Commercial shipments usually average 4 to 6 per cent rotenone "resins," but selected strains materially raise this average. Rotenone preparations may be utilized in powdered form (pulverized root is mixed with talc or clay to about 1 per cent active ingredient), as wettable powder emulsions, or as an extract dissolved in solvents such as acetone, pine oil, xylol, or naphthalenes. Rotenone should not be mixed with lime or alkaline materials, which cause its deterioration.

Thus rotenone, first isolated from *Derris* in 1902, and found in *Lonchocarpus* in 1926, has come to have immense practical value. For truck and canning crops it may be used to control many serious pests without hazard to the consumer and without residue tolerance problems. It serves also in livestock dips, killing ticks, lice, fleas, and other pests without harming the mammals. Mexican bean beetle, wooly apple aphid, European corn borer, pea aphid, housefly, mosquito, and cockroach, are all combatted by rotenone, although it is reasonably safe for honeybees. Before World War II, at least 50 per cent of all production came from the Far East (*Derris*), but several thousand tons of crude and powdered *Lonchocarpus* roots have been imported by the United States annually since then, mostly from Peru.

Pyrethrum. Daisylike flowers of certain species of *Chrysanthemum*, Compositae, yield pyrethrum. It is purported that the insecticidal value of pyrethrum was known to the Chinese in the first century, and the Persians certainly used it four centuries ago. In the 1800's Dalmatia (Yugoslavia) was center of the industry, continuing so until World War I, when Japan assumed leadership, Japan in turn yielding to Africa during World War II. In the 1960's pyrethrum growing was introduced into the New Guinea highlands as a particularly suitable self-help industry.

The pyrethrum daisies are perennial herbs with dissected leaves, bearing in spring and summer about fifteen white to crimson

Pyrethrum flowers are ground into a fine powder and weighed before further processing. New Guinea. (Courtesy Australian News & Information Bureau.)

Tanks in which ground pyrethrum flowers are mixed with petroleum ether to extract pyrethrum. New Guinea. (Courtesy Australian News & Information Bureau.)

flower heads on long, erect stems. These flower heads, the mature ovaries especially, are the source of commercial pyrethrum. "Persian insect powder" is derived from *C. coccineum* (*C. roseum*), basis for the industry in New Guinea. "Dalmatian powder" comes from *C. cinerariaefolium*, native to southeastern Europe and now widely cultivated throughout the world, especially in Japan and Africa. In the New World some is grown in California and Brazil.

Pyrethrum is usually sown in seedbeds in late summer or early spring, or is propagated by crown separation. In April or May the plants may be transferred from seedbed to field, being spaced about 2 feet apart. Good drainage and abundant sun are required. Usually one season is needed for the plants to become established from seed, after which yields of flower heads of 400 to 500 lb to the acre can be expected for about 4 years (rotation is then advisable). In most parts of the

world the heads are picked by hand when just beginning to expand. The picker gathers 20–30 lb of flowers per day. The heads are sun-dried or are dried under artificial heat at temperatures less than 160°F to yield up to 3 per cent crude pyrethrum insecticide. The active principles are pyrethrins and volatile oils. Pyrethrum serves as a contact insecticide and in parasiticide lotions. It may be used as a dust, or as an emulsion in combination with a suitable emulsifier or spreader. Increasingly, preliminary extraction is made at place of production, to save cost of shipping large bales of dried flowers. About 20 pounds of pulverized dry flowers yield a pound of crude extract after leaching with petroleum ether or naphtha, and concentration by distillation.

Pyrethrum is especially used in aerosols. Like rotenone, this insecticide has come into wide use only within recent decades. The United States imports several thousand tons of pyrethrum annually. World production runs over 10,000 tons. Because of low labor costs Africa is the main producing center for pyrethrum, principally from the Rift Valley of Kenya and Tanzania. The species grows best at higher elevations, from 5,000–10,000 feet.

Partial listing of other plants capable of yielding insecticidal substances.

Amorpha fruticosa, Leguminosae—resinous pustules of the pod of this species contain a glycoside called amorphin ($C_{33}H_{40}O_{16}$), which acts as a contact and stomachic insecticide and as a repellent.

Anabasis aphylla, Chenopodiaceae—alkaloids, particularly anabasine (an isomer of nicotine), can be obtained from leaves of this plant indigenous to the steppes of the Caspian region.

Croton tiglium, Euphorbiaceae—the resinous oil from seeds of this shrub has found use by the Chinese as an insecticide.

Duboisia hopwoodii, Solanaceae—nornicotine, more toxic to certain insects than nicotine, is found in the dried leaves of this Australian plant, as well as in various kinds of tobacco.

Haplophyton cimicidum, Apocynaceae—the dried leaves of this "cockroach plant" of Mexico have been used since time immemorial for killing flies, fleas, roaches, lice, and other pests. It is used both as stomach and contact poison, the crude "alkaloid" reportedly being as toxic against certain insects as is pyrethrum.

Heliopsis longipes, Compositae—roots of this Mexican plant yield an amide ($C_{14}H_{23}ON$) which has paralyzing action on certain insects.

Milletia pachycarpa, Leguminosae—both seeds and root of this Chinese plant yield an insecticide of the rotenone type.

Phellodendron amurense, Rutaceae—oil from the seed of this East Asiatic tree is as toxic to certain insects as is pyrethrum.

Quassia amara, Simaroubaceae—the water-soluble extract, quassin, from wood of this and other simaroubaceous trees of tropical America serves to control some types of aphids. It can be absorbed from soil solution by growing plants to "immunize" them to certain kinds of insect attack.

Randia dumetorum and R. nilotica, Rubiaceae —the former species is Indian (Ceylon) and the latter African. The root of both, when powdered, yields a product of some insecticidal value.

Rhododendron molle, Ericaceae—flowers of this Chinese species, when powdered, show mild insecticidal effects.

Ryania speciosa, Flacourtiaceae—alkaloids extracted from the stems and roots of this Mexican plant are toxic to many kinds of insects and to rats.

Schoenocaulon officinale, Liliaceae—seed of this American plant has served as an insecticide since the sixteenth century, mainly for destruction of lice.

Sesamum indicum, Pedaliaceae—oil from *Sesamum* seed is a powerful synergist with pyrethrum.

Tephrosia vogelii, Leguminosae—this tropical African species can be cropped as an annual in the West Indies and parts of the United States, yielding leaf abundantly, from which is extracted a type of rotenone. Rotenone recovery is better from fresh leaf than from dry, necessitating local processing. It is hoped that strains can be bred to yield as much as 5 per cent rotenoid, providing a commercially feasible crop adapted to mechanization.

Tripterygium wilfordii, Celastraceae—this is the "thunder-god vine," cultivated in China, where the pulverized roots are widely used for control of insects infesting vegetables.

Zanthoxylum clava-herculis, Rutaceae—a solvent extract from the bark of this shrub of the southeastern United States is an effective synergist with pyrethrum.

Growth Regulants

A practical outgrowth from pure research on plant growth substances are the hormonal growth regulants. Particularly outstanding is 2,4-D (2,4-dichlorophenoxyacetic acid, as salts, esters, and amides), typifying a group of selective herbicides that destroy certain plants (most dicotyledons) while not affecting others (most monocotyledons) at the same concentration. Thus 2,4-D is the basis for the familiar lawn sprays which eliminate dandelions, plantains, and many other weeds without injuring the grass. The same substance sprayed by plane on a wheat field represses the broadleaf weeds but leaves the grain unharmed, its yield increased because competition from weeds is lessened.

About the same time that 2,4-D was developed, so were "synthetic auxins," used to increase set of fruit, especially with greenhouse tomatoes and snapbeans. In the 1950's other new growth regulators were discovered, notably the gibberellins and the cytokinins. Later came the abscisins, the morphactins—and the end is not in sight. There are chemical variations to these growth-regulating substances: for example, about two dozen gibberellins have been isolated, and scores of different higher plant species yield cytokinin from structures so diverse as fruit, seed, root, and plant sap. There is potential for both good and evil in these developments: the same type of regulants that promote rooting of horticultural cuttings, control windfall and ripening of fruits, and regulate size and shape of plants and plant structures also can be used for biological warfare that would destroy crops of the enemy!

Most growth regulants are now rather easily and inexpensively synthesized. Some are still more economically derived from plants than they can be produced synthetically, especially fermentation products from fungi. Commonplace today, in addition to the auxins and phenoxy herbicides, are maleic hydrazide, CCC or cycocel, B-Nine or alar, and other growth-retarding compounds. Their effect is contrary to the growth-promoting properties of gibberellin; they are used to shorten internodes, thicken stems, improve color, and regulate flowering and fruiting. By 1966 the USDA had given label clearance for thirty-four different growth-regulating chemicals, all stemming from the original investigations of auxins found in growing stem tips.

Auxins. Investigations during the 1940's identified such substances as indolebutyric acid, naphthaleneacetic acid, and similar substances which control gross growth of the stem in remarkably small concentration (only a few parts per million). These growth regulants are found in a number of natural materials as well as stem tips, but were soon inexpensively synthesized from commonplace chemicals. Rooting compounds, fruit-setting and fruit-thinning sprays, and substances that prevent drop of fruit were soon on the market commercially. Subsequent development of the other classes of growth regulators has permitted the combining of effects, to permit sophisticated thinning of fruits, earlier flowering and fruiting of diverse crops, and even control of plant shapes (for sales appeal, and so as to make most effective use of photosynthesis).

The phenoxies, **2,4-D, 2,4,5-T, 2,4,5-TP, etc.** 2,4-D has proved to be one of the most remarkable growth regulators developed. As early as 1945, 1 million pounds were produced annually in the United States, and over 50 times this much in the 1960's. The mode of action of the phenoxies permitted an entirely new technique in plant control. Prior to 2,4-D, herbicides relied upon caustic or poisoning properties, killing only the plant tissue contacted. Hormonal herbicides, on the other hand, cause the plant to decline from physiological upset. Respiration and other phases of catabolic metabolism are increased, as is usually cell proliferation. The normal cell arrangements and conductive channels are disrupted and the food reserves exhausted. The plant curls, and becomes distorted; eventually, if given sufficient treatment, the whole plant dies (not just the parts treated). Agriculturally the phenoxies may be used as either spray or dust. The herbicides are usually applied directly to the vegetative parts of the plant to be killed, but in the case of woody plants sometimes to wounds opened on the trunk. With most plants the seedling stage is the most susceptible, although there also occur genetical, seasonal, and temporal differences in susceptibility.

The manufacture of 2,4-D is by the chlorination of phenol, which yields a mixture of chlorinated compounds from which the compound with chlorine in the 2,4 position is readily separated by distillation. This compound is made into the sodium salt and combined with chloracetic acid to give 2,4-D. It is inexpensive, without offensive odor, is reasonably nontoxic and noncorrosive; it is herbicidally effective in minute quantities (generally used about 1 lb per acre for broadleaf-weed control). When first commercially available in 1945, it cost $125 per lb, but within a year its price had dropped to $3. Since then 2,4-D has become a byword for the home gardener as well as professional horticulturists and agriculturalists. In very light dosage it is often a growth stimulant, similar to the auxins. And its great utility has been incentive for the discovery of many additional systemics and selective herbicides such as silvex (2,4,5-TP), dicamba, the uracils, the triazines, and many others.

The Gibberellins. A disease of rice caused by the fungus *Gibberella fujikuroi*, Ascomycetae, has long been known to cause excessive lengthening and subsequent toppling of rice plants in the Orient. Japanese researchers extracted and chemically identified the substance in the 1950's, giving it the name "gibberellin" (as noted, there are numerous chemical variations). In minute quantities the gibberellins, usually applied as a spray, cause marked stem elongation in many species, overcoming both physiologic and genetic dwarfing controls. Gibberellin can often substitute for cold and light requirements in encouraging the flowering and development of plants, and can promote fruit setting, fruit enlargement, elimination of seeds, stimulation of sprouting, the overcoming of dormancy, and so on. Practically, gibberellin has revolutionized the growing of Thompson seedless grapes, substituting for hand thinning, increasing the size of fruit, and helping prevent fruit disease. Much of the navel orange crop is sprayed, to keep the fruit rind juvenile and postpone its deterioration. Gibberellin has also been used in the production of hybrid cucumber seed, to break dormancy with seed potatoes and lettuce seed, and to delay maturity with many fruits and crops. It increases yields of hops and sugar cane, reduces cold requirement of rhubarb, improves the malting quality of barley, and increases fruit set with a number of crops.

Other Growth Regulants. By early 1971 the cytokinins received only experimental release. But they do seem useful for prolonging the storage life of green vegetables, suppressing respiratory activity, and holding the commodities green for several days longer than would be normal. At least one of the cytokinins, zeatin, occurring naturally in corn, has been synthesized. Maleic hydrazide is widely available, used to reduce the need for roadside mowing and tree trimming under powerlines. As a postharvest foliar spray it checks sprouting of onions and potatoes and it has been utilized to control sucker growth in tobacco. Cycocel, alar, and phosphon have been used extensively on ornamentals to regulate height. Cycocel can increase the yield of wheat by encouraging tillering while preventing lodging (because of shorter, thicker stems). Cycocel and alar are useful for lengthening the life of cut flowers, for increasing cold resistance in cabbage, and for promotion of fruit set on a number of crops. Alar sprays delay deterioration of mushrooms and harvested lettuce; application at proper concentration to fruit trees reduces shoot growth, promotes flower formation the following season, often delays spring flowering (and thus is protective against frost), and has physiological influence that is often beneficial on a number of fruit crops. Several regulants, including some of the synthesized herbicides, increase protein content of food plants, potentially useful in lands of inadequate dietary balance.

Tobacco—Insecticide, Medicinal, Masticatory, and Fumitory

One of America's gifts to the Old World, tobacco, was introduced there first for its supposed medicinal virtues. In the seventeenth and eighteenth centuries it was rec-

The tobacco plant. (Courtesy Chicago Natural History Museum.)

ommended as a panacea, reportedly capable of curing dozens of illnesses including the common cold, halitosis, toothache, rheumatism, indigestion, blood poisoning, cancer, hydrophobia, and numerous other ailments. In short order, however, it became the world's foremost fumitory, and in some parts of the globe a masticatory to rival coca and betel. Then, as the war against insects to insure greater and more certain crop production captured world concern, the alkaloid nicotine was extracted, or the tobacco leaf itself was powdered, to yield one of the three most important insecticides known from plant sources. Increasingly grown, tobacco became a source of pleasure to millions as a fumitory, its most important use eventually. At the same time it fouls the air of public places, wastes untold productive effort, and is an habituating narcotic linked to increased incidence of lung cancer. Certainly tobacco's importance merits major attention in a chapter devoted to extractives!

Nicotiana, the tobacco genus. Modern tobacco is largely derived from the species *Nicotiana tabacum*, Solanaceae, although *N. rustica* appears to have been the first species introduced into Europe from America. Both species are apparently tetraploids ($n = 24$). *N. rustica* is native to North America, where it was much grown by the Indians before the arrival of Columbus. *N. tabacum* appears to have been indigenous to tropical America, but is not known with certainty in the wild. A review of the genus by Goodspeed indicates that *N. tabacum* may have originated as a chance cross between ancestral forms of *N. sylvestris* and some distant relative of *N. otophora*. It appears as though perhaps the species is relatively recent, having been brought quickly into domestication after formation as an amphidiploid, possibly in middle South America. *N. tabacum* and *N. rustica* together, of the more than sixty species of *Nicotiana*, yield the

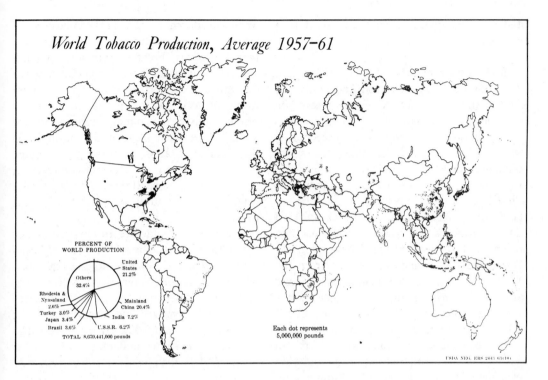

World Tobacco Production, Average 1957-61

PERCENT OF
WORLD PRODUCTION

United States 21.2%

Others 32.4%

Rhodesia & Nyasaland 2.6%

Turkey 3.0%

Japan 3.4%

Brazil 3.6%

Mainland China 20.4%

India 7.2%

U.S.S.R. 6.2%

TOTAL 8,659,441,000 pounds

Each dot represents 5,000,000 pounds

USDA NEG. ERS 2445 65(10)

astounding total of more than 4 million tons of tobacco leaf, the annual world production. *N. tabacum* is cultivated throughout the world; *N. rustica* to a lesser extent, primarily in southern Europe and the Near East.

The tobacco plant is a coarse, glandular-pubescent perennial (usually grown as an annual), with large, alternate leaves and a terminal panicle of five-parted, tubular flowers. Stamens are borne on the corolla tube, and the many ovules yield numerous minute seeds. The leaf is the valuable portion, but possesses a rank, harsh odor and taste until it is fermented, a process of curing and aging that gives tobacco leaf its pleasant aroma. Selected strains of this once-tropical plant can be cultivated as annuals as far north as 60°N and as far south as 40°S. North America and Asia each produce over one-fourth of the world's tobacco, Europe (including the U.S.S.R.) nearly 20 per cent. The chief producing countries are the United States (over 1 million tons), China and India

(nearly a half-million tons each), followed by Brazil, Japan, U.S.S.R., Turkey, Greece, and Rhodesia all closely bunched.

The alkaloid nicotine ($C_{10}H_{14}N_2$) is the most characteristic constituent of tobacco, although nornicotine ($C_9H_{12}N_2$) is more abundant in some strains. Nicotine was first isolated in 1828. It volatilizes easily with steam (heat) to give the characteristic irritating vapor. It is lethal to most insects as a contact poison. Tobacco smoke contains the common gases nitrogen, oxygen, and carbon dioxide, and in addition small quantities of the poisonous carbon monoxide, hydrogen sulfide, hydrocyanic acid, and ammonia. Among the liquids and solids in tobacco smoke are nicotine and its derivatives; pyridine compounds; resins; essential oils; methyl alcohol; acetone; formic, butyric, and acetic acids; and phenols. Nicotine is the most important physiological factor, being responsible for the narcotic and soothing qualities of tobacco. Smoking and chewing tobacco usually contain 1 to 3 per cent of this alkaloid.

Brief history of tobacco use. It is certain that tobacco was much used and cultivated by the Indians of the Americas before the Conquest. It was considered as of a semi-sacred nature. On ceremonial occasions it was smoked, chewed, or taken as snuff, often in association with other herbs or barks. In 1558 Francisco Fernandez, physician to Phillip II of Spain, sent tobacco to the homeland from Mexico, as a medicinal. In 1565 Sir John Hawkins carried Florida tobacco to England. By then the Spanish were already employing cigars made of tobacco from the West Indies. About 1579 Jean Nicot, French ambassador to Portugal in whose honor Linnaeus named *Nicotiana*, introduced tobacco to the court of France. John Rolfe initiated commercial tobacco culture in Virginia in 1612, and thus began the ruination of some of the best lands in the New World. From then on, growing and export of tobacco were to constitute a chief means of sustenance for the colonies, and tobacco remained an important item of commerce for the new nation that sprang from them.

Of the ways of smoking tobacco, the Spanish generally favored the cigar. The English, however, adopted the gentlemanly custom of pipe smoking popularized by Sir Walter Raleigh. The most elaborate form of pipe smoking, too extravagant for English decorum—use of the ornate "hookah" or waterpipe—was widely utilized in the Near East. The French were the chief takers of tobacco as an inhaled powder—snuff—a practice accepted elsewhere in a minor way.* But it was not until western European armies

were introduced to the Turkish custom of smoking cigarettes during the Crimean War that cigarettes became generally popular and accepted in the West. Since that time the cigarette has far surpassed in favor all other forms of tobacco and ranks today as the leading tobacco product, accounting for about 83 per cent of tobacco consumed in the United States. Interestingly enough, American cigarette manufacturers consider it essential to blend domestic tobacco with smaller-leaved, untopped, aromatic Turkish tobaccos.

Cigarette manufacture follows essentially the following procedure. The properly aged tobacco goes to a conditioning chamber, where it is uniformly remoistened. It is then "stemmed," the coarse petioles and large veins being removed by machine and the leaf cut into "strip." The strip goes through an ordering machine, undergoes some further aging, and is then ready for blending with similarly conditioned Turkish tobaccos. Humectants (moisture-holding substances) such as glycerine, diethylene glycol, or concentrated apple juice are applied in the proportion of 2 to 4 per cent. Flavorings, including sugars, licorice (mostly for chew tobacco), coumarin, rum, menthol, and the like, are also added. Domestic and Turkish tobaccos are thoroughly mixed in the rotatory blending drums and sent to the high-speed cutting machines for shredding. Shredded tobacco is given a final mixing or blending before passing to the "making" machine, where the tobacco is mechanically rolled in the proper type of cigarette paper. The resultant slender strand is cut to the proper length, provided a filter in many brands, and the cigarettes automatically packaged

---

* An 1840 London testimonial to snuff reportedly read as follows: "Sir—I was born stone-blind: and so continued up to my seventeenth year; when upon taking one small pinch of your infallible and miraculous regenerating, penetrative sight-perpetrator, my eyes opened, strong and clear as those of Argus. I have ever since taken about a pound and three-quarters a day of my inanimate animator, or second parent; am now 96, can read the smallest type without glasses by moonlight and drink barrels of the most potent beverages without the dream of a headache!"

at the rate of thousands per minute. The manufacture of cigars, chewing tobacco, pipe tobacco, and snuff is somewhat less elaborate.

TOBACCO CULTURE AND CURING. Tobacco growing, being practiced the world over in many climates and on diverse soils by many races of men, has naturally become locally specialized. Many of the named grades depend upon the local cultivation techniques and region where produced for their characteristics. Nevertheless, certain broad principles of tobacco culture apply in all parts of the globe. For one thing, tobacco is started in seedbeds and transplanted to fields as seedlings. This is necessary because of the fineness of the seed and its inability to produce many surviving seedlings in competition with weeds or under harsh environmental conditions. The seedbed is usually sterilized, in technologically less-advanced areas often by a wood fire built over the soil, but increasingly with chemicals such as methyl bromide. Tobacco is quite susceptible to a wide variety of diseases and nematodes, being in this respect one of the "touchiest" of the important market plants: hence the need for soil fumigation and subsequent disease control. The minute seeds are mixed with sieved soil, sand, ashes, or cornmeal, and then broadcast on the seedbed; or they may be dispersed in water and spread with a sprinkling can. One ounce of seeds suffices to plant over 2,000 square feet of seedbed, providing 60 seedlings per sq ft. Hybrid seed is increasingly being used in the burley belt of the United States, with over three-fourths of the crop planted to hybrid tobacco. Hybrid seed is produced by hand pollination of a male–sterile line; the resulting seed is worth more than its weight in gold. Hybrid seed is said to give a more uniform, quicker-starting, and higher-yielding crop.

Seeds may be started in hotbeds, cold-frames, or, in warmer climates, in beds covered with cloth or plastic. The young seedlings are thus protected, with conservation of heat and moisture. Regular fungicidal treatment controls disease, and no smoking is allowed nearby lest tobacco virus by chance be introduced. In 6 to 10 weeks the seedlings are ready for transplanting ("pulling") to the fields. Transplanting may be done by hand, but in the United States is often mechanically accomplished by a planter that opens a furrow into which the seedling is deposited, closes the soil about the seedling, and applies a measured amount of water, all automatically.

Another requirement for tobacco growing, inasmuch as the plant is a heavy feeder, is a well-drained soil rich in potash and phosphorus but usually devoid of high concentrations of nitrogen. The soil must be clean-cultivated, a requirement that causes tobacco soils, like those of corn, to be subject to ready erosion. Thus tobacco should not be grown on sloping land, and the soil should be amply fertilized.

A general practice in tobacco culture is topping. Topping removes the terminal growing tip, preventing flowering and seeding and redirecting food that might be so consumed into the leaves. In conjunction with topping, de-suckering of axillary shoots must be practiced, or the foods saved by the first practice will simply go into these shoots and fail to stimulate leaf development. Chemical regulation of suckering is increasingly practiced, using regulants such as maleic hydrazide.

Harvesting may be accomplished in either of two ways. The entire plant may be cut and speared on a stick for hanging in the curing shed, or the leaves may be picked by hand individually as they mature and strung in separate bundles over sticks for curing. With the former method only a portion of the leaves can be at ideal maturity, but the labor saving is substantial; in the latter case best-quality leaf is obtained but at a greater expenditure for labor. Curing is accomplished by one of three systems: air-curing, flue-curing, or (less common) fire-curing. Air-curing is a slow natural process of preliminary fermentation in well-ventilated barns, with application of heat only under unusual circumstances (e.g., wet weather). Flue-curing is a quicker fermentation (usually in 84 to 96 hours) in small barns or sheds with graduated artificial heat from carefully regulated furnaces. Fire-curing involves smoking from slow fires, usually without material rise in temperature. Occasionally sun-curing—drying in the open on special scaffolds—is practiced, particularly in the production of oriental and Turkish tobaccos. During curing, water content of the tobacco leaf is reduced from 80 to 20 per cent, with some loss as well of dry matter, particularly starch (which is converted to sugar) and solubilization of proteinaceous substance. Five pounds of fresh leaf is needed to produce about 1 pound of cured tobacco.

After curing, hand sorting and tying of similar-grade leaves into "hands" takes place. In this form the tobacco is usually marketed, either by personal agreement or at the tobacco auctions. Thereafter the tobacco is aged, undergoing a more complete fermentation that develops the final aroma.

This fermentation usually takes place in large warehouses, with the tobacco properly moistened and packed in huge hogsheads or other boxings, and may continue for 6 months to 4 years.

It is not easy to make rhyme or reason of the classification of tobacco types: the names are based largely on custom. Usually broad groupings are made on the basis of strain of seed planted, method of curing and growing, locality where produced, color, and use to which put. Thus there occur such designations as "white burley," "Maryland," "shade-grown," "piedmont," "mahogany," "filler," "perique," and multitudes of others. Individual farmers generally distinguish between "leaf" and "lugs," the latter consisting of leaves from the lower half of the plant; there are also many finer designations. The U. S. Department of Agriculture established a system of classification broken down into classes, types, and grade groups. According to this classification there are seven classes (see Table 12-3): flue-cured, fire-cured, air-cured, cigar-filler, cigar-binder, cigar-wrapper, and miscellaneous. Flue-cured tobacco accounts for more than half of domestic production, mostly used for cigarettes, and the chief type that is exported. Cigar-wrapper types are the most difficult

A floor of Burley tobacco ready for sale at auction. (Courtesy USDA.)

Table 12-3

Volume of Production, Domestic Disappearance and Exports, and Domestic Usages
in Manufactured Tobacco Products, by Kinds in Mid-1960's

Class	Typical Production		Disappearance		Usage in Products	
	Quantity	Percentage of Total	Domestic	Exports	Principal	Other
	Millions of Pounds	*Percentage*	*Percentage*	*Percentage*		
Flue-cured	1,225	59.2	62	38	Cigarettes	Smoking, chewing
Fire-cured	50	2.4	54	46	Snuff	Chewing, strong cigars
Air-cured:						
Burley	600	28.9	90	10	Cigarettes	Smoking, chewing
Maryland	35	1.7	64	36	Cigarettes	Cigar filler
Dark	22	1.1	74	26	Chewing	Smoking, snuff, cigar filler
Cigar-filler	90	4.3	99	1	Cigar filler	Scrap chewing
Cigar-binder	30	1.5	90	10	Scrap chewing	Cigar binder
Cigar-wrapper	18	0.9	70	30	Cigar wrapper	—
Total	2,070	100.0	73	27		

From U. S. Department of Agriculture, Consumer and Marketing Service, Washington, D.C., Miscellaneous Publication No. 867, 1966.

to produce, requiring elastic leaves free of blemish; they are grown under the protection of cloth screening ("shade-grown").

Outside the United States the best-known producing countries include: Puerto Rico, cigar tobaccos; the Philippines, filler-leaf cigar tobacco; Canada, flue-cured tobaccos; Cuba, outstanding cigar tobaccos; Sumatra, cigar-wrapper leaf; and Greece and Turkey, Turkish and oriental cigarette tobaccos. In each case varying local production methods are followed, with hand labor, always high in tobacco production, typically playing a greater part than in the United States.

SUGGESTED SUPPLEMENTARY REFERENCES

Akehurst, B. C., *Tobacco*, Humanities Press, N. Y., and Longmans, London, 1969.

Barron, F., M. E. Jarvik, and S. Bunnell, "The Hallucinogenic Drugs," *Scientific American* **210,** 29–37 (April, 1964).

Brooks, J. E., *The Mighty Leaf*, Little, Brown Co., Boston, 1952.

Garner, W. W., *The Production of Tobacco*, Blackiston, N. Y., 1951.

Goodspeed, T. H., *The Genus Nicotiana*, Chronica Botanica, Waltham, Mass., 1954.

Henry, T. A., *The Plant Alkaloids*, Blackiston Co., Philadelphia, 1949.

Hoffer, A., and H. Osmund, *The Hallucinogens*, Academic Press, N. Y., 1967.

Hofmann, A., in A. Burger, *Chemical Constitution and Pharmacodynamic Action*, M. Dekker, N. Y., 1968.

Jackson, B. P., and D. W. Snowden, *Powdered Vegetable Drugs*, American Elsevier, N. Y., 1968.

Jacobson, M., *Insecticides From Plants*, Agriculture Handbook 154, USDA, Washington, D. C., 1958; catalogue of hundreds of species.

Kaplan, J., *Marijuana—The New Prohibition*, World, N. Y., 1970.

Kingsbury, J. M., *Poisonous Plants of the United States and Canada*, Prentice-Hall Inc., Englewood Cliffs, N. J., 1964.

Krochmal, A., R. S. Walters, and R. M. Doughty, *A Guide to Medicinal Plants of Appalachia*, USDA Forest Service Research Paper NE 138, Washington, D. C., 1969.

Lewin, L., *Phantastica: Narcotics and Stimulating Drugs, Their Use and Abuse*, E. P. Dutton Co., N. Y., 1964.

Lloydia, a quarterly journal of pharmacognosy: various items.

Muenscher, W. C., *Poisonous Plants of the United States*, Macmillan Co., N. Y., 1961.

Raffaut, R. F., *A Handbook of Alkaloids, and Alkaloid-Containing Plants*, John Wiley and Sons, N. Y., 1970.

Schmutz, E. M., B. N. Freeman, and R. E. Reed, *Livestock-Poisoning Plants of Arizona*, Univ. of Arizona Press, Tucson, 1968.

Schultes, R. E., "Hallucinogens of Plant Origin," *Science* **163**, 245–54 (1969).

Taylor, N., *Plant Drugs that Changed the World*, Dodd, Mead, N. Y., 1965.

USDA, *Tobacco in the United States*, Misc. Publ. 867, Consumer and Marketing Service, Washington, D. C., 1966.

Usdin, E., and D. H. Efron, *Psychoactive Drugs and Related Compounds*, U. S. Public Health Publ. No. 1589, Washington, D. C., 1967.

Vogel, V. J., *American Indian Medicine*, Univ. of Oklahoma Press, Norman, 1970.

Von Euler, U. S. (ed.), *Tobacco Alkaloids and Related Compounds*, Macmillan Co., N. Y., 1965.

Willaman, J. J., and B. G. Schubert, *Alkaloid-Bearing Plants and Their Contained Alkaloids*, USDA Techn. Bull. 1234, Washington, D. C., 1961.

Williams, L. O., *Drug and Condiment Plants*, USDA Agriculture Handbook 172, Washington, D. C., 1960.

Woodson, R. E., H. W. Youngken, E. Schlitter, and J. A. Schneider, *Rauwolfia: Botany, Pharmacognosy, Chemistry, and Pharmacology*, Little, Brown, N. Y., 1957.

Various items in the periodicals *Economic Botany*, *Natural History*, *Science*, *Scientific American*, and *World Crops*.

CHAPTER 13

Vegetable Oils, Fats, and Waxes

VEGEBABLE OILS, FATS, AND WAXES ARE A far more important class of extractions from the standpoint of volumes and values than the tannins, dyes, essences, and flavors, discussed in previous chapters. The production of prominent plant oils is frequently on the order of hundreds of thousands and even millions of tons annually, and oils, fats, and waxes are necessary in one way or another to a variety of the great industries of the world. Within the limits of the world's arable lands, plant sources potentially offer almost illimitable supplies of these essential items, for oils, fats, and waxes are common in many plant parts of many plant species, and can be readily produced by agricultural means when demand and prices make this profitable. Most of the oils used today for margarine, paints, soaps, and similar familiar products come from cultivated plant sources, such as soybean, cotton, flax, castorbean, tung, and olive. A few important wild plant sources still exist in the tropics, notably the various oil- and wax-yielding palms.

Chemically oils, fats, and waxes are very similar. Waxes are fatty-acid esters of monohydric alcohols. Waxes are impervious to the passage of water, and usually serve plants as a protective coat (warding off desiccation) on the surface of leaves or stem. Oils and fats are the glycerides (trihydric) of similar organic acids. The distinction between oil and fat is largely a physical one: fats are solid or semisolid at ordinary temperatures whereas oils are fluid. Quite obviously an "oil" at one climate or season may be a "fat" at another. Oils are subdivided, according to their ability to oxidize into a tough film, into nondrying or drying types; or according to their usage into groups such as edible, soap, industrial, and miscellaneous. All vegetable oils, fats, and waxes seem to be synthesized by plants from the elementary carbohydrates of photosynthesis.

The utility of oils, fats, and waxes to the plant itself is less obscure than in the case of tannins, dyes, and essential oils. It is well known in plant and animal metabolism that fats and oils constitute an abundant and convenient reservoir of stored energy. They contain about twice as many calories per gram as either carbohydrates or proteins. Fats and oils are found abundantly in seeds, which must provide energy for early growth of the seedling. Man's most important plant oils come from seeds. As a general rule seeds rich in oil are also rich in protein, a valuable correlation in choosing and developing food

Table 13-1

Oil Content of Seeds by Family (range and mean shown for the samples analyzed in each family)

Family	No. of Samples	Oil Content %
Myristicaceae	1	
Thymelaeaceae	1	
Olacaceae	1	
Simarubaceae	6	
Theaceae	2	
Symplocaceae	1	
Magnoliaceae	2	
Staphyleaceae	1	
Oxalidaceae	1	
Martyniaceae	5	
Styracaceae	2	
Lauraceae	3	
Buxaceae	2	
Sapindaceae	10	
Calycanthaceae	2	
Cucurbitaceae	59	
Casuarinaceae	1	
Meliaceae	3	
Bombacaceae	3	
Araliaceae	3	
Campanulaceae	5	
Lardizabalaceae	2	
Apocynaceae	6	
Loasaceae	3	
Papaveraceae	18	
Euphorbiaceae	37	
Cornaceae	5	
Taxaceae	3	
Saxifragaceae	3	
Ericaceae	5	
Celastraceae	3	
Resedaceae	2	
Guttiferae	2	

Family	No. of Samples	Oil Content %
Oleaceae	7	
Aceraceae	6	
Ulmaceae	5	
Liliaceae	40	
Rutaceae	9	
Verbenaceae	9	
Portulacaceae	1	
Umbelliferae	80	
Fouquieriaceae	1	
Aquifoliaceae	3	
Malvaceae	81	
Elaeagnaceae	2	
Primulaceae	1	
Fagaceae	5	
Menispermaceae	2	
Betulaceae	5	
Zygophyllaceae	5	
Lythraceae	6	
Tiliaceae	3	
Sapotaceae	2	
Loganiaceae	2	
Amaryllidaceae	7	
Cochlospermaceae	2	
Geraniaceae	3	
Berberidaceae	1	
Anacardiaceae	19	
Polygalaceae	1	
Iridaceae	5	
Pittosporaceae	2	
Phytolaccaceae	1	
Rubiaceae	7	
Juglandaceae	1	
Theophrastaceae	3	

Left chart families:

Family	Count
Cruciferae	130
Capparaceae	15
Sterculiaceae	5
Rhamnaceae	10
Moraceae	6
Asclepiadaceae	10
Pinaceae	11
Caprifoliaceae	7
Flacourtiaceae	2
Linaceae	5
Bignoniaceae	13
Valerianaceae	3
Dilleniaceae	2
Dipsacaceae	3
Labiatae	55
Palmae	18
Gentianaceae	1
Compositae	382
Boraginaceae	12
Myricaceae	3
Ranunculaceae	39
Onagraceae	21
Limnanthaceae	6
Urticaceae	1
Scrophulariaceae	34
Vitaceae	6
Santalaceae	1
Polemoniaceae	6
Malpighiaceae	2
Aristolochiaceae	1
Hamamelidaceae	2
Solanaceae	31
Alismaceae	1
Cactaceae	3
Rosaceae	38
Balsaminaceae	1

Oil Content %

0 10 20 30 40 50 60 70 80

Right chart families:

Family	Count
Tamaricaceae	2
Trochodendraceae	1
Convolvulaceae	12
Tropaeolaceae	1
Sabiaceae	1
Hydrophyllaceae	10
Amaranthaceae	9
Cyperaceae	12
Cunoniaceae	1
Aizoaceae	6
Nyssaceae	1
Plumbaginaceae	4
Plantaginaceae	11
Burseraceae	5
Chenopodiaceae	19
Platanaceae	2
Erythroxylaceae	1
Caryophyllaceae	21
Gramineae	129
Leguminosae	704
Nyctaginaceae	2
Hippocastanaceae	2
Juncaceae	1
Bixaceae	2
Punicaceae	1
Cistaceae	3
Ginkgoaceae	1
Polygonaceae	30
Ebenaceae	3
Araceae	1
Myrtaceae	4
Acanthaceae	2
Combretaceae	1
Cannaceae	3
Myrsinaceae	1

Oil Content %

0 10 20 30 40 50 60 70 80

From Q. Jones and F. R. Earle, *Economic Botany* **20**, 129 (1966).

crops. Starchy seeds tend to be of larger size than those rich in oil, and contain less oil and protein. Among the oil seeds, however, the larger types are generally richer in oil, the smaller in protein.

Table 13-1 shows some of the 160 "oil seed" families investigated by the USDA as potential sources of new farm crops, and the general range of oil content in seeds of the species tested. Note that oil content is frequently above 50 per cent. Of the families that are well sampled, the Simarubaceae, Sapindaceae, Cucurbitaceae, Papavaraceae, Euphorbiaceae, and Cruciferae stand out. High in protein as well as oil are notably the Cucurbitaceae and Leguminosae. In spite of their great importance, the history of oils, fats, and waxes is far less spectacular than that of dyes, essences, and spices. Because of their abundance, universality, and interchangeability, they have been largely taken for granted. Seldom have expeditions charted the unknown in their search, or have imperialists threatened far corners of the globe to secure them. Vegetable oils and waxes have attained their greatest importance in comparatively recent times, with

the development of industrial economies and concentrated, urban populations. There is every indication that plant oils will continue to play an increasingly important part in man's activities, finding increased use in food products, plastics, and various manufactures.

Fats and Oils

Fats and oils are believed to be formed in plants by the synthesis of fatty acids from carbohydrates and the combination of these through enzymatic action with glycerin to form triglycerides. Triglycerides are colorless and practically tasteless, so that taste or color in an oil is largely due to the presence of traces of essential oils or pigments. Fats and oils also contain small percentages of sterols, phosphatides, lecithins, vitamins, and like substances. Most fats and oils are unstable and become rancid when stored for considerable lengths of time, particularly at high temperatures and with free access of air. Rancidity is due to breakdown of the glycerides into various aldehydes, ketones, and the like.

Vegetable oils are seldom a single fatty

Table 13-2

Frequency of the Seven Most Important Fatty Acids in Several Representative Vegetable Oils and Two Animal Fats

Fat or Oil	Percentages of Acids						
	Lauric	Myristic	Palmitic	Stearic	Oleic	Linoleic	Linolenic
Coconut	45	20	5	3	6		
Palm kernel	55	12	6	4	10		
Beef tallow		2	29	25	44		
Lard			25	15	50	10	
Olive			15		75	10	
Peanut			9	6	51	26	
Cottonseed			23		32	45	
Maize			6	2	44	48	
Linseed		3	6			74	17
Soybean			11	2	20	64	3

Adapted from Alsberg and Taylor, *The Fats and Oils: A General Review*, Stanford, Cal., Stanford Univ. Press, 1928.

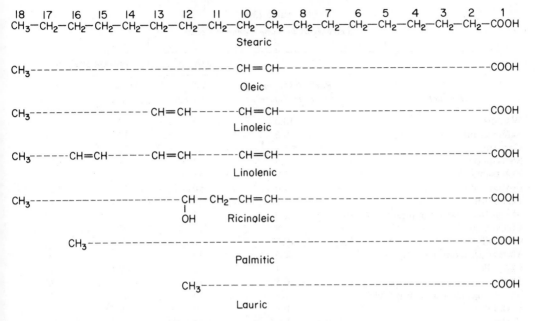

Figure 13-1 Most common fatty acids of seed oils used for nonfood purposes. (After Maclay, Matchett, and Pollack, *Economic Botany* **17**, 26 (1963).)

acid, but instead a mixture of several. Characteristics of the oils depend upon the relative abundance of the component fatty acids (Table 13-2). These can often be altered chemically, as by hydrogenation; to a high degree vegetable oils are interchangeable or substitutable. In this sort of competitive atmosphere an oil must be in constant supply at a stable price in order to command much of a commercial following.

An important consideration about a fatty acid, and therefore the oil containing it, is its degree of saturation. The more the double bonds (see the sequence from saturated stearic through progressively less saturated

oleic, linoleic, and linolenic, Fig. 13-1), the more prone is the oil to oxidize as a waterproof film (i.e., be a "drying oil"). A familiar categorization of vegetable oils is as follows:

Nondrying oils—largely glycerides of saturated and oleic acids, with little or no linoleic or linolenic acid present, remaining fluid for prolonged periods upon exposure to air, and having a low (less than 100) iodine number*; they are found notably in tropical plants. Included are such oils as palm, peanut, olive, castor, rape, and almond.

Semidrying oils—intermediate between the two other categories, without appreciable

---

* The ability of an oil to absorb iodine correlates highly with its drying properties, so that the higher the iodine number the more capable of oxidation to an elastic film is any given oil. One of the fundamental statistics for any commercial plant oil is its iodine number. Iodine numbers for some common vegetable oils are: linseed, 173–201; tung, about 171; soybean, 137–143; sunflower, 119–135; corn, 111–130; cottonseed, 108–110; sesame, 103–108; rapeseed, 94–102; peanut, 83–100; olive, 79–88; palm, 51–57; cacao butter, 32–41; palm kernel, 13–17; coconut, 8–10. By way of comparison, iodine numbers for some animal oils are: menhaden, 139–173; whale, 121–147; lard, 46–70; beef tallow, 38–46. [Adapted from Alsberg and Taylor, *Fats and Oils: a General Review*, Stanford Univ. Press, Stanford, Calif., 1928.]

Table 13-3

Fatty Acids in Some Common Foods

Product	Ratio of Polyunsaturates to Saturates	Fatty Acids as Percentage of Total Fat		
		Poly-unsaturates	Mono-unsaturates	Saturates
Walnuts	10.0	72	17	7
Safflower oil	9.0	72	15	8
Corn oil	4.8	53	32	11
Soybean oil	3.9	59	20	15
Fish (salmon)	3.5	53	25	15
Cottonseed oil	2.0	50	21	25
Peanut oil	1.6	29	47	18
Margarine (containing liquid corn oil)	1.5	27	51	18
Chicken fat	0.8	26	38	32
Olive oil	0.7	8	76	11
Margarines (conventional)	0.5	13	57	26
Egg yolk	0.4	12	49	32
Lard	0.3	11	46	38
Shortenings (animal + vegetable)	0.3	12	41	43
Pork fat	0.2	9	49	38
Butter	0.07	4	35	55
Lamb fat	0.07	4	36	56
Beef fat	0.06	3	44	48
Chocolate	0.04	2	37	56
Coconut oil	—	Trace	7	86

From A. L. Elder and D. M. Rathmann, *Economic Botany* **16**, 199 (1962).

linolenic acid but with considerable amounts of linoleic and more saturated acids, drying slowly and at elevated temperatures, and with an iodine number 100–130. Included in this category are such oils as cottonseed, sunflower, sesame, croton, and maize.

Drying oils—high in glycerides of the unsaturated type, particularly linoleic and linolenic, but low in oleic compounds; drying or absorbing oxygen readily from air to form an elastic film; and having a high (over-130) iodine number. These oils are found notably in temperate plants. Included are such oils as linseed, soybean, tung, hempseed, nut, poppy, and safflower.

In general nutrition, lack of the proper fatty acids can result in deficiency symptoms marked by skin eczemas and other disorders. Linoleic oil is converted by the organism into the essential arachidonic acid (which seldom occurs in vegetable oils), and should be a major element in food oils. Linolenic oil is partially sufficient. Traditionally, however, edible oils have been chosen mostly for flavor and availability rather than fatty-acid content. Table 13-3 lists several common food oils, indicating their degree of saturation. Unsaturated types are less apt to cause high cholesterol levels in the blood; table spreads and cooking oils predominating in unsaturated fatty acids have gained favor in the campaign to combat heart disease.

Extraction methods. Oils and fats are contained as insoluble droplets or deposits within the cells of plant tissue. They must be freed for use by man by breaking and crushing the cells, thus permitting the escape of oil or its dissolution by suitable solvents. Most oils are derived from seeds.

Pressing olives in North Africa by a primitive means of expression. (Courtesy Chicago Natural History Museum.)

The harvested seeds are thoroughly winnowed and freed from remaining impurities by screening and passage over magnets. Decorticating machines crack the seed coats where necessary and remove the husks from the kernels, the final separation usually being accomplished by air-blast or sieving or, less frequently, by flotation. The kernels are next crushed between millstones or rollers, to a varying degree, depending upon the seed in question and method to be followed in further processing.

The methods of fixed oil extraction are not greatly different from those described for essential oils. They involve expression, solvent extraction, boiling or centrifuging, or even bacterial fermentation. Expression is the most important of these methods, and may be of either the hot or the cold type. Hot expression is the more common, and involves precooking of the kernel, usually with steam, to facilitate oil flow. Cold expression involves mere pressing of the kernel without preliminary treatment. In either case the expressed oil drains to settling tanks, where the clear oil at the top is withdrawn from the "press foots." The withdrawn oil may be passed through filter presses, and, if needed, clarified by heating (to coagulate albuminoids, for example) or by agitation with caustic soda (to saponify free fatty acids). In some cases bleaching or deodorization are also practiced. The press cake left in the hydraulic press or expeller is used in cattle feeds, and in some instances is almost as important a product as is the expressed oil.

Solvent extraction is the only other important method of procuring fixed oils. In this method particular attention must be paid to preliminary grinding, for quite obviously the solvent must have access to all oil-bearing cells to function economically, and yet must not be impeded in flowing by a compacted mush of too finely pulverized material. A number of solvents can be used, including gasoline, benzene, various chlorinated ethylenes, carbon disulfide, etc. The solvent is usually evaporated off at temperatures of 60° to 110°C, leaving as a residue the extracted oil. Many different milling procedures and types of equipment, which cannot be discussed here, are of course available; the latter vary from continuous to batch extractors. Solvent extraction is often practiced on the press cake remaining after expression, to obtain the last small percentage of oil.

Saponification and soap making. Saponification of fats and oils serves to form soaps. It is a hydrolytic action in which the triglyceride is split into three fatty-acid molecules and one molecule of glycerin. Combination of the fatty acids with a metal yields the soap. Hard soaps are formed with sodium, soft soaps with potassium; rarely salts of other metals, such as lead (medicinal soaps), zinc (ointment soaps), and aluminum (waterproofing soaps), are utilized. Soaps may be made in one operation or in two. In the former case the fat or oil is treated with the appropriate amount of caustic soda (to make soap and glycerin), to which is added common salt (to form brine), in which soap is insoluble. The mixture separates into an upper layer of pure soap and a lower layer of brine and glycerin. In the two-operation method the fat is first chemically separated into glycerin and fatty acid; to the latter is then added caustic soda to make the soap. The number of milligrams of potassium hydroxide required to convert 1 gram of fat completely into glycerin and potassium soap is known as the saponification number. It serves particularly as an index of the solubility of the soap in water: the higher the number the greater the solubility.†

Soaps were apparently unknown to the ancients, although natural substances from certain barks are used by primitive peoples. The first deliberate making of soap seems to have been by the Gauls about a century after they were conquered by Rome. Goat tallow (the fat) was boiled with beechwood ashes (the alkali) to make a combination pomade and hair dye, which, it was soon found, strangely made suds in hot water that were capable of cleaning skin, clothing, and utensils with a rapidity never before known. Later, during the Dark Ages, unknown benefactors of the future soap industry of southern France discovered the advisability of substituting sodium alkali for the customary potassium of beechwood ashes and of adding rosin as a hardening ingredient. About this time olive oil came to be substituted for goat tallow. But soap making remained a home operation until well past the middle of the nineteenth century. Then discovery of the chemical reactions involved in soap making, coupled with the modern availability of pure caustic soda, paved the way for the tremendous soap and detergent industry. Synthetic detergents and other surface-active agents require chemical modification of fatty-acid derivatives, especially sulfonation. Coconut is the most used oil in the soap and detergent industry. The by-product of soap making, glycerin, has proved useful for many industrial purposes.

Hydrogenation and cooking fats. Hydrogen can be added to unsaturated fats and oils to bring them to saturation or to a higher degree

---

† Some comparative saponification numbers are: coconut: 246–260; palm kernel, 242–250; palm, 196–205; peanut, 190–196; cottonseed, 193–195; linseed, 192–195; soybean, 193; olive, 185–196; corn, 188–193; rapeseed, 170–179. These compare to the saponification numbers of certain animal fats as follows: butter, 220–233; beef tallow, 193–200; lard, 195. [After Alsberg and Taylor, *op. cit.*]

of saturation. This process is hydrogenation, by which, for example, oleic acid can be changed to stearic acid. Unfortunately, chemists cannot quite as easily accomplish the reverse, whereby the more valuable drying oils can be cheaply derived from the less valuable nondrying types. In hydrogenating unsaturated oils, the melting point is typically raised, so that oils of one sort become capable of substitution for others. Thus oils may be inexpensively transformed into fats. By this means the much-used vegetable shortenings are made. Commercial hydrogenation is based on the principle of adding a catalyst, usually finely divided nickel or a nickel compound, to the dry oil and heating it in a closed chamber into which hydrogen gas is forced under pressure. Later, after the oil attains the desired degree of saturation, the nickel is filtered from the newly created fat.

Margarine. Discovery of hydrogenation and the techniques of refining vegetable oils opened the way for the rise of the margarine industry. Originally, at its invention as "synthetic butter" in 1870, margarine was a weird mixture of beef fat digested with pepsin, plus casein, udder extract, and sodium bicarbonate. In the next three decades improvements were made, but the big steps forward were Wesson's discovery of how to remove objectionable taste and odor from cottonseed oil and development of the ability to hydrogenate this oil to semisolid form. A path was then open to successful utilization of a number of vegetable oils, including coconut, palm, cottonseed, and soybean. Margarine ordinarily consists of about 80 per cent refined and hydrogenated vegetable oil, some 16 per cent cultured skim milk, and small amounts of a glycerin derivative, lecithin, salt, vitamin A, and perhaps benzoate of soda.

Linoleum. Discovery of how to make linoleum occurred in 1863, when the English

inventor Frederick Walton, meditating over the scum from an open paint pot, realized he had in it a satisfactory binder for pulverized cork, until then bound with rubber in an older type of floor covering. He named the product linoleum from the Latin *linum* (flax) and *oleum* (oil). Linoleum still utilizes a felt or burlap backing, oxidized linseed oil, ground cork or wood flour, small proportions of resins, and various pigments. The linseed oil is oxidized with driers and resins in a large tank into which oxygen is forced. The oxidized product is cured, then mixed with ground cork and pigments and pressed onto the burlap or felt backing. The resulting sheets, rolled to proper thinness, are finally baked for several days. The tough, elastic, and practically waterproof qualities of the oxidized linseed oil give to linoleum much of its usefulness. In advanced economies linoleum and oilcloth have been superceded by various plastic materials for floor coverings, many of which, however, utilize vegetable oils in their manufacture.

Paints. The oil paint industry demands large quantities of drying oils, linseed, soybean, and tung being among the most used. House paints are basically such oils into which, after clarification, coloring pigments have been ground. Varnishes are generally made from highly refined linseed or treated tung oils which bleach at varnish-cooking temperatures of about 600°F, and to which resins are added. Enamels utilize these oils or perilla and other minor oils. Tung and soybean oil blends have proved effective in making enduring paints. One refrigerator finish calls for phthalic anhydride, glycerin, and soybean oil, which, after proper cooking, are dissolved in a solvent and the pigment added. Water-soluble "latex" paints have supplanted oil paints for many purposes, but these, too, may employ vegetable oils in their manufacture. Soybean and tall oil are the chief oils used in alkyd paint formulations.

Oils for other purposes. Scores of millions of pounds of epoxidized vegetable oils are used in plasticizer manufacture, especially for polyvinyl chloride; castor, palm, and coconut oils are much utilized. As lubricants, vegetable oils have been largely replaced by mineral oils, but they still find some use in lubricating greases, and as cutting oils for machine tools. Only nondrying oils are suitable for these purposes, for other types, of course, would become sticky and gummy. Candle making today, in contrast to the heyday of whaling, when spermaceti was abundant, relies largely upon free fatty acids, particularly stearic and palmitic, which may on occasion come from plant sources. Glycerin, derived from saponification of vegetable fats and oils, serves to make nitroglycerin, antifreeze, cellophane, plastics, and film. It also finds use as a component of certain pharmaceuticals. Various fatty acids and oils are used for hot-dip treatment of metals, in making synthetic rubber, buffing compounds, insulation coatings, adhesives, printing inks, insecticides; in processing leather, rubber, and textiles; and in many other ways.

The Important Oil Plants

Nondrying types. Of the several score of nondrying oils castor, olive, palm, and peanut oils stand out.

CASTOR. One of the world's important industrial oils, castor oil, is obtained from the seeds of *Ricinus communis*, of the Euphorbiaceae, a widely adapted herbaceous shrub probably indigenous to Africa and today found wild and in cultivation throughout the world. Chromosome number is $2n = 20$, and tetraploid cultivars have been bred in Africa. The plant is generally treated as an annual, although there are perennial types in the tropics; it is frequently planted as a summer ornamental in temperate regions. It attains a height of several meters, has coarse stems, large, palmately lobed leaves, inconspicuous flowers, and spiny capsules containing large, mottled seeds. Oil content of the seeds ranges from 35 to 55 per cent. Production is chiefly from Brazil, India, Thailand, Manchuria, and Mexico, where cultivation is usually a local family operation. World production exceeds a half-million tons annually, with the United States as the largest importer. Harvest is ordinarily by hand, and hence production is an enterprise of cheaplabor areas. However, dwarf strains of castorbean have been bred that hold their seeds and are amenable to machine harvest. With these, planting, cultivation, and harvest can be adapted to mechanized agriculture, with yields sometimes more than 1 ton per acre.

The oil is obtained from the seed chiefly by cold expression. The press cake left is

Harvesting a field of Cimarron castorbeans mechanically. (Courtesy USDA.)

DRY FILLED **SEGMENTS**

CLEAN SEEDS **HULLS**

SPIKE **BROKEN AND DECORTICATED** **LIGHT AND " POPS"**

Castorbean inflorescence and seeds, source of a premier industrial oil. (Courtesy USDA.)

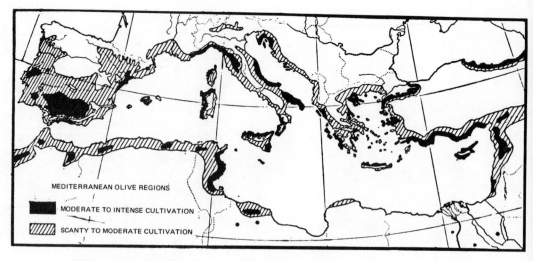

Figure 13-2   Olive producing areas of the Mediterranean region. About 99% of the world's olive crop is produced in area shown by map. (Courtesy *Economic Botany*.)

ordinarily ground and subjected to solvent extraction. The ultimate residues or extractions contain a toxic alkaloid, ricin, and are not usable as edible oils or stock food without further treatment. Castor pomace is much used as a fertilizer in India. Castor oil was used by the ancient Egyptians as an illuminant, but today finds greatest use as an industrial raw material. It is used as a lubricant; in manufacture of some soaps, synthetic rubber, imitation leather, linoleum, oilcloth, plastics, typewriter ink, baking finishes; and as an ingredient of nylon. Dehydrated castor oil, derived by heating the oil with suitable catalysts, serves as a drying oil used in mixture with more expensive oils in the paint industry. Sulfonated castor oils are used as a dyeing aid, particularly with alizarin; as a medicinal; in plasticizers; and in finishing fabrics and leather.

OLIVE. The ancient and valuable olive oil is obtained from the common olive tree, *Olea europea*, of the Oleaceae, of which there are hundreds of cultivars, some developed solely for oil extraction. The species is a small evergreen tree native to western Asia, extensively grown in the Mediterranean basin

(Fig. 13-2) since the beginning of historical times, and more recently in Australia, South Africa, Mexico, California, and to a slight extent in South America. Propagation of select strains is by cuttings, layering, or budding. Trees require 10 or more years to reach a profitable bearing state, and domestic strains are frequently grafted onto wild stock in the Mediterranean area. The tendency toward alternate years of heavy bearing is counteracted in California by hormonal sprays. With olives grown for oil, thinning to control fruit size is not so important. Olives contain 14 to 40 per cent oil. Many orchards are small, local enterprises. World production runs between 1 and 2 million tons annually, with Spain, Italy, Greece, Turkey, and North Africa the leading producing centers. Several European countries and the United States are the chief importers. Quite obviously olive growing is a long-term project, and production cannot be increased suddenly with demand, as can be done in the case of annual crops.

Harvest begins in late autumn, just before the fruit reaches maturity. In harvesting oil olives, a cloth is usually spread beneath the trees and the fruits are knocked onto it from the trees with poles or rakes. Ideally the gathered fruit should be processed immediately, but this is possible only at the larger, better-equipped mills. Extraction is principally by expression after the olives have been freed of foreign matter. Usually two to four pressings are made, the first yielding the premier-quality, greenish-yellow "virgin" oil, which can be utilized without refining. Equipment and techniques vary from the archaic to the most modern, but in the more up-to-date mills, the first pressing does not crush the seed. After the second pressing the cake is solvent-treated, or pulverized under water and the last traces of oil separated by flotation. Press cake, before solvent action, may contain as little as 4 per cent to as much as 20 per cent oil. Other less-used techniques involve grating or beating followed by centrifuging or roller crushing, with heating of the pomace in water. Final residue may be used as a mulch in the orchards, as a cattle feed, or even as a fuel for the boilers. The expressed oil is separated from the aqueous liquor by settling or centrifuging, and is further filtered, refined, and blended as needed. Highest-grade olive oil finds use primarily as a salad or cooking oil; later pressings or solvent-extracted portions are used to make soaps. Some oil is used as a lubricant and in medicinals.

PALM. If under the heading of palm oil is included all oil derived from members of the Palmae, the palm family, embracing the coconut (*Cocos nucifera*), the African oil palm (*Elaeis guineensis*), the babassú and other South American palms (principally *Orbignya oleifera*, *O. martiana*, *Cocos coronata*, and species of *Attalea* and *Astrocaryum*), production is equal to all other sources of nondrying oil combined. Marketed production exceeds 4 million tons annually; and in many undeveloped tropical areas more than half of the total production is consumed locally, never reaching the market, and thus is not included in the statistics of annual production. Palms are essentially tropical plants, with slender, unbranched trunks, very large fan-shaped or pinnate leaves, large inflorescences of minute flowers, and enormous bunches of a variety of fruits commonly rich in oils. Indeed, palms may be considered the most important family of plants in the tropical economy, yielding not only edible fruits and oils but thatching, fuel, construction material, fibers, waxes, vegetable ivory, and even fishhooks (from trunk spines). One might safely say that the palms are to the tropics what grasses (including cereals) are to temperate localities.

Yet in spite of their importance palms are not widely cultivated (except as ornamentals in landscaped areas), economic reliance often being placed on the ubiquitous wild plants in undeveloped regions. Consequently production of oils from palms is apt to be crude and wasteful, without proper supervision, efficient machinery, or progressive techniques.

Extraction of the kernel from a coconut for processing into copra. Malaysia. (Courtesy USDA.)

Many importing countries still ship un-processed or only partially extracted palm nuts to domestic mills for oil extraction.

Palm oils, in the broad sense, may be extracted either from the pericarp (meso-carp) or from the endosperm of the seed. The oils derived from the first source are the "pulp oils," those from the latter the "kernel oils." In some cases both types are derived from the same fruit; in others, a palm fruit yields either pulp or kernel oil alone. All kernel oils are very similar—high in lauric acid and with high saponification numbers. Examples are the commercial "coconut oil" and "palm kernel oil." These find greatest use in the making of soaps, and yield glycerin as a by-product. Pulp oils vary, in some cases resembling kernel oil and in others olive oil. Oil palm pulp oil plays an important role in the manufacture of tin plate (the clean metal surface is protected with oil before application of the plating) and of candles. Various oils from palms are used in food delicacies, margarines, lubricants, fuel, cosmetics, and dentifrices; the press cake serves as stock feed, fertilizer, and fuel.

Most palm fruits are simply picked from the ground where they fall when ripe. Babassú palms (*Orbignya speciosa*) in the state of Maranhão, Brazil, for example, which are numbered in the millions, are a great blessing to the inhabitants who eke out

an existence gathering them. But gathering the palm fruits only partly solves the problem. Cracking the tough husks is even more laborious. Only to a limited extent has it been possible to use machinery to accom-plish cracking and kernel removal, so that over much of the tropical world this is still a hand operation.

The world's most important palm, and one of man's first oil plants, is the coconut. It is presumably native to the Old World tropics, but has been scattered by sea cur-rents throughout the coastal areas of all oceans. It has, moreover, been widely planted within the last century, and most production today comes from planted but largely untended groves in the Far East. The graceful, leaning trees, up to 100 feet tall and with picturesque crowns of about twenty pinnate leaves, may in certain cases produce 300 or 400 coconuts per tree annually, although the average may be only one-tenth this much. World production, in terms of copra, is over 3 million tons annually. The outer husk of the coconut fruit is removed, to be discarded or used as fiber (*see* Coir, Chapter 7), the "shell" cracked, and the inner "meat" usually smoke- or sun-dried to yield the oily copra of com-merce. Splitting of the coconuts is either by machine or by hand. Copra is hot-expressed, either in the producing country or by the importer, to yield a coconut oil that is sub-

Palm seeds for oil. Belem, Brazil. (Courtesy Chicago Natural History Museum.)

Seed clusters of African oil palm (*Elaeis guineensis*) about 4 years old. Peru. (Courtesy UDSA.)

sequently refined. The press cake is sold as stock feed. The Philippines and East Indies are the leading production centers, the United States the biggest importer.

The fruit of the African oil palm, abundant in the wild in the coastal areas of West Africa and cultivated in the Congo, East Indies, and tropical America, is like a small coconut. The outer husk or pericarp, which is 30 to 70 per cent oil, yields palm oil proper; the nut yields the saturated palm kernel oil of commerce, which is similar to coconut oil. Expression of the kernel is seldom carried out in the producing region, although the pulp oil has been obtained locally by crude boiling in water, usually following fermentation in holes in the ground.

In more progressive, capitalized areas there has been a trend toward steam distillation, followed by centrifuging and even hydraulic expression. The kernels, after shipment to industrialized consuming countries, are either expressed or solvent-treated. The largest producing area is Africa (Nigeria, in particular). In Indonesia special efforts were made under Dutch hegemony to develop high-yielding cultivars as a companion crop for rubber plantations. World production averages better than 1 million tons annually.

All in all the palm family, found mostly wild in the tropics, and with scores of species already utilized for oil production and as many more potentially useful but as yet untried, will continue to yield sizable amounts of commercial oils for decades to come. In fact, northern Brazil, the world's richest storehouse in quantity and variety of palms, has been only partially exploited so far for its palm oils. Some of the better-known species are listed among "Some other nondrying oils" at the end of this section.

PEANUT. The peanut, the seed of *Arachis hypogaea*, of the Leguminosae, probably is indigenous to Brazil but today is cultivated in most tropical, subtropical, and even temperate regions, particularly in India, China, west Africa, and the southeastern United States. Europe is the chief importing and processing region. In India the oil is hydrogenated to make ghee. In the United States over half of the crop is ground for peanut butter. After World War II the British undertook a gigantic peanut-growing program in East Africa, which, however, proved somewhat short of a complete success.

Arachis is a small genus of about twenty species, the commercially important *A. hypogaea* having a diploid chromosome complement of 40. The peanut, also known as groundnut, earthnut, goober, or pinder, is one of nature's more peculiar plants, in that the air-borne flowers, after pollination, are

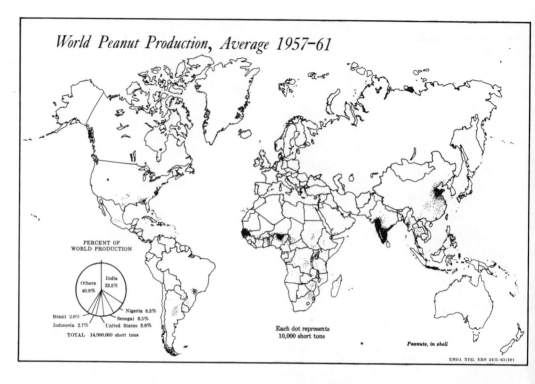

World Peanut Production, Average 1957-61

PERCENT OF
WORLD PRODUCTION

Others 40.8%
India 33.5%
Nigeria 8.3%
Senegal 6.5%
United States 5.6%
Brazil 2.6%
Indonesia 2.7%
TOTAL 14,900,000 short tons

Each dot represents
10,000 short tons

Peanuts, in shell

USDA NEG. ERS 2431-63(10)

Virginia runner peanuts on freshly pulled vine. (Courtesy USDA.)

thrust on elongating gynoecia beneath the soil, to mature there into the characteristic shelled peanut. The plants are branched, trailing annuals, with quadrifoliolate leaves, large stipules, and inconspicuous flowers. A multitude of varieties are grown—in the United States, principally the small-podded Spanish variety, or the large-podded Virginia Runner or Virginia Bunch. Peanuts thrive best in light, sandy soils. They are planted, cultivated, and harvested by hand throughout most of the world, but handled by machine (see Fig. 13-3) in the United States with average yields per acre of nearly 2 tons (in the shell). Most oil extraction takes place in Europe, chiefly with Indian and African peanuts, but India and other tropical countries are coming more and more to establish their own extraction mills. In the United States only 15 to 20 per cent of the crop is normally expressed, depending upon the relative demand for peanuts in candy, peanut butter, and salted or roasted peanuts. Before

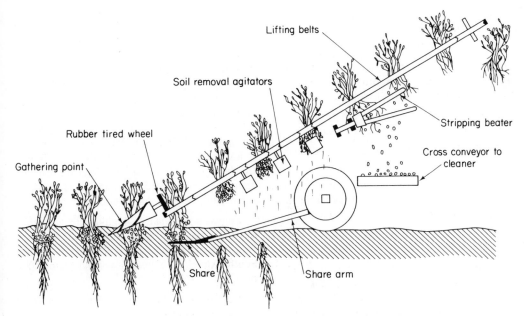

Figure 13-3 Diagram of 'once-over' harvester developed for harvesting the peanut crop in East Africa. (After J. C. Hawkins, *World Crops* **14**, 181 (1962).)

crushing, the nuts are cleaned and shelled by machine. Shelled nuts are 36 to 45 per cent oil. The same mills which crush cottonseed in the United States can also be used to extract peanut oil. Both hydraulic presses and expellers are used. Hot expression is usually practiced in the United States, but both solvent extraction and cold expression (followed by hot expression) are used to some extent in Europe.

Peanut oil finds use in margarine and shortenings, as a salad or cooking oil, in soaps, and for various industrial purposes. It is high in oleic, modest in linoleic, and low in certain other fatty-acid components. The press cake, rich in protein and containing 6 to 8 per cent oil, is a valuable livestock feed. Since the peanut is both a high-yielding crop and an annual one adaptable to market fluctuations and agricultural conditions, its future as an important oil source seems assured. World production of peanuts totals about 16 million tons annually, largely in India, China, and Nigeria.

Some other nondrying oils.

Acorn—from seed of species of *Quercus*, Fagaceae; North America.

Allanblackia—from seed of species of *Allanblackia*, Guttiferae; Africa.

Almond—from kernels of bitter and sweet almonds, *Prunus amygdalus*, Rosaceae; Mediterranean area, California.

Andiroba or carapa—from seeds of species of *Carapa*, Meliaceae; tropical America and West Africa. *C. guianensis* of the lower Amazon yields large triangular seeds from which is expressed a bitter fat rich in myristic and palmitic esters, sometimes used to make soap.

Avocado—from fleshy portion of fruit of *Persea americana*, Lauraceae; tropical America.

Batiputa—oil from fruits of the Brazilian shrub, *Ouratea parviflora*, Ochnaceae.

Ben or moringa—from seeds of *Moringa oleifera*, Moringaceae; Mediterranean area (indigenous), West Indies.

Borneo tallow—from species of several genera of the Dipterocarpaceae; Far East.

Cacao butter—from seeds of the cocoa tree (*see* Chocolate, Chapter 20), *Theobroma cacao*, Sterculiaceae; world tropics.

Cashew—from kernels of *Anacardium occidentale*, Anacardiaceae; South America, and introduced throughout world tropics.

Chinese vegetable tallow—from fruit of *Stillingia sebifera*, Euphorbiaceae; Far East.

Chufa—from rhizomes of *Cyperus esculentus*, Cyperaceae; southern Europe, Africa.

Cuphea—the seeds of *C. carthagenensis* contain about 33 per cent of an oil largely of lauric composition, possibly substituting for coconut oil.

Curuá—a kernel oil from *Orbignya* (*Attalea*) *spectabilis*, Palmae; Amazonian Brazil.

Flacourtiaceae—numerous genera (e.g., *Hydnocarpus*, *Oncoba*, *Carpotroche*, *Lindackeria*, *Mayna*, etc.) of this family yield useful seed oils, including the famous chaulmoogra oil formerly used to treat leprosy; world tropics.

Hazel or filbert—from nuts of *Corylus avellana*, Betulaceae; Europe.

Jacassar or kussum—from seeds of *Schleichera trijuga*, Sapindaceae; Far East.

Japan tallow—from berries of species of *Rhus*, Anacardiaceae; Orient.

Kapok—from seed of *Ceiba pentandra*, Bombacaceae; Far East and South America. A by-product of kapok fiber industry.

Lesquerella—several, annual, biennial, and perennial herbs of this genus of the Cruciferae abundant in the arid regions of western North America yield seeds having up to 30 per cent hydroxylated fatty-acid esters that might extend the usefulness of castor oil.

Mafura—from fruits of *Trichilia emetica*, Meliaceae; East Africa.

Murumuru—oil from the fruit of *Astrocaryum murumuru*, Palmae; Amazonian Brazil.

Myristicaceae—from seeds and aril of *Myristica fragrans* (nutmeg butter) and various species of *Virola* (Ucuhuba butter, Brazil; Ochoco butter, West Africa; etc.).

Ouricuri—kernel oil from *Cocos* (*Syagrus*) *coronata*, abundant in eastern Brazil; species is more important for leaf wax.

Paraguay cocopalm or Mbocaya—both kernel and pulp oils from *Acrocomia totai*, Palmae, abundant wild in central and eastern Paraguay and extending into contiguous Brazil, Argentina, and Bolivia. The fruit is a better source of kernel oil but a poorer source of pulp oil than African oil palm (*Elaeis*), but inadequate means for quick processing results in inferior oil with much free fatty acid. Some 2 million tons are produced annually, mainly used for soap. The fruit is also eaten, both by man and livestock.

Patauá or Batauá—a high-quality oil, about three-fourths oleic glycerides, from the outer fruit pulp of *Jessenia bataua*, Palmae, of northern South America, almost indistinguishable from olive oil.

Pili nut—from kernels of species of *Canarium*, Burseraceae; Far East.

Pracaxi—seed oil from species of *Pentaclethra*, Leguminosae, of which there is some production in Amazonia, primarily for soap.

Sapotaceae—Djave or Adjab butter is from kernels of *Mimusops djave;* West Africa. Illipé is from species of *Bassia*, indigenous to India. Other sapotaceous trees yield Shea butter (Africa), Katio oil (Borneo), Phulwara butter (India), Siak fat (Malaya), and so on.

Sawarri or suari, piqui—from seeds of species of *Caryocar*, Caryocaraceae; South America. Sir Henry Wickham placed as much emphasis on *C. villosum* as upon *Hevea* when sending seeds from Brazil to England in 1876 for use in the British colonies, but planting in Malaya never reached the importance of rubber. *Caryocar* fruit yields both pericarp and kernel oils similar to palm oil (nearly 50 per cent each palmitic and oleic fatty-acid esters), and would seem an excellent source of edible oils from a tropical crop.

Soapnut—*Sapindus trifoliatus*, Sapindaceae, of India, is a source of seeds rich in oleic components; it is cultivated in Bengal for its saponin-rich fruit that is used as a detergent. The seeds contain about 15 per cent oil high in oleic and arachidic acids, useful in soap making.

Teaseed—from seed of *Thea sasanqua*, Theaceae; Orient.

Tucum—oil from the fruit of *Astrocaryum tucuma*, Palmae, of the Amazon valley.

Ximenia—from seed of *Ximenia americana*, Olacaceae; oil from southern India.

Semidrying types. Four semidrying vegetable oils are of prime importance: corn (or maize), cottonseed, sunflower, and tall oil. The last, a by-product of kraft paper making from southern yellow pine, was discussed under forest products, and is merely mentioned here because its volume is appreciable and it serves as an inexpensive industrial raw material. The oil consists of about equal parts oleic and linoleic fatty acids, and is used chiefly for protective coatings (including alkyd paints), but also for soap–detergents and as a chemical intermediate. Of lesser but still considerable importance are rapeseed and sesame. Soybean oil is often considered a semidrying type, but, following Jamieson (see References at the end of chapter), it is listed here among the drying oils.

CORN OR MAIZE. Little need be said of maize or Indian corn, *Zea mays*, of the Gramineae, since it will be discussed later as one of the world's most important cereal plants. The species is believed to have originated in the American tropics, the modern form after introgression with *Euchlaena* (itself possibly a hybrid of ancient corn and *Tripsacum*). Today it is cultivated as an annual crop the world over in innumerable cultivars and hybrids. It is planted, cultivated, and harvested by hand in less-advanced tropical areas, but by the most efficient mechanical means in North America. Only a small portion of the world's tremendous corn crop goes into the making of corn oil, which is, as a matter of fact, little more than an important by-product of the manufacture of cornstarch and other corn products.

An individual corn grain, botanically a fruit, consists of two parts: the small, oily "germ" or embryo, constituting 6 to 13 per cent of the grain, and the more abundant starchy endosperm. The common milling method is wet milling: the grain is first soaked in a weak sulfurous-acid solution which swells and softens it, dissolving out certain carbohydrates and proteins. The resulting solution is "corn steep liquor," important in the production of antibiotics. The softened grain is then cracked in an attrition mill, between paired, revolving, toothed discs, which free the germ. The loose, cracked grain is sent to flotation tanks where the oily germ (50 per cent oil) is floated off as a layer above the water, while the denser endosperm sinks to the bottom, to be retrieved and further processed to yield corn starch. The germ is expressed in expellers, or sometimes solvent-treated to yield corn oil. This oil, after refining, is used principally as a cooking and salad oil. In composition it consists largely of oleic and linoleic glycerides.

COTTONSEED. Prior to the American Civil War and the advent of the cottonseed oil mill, the seed from cotton, *Gossypium sp.*, of the Malvaceae, was considered of

Cotton seed after ginning. (Courtesy Chicago Natural History Museum.)

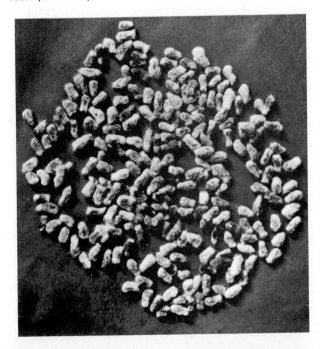

little value and was usually thrown away. Today such seed, a by-product of the cotton fiber industry, is the basis of an important oil and stock feed industry. Eighty per cent of cottonseed goes into the production of oil, feeds, fertilizers, and cellulosics, which often approach in value the fiber itself. No part of the cottonseed is wasted. The linters or short seed fuzz, which cannot be ginned, are first removed; they serve as a cellulose source (see Rayon, Chapter 5). The hulls as well as the linters find use in the manufacture of cellulosics. The hulls, which contain considerable potassium, were formerly burned and the ash used as fertilizer, but more recently have been recombined with the press cake as roughage in stock feeds. Cottonseed consists of about 55 to 60 per cent hulls and 40 to 45 per cent kernels or "meats." The kernels are about 35 per cent oil and 10 per cent protein.

Cottonseed received at the processing mill is first thoroughly cleaned (by screening, blowing, passing over magnets). Removal of the lint follows, in de-linters which operate on the principle of the cotton gin but which have finer teeth in their saws. The de-linted seeds are machine-cracked, much as are corn grains, and the hulls largely removed by screening (in shakers) and suction. The kernels are roller-crushed, then cooked (cooking expels moisture, detoxifies the oil and makes it fluid, coagulates albumen) prior to treatment by hot expression in hydraulic presses or expellers. Sometimes whole (undecorticated) seeds are expressed, a procedure that circumvents many preliminary steps. Solvent extraction is more economical and efficient, however, and has been increasingly used in the United States. Cottonseed oil consists chiefly of linoleic, oleic, and palmitic glycerides.

Crude cottonseed oil is ordinarily shipped in tank cars to refineries, where the refining process follows the usual procedures of filtering, saponification of free acids, bleach-ing, deodorization, and so on. The refined oil is used to some extent as a cooking and salad oil or for soap, but the greater part is hydrogenated to produce margarine and shortenings. Nearly 2 million tons of cotton-seed oil are produced annually, chiefly by the United States. World production of cotton-seed runs over 21 million tons annually. Utilization is chiefly by the United States and the European countries. Production of cottonseed oil is of course contingent upon the demand for cotton fiber. The more than 4 million tons of protein represented by cottonseed is mainly used in cattle feeds, although limited amounts can be used to fortify human foods in regions of protein deficiency (e.g., Incaparina, for Latin America). Fortunately, the toxic gossypol found in glanded cultivars of cotton is largely inactivated by processing. Cotton breeders, however, have developed glandless cultivars which promise to be equally as productive of fiber as are the glanded cottons, but lacking gossypol in the seed.

SUNFLOWER. The Soviet Union is by far the largest producer of sunflower seed, from *Helianthus annuus*, Compositae. Over half of the nearly 8 million tons of world production comes from the U.S.S.R., with sufficient quality and attractive-enough price to take markets away from competing oils such as soybean, especially in Europe. Other prominent producing countries are Rumania, Argentina, and Bulgaria; although still a minor oil in North America, it is gaining importance as a crop in the South and Plains states. The sunflower is native to North America, but high-oil cultivars were developed in eastern Europe.

The cultivated form of the sunflower is not known in the truly wild state. In pre-Colombian times the American Indians were already growing sunflowers for the edible seeds, possibly derived from the subspecies *jaegeri* of the American Southwest. Introgression has apparently occurred with other annual species and varieties. For a number of decades sunflower has been a crop of minor interest for birdseed, and, toasted, even as a human food. Such seeds normally contain up to 30 per cent oil, but the newer cultivars brought from Russia not only outyield the former types, but contain up to 50 per cent oil. One hundred pounds of seed yield about 40 pounds of oil, which is about 68 per cent linoleic glycerides. There are over 100 species of *Helianthus*, with many cultivars of *H. annuus* (diploid chromosome count 34). Giant cultivars grow as tall as 15 feet, with heads 20 inches in diameter, but dwarf types growing only a few feet not only have highest oil content but are amenable to mechanical harvest with adapted combines. The species is relatively tolerant of drought, and fits nicely into rotation schemes in small grain-farming areas. Per acre yields run a half-ton or more.

Sunflower oil is mostly utilized for edible purposes. It is said to be much like safflower, more stable than soybean. It has proved excellent for frying potato chips. Excess capacity at cottonseed oil mills can be utilized in the United States for extracting sunflower oil. Residues are 35 to 50 per cent protein, suitable for stock feed (but, being low in lysine, not of as much value as is soybean meal). The oil, in addition to being utilized as a shortening and in the manufacture of margarine, is blended with linseed in paints, and in some areas used as a lubricant or illuminant. Unprocessed seed is fed to poultry, caged birds, and livestock, and the whole plant may be used for silage or as a green manure. The crop is planted from seed, grown the same as small grain.

Some other semidrying oils

Apricot kernel—from *Prunus armeniaca*, Rosaceae. Apricot pits are derived in considerable quantity from the manufacture of dried apricots; the kernels contain 40 to 45 per cent oil, mostly of the oleic type. Cherry (*P. cerasus*),

peach (*P. persica*), and plum (*P. domestica*) yield similar kernel oils.

Beechnut—from *Fagus sylvatica*, Fagaceae; expressed in Europe.

Brazilnut—from *Bertholletia excelsa*, Lecythidaceae; expressed in Brazil.

Citrus-seed—from orange, grapefruit, lemon, and lime seeds, *Citrus sp.*, Rutaceae; at canning or juice factories.

Croton—from *Croton tiglium*, Euphorbiaceae; southern Asia.

Fennel—*Foeniculum vulgare*, Umbelliferae, has already been mentioned as a source of essential oil. Excellent fixed oil yield from the seed has been obtained in experimental work, well over 1 half-ton per acre. Up to 26 per cent oil is obtained from the seed, about 60 per cent of this petroselinic glyceride. This can be cleaved to yield lauric and adipic acids. Lauric acid is widely used to make surface-active materials and plasticizers. Adipic acid is used in nylon, plasticizers, urethane foams, elastomers, lubricants, certain foods, and pharmaceuticals.

Pine—from seed of *Pinus monophylla* and other species, Pinaceae; western United States, Europe, India.

Rape—from seeds of species of *Brassica*, Cruciferae; chief producing areas in order of importance are China, India, Europe, and Canada. The United States is the chief importer of the oil; Japan, Italy, Germany, and Great Britain are the chief importers of seed. World production of seed runs several million tons annually, most of which is consumed locally. The seeds contain 30 to 45 per cent oil, which consists largely of glycerides of erucic, oleic, and linoleic acids. The extracted oil is used as an edible oil, lubricant, fuel, and in synthetic rubber. It is obtained by expression or solvent extraction.

Rice—from hulls of *Oryza sativa*, Gramineae; expressed and solvent-extracted in Japan. Minor quantities of similar grass oils are obtained from rye, oats, millet, and wheat.

Sesame—from seed of the coarse, herbaceous annual *Sesamum indicum*, Pedaliaceae. This oil is also known as benne, teel, and gingili. About $1\frac{1}{2}$ million tons of seed are produced annually, mostly in China and India and to a lesser extent the Sudan and Mexico. The Mediterranean countries, Japan, and the United States are the chief importers. The plant, probably native to Africa, has been grown for seed since ancient times in Asia, for use on bakery goods and food delicacies. The seeds contain 45 to 55 per cent oil, largely oleic and linoleic glycerides. The oil, extracted by expression, serves as a salad or cooking oil, and in making margarine, shortenings, and soap. Diploid chromosome number is 26, and crossing with certain of the thirty-four other species of the genus shows only occasional fertility.

Tomato—from waste seed of catsup and soup making, from *Lycopersicum esculentum*, Solanaceae; chiefly in Italy.

Umbelliferae—from seed of various genera of this family (parsley, celery, anise, fennel, dill, carrot, coriander, cumin, chervil, caraway); solvent-extracted or expressed.

Drying types. Some of the most expensive vegetable oils are drying oils, notably linseed, soybean, and tung; and of fair impor-

tance oiticica, perilla, and safflower. With trends away from oil paint to latex types, demand from the paint industry has been decreasing, although demand for edible unsaturated oils has been increasing.

LINSEED OR FLAX. The flax plant, *Linum usitatissimum*, of the Linaceae, was discussed in the chapter on fibers. Differing cultivars are generally planted for oil seed than for stem fiber. The most important growing areas for seed flax are the United States, Argentina, Canada, India, and the U.S.S.R., altogether accounting for about 85 per cent of world production. The United States and various European countries are the chief consumers. World production well exceeds 3 million tons annually. The seed contains 32 to 43 per cent oil, mainly linolenic, oleic, and linoleic glycerides. Most linseed oil was traditionally consumed by the paint, oilcloth, linoleum, and protective-coating industries, most of which have been decreasing their demand for linseed oil in recent years as plastics and synthetics assumed greater

Combine harvesting flax seed in North Dakota. (Courtesy USDA.)

importance. But linseed oil is still highly recommended as a sealer and surface coating for decorative stone and slate, floors, paneling, furniture, wooden handles, and sporting equipment. A promising large-scale usage is for treating concrete highways and bridges, to prevent flaking and scaling that is encouraged by winter salting. Newly poured concrete sprayed with a linseed oil emulsion cures better and its surface is maintained longer. The oil can also be used in soap and leather production, and somewhat for edible purposes. It may be obtained by either hot or cold expression, after the usual preparatory cleaning processes. Like cottonseed, flax contains a toxic glucoside; this is automatically detoxicated, however, in heating, either in hot pressing or in refining. Refined linseed oil will keep almost indefinitely in sealed containers. The press cake is valuable as stock feed.

SAFFLOWER. Safflower, *Carthamus tinctorius*, Compositae, diploid chromosome number 24, has been recognized as a dye plant since ancient times, but only recently has become prominent as a source of an unsaturated oil. The species is known only from cultivation, and may have originated from *C. lunatus*, *C. flavescens*, *C. palaestinus*, or *C. oxyacantha* of northeastern Africa and Asia Minor. Safflower was first cultivated in Egypt, somewhat later in China, and was introduced into Mexico by the Spanish explorers.

Safflower oil has become increasingly used as an edible oil not conducive to cholesterol build-up in the blood. It is rich in linoleic components, but perhaps reflecting its mixed heredity has proved extremely variable upon chemical analysis. California studies on safflower procured from many parts of the world show variations in iodine number from as low as 87 to as high as 149, in proportion of linoleic acid from about 9 to 85 per cent, and of oleic acid from 9 to 87 per cent. Obviously there is ample latitude for selection of superior, tailor-made cultivars. Plant breeders have used a recessive genotype for the female parent in producing hybrid cultivars having a high oil content.

Safflower oil is used principally in the production of margarine, salad oils, mayonnaise, shortening, and other food products. The meal that is a residue after oil extraction is a good protein supplement for livestock. The oil is also used in paints and varnishes because of its nonyellowing characteristics. Up to 40 per cent oil is contained in safflower seed, and seed meal is up to 50 per cent protein.

The safflower is well adapted to drier climates and is salt-tolerant; in the United States it is planted especially in the Southwest and the northern Great Plains. The crop generally does well where barley is grown, and is handled in much the same fashion. However, safflower seed is twice as valuable as is barley grain, and if yields are comparable, is the more remunerative crop. The crop is planted in early winter in Arizona and California, in mid-spring in the Plains states. Plantings may be broadcast (sometimes from the air), drilled (6–14 inches spacing for solid stands), or planted in rows (especially where cultivation is needed for weeds, and under irrigation). Safflower is let stand to dry thoroughly before combining, since neither the plant lodges nor the seeds shatter easily.

SOYBEAN. The soybean, *Glycine max*, of the Leguminosae, known in hundreds of cultivars, is an ancient cultivated plant of eastern Asia. The Chinese Emperor Sheng Nung is said to have mentioned it in a publication of 2838 B.C. (probably in error), but there is no record of its first domestication nor archeological findings of the Neolithic period in northern China, where the soybean emerged as a cultivated crop in the eleventh century B.C. It was considered one of the five "sacred grains" of the ancient Chinese. The German botanist Kaempfer first brought soybeans to Europe in the seventeenth

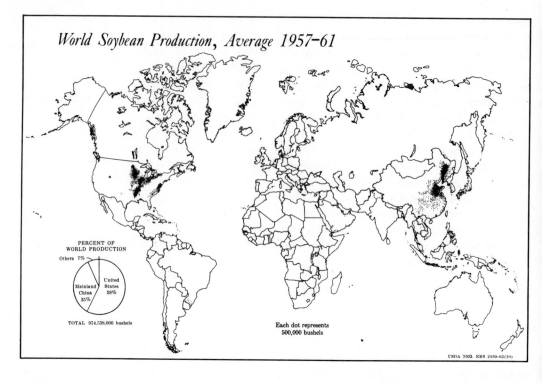

World Soybean Production, Average 1957-61

PERCENT OF WORLD PRODUCTION

Others 7%

Mainland China 35%

United States 58%

TOTAL 974,538,000 bushels

Each dot represents 500,000 bushels

USDA NEG. ERS 2430–63(10)

century, and they were introduced by chance into the United States in 1804. Scant attention was given the soybean in North America until the 1930's, whereupon its rise to become the second most important cash crop in the Midwest was phenomenal. It has been used throughout the world as pasture, cover crop, green manure, for the edible food, and recently on a large scale for the seed oil and protein cake. Production worldwide now exceeds 30 million tons annually, of which the United States contributes about two-thirds and China nearly a third, with relatively insignificant amounts from other countries.

As a "cinderella crop" in the United States, time enough has not passed to answer many important questions about the soybean. Its response to fertilization is uncertain; at least those cultivars that are strongly nodulating show little response to nitrogen, although the crop is a heavy feeder. Newer, more remunerative cultivars are continuously being bred, in order to lower costs and enable the

soybean to compete more successfully for the world oil market. Depending upon local circumstances, sometimes row planting, sometimes broadcast seeding, proves more remunerative, and the most desirable row spacing is not always clear. At least for more northerly regions a growth regulant (TIBA) is offered that makes the plant more compact and pyramidal, usually with increased yield. The cultivated soybean is an erect annual up to 3 feet tall, with trifoliolate leaves and small flowers borne in clusters. The hairy pods contain only a few seeds, some cultivars maturing quickly and others requiring 150 days or more. In the United States yields average around 25 bushels per acre (over 30 on good farmland, such as in Illinois), with adaptability and climatic requirements about the same as for corn. Weed control is increasingly preventive with herbicides, and harvest is by combine. The diploid chromosome number of soybean is 40.

J. W. Purseglove (Tropical Crops, New

York, John Wiley & Sons, 1968) believes *G. max* to be derived from *G. ussuriensis*, a prostrate legume of eastern Asia, possibly involving hybridization with other species such as *G. tomentosa* and *G. gracilis*. The nitrogen-fixing bacterium, *Rhizobium japonicum*, is said to be specific to the soybean. Seed inoculation is recommended, especially if the crop is planted in a soil not previously having grown soybeans. The species is a short-day plant, although selected cultivars have now adapted the species to many regions. The plants are self-fertile, and mostly self-pollinated.

The soybean seed is 13 to 25 per cent oil, 30 to 50 per cent protein, and 14 to 24 per cent carbohydrate. After solvent extraction of the oil, the meal (used primarily as stock feed) is nearly 50 per cent protein. In many parts of the world the meal is used to fortify human food, such as in Israel where 10 per cent soybean flour is included with wheat flour in

Soybean plant and shelled beans. (Courtesy USDA.)

baking bread. Techniques have been developed to use soybean meal in place of meat, with the cooked food practically indistinguishable from that made from fresh meat. Depending upon market conditions now the oil, now the protein meal, is the more important and in greater demand. The oil is used extensively in the Far East as a food, and in the West as a salad oil and especially for the manufacture of margarine and shortening. Industrially the oil goes into paints, linoleum, printing inks, soap, insecticides, stabilizing agents for food, cosmetics, leather, plastics, and so on. The oil is about 50 per cent linoleic, 30 per cent oleic, and 7 per cent linolenic components. In some parts of the world extraction is very primitive, and in other areas quite advanced. Modern crushing mills usually extract the oil with continuous screw-type expellers, although solvent extraction (which retrieves more oil) is increasingly used (said to extract about 50 pounds more oil from 1 ton of seed). Less than 2 per cent oil generally remains in the meal after solvent extraction, and many times this much with crude expression.

TUNG. Tung oil is produced from several species of *Aleurites*, Euphorbiaceae, diploid chromosome complement 22. *A. montana* is most grown in the tropics, usually at higher elevations, *A. fordii* in subtropical areas including central China and the Gulf Coast of the southeastern United States. Of minor importance is *A. trisperma* of the Philippines, used locally there for oil production, and *A. moluccana*, the candlenut tree of Malaysia, which yields an inferior oil. The two commercial species are both native to China, but now widely distributed around the world. *A. fordii* was first introduced into the United States in 1904, from seed sent by the counsel-general at Hankow, China. It offers the more easily hulled seed amenable to mechanical processing.

Tung trees grow rapidly in suitable environment, yielding a few oil nuts within 3 years. The trees are handsome, with dark green, ovate leaves, and clusters of whitish flowers produced terminally early in spring. The fruit is a spherical, angled drupe about 6 centimeters in diameter containing three to five seeds. To some extent the fruit dehisces on the tree, but more often falls to the ground while still containing the nuts. These are typically gathered by hand. To avoid excessive moisture content, the fruit is generally left to dry on the ground under the trees for several weeks. Because of the high labor costs in gathering fruit, tung orchards in the United States have had difficulty in competing in the world oil market, and the industry has been declining (even though protected by government subsidy). In the United States the nuts are mechanically hulled, ground, preheated, and expressed in a continuous screw press. The oil is filtered, and the press cake often subjected to further expression or solvent extraction to retrieve the remaining oil. The fruit yields 15 to 20 per cent oil, 40 to 45 per cent meal.

Tung oil is three-fourths or more eleostearic fatty acid, with smaller amounts of oleic, linoleic, and palmitic glycerides. Its iodine number is 163. Eleostearic acid is isomeric with linolenic acid, and taken internally is a strong purgative. In China it has long been used in certain medicinals, although more commonly employed for waterproofing various materials and as a component of caulk and mortar. Its ability to polymerize to a tough coating has made it especially valued in paints, varnishes, linoleum, oilcloth, and printing inks, where a reactive drying oil is needed. The press cake is used chiefly as a fertilizer since it is toxic as a stock feed unless especially treated. At the peak of oil paint popularity, over 50,000 tons of tung oil were used annually in the United States paint industry alone, about three-fourths of which was usually imported. World production of tung oil generally runs over 100,000 tons annually, about two-thirds of

it from China (relatively small amounts from Argentina and the United States). Seedlings are in full production at 10 years of age. In many parts of the world superior clones are bud-grafted onto open-pollinated seedlings, such as in the managed plantations of eastern Africa. In the United States the five named varieties released by the USDA come sufficiently true as seedlings so that they are usually planted from seed. Tung is relatively intolerant of competing vegetation, and highest yields are achieved in orchards that are clean-cultivated. The tree responds well, too, to fertilization. Under good growing conditions oil yields of nearly a half-ton per acre should prove possible.

Some other drying oils.

Candlenut—name for *Aleurites moluccana*, Euphorbiaceae, one of the tung oils. Other species of *Aleurites* are known as Abrasin, Japanese tung, and Bagilumbang.

Cedarnut—from *Pinus cembra*, Pinaceae; Alps to Siberia.

Comandra—*C. pallida*, Santalaceae, bastard toadflax, yields a seed oil that is nearly half ximenynic acid, an acetylenic oil having strong drying properties.

Crambe—*C. abyssinica*, Cruciferae, sometimes called colewort or Abyssinian kale, is a widely adapted cool-season crop beginning to find favor as an oil source. Yields of more than 1 ton per acre of seed have been obtained experimentally under irrigation. The fruit is about 32 per cent oil, approximately 60 per cent of which is erucic-acid glycerides. The seed meal is high in protein, but contains toxic glucosides which must be removed to make it palatable and nutritious. Crambe oil has proved excellent as a lubricant in the continuous casting of steel, a use of considerable potential; it is also useful as a chemical raw material and rubber additive, and in the formulation of waxes, plastics, and synthetic fibers (nylon, 1313). Purdue University has introduced the cultivar Prophet, which has yielded 1 ton of seed per acre in central Indiana.

Dimorphotheca—seed of the cape marigold, *D. sinuata*, Compositae, an ornamental plant of South Africa, yields dimorphecolic fatty acid. The seed is approximately 34 per cent oil, which in turn is about 50 per cent of the unique glyceride. It reacts chemically to produce a wide variety of materials useful in surface coatings, plastic foams, lubricants, surfactants, and industrial chemicals. The structure of the oil is similar to dehydrated castor, isomerized safflower, and tung oils. The seed meal is relatively deficient in lysine and methionine, but otherwise is a suitable source of amino acids. Several related genera of the Calenduleae tribe, mostly native to South Africa, yield similar oil, including *Castalis*, *Osteospermum*, *Gibbaria*, and *Chrysanthemoides*.

Euphorbia—from several species of *Euphorbia*, Euphorbiaceae. *E. lagascae* contains a seed oil rich in vernolic acid, an epoxidized type that can substitute for the more expensively altered vegetable oils used in plasticizers. The seed contains up to 50 per cent oil, about 60 per cent of which is epoxyoleic.

Grapeseed—from expression of seeds of *Vitis vinifera*, Vitaceae; a by-product of the wine and raisin industries; Europe, Argentina, North Africa, California.

Hempseed—from *Cannabis sativa*, Moraceae; particularly Asia and Europe; used as edible and paint oil.

Lallemantia—from *Lallemantia iberica*, Labiatae; southeastern Europe and central Asia.

Mercurialis—from a few species of *Mercurialis*, Euphorbiaceae; Europe.

Nigerseed—from *Guizota abyssinica*, Compositae; Africa, India. Cold- or hot-pressed for edible oil, soaps, paints, and illuminants. India produces 75,000 tons annually in the hill country.

Oiticica—important kernel oil from seed of *Licania rigida*, Rosaceae; Brazil. A significant industry has been built up around procurement of this seed in arid northeastern Brazil. The seed, from a semi-evergreen wild tree, contains 55 to 62 per cent oil. A single seed a few centimeters long is enclosed in a friable husk. Seeds are usually collected from December to April. The oil is extracted in hydraulic presses or expellers, or in some instances by solvent extraction.

It is a strong drying oil similar to tung, solidifying completely on exposure to sunlight or air. A number of mills in Brazil produce up to 25,000 tons of the oil annually. It has many industrial uses, particularly for paints, linoleum, printing inks, moisture-proofing, auto brake-bands, fiberboards, metallic soaps, and certain rubber products.

Perilla—an important oil, rivaling tung, from seeds of *Perilla frutescens* and its varieties, Labiatae; northern India, China, Japan, Soviet Union. The plant is a branched annual, 3 to 5 feet high. Seed yields up to 1,500 pounds per acre have been obtained. Difficulty is encountered in harvesting the seed, which tends to drop quickly upon ripening and does not all ripen at the same time. Oil content of the seed averages about 38 per cent, and is mostly of linoleic and linolenic composition. The oil is extracted by expression or solvent action, usually from roasted seed. It is used as an edible oil and for oiling paper in Asia; and for paints, varnishes, printer's ink, linoleum, and waterproofing. World production amounts at times to 200 thousand tons, almost entirely from Asia. Japan and the United States are the largest importers.

Poli—from *Carthamus oryacantha*, Compositae; India.

Poppyseed—from *Papaver somniferum*, Papaveraceae; Europe, Asia, and elsewhere. Seeds of the opium poppy contain up to 50 per cent oil. They are usually cold- or hot-expressed.

Rubberseed—from *Hevea brasiliensis*, Euphorbiaceae, the Pará rubber tree, grown throughout the world tropics. The oil is extracted by expression or solvent action, and is used chiefly for soap and paints.

Vernonia—the seed of *Vernonia anthelmintica*, Compositae, is rich in vernolic acid, mentioned under *Euphorbia*. The seed contains 22–28 per cent oil, about three-fourths of which is of vernolic esters. The seed is said to be processable by the same continuous solvent extraction techniques used for soybeans.

Waxes

Vegetable waxes have been largely obtained from exotic and distant sources in regions little known in the great consuming centers. Today they play a minor role in the world economy—in comparison to mineral, animal, and synthetic waxes—accounting for only 3 per cent of world consumption (the United States alone uses over a half-million tons of wax annually, mostly from petroleum sources). Yet the king of all waxes, carnauba, comes from a stately palm in the inhospitable climate of the northeastern states of Brazil. It is esteemed for polishes and carbon paper. Vegetable waxes are also used in paper and textile sizing, candles, coatings, and leather goods.

Like beeswax, man's first wax and his standard of definition, vegetable waxes are fatty-acid esters of monohydroxy alcohols. The fatty acids usually have about 30 carbons to the chain. Vegetable waxes contain a wide range of organic compounds, and may differ chemically for different parts of the same plant, or with season, locale, and age. Wax occurs in almost all plants in greater or lesser abundance, as a protective coating on the epidermis to prevent desiccation, oxidation, pest attack, and undue abrasion. It may also shield the plant from excessive ultraviolet radiation. Leaves are the most important source of vegetable waxes. The cuticle is overlaid with an epicuticular wax, the commercial product: this may constitute as much as 4 per cent of the green leaf, 15 per cent of the dry.

Commercial preference for one wax over another is largely a question of the quantity obtainable. In most plants the quantity of wax is small and hence cannot be extracted economically. On the other hand, the plants that yield the most wax usually grow in inhospitable arid environments where the heavy wax affords needed protection against a hostile climate. Competition from cheaper synthetic and petroleum waxes (paraffin) has largely limited use of vegetable waxes to a few special types, such as the comparatively expensive but unequaled carnauba. The total annual production of

vegetable waxes is only about 20,000 tons, a small amount in comparison to the large tonnages of vegetable oils produced. Yet demand exists for more high-quality vegetable wax if such can be had at a moderate price. Research is under way to obtain good vegetable waxes from diverse agricultural crops or plant wastes.

CARNAUBA. The slow-growing carnauba palm, *Copernicia cerifera*, rears its crown of fan-shaped leaves to the drying winds of Ceará and Piaui, Brazil. The fiercer the scorching winds and the worse the drought of the dry season, the heavier seems to be the wax secretion on the leaves. It is then that the carnauba, named by Humboldt "tree of life," offers succor to a wasting economy.

The carnauba palms are found wild in tremendous number in the northern part of the states of Piaui and Ceará, Brazil. They prefer moist soil, and are commonly found in lowland areas with a high water table. In the last half-century, plantings of the tree have also been made. At one time the big leaves of carnauba were cut indiscriminately by uneducated workers, but more recently harvest has been restricted by law so that excesses will not "kill the goose that lays the golden egg." Only a certain number of leaves are allowed to be removed from any tree in a given year, in two cuttings of ten to twenty-five leaves each during the dry season.

Leaves are cut by machete where possible, by long pole saws or knives and with the help of ladders where the tree towers too high. Cut leaves are taken to drying sheds, where they are usually stripped (shredded) by hand or by machine, and then are allowed to lie for 3 to 5 days in an enclosure. As the leaf fragments wither and dry, the layer of wax on the surface becomes loose and falls as a whitish powder. Laborers flail the dry leaf parts until all the wax falls away. The leaf parts are discarded or, in managed plantations, returned to the palms as a mulch, and the loose wax is swept into containers. Next it is melted, strained, and molded for marketing.

The wax is largely myricyl cerotate, never economically synthesized and unique in its hardness, high melting point, ability to take a lasting polish, and compatability. Most crude wax goes to the United States, the chief importer of carnauba, for final refining (remelting in water) and compounding. Wax derived from the erect center leaves ("olho") is usually yellow; that from older, outer leaves of the crown is brownish or grayish and is less esteemed than the yellow.

Production of carnauba wax amounts to several thousand tons annually, from about 70 million trees that contribute an average of 6 ounces each. Carnauba wax finds use in polishes, lubricants, floor and automobile waxes, insulating materials, carbon paper, chalk, matches, phonograph records, plastics, cosmetics, protective coatings, films, and many other products. It is the hardest, highest-melting, natural commercial wax; and it is often added to other waxes to increase their melting point, hardness, toughness, and luster, and to decrease stickiness, plasticity, and crystallizing tendencies.

CANDELILLA. In desertlike northern Mexico, south and west of the Big Bend country (Texas), grows the largely leafless candelilla shrub, *Euphorbia antisyphilitica*, of the Euphorbiaceae. It is 1 to 6 per cent wax by weight. *Pedilanthus pavonis*, Euphorbiaceae, is reported as also being used under the name candelilla. The wax is found as a thin film on the stems, giving them a whitish cast. Seeds are produced in May to July; seeding is solely by natural means. Habitation by man of the harsh candelilla country is difficult, and it is unlikely that candelilla will ever become a cultivated plant. Wild plants have been harvested intermittently for a half-century, with what seasonal labor is available at the peak of the dry season, when wax content of the candelilla is at its highest.

Canadelilla plants are pulled by hand and sun-dried for a few days. The wax is ordinarily obtained by boiling the plants in acidified water, the melted wax being skimmed from the surface of the cauldron. About 8 pounds of sulfuric acid are added to the brew for each 100 pounds of candelilla. The wax is allowed to harden into crude cakes, which are cut to convenient size for taking to market. Portable apparatus is used, since supplies of candelilla in any one locality are soon exhausted. The crudely cast wax is again refined at commercial centers, since it contains much debris. The final product finds use in most of the same ways as carnauba, and is mostly utilized in blends. It is especially used for candles, the initial reason for its exploitation in Mexico. Production is limited to several thousand tons annually, since only wild plants are exploited and the stands fast becoming exhausted. Mexico has taken measures to prevent excessive harvest. Most of the production is consumed by the United States.

Some other wax plants.

Bayberry or myrtle—tallow covering the berry of species of *Myrica*, Myricaceae; North America, South America, and South Africa; mostly used for candles.

Carandá or caranday—from *Copernicia australis*, Palmae; Brazil and Paraguay. The wax is similar to carnauba in quality. An immense reserve of caranday exists in the relatively uninhabited Gran Chaco, but wax production there is still little developed.

Caussu—the large leaves of *Calathea lutea*, Marantaceae, yield a commercial wax. The species is a perennial herb frequent in the lowlands of Latin America. Extraction is much as with carnauba leaf.

Cotton—a wax can be solvent-extracted from cotton fibers, *Gossypium sp.*, Malvaceae. Similar waxes can be extracted from hemp, flax, and other fibers.

Douglas fir—a wax can be derived from the bark of *Pseudotsuga menzeisii*, Pinaceae. Waste bark from logging operations is utilized, yielding other products from the solvent extraction (e.g., tannins as well as wax).

Esparto—esparto grass, *Stipa tenacissima*, Gramineae, of North Africa, has been mentioned as a fiber source used for paper making in Britain. The wild grass has long been important in the economy of the Barbary Coast. It has been used for weaving (sandals, hats, brooms, carpets, and the like, for local use); rope; fuel; stock feed; and similar uses. Almost the entire crop is pulled by hand and transported by camels to the ports. Wax is extracted from the grass by mechanical cleaning, before the grass is pulped. The impure wax is purified by solvent dissolution and subsequent distillation. It finds much the same uses as other vegetable waxes, and is not greatly inferior to carnauba in quality. A few hundred tons are produced annually.

Jojoba or goatnut—a liquid wax (monohydric esters of long-chain acids rather than triglycerides —see Table 13-4) can be obtained from expression of the fruit and the seed of *Simmondsia californica*, Buxaceae, native to the dry foothills of California– Arizona and northern Mexico. It may be altered to the solid state by hydrogenation. Attempts have been made to introduce jojoba into cultivation as a new crop for the southwestern United States, with vacuum machine pickup of fallen seeds to eliminate costly hand gathering. Yield is about 2 pounds of seed annually per shrub.

Limnanthes—species of *Limnanthes*, Limnanthaceae, are familiar spring annuals in the Pacific states of the United States. Seed yields of about 1 ton per acre have been obtained experimentally, but cultivars suitable to agricultural growing and economical harvest remain to be developed. Fatty acids from the seed are 95 per cent longer than an 18-carbon length, yielding a liquid wax similar to jojoba. Seed of limnanthes contains 20–33 per cent oil and 15–25 per cent protein.

Ouricuri or licuri—from *Cocos* (*Syagrus*) *coronata*, Palmae, already listed as an oil source (from fruit). The leaves are cut and the waxy coating scraped free by hand (the wax does not separate upon drying as does carnauba wax). The resultant powder is cleaned, melted, and cast prior to marketing. Ouricuri often substitutes for carnauba. Wild stands of this small palm are frequent in northeastern Brazil.

Table 13-4

Chemical Composition of Jojoba Oil

Compound	Percentage
Saturated acids (various C_{20} to C_{26})	1.64
Palmitoleic acid, $CH_3(CH_2)_7CH:CH-(CH_2)_5COOH$	0.24
Oleic acid, $CH_3(CH_2)_7CH:CH(CH_2)_7-COOH$	0.66
Eicosenoic acid, $CH_3(CH_2)_7CH:CH-(CH_2)_9COOH$	30.30
Docosenoic acid, $CH_3(CH_2)_7CH:CH-(CH_2)_{11}COOH$	14.20
Eicosenol, $CH_3(CH_2)_7CH:CH(CH_2)_9-CH_2OH$	14.60
Docosenol, $CH_3(CH_2)_7CH:CH-(CH_2)_{11}CH_2OH$	33.70
Hexacosenol,* $C_{26}H_{51}OH$	2.00

* Hexacosenol is known to contain one double bond but its position has not been established.

From N. B. Knoepfler and H.L.E. Vix, *Agricultural and Food Chemistry* **6**, 119 (1958).

Palms—in addition to the specific palm genera listed above, others are potential sources of leaf wax, such as *Ceroxylon andicola* of the Andes, and *Raphia pedunculata* of Madagascar.

Sisal and henequen—wastes left in fiber production from these two famous hard-fiber plants, species of *Agave*, Amaryllidaceae, provide a commercial source of wax.

Sorghum—*Sorghum bicolor*, Gramineae, yields a useful wax derived from the hull of the grain. It is removed by solvent extraction.

Sugar cane—sugar cane, *Saccharum officinarum*, Gramineae, yields commercial wax from the clarifier muds in sugar refining. Either solvent extraction or vacuum distillation can be employed. Many thousands of tons of such wax are potentially available as a by-product from the world's sugar mills. The wax is chemically complex, containing about one-third free acids and alcohols as well as mixed glycerides.

The Future of the Vegetable Oil and Wax Industries

The vegetable oil industry, massive and dedicated to low costs, seems anything but exciting. Its major edible and industrial oils, entering commerce in millions of tons annually, would seem firmly entrenched and unchallengeable. Not necessarily so! First of all, most of these oils are substitutable, and changes in price, supply, or even the political situation (e.g., tariffs) can upset markets in a hurry. In the early 1960's soybean oil produced in huge amounts under mechanized growing in the North American corn belt would have seemed unstoppable. However, its star rising almost equally as fast, sunflower oil from eastern Europe, more economical, and perhaps aided by Common Market political considerations, unseated soybean oil in Europe before 1970.

Secondly, with vegetable oils so ubiquitous in the plant kingdom, no source has a corner on the market. It is estimated that only 1 per cent of the plant species known are presently applied for human use; what might the possibilities be in the remaining 99 per cent? So far as vegetable oils are concerned, only in recent years has systematic screening of the possibilities been undertaken. Improved chemical technology not only makes it possible to better analyze usefulness of oils, but is discovering unique types previously not even known. There seems little doubt that vegetable oils fitted to specialized purposes will become increasingly common—especially since active programs are underway to find new crops adapted to agriculture. In addition to the familiar, widely used fatty-acid glycerides—oleic, linoleic, linolenic, palmitic, and stearic—other glycerides such as the eleostearic that makes tung oil unique, and the ricinoleic characterizing castor oil, can be expected to become important market items. Indeed, chemists are continuously dicovering new fatty acids, with unanticipated positions of unsaturation, and new isomers having unique physical and chemical properties. It is only within recent decades that epoxy fatty acids were discovered in natural oils; what may be the new discoveries just

around the corner? The future of the vegetable oil industry, which is rapidly growing in chemical technology and changing in other ways, is up for grabs.

It is true, however, that a world growing increasingly hungry is likely to demand all of the edible oils now produced, as well as new additions. Of the more than 20 million tons of vegetable oils produced annually in the 1960's, the greatest share was for edibles, only about 10 per cent being directed to industrial purposes. About a quarter of the world's edible oils came from the two legume seeds, soybean and peanut. One-quarter of the total tonnage of edible oils was produced in the United States. Corn oil is prominent there, along with soybean, but to an extent is a by-product of cornstarch production, just as cottonseed oil is a by-product of fiber production. But especially in the production of legume oils does a proteinaceous seed cake result, valuable as an animal feed. More and more these proteinaceous meals are being utilized to enrich human food, too, especially in regions of dietary imbalance.

But no oil need feel secure in its market. Since mid-century soaps once relying upon vegetable oils for 96 per cent of their raw material have given way to detergents utilizing only 19 per cent oil. In the United States, at least, vegetable shortenings have rather thoroughly displaced lard as a cooking fat. Margarine from hydrogenated vegetable oil has gradually surpassed butter in usage, mostly for reasons of economy. In paint, drying oils have lost position to synthetic resins much used in latex paints. There is great give-and-take in the market. On the whole, vegetable oils have maintained their position well, encouraged by economies from the new technology. For example, improved extraction techniques have been developed, such as the continuous screw press and effective solvent extraction methods. Yes, vegetable oils and fats seem destined to be used in astonishing quantities, both for foods and as industrial raw materials.

The situation with vegetable waxes is not so certain, Every indication points to a continued demand for quality waxes, but whether vegetable waxes can hold their own in competition with the synthetic and mineral types depends upon many factors, not the least of which are labor costs for gathering these substances when not agriculturally produced. Efforts are constantly being made to find a substitute for carnauba that equals its high quality; were this to be achieved, vegetable waxes, like vegetable dyes, would become largely of historical interest only. It is unlikely that the present dominant vegetable waxes will be able to improve their position materially, because of inefficient production from wild rather than cultivated sources. Waxes from abundant, concentrated plant wastes, such as sugar cane or grain hulls, would seem to offer greater potential for growth.

SUGGESTED SUPPLEMENTARY REFERENCES

Books

Bailey, A. E., *Industrial Oil and Fat Products*, Interscience, N. Y., 1951.

Berg, G. L. (ed.), "New Comprehensive Manual Modern Soybean Production," *Farm Technology* **23**, Meister Pub. Co., Willoughby, Ohio, 1967.

Eckey, E. W., *Vegetable Fats and Oils*, ACS Monograph 123, Reinhold, N. Y., 1954.

Fremond, Yan, Robert Ziller, and Mide Lamothe, *The Coconut Palm*, International Potash Inst., Berne, Switzerland, 1968.

Hilditch, T. P., and P. N. Williams, *The Chemical Constitution of Natural Fats*, 4th ed., Wiley, N. Y., 1964.

Jamieson, G. S., *Vegetable Fats and Oils*, Reinhold, N. Y., 1943.

Kirschenbauer, H. G., *Fats and Oils, An Outline of Their Chemistry and Technology*, 2nd ed., Reinhold, N. Y., 1960.

Norman, A. G. (ed.), *The Soybean—Genetics, Breeding, Physiology, Nutrition, Management*, Academic Press, N. Y., 1963.

Sauer, J. D., *Plants and Man on The Seychelles Coast*, Univ. of Wisconsin Press, Madison, Wisc., 1967.

Scott, W. O., and S. R. Aldrich, *Modern Soybean Production*, Farm Quarterly, Cincinnati, 1970.

USDA, *Growing Safflower*, Farmer's Bull. 2133, Washington, D. C., 1966.

USDA Yearbook, *Seeds*, Washington, D. C., 1961.

Weiss, E. A., *Castor, Sesame and Safflower*, World Crop Series, Leonard Hill, London, 1971.

Wrath, Albin H., *The Chemistry and Technology of Waxes*, 2nd ed., Reinhold, N. Y., 1956.

Woodroof, J. G., *Peanuts—Production, Processing, Products*, AVI Pub. Co., Westport, Conn., 1966.

——————, *Coconuts: Production, Processing, Products*, AVI Pub. Co., Westport, Conn., 1970.

Periodicals

American Oil Chemists' Society (journal): various issues.

Economic Botany (quarterly periodical): various issues.

Farm Quarterly (periodical): various issues.

World Crops (periodical, London): various issues.

CHAPTER 14

Carbohydrate Extractives:
Sugars and Starches

THIS CHAPTER IS TRANSITIONAL FROM PLANT derivatives that are appreciably processed (the extractives much used industrially) toward food plants (which are mainly consumed with mere grinding or cooking). There are no clear-cut boundaries between carbohydrate extractives and carbohydrate-containing food and drink. Extractives are usually prepared through fairly elaborate technologies, but more often than not the derivatives are reincorporated into prepared foods. Thus a half-billion tons of extracted sugar annually nourishes millions of human beings directly, usually as an additive making other food and drink more palatable, but also is used for fermentations and as an industrial raw material. Extracted starch has even wider industrial applications.

Sugars are the interim, soluble form of food in living cells, and starch the common form in which food is stored in the cell. As such starch is an important concentration of energy, both for metabolism of the plant and, extracted, for man's manifold uses. Being soluble, sugar is typically extracted as an expressed juice, while starch grains are mechanically separated after fragmentation of the cell. Both sugar and starch go into a wide variety of foodstuffs. Starch is an inexpensive textile and paper sizing; nearly a quarter-million tons are used in the United States annually for stiffening textile warp, and another half-million for coating paper (each ton of paper is said to employ about 27 pounds of starch on the average). Starch has been yielding some ground industrially to synthetics such as the polyvinyls, but its ready abundance and inexpensiveness favor it as a raw material. A corn cultivar has been bred to yield high-amylose starch economically, especially well adapted to paper sizing and binding (forming as it does an especially strong, grease-resistant film). Amylose is chemically identical with cellulose except for the manner in which the polymer units are linked. Starch is a raw material for manufacture of industrial glycosides, used as humectants in sweets and pastries, and in the manufacture of alkyd resins for paint. Glycosides also have some application in making biodegradable detergents, and in the manufacture of insulating foams and urethane plastics. It is evident that we have here a plant extractive widely useful industrially as well as in foods.

The carbohydrate molecule is basically less complicated than are the fats and oils, proteins, alkaloids, etc. which have been the objective of the extractions discussed in other chapters. Carbohydrates contain carbon, hydrogen, and oxygen in approximately the ratio of $1:2:1$ (e.g., glucose, $C_6H_{12}O_6$; sucrose, $C_{12}H_{22}O_{11}$; starch, $[C_6H_{10}O_5]_n$).

The comparatively high proportion of oxygen makes carbohydrates a less efficient source of energy than are fats and oils, since less oxidation (whereby energy is released) is possible. Carbohydrates are believed to be the initial photosynthetic product, and the building blocks for organic synthesis that yields the more complicated molecules. In this sense the energy packed into a simple sugar represents all plant substance. In this chapter we shall confine our attention to the two most widely useful carbohydrate extractives, sugar and starch, leaving for the section on food plants an appreciation of carbohydrates as a nutritional component.

Sugars

There are many sugars that have some utility in man's complicated world. These include among others glucose (karo syrups), lactose (bacteriological broths), levulose (diabetic sugar), dextrose, mannose, and others. But far exceeding all others in importance is sucrose, common cane or beet sugar, which is extracted annually in amounts reckoned in excess of 60 million tons. The per capita consumption for the world averages over 20 pounds per year, for the United States over 100 pounds. This important carbohydrate commonly occurs in small amounts in many plant species and plant parts, from onions to apples. In fact insoluble substances such as starch may be changed to soluble sugar form to be moved from location to location in the plant and then recombined into the insoluble form. But sugars seldom occur abundantly enough in locations from which they can be extracted to merit large-scale commercial attention. In only two plant species, sugar cane and sugar beet, has man found an economical source of sugar.

In Biblical times honey, made by the honeybee from the sugary nectar of flowers,

Harvesting sugar cane on a plantation in Brazil. A papaya or "mamao" is growing at the right. (Courtesy Chicago Natural History Museum.)

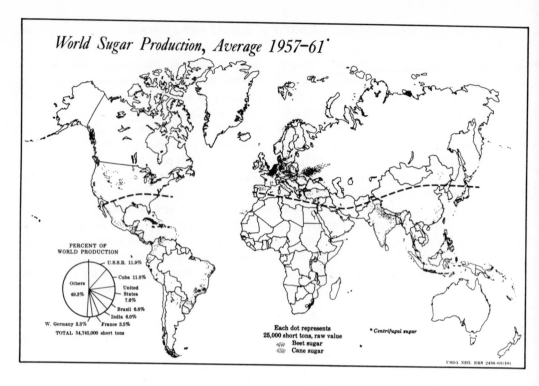

World Sugar Production, Average 1957–61*

PERCENT OF
WORLD PRODUCTION

U.S.S.R. 11.9%
Cuba 11.8%
United States 7.6%
Brazil 6.6%
India 6.0%
France 3.5%
W. Germany 3.3%
Others 49.3%

TOTAL 54,745,000 short tons

Each dot represents
25,000 short tons, raw value
Beet sugar
Cane sugar

* Centrifugal sugar

USDA NEG. ERS 2436-63(10)

constituted the only sweetening. The prosperous land was one "flowing with milk and honey." By the time of the Roman Pliny, man had learned to extract, although inefficiently, "a kind of honey" from bamboos. Thus began what was to become a tremendous world industry. Sugar then and throughout the Middle Ages was not common; it remained costly, used only as a medicinal or as a luxury of kings. It was ordinarily sold in the apothecary shops, along with various herbs. Sugar was apparently introduced to Europe through the overland caravan trade from India. The inhospitable hills of Afghanistan and the bleak wastes of the caravan route seem to have effectively prevented introduction of the living "Indian honey-bearing reed" into Asia Minor for some centuries. But at some time during the Middle Ages and the epoch of Arabic hegemony, sugar cane was widely introduced into the Near East and North Africa. The Arabs of this period seem to have made the first scientific attempts at sugar extraction. Then, in the fifteenth and sixteenth centuries, the Portuguese and Spanish, world colonizers and merchants, introduced sugar cane throughout the Old and New World tropics. Its culture rapidly expanded, and sugar quickly became one of the most demanded imports of Europe. In seeking independence of the tropical source (sugar cane), German experimenters undertook investigations concerned with the possible extraction of sugar from the beet, in which it had been noted as early as 1590 that considerable quantities were contained. Since that time the world has relied for its sugar on the tropical cane together with the temperate beet. Both yield nearly pure sucrose and hence are interchangeable in use.

The U.S.S.R., United States, Cuba, Brazil, India, Germany, France, Mexico, China, Australia, and the Philippines rank roughly in that order in production of sucrose. About half the sugar consumed in the United States is used for fruit and vegetable packing.

Sugar availability and sugar prices have had a chaotic history, ranging (originally) from a luxury item for the wealthy only, to an abundant commodity worth as little as a cent or two per pound. So important is sugar production to the economy of many tropical countries that from time to time quotas and marketing agreements have been entered into. As the major consumer of sucrose, the United States has chiefly shouldered the responsibility for supporting sugar prices. Sugar quotas granted by the United States have varied from time to time, depending upon political as well as economic considerations. The Sugar Act of 1965 reflected political disaffection with Cuba, formerly one of the major suppliers of sugar for the United States. The 1965 law allotted a base of 3 million tons per year (approximately 30 per cent of the nation's needs) to the domestic beet industry. Domestic cane growers (with an increased allotment at the expense of Cuba) received a base of 1.1 million tons; the remaining 6.2-million-ton allotment was spread among Hawaii, Puerto Rico, and various foreign sources. Through most of the 1960's the Soviet Union accepted Cuban production, in support of a Castro regime at political odds with the United States.

Cane sugar. There are a number of species of the sugar cane grass, *Saccharum*, Gramineae, with the commercially important canes credited to *S. officinarum*, a polyploid with a diploid chromosome count presumably 80 (cytogenetic studies sponsored by the Hawaiian Sugar Planters Association show some clones to be stable, and others variable with diploid counts ranging from 51 to 185). Actually, commercial cultivars are hybrids with three or four *Saccharum* species in their ancestry, and are generally aneuploid. It is believed that most of the modern cultivars derive from the *S. officinarum* "noble canes," probably native to New Guinea, which crossed naturally with other species. The primitive chewing canes were doubtless improved by selection before the science of plant breeding developed, then by intraspecific crop

breeding, and finally interspecific crossing that has yielded the prime clones of the modern sugar plantation. Sugar cane breeding was undertaken in Southeast Asia, particularly Java, as early as 1885. There, crosses and back-crosses involving *S. spontaneum* and *S. sinense* yielded a markedly superior clone (P.O.J. 2878) in 1921 that was to change sugar cane growing all over the world. Other improved cultivars have been bred in Hawaii and elsewhere, with *S. robustum* entering the pedigree. Rise in sugar yield has been amazing. Average annual yields of 2 tons per hectare in Java in 1840 were increased to 10 tons by 1910 and to 20 tons by 1940. Of course some of the gain resulted from improved growing technology, as well as because of breeding for higher yield and disease resistance.

Sugar cane is a plant of the humid tropical lowlands. Whether introduced from Indonesia or native, sugar cane was grown in India in ancient times. As Western trade with the

Propagation of sugar cane is by stem section, new plants sprouting from the nodes, (as revealed in this cut-away). (Courtesy USDA.)

Far East developed, sugar was brought along the ancient trade routes to the Mediterranean, where some of the primitive canes are still grown as a minor crop for chewing. Sugar cane was cultivated in the Canary Islands before 1493, and supposedly Columbus took starts from there to the New World. By the early 1500's sugar was widely planted in the West Indies, and it was spread to Mexico by 1520, to Brazil by 1532, and to Peru by 1533. In the West Indies many of the islands such as the Virgin Islands were eventually completely denuded of native forest and planted entirely to sugar cane. Sugar cane was raised in the French colony of Louisiana, in what is now the continental United States, as early as 1753. The high-yielding cultivars from Java did not supplant low-yielding cane in the New World until the late nineteenth century.

A little cane is still grown in the Mediterranean area, especially southern Spain. As noted earlier, there has been increased interest in cane growing in the southern United States, especially Florida and Louisiana, as imports from Cuba ceased. A number of countries in Africa produce fair amounts of sugar, and a great deal is grown in South America, especially in Brazil. Other major world producing regions are the islands of the Caribbean and coastal Mexico, islands of the central Pacific (especially Hawaii and the Philippines), and southern Asia. World production of cut cane approaches a half-billion tons annually, and of extracted sugar nearly 60 million tons.

Cane is a coarse grass, rooting from the nodes and reaching heights as tall as 20 feet. It roots easily from cuttings, the usual method of propagation. Until comparatively recently sugar cane was crudely planted and tended by hand, and still is today on small local holdings where efficiency is not at a premium. A section of cane about three nodes in length is simply stuck in a hole dug with a stick. Competing weeds are kept somewhat in check by any feasible procedure. Modern sugar plantations, however, that must compete in the world markets can hardly afford such tedious methods. Instead, planting machines open a furrow, drop the stem sections in, and then cover them at intervals. Mechanical cultivation, herbicide treatment, fertilizing, and flaming may precede and follow planting.

Usually 1 to $1\frac{1}{2}$ years are required from the time the cane is planted until it is harvested. Harvest was once entirely by hand, but on modern plantations it is done by gigantic cutting machines—a sort of combine for sugar cane. Some machines cut away tops and foliage, and in other cases the cane is flamed for defoliation before harvest. Many machines cut the cane into sections as it is harvested, and load the sections automatically into trucks accompanying the harvester in the field. In very rich soils, good production will continue from the rhizomes of one planting for as long as 20 years. At the other extreme, plantings must be handled as an annual crop, a system required by law in Java. The usual productive period is some-

Mature sugar cane in Louisiana ready for harvest. (Courtesy USDA.)

where between the two extremes. On the average, replanting is perhaps undertaken each 2–3 years.

Harvested cane, containing about 12 per cent sucrose, is brought to centralized locations or factories for extraction. Cane yield can run as high as 100 tons per acre under ideal growing conditions. Extraction is almost invariably an expression process, involving at its simplest the running of the cane between two hand-turned, fluted, wooden rollers, which squeeze out a goodly portion of the juice. Where more elaborate equipment is available, the cane stems are run between sets of three steel rollers, either with or without preliminary shredding. Three or even four such expressions are performed, each subsequent expression involving rollers set more closely together than for the preceding expression. After each expression the cane may be wetted to facilitate nearly complete extraction of all sugar present. The residue of fibrous stalks, bagasse, has in the past been of little value. Recently, however, it has been compounded into fiberboard or used as an ingredient in paper. More commonly it has served as an inefficient fuel for the boilers of the extraction factory.

The expressed juice undergoes evaporation—in primitive areas by boiling in open cauldrons, with or without addition of wood ashes (alkali) to aid in clarification. The dark, sticky sugar resulting contains sucrose, glucose, and small percentages of gum, protein, organic acids, pectin, ash, minerals, soils, pigments, and the like. As such it is a more complete and nourishing food than the highly refined white sugar. It is often a staple in the diet of certain peoples, as is the "rapadura" in rural parts of northeastern Brazil. Modern mills, utilizing evaporation chambers or vacuum pans, add chemicals such as sulfur dioxide, lime, sodium carbonate, phosphoric acid, etc. to precipitate out from cane juice nonsugars that are not coagulated by heat (clarification). The treated solution is passed through filter presses (defecation) and then concentrated to a density that will cause sucrose to crystallize out from the thick syrup or "massecuite." Centrifuges separate the liquid portion, molasses, from the sugar crystals. The molasses may be subjected to additional crystallizations. It is eventually sold for human consumption, as an ingredient of cattle feeds, or for fermentation into rum, ethyl alcohol, and vinegar. The raw sugar, about 96 per cent sucrose, is subjected to further refining, often in the consuming countries. This refining involves repeated washings and recrystallizations, and decolorization with carbon. The refined sugar is eventually packaged in loaf, lump, granular, or powdered form, almost completely free of its healthful "impurities."

Changes and the trend toward mechanization affecting agriculture generally have not spared the cane sugar industry. In spite of domestic allotments and price support, sugar production in the United States has had to become highly mechanized in order to remain competitive. In Hawaii the 30 man-hours required to produce 1 ton of raw sugar at mid-century have been reduced to about half this amount (and at the same time yields have increased to over 9 tons per acre with improved cultivars, and greater use of fertilizers and pesticides). Technology has improved the centrifugal separation of sugar from molasses, and ion-exchange methods (utilizing synthetic resins) are sometimes used for production of specialty sugars. It has been suggested that the juice from sugar cane be used as a growth medium for proteinaceous fungi, to supply more protein food in parts of the world where this is deficient. It is said that the 40 tons of cane produced annually per acre could be transformed into nearly 1,300 pounds of fungal protein, a protein yield higher than is possible from soybeans or other proteinaceous crops. Onset of flowering in cane, partly a response to night length, reduces sugar yields approximately 15 per cent; chemical sprays such as of diquat are increasingly used to inhibit flowering, increasing yield and simplifying harvesting schedules. Of course the worldwide search goes on for cultivars well adapted to local soils, resistant to local diseases, and responding well to fertilizer and local environmental stimuli.

Beet sugar. A selected line of the common beet, *Beta vulgaris*, Chenopodiaceae, native to northern Europe, was developed as a sugar source primarily in Germany. The species is presumably derived from the wild beet of northern Europe, *Beta maritima*. About two-fifths of the world's sugar, more than 23 million tons, now derives from the sugar beet. For centuries the common comestible beet was known to contain more than ordinary

Contemporary cartoon satirizing Napoleon's attempt to establish a sugar-beet industry. (Courtesy *Scientific Monthly*.)

The sugar beet. (Courtesy Chicago Natural History Museum.)

amounts of sugar, but not until 1774 did the chemist Marggraf in Berlin, investigating potential sugar plants, demonstrate the utility of the sugar beet. The species became especially valuable when selection and breeding progressively increased the average 2 to 4 per cent sugar content to 15 to 20 per cent. It was Marggraf's practical student, Achard, who first bred large-size, sugar-rich beets from the motley stocks cultivated by Silesian peasants. In the early years of the 1800's, Napoleon encouraged the newborn sugar beet industry, and was ridiculed for his efforts. Nevertheless, as cane sugar imports were cut off by Napoleon's edict forbidding trade with England, the beet sugar industry became firmly established. Destructive wars and renewed imports of cane sugar after 1815 disrupted the industry, but by the early twentieth century it had again revived sufficiently so that European production was reckoned in the millions of tons. Sugar beet growing became established in the United States only after 1875, but it ranks today as one of the foremost agricultural pursuits. Plant breeders have persistently raised the sugar beet's productivity and resistance to disease, and more efficient growing and handling have been continually achieved. As a result the sugar beet is able to compete well with sugar cane.

Commercial growing of the sugar beet is confined to the Temperate Zone. It is carried on in all continents except Africa. Northern Europe (especially Germany, France, Poland, and Czechoslovakia), the U.S.S.R., and the United States are the chief producers. In the United States a score of states from Maine to California (prominently Colorado, Idaho, and Michigan) produce about 3 million tons of beet sugar annually. About 15 per cent of the average beet is sucrose. The residue from sucrose extraction (pulp plus tops) finds use as a valuable livestock feed, fertilizer, and source of pectin. The pulp is said to contain as much as 20 per cent pectin. This pectin can be extracted by treatment with dilute acid.

Sugar beet seed is sown in rows in early spring, in carefully prepared seedbeds. A close stand of seedlings must be thinned, once a hand operation but becoming an operation by machines equipped with a rotating propellerlike blade. Introduction of monogerm seed (and precision planters) promises to eliminate the need for thinning where mechanized agriculture prevails. Mechanical cultivation (one device uses an electric eye to distinguish beet from weed) or herbicide treatment serves to keep weeds in check, and by mid-autumn the mature crop is ready for harvest. Mature beets have large, conical, whitish roots weighing 1 to 5 pounds, with two broad longitudinal belts of small rootlets and a crown of glossy, crinkled leaves. The plant is biennial, and stock retained for seed flowers the subsequent year. In the United States beets for seed are generally planted in autumn, left in the ground over the winter, and the seed harvested the following summer.

Traditionally sugar beets were gathered and topped by hand, but mechanical devices for accomplishing this have been developed, and in the United States nearly all of the crop is mechanically harvested. One type of machine loosens the beets with plowshares, then clasps the leaves and elevates the entire plant to the topping mechanism. Another utilizes a gigantic spiked wheel to lift the beets. It is thus possible to effect entirely by mechanical means digging, topping, loading on trucks, and windrowing of tops. Increasing efficiency may be expected as tailor-made, high-yielding beet strains are continually developed, and mechanical planters and harvesters are further perfected.

Sugar is extracted from harvested beets by a diffusion process preceding a recovery expression. The beet is first thoroughly washed, then shredded. The shreds are soaked in tanks of circulating hot water, permitting diffusion of the soluble sugar into the water solvent. Ninety-seven per cent of the sugar is so removed. The extracted beets are discharged to a press, where the remaining sugary water is pressed out. The pulp cake is utilized as previously indicated. The sugar solution is treated essentially as is cane juice. The clarification process involves application of lime, carbon dioxide, and eventually sulphur dioxide, with several filterings to remove sediment. First, second, and third runs of massecuite, subsequent to centrifuging and final refining, are carried out on the mechanical principles described for cane sugar.

Although all sucrose, from whatever source, is identical chemically, beet sugar was long regarded as "inferior" to cane sugar, and sold at a price disadvantage. Today sugar from all sources is regarded as equally valuable, and the sugar industry worries less about internal competition than competition from synthetic sweeteners such as saccharin much used in dietetic foods. Technical advances offer hope of additional economies in the beet sugar industry, and a stronger market position. Attempts are being made to harvest beets at the time of maximum sugar content. Some years ago the Dutch government spon-

Sugar extraction factory in Colorado for processing sugar beets. (Courtesy USDA.)

sored research at a phytotron in California directed toward determining at just what combination of environmental factors (day length, temperature, etc.) sucrose content would be optimum in the beet, in order to schedule harvest nationally for the appropriate instant! In some sections of the United States the farmer is paid for his beets on the basis of a chemical analysis showing sugar content, rather than on weight alone. Processers are attempting to find ways for better holding beets (in which the sugar quality deteriorates if held too long), and for avoiding the seasonality of an industry in which harvest and processing is generally accomplished within 120 days. Continuous-automated systems have been developed to replace the smaller batch operations, in turn calling for larger processing factories and increased capital demands. At least in the United States trends in the beet sugar industry are toward fewer, larger, well-financed corporations capable of supporting automation and cost-cutting research (reconstitution of limewater used in clarification is already being practiced in some areas, avoiding much of the hauling expense for new lime).

Maple sugar. The gathering of maple sap for sugar and syrup cannot compare in importance with the cane and beet sugar industries, but in picturesqueness it exceeds both. Gathering sugar from the sugar or hard maple tree, *Acer saccharum,* Aceraceae, was a trick of the American Indian, unknown to the white man until colonization of lower Canada and New England was well under way. The first extant record of white man's acceptance of woodscraft sugaring is dated 1673. Indian custom was simply to hack wounds in the sugar maple trunk in the early spring and collect the sap that dripped out in crude wooden or pottery vessels. This sap was concentrated by boiling (accomplished by dropping hot rocks into the sap, for containers resistant to fire for long periods were not to be had until the white man brought copper and iron), or by freezing the sap and removing each day the upper layer of ice that formed. The early settler simply adapted his means to Indian methods, eventually substituting drilling for hacking and using sumac and later metal spiles to conduct the sap drip to metal pails. Boiling-down was first done in iron kettles over an open fire, and later in special sugar houses.

The sugar maple is a common forest tree of the northeastern United States. As early as February or as late as April, depending upon season and climatic zone—whenever warm, sunny days follow cool, crisp nights—the sap of this maple, 95 per cent water, begins to flow. Then commences day-long activity in the "sugar bush," as the maple stands are called. The sugary sap apparently results from conversion of starches accumulated during the previous growing season into sugars during the winter, mostly in ray cells. The 2 to 6 per cent sugar content is largely sucrose. Sap flows for an average of 34 days (the shortest run on record is 9 days, the longest 57).

When trial has shown that the sap is flowing, holes about 2 inches deep are drilled into the sapwood, one to four to the tree, depending upon tree size. Today power drills are usually employed. A spile is inserted into the hole and a pail hung below it. The sap drips down the spile into the pail at a rate of sometimes better than 1 drop per second. Collection of the accumulating sap is a daily operation, formerly practiced by men wearing a sap-yoke with pails hung from its ends; in more modern times, by sleds,

Operating maple "sugar bush" in Vermont in mid-March. (Courtesy USDA.)

wagons, or power equipment driven into the woods or by gravity piping from the bush.

The sugar house is the scene of final processing. A shallow-pan evaporator, as much as 20 or 30 feet long, usually wood-fired at one end, receives the sap. The sap is here boiled down to a specified thickness, as indicated by temperature and density, and drawn off for filtering to remove precipitated limy materials. The resulting product is the famed maple syrup, weighing 11 pounds to the gallon. Further careful boiling produces a thick paste, which congeals and crystallizes when cooled to become the traditional New England confection, maple sugar.

Today maple sugar and syrup are expensive treats, a luxury. It was not always so. The American Indian, having no other sugar save occasional honey, cherished this product of the abundant maple as a sweetening. Similarly in early colonial days before the import and introduction of sugar cane, maple sugar was much a necessity in the colonists' limited fare. Production continuously increased, until about 1860, when cane sugar began to usurp maple sugar's place in the New England larder and decreasing abundance of forest progressively restricted production. But many thousands of gallons of syrup are still produced annually from New England to Missouri, often adulterated with cane sugar, however. A hint of the old romance of maple sugaring still lingers from Ohio to New England, as the first warm days of spring bring thousands of hardy and independent farmers into their sugar bush.

Most of the sugar and syrup made is sold to large syrup companies. The remainder is consumed locally or marketed independently. Strangely enough, the tobacco industry is one of the largest consumers of maple products, for flavoring tobacco. Annual production amounts to several million gallons (of syrup or its equivalent), almost entirely from wild trees. Deforestation and general "decline" of the sugar maple suggests that maple sugar will continue to become progressively less important as a sugar commodity.

Other sugar sources. Considering the many plant sources for vegetable oils, it is indeed strange that so few exist for sugar. In addition to the three previously discussed, the following have had minor importance.

PALM SUGAR OR JAGGERY is obtained from the sap of several palms, including species of *Phoenix*, *Borassus*, *Arenga*, and *Cocos*. The palm sugar industry of India and the Eastern tropics is an old one. The sugary sap is obtained by severing the expanding inflorescence and affixing some suitable receptacle for the vigorously rising sap. The cut or bruised peduncle will continue to yield sap or "toddy" (so-called when allowed to ferment) for a number of weeks or even months, at a rate of as much as 1 gallon per day. Collection commonly starts in November or December and lasts until summer. The gathered juice is boiled to syrup and sugar much as is maple sap. The Indian *Phoenix sylvestris* is typically incised or boxed on the upper trunk to yield sap. Each box is about 4 inches deep; on alternate years the box is made on the side opposite from that cut the previous year. Each second year the new box is hewed on new growth above the old box; thus palms tapped for several years have a curious zigzag appearance. The sugary sap, called neera, is about 10 per cent sucrose. It is generally boiled locally and crudely centrifuged to yield unrefined lump sugar called gur.

SORGHUM is derived from *Sorghum bicolor* var. *saccharatum*, a maizelike plant of the Gramineae, known in many horticultural strains. Other sorghum cultivars are better known for grain, forage, silage, and broom materials. The limited syrup acreage is in the Midwest and South of the United States. The syrup is mostly produced and consumed locally. Sorgo or sweet sorghum for syrup is planted like corn. Extraction and concentration of the juice involves simple, farm–home procedures, in principle similar to those used to derive massecuite in the cane sugar industry. The process is seldom carried beyond the syrup state.

PLANT NECTAR is well known as a source of sucrose, glucose, and levulose for the honeybee, which partially digests it and compounds it into flavorful honey. Honey is about three-fourths fructose and glucose, with minor amounts of proteins, mineral salts, and water. The flavor of honey largely depends upon the species of plants the honeybees visit. Flowers containing substantial amounts of essential oils usually impart characteristic flavors. Nectars from some African and Asiatic flowers have been reported to be consumed directly by man. These include the Madhuca tree, *Madhuca latifolia*, Sapotaceae, India; the honey flower, *Melianthus major*, Melianthaceae; and the Boer honey pots, *Protea mellifera* and *P. cynaroides*, Proteaceae, South Africa.

STARCH HYDROLYSIS. In Japan and elsewhere starch, from rice, millet, other cereals or potatoes, is hydrolized to sugar enzymatically (with malt), much as in making wort in brewing. Dextrin and maltose are the typical sugars so derived, and are collected by evaporation following hydrolysis. In North America a glucose syrup (Karo) is made by hydrolyzing corn starch with dilute acids. Fructose, used by diabetics, is made from the polysaccharide inulin, which may be derived from the tubers of the dahlia (*Dahlia pinnata*), the Jerusalem artichoke (*Helianthus tuberosus*), and other species.

Starches

Starch, even more so than sugar, is a common and readily extracted reserve food in various plant parts. Starch is almost invariably present in plants consumed for food. Being ordinarily insoluble, starch is commercially extracted by physical separation, for a multitude of industrial and food uses. The starch-bearing cells are by some means ruptured, and the abundant starch grains floated free or separated by centrifuging. There are several different types of starch —differing in both the chemical linkages and the physical nature of the grains. It is usually possible to identify the source of a starch grain by microscopic examination.

Starch can be extracted from practically any cereal grain, tuber, or other plant storage organ. Commercially two sources in particular have been exploited: the temperate *Zea mays* (maize or corn) and the tropical *Manihot esculenta* (locally called manioc, mandioca, cassava, yuca, tapioca, and so on). The former yields over 1 million tons of cornstarch annually, and another $1\frac{1}{2}$ million tons are chemically converted (into syrup, sugar, dextrin, etc.). Extracted starch is encountered in modern economy practically everywhere—in various foods such as puddings, candies, chewing gum, bakery products, and the like; in adhesives, laundry starches, industrial sizings, pharmaceuticals, oil-well drilling compounds; and (by hydrolysis as corn sugar) in a number of products from confections to leather tanning, in syrups for uses from infant feeding to tobacco flavoring, and in dextrins for envelopes, stamps, and other uses.

Cornstarch. The extraction of corn oil from the embryo of the seed of maize, *Zea mays*, Gramineae, has been discussed in a previous chapter. To effect separation of the germ the cleaned corn grains are steeped for a number of hours in a weak sulfurous acid solution, which softens the grain and dissolves out certain soluble substances (the valuable corn steep liquor). The grain is then cracked in special mills

Corn grain, the chief source of commercial starch in the United States. (Courtesy USDA.)

and the germ, lighter than water, is floated off in flotation tanks. The heavy sediment of the flotation tank constitutes the remainder of the corn grain—starch and gluten (protein) from the endosperm, and the various seed coats or hulls. This mass is finely pulverized in burr mills, and, in wet-process milling, separation of the various components is effected as follows. The coarse hulls are first removed by washing the pulverized endosperm through fine mesh cloth on rotating reels, and finer hulls by subsequent sifting in succeeding shakers. The suspension of gluten and starch that washes through was formerly separated in elongated starch tables, shallow channels into which the mixture is slowly fed. Most of the starch settles out from the slow water as a white sediment, while the gluten washes to the end of the channel. The gluten is usually recombined with the hulls for cattle feeds, or occasionally finds chemurgic use in plastics or synthetic fibers. More economical separation is achieved by centrifuging, now commonly employed. By whatever means of separation, the starch is washed in a strong stream of water to suction pads which pick it up from "solution" and subject it to further washing with a spray of clean water. The starch is then scraped from the suction pad, dried, and pulverized, ready for market.

A portion of the starch produced in the United States is hydrolized with weak acid, to form glucose syrup or corn sugar. This hydrolysis is a controlled digestion process, similar to the chemical breakdown starch undergoes when taken into the human digestive tract. By altering conditions or by use of special enzymes, a corn syrup or corn sugar derived from corn syrup can be made to contain specified types and percentages of simple sugars, dextrins, and the like, according to the use to which they are to be put. Most starch, however, is consumed directly. The finer grades of cornstarch are utilized for human food; cheaper grades find industrial use as sizing, laundry starch, hydrolysis stock, and the like. Paradoxically, as with sugar, man refines starch so highly that the most healthful mineral and protein portions go into stock feed or for other uses. From the nutritional standpoint, the cattle get the best of it.

Other starch sources. MANIHOT. The growing and use of the tropical cassava, manioc, or yuca, *Manihot esculenta*, Euphorbiaceae, as a food plant is discussed later in this book. In the Philippines, East

Indies, and the Caribbean tropics it is also much cultivated as a source of crude flour from which tapioca and a high-amylopectin starch are extracted. Principles involved in extraction are essentially the same as those described for cornstarch. Extraction is carried out crudely in the Far East or more efficiently in the consuming country. Most of the cassava starch used in the United States is imported as the crude flour obtained from the grinding of the fleshy roots, and the starch is separated by appropriate mechanical means. It finds wide industrial use in sizings and a variety of mucilages. The edible tapioca is made by heating cassava starch until the grains swell, partly hydrolize, and aggregate into the characteristic tapioca pearls.

POTATO. The common white or "Irish" potato, *Solanum tuberosum*, Solanaceae, contains an abundance of starch. In Europe, especially, it takes the place of corn in the production of commercial starch. Some potato starch is produced in the United States from culls, left unsalable in amounts of several thousand tons a year in the potato-raising states. The potatoes are washed, pulverized, and strained. Thereafter the starch is isolated as in the cornstarch industry. The product finds a market primarily as a cotton warp sizing or as raw material for industrial fermentation. Of course the waste pulp is used in cattle feeds.

WHEAT. Since ancient times, a starch extract has been derived from wheat, *Triticum sp.*, Gramineae.

RICE. Imperfect rice grains, *Oryza sativa*, Gramineae, serve as a starch source. The starch is freed after softening of the grain by a series of sievings.

ARROWROOT. The tuberous rhizomes of several Scitaminaceous plants, including *Maranta arundinacea*, Marantaceae; *Canna edulis*, Cannaceae; and *Curcuma angustifolia*, Zingiberaceae, yield arrowroot starch. Extraction follows the principle of cleaning, crushing, washing the starch free, and

centrifuging or allowing it to settle out. Arrowroot is said to be easy to digest, and is used in foods for invalids and children.

SAGO. Sago starch is extracted from the stems and possibly the roots of certain palms and cycads. The palm *Metroxylon sagu* of Malaya and the East Indies yields the most common sago starch from the pithy portion of the stem. Extraction involves the usual pulverization, straining, and sedimentation procedures.

The Future of the Carbohydrate Extractive Industries

Extracted plant carbohydrates can be expected to hold their own in the battle for markets, in a world that must increasingly rely upon renewable crop materials as forest and mineral resources approach exhaustion. In terms of energy per acre there are few more remunerative crops than sugar cane, and there is ample tropical environment where it can be grown. Sugar beets, of course, must face the competition of other crops for the most remunerative use of the land where beets are now grown; the future here is more problematic, but, as was noted in the discussion of the sugar beet, trends toward mechanization and higher capital investment augur well for keeping the sugar beet competitive at least in the short run. Although sugars of various sorts can be secured from many plants and plant parts, it is unlikely that any new crop can break into the market (except locally) in face of a sugar industry efficiency that yields sucrose on the international market at just a few cents per pound.

The situation with starches is not so cut-and-dried. Many are the potential sources of plant starch, local availability and economy in production (often from cull products grown for other purposes) generally determining which may be used. It is natural that a sizable cornstarch industry would have developed in places such as the United States, where corn is a dominant crop, and potato or cassava starches in places such as eastern Europe and the tropics where these root crops dominate. To a great degree, the extent of starch extraction correlates very closely with that of food production, the subject of the next part of this book. The cereals and root crops that are high-yielding in starchy food are also those adaptable to starch extraction. To an extent starch battles with synthetics for industrial markets, such as for textile and paper sizing. Still, starch produced from annual crops is an abundant and economical raw material, and it is difficult to imagine it being abandoned in favor of other substances that must be more expensively produced from exhaustible natural resources.

The chemist has just begun to manipulate the carbohydrate molecule. It was noted that derivatives from starch can be made into urethane foam, alkyd paint, and certain types of detergents. Methyl glucoside has been manufactured commercially by treating a slurry of corn sugar with methanol, to provide the base for an insulating foam. Similar applications of starch will doubtless develop, and already the demand for a special type of starch (high-amylose) has spurred the breeding of a special corn cultivar to furnish the distinctive raw material. Possibilities of this nature would seem unlimited, held back only by economic forces that govern the cost of competing raw materials. It is reassuring that mankind has "in reserve" an inexpensive, renewable source of energy in the several plant species that yield carbohydrate extractives abundantly. In an overpopulated world these will doubtless be much called upon.

SUGGESTED SUPPLEMENTARY REFERENCES

Barnes, A. C., *The Sugar Cane*, World Crops Books, Leonard Hill, London, 1964, and Interscience, N. Y., 1965.

Brautlecht, C. A., *Starch, Its Sources, Production and Uses*, Reinhold Pub. Co., N. Y., 1953.

Deerr, N., *The History of Sugar*, Chapman and Hall, London, 1949.

Hughes, C. G., E. V. Abbott, and C. A. Wismer, *Sugar-cane Diseases of the World*, American Elsevier, N. Y., Vol. I, 1961, Vol. II, 1963.

Humbert, R. P., *The Growing of Sugar Cane*, American Elsevier, N. Y., 1968.

King, N. J., R. W. Mungomery, and C. G. Hughes, *Manual of Cane Growing*, American Elsevier, N. Y., 1965.

McGinnis, R. A. (ed.), *Beet-sugar Technology*, Reinhold Pub. Co., N. Y., 1951.

Nearing, H., and S. Nearing, *The Maple Sugar Book*, John Day Co., N. Y., 1950.

Pigman, W., *The Carbohydrates—Chemistry, Biochemistry, Physiology*, Academic Press, N. Y., 1957.

USDA Yearbooks, Washington, D. C.: appropriate sections especially of such issues as *Farmer's World*, 1964; *Power to Produce*, 1960; and *Crops in Peace and War*, 1950–51.

See also the Supplementary References in Chapters 16 (cereals), and Chapter 18 (especially with reference to root crops).

PART IV

Plants and Plant Parts Used Primarily for Food and Beverage

FOOD PLANTS ABOVE ALL OTHERS, ARE ESSENTIAL FOR LIFE: THEIR SELECTION and refinement have proceeded as civilization has evolved. In a world increasingly hungry they become increasingly important. Tremendous world commerce takes place with the major foodstuffs, but equally they are the concern of every man from sophisticated gourmet to subsistence farmer. Plants are also "food" for our esthetic being. And they often serve in ceremonial ways, or for purposes not reviewed in previous sections.

CHAPTER 15

Food Plants

THE INTRODUCTORY CHAPTERS EMPHASIZED the importance of plants—mainly food plants—in the development and very existence of man. Many of the important species in use today were domesticated so long ago that we can only guess at the wild forms and loci of domestication. Think of the many plant products eaten or otherwise used daily that are derived from the important agricultural crops of the world. The chances are that all were domesticated before the first chronicles were ever chiseled on stone. Of the hundreds of important economic plants, only a few come to mind as having a more-or-less precise history of domestication— rubber, quinine, a few nut and fruit trees, and some ornamentals. One can only admire the perspicuity of our ancestors who first planted certain wild annuals, saw them through to maturity, and saved necessary seed for the next crop. Without agriculture, man would have remained an infrequent nomad, devoid of the material and intellectual empire that is his today.

World domestication centers. Vavilov and his followers have mapped out several world centers for domestication, as was mentioned in Chapter 2. Although domestication of food plants may actually have been more diffuse than use of the word "center" implies, Vavilov's concepts suggest a few generalizations. Each center is a region of limited rainfall and mild climate. Neither the humid tropics nor the cold temperate regions have apparently contributed many food plants of world importance. The putative centers of domestication are all more or less mountainous. Several important domesticated plants seem to have had multiple points of origin, one cultivar being derived at one time and in one place and another in an entirely different locality at presumably a widely different date. Thus soft wheats presumably came from southwestern Asia, whereas hard wheat originated in the Mediterranean area. Vavilov further contended that weed "hangers-on" of the primary cultivated crops often became equally or more important in environments inhibiting in one way or another the primary crop, and thus essentially domesticated themselves. Rye and oats may have developed thus from weed species in wheat fields.

Vavilov's chief domestication centers can be assigned continental location roughly as follows: (1) *Southwestern Asia* (Afghanistan through Asia Minor), the most important Old World center, where probably were first domesticated soft and club wheats, rye, buckwheat, broadbeans, peas, lentils, carrots, turnips, apples, pears, plums, sweet cherries, almonds, and pomegranates; (2) *Southeastern Asia* (the mountains of Tibet and China north and east of the Himalayas), where presumably were domesticated forms

Table 15-1

New World and Old World Food Plants before 1492

New World			Old World		
maize	pumpkin	wheat	onion	wine grape
manioc	peanut	rye	garlic	apricot
lima bean	chayote	barley	spinach	peach
potato	papaya	oats	eggplant	olive
sweet potato	avocado	millet	lettuce	fig
kidney and	pineapple	sorghum	endive	almond
other beans	custard apple	rice	celery	quince
tomato	soursop	buckwheat	asparagus	pomegranate
green pepper	cherimoya	turnip	peas, various	watermelon
Jerusalem	guava	cabbage	soybean	cucumber
artichoke	chocolate	rutabaga	broadbean	banana
sunflower	cashew	chard	dasheen	orange
quinoa	sapote	mustard	yam	lemon
squash		radish	apple	lime
		beet	pear	date
		parsnip	plum	mango
		carrot	cherry	clover
		breadfruit	alfalfa	
		mangosteen	bluegrass	

of oats and barley, soybeans, certain millets, yams, citrus fruits, peaches, and various vegetables; (3) *Northeastern India–Malaysia* (southeast of the Himalayas), where presumably were domesticated rice, sugar cane, bananas, and breadfruit; (4) *North Africa* (especially Abyssinia) where presumably were domesticated hard wheats, oats, barley, certain millets, and coffee; (5) *The Mediterranean Basin*, where presumably were domesticated the olive and fig; and (6) *The New World* (Peruvian area, and secondarily southern Mexico), where were domesticated the New World food plants listed in Table 15-1.

In each ancient culture center, certain food plants were basic. In Mexico and southward, maize was the most important single food. In the middle Andes (parts of Bolivia and Peru), the potato was pre-eminent, although beans, maize, and other foodstuffs were of considerable secondary importance. In the Mediterranean basin, Asia Minor, and southwestern Asia the

cereals—wheat, barley, rye, and oats—were most important, while India and China were undoubtedly the domain of rice and soybean.

Practically no important food plants or domestic animals were common to both hemispheres, the coconut, certain gourds, and the dog perhaps offering exceptions. But in the era of expansion and colonization following the Middle Ages, food plants were spread by man from one corner of the earth to another, and ultimate sources were thus confused: for example, the French thought maize to have come from Turkey, naming it *blé de Turquie;* it probably did, but only after it had first reached there from the New World. Similarly, the Incan "Irish" potato reached North America via Europe. Other chapters have repeatedly referred to Spanish, Portuguese, Dutch, and English worldwide trade in search of spices, aromatics, drugs, sugar, and other food products. Never before this era of travel had man been so adept in altering the ecology of the globe, and today we grow indiscriminately Old World plants

Digging with the foot-plow for planting potatoes in the high Andes. (Courtesy Chicago Natural History Museum.)

Indians gathering cowlily fruits. (Courtesy Chicago Natural History Museum.)

in the New World and New World plants in the Old, side by side with those native to each area.

Food plants of early cultures. The probable progression from a roaming and hunting to a sedentary society employing agriculture was reviewed briefly in Chapter 2 for two of the more thoroughly investigated parts of the world—the Zagros Mountains east of the Tigris–Euphrates, and the Tehuacan Valley of southern Mexico. Evidence from many disciplines provides plausible documentation of the domestications that made these among the first great centers of civilization.

Of course other localities were inhabited too; it appears that in the tropics especially such acculturation as was achieved depended upon a system of shifting agriculture. "Slash-and-burn" land preparation is still practiced in less populated tropical lands, and bears such names as ladang in the Orient and milpa in Central America. Forest is killed, then burned in the dry season. When only primitive tools were available girdling must have been the chief means for killing large trees. Ashes from the burned forest have temporary effect as a fertilizer, stimulating first-season growth of the crop. In primitive societies the crop is usually planted by poking holes into soft soil as the rainy season begins. Shifting cultivation cannot be discussed in detail here, but obviously the system is appropriate only for lightly populated lands. The burned-over forest remains fertile and sufficiently free of brush for only a few years at most, whereupon the process must be repeated elsewhere. It has been surmised that major civilizations such as that of the Mayans in Guatemala may have disappeared because shifting cultivation exhausted all the nearby land that supported the advanced culture. One can agree with H. H. Bartlett (*Man's Role in Changing the Face of the Earth*), when he writes:

The rise of man from brute to savage, from

savage to barbarian, and from barbarian to the semi-civilized man of today has come at great expense to the resources of the world as it was, let us say, before man's mastery of fire and tools enabled him to dominate the earth.

CHINA. The Western world hears relatively little about China, the center of early Oriental civilization. Kwang-Chih Chang, in *Archaeology of Ancient China* (*Science* **162**, 521 Nov. 1, 1968), after discussing the various stages of incipient agriculture, states the following:

This Neolithic culture centering on the domestication and cultivation of foxtail millet is also the earliest identifiable manifestation of a Chinese culture. In contrast to other contemporary traditions, such as in the Near East and Nuclear America, the Chinese tradition in Neolithic times is characterized by the following traits and complexes. A large variety of plants was cultivated and used for food or fabrics, or both. In addition to the foxtail millet, plants cultivated for food included the broomcorn millet (*Panicum miliaceum*), many vegetables of the genus *Brassica*, and possibly sorghum, and the soybean; those for fabric included at least one type of hemp. The fiber of plants was made into cordage and cloth, and basketry and matting were also highly developed. The silkworm (*Bombyx mori*) was raised for silk. Dog and pig were the most prominent domestic animals.

SOUTHEAST ASIA. Archeological findings dated about 7000 B.C. in Thailand reveal species of almond (*Prunus*), *Terminalia*, betel (*Areca*), bean (both *Vicia* and *Phaseolus*), pea (*Pisum*), *Raphia*, bottlegourd (*Lagenaria*), and Chinese waterchestnut (*Trapa*). A bit later in the sequence are *Canarium*, cucumber (*Cucumis*), *Piper*, butternut (*Madhuca*), candlenut (*Aleurites*), bamboo, and other useful plants. Presumably shifting cultivation permitted development of civilization in this forested region.

MESOPOTAMIA. The wonderful variety of habitat in the Zagros Mountains west of the Fertile Crescent has yielded a wealth of cultivars that eventually sustained sedentary cultures in the dry lowlands, under irriga-

Gathering lotus for food. Stems, seeds, and young leaves are edible. (Courtesy USDA.)

tion. Here shifting cultivation was less a requisite than with most early civilizations. Wild wheat and barley were intensively collected and eventually brought into cultivation, also alfalfa and several small-seeded legumes. Wild barley was transformed into a free-threshing type before 7000 B.C., and a free-threshing hexaploid wheat was cultivated by 6000 B.C. Oil-rich foods seem to have been obtained by gathering wild fruits, such as acorns, pistachios, hazelnuts, and olives, but there was domestication of weed grasses (e.g., Italian millet, presumably derived from the wild green millet, *Setaria viridis*) and cultivated flax (derived from the wild *Linum bienne*).

PREHISTORIC EUROPE. About 8000 B.C. the ice age was drawing to a close in Europe, and northern Europe was recolonized by hunting peoples. Before 3000 B.C. food-producing cultures had arisen in many locations, especially around lake margins and in the lowlands near the sea. It has been demonstrated that a type of shifting cultivation could have been undertaken (and evidence further suggests it was) in the deciduous forests of Europe using only the stone tools of appropriate archeological age. Oats and rye were grown for a few years, after which the land was abandoned, returning to birch and alder. After 20 to 30 years young forest might again be cleared by slash-and-burn techniques. Within three or four millennia many crops had been brought into into domestication; familiar species cultivated in Charlemagne's empire are listed in Table 15-2.

Flail threshing grain in rural Europe. (Courtesy Chicago Natural History Museum.)

Table 15-2

Species Listed in Capitulare de Villis vel Curtis Imperialibus of A.D. 812 ("Directions for the Administration of Imperial Courts or Estates"), Reign of Charlemagne

Agrimonia eupatoria	Agrimony	Armoracia rusticana (=*Armoracia lapathifolia*)	
Agrostemma githago	Corn cockle		Horseradish
Alcea rosea (=*Althaea rosea*)	Hollyhock	*Artemisia abrotanum*	Southernwood
Allium ascalonicum	Shallot	*Artemisia dracunculus*	Tarragon
Allium cepa	Onion	Arum dracunculus (=*Dracunculus vulgare*)	Dragon
Allium fistulosum	Onion, Welsh or Spanish	*Asarum europaeum*	Ginger, wild
Allium porrum	Leek	Athamanta meum (=*Athamanta sp.?*)	Athamanta
Allium sativum	Garlic	*Atriplex hortensis*	Orach
Allium schoenoprasum	Chive	*Avena sativa*	Oat
Althaea officinalis	Marsh mallow	Balsamita vulgaris (=*Chrysanthemum balsamita*)	
Althaea rosea	Hollyhock		Costmary
Ammi majus	Herb William	*Beta vulgaris*	Beet
Amygdalus communis (=*Prunus amygdalus*)	Almond	Betonica officinalis (=*Stachys officinalis*)	Betony
Amygdalus persica (=*Prunus persica*)	Peach	Blitum capitatum (=*Chenopodium capitatum*)	Blite
Anethum foeniculum (=*Foeniculum vulgare*)	Fennel	*Brassica caulorapa*	Kohlrabi
Anethum graveolens	Dill	Brassica eruca (=*Brassica sativa*)	Roquette
Anthriscus cerefolium	Chervil; Chervil, salad	*Brassica napus*	Colza; Rape; Turnip
Apium graveolens var. *dulce*	Celery	*Brassica nigra*	Mustard, black
Apium petroselinum (=*Petroselinum crispum*)		*Brassica oleracea* var. *capitata*	Cabbage
	Parsley	Brassica oleracea var. caulorapa (=*Brassica caulo-*	
Arctium lappa	Butter bur	*rapa*)	Kohlrabi
Armoracia lapathifolia	Horseradish	*Brassica rapa*	Colza, Kohlrabi; Rape; Turnip

390

Table 15-2 (cont'd)

Species Listed in Capitulare de Villis vel Curtis Imperialibus of A.D. 812 ("Directions for the Administration of Imperial Courts or Estates"), Reign of Charlemagne

Cannabis sativa	Hemp	*Lactuca sativa*	Lettuce
Carum carvi	Caraway	*Laurus nobilis*	Sweet bay
Castanea sativa	Chestnut, Spanish or Eurasian	*Lens esculenia* (=*Lens esculenta*)	Lentil
Castanea vesca (=*Castanea sativa*)		*Lens esculenta*	Lentil
	Chestnut, Spanish or Eurasian	*Lepidium sativum*	Cress, garden
Centaurium species	Centaury	*Levisticum officinale*	Lovage
Chenopodium capitatum	Blite	*Lilium candidum*	Lily, white
Chrysanthemum balsamita	Costmary	*Linum usitatissimum*	Flax, linen
Chrysanthemum parthenium	Feverfew	*Malus sylvestris*	Apple
Cicer arietinum	Chickpea	*Malva rotundifolia* (=*Malva sp.*?)	Mallow, running
Cichorium endivia	Endive	*Malva sylvestris*	Mallow, European
Cichorium intybus	Chicory	*Matricaria parthenium* (= ?)	Feverfew
Citrullus colocynthis (= ?)	Bitter-apple	*Mentha aquatica*	Mint, water
Coriandrum sativum	Coriander	*Mentha crispa* (=*Mentha spicata*)	Spearmint
Corylus species	Hazelnut	*Mentha gentilis*	Mint, garden
Corylus avellana	Hazelnut, European	*Mentha pulegium*	Pennyroyal
Costus hortorum (= ?)	Costmary	*Mentha rotundifolia*	Mint, apple
Cucumis colocynthis (= ?)	Bitter-apple	*Mentha spicata*	Spearmint
Cucumis melo	Melon	*Mentha sylvestris*	Mint, wood-
Cucumis sativus	Cucumber	*Mespilus germanica*	Medlar
Cucurbita lagenaria (=*Cucurbita pepo* var. *ovifera*)		*Momordica elaterium* (=*Ecballium elaterium*)	
	Bottlegourd		Cucumber, squirting-
Cucurbita pepo	Pumpkin	*Morus nigra*	Mulberry
Cucurbita pepo var. *ovifera*	Bottlegourd	*Nepeta cataria*	Catmint (catnip)
Cuminum cyminum	Cumin	*Nigella damascena*	Love-in-a-mist
Cydonia oblonga	Quince	*Nigella sativa*	Fennel flower
Cydonia vulgaris (=*Cydonia oblonga*)	Quince	*Origanum dictamnus*	Dittany of Crete
Cynara cardunculus	Cardoon	*Panicum miliaceum*	Millet, common or proso
Daucus carota	Carrot	*Papaver somniferum*	Poppy, opium
Dictamnus albus	Dittany	*Pastinaca sativa*	Parsnip
Dipsacus fullonum	Fuller's teasel	*Petroselinum crispum*	Parsley
Dracunculus vulgare	Dragon	*Phaseolus vulgaris*	Bean, kidney
Ecballium elaterium	Squirting-cucumber	*Pimpinella anisum*	Anise
Eruca sativa	Roquette	*Pinus cembra*	Pine, Swiss stone
Erythraea centaurium (=*Centaurium sp.*)	Centaury	*Pisum arvense* (=*Pisum sativum* var. *arvense*)	
Euphorbia lathyris (=*Euphorbia lathyrus*)			Pea, field
	Caper, wild	*Pisum sativum*	Pea, garden
Euphorbia lathyrus	Caper, wild	*Pisum sativum* var. *arvense*	Pea, field
Fagus species	Beechnut	*Prunus amygdalus*	Almond
Ficus carica	Fig, common	*Prunus cerasus*	Cherry, sour
Foeniculum vulgare	Fennel	*Prunus domestica*	Plum
Gentiana centaurium (= ?)	Gentian	*Prunus persica*	Peach
Gladiolus communis	Cornflag	*Pyrus communis*	Pear
Heliotropium europaeum	Heliotrope	*Pyrus malus* (=*Malus sylvestris*)	Apple
Hibiscus syriacus	French tree-mallow	*Quercus* species	Acorn
Hordeum vulgare	Barley	*Raphanus sativus*	Radish; Radish, long; Rave
Iris florentina (=*Iris germanica* var. *florentina*) Orris		*Rosa centifolia*	Rose
Iris germanica	Iris, "German"	*Rosmarinus officinalis*	Rosemary
Iris germanica var. *florentina*	Orris	*Rubia tinctorum*	Madder
Isatis sativa (=*Isatis tinctoria*)	Woad, dyer's	*Rumex acetosa*	Sorrel, garden
Isatis tinctoria	Woad, dyer's	*Ruta graveolens*	Rue
Juglans species	Walnut	*Salvia officinalis*	Sage
Juglans regia	Walnut, English	*Salvia sclarea*	Clary
Juniperus sabina	Savin	*Satureja hortensis*	Savory, summer

Table 15-2 (cont'd)

Species Listed in *Capitulare de Villis vel Curtis Imperialibus* of A.D. 812 (*"Directions for the Administration of Imperial Courts or Estates"*), Reign of Charlemagne

Scandix cerefolium (=*Anthriscus cerefolium*) Chervil		*Tanacetum vulgare*	Tansy
Scilla maritima (=*Urginea maritima*)	Sea-onion	*Trigonella foenum-graecum*	Fenugreek
Secale cereale	Rye	*Triticum aestivum*	Wheat
Sempervivum tectorum	Houseleek	Triticum sativum (=*Triticum aestivum*)	Wheat
Seseli tortuosum (=*Seseli* sp.?)	French heartwort	*Triticum spelta*	Spelt
Setaria italica var. *stramineo-fructa*	Millet, German	*Tussilago farfara*	Coltsfoot
Sinapis nigra (=*Brassica nigra*)	Mustard, black	Tussilago petasites (=*Tussilago farfara*)	Coltsfoot
Smyrnium olusatrum	Alexanders	*Urginea maritima*	Sea-onion
Sorbus domestica	Service-tree	*Vicia faba*	Bean, broad
Stachys officinalis	Betony	*Vicia narbonensis*	Vetch
Tanacetum balsamita (=*Chrysanthemum balsamita*) Costmary		*Vitis vinifera*	Grapes

From John Asch, comp., *Garden Journal* **18**, 147 (Oct. 1968).

TROPICAL AFRICA. When David Livingstone first explored Africa in the mid-nineteenth century, he found a system of shifting cultivation probably unchanged for millennia. Ecologically the system is well suited to a tropical climate where land held cleared turns to intractable grass. Intensive cropping is practiced for 1–3 years, perhaps 4 years in the savanna woodlands, before the land is turned back to fallow and forest regeneration. Yams, melons, calabash—and since their introduction beans, cassava, pumpkins, and corn—are planted in the burned-over area with onset of the rainy season, profiting from the fertility of the wood ashes.

MESOAMERICA. In southern Mexico and Central America incipient cultivation seems to have begun between eight and seven millennia B.C., with small black beans (*Phaseolus*) and squash seed (*Cucurbita*) planted around habitations to supplement acorns, peanuts, and other wild fruits and seeds. Progressively, primitive corn, chili peppers, avocados, amaranths, tepary beans, zapotes, and various squashes were domesticated. By 1000 B.C. the basic crosses with *Euchlaena* that yielded modern corn appear to have occurred, perhaps the greatest food plant advancement in the New World. The relatively arid climate of the Oaxaca and Tehuacan valleys, surrounded by more pluvial highlands, offers, like the Fertile Crescent in Asia Minor, wide variation of habitat ideal for domestication and establishment of sedentary cultures without the need for excessive shifting cultivation. In areas such as these the most prominent early civilizations seem to have been born.

NORTH AMERICAN INDIAN. Many North American Indian tribes carried on an agriculture of greater or less complexity, but even these relied to some extent upon wild plants; and other, hunting tribes depended entirely upon wild plants for their limited vegetable fare. The plants cultivated were chiefly maize, beans, pumpkins, squashes, and sunflowers. In addition to these there were nuts and fruits from many wild trees and shrubs, greens and vegetables from wild herbs, bulbs and tubers for use in making meal or seasonings, and seed of wild grasses for grinding to flour. Between 1,000 and 2,000 wild species are known to have been used by the Indians as food plants.

Acorns from various oaks, *Quercus*, were one of the most important native foods of the North American forest Indians. The acorns were pulverized, then leached in streams for several days to remove the bitter taste. Other much-used "nuts" included: seeds of the aquatic American lotus, *Nelumbo*

An assortment of New World fruits and vegetables. (Courtesy Chicago Natural History Museum.)

An assortment of Old World fruits and vegetables. (Courtesy Chicago Natural History Museum.)

lutea, piñon nuts from *Pinus edulis*, beechnuts from *Fagus grandifolia*, hickory nuts from *Carya*, black walnuts and butternuts from *Juglans*, hazelnuts from *Corylus*, and the like. Fresh fruits used raw, cooked, dried, or crushed included: serviceberry (*Amelanchier canadensis*); blueberries (*Vaccinium*); *Smilax herbacea*; blackberries and raspberries (*Rubus*); currants and gooseberries (*Ribes*);

strawberries (*Fragaria*); various cactus fruits; wild plums and cherries (*Prunus*); hawthorns (*Crataegus*); crabapples (*Malus*); persimmons (*Diospyros*); mesquite (*Prosopis*). Indians also ate, raw or cooked, the flowers of many species, including the redbud (*Cercis canadensis*); milkweed buds (*Asclepias*); and Spanish bayonet (*Yucca*). For greens or potherbs almost anything tender and not fibrous was eaten, raw or, more commonly, cooked. Included were milkweeds, clover, *Heracleum, Oxalis, Caltha, Montia, Phytolacca*, and so on. Wild "grains" consumed included: wild rice (*Zizania aquatica*), now available on the market; cane grass (*Phragmites communis*); and seeds of goosefoot, amaranth, *Loasa*, mints, and composites, often mixed with corn meal. Roots and rhizomes used, particularly for winter fare, included: wild onion (*Allium*); mariposa (*Calochortus*); camass (*Camassia esculenta*); spring beauty (*Claytonia virginica*); *Trillium*; trout lily (*Erythronium*); Jack-in-the-pulpit (*Arisaema triphyllum*); arrowhead (*Sagittaria*); cattails (*Typha*); wild sweet-potato (*Ipomoea pandurata*); bracken fern (*Pteridium aquilinum*); and many others.

JAPAN. The soybean, the most important food legume in Asia, has been cultivated since before recorded history. The beans are eaten freshly boiled, made into bean curd (tofu), or mashed and fermented along with malt and salt (miso). The tubers of swamp plants such as *Sagittaria* and *Eleocharis* are consumed both raw and cooked. Taro or dasheen, *Colocasia esculenta*, is popular in warmer sections, as is fresh ginger rhizome; lily bulbs are eaten, too. The young leaves of *Hosta* serve as greens. Both tubers and seeds of *Nelumbium* are eaten, as are the seeds of the waterchestnut, *Trapa*. Young bamboo shoots, fiddleheads of the bracken fern (*Pteridium*), and young petioles of burdock (*Arctium lappa*) are other unusual delicacies.

Modern Food Production. Agriculture in the last third of the twentieth century is in a rapid stage of transition and is exhibiting many contrasts. In some parts of the world, especially the so-called "underdeveloped" countries, food production is still of a subsistence nature, primitively carried on with

The old way of land preparation in the United States, still practiced in a few remote areas. Hours spent are not very remunerative, although crop growth can be adequate. (Courtesy USDA.)

The usual way of preparing land in the United States, a more economical use of man-hours than when less mechanized. In technically advanced nations man-hours are the most costly agricultural input.

few or no mechanical aids. At the other extreme, in the technologically advanced nations, capital and machines have increasingly been substituted for hand labor (which is one of the most costly inputs in farming there). Mechanization can be efficient only where enough land is available and population pressures are reasonable; output per acre is usually higher where land is meticulously tended by hand (as in parts of the Orient) than where machines do the job. Fertilizers, pesticides, drainage, and other techniques associated with mechanization help marginal lands, but nothing equals customized human care to extract the full potential from small holdings. With the world running short of good farming land, it is problematic how much farther mechanization can increase per capita production. Nevertheless, the record in the last half-century is certainly impressive (see Fig. 15-1).

And certainly in the future the pressures on agriculture will not lessen, what with a large part of the world undernourished today. Already the predictions of Krantz and Hills are manifest—that farming will shift from brawn to brain, from a craft to a science, production orientation to profit orientation, from cash to perpetual credit, from home-grown feed to purchased feed, from farm-produced fertilizer and horsepower to machinery and high-analysis chemical fertilizer, and from worker to manager. The trend is toward professionalized management of larger, more sophisticated units, in which biological and mechanical technology and capital substitute for labor. Figure 15-2 tells the story of how the economy is sustained in an industrialized nation.

Although there is little hope for any part of the world unless population control can be achieved, food insufficiency is an immedi-

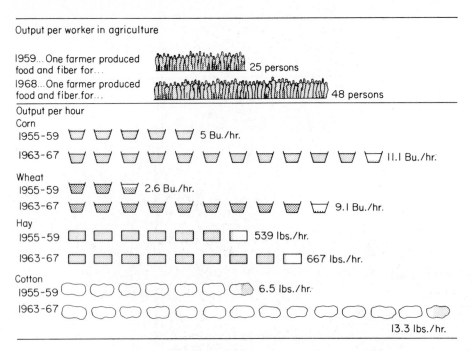

Output per worker in agriculture

1959...One farmer produced food and fiber for... 25 persons

1968...One farmer produced food and fiber for... 48 persons

Output per hour

Corn
1955-59 5 Bu./hr.
1963-67 11.1 Bu./hr.

Wheat
1955-59 2.6 Bu./hr.
1963-67 9.1 Bu./hr.

Hay
1955-59 539 lbs./hr.
1963-67 667 lbs./hr.

Cotton
1955-59 6.5 lbs./hr.
1963-67 13.3 lbs./hr.

Figure 15-1 The bounty of mechanization in industrialized societies.

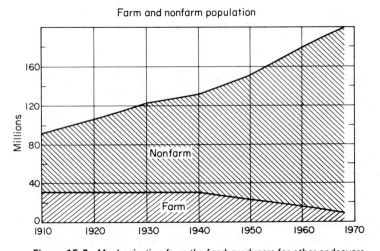

Farm and nonfarm population

Figure 15-2 Mechanization frees the food-producers for other endeavors.

ate problem in heavily populated lands such as those of the Orient (Fig. 15-3). The less developed countries even lack exports enough to exchange for imported food "borrowed" from other less heavily taxed parts of the world. Some regions, still moderately populated, such as central Africa, lack the facilities for materially increasing food production; there is insufficient education, technical training, transportation and distribution, and aids such as fertilizer and pesticides. So far there has been relatively little research directed toward discovering high-yielding, adapted cultivars, either. New, useful food plants are needed, especially of types providing good nutritional balance.

World agricultural production

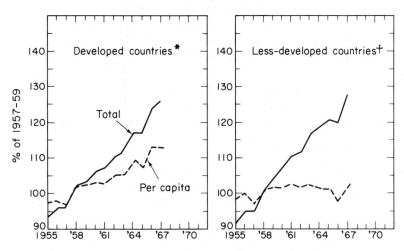

*North America, Europe, USSR, Japan, Republic of South Africa, Australia and New Zealand

†Latin America, Asia(except Japan and Communist Asia), Africa (except Republic of South Africa)

Figure 15-3 The crux of the food problem is that in many lands gains in productivity are offset by increased population: productivity cannot increase indefinitely.

Food plants rich in protein are especially needed for populations subsisting mainly on carbohydrates, to avoid further increase in kwashiorkor disease (protein deficiency). In addition to finding new locally adapted food species, the plant breeder can provide new cultivars of present crops having higher protein content. Corn belt crops in the United States exhibit quite a range of variation, with maize showing 8.3–13.4 per cent protein, sorghum 10.6–13.2 per cent, oats 7.4–22.6 per cent, rye 13.0–15.4 per cent, and soft wheat 11.6–15.6 per cent. These measurements are of feed grains, and obviously there may be great differences in nutritional value within a single feedstuff.

The many problems involved in providing a proper and sufficient diet for mankind are not only quantitative, but also relate to human habits and preferences. Many ill-nourished people refuse to eat strange foods even if doing so would correct their dietary deficiency. Some people accustomed to rice will not eat something so similar as wheat, even when at the brink of starvation. But progress is being made. Today soybeans can be processed so as to be indistinguishable from meat; this is perhaps a foretaste of the day when vegetable protein must substitute for less efficient animal protein (there is great energy sacrifice in turning plant protein into meat; beefsteak protein will always be more costly than the plant substance cattle consume). Let us hope that human procreation is brought under control before the world must resort to the even more efficient (but less palatable) production of protein by eating yeasts and algae rather than steaks and chops.

AFRICA. African agriculture is still on the whole primitive; in tropical areas much of it still by slash-and-burn technique. As populations increase, the fallow between cropping cycles must shorten; the soil then worsens and yields decline drastically. On the favorable side, much of Africa is still lightly populated, and tropical areas have ample rainfall. In southern Africa, on central African plantations, and to some extent along the Mediterranean coast, crop production is as advanced as anywhere in the world. Elsewhere subsistence farming prevails, on small native plots or tribal lands according to ancient custom. Ecologically this fits the generally poor soils well, but it is not sufficient to maintain an acceptable standard-of-living today. The greatest need is for improved education. Intensive research to determine what agricultural practices are of value locally would be a logical second step. Sponsored investigations are underway toward increasing the output and quality of food crops such as maize, soybeans, peanuts, cassava, yams, and grain sorghum. Rural development programs have been launched in the populated coastal belt of western Africa endeavoring to expand participation in the growing of oil palm, rubber, coffee, cocoa, citrus and other fruit trees, and mahogany. As yet little fertilization is practiced, nor, except on plantations, advantage taken of improved technology.

In arid North Africa nomadic animal tending is the rule. Cereals, particularly hard and soft wheats and barley, are the chief food crops planted along the Mediterranean coast and in the Nile valley. Subtropical trees and vines (olive, dates, citrus, and the like), and truck crops, receive some attention. Through much of central Africa native plantings are of a mixed nature—cereals with legumes, maize with sorghum, yams with okra, and so on. In tropical regions export crops dominate the agriculture—palm oil, cacao, bananas, peanuts, coffee, etc.—but nationalism in recent years has reduced the incidence of foreign-controlled plantations. Foods for local consumption are mainly minor cereals and starchy roots (maize and yams near the coast; sorghum, finger millet,

cassava, and sweetpotatoes inland; rice along rivers). Eggplants, tomatoes, beans, plantains, pineapples, pawpaws, peppers, and suchlike are frequent. Southern Africa is industrialized, the farming efficient. Because of the semiarid climate agriculture tends to be pastoral; the chief crops are cereals, legumes, and forages.

ASIA. It is impossible uniformly to characterize so vast a continent as Asia. In Indonesia and Malaysia dasheen, manioc, rice, and bananas reign as local foods. In China and Japan barley, rice, wheat, soybean, maize, and vegetables are common. India grows a diversity of domesticated plants, with rice, maize, cassava, wheat, millet, sugar cane, and many subtropical fruits important. Western and northern China is largely a land of nomadic herdsmen and, like Siberia, has little managed agriculture. The chernozem plains of southwestern Asia represent some of the most important cereal lands of the world, with wheat and other small grains the staples.

In most of Asia agriculture is intensive, but primitively carried on. Soil tillage is still largely by hand or with bullock. So concentrated are the populations of India, Indonesia, China, and Japan that the most effective possible use must be made of what arable land there is. For example, the Chinese are noted for trudging many miles to carry sewage to the small home plot—a source of fertilizer that in the United States is for the most part flushed into the rivers. In most regions agronomic learning, tools, and locally adapted high-yielding crop cultivars are just becoming available (Japan is the exception), so that the land yields at somewhat less than capacity.

Especially significant for southeastern Asia is establishment of the International Rice Research Institute in the Philippines, sponsored by the Ford and Rockefeller Foundations. In spite of the reliance of nearly 2 billion people upon rice as their basic food, rather little had been done to produce higher-yielding cultivars that would not lodge when heavily fertilized. Since 1962, when the Institute was opened, a better understanding of rice and a significant improvement in cultivars has been achieved; many of the tropical Asian countries are approaching rice yields heretofore occurring only in the advanced farming of Japan. Development of stiff-straw rices yielding nearly double what the local rices produce promises to forestall world famine for at least a few years!

With little new land to cultivate, India has been sorely pressed to provide even a substandard diet for its increasing population. There has been some progress in improving food production through the use of fertilizer, pest control, and improved technologies, but each gain is immediately offset by still greater population. Average per capita calorie intake per day is less than 2,000 in India, compared to an average of around 3,000 in North America and Europe. Under prevailing circumstances there seems little chance that this can be much improved, although newer cultivars and improved technologies should help reduce the severity of the periodical famines. High-yielding cultivars of wheat, rice, and maize, developed in other parts of the world, have already been brought into widespread usage. West Pakistan has been made nearly self-sufficient in wheat, its major crop, by using cultivars and methods developed at the International Center for Maize and Wheat Improvement in Mexico.

With the aid of outside financial support seldom available to undeveloped countries, Israel has demonstrated the outstanding achievements possible in scientific food production using modern methods and irrigation. An arid and abused land has been made to "flow with milk and honey." Unfortunately similar expansion in the nearby Arab countries is limited for lack of water as well as technical skill. The imbalance of food in the Middle East as a whole remains critical. Research directed at discovering suitable cultivars of sugar cane, maize, sorghum, sesame, peanuts, soybeans, eggplants, and tomatoes, as well as adapted strains of Temperate Zone food plants, is, however, being pursued. The main needs are land reclamation, and education for illiterates. Where irrigation is possible, reclaimed land can grow citrus fruits, dates, olives, papayas, field and vegetable crops, and cereals. Much of the population, however, leads a nomadic life, depending upon goats and other livestock to eke out existence on land badly overgrazed. For many the food supply is at starvation levels, little better than in India.

AUSTRALIA. Australia, 40 per cent desert and almost another 30 per cent suited only to pasturage, can support a large population only in certain coastal areas. Perhaps no more than 1 per cent of the total land area is cultivated, and this is planted mostly in wheat. Most agricultural income is derived from pastoral pursuits, particularly the production of wool, on natural range. Where agriculture is practiced, modern methods and mechanized means are used, as in North America.

EUROPE. Europe has exerted inordinate influence upon world agriculture. First, the great colonial powers of Europe—especially Spain, Portugal, Holland, France, and England—have dictated the export crops to be grown over much of Africa, the Far East, and South America. Second, as the world's greatest consuming center of raw materials, and one able to afford needed supplies, Europe largely created world demand. And third, Europe superimposed the diversification of podzol "felled forest" agriculture on adaptable lands elsewhere, her own population, which eventually colonized most of the New World and other continents, having been nurtured by podzol-type soils.

Within Europe itself the status of agriculture ranges from excellent to poor.

Peasant holdings of the back country are still operated in a rather unscientific fashion, but in northern Europe the most intelligent and effective agriculture in the world is the rule. By and large, agriculture in Europe is not nearly so mechanized as in the United States, but the necessary implements and marketing systems for efficient work are available, as they are not in much of Africa and Asia. Heavy populations over many centuries have also taught Europe the importance of conservation and efficient care of the land—procedures only beginning to be practiced in the New World.

As regards agricultural commodities Europe exists on a deficit basis, needing some imports to support her heavy population. Yet she does produce some four-fifths of her food, and nearly half of her land area is arable. Chief crops raised are potatoes (largely in northern Europe), wheat (largely in Spain, France, and the Danube basin), oats (especially in France, Germany, and Poland), rye (especially in Germany and Poland), barley (especially in central and southern Europe), and maize (especially in Rumania, Yugoslavia, and southern Russia). Fresh fruits and nuts, legumes, sugar beets, and forage for the dairy industry are also very important, especially in western and northern Europe. Olives are much grown in the Mediterranean area. The soils and climate of Europe encourage, on the whole, diversified agriculture. The diet of northern Europe is less than half cereals and potatoes; animal and dairy products, fats and oils, sugar, fruits, and vegetables make up almost as large a proportion. But in the Balkan area cereals and potatoes account for about three-fourths of the diet.

NORTH AMERICA. From Panama through Mexico, and in the West Indies, agriculture is mostly a matter of small home plots worked by hand in a semi-subsistence fashion. Exceptions are the larger fincas and plantations, sometimes operated by foreign interests. On these may be found modern mechanized equipment and adherence to conservation practices. The principle plantation crops are bananas and sugar, grown mostly for export. Crops of the small home plots include maize, beans, squash, peppers, tomatoes, bananas, and in suitable environments potatoes, rice, citrus and other fruits, and various vegetables.

Food products of the San Salvador market. Forty-six of the most popular locally grown items are shown here. (Courtesy USDA.)

Plowing land with a yoke of oxen and a wood plow in El Salvador. (Courtesy USDA.)

Mexico has been in the enviable position of raising its agricultural production by over 4 per cent each year in recent decades, becoming essentially self-sufficient in wheat and doubling yields of corn and beans. Still, most of Latin America is barely self-sufficient in food production, and countries with large exports such as Cuba (sugar) must compensate by importing basic foods.

The United States (see Table 15-3) and Canada offer a spectacular example in world agriculture of how ever-increasing mechanization has compensated for high manual-labor costs and permitted farmers of this continent to realize a high standard-of-living. Tractors, combines (see Fig. 15-4), mechanical pickers, and ready means of transportation make the efforts of one man more productive than those of multitudes in less mechanized countries. Several agricultural "belts" corresponding more or less to climatic zones exist, notably the wheat belt of the somewhat arid Great Plains, from central Canada to Texas; the corn belt of the moister and warmer Midwest, centering in Iowa and Illinois; the diversified crop lands of the East and Far West; and the citrus groves of Florida and California. Rice is grown in the lower middle Mississippi valley, peanuts throughout the South, soybeans in the Midwest, field beans in the North and West, peas in the North, and various truck crops over the entire area. Varying forage plants are also grown throughout the area, and many of the primary crops raised are utilized industrially rather than directly as human food. The United States and Canada are among the world's greatest food surplus countries.

SOUTH AMERICA. Latitudinally, South America extends from about 12° north of the equator almost to the Antarctic Circle;

Figure 15-4 How a combine is constructed. (Courtesy John Deere & Co.)

This cutaway view shows how grain and straw are handled from cutter bar on through separator.

The reel (1), divides grain and holds it to cutter bar (2), until cut. The auger (3) carries grain from both ends of platform to center of auger (4). Retracting fingers in auger beater (4), take material and feed it to feeder beater (5). Feeder beater (5), moves grain to feeder conveyor chain (6). The chain (6), delivers grain to rasp-bar or spike-tooth cylinder (7).

As grain travels between cylinder (7), and concave (8), over grate fingers (9), and back against separating beater (10), the greater part of separating takes place. Separating beater (10) strips straw from the cylinder (7), deflects grain through finger grates (9), and passes straw onto the straw walkers (11).

Most of the grain falls through concave grate (8) and fingers (9) onto grain conveyor (12). Straw and remaining loose grain are passed along to the straw walkers (11). Curtain (13) keeps grain from being thrown over. On its outward movement, straw is agitated by straw walkers (11). The remaining grain falls through openings in walkers and flows back through straw walker grain return pans (14) onto auxiliary chaffer (15). Straw is dropped off end of the straw walkers and out separator. The straw can be spread by straw spreader (special equipment) or broken up by straw chopper (special equipment).

After grain and chaff leave conveyor, (12) a blast of air from fan (16), through adjustable windboards (17), is directed against auxiliary chaffer (15), chaffer (18), chaffer extension (19), and sieve (20). The air blast, with aid of sieve agitation, blows chaff away and moves tailings to tailings auger (21). The tailings auger (21) carries tailings to tailings elevator (22), which conveys them through cross-auger (23), to center of cylinder (7), for rethreshing.

Clean grain after dropping through auxiliary chaffer (15), chaffer (18), chaffer extension (19), and sieve (20), is carried by clean grain auger (24), to elevator (25). Elevator (25) delivers clean grain to tank loading auger (26). The loading auger (26) distributes grain evenly to grain tank (27). Grain tank unloading auger is (28).

The combine is one of the revolutionary labor-saving inventions in agriculture. Until late in the 19th century, harvesting and threshing of grain were separate operations, involving considerable handling. A successful combined harvester and thresher, known today as the combine, became commercially available about 1880, but was widely used in the Midwestern grain belts only after World War I. Immense self-propelled combines now harvest almost all seed foods grown in the United States. Basically a combine consists of the following operational units:

Header. This cuts the grain and conveys it to the thresher. It usually consists of reel, sickle, and conveyor. The rotating reel pushes the grain over and into the cutter bar, from which it falls onto the conveyor belt leading to the thresher.

Thresher. In the thresher the grain heads are forced between a high-speed rotating cylinder (equipped with rasp-bards, spike teeth, or the like) and stationary "concaves" similarly equipped. The beating action and abrasion free the seed from hulls or pods.

Separator. The seeds, chaff, and small straw fragments fall through a grating in the separator as the straw is conducted off by beaters and conveyors. Vibration screenings and air blasts serve to remove chaff, unthreshed seed heads, and other trash from the grain.

Cleaners. Threshed grain, along with weed seed and other foreign material not removed by the chaffer, goes through other agitation screenings, sievings, and air blast action, adjusted to the particular grain and prevailing conditions. The clean grain is conducted to suitable storage or blown directly into trucks.

Table 15-3
Principal U.S. Crops

Crop	Acreage and Production in Thousands, 1968	Crop	Acreage and Production in Thousands, 1968
Corn for Grain		Sweet Corn for Processing	
Acreage	55,886	Acreage	514.7
Yield per acre, bu	83.0	Yield per acre, tons	4.87
Production, bu	4,636,456	Production, tons	2,508.5
Soybeans for Beans		Tomatoes for Processing	
Acreage	40,949	Acreage	373.0
Yield per acre, bu	26.4	Yield per acre, tons	17.3
Production, bu	1,079,627	Production, tons	6,448.6
Old soybeans on farms, bu	55,631	Summer Snapbeans*	
Winter Wheat		Acreage	23.2
Acreage	43,005	Yield per acre, cwt	37
Yield per acre, bu	29.1	Production, cwt	856.0
Production, bu	1,251,537	Mid-summer Cantaloupes*	
All Spring Wheat		Acreage	54.6
Production, bu	345,062	Yield per acre, cwt	138
All Wheat		Production, cwt	7,542.0
Production, bu	1,596,599	Late-summer Watermelon*	
Oats		Acreage	25.2
Acreage	17,765	Yield per acre, cwt	126
Yield per acre, bu	52.6	Production, cwt	3,169.0
Production, bu	934,424	Late-summer Cabbage†	
All Hay		Acreage	15.5
Acreage	63,567	Yield per acre, cwt	204
Yield per acre, tons	1.95	Production, cwt	3,160.0
Production, tons	123,827		

* Fresh market only.
† Includes fresh market and processing.

altitudinally, from the equatorial lowlands of Amazonian Brazil to the high Andes along South America's west coast. Obviously, habitats are diverse and extreme, and concomitantly the agriculture. In the tropical lowlands manioc is usually the staple food. It is accompanied by beans, pumpkins, tomatoes, bananas, pineapples, and other tropical fruits, and, where feasible, by maize and rice. Cacao is much grown on the east coast of Brazil, as is coffee in the south. Southern South America, including the plains area of southern Brazil, Uruguay, and Argentina, is largely pastoral, but produces export quantities of maize, wheat, and other cereals. Andean South America cultivates several food plants unknown in any other part of the world, such as quinoa, oca, anyu, and yautia, but also relies heavily on the well-known potatoes, beans, maize, and vegetables or fruits already listed for the lowlands and for Central America.

In southern South America some mechanical and scientific progress in agriculture has been made, but over most of the continent hand labor with crude tools is the sole means of food production. Much of the area is wilderness; and of the cultivated land, title is often in the hands of absentee landlords. Many agricultural workers are ill-educated and have little interest in conserving the land. Forest is cleared or burned where possible, and manioc or other foods planted and cared for more or less inefficiently. On the whole, tropical America imports more food than it produces, although temperate South America is an area of surplus.

The future of world agriculture. The old Malthusian hypothesis that population will eventually outstrip food supply is now a reality facing mankind. Must famine overtake the world? The future does look bleak unless man soon regulates his numbers, and avoids wasteful exhaustion of his resources. Improved agricultural technology can at best buy but a decade or so before food production reaches a ceiling, imposed by the maxi-

mum possible photosynthesis, on all of the land suitable for effective crop production. Before such a ceiling is reached, of course the law of diminishing returns will set in: food will become more expensive, producible in sufficient quantities only by "extravagant" measures requiring a high level of inputs (e.g., expensive soil improvement, irrigation, heavy fertilization, and expert technical skills). Of one thing we can be certain: there will be no dearth of demand for almost any efficiently produced food. The plants that must largely support the demand are the subjects of the next several chapters.

As Fig. 15-3 shows, the immediate problem of insufficient food is more critical in less developed countries than in developed ones, where output for the moment keeps

Agriculture adapted to the land. Hilly land is kept in grass to avoid erosion, and a balance struck for enjoyable living where population pressure is not yet too great. (Courtesy USDA.)

Mechanized agriculture. With the aid of elaborate equipment one man cultivates, fertilizes, seeds, and applies preventive pesticide over several acres per hour.

pace or exceeds the population growth. Even where sufficient food can be produced in developed lands to feed people of under-developed regions, many difficulties are experienced in achieving food distribution, satisfactory payment for production (if any), and a willingness by the recipients to work toward self-sufficiency. International programs for food distribution are not only costly, but can turn into a continuing drain on production that eventually even the surplus countries will themselves need. More-over, as humanitarian as international food aid is, in many cases it merely postpones correction of imbalances that eventually must be corrected.

In technologically advanced parts of the world such as North America and Europe the pace of agricultural change has been truly astounding. But this is "only the begin-ning." Large, corporate farming will in-creasingly provide most foodstuffs; con-currently, some small-scale, "inefficient" food production by part-time "farmers" who derive income from other sources can be anticipated. The latter fill a need for fresh and seasonal produce, of a quality not obtainable through mass marketing—some-thing like the family garden, and subsistence horticulture in less advanced agricultural countries. However, the basic business of feeding the nation falls the lot of the "big" farmer able to capitalize extensive lands and labor-saving equipment. Corporate rather than family farming may become the rule. This agribusiness will be professionally managed and financed as are other businesses. A systems approach will be used. Cultivars for the particular soil and end use, time of planting, quantity of fertilizer and pesticides, harvesting schedule, up-to-the-minute assess-ment of the market, etc. may be scheduled by computer. Specialists never setting foot in the field may be called upon to recom-mend programs (such as preventive weed and insect control based upon known cycles of occurrence) just as the services of doctors or lawyers are engaged.

In less advanced parts of the world the means for the preservation of food accu-mulated during times of surplus are badly needed. Without electricity, freezers and canning facilities are not possible, and food cannot be held for any length of time under tropical conditions. Perhaps the answer lies in government dehydration facilities, or in radiation preservation. Food quality is another problem: at least 20 per cent of the population in underdeveloped nations is undernourished, and 60 per cent malnourished (see Fig. 15-5). Attempts have been made to enrich foods for these peoples, by fortification with vitamins and amino acids (lysine), or the preparation of special high-protein supplements. There are many sociological barriers still to be overcome, however, such as traditions as to just which foods are considered comestible.

Choice of food may not be so wide in the future as it is now; trying to support 7 billion human beings by the year 2,000 on the same land that supported only half that number in 1968 (and this often inadequately)

is bound to call for some belt-tightening. Not only will the demand for efficiency rule out foods of low nutritive value for many, but decisions will have to be made about whether the total environment is to be turned essentially into a food factory. The loss in human and esthetic values can be great in a land devoted to sustenance without regard for wilderness, spaciousness, and other intangibles that many people cherish more than life itself. Whatever the outcome, those food and ornamental plants that are the subject of the next six chapters have carried mankind and his civilization this far: they bid fair to be of even greater importance in the years immediately ahead. If only world population can be contained, there is ample wherewithall here, and in selections still to be made from the wild, to support civilization with dignity if not opulence.

Figure 15-5 Population and food. (Courtesy Food and Agriculture Organization.)

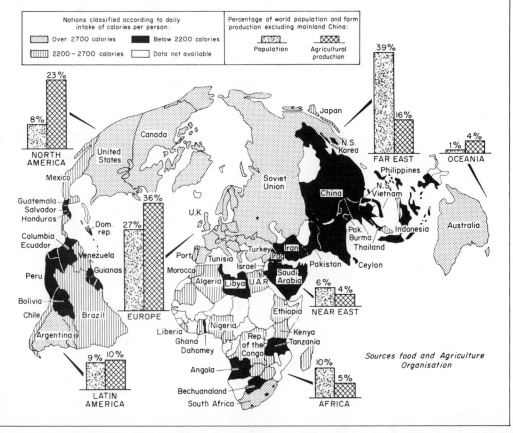

SUGGESTED SUPPLEMENTARY REFERENCES

Altschul, A. M. (ed.), *Processed Plant Protein Foodstuffs*, Academic Press, N. Y., 1958.

——————, *World Protein Resources, Advances in Chemistry No. 57*, American Chemical Society, Washington, D. C., 1965.

——————, *Proteins-Their Chemistry and Politics*, Basic Books, Inc., N. Y., 1965.

American Society of Agronomy, *World Population and Food Supplies 1980*, Madison, Wisc., 1965.

Brown, L. R., *Seeds of Change: The Green Revolution and Development in the 1970's*, Praeger, N. Y., 1970.

Delorit, R. J. and H. L. Ahlgren, *Crop Production*, Prentice-Hall Inc., Englewood Cliffs, N. J., 3rd ed., 1967.

Desrosier, N. W., *Attack on Starvation*, AVI Pub. Co., Westport, Conn., 1961.

Dijkman, M. J., *Tropical and Subtropical Agriculture*, Macmillan, N. Y., 1961.

Harrison, S. G., G. B. Masefield, and M. Wallis, *Oxford Book of Food Plants*, Oxford Univ. Press, N. Y., 1969.

Irvine, F. R., *West African Agriculture*, Oxford Univ. Press, London, 1970.

Leiner, I., *Toxic Constituents of Plant Foodstuffs*, Academic Press, N. Y., 1969.

Padfield, H., and W. E. Martin, *Farmers, Workers and Machines: Technological and Social Change in Farm Industries of Arizona*, Univ. of Arizona Press, Tuscon, 1965.

Peterson, M. S., and D. K. Tressler, *Food Technology the World Over*, AVI Pub. Co., Westport, Conn., Vol. I, 1963, Vol. II, 1965.

Stakman, E. C., R. Bradfield, and P. C. Mangelsdorf, *Campaigns Against Hunger*, Belknap Press of Harvard Univ. Press, Cambridge, Mass., 1967.

USDA Yearbook, *Food*, Washington, D. C., 1959.

Webster, C. C., and P. N. Wilson, *Agriculture in the Tropics*, Longmans, London, 1966.

Wharton, C. R., Jr. (ed.), *Subsistence Agriculture and Economic Development*, Aldine Pub. Co., Chicago, 1969.

Wrigley, G., *Tropical Agriculture*, Faber & Faber, London, 1969.

Yanofsky, E., *Food Plants of the North American Indians*, U.S.D.A. Misc. Publ. 237, Washington, D. C., 1936.

CHAPTER **16**

The Cereals

NO GROUP OF FOOD PLANTS HAS BEEN SO important to mankind as have the cereal grains. Cereals are the staff of life to most Temperate Zone and many tropical peoples. Without them civilization could not exist nor the modern economy function. Cereals were nearly indispensable to budding civilization, and in a modern world they provide the main sustenance for more than 2 billion people. Their production runs to more than 1 billion tons annually.

The true cereals come entirely from the

A young farmer poses proudly with his invaluable wheat in a field near the village of Varanasi, Uttar Pradesh, India. Wheat is one of the three major cereals supporting civilization. (Courtesy USDA.)

grass family, the Gramineae, although pseudo-cereals such as buckwheat (*Fagopyrum*, Polygonaceae) and amaranths (*Amaranthus*, Amaranthaceae) having a different fruit morphology are sometimes listed with them. The grain constitutes a fruit in which the dry ovary wall closely invests and fuses with a single seed and its coats. Grains usually occur in rather concentrated clusters, which in most parts of the world are today mechanically harvested, the individual seeds freed by a variety of ingenious machines such as the combine described in the previous chapter. Grains are easily handled and kept, a circumstance that makes them pre-eminent among foods that can be stockpiled. They possess a concentrated store of food material, largely carbohydrate, but containing some protein, oil, and vitamins. Foods with higher water content, such as potatoes, manioc,

or various vegetables, cannot be stored and handled like grain without special care and drying. All societies have had granaries, and primitive peoples propitiated the gods with gifts of grain. The word "cereal" itself stems from Ceres, the goddess considered the giver-of-grain by the ancient Greeks; to her were made offerings of wheat and barley.

The chief cereals are maize or corn, wheat, rice, oats, barley, rye, sorghum, and millets. The first three are by far the most important. Of these corn is the premier cereal of the New World, particularly in the warm temperate (summer) and subtropical climates; wheat is a contribution from western Asia and Europe to temperate climates where reduced rainfall and cold often limit the usefulness of maize; rice is the pre-eminent tropical cereal, the basic food source for the greatest portion of the world's peoples. The more than

Planting rice in the Philippines. (Courtesy Chicago Natural History Museum.)

Grinding maize for tortillas. (Courtesy Chicago Natural History Museum.)

1 billion tons of cereal produced annually throughout the world are grown mainly in the Orient (rice), North America (maize and wheat), and Europe (wheat and other small grains). Cereal production in Africa is low by comparison with other continents, but, interestingly, utilizes a much higher proportion of unusual and minor cereals such as various millets and sorghums.

Maize, corn, or Indian corn, Zea mays. A great deal of light has been shed upon the origin of maize from archeological findings during the 1960's. Especially have investigations in the Tehuacan Valley of Mexico supplemented previous surmises based only upon genetic research. A primitive maize grew and probably was cultivated in southern Mexico seven millennia B.C. By 1500 B.C. agriculture was well established there, and in the next thousand years reliance upon wild plants and game nearly ceased. Improved maize was by then a chief crop. Change from the "wild" form to a more modern type apparently involved hybridization and introgression with teosinte about 1500 B.C. Perhaps because of the competitive pressure of these newer forms and man-caused changes in the habitat, wild maize disappeared from the record before the birth of Christ and is presumed extinct. While it is uncertain whether similar domestications may have occurred elsewhere than in southern Mexico, archeological discoveries almost as ancient are found in South America, another postulated locus for domestication. It may be that early races of maize were introduced into South America from Mexico, and a secondary evolutionary center established in the northern Andean foothills (where "flint" genes were picked up by maize).

411

Modern maize growing in the United States. (Courtesy USDA.)

Maize pollen 80,000 years old has been recovered in soil cores near Mexico City, so certainly the wild ancestral plant was indigenous to this area. It has been suggested that cultivated maize originated through natural crossing with other genera, perhaps gamagrass (*Tripsacum dactyloides*), to form teosinte (*Euchlaena mexicana*), followed by back-crossing to the latter. *Tripsacum* itself may have come from the crossing of primitive *Zea* with *Manisurus*. If the center of dispersal of the modern races of maize is truly southern Mexico, it appears that distribution occurred into South America thousands of years ago, with the strains evolving there being spread northward through the Caribbean islands and eventually into eastern North

America. At the same time, Mexican races of maize were taken northward into the western United States, and then northeastward as far as New England where the early European colonists first became acquainted with the species—a species destined to yield more wealth than all the treasure envisioned by the early explorers.

North America remains the center for maize production. The crop is intensively managed there (see Table 16-1), and its growing mechanized, to make it the foremost seed and industrial grain. Only limited quantities of corn are consumed directly as human food in the United States—as fresh sweet corn ("roasting ears"), for canned corn, and ground into cornmeal for baking. Nearly 90 per cent

412

Maize taken from ancient Peruvian graves. (Courtesy Chicago Natural History Museum.)

is fed to livestock. North America accounts for about half of the world production of corn grain, which totals approximately a quarter-billion metric tons worldwide. Brazil, Mexico, Rumania, Yugoslavia, Argentina, India, and South Africa are other prominent corn growing areas, although not nearly matching the United States in quantity, nor usually yields.

At the time of the New World discovery, maize was the chief foodstuff with most of the Indians. By then many selections had been made of this variable and genetically plastic crop, equipping it for a variety of habitats, and developing such diverse types as popcorn, northern flint corn, and southeastern dent corn (ancestral to most corn belt cultivars of today). In forested areas there was limited planting by slash-and-burn techniques, in drier habitat seasonal cultivation or growing under irrigation. Early historians remarked that the Mohawk Iroquois women soaked corn seed for a few days before planting a few seeds to a hill, often accompanied by the burying of fish residues as a starter fertilizer. Various rituals were carried out to ward off evil spirits, and human sentries posted to discourage pilfering birds and animals. Often other crops, such as squash and beans, were interplanted with the corn. Of course weeding

Samples of modern maize. Shown from left to right, top and bottom, are ears of pop, flint, flint, dent, flour, and sweet corn. (Courtesy Chicago Natural History Museum.)

Table 16-1

Progressive Increase in Yields of Maize in the United States through Breeding and Improved Technology, Colonial Times to Present

Period	Average Yield, Bushels per Acre
Colonial	few
1930's	25
Early 1960's	64
Central Illinois, 1965	90
Good farmland, 1970's (projected)	125–180

and such cultivation as was practiced was carried out entirely by hand, using sticks as crude tools.

Maize was gradually distributed to other parts of the world, with varying degrees of acceptance. In Europe maize has come to be grown and utilized much as in North America. But the Maoris of New Zealand, who have accepted only three introduced food plants, one of which is maize (introduced in 1772), adopted the custom of soaking corn ears in water for 6 months or more before consuming the then thoroughly rotted grain. In parts of Peru the custom is similar, involving a soaking and fermentation before consumption. The slimy, strong-smelling kernels are crushed, boiled to make a gruel, then fried, or eaten warm with milk and sugar.

In most of the Orient maize is grown only where the land is not suitable for rice. The grain is somewhat used for feed, or ground to make puddings for human consumption, as well as being eaten fresh. In India corn flour may be used, or the grains parched on beds of hot sand to improve keeping quality. The American Indians frequently parched or popped corn before storing the grain in pits or small caves dug in canyon walls. Soaked in wood-ash lye, maize keeps well as hominy, a way of preserving the crop adopted by early colonists in North America from the Indians. In drier climates the corn grain preserves well without special treatment. The chief food for many American Indians, such grain was ground into meal on a simple mortar and pestle fashioned from stones. One reason the

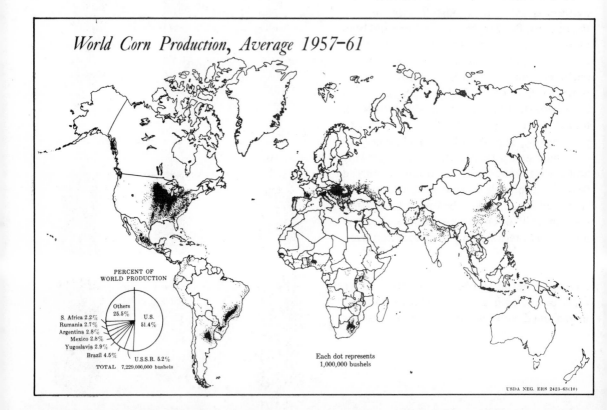

World Corn Production, Average 1957-61

PERCENT OF WORLD PRODUCTION

Others 25.5%
U.S. 51.4%
S. Africa 2.2%
Rumania 2.7%
Argentina 2.8%
Mexico 2.8%
Yugoslavia 2.9%
Brazil 4.5%
U.S.S.R. 5.2%
TOTAL 7,229,000,000 bushels

Each dot represents 1,000,000 bushels

USDA NEG. ERS 2423-63(10)

teeth of some Indians were worn down nearly to the gum line was the continual abrasion from sand in the cornmeal prepared by grinding between stones.

The modern cultivars of maize scarcely resemble the ancestral type. Mangelsdorf has been able to reconstruct maize closely resembling the wild type as found in archeological sites by crossing and back-crossing primitive podcorns and popcorns. Modern corn, with closely investing husks, and grain which strongly adheres to the cob, could hardly perpetuate itself in nature. It is entirely dependent upon man, perpetuated through planting into cared-for seedbeds. In technologically advanced parts of the world highly specialized tools—including cultivators, tailored fertilizers, precise planters, various pest controls, picker–shellers, combines, drying equipment, and so on—have been developed for the cropping of this now highly specialized species. As we shall see, there is also a highly specialized industry for the production of hybrid seed, now almost exclusively sown in the United States.

Among major crops maize is an extreme example showing interdependent evolution of technology and biological adaptation. Neither the crop nor the special technology could be effective one without the other; working together, however, yields have been increased above 200 bushels per acre, and the limit is not yet in sight. The breeder may be called upon to develop morphologically still different cultivars for mechanized growing, perhaps even harking back to ancient maize in having multiple seed-bearing inflorescences (the cultivar bred specifically for combining rather than picking of ears). Perhaps the remarkable genetic flexibility of maize will be utilized to produce low-growing types adapted to high field populations and short seasons, instead of traditional "tall corn" with but one or a few large ears per plant. Surely cultivars tailored for nutrient content can be anticipated. Already one rich in amylose has

become commercially important (see Chapter 14 on carbohydrate extractives). Other mutants show altered protein composition, and cultivars richer in lysine and tryptophan (in which maize is normally relatively low) can lend increased feeding value to corn.

Corn is noteworthy as the proving ground for heterosis, the basis of the hybrid corn industry. Hybrid vigor from crossing inbred lines has not only increased corn belt productivity greatly, but has provided a means of retrieving profit from the invention of a new hybrid; hybrid seed will not come true to type, so requires purchase each year from the originator who controls the inbred lines. Moreover, being monoecious, corn is simple to cross; all that is needed is to remove the tassel (bearing the male flowers) in order to assure that its seed is pollinated by another strain planted adjacent to it. Since the early days of the hybrid corn industry, when crossing was assured through de-tasseling, hybridizers have discovered male–sterility factors which cause a plant to be sterile to its own pollen. It will, however, produce seed with another line that carries the fertility-restoring factor (which in many crops is carried in the cytoplasm). Select, inbred lines, when crossed, of course provide first-generation progeny of great potentiality. Inbred suitability for such crossing is determined today by sophisticated computations involving crosses to a tester line, although originally it was a trial-and-error procedure. The crossing of two suitable inbred lines is known as a single cross. Although the seed thus produced carries all the potentiality that heterosis confers, it is not abundant, being borne on a relatively weak inbred plant. Thus a double-cross method of seed production is normally used. Progeny from the crossing of lines A and B are in turn crossed with progeny from lines C and D, to provide both heterotic vigor and high yields. Sometimes only three lines are used, the so-called triple cross, with the progeny of lines A × B pollinated

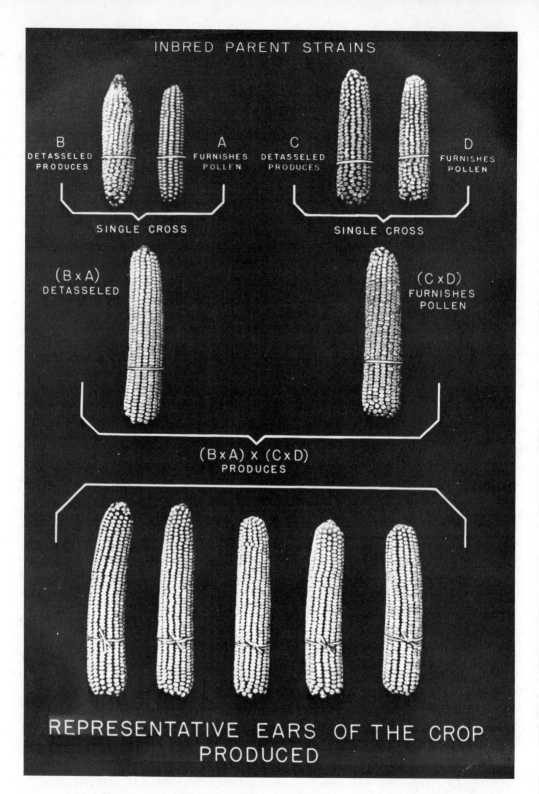

Method of producing double-cross hybrid seed corn and representative ears of the crop produced from hybrid seed. (Courtesy USDA.)

by a single third inbred. In any of these cases the essentially homozygous condition of the inbred becomes heterozygous in the hybrid seed offered for sale, a requisite for heterosis. Of course seed collected from a crop planted from hybrid seed undergoes genetic reassortment, and will yield a variable population if in turn sown. Many breeders feel that chances of significant gain through isolation and crossing of more new inbred lines is now limited, and that mass selection of mixed populations may show greater promise for yield enhancement in the future.

There are several categories of corn in corn belt farming. The *flint corns* are much like those found being used by the Indians of the northeastern United States, the kernels consisting of a hard starch difficult to grind. The *dent corns*, apparently descended from types originally found in the southeastern United States, have an indentation (the "dent") on the kernel and soft starch internally with a hard capping; dent corns are especially high-yielding and much used for modern cultivars. *Flour corns* have kernels entirely of soft starch, and were popular with the American Indians of the Southwest, since they grind more easily by hand. For the home garden and fresh food there are many varieties of *sweet corn*, the kernels rich in sugar; they are consumed boiled or roasted in the green stage. Kernels of *popcorn* burst upon heating, a delicacy in the modern day, but an especially useful type with primitive peoples since the grains became comestible upon heating without the need for grinding. These are the main commercial groupings but, of course, scattered throughout the world are many local cultivars, including primitive podcorns in which each grain bears a husk. Some will no doubt contribute germplasm to maize-of-the-future.

In North America, and to a lesser extent in other parts of the world, most of the corn crop is fed to cattle and hogs, only 10 per cent or so being used for industrial purposes

and for direct human consumption. Chapter 14 reviewed cornstarch, the chief industrial derivative, which in turn may be hydrolized or fermented into syrups, adhesive polymers, and so on. The approximately 10 per cent of protein in the grain is valuable for animal feeds, and that removed in the initial steeping (corn steep liquor) is useful for the production of antibiotics. The oily embryo contains up to 13 per cent oil, which upon crushing or solvent extraction yields about 50 per cent of its weight in corn oil. Residues and hulls removed in processing are usually reincorporated into cattle feeds.

Wheat. Wheat, mainly *Triticum aestivum* (= *T. sativum* = *T. vulgare*), the second of the three great cereals and one of the two bread cereals (the other being rye), feeds Temperate Zone peoples of five continents. It was introduced into Mexico about 1520 by the Spaniards, and grown on islands off the Massachusetts coast by 1602. The American wheat belt is today nearly as important as is the corn belt, production in the United States totaling more than 30 million tons annually, principally from the Ohio valley, the Prairie States and eastern Oregon and Washington. World production is over a quarter-billion tons, mostly from the U.S.S.R., the midwestern United States, Canada, central Europe, Turkey, Argentina, northeastern China, southeastern Australia, and northwestern India.

Wheat is typically a crop of cool climates with moderate rainfall. A combination of high temperature and high humidity is almost invariably fatal to the crop, and limits its usefulness in the tropics. The world's greatest harvests come from grassland soils and climates. Wheat planted in autumn is remarkably winter-hardy, except under the most severe conditions when a snow blanket is lacking. Plants tiller well, giving many stalks and hence many seed heads, in cool spring weather, early in the plant's growth. Wheat

Mechanized harvesting of wheat with combine.

This Punjab farm laborer harvests wheat with small hand sickle. Practically all Indian wheat is harvested by hand. (Courtesy USDA.)

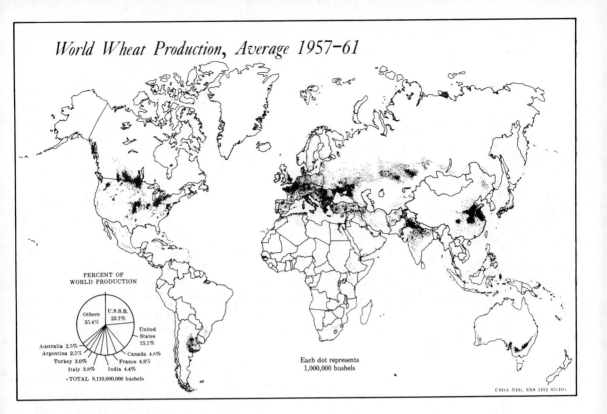

World Wheat Production, Average 1957–61

PERCENT OF
WORLD PRODUCTION

Others 35.4%
U.S.S.R. 23.7%
United States 15.1%
Australia 2.5%
Argentina 2.5%
Turkey 3.0%
Italy 3.8%
India 4.4%
Canada 4.8%
France 4.8%
·TOTAL 8,110,000,000 bushels

Each dot represents
1,000,000 bushels

USDA NEG. ERS 2412 63(10)

grown in humid regions usually produces soft and starchy grain, giving a "weak" flour.

When man first began to cultivate wheat many thousands of years ago, presumably in the Mediterranean area,* other considerations than that of mere yield were to guide his selections. Wheats were destined to become the staple food in many diverse climates, where they were subject to attack by many different diseases and were harvested by various means. Varieties to meet these varying conditions were inevitably selected. The modern plant breeder has had his hands full maintaining suitable strains in the continuing fight for increased production. Particularly is this true with regard to resistance to wheat rust, *Puccinia graminis,* a Basidiomycete. Many are the wheat strains once resistant to this rust, abandoned

as mutating rust strains again attacked the plant. Of the hundreds of wheat cultivars used today, only a few will likely be important in the agricultural economy of tomorrow.

The plant called wheat is a complex combination of the several uncertain species comprising the genus *Triticum.* Two of these (*T. monococcum,* einkorn; *T. aegilopoides,* wild wheat) are diploid ($2n = 14$), about nine are tetraploid (*T. dicoccoides,* wild emmer; *T. dicoccum,* emmer; *T. durum,* durum or macaroni wheat; *T. pyramidale,* Egyptian cone wheat; *T. polonicum,* Polish wheat; *T. turgidum,* poulard or rivet wheat; *T. orientale,* Khorasan wheat; *T. antiquorum;* and *T. timopheevi*), and four are hexaploid (*T. aestivum = T. vulgare = T. sativum,* bread wheat; *T. compactum,* club wheat; *T. spelta,* spelt; *T.*

---

* Bread wheat is known to have been grown in the Nile valley by 5000 B.C.; in the Euphrates and Indus valleys by 4000 B.C.; in China by 2500 B.C.; in England by 2000 B.C. Presumably this evidence indicates diffusion from Mediterranean centers of domestication, and corresponds well with the spread of civilization as evidenced, for example, by the use of wheel or cart. Vavilov believed soft wheats to have originated near the mountains of Afghanistan, durum wheats from North Africa, and einkorn from Asia Minor.

Ancient wheat gathered in an archeological site in Iraq by the Field Museum–Oxford University expedition. (Courtesy Chicago Natural History Museum.)

sphaerococcum, Indian dwarf wheat). Apparently there has been natural crossing in the wild between the Hordeaceous *Agropyron, Triticum,* and *Aegilops;* and the modern wheats are believed to contain genes from species of these genera. The tetraploid wild wheat, *T. dicoccoides,* may have resulted from a cross of *T. aegilops* with an unknown species of *Agropyron,* and in turn may have been an ancestor of spelt wheat (by an *Aegilops* cross?) and *T. orientale. T. aegilops* may likewise have given rise to *T. monococcum,* which in crossing with *Agropyron triticeum* yielded the tetraploid *T. antiquorum,* perhaps an ancestor of modern tetraploid wheats and part-ancestor (by another *Aegilops* cross?) of hexaploid wheats. The hypothetical lineage of several wheat species is given in Fig. 16-1.

Einkorn wheat, *T. monococcum,* is unusual in that it produces a single seed in

Figure 16-1   Tentative derivation and chronology of wheat.

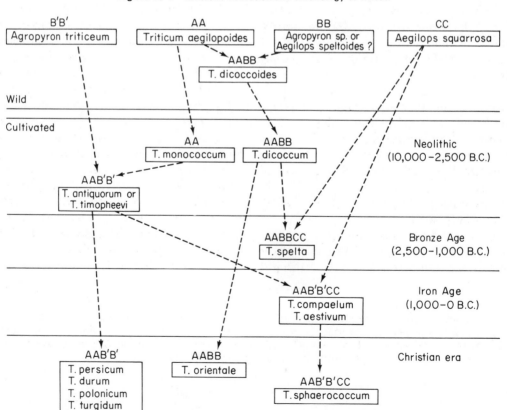

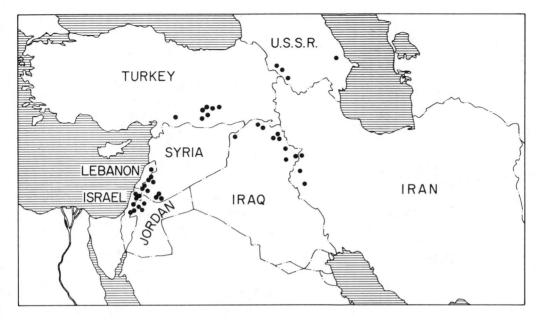

Figure 16-2 Distribution of known and reasonably certain sites of wild emmer wheat. (After Harlan and Zohary, *Science* **153**, 1078 (1966).)

each spikelet. It yields a scanty harvest, but will grow in poor, rocky ground and is still cultivated to a minor extent in mountainous areas of southern Europe. It has been found in Swiss lake dwellings of the Stone Age and in the Jarmo excavations of Iraq dating back nine millenia. Emmer, *T. dicoccum* (Fig. 16-2) was cultivated by the ancient Mediterranean peoples, and has also been identified in ancient archeological diggings. It is still cultivated to some extent in mountainous Europe and to a very limited extent elsewhere (e.g., the Dakotas, as a livestock feed). Spelt, *T. spelta*, like emmer and einkorn, retains its glumes when threshed, and finds minor use only as stock feed. It is said to be cultivated still in northern Spain. Polish wheat, *T. polonicum*, curiously not indigenous to Poland, is cultivated in southern Europe and North Africa. The grain, of no value for milling, falls readily from the head. Poulard (rivet or English) wheat, *T. turgidum*, does not yield heavily and is of only slight economic importance. Club wheat, *T. compactum*, has been most grown in Chile, the Far West in the United States, and India.

The spring-sown durum, *T. durum*, finds considerable use as a semolina or macaroni wheat, because of its high gluten content. It is grown principally in Spain, Algeria, the Ukraine, India, and in the northern wheat belt of the United States. Bread wheat (*T. aestivum* = *sativum* = *vulgare*) occurs in innumerable varieties, among which are many famous names (Marquis, Red Fife, and so on). This is the chief commercial species today, but it is one not known in the wild.

Bread wheats are separated into two primary groups on the basis of their mode of cultivation. These groups are termed spring wheats and winter wheats. Spring wheats require at least a 90-day growing season, and are usually sown in March for harvest in autumn. They are grown in colder regions, such as the northern wheat belt of the United States and Canada to the Arctic Circle. Only barley, among cereals, can be grown in colder climates than can spring wheat. Winter wheats may be cultivated where autumnal rains are favorable and winter climates less severe. In North America they are planted in the southern wheat belt in

Seventeen different grasses and wheats. 1, 2, 3, wild grasses related to wheat; 4, Einkorn; 5, Emmer; 6, Spelt; 7, Polish wheat; 8, Poulard; 9, Club wheat; 10, Durum; 11, Turkey wheat, a hard, winter, bearded type; 12, Wilhelmina, a soft, winter wheat; 13, Pacific bluestem, a soft, spring wheat; 14, Dicklow, a soft, spring wheat; 15, Marquis, a hard, red spring wheat; 16, Red Fife, a northern, hard, spring wheat; 17, Kitchener, a hard, spring wheat. (Courtesy Chicago Natural History Museum.)

September for harvest the following July. The hard red and durum wheats are mostly grown in the spring wheat belt, and soft or hard red winter wheats in the winter wheat belt. Yield of wheat in Kansas is about 25 bushels per acre on the average, but on intensively managed lands and certainly with hybrid cultivars can exceed 50 bushels.

Spring wheat must be planted as early as the soil can be worked, to obtain the necessary length of growing season. In the earliest recorded cultivation of wheat, the grain was broadcast by hand, thickly enough to discourage weed growth, as it still is in a few parts of the modern world. Perhaps cattle were used to trample the seed into the roughly cleared or burned-over soil. Harvest was accomplished with bone sickles wielded by man or woman, and the cut stalks were bound into bundles. Threshing consisted of trampling, beating, or flailing the grain free from the straw and hulls. The chaff was removed by winnowing—tossing the grain into the air to let the lighter hulls blow away. These very principles have been incorporated into the modern combine. In the United States wheat is almost entirely combine-harvested. In early summer many combines and combine crews start in Texas and Oklahoma, working on the winter wheat crop, and gradually move northward as the wheat matures, eventually in autumn harvesting the spring wheat of the northern United States and Canada.

The wheat plant is an annual, probably descended from a perennial. From the base of the plant several tillers (lateral branches) grow, each characteristically bearing a seed head. Varieties with strong stems and abundant adventitious roots are preferred, since they will not lodge (fall over) as readily when heavily fertilized. Development of the Gaines variety for the Pacific Northwest, from the dwarf Japanese "Norin-10" germplasm, was a significant breakthrough for increasing wheat yields during the 1960's: 100 or more bushels per acre are obtained from this non-lodging cultivar when moisture is not limiting. Norin-10 heredity also sparked the "Green Revolution" resulting from wheats bred in Mexico for which N. E. Borlaug received Nobel Peace Prize recognition in 1971. The inflorescence in wheat is a terminal, compound, distichous spike. In most cultivars the axis is not brittle, a condition that in spelt, emmer, and other wheats causes the spikelets to disarticulate. The flowers are usually perfect and self-pollinated, one to five to each spikelet.

The wheat grain, like the corn grain, consists of a small, oily embryo or "germ" at its lower end, with the single cotyledon applied against the starchy endosperm, and this in turn surrounded by the proteinaceous aleurone cells and finally the husks (remains of nucellus and integuments and pericarp). The pericarp and aleurone become bran upon milling. The pericarp amounts to about 5 per cent of the grain, the aleurone layer 3 to 4 per cent, the endosperm 82 to 86 per cent, and the germ about 6 per cent. One analysis of wheat gives 14.2 per cent protein, 2.4 per cent fat, 78.7 per cent carbohydrate, and the remainder mostly mineral matter and moisture. Hard wheats generally average 13 to 16 per cent protein, soft wheats 8 to 11 per cent. Wheat is higher in protein by a slight percentage (and a little lower in fat and carbohydrate) than is maize. Because of the two binding proteins, glutenin and gliadin, wheat flour is suitable for making bread, whereas maize flour is not. "Stronger" wheat flours are those that contain the higher percentage of protein resulting in a stickier or more tenacious dough for baking.

Market classification of wheat does not necessarily bear any relationship to botanical groupings. **Hard Red Spring** is essentially a bread wheat, used for quality flour. The kernels are short, thick, asymmetrical, hard,

A farmer of Uttar Pradesh, India, winnows wheat by letting the wind blow the chaff away from the pile of grain, after the grain is trampled by bullocks. (Courtesy USDA.)

flinty, rich in protein, and with an "elastic" gluten permitting swelling in baking without breakage of crust. In the United States such wheat is grown chiefly in the northern wheat belt. **Hard Red Winter** is similar to the preceding. The kernels are long and narrow, symmetrical. In the United States it is grown across the continent south of the colder regions, Kansas being the leading state. **Soft Red Winter** is a comparatively high-starch–low-protein grain, giving an inelastic dough used in pastry products rather than in bread. The kernel is long and wide. This type of wheat is cheaper than the preceding and may be used as livestock feed. In the United States it is grown chiefly in Ohio, Indiana, and Illinois. **Durum** is a heavy-bearded, pointed-kernel, amber-colored wheat, grown like hard red spring wheat, moderately starchy (depending upon weather conditions), and much used for macaroni products. In the United States it is grown only in the Dakotas and Minnesota. **Red Durum** is grown in the same area as, and is similar to, durum, but used primarily for livestock feed. **White** are unpigmented spring and winter varieties, "soft" or starchy and poor in protein and gluten, used for pastries and breakfast foods. In the United States such varieties are grown principally in the Far West and Northeast.

The milling of wheat for flour, like the milling of any product, involves a preliminary cleaning during which weed seed or foreign matter is removed. The grains are moistened or "tempered," a process that "toughens" the bran and prevents its fragmentation. The grains are then run between a series of corrugated rollers to break or crush the seed. The bran and flattened embryos* are screened off from the endosperm in shakers, for use in cattle feed, and the endosperm fragments or middlings are further purified in blowers. The middlings, nearly bran-free, are then pulverized between smooth rollers to become flour, which may be further treated by bleaching. Highest-quality flour is known as patent flour. The flour yield in milling is about 70 per cent: a 60 pound bushel of wheat should give about 42 pounds of flour. Modern roller milling is in striking contrast to the cumbersome methods of earlier times, when the pestle and mortar were used to beat or bruise the grain into a coarse flour. Finer flours were prepared by repeated sievings or grindings in a hand mill. The old flour mill consisted of a grooved upper rotating stone which turned against another fixed stone, the grain being fed into a central cavity. As the grain worked out the radiating grooves between the millstones, it was pulverized to flour.

In the making of bread, one of the staple items of the diet of the world, flour from wheat is supplemented by another ancient domesticated plant, yeast. Bread has been used for so long that it has come to be more or less synonymous with food. In Biblical times loaves from wheat and barley were the principal item of diet. St. Jerome wrote in A.D. 388, "He is rich who doesn't want for bread." Some bread 4,000 years old has even been recovered from an Egyptian tomb. Bread is made by adding yeast to glutenous wheat dough consisting of moistened flour in combination with small quantities of fat, sugar, milk, and salt. The yeast ferments carbohydrate materials, freeing carbon dioxide. As carbon dioxide bubbles form, they become trapped in the sticky dough, causing it to rise and become as it sets in baking (in which the gluten is coagulated) the familiar light, porous loaf. Various baking powders can be substituted for yeast to release the carbon dioxide essential in baking.

The commonest uses of wheat are for the familiar foods made from flour—bread, rolls, biscuits, muffins, waffles, pies, cookies, macaroni, spaghetti, noodles, crackers, dumplings, and the like. Breakfast foods and beer or other alcoholic fermentations also take large quantities of wheat. The bran left over from milling amounts to about a quarter of United States wheat production, and finds important use as a high-grade animal feed. Wheat straw finds use as a stuffing or weaving material, and as a mulch. Wheat harvested in an immature state, the "milk stage," can be used as a nutritious and palatable forage. Usually, however, whole wheat is too valuable to be used as feed. Before maize was known to the Old World, wheat starch sized the paper and starched the ruffles of the Renaissance cavaliers. In recent decades maize has monopolized 98 per cent of the starch business, but the

---

* The oily embryo is not pulverized in roller milling, but flattens into a relatively large lump that can be readily screened off. To produce the white flour demanded by the more "progressive" people involves the same irrational procedures required for the production of pure sugar, white rice, and the like—the most nutritious portions of the grain are discarded, to be fed to cattle. Thus the embryo, high in fats and vitamins and natural balance for the starchy endosperm, and the mineral- and protein-rich outer grain coatings are rejected. Of course there is a practical reason for eliminating the germ from the flour—else the keeping quality of the flour would not be as great. But the stone-ground wheat flour of the colonists retained 60 per cent of the original thiamine; modern roller-milled white flour retains only some 6 to 16 per cent. For this reason some breads are "enriched," by the addition of minerals, vitamins, and soybean flour.

wet milling of wheat, which washes the starch from gluten dough, yields special-purpose starches. The gluten is used in adhesives, emulsifiers, polishes, meat flavors, synthetic vitamins, and the like.

By the early 1970's plant breeders had devised means for producing hybrid wheat, attempting to take advantage of heterosis the same as with maize; seed, however, was not yet commercially abundant nor proved economically worthwhile under all circumstances. Because of extra costs in production of hybrid wheat seed, it must sell for several times the price per pound of conventional cultivars. In most wheat-producing areas it would have to increase productivity 25 per cent or more in order to justify its cost, not always a certainty where lack of rainfall or other hazards can limit production.

Potentiality for hybrid wheat began with a discovery by Japanese scientists in 1951 of male sterility in wheat developed from a cross with wild *Aegilops*. However, no fertility restorer could be found suited to American cultivars, so similar crosses were investigated in the United States looking toward development of a male–sterile line. The most promising line was developed by crossing *T. timopheevi* with Bison bread wheat, in Kansas. Shortly thereafter fertility-restoration factors were also transferred from *T. timopheevi* to various bread wheat lines. The stage was then set for producing a hybrid wheat by mixed plantings of several rows of a male–sterile line, with occasional rows of a pollinating line carrying the fertility restorer. Tests of hybrid cultivars planted under favorable growing conditions have generally shown yield increases of 20–40 per cent as compared to the parent varieties. A Bison × Scout cross in Kansas has yielded 57 bushels per acre in an experimental planting. There may be a beginning in this of a great new hybrid-seed industry analogous to that for hybrid corn, along the lines already proved commercially feasible for hybrid sorghum, hybrid onions,

and other crops which normally have perfect flowers. But obviously many problems and expensive procedures are involved, such as developing and maintaining the male–sterile and restorer lines, then finding suitable "inbred" lines for crossing such that heterosis is material and flowering time (and pollen transfer) feasible in field plantings.

TRITICALE. Of considerable interest are the hybrids of *Triticum aestivum* and *T. turgidum* with rye, *Secale cereale*. The sterile hybrids from such crosses can be made fertile by doubling the chromosomes of the hybrid by various techniques. The hybrid has been named *Triticale*. The early crosses were with hexaploid wheat, the hybrid containing 56 somatic chromosomes, the earliest of such crosses dating back to 1875. More recently there has been interest in crosses with tetraploid wheat, giving amphiploids with 42 somatic chromosomes. By the beginning of the 1970's none of the *Triticale* crosses had yet proven competitive with wheat, but there was promise that further breeding may increase yields and quality sufficiently to establish *Triticale* as a commercial grain. Reports from the Soviet Union indicate that *Triticale* has a higher protein content and is more winter-hardy than wheat, but most tests have shown yields to be less than with wheat and lodging a fairly serious fault. *Triticale* grain is larger than wheat, but less abundant on the spike.

Rice. Rice, *Oryza sativa*, the third of the three great cereal plants, is presumably of Far Eastern origin, although there have been advocates of the view for an African center of domestication. Over half the world's population subsists wholly or partially upon rice. Thousands of strains are known, cultivated or escaped, in India and China. Domestication must have first taken place very long ago, for rice is known to have been grown in China nearly 3,000 years before Christ. Rice has always been primarily a crop

Rice fields in Thailand as they appear immediately after planting. (Courtesy USDA.)

of the Orient, where even today over 90 per cent of the world's crop is produced and consumed. Rice was introduced into the American colonies in 1647, and became an important crop in the Carolinas before the end of that century. Rice production in the United States has remained modest in comparison with that of wheat and maize, and constitutes only about 1 per cent of the world crop. The South, chiefly Arkansas, Louisiana, Texas, and California, produces more rice than the United States can consume. Efficiency resulting from mechanization has enabled domestic growers to compete with oriental rice raised by cheap labor.

In the Far East rice is usually the mainstay of the diet. Annual consumption of 400 pounds per person is not uncommon. In contrast North Americans consume a mere 8 pounds per person annually. Rice is the only one of the great cereals that is largely consumed by man directly as harvested (after dehulling and often polishing, of course). Maize, it will be recalled, is to a great extent fed to animals, to be ultimately consumed as meat or dairy produce; wheat is mostly milled to flour, then baked; but rice is usually simply boiled for direct consumption.

The rice plant is a variable annual of from thirteen to twenty-three species and many strains, with a chromosome complement of $2n = 24$ (several American cultivars are tetraploid, and their triploid hybrids are fertile). The cultivated rices are grouped as

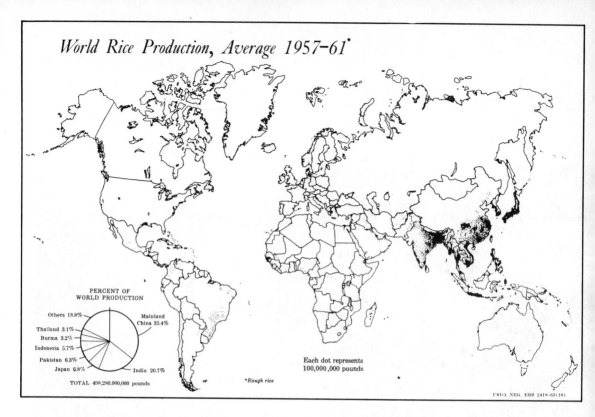

World Rice Production, Average 1957–61*

PERCENT OF
WORLD PRODUCTION

Others 18.8%
Thailand 3.1%
Burma 3.2%
Indonesia 5.7%
Pakistan 6.3%
Japan 6.8%
Mainland China 35.4%
India 20.7%

TOTAL 498,280,900,000 pounds

Each dot represents
100,000,000 pounds

*Rough rice

USDA NEG. ERS 2418–63(10)

O. sativa, a catchall complex conveniently divided into *indica* cultivars of the tropics (long-grained, weak-strawed, relatively light-yielding) and *japonica* cultivars of the sub-tropics (shorter-grained, stiff-strawed, heavy-yielding). Asiatic forms have been considered *fatua* or *rufipogon* subspecies, American forms as *perennis* (a term of doubtful validity), and African forms as *barthii*. In some classifications a *fatua* or *rufipogon* series and a *perennis* complex have been proposed. Within the perennis complex the American forms would be designated *cubensis,* the Asiatic form *balunga,* and the African ones as *barthii.* The *fatua* types seem to occur in the wild as annuals, in shallow swamps, and the *perennis* types (at least the Asiatic *balunga*) in deep swamps. African rice is reported to spread by rhizomes, although it is not clear whether this is an inherent characteristic or a partial response to the environment. Under tropical conditions plants may be perennial to a degree, perpetuated perhaps by rhizomes, and by new tillers from the ground. In common with other cereals rice has a root system of mostly secondary origin. Stems are typically about 10 nodes tall, but may have as many as 20 nodes, with the internodes fully or partly covered by the leaf sheath. Tillering usually occurs from the lower nodes, the tillers being either few or many. The apical leaf is broader than the others and is known as the "flag," becoming almost vertical as the terminal panicle emerges. In most cultivars the spikelets are solitary, and each ordinarily bears a single, self-pollinated flower.

Rice demands relatively warm temperatures and abundant moisture. Level, fairly heavy soils with a more-or-less impervious subsoil, and tropical or subtropical climates, commonly meet these demands. The usual lowland rice is grown on land submerged for 60 to 90 days during the growing season. Thereafter the water is drained away, if possible, to facilitate harvest. In the United States the rice crop is fertilized and drill-planted on drained land in spring just as any other cereal, and then, when the seedlings are a few inches tall, the fields are irrigated. Especially in California seeding of irrigated

428

land may be achieved by airplane. About two weeks before harvest the fields are allowed to dry, and combines used to gather the crop. Harvest is best when the moisture content of the rice grains is from 23 to 28 per cent. Upland rice can be grown without irrigation, which costs several dollars per acre for the 30 inches of water needed each growing season, but its yields are less than those given by lowland rice. Less than 10 per cent of the world's crop is upland rice.

Oriental methods of growing rice (called *paddy* there) are in sharp contrast to the mechanized production in the United States. In parts of the East, where there are perhaps 2,000 inhabitants per square mile, all the original forest has been cleared for rice growing. Rainfall and river water, the latter sometimes hand dipped or lifted on a human-powered water-lift, are used to flood the laboriously prepared fields. Before flooding, the land is scratched with a simple pointed plow, typically drawn by water buffalo, and dung or sewage, if available, is applied. The ground may be smoothed by dragging a log over it. Dikes are repaired; and at the appropriate season rains begin to fill the "paddies." Rice seed may then be sown by broadcasting, and perhaps be trampled into the mud. In most localities, however, the rice is sprouted in flooded seedbeds, and transplanted by hand to the soggy fields at the appropriate stage. During the growing season men and women wade into the fields to pull up the weeds by hand. As the grain matures, the fields are drained and the seed heads cut

Planting rice in Thailand with water buffalo. (Courtesy USDA.)

with hand tools. Japan practices the most progressive rice growing in the Orient. Yields average nearly 2 tons per acre, and range as high as $4\frac{1}{2}$ tons. Fertilization is as intensive as on North American corn land. A complete fertilizer is generally incorporated prior to flooding and hand planting of the seedling rice plants. Another fertilization is undertaken just before the panicles form, to increase yields. The fields are drained for harvesting and threshing, typically accomplished with small portable machines brought to the site. A green manure such as vetch may then be planted and later plowed down. Or the paddy fields may be used for a winter crop such as barley, wheat, or rape.

Rice grains may be classified as glutenous or nonglutenous. The latter type possesses an endosperm starch that is hard and "flinty" (transports and mills well), and is not sticky when cooked (since it lacks the dextrins present in glutenous types). These nongluten-ous types are the most important. They re-

An Indonesian farmer fertilizing his rice crop. (Courtesy USDA.)

Fertilization of rice by plane in the United States. (Courtesy USDA.)

430

quire a relatively long growing season and so are mainly produced in tropical climates. Rices are also classified according to kernel shape, into short (about 5.5-millimeter), medium (about 6.6-mm), and long (7- to 8-mm) types, with the long-grain forms being most esteemed. The rice grain is similar to that of wheat, having a small embryo and a single cotyledon applied to the starchy endosperm, surrounded by aleurone cells and the fused hulls.

Processing the crop involves threshing, in which the grain is separated from the seed head and chaff, and the nutritious seed is freed from surrounding hulls or bran. In the United States this is done by combines, and is followed by polishing in rice mills. Rice hulls find use for furfural and other chemurgic products. In the Orient threshing is usually accomplished by gently beating, abrading, or flailing the collected rice heads, or by treading barefoot upon them. Grain so treated is winnowed free of chaff by tossing it into the air above a sheet or mat. In Afghanistan green rice is mixed with heated sand to harden and crisp the hulls, which are then separated by crude milling on water-powered devices that pound the grain against the soil. Screening and winnowing follow. In Ecuador much of the rice is fermented by massing it on the floor while damp, and covering it with a tarpaulin for a few days. When dried it cooks more quickly than unfermented rice in the high altitides of the Andes. The brown rice obtained by "primitive" processing still possesses the nutritious outer proteinaceous and mineral layers. It is fortunate that most Eastern peoples have not adopted the Western custom of polishing rice to a creamy whiteness, and in doing so eliminating much of the nutritional value. In the United States most rice undergoes a polishing process, during which the outer, colored, protein-rich parts (12 per cent protein) are abraded away, to become cattle food, and the starchy white endosperm alone is left for human consump-

A woman threshes rice with a finger stripper in Japan. (Courtesy USDA.)

A wooden rice mill for preparing grain in the eastern Tropics. (Courtesy USDA.)

431

tion. Polished rice is approximately 92 per cent carbohydrate and has only 2 per cent additional material of nutritional value.

Annual yields of rice per acre greatly exceed those of wheat; yet because of specialized growing conditions and consequent demand for labor, rice is an "expensive" crop, at least in terms of human effort, compared to "cheaper" wheat. Rice is today grown chiefly in the moist tropical areas of India and southeast Asia. There is notable production in equatorial Africa and the American tropics. World production of rice exceeds a quarter-billion tons annually, surpassing wheat slightly. Three-fifths of the crop is grown in India–Pakistan, China, and Japan, and most of it is consumed domestically. The International Rice Research Institute has been instrumental in developing improved rice cultivars and

Table 16-2

Yield of Rice in Asia, 1961/62–1963/64 Average Compared with That Ten Years Earlier

Regions and Countries	Yield (tons per hectare)		Per- centage Change
	1951/52 to 1953/54	1961/62 to 1963/64	
Northeast Asia			
Japan	4.00	5.14	28.5
Korea (Rep. of)	2.45	3.08	25.7
Taiwan	2.48	3.34	34.7
Total	3.44	4.42	28.5
Southeast Asia			
Burma	1.42	1.59	12.0
Cambodia	1.00	1.07	7.0
Indonesia	1.65	1.80	9.1
Malaysia	1.32	2.07	56.8
Philippines	1.18	1.24	5.1
Thailand	1.32	1.52	15.2
Total	1.39	1.55	11.5
South Asia	1.23	1.53	24.4

Adapted from S. C. Hsieh and V. W. Ruttan, Stanford Univ. Food Research Institute, *Studies* **VII**, No. 3 (1967).

releasing them widely. IR-8 especially (a short-strawed, high-yielding hybrid adapted to heavy fertilization) has increased rice production significantly. Table 16-2 shows yield increases achieved in various countries of eastern Asia in the years between 1951 and 1964, with the trend still continuing.

Oats. Oats, species of the genus *Avena*, like maize are mostly locally consumed as feed, and thus for the most part enter the human diet, if at all, as meat or dairy produce. Oats are among the most widely grown cereals of North America and Europe, being better adapted to a wide range of soil types, cultural techniques, and climatic extremes than are most other cereals. Oats are probably of polyphyletic origin, domestication usually being considered to have centered in northern Africa, the Near East, and temperate Russia—fairly recently (in the archeological sense), perhaps as late as 2500 B.C. The common cultivated species, *A. sativa*, is not known in the wild, although it is frequently encountered as an escape. Other species of the genus are common in the wild, and a number of them, perhaps even *A. sativa*, may have spread widely as weeds in fields of other grains, especially barley.

Botanical classification of oat species is confused. The common cultivated oat, *A. sativa*, is believed by many to have sprung from *A. fatua*, the common wild oat; other commercial varieties probably arose from *A. sterilis*, the red wild oat. All three of these species are hexaploid ($2n = 42$), as are also *A. byzantina* (sometimes considered a subspecies of *A. sterilis*); and *A. nuda*, the naked oat. Tetraploid oats ($2n = 28$) include *A. barbata*, the slender oat, and *A. abyssinica*. Diploid oats ($2n = 14$) include *A. brevis*, the short oat; *A. strigosa*, the sand oat; *A. wiestii*, the desert oat; and *A. nudibrevis*. The diploid oats are of little commercial importance, although they are cultivated to a limited extent in parts of Europe. The "A" genome of the diploid *A. strigosa* group is found also in the tetraploids and hexaploids. Of the tetraploid species, *A. barbata* is a tall, weak range grass of the western United States; and *A. abyssinica* is said to be grown for forage to a limited extent in the desert areas of northern Africa. *A. fatua* is a troublesome weed in oat fields. *A. sterilis* (and its varieties) is similar to *A. fatua*, but with larger spikelets. *A. byzantina* is putatively a descendant of *A. sterilis*, and is cultivated in several varieties, particularly in the warmer portions of the oat range. In *A. nuda* the lemma and palea are incompletely fused to the grain, yielding a naked seed easily upon threshing. Nevertheless, this species is not much cultivated, primarily because of its inferior yield. *A. sativa* (and its varieties) is the most important oat species; to it most of the modern cultivated strains belong.

Oats are essentially a crop of moist-temperate regions. Like wheat, they are broadly classified into spring or winter types, depending upon the season of planting. In the United States winter oats can seldom survive north of central Missouri, and most domestic production comes from spring oats of the northern states. Disease prevents oats from prospering in the humid Southeast. In the United States yields average about 1 ton per acre, but in northern Europe (Netherlands, Denmark) yields approach double this amount.

In the oat plant, as in other cereals, the main root system springs adventitiously from ground-level nodes. Stems have four to eight nodes, with tillers developing from the base in early growth. Lodging of stems is common, a fact that attests to their weakness, even though they are generally thicker than wheat stems. The leaf may be dark or light, and the leaf sheath closely envelops the internode. The inflorescence is a branched panicle, sometimes expanded and sometimes condensed and one-sided, with the ultimate branchlets terminating in glume-invested spikelets of one to seven flowers (of which the upper ones are

Close-up of oat grain. (Courtesy USDA.)

often staminate or infertile). Lemmas are either awned or awnless, depending upon variety, and with the paleas constitute a higher percentage (crude fiber) of the threshed grain than is usual in other cereals. The florets are ordinarily self-pollinated.

The oat grain is fairly rich in protein and fat, and constitutes a valuable foodstuff. One analysis breaks the oat grain down as follows: 13.8 per cent protein, 4.3 per cent fat, 66.4 per cent carbohydrate, and 12.2 per cent crude fiber (hulls—the lemmas and paleas). Selections of *A. sterilis* made in Israel have yielded as high as 30 per cent protein, and promise to be useful in breeding programs for higher-yielding, disease-resistant cultivars. Some 3 or 4 per cent of the crop goes into food for human use, mostly as rolled oats.

Combine-harvesting of oats in California. (Courtesy USDA.)

The protein is not of the glutenin type, so that oat flour has no value for breads. It may, however, be used for cakes, biscuits, and breakfast foods. Oatmeal is prepared by coarse grinding, rolled oats by running the dehulled seed (groats) between rollers in a steam chamber that softens the grain. In animal husbandry oats may be pastured or they may be cut for hay before maturity. Oat straw finds use as an emergency feed and as animal bedding. The value of the straw has caused some farmers to use the binder and thresher for oat harvesting, in preference to the combine (which scatters the fragmented straw over the fields). Oats were formerly much used as a "nurse crop" seeded with alfalfa, yielding a grain harvest the following spring while the forage was establishing.

Oats were introduced into North America in 1602, for plantings made in the Elizabeth Islands off the coast of Massachusetts. The United States is now usually the world's foremost producers of oats, although demand has been decreasing ever since machinery replaced the horse. Production is chiefly from the Wisconsin–Minnesota–Dakotas region. New stiff-strawed cultivars (such as Holden and Portal) adapted to heavy fertilization have been developed for this region. Multiline plantings are coming to be more used in the humid Northeast. Other oat-production centers are Russia, northern Europe, and Canada. World production is about 50 million tons annually, almost exclusively from temperate North America and Europe.

Barley. Barley, a species–complex of the genus *Hordeum* with about twenty species, is another versatile cereal much like oats in this respect. It can adapt itself well to high altitudes, cold weather, and a short season such as may be caused by a summer drought. Hot and humid weather, however, is as inimical to barley as it is to wheat. Barley is one of the oldest of cultivated cereals, having been used as food for man and beast in ancient

434

Egypt, and in China around 2800 B.C. It was used as a staple in southern Europe from time immemorial, until supplanted by wheat (after discovery of yeast bread) and by rye in Christian times. Presumably it was first domesticated in the Near East (Abyssinia and Tibet have also been considered as possible loci of domestication), where the wild barleys *H. spontaneum* and *H. maritimum* can still be found. Barley is typically a special-purpose grain rather than a general-market crop, finding its greatest use as animal feed and for malt. Minor quantities serve as human food, being consumed like rice, mostly in the Far East.

Cultivated species of barley, presumably descended from *H. spontaneum* (which crosses readily with them), include *Hordeum vulgare* (= *H. sativum* = *H. hexastichon*), a six-row type, with all lemmas awned or hooded, the usual commercial species; *H. intermedium*, a six-row species without awns or hoods, little used because of inferior yields; *H. distichon*, a two-row type with lateral spikelets reduced but more or less complete; and *H. deficiens*, a two-row species with lateral spikelets reduced to glumes only. All four are diploid ($2n = 14$), and most authorities lump them together as a single cultigen species, *H. vulgare*. Tetraploid species include *H. jubatum*, *H. marinum*, and *H. bulbosum*. In the United States the six-row types have been preferred to the two-row cultivars, and are more generally grown. Archeological findings indicate that the earliest cultivated barleys were of the two-row types, little different from wild *H. spontaneum* (Fig. 16-3). Like oats and wheat, barley is classified agronomically into spring and winter types, on the basis of planting season. Spring barleys can be planted farther north than probably any other cereal, maturing in only 2–3 months. Greatest production in the United States is from spring barley, grown mostly in the north-central states and in California. There, much of the crop serves as pasture and cover crop.

Grainheads of barley. (Courtesy USDA.)

The barley plant is ordinarily an annual, developing a heavy adventitious root system about the time of tillering. Tillers are usually fewer than in wheat, and commonly five to eight nodes tall (at times up to thirteen nodes). The inflorescence is a cylindrical spike, in which the tough, flattened rachis does not disarticulate. At each rachial node are borne three one-flower spikelets (thus basically there are six rows of spikelets, three to each side from consecutive nodes), of which the outer spikelets may be infertile or more or less aborted (two-row barley). Glumes and lemmas are typically awned (bearded barleys). Cross-pollination of the floret rarely occurs. The mature grain is frequently highly colored, owing to pigmented aleurone layers.

The barley grain is not greatly different from that of wheat, consisting of typical outer hulls (lemma and palea) fused with the pericarp, surrounding the abundant

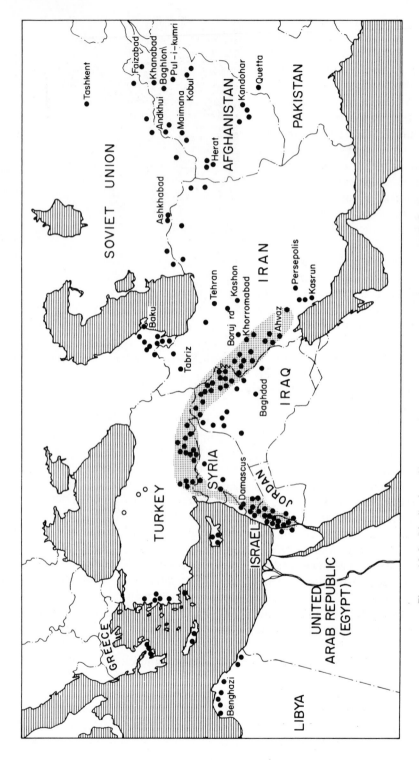

Figure 16-3 Distribution of known and reasonably certain sites of wild barley. Massive stands in fairly primary habitats may occur within the shaded area. Elsewhere, wild barley may be abundant, but confined to highly disturbed habitats. (After Harlan and Zohary, *Science* **153**, 1075 (1966).)

endosperm (with its outer aleurone layer), and the smaller embryo. Depending upon the environment in which it is grown, the grain may contain greater or lesser amounts of protein (aleurone cells). In the United States, newer nitrogenous soils of the western plains, and the dry conditions prevailing there at heading time, give a "flintier" barley, higher in protein, than do soils and climates farther east. Such barley is not in favor for malt production, the "softer" starchy barley being preferred for its greater malting facility and yield. This situation is in contrast to that existing in wheat production, where high protein content ("strong" wheat) usually commands a commercial advantage. One analysis of barley grain breaks it down as follows: 13.4 per cent protein, 1.9 per cent fat, 76.1 per cent carbohydrate, and 5.6 per cent crude fiber.

Early seeding is preferred for barley, which should be well matured before harvest. In technologically less advanced parts of the world it may be hand gathered, or even where combined, dried in windrows before threshing to reduce grain moisture to 15 per cent or less for best keeping. Yields for the United States as a whole average a little less than 1 ton per acre, but individual yields more than double this have been recorded in Denmark. Test plantings of the hybrid cultivar, Hembar, in Arizona have yielded over 4 tons per acre.

Barley finds considerable use as a substitute for maize in animal feeding where climatic conditions (cool, dry summer) favor it as a feedstuff. It is regarded as nearly as efficient a feed as is maize. In some localities barley is rolled or partly crushed before it is fed to the stock. Over 30 per cent of the crop is germinated for malt, consumed largely by the brewing industry, but also used for breakfast foods, malt syrups, confections, and the like. Broken grains will not, of course, germinate, and do not therefore qualify as malt grain, for which a premium is usually paid. Thus care in threshing and handling can

greatly affect the sale price of malt barley. Good malting quality can be tested for in the laboratory, where protein content and diastatic potential are determined. The making of malt is a relatively simple process. Grain is steeped, then germinated under controlled humidity, temperature, and atmosphere. The sprouting barley develops a high content of amylase and other enzymes, which are valued for digestion of starch adjuncts in the brewing of beers and liquors. Germination is stopped at the stage of maximum enzyme content, by heating and drying the sprouting grain. In the brewing process, malt is mixed with other grains to provide the mash; malt enzymes reduce the starch to simple sugars in the resulting wort, which is later fermented by yeast to make a beer.

Pearl barley, which amounts to about 3 per cent of the total production in the United States and is used mostly for soups, is prepared by a dehulling process accomplished by mechanical abrasion, similar to the process used for oats. In northern Europe and parts of Asia barley is more commonly consumed directly as human food than it is in the United States. Very minor quantities of barley are made into flour, for use by invalids and babies. This nonglutenous flour does not have good bread-making qualities, but was nevertheless used to make bread before the more palatable wheat flour became common.

The chief production regions for barley are the north-central states of the United States and adjacent Canada, the U.S.S.R., central and northwestern Europe, northern Africa and Spain, north-central India, and northeastern China and Japan. World production amounts to more than 100 million tons annually, with the U.S.S.R. and China probably the leading producers, followed by the United States, northern Europe, Turkey, India, and Morocco. The Mediterranean area, China, Japan, and India produce chiefly winter barley. The United States production amounts to some 9 million tons annually. Leading producing states for spring barley are the Dakotas, California, Montana, and Minnesota; most winter barley is grown in the southeastern states, for autumn and spring pasture and as a cover crop. Male–sterile lines carrying a trisomic chromosome have been developed through the crossing of several species, and hybrid barley is becoming available commercially.

Rye. The commercial rye plant, *Secale cereale*, is an important crop in northern Europe, but scarcely elsewhere. It is a diploid species ($2n = 14$), as are apparently also the other species of the genus. The various species will hybridize readily (as will also wheat with rye). *S. cereale* has not been authenticated in the wild, although it is common as a naturalized escape. Possibly cultivated rye was domesticated from the wild rye, *S. montanum*, of the Mediterranean region, although some students regard *S. anatolicum* of southwestern Asia as a more likely progenitor. Intermediate weed forms between the cultivated and the wild species are not uncommon. Quite possibly domesticated rye originated as a weed among barley and wheat. The species is believed to be of more recent origin than most cereals, for besides its similarity to wild forms there are no traces of it among Egyptian ruins or Swiss lake dwellings. It was, however, known to the Greeks and Romans.

Rye is typically a plant of cool, nonhumid climates, and can stand cold winter temperatures better than most other cereals can. No other winter cereal can be grown as far north. It is a fairly reliable crop on almost any soil, although it naturally does better on fertile land. Because of its ability to grow on soils too poor for other crops, it has at times been termed "poverty grain." Both spring and winter types are known; winter rye is the more important. Yields in

the United States amount to about 0.6 ton per acre, although on good soils and especially in northern Europe yields double this or more are not uncommon.

The rye plant is an annual under cultivation, although tending at times to maintain itself as a perennial by sprouting anew from the stubble. The abundant adventitious root sytem is one of the most profuse among cereals. Stems are slender and tough, four to six nodes tall, and often of a more or less purplish color. Both stem and leaves bear a waxy epidermal coating. The inflorescence is typically an elongate spike with about thirty three-flowered spikelets. Of the three florets, the uppermost is usually abortive. Lemmas are typically awned, and the grain is thus "bearded." Flowers are ordinarily self-sterile, so that cross-pollination is usual.

The mature grain is more slender than that of wheat, usually grayish yellow, free from the lemma and palea. The aleurone layer is confined to a single layer of cells. One analysis of the grain breaks it down as follows: 13.4 per cent protein, 1.8 per cent fat, 80.2 per cent carbohydrate, and 2.3 per cent fiber. The protein is partly glutenous (contains gliadin but not glutenin), so that rye commonly serves as a bread grain. A large portion of the European population lived for centuries mainly on "schwarzbrot," the black, soggy, rather bitter bread still not infrequent in rural Germany, Poland, and Russia, whose main ingredient is rye flour. In the United States rye bread is made of a combination of wheat flour and rye flour.

Today the major portion of the rye crop goes into stock feed. The plant is also much used for hay and pasturage, cover crop, and green manure. The long, fine, highly resistant straw finds service for animal bedding, stuffing, thatching, and paper manufacture. The grain may be fermented to yield alcoholic beverages or industrial alcohol. Ergot (see the section on drugs in Chapter 12) is a parasite of the rye plant.

Russia leads the world in the production of rye, followed by adjacent northern European countries. Total world production, about 95 per cent of which comes from Europe, amounts to some 35 million tons. The United States grows about 1 million tons of rye annually, in all parts of the country but most heavily in the Dakotas and Nebraska. About two-thirds of this crop goes into stock feed.

Sorghum. Sorghum has been a widely cultivated crop in Asia and Africa for four millenia. It is said to have been a principal item in the diet in the ancient city of Ninevah on the Tigris. The genus *Sorghum* embraces several types of economic plants. There are four broad groupings: (1) forage or syrup sorghums or sorgos, with sweet stalks; (2) grain sorghums, without sweet stalks (Kaffir, Durra, Milo, etc.); (3) broomcorns, grown for the umbelliform inflorescence; and (4) grass sorghums, for pasture, hay, and silage (Sudangrass, etc.). All these types can be most conveniently lumped as the many-faceted species, *Sorghum bicolor* (*S. vulgare*). Some authorities recognize three morphologically distinct varieties that exhibit allopatric distribution and various genetic isolating mechanisms: *arundinaceum* ("bicolor" complex), *aethiopicum* (guinea complex), and *verticilliflorum* (kaffir–durra–nervosum complex). Sorghum attains greatest importance in man's life in Africa, but the species is also grown and harvested in India, China, Manchuria, the United States, and to a lesser degree elsewhere. Grain sorghums are known in innumerable cultivars. All are annuals. The many varieties—"species" under some systems of classification—cross readily, and all have a chromosome complement of $2n = 20$. The closely related Johnsongrass, *S. halapense*, is a perennial weed of cultivated areas or useful forage plant, depending on circumstances. It is similar to Sudangrass (*S. bicolor* var. *sudanensis*), but has a $2n = 40$

Grain sorghum in the United States. (Courtesy USDA.)

chromosome complement, and rarely crosses with the common sorghums. Sorghums are probably indigenous to Africa, but cultivated species may have originated in Asia as well. Sorghum was known in Egypt prior to 2200 B.C., and was introduced into the United States about the middle of the nineteenth century.

Perhaps one reason for sorghum's great resistance to drought is its extensive root system, greater than that of maize, in contrast to its modest leaf area. Stems commonly have about ten nodes, with a lateral bud at each, but only the two or three lowermost ordinarily develop into tillers. Originally strains introduced into the United States were tall, but breeding programs have reduced the internode length yielding dwarf varieties suitable for combine harvesting. Almost all planting is now to hybrid cultivars, the seed produced through crossings made possible by development of male–sterile lines. The inflorescence is usually a compact panicle, with many sessile (perfect) and pedicellate (usually staminate) spikelets containing a single functional floret each. The lemma is awned. Self-pollination is the rule. Grains are small in comparison to those of maize, but have a similar starchy endosperm. Composition of the grain is, on the average, 12 per cent protein, 3 per cent fat, and 70 per cent carbohydrate.

Sorghum is a short-day plant, maturing as the length of the day decreases. It is remarkably drought-resistant, and prefers warm climates. It is typically grown where climates are too dry or hot for maize, and as a quick substitute where failure threatens another crop (as with weevil attack on alfalfa in the northeastern United States).

Sorghum is grown as a grain for human consumption mostly in Africa and Asia. There it is used just as soon as it is grown; for without milling (to remove the embryo) flour would become rancid. The pulverized grain is usually consumed as mush (in Africa) or in tortilla-like cakes (in Asia). The species sometimes serves also as cover crop or green manure, and the stems for fuel or weaving (in Africa and Asia). In the United States about one-fourth the sorghum acreage is devoted to forage or silage, and three-fourths to grain (which is later incorporated into poultry or stock feeds). Grain production in the United States amounts to over 20 million tons annually. Both "waxy" and "sweet" varieties have been developed, the former finding use to some extent as a substitute for manioc. Industrial processes utilizing maize can utilize sorghum grain for the same purposes. Moreover, a wax similar to carnauba is reportedly obtainable from the hulls. Production of sorghum centers in the Great Plains area, from the Gulf to the Dakotas. As much as 2 or 3 tons of sorghum grain per acre is possible. World production of grain sorghum is approximately 35 million tons annually, largely from China, India, and central Africa.

Millets. A number of small-seeded, annual grasses, used both as grain and for forage, are classed as millets. Uphof lists a dozen millets in seven different genera as worthy of mention in his *Dictionary of Economic Plants.* None are of major importance in the Western world, but many form a staple in the diet of some Asiatic and African peoples, particularly in that of the poorer classes: millets have been termed the "poor man's cereal." Millet grain is usually not distinguished from sorghum grain in production statistics. World production of both together approaches 80 million tons annually, of which about 15 million is certainly millet and 35 million certainly sorghum, the remaining 30 million being possibly either. Millets are grown principally in the same areas as is sorghum, typically in regions of limited rainfall. Often they are grown as mixed crops with legumes, maturing quickly (in 3 or 4 months). India, Pakistan, and Africa just south of the Sahara account for most of the world production. In India, where millet is especially important, the grain is considered to have been cultivated and used for bread since prehistoric times. The importance of millet as human food is demonstrated by the fact that over 80 million acres in India are devoted to millet growing, yielding over 12 million tons of grain. In the United States millets are incorporated to a limited extent in stock feed and birdseed, but are not consumed directly as human food.

Because millets are adapted to a wide range of soils in semi-arid climates, they are often utilized for emergency hay or catch crops. New pearl millet cultivars have been developed in recent years that are useful as forage plants in the humid Southeast of the United States as well as for grain. In the United States millet is mechanically planted the same as small grain, and seed combine-harvested. In

Millet grown in India. (Courtesy The Rockefeller Foundation.)

much of the world, however, millets are primitively grown with little cultivation and almost no mechanical handling. The crop is cut by hand with sickles, and carried to communal threshing grounds where the grain may be "threshed" by treading and winnowing. Under primitive conditions, on poor soil, and with scanty rainfall, few crops can provide so much sustenance as does millet. About 85 per cent of world production is consumed as human food, even though the grain is nonglutenous and unsuitable for making typical bread. In southeastern Europe a flat bread or a porridge may be made, or the millet may be fermented into a beverage. In western China, India, Pakistan, and the southern U.S.S.R., millet may be consumed as an entire grain, or the grain may be ground into flour, or even sprouted before consumption. Millet is said to be somewhat lower in protein content than wheat, but richer in fat, and about the equal of rice in nutritive quality.

A few of the more familiar millets are listed below:

FOXTAIL OR ITALIAN MILLET, *Setaria italica.* The inflorescence is a loose, simple spike. There are five or six different types, in-cluding the common, Hungarian, Aino, German, Siberian, and Golden Wonder millets. The species supposedly originated in India, but is principally cultivated today in the Near East and China. It is highly drought-resistant.

PEARL MILLET, *Pennisetum glaucum.* Pearl millet is the second most important millet of India after sorghum, being grown throughout the country. The inflorescence is a dense, simple spike. Pearl millet is grown as a rainy-season crop in India and the Sudan, and yields a very nutritious flour, with minimum demands for soil and moisture. Hybrid seed from select parent cultivars is now available, used in India for food grain, and for forage in the United States. Growth of hybrid pearl millet in the United States has been made economical by the discovery of male sterility in the species.

FINGER MILLET OR RAGI, *Eleusine coracana.* Finger millet is an important millet of southern India and the principle food staple for people in southern Sudan and northern Uganda. The species has dense, digitate spikes, and may be derived from *E. indica.* In contrast to most millets, finger millet requires coolish weather and adequate rainfall

441

Threshing millet in Nepal. (Courtesy USDA.)

to mature; even then per-acre yields are not great.

PROSO MILLET, *Panicum miliaceum*. The inflorescence is an unawned, drooping panicle. This species is extensively grown in Soviet Union and central Asia, and is probably the ancient millet of the Chinese and of the Swiss lake dwellers, perhaps antedating wheat in cultivation. The grain is enclosed in a hard, shiny lemma and palea.

JAPANESE BARNYARD OR SANWA MILLET, *Echinochloa frumentacea*. This millet has a short, awned raceme. The grain is firmly enclosed in the hardened lemma and palea. The species may be derived from the widely distributed *E. (Panicum) crusgalli*, which is also cultivated to some extent in the Far East.

TEFF, *Eragrostis abyssinica*. Teff is an important food millet of Ethiopia and the eastern African highlands, in several cultivars (some with white and some with black grain). The plant is used as a forage as well as for food grain, and the straw is mixed with mud for constructing huts. Women grind the grain on flat stones to make a flour, which is baked into large, flat pancakes called *injara*, the basic food of the Amharas. After early morning prayers and chores, a typical Amhara husband returns to the house for breakfast of injara and hot pepper sauce. Teff is planted to fields cultivated by oxen with a crude plow to break the soil, followed by hand sowing of teff seed and another plowing to cover the seed. Harvesting is painfully slow with small hand sickles.

OTHER FINGER MILLETS, *Digitaria spp*. Acha, *D. exilis*, is a millet much grown on the upland plateau in central Nigeria, Africa.

442

Yields are poor, but the seed is highly nutritious and palatable. Other species, such as *D. sanguinalis* (crabgrass) and *D. iburua*, also yield comestible seed.

Some other cereals.

WILD RICE, *Zizania aquatica*. Wild rice is a nutritious grain which grows wild in the shallow lakes of the northern United States (Minnesota, especially) and southern Canada. It was an important staple for the Chippewa Indians; archeological findings from prehistoric times indicate that gathered grain was charred on fire hearths, after which it was threshed by treading in large earthen bowls termed "jig pots." Until recent regulations were established in Minnesota to conserve wild rice, the Indians occupied favorable sites seasonally for harvesting the wild grain and preparing it for storage by drying it on mats in the sun, or parching it in a kettle over an open fire (then threshing and winnowing to remove the husks and chaff). Wild rice was an important Indian food when the Northwest Territory was first explored, and has since become an expensive gourmet delicacy.

The Indians harvested the grain by paddling canoes among the ripening seed heads in September, flailing the grain into the canoe bottom from which it was later gathered. Wild rice is still so gathered on the Chippewa lands in northern Minnesota, chief source of United States production. On private lands and in Canada there has been some mechanization, with cutting and gathering machines mounted on barge-like boats that work through the wild stands. It is feared that permitting efficient mechanical techniques on the Chippewa reservation would soon exhaust the wild stands and put an end to this small but profitable enterprise for the Indians. Already the wild stands are much dissipated, and annually less than 2,000 tons of wild rice come to market these days, as a specialty food to be sold at prices tenfold that of other cereals. About 60 per cent of production is from Minnesota. The grain, much of it still gathered from wild stands, is usually full of wet foliage and lake debris; it shrinks to about 40 per cent of its harvested weight during processing.

A recent taxonomic revision of wild rice established three varieties, of which the one growing in northern Minnesota and southern Canada, responsible for most of the commercial harvest, is termed *Z. aquatica interior*. The other varieties grow slightly above tide level along the Atlantic coast from Canada to Florida, and are important as wildlife food but not harvested by man. Early attempts at cultivating wild rice were unsuccessful, partly for unknown causes, and partly because it is difficult to maintain proper water levels which seem critical for the establishment and growth of the wild plants. Since the grain ripens unevenly, the flailing operations involved in canoe harvesting permits sufficient seed to fall back into the lakes to replenish the stand. The species is an annual, and its perpetuation depends upon natural reseeding. Since the mid-1950's wild rice has been increasingly grown in artificial paddies constructed with 8-foot dikes. Flooded paddies are seeded from boat or plane, and drained for combine harvesting. Yields run about 300 pounds per acre compared to about 40 pounds for uncultivated stands. Development of nonshattering cultivars should increase yields seven- to tenfold. But the special conditions required will always make wild rice a costly crop to grow, not remunerative as a nutritional source in competition with dry-land grains.

The nutritive quality of wild rice is on a par with unpolished rice or whole wheat. Although the hulls are removed—sometimes still by treading and hand winnowing, but increasingly by small fanning mills installed by the Indians—the grain usually retains the nutritious outer aleurone layers. Typically green rice is sold to processors sacked directly from the collecting canoes. It is sun-dried for a couple of days, turning brown, then parched in tumbling drums. The heat turns the grains a rich black. Screening, dehulling by machine, and removal of chaff by suction devices follows. Gravity tables effect a final sorting according to grain size and condition.

WILD OATS, *Avena sterilis*. The USDA has screened thousands of wild oat introductions from various parts of the world, and found a type from the Mediterranean region yielding as much as 30 per cent protein (compared to around 19 per cent for the best commercial oats grown in the United States). Some strains of wild oats have large seeds, and exhibit superior disease resistance, raising the hope that plant breeding may yield a wild oat cultivar competitive with other cereals, and especially useful as a high-protein source. The tendency for seed shatter will have to be eliminated, but researchers estimate that yields as high as 150 bushels per acre of 25 per cent protein grain are theoretically possible. Other wild species, such as *A. magna*, a tetraploid, may contribute useful characteristics, although difficulties are encountered in crossing *A. magna* with the hexaploid *A. sterilis*.

JOB'S TEARS, *Coix lachryma-jobi*. The Job's tears grass or Adlay is native to southeastern Asia, but has been introduced and cultivated throughout the world tropics. The panicle bears attractive, shiny white grains, fancifully resembling tears and often used ornamentally as beads. These serve as a food item for poorer peoples, particularly in the Far East. The hulled grains can be parched, boiled like rice, or milled into a flour for making a type of bread. Reportedly the grain has a higher protein-to-carbohydrate ratio than most cereals.

OTHER WILD GRASSES. Many grasses yield small quantities of seed which are locally useful as food. In most cases, however, the grain is too small, processing too difficult, yields too scant, or cultivation too precarious to arouse widespread interest in the species as a cultivated food plant. Such is the wealth of the grass family, however, that candidate species for domestication can be found in all parts of the globe. To some extent mannagrass (*Glyceria fluitans*), Guineagrass (*Panicum maximum*), Fonio (*Paspalum longiflorum*), Burgu (*Panicum burgii*), reedgrass or canegrass (*Phragmites communis*), *Echinochloa stagnina*, and many other species have been utilized for their seed grains.

SUGGESTED SUPPLEMENTARY REFERENCES

Aufhammer, E. G., P. Bergal, A. Hagberg, F. R. Horne, and H. van Veldhuizen, *Barley Varieties*, 2nd ed., American Elsevier, N. Y., 1958.

Chandler, Robert F. (ed.), *Rice Genetics and Cytogenetics*, Proceedings of symposium held at International Rice Research Inst., Los Banos, Laguna, Philippines; Am. Elsevier Pub. Co., N. Y., 1964.

Coffman, F. A. (ed.), *Oats and Oat Improvement*, American Society of Agronomy, Agronomy Monograph 8, Madison, Wisc., 1961.

Doggett, H., *Sorghum*, Longmans, Green and Co., London, 1970.

Dore, W. G., *Wild Rice*, Pub. 1393, Research Branch, Canada Dept. of Agriculture, Ottawa, 1969.

George Washington Univ., Bibliography of World Literature, Scarecrow Press, Metuchen, N. J.: *The Millets, 1930–1963*, 1967; *Sorghum, 1930–1963*, 1967; *Rice*, 1963.

Grist, D. H., *Rice*, 4th ed., Longmans, London, 1965.

Hitchcock, A. S., *Manual of the Grasses of the United States*, 2nd ed., rev. A. Chase, USDA Misc. Publ. 200, Washington, D. C., 1950.

Inglett, G. E. (ed.), *Corn: Culture, Processing, Products*, AVI Pub. Co., Westport, Conn., 1970.

Kent, N. L., *Technology of Cereals*, Pergammon Press, Elmsford, N. Y., 1966.

Leonard, W. H., and J. H. Martin, *Cereal Crops*, Macmillan, N. Y., 1963.

Matz, S. A. (ed.), *Cereal Science*, AVI Pub., Co., Westport, Conn., 1969.

——————, *Cereal Technology*, AVI Pub. Co., Westport, Conn., 1970.

Milthorpe, F. L., and J. D. Ivins (eds.), *The Growth of Cereals and Grasses*, Butterworths, London, 1966.

Peterson, R. F., *Wheat*, Interscience, Wiley, N. Y., 1963.

Pierre, W. H., S. R. Aldrich, and W. P. Martin, *Advances in Corn Production*, Iowa State Univ. Press, Ames, 1966.

Quisenberry, K. S. and L. P. Reitz, (eds.), *Wheat and Wheat Improvement*, American Society of Agronomy, Agronomy Monograph 13, Madison, Wisc., 1967.

Wall, J. S., and W. M. Ross (eds.), *Sorghum: Production and Utilization*, AVI Pub. Co., Westport, Conn., 1970.

CHAPTER 17

Other Food Seeds
and Forages

NATURE PROVIDES THE EMBRYO, THE YOUNG plant-to-be, with an abundance of concentrated food stored in the seed for its "start in life." Man appropriates this food from the plant world to sustain his civilization. The previous chapter discussed the world's most useful food seeds, the cereal grains. Cereal plants are readily adapted to domestication and agriculture, and their seeds to easy harvesting and storage for protracted periods without deteriorating. But plant families other than the Gramineae have contributed some almost equally important species to man's vital needs. Of especial importance is the family of the Leguminosae, with its contribution of beans, peas, soybeans, and many pasture plants. The legumes almost alone among food plants give high-protein seeds, and thus provide something of a natural balance to the carbohydrate-rich cereals. Not only do the legumes provide nitrogenous molecules for man and his domestic animals directly; they also, bearing as they usually do the vital nitrifying bacterial symbionts in root nodules, build up the soil so the other plants and eventually animals may satisfy protein needs. In the web of life legumes are certainly as important as the grasses.

As world populations increase more attention will doubtless be given certain seeds, and proteinaceous leaf that is in abundant supply but little used at present. Cottonseed and citrus seed, for example, come to mind. True, edible and industrial oils are extracted from such seeds; but it should prove possible, with present-day research and technology, to free a seed such as cottonseed of disagreeable or toxic principles, and permit a higher level of usage as human food. Likewise, there seems little question but that the foliage of fast-growing tropical plants could be made available for the nourishment of livestock, especially "waste" such as cassava leaf (a by-product of the growing of the plant for its fleshy root). Cassava is low only in methionine among amino acids essential to human beings (see Table 17-1). Also, there seems little question but that many additional tree nuts remain to be selected and brought into cultivation, especially in the tropics where natural conditions are suitable for a tree crop but not for annual cultivation. Large seeds from trees are by-and-large a more nourishing food than the more generally consumed tree fruits.

In this chapter attention will first be given to certain "pseudo-cereals"—plants whose hard seeds resemble grains, and are

Table 17-1

Amino Acid Profile for Cassava Leaf

Amino Acid	Average Percentage for Jamaican Samples
Alanine	5.98
Arginine	5.28
Aspartic Acid	10.14
Cystine	1.37
Glutamic Acid	10.22
Glycine	5.39
Histidine	2.23
Isoleucine	5.01
Leucine	8.89
Lysine	7.20
Methionine	1.65
Phenylalanine	5.82
Proline	4.64
Serine	5.16
Threonine	4.92
Tryptophane	1.47
Tyrosine	4.18
Valine	5.73

Adapted from D. J. Rogers and M. Milner, *Economic Botany* **17**, 215 (1963).

similarly treated. Legumes as food seed plants will next be considered; and then other food seeds, many commonly known as nuts. Finally the part both legumes and grass plants play as forage—for the production of animal protein to be consumed by man as meat and milk—will be briefly discussed.

The Seeds of Annuals other than Grasses with Grainlike Seeds: the Pseudo-Cereals

Buckwheat. Buckwheat, *Fagopyrum esculentum*, Polygonaceae, is related to the rank docks (*Rumex*) and smartweeds (*Polygonum*). Like them it is adapted to poor soils, more so than the true cereals. Buckwheat is a crop of cool, moist climates, and accommodates to high elevation and a short growing season. It is singularly free of disease. Buckwheat is cultivated mostly in the northeastern United States, northwestern France, and eastern Europe. It probably originated in Central Asia (perhaps western China), seemingly fairly recently in comparison to the

Principal types of dry colored beans. Legume seeds complement cereal grains in human diets throughout the world and are a richer source of protein. (Courtesy USDA.)

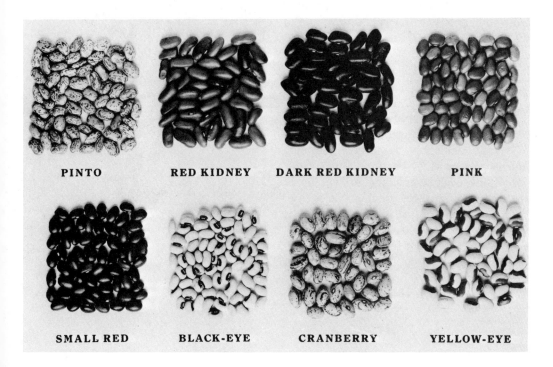

PINTO   RED KIDNEY   DARK RED KIDNEY   PINK

SMALL RED   BLACK-EYE   CRANBERRY   YELLOW-EYE

cereals. It was introduced into Europe in the Middle Ages, and into North America by the early colonists.

In the United States most buckwheat seed is used for stock feed, but in Europe, particularly parts of Poland and Russia, it is a basic food item in the peoples' diet. There a starchy flour obtained by milling the "grain" is consumed as porridge and soup. In the United States buckwheat flour is used chiefly to make pancakes. Buckwheat makes an excellent green manure, bee plant, and catch crop. The foliage has been utilized as a source of rutin, which is used medicinally to control hemorrhaging.

The buckwheat plant is a rapid-growing, succulent annual which branches profusely and soon smothers competing vegetation if given a good start in a well-prepared seedbed. It has a diploid chromosome complement of 16, and tetraploids have been induced. The leaves are alternate, lanceolate, and with the stipular sheath characteristic of the Polygonaceae. The inflorescences are axillary and terminal cymes of perfect but dimorphic flowers, blooming from as little as 12 weeks after planting until frost. The flowers of *F. esculentum* are largely self-sterile, and require cross-pollination. Characteristic three-angled seeds, the buckwheat grains, are matured abundantly unless hot weather causes blasting of the flowers. The dicotyledonous seeds with appreciable endosperm contain about 11 per cent protein, 2 per cent fat, 11 per cent digestible carbohydrate. The gluten content, although low, is reportedly sufficient to permit making of yeast bread. Milling follows the general practices common for wheat, involving a cracking to loosen the seed coats, sieving, and grinding. Yields from buckwheat grain consist of 60 to 75 per cent flour, 4 to 18 per cent middlings (embryo), and 18 to 26 per cent hulls.

The usual small-grain combines gather the crop in the United States. Yields range up to 40 bushels per acre. United States pro-

duction amounts to less than 30,000 tons, mostly from New York, Pennsylvania, Michigan, and Wisconsin. World production probably exceeds 3 million tons annually, mostly from the U.S.S.R. Tartary buckwheat, *F. tataricum*, is somewhat grown for poultry feed. Grain of the related *Polygonum nodosum* was harvested by Iron Age man in northern Europe, but is not today cultivated.

Quinoa. Quinoa, Hupa, Dahue, and various other Indian names refer to the seed of *Chenopodium quinoa*, Chenopodiaceae. The species is a tetraploid ($2n = 36$) little known outside the higher Andes of tropical South America. Yet there it constitutes, just as it did in Incan times, a staple in the diet of some Indian-blooded peoples of Peru, Bolivia,

Quinoa, *Chenopodium quinoa*. (Courtesy Chicago Natural History Museum.)

and Ecuador. The plant is indigenous and still wild in this region, and was presumably first domesticated in the vicinity of Lake Titicaca long before the coming of the *conquistadores*. Indeed, the Incans, in whose empire quinoa was more widely grown than any other plant save the potato, are said to have regarded the species as a sacred plant, and to have ceremoniously opened the first furrow for its growing season with a golden implement.

Quinoa is seeded in rows or broadcast in autumn and crudely covered with soil by dragging thorny branches over the field or by rude harrowing. Germination is rapid, the seedlings emerging in as little as 1 week after planting. Quinoa plants are tolerant of light frost and scanty rainfall. They are thinned and weeded by hand, and mature a heavy head of seed in 5 or 6 months. Then the whole plant is cut and left to dry in the sun for several days. Threshing is accomplished by flailing with a stick, and the chaff is separated by winnowing. Average yield is less than 1 ton of grain per hectare, although triple this is possible under favorable circumstances.

The quinoa plant is quite plastic and is known in many local "varieties," distinguished largely by color. The species much resembles its North American relative, lamb's quarters (*C. album*). It is a shallow-rooted, hollow-stemmed, herbaceous annual 1 to 2 meters tall at maturity, with alternate, marginally lobed leaves rich in calcium oxalate, and inconspicuous, perfect flowers. The quinoa seed, in addition to its direct consumption as human food, serves as a poultry and livestock feed and source of an alcoholic beverage. Tender stems and leaves of this plant can be used as salad greens. The seed can be eaten boiled or toasted, but most is ground into flour for tortillas or porridge (or is mixed with wheat flour for bread). The grain may also be fermented for chicha (see Chapter 20, "Beverage Plants").

It is about 55 per cent carbohydrate, 15 per cent protein, and 4 per cent fat.

C. nuttaliae, Huauzontle, another tetraploid grain chenopod of Mexico, is similar to quinoa. It was much grown in central Mexico before the Conquest, but is now gradually passing out of cultivation.

Cañihua. Cañihua or cañigua, *Chenopodium pallidicaule*, Chenopodiaceae, is similar to its close relative, quinoa, but is diploid ($2n = 18$) and somewhat shorter. It is cultivated to a limited extent in the same area for the same uses as quinoa, but is adapted to even higher elevations because it has a shorter growing season (135–145 days for cañihua as against 165–175 for quinoa). Cañihua ash, high in calcium, is used by coca chewers in Bolivia.

Amaranth. Several species of *Amaranthus*, Amaranthaceae, have yielded food seeds to primitive peoples, although none have become important as modern-day cultivated plants. Jataco, achita, or quihuicha, *A. caudatus*, is one such species of the South American Andes, and can be cultivated and used in much the same way as quinoa (a name sometimes used for this seed, too). *A. hybridus* is also grown in the Andes. *A. leucocarpus* is a Mexican species reportedly used for food and ceremonial purposes in ancient times. It is said that eighteen imperial granaries were filled by the Aztecs with the small seed of this species. *A. cruentus*, an Asiatic species, is reported to be cultivated in Asia and in Europe; it has been naturalized in the New World. *A. frumentaceus*, presumably native to India, has been cultivated there for food among the wilder hill tribes.

Cucurbitaceous seeds. Squashes, watermelons, and other members of the Cucurbitaceae will be discussed later among fruit foods. Their seeds were commonly parched and eaten by primitive peoples. The seeds are relatively large and rich in oils and protein.

Sunflower. The sunflower, *Helianthus annuus*, Compositae, today grown chiefly as an oil plant (see under "Semidrying oils" in Chapter 13), should also be listed among food seeds. The species was a treasured plant of the American Indians, widely spread, and cultivated from Mexico northward. Typically the seeds were parched and then ground into meal.

Many other species listed among the oil plants in Chapter 13 (e.g., hemp, sesame, and others) also have provided or can provide seed for direct consumption.

Legumes or Pulses

Legume seeds, sometimes termed "grain legumes," are second only to the cereals as a source of animal and human nutriment. From the qualitative standpoint they perhaps rank first among food plants. Note in Table 17-2 the comparatively greater nutritional richness of legume seeds, generally 25 per cent protein or more. Legumes have two or three times the protein content of most cereal grains, and an even higher proportion than the starchy root crops. Legume seeds are especially important as a proteinaceous complement to cassava and plantains in the primitive, tropical areas where diet fare is limited and quantity often inadequate. The kwashiorkor disease so frequent in Africa, due to protein deficiency, can generally be overcome with an improved diet that includes legumes (or fish and animal proteins, where these can be obtained). Not that the protein balance of legume seeds is perfect for human needs. But legumes do supply lysine, the essential amino acid that is limited in cereals such as maize. Many authorities conclude

Table 17-2

*The Composition of Some Legumes and Other Foods**

Food	Calories per 100 g	Percentage of: Protein	Fat	Carbo-hydrate	Vitamin A I.U. per 100 g	Thiamin	Ribo-flavin	Nicotinic Acid	Ascorbic Acid
Millet, finger, meal	332	5.5	0.8	76	trace	0.15	0.07	0.8	0
Maize, meal, 96% extr.	362	9.5	4.0	72	trace†	0.30	0.13	1.5	0
Rice, lightly milled	354	8.0	1.5	77	trace	0.25	0.05	2.0	0
Sorghum flour	353	10.0	2.5	73	trace	0.40	0.10	3.0	0
Cassava, fresh	153	0.7	0.2	37	trace	0.07	0.03	0.7	30
Yam, fresh	104	2.0	0.2	24	20	0.10	0.03	0.4	10
Bambarra groundnut, fresh	367	18.0	6.0	60	trace	0.30	0.10	2.0	trace
Cowpea	340	22.0	1.5	60	20	0.90	0.15	2.0	trace
Groundnut, dried	579	27.0	45.0	17	trace	0.90	0.15	17.0	trace
Soybean	382	35.0	18.0	20	trace	1.10	0.30	2.0	trace
Fish, sea, lean, fillet	73	17.0	0.5	—	trace	0.05	0.10	2.5	trace
Beef, lean	202	19.0	14.0	—	trace	0.10	0.20	5.0	trace
Eggs, hens'	158	13.0	11.5	0.5	1000	0.12	0.35	0.1	trace

* Figures relate to edible portion.

† Yellow maize 150. I.U.

From W. R. Stanton et al., *Grain Legumes in Africa*, Food and Agriculture Organization of the United Nations, Rome, 1966.

that it is more economical to supply necessary lysine from a legume than to attempt introduction of high-lysine maize strains that do not yield so abundantly as the conventional cultivars. Legume protein is not so properly proportioned as animal protein in amino acids needed in the human diet, often being deficient in some and over-rich in others. But where a population faces starvation or suffers from a poor diet, legume seeds help greatly, substituting nicely for the more expensive animal proteins.

Increasing emphasis can be expected throughout the world on the growing of legumes for their seeds. In the search for new cultivars most genera have received little attention compared to the cereal grains. Indeed, archeological and historical knowledge of legumes is quite limited for so important a group. Many of the beans were domesticated as long ago as the cereals, and they have been almost as basic for the development of civilization. All told, over 40 million tons of the familiar legume seeds come to market each year, and many more millions of tons are consumed locally and by wildlife, especially of the less prominent species that fail to enter the statistics.

Beans. The word *bean* denotes a great number of forms, species, and even genera of legumes, the seeds or pods of which constitute some of mankind's most important foods. It is necessary to specify the particular kind of bean.

The broadbeans (*Vicia faba*), soybeans (*Glycine max*), and a few other types were well known in the Old World before the time of Columbus, but the choice food beans of the world, the large-seeded species of *Phaseolus*, were the exclusive ward of the New World Indians. *Phaseolus vulgaris* was probably domesticated in central Mexico, where today the cultivated forms grade imperceptibly into escapes and wild forms. Archeological discoveries dating back more than five millenia B.C. show beans to have been under at least incipient cultivation then. By 1492 lima beans, scarlet runner beans, string beans, shell beans, white beans, black beans, pea beans, black-eyed beans, and kidney beans—all *Phaseolus*—were spread

Green or snap beans are eaten in the pod: left to mature, dry beans (seeds) would result. (Courtesy USDA.)

Mechanical bean pickers work in relays in this Wisconsin snap bean field. Eight tons per acre have been obtained with well fertilized snap beans. (Courtesy USDA.)

from Peru to New England. The invading white man quickly recognized their usefulness, and very soon these nutritious, protein-rich seeds were providing a livable diet on long ocean voyages or for pioneering expeditions. The world now had a proteinaceous food as easily stored, handled, and transported as were the cereal grains.

Beans are warm-season annuals, sensitive to temperature extremes and requiring a modest amount of moisture. They are, however, quite tolerant of soil types. Over much of the world they are planted, tended, and ultimately harvested by hand. In the United States commercial plantings are generally mechanically handled. Harvest of dry beans involves cutting and threshing or combining. Yield of field-grown dry beans in the United States is a little under 1 ton per acre. Green beans are picked by hand and marketed locally. World dry-bean production (exclu-

sive of broadbeans) is about 10 million tons annually. Brazil and India are the largest producers, followed by the United States, Mexico, southern Europe, and Japan. Michigan, California, and New York (green beans) lead in production of edible beans in the United States, and green beans are grown as a garden plant in all sections.

The genus *Phaseolus* embraces both annual and perennial characteristics. Roots of tropical species often tend to be perennial, although the stems are typically annual and herbaceous. The leaves are pinnately trifoliolate, stipulate, and stipellate. Inflorescences are axillary racemes, bearing white or yellow to purplish, bibracteate, perfect flowers. The fruit or legume is usually fairly large, containing a few to several seeds, the commercial beans.

Some 100 or more species occur in both the New and Old Worlds. The Asiatic

Phaseolus aconitifolius, P. angularis, P. aureus, P. calcaratus, and *P. mungo* possess slender cylindrical pods with small seeds and have yellow flowers. The American species, *P. acutifolius, P. coccineus, P. lunatus,* and *P. vulgaris* possess more-or-less flattened pods with large seeds and have white to purple flowers. Included within these species, particularly *P. vulgaris,* are the most important beans of the world.

KIDNEY, FIELD, GARDEN, OR HARICOT BEAN; FRIJOL; *P. vulgaris.* This is the most widely cultivated of the beans, and is known in both "bush" and "vine" or "pole" forms. It yields the familiar "green," "string," "snap," or "wax" beans when the entire pod is eaten immature, or the many familiar types of dry beans (kidney, pea bean, pinto, great northern, marrow, yellow-eye, and others) when the mature dry seeds are shelled from the pod. This very variable species is native to tropical America, the indigenous forms having smallish dark seeds, from which presumably cultivars have repeatedly been selected. Compared to maize, *P. vulgaris* has shown very little evolutionary change under domestication. It occurs as an escape and hybridizes sufficiently with wild beans to cause a great deal of confusion in classifying it. It is known to have been in cultivation under irrigation in Mexico four millenia B.C. Protein content in various strains ranges from 17–37 per cent, with an average of around 25 per cent (greatly influenced by growing conditions).

LIMA OR BUTTER BEAN, *P. lunatus* (including *P. limensis = P. lunatus* var. *microcarpus,* the large-seeded South American forms). Another confusing delimitation of species and uncertain history of origin is found with the lima beans, "aristocrats of the bean family." It is usually presumed that the many "wild" plants of this species found today in the American tropics are naturalized escapes rather than indigenous. The earliest lima beans in archeological diggings are from Peru, dated about 3800 B.C. Three subgroup-

ings with numerous cultivars are recognized, including the sieva beans—small, flat, colored-seeded forms, sometimes regarded narrowly as the species—and "large-flat," and "potato" types, later-ripening and more susceptible to cold than are the sievas. West Indian forms are often round-seeded. Archeological records show sieva types to have been domesticated anciently in North America, large-seeded types in Peru. Both "pole" and "bush" types are grown, the bush forms having been developed from sports of the pole type. The pods and seeds of lima beans are broader and flatter than those of kidney beans. In the tropics lima beans are generally perennials; in the United States, the country of their greatest cultivation, they are treated as annuals. The beans contain adequate amounts of protein, fat, carbohydrate, and vitamins, and may serve as a delightful substitute for meat. Lima beans have been found in prehistoric graves of Peru and Brazil, and had already been spread from middle South America to New England by 1492. United States production is today mostly from California.

TEPARY, FREEMAN, OR ESCOMITE BEAN, *P. acutifolius* var. *latifolius.* The nearly round, small, white to bluish-black seeds of this species are the tepary beans of western Mexico and the southwestern United States. They were much cultivated by Mexican Indians in prehistoric times, dating back 5,000 years in archeological findings at Tehuacan. They have been somewhat used as a drought-resistant species for hot, arid climates in modern times.

SCARLET RUNNER BEAN, *P. coccineus* (*P. multiflorus*). This species is a perennial with thickened, tuberous roots, indigenous to the cool, humid uplands of southern Mexico and Guatemala. It is normally cultivated as an annual. In the United States it is often grown for ornamental purposes, but is a common garden bean in Europe.

BLACK GRAM OR URD BEAN, *P. mungo.* This is a pubescent species extensively grown in India and the East. It is probably native to India.

MUNG, GOLDEN GRAM, OR GREEN GRAM, *P. aureus.* This species is related to the preceding one, and is widely cultivated in India. The Chinese use it by germinating the seeds and consuming the "bean sprouts" when they are about 4 days old. Decoctions of the seeds are also reportedly used medicinally in treatment of colic, and sometimes the sprouts are candied.

ADZUKI BEAN, *P. angularis.* This bushy annual is widely grown in China and Japan, ranking after the soybeans.

RICE BEAN, *P. calcaratus.* This species, native to southeastern Asia, is cultivated to a limited extent in China, Japan, India, and the Philippines.

MOTH OR MAT BEAN, *P. aconitifolius.* This is a prostrate species with lobed terminal leaflets, widely cultivated in India; presumably recently domesticated. Both the small seeds and green pods are eaten.

Broad, Windsor, Horse, or Scotch beans. The broadbean, *Vicia faba*, is probably native to northern Africa and the Near East, and closely resembles wild *V. pliniana* of Algeria. It has been under domestication for some 4,000 years, and was well known to the ancient Egyptians and to the Greeks. The species is an erect, vigorous annual up to 2 meters tall, with four-angled stems and pinnate leaves of two to six entire leaflets lacking tendrils. Large, thick pods are borne on axillary racemes, and contain seeds variable in size and shape. The diploid chromosome number is 12 or 14. Three subspecies are recognized, *paucijuga* of India, *eu-faba* (small-seeded varieties of Europe and Asia), and *major* (the typical culinary type). Other species of the genus are also cultivated, the untendrilled wild bitter vetch (*V. ervilia*) being grown for forage in southeastern Europe and western Asia. *V. narbonensis*, *V. villosa*, *V. sativa*, *V. angustifolia*, *V.*

atropurpurea, and *V. cracca*, all tendril-bearing, are also sometimes cultivated, almost entirely for pasturage, hay, and silage. The broadbean was the only common edible bean in the Old World before 1492, and it is still cultivated there to the extent of over 1 million tons annually, mostly in Italy. Yields often exceed 1 ton per acre. It is scarcely grown at all in the United States, but to a limited extent in Mexico and Brazil. Production of forage vetch seed in the United States, however, amounts to many thousands of tons annually. Broadbean is used as food for livestock as well as man. The bean is about 25 per cent protein, 49 per cent carbohydrate, with less than 2 per cent oil.

Peas. The pea, *Pisum sativum*, was probably domesticated in central Asia, with possibly secondary developments in the Near East and North Africa. In these areas a few wild species similar to the cultivated pea and able to hybridize with it are found (e.g., *P. arvense* and *P. elatius*), although *P. sativum* itself is not known in the wild. Remains of peas have been found in the Swiss lake dwellings of the Bronze Age, and the plant was cultivated by the Greeks and Romans. Until the late Middle Ages there is no hint of peas being eaten "green." But by the 1500's, in France especially, the more sugary green or garden peas had become the fashion—a fashion that has persisted into modern times in the Western world. In the United States dried peas, the "split peas" much used for soup, are consumed less than are green peas. The United States production of green peas (*P. sativum* var. *hortense*), amounts to about a half-million tons, some 95 per cent of which is processed (canned or frozen). Fresh peas come mainly from California, Colorado, and New York, processed peas from Wisconsin, Washington, Oregon, and Minnesota. World production of dry or field peas (*P. sativum* var. *arvense*) amounts to nearly 15 million tons annually, less than

300,000 tons from the United States and more than ten times this in the U.S.S.R., China, and India, the latter three being by far the leading producers. Much of the production of dry or field peas is used in stock feed. Dry peas are about 23 per cent protein, 59 per cent carbohydrate, and only 1 per cent oil.

Pea plants are glaucous, climbing, or trailing annuals adapted to a moderately cool growing season and a fair amount of rainfall. The leaves have two to six leaflets and terminate in tendrils. Stipules are large and leaflike, the flowers modest, papilionaceous, and with a diadelphous staminal tube. The diploid chromosome number is 14. Both "bush" and "vine" forms of the species are available. Peas are grown in spring or summer, particularly in Wisconsin and Ontario in North America, northern Europe, and northern China in Eurasia. Dry peas are easily cut and threshed, but invention of the "viner," a threshing machine capable of removing green peas from vines and pods, permitted development of the pea processing industry. Cultivars have been bred that produce mature pods all about the same time. Pea plants sometimes find use as green manure, or as hay in combination with grasses. They are especially renowned as the organism of Gregor Mendel's epochal experiments that ushered in the science of genetics.

Chick or gram peas or garbanzos. The western Asiatic genus *Cicer*, including a dozen or so species, has contributed to man's larder the chick pea, *C. arietinum*, used both for direct consumption of the seed and as forage. The chick pea is not known definitely from the wild, but escapes are not uncommon in eastern Mediterranean lands. The plant is a small, herbaceous, glandular-pubescent annual, with bipinnate leaves (which exude malic and oxalic acids), large, serrate stipules, and nine to fifteen leaflets. The pods, sometimes eaten with the seeds, are short, bearing only one or two wrinkled and pointed seeds of various colors. These sometimes serve as a coffee substitute. The diploid chromosome number is 16. The species is grown principally in India and is well adapted to cool, semiarid conditions. In India seeds are eaten cooked as dhal, or a flour may be milled for use in confections. The seeds are about 17 per cent protein, 5 per cent oil and 61 per cent carbohydrate. The chick pea is an important food for the peoples of northern Africa, Spain, and tropical America as well as India. World production reaches about 7 million tons annually, 90 per cent of it from India–Pakistan. United States production is insignificant. Yields seldom average as much as a half-ton per acre.

Cowpeas. "Peas" in the deep South of the United States invariably refers to cowpeas or black-eyed peas, *Vigna sinensis* (often merged into the wild cowpea species, *V. unguiculata;* in Africa the two "species" exhibit a range of intermediate forms). Diploid chromosome number is reported as 22 and 24. Actually this species is more closely related to beans than to peas. Most species of *Vigna* are erect or twining continuously growing herbs or subshrubs of tropical or subtropical origin. The leaves are pinnately trifoliolate. In one strain—the asparagus or yard-long bean (sometimes considered the separate species *V. sesquipedalis*)—the somewhat inflated pods exceed 3 decimeters in length. In the catjang strain or catjang pea (sometimes considered the species *V. catjang*), rarely cultivated, the pods are only about 1 decimeter long. Pods of the common cowpea are generally 2 to 3 decimeters in length.

The cowpea is probably native to central Africa, and is the most widely grown cultigen of African origin. It had been spread as far east as India in Sanskritic times and was well known to the ancient Greeks and Romans. The cowpea was introduced into

Cowpeas, source of edible beans and an important forage in many parts of the world. (Courtesy USDA.)

the West Indies and thence into the Carolinas before the early 1700's. In North America it is perhaps grown more as a green manure and forage plant than for the blackeye pea dishes omnipresent in the South. The species grows well on almost any soil, but cannot stand cold or frost. In Africa cowpea is generally broadcast into mixed plantings with other crops, the thinnings used as potherbs. In the United States crops are generally rotated with cotton or maize, or the cowpea is drilled with sorghum for forage. Leading producing states of the United States are Texas, Georgia, Oklahoma, and the Carolinas. United States production is about 30,000 tons annually. No statistics are available for world production, but this must reach many hundreds of thousands of tons, for the species is much grown in Africa, China, and India. In Africa the dried seed may be ground into meal, or the seeds (and immature pods) eaten fresh, or frozen or canned. The seed is approximately 23 per cent protein, 1 per cent oil, and 57 per cent carbohydrate. Yields with modern agricultural techniques often exceed 1 ton per acre, but under primitive circumstances may be only a quarter-ton.

Lentils. The lentil, *Lens esculenta* (*L. culinaris*), is one of the oldest of legumes, and also one of the most nutritious. Large-seeded (subspecies *macrospermae*) and small-seeded (subspecies *microspermae*) groups of cultivars are recognized. The lentil is believed to be indigenous to southwestern Asia; it was introduced into Greece and Egypt before Biblical times, and was taken as far east as China. Wild species putatively ancestral occur in Turkey. Diploid chromosome number is 14. The plant is a small pubescent annual, with pinnate leaves bearing four to fourteen leaflets, tendrilled terminally. Pods are short and usually contain only two lens-shaped, pealike seeds. The seeds are commonly made into a soup or porridge, and constitute an excellent meat substitute. In North America lentils are grown chiefly in the Palouse area of Washington–Idaho, where some 40,000 tons are generally produced annually by 3,000 growers. Yields are about a half-ton per acre. Plantings are generally in rotation with wheat, the vines permitted to dry completely before combining and final cleaning at local processing plants. Lentils constitute an important food in India–Pakistan, southwestern Asia, northern Africa, and southern Europe. World production is slightly more than 1 million tons annually. The seed is approximately 25 per cent protein, 1 per cent oil, and 56 per cent carbohydrate.

Soybean. The soybean, *Glycine max*, was considered as an oil source in Chapter 13, to which the reader is referred for a discussion of this important species. Since ancient times soybeans have been consumed directly as food in Asia, and "edible soybeans" are coming to be more widely grown in the United States as a garden plant. They constitute a very nutritious food, and may be eaten "green" (boiled or baked), roasted as an appetizer, or made into flour that is used with other flours to make breakfast food, bread, meat substitutes, and the like. In Japan the soybean has long been used fermented to make miso, and for flavorings such as soy sauce; the soybean thus furnishes a nice complement to the starchy rice staple. Miso is made by inoculating trays of rice with *Aspergillus oryzae*, and leaving them until they mold abundantly. Then a ground preparation of cooked soybeans and salt is mixed in, the mass allowed to ferment and age for several weeks before being ground into a paste. Per capita daily consumption of miso in Japan is about 30 grams (slightly over 1 ounce). As was noted in the discussion of soybean oil, world production of soybeans exceeds 30 million tons annually, almost two-thirds from the United States. Use of soybean flour, about half protein, has been suggested for improving the diet of undernourished populations such as in India, where the crop is still little known.

Peanuts. The peanut or groundnut, *Arachis hypogaea*, an ancient gift from South America for making North American baseball games more enjoyable to watch, was also discussed in Chapter 13. As a food it constitutes a national favorite in the United States as appetizer and delicacy, and for making peanut butter. Peanuts are also used as stock feed, particularly for hogs (with the hogs doing the harvesting), and the tops

The soybean, much consumed directly in the Orient. Garden varieties are available in the United States. (Courtesy USDA.)

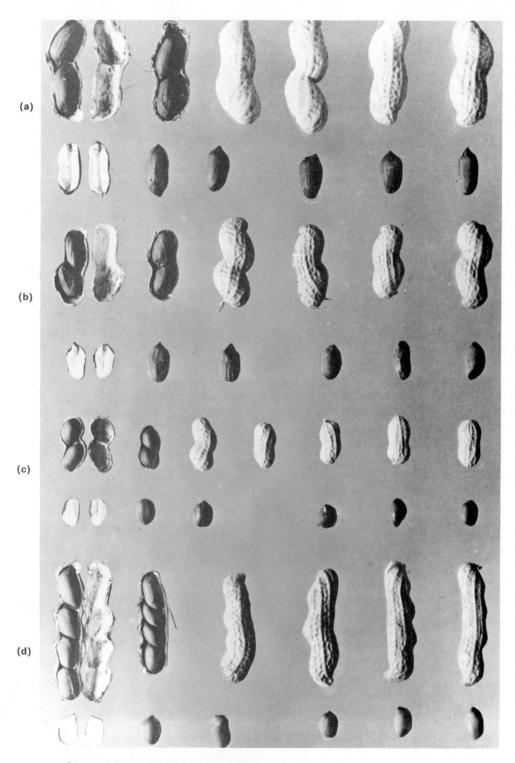

Commercial types of peanuts: a, Virginia Bunch, or Virginia Runner; b, African, or North Carolina; c, Spanish; d, Tennessee Red. (Courtesy USDA.)

458

Mechanically digging peanuts on a Georgia farm. (Courtesy USDA.)

serve as hay. Peanuts are about 30 per cent protein, 48 per cent oil, and 12 per cent carbohydrate.

The United States produces about 1 million tons of peanuts annually, chiefly from Georgia, North Carolina, Texas, Alabama, and Virginia. Over half of this goes into peanut butter. The familiar large-kernel peanuts come mainly from the Virginia-Carolina region; the smaller "Spanish" nuts from the deep South, where a goodly portion serves as stock feed. India, West Africa, and China are the leading groundnut-producing areas, responsible for more than half of world production that reaches about 15 million tons annually.

Other Legumes of Lesser Importance

Herbaceous species. PIGEON PEAS, CAJAN BEANS, OR CONGO BEANS, *Cajanus cajan* (*C. indicus*). The pigeon pea is a branched, pubescent, trifoliolate-leaved shrub up to 3 meters high, widely grown in tropical and subtropical countries, particularly India, equatorial Africa, the East Indies, and the West Indies. It is probably native to Africa, where it is often found naturalized or wild; certainly it was cultivated in Egypt at least four millenia ago. The diploid chromosome count is 22. The species is usually grown where rainfall is less than that best suited to cowpeas: it is deep-rooted and resistant to drought. Pigeon peas are eaten either as the green pods or as ripe seeds; in India they are made into dhal. In Africa they are generally interplanted with cereals or root crops, or used in rotation to restore run-down fields. The pigeon pea is also much used as a forage.

JACK BEAN, GOTANI, OR HORSE BEAN, *Canavalia ensiformis*. The erect jack bean is native to tropical America. Fragments dating to 3000 B.C. have been recovered from archeological sites in Mexico. The young pods are used as a vegetable; the herbage serves as a forage and green manure. Immature jack beans are eaten like lima beans. The species is cultivated in almost all tropical countries. Diploid chromosome count is 22.

SWORD BEAN, *Canavalia gladiata*. The twining sword bean is probably native to India and is widely cultivated throughout the East. It may have been derived from *C. virosa*. As with the jack bean, the young pods are eaten as a vegetable.

GUAR OR CLUSTER BEAN, *Cyamopsis tetragonaloba*. This plant was discussed in Chapter 9 as a gum plant. It is used as a forage and green manure, and in India the young pods serve as food for humans.

LABLAB, BONAVIST, OR HYACINTH BEAN, *Dolichos lablab* (*Lablab niger*). The lablab, one of about fifty species in the genus, is a woody, perennial, trifoliolate climber, bearing abundant pods. It is reported to be native to India, and has been in cultivation since ancient times. It is drought-resistant and adapted to well-drained soils. Both pods and seeds are eaten, although the plant serves more for cover crop, hay, forage, and as an ornamental than for human food. The species, although perennial, is treated as an annual in cultivation. Diploid chromosome count is reported as 22 and 24.

HORSEGRAM, *Dolichos uniflorus*. The horsegram is a poor man's pulse of southern India. The seeds are generally parched, and then boiled or fried. The seeds are smaller than those of the lablab, and as the name suggests serve as food for cattle and horses. Diploid chromosome count is 24.

GEOCARPA GROUNDNUT, *Kerstingiella geocarpa*. This species is another annual African groundnut, related to the Bambarra groundnut, but only locally cultivated. Like the peanut, it matures the fruit underground.

GRASS PEA, CHICKLING PEA, OR KASARI, *Lathyrus sativus*. Grass pea is native to southern Europe or western Asia, and is now especially grown in India where it often occurs as a weed in the barley fields. Although mostly grown for fodder, the seed is the least expensive pulse in India, and often eaten by the poorer classes, especially in times of famine. The species is somewhat grown in the northern Sudan, too. Diploid chromosome count is 14.

WINGED PEA, *Lotus tetragonolobus* (*Tetragonolobus purpureus*). Both the seeds and the pods of this southern European species are edible. It is a small, spreading, hirsute annual with trifoliolate leaves.

YAM BEAN, *Pachyrhizus erosus*. The young pods are sometimes eaten like green beans, although species of this genus may be more important for their tubers, which are eaten both raw and cooked. It is now widely planted in the oriental tropics. Diploid chromosome count is 22.

GOA BEAN, *Psophocarpus tetragonolobus*. Five species of this tuberous-rooted genus are grown principally in the African and Asian tropics. Young pods are used as a vegetable, and the ripe seeds are said to be eaten after parching in Java.

KUDZU, *Pueraria thunbergiana*. The kudzu bean has been much used as a cover crop and forage plant in the southern United States, but the seeds can be used for human food. Probably native to Japan, kudzu is a rank, perennial vine with large trifoliolate leaves and racemes of purple flowers.

VELVET BEANS, *Stizolobium*. Several species of this genus of rank, twining annuals, native to Asia, are cultivated. Although mostly used for pasture, hay, and green manure purposes, pulverized beans can be utilized both as a stock and human food.

BAMBARRA GROUNDNUT, *Voandzeia subterranea*. The species is found wild in the district of Bambarra on the upper Niger in Africa, where the plant is also cultivated for edible seeds eaten in the immature state, or the dried seeds roasted and ground into flour. It was introduced into other parts of Africa, Brazil, and the Orient early in the seventeenth century. It is an annual herb with creeping stems that root at the nodes, the leaves pinnately trifoliolate. Like the peanut the pod matures underground, and in Africa entire plants are pulled up for harvest. Yields are approximately a quarter-ton per acre of dried seeds. Diploid chromosome count is 22.

Tree species. The tree legumes have never attained great importance as sources of human food, although the species locally or potentially useful are legion. Particularly in the tropics the fallen seeds from leguminous trees serve as incidental sustenance to animal life, and occasionally to man. One important advantage rests with tree legumes as a food source, in comparison to annual crops— they can be utilized on slopes and with soils unsuited to cultivation. By planting tree species it is possible to make much unproductive land at least partially useful to agriculture. Only a few of the many useful tree legumes can be mentioned here.

CAROB, *Ceratonia siliqua.* Reference to this small, evergreen Mediterranean tree has been made in a previous chapter dealing with gums. The pods and seeds are also much used as cattle food.

TONKA BEAN, *Dipteryx odorata.* The seeds of tonka bean are rich in coumarin, have been used as a substitute for vanilla flavoring more than directly for food. The species is indigenous to the forests of South America.

HONEY LOCUST, *Gleditsia triacanthos.* This common ornamental and forest tree of the North American hardwood forest belt produces very large pods rich in sugar.

LOCUST BEANS, *Parkia filicoidea, P. biglobosa.* The large, edible pods of these African trees are consumed for the pulp; the seeds are roasted and consumed like coffee, or are boiled, pulverized, and fermented a few days to produce a highly nutritious protein- and fat-rich food similar to cheese. Most of the crop comes from the wild.

RAIN TREE, *Pithecolobium saman.* This huge tropical tree is much cultivated throughout the world tropics. The pod is useful as a stock feed.

MESQUITE, ALGARROBA, OR KEAWE, *Prosopis spp.* This genus, consisting of one or more uncertain species ranging from the arid southwestern United States to Chile, can produce heavy yields of legumes in arid climates unsuitable for the usual crops. It grows rapidly in barren areas to yield an important stock feed. The flowers are an excellent honey source.

GALLITO, *Sesbania grandiflora.* The young pods of this small tree are eaten lightly steamed or boiled in the Caribbean area.

Seeds of Nonleguminous Woody Perennials

This section will be devoted largely to the seeds or fruits known as nuts consumed for food by man and other animals. None are of first importance as world foods. Most frequently they are regarded as delicacies, to be consumed as sweetmeats too expensive for general consumption. Nuts can be handled and stored almost as well as can legumes and cereals, but they cannot be harvested as well by machine. The majority are today cultivated, but some, such as Brazil-nuts and piñon nuts, are still collected from wild trees. In cultivating nut trees, good use can be made of land of inferior agricultural possibilities. As a class of foods the nuts are somewhat different from either the cereals or legumes, being, as a rule, a concentrated source of oil rather than of carbohydrate or protein. Those nuts discussed with oils in Chapter 13 will not be treated further here.

Rather arbitrarily, any dry seed from woody plants will bear mention in this section, including in addition to the true botanical nuts (dry, indehiscent, one-seeded fruits) such "nuts" as the almond and the walnut, which are the seeds of drupes.

Almonds. Strangely, the almond, seed of *Prunus amygdalus* (*P. communis*), Rosaceae, comes from a peachlike fruit, the outer flesh of which, however, is not sweet and succulent as it is in the peach. In the peach, we consume the flesh and throw away the "pit"; in the almond, we keep the pit and discard the thin, fibrous flesh. The almond pit is usually naturally ejected from the splitting "hull" as the fruit ripens. This pit is cracked or the nut marketed whole. The almond has a rather high protein content (21 per cent).

The almond is a small, graceful tree, with slender, alternate leaves and very colorful spring blossoms. Two varieties of almond occur, the sweet (var. *dulcis*), which is the usual comestible type grown in the United States and southern Europe, and the bitter (var. *amara*), "poisonous" because of prussic acid, but from which, in southern Europe, oil of bitter almonds is extracted, the prussic acid being eliminated in processing. Almonds are propagated by slips and by budding onto peach, plum, or almond rootstock. They are well adapted only to subtropical, nearly frost-free climates similar to that of their native Mediterranean region. The trees bear before they are 7 years old, the fruit ripening in the autumn. The varieties are usually self-sterile. When the trees become established, yields of a half-ton of nuts per acre are not uncommon. The nuts are increasingly harvested by mechanical tree shakers. California is the leading almond state, producing almost 95 per cent of the approximately 80,000 tons annual production

An almond orchard. (Courtesy Chicago Natural History Museum.)

in the United States. Almond seeds were first brought there by the Franciscan Fathers in 1843, for planting at the missions. World production approximates a quarter-million tons annually, chiefly from Spain, Italy, Australia, and South Africa, in addition to the United States. The sweet almond has remained since Biblical times one of the world's most popular nuts.

Brazilnuts. Brazilnut, creamnut, niggertoe, or *castanha do Pará* (chestnut of Pará), *Bertholletia excelsa*, Lecythidaceae, constitutes an important native food and export item from the Amazon valley. The Brazilian sapucaia or paradise nut, *Lecythis sp.* (*L. usitata* and *L. ollaria* are said to be the species most commonly reaching market), of the same family, is similar in most respects to the Brazilnut. *B. excelsa* is one of the most magnificent trees of the rainforest, often towering 50 meters or higher with a straight, branchless bole. Twelve to twenty seeds or nuts are borne in large, urnlike, ligneous fruits, the "monkey pots," and individually are about the size and shape of a segment of orange. The fruits are called "monkey pot" because monkeys are said to stick a paw into the large ligneous fruit gnawed open by rodents, grabbing a handful of nuts and becoming "trapped" because they are unable to withdraw the full fist. The nuts are rich in oils (up to 70 per cent). The seeds, shed from the tree, are collected in the upper Amazon as a part-time occupation. Collectors often wear padding on their heads lest they be felled by a falling monkey pot "bomb." The trees are scattered over the better-drained sites throughout the Amazon basin. Little attempt is made to cultivate them in their native area, and elsewhere they have survived poorly. The large white blossoms appear almost a year in advance of fruiting, and furnish a preliminary indication of what yields will be forthcoming.

As early as 1633 Netherlands traders sent

Branch of *Lecythis usitata* with fruits. The seeds are the sapucaia nuts of commerce. (Courtesy Chicago Natural History Museum.)

Brazilnuts to Europe, where for centuries they were mechanically shelled and then re-exported the world over, even back to Brazil. Shortly after World War I a domestic shelling industry was established in Brazil, and export of both shelled and unshelled nuts has increased until about 40,000 tons are now exported annually, mostly to the United States and Europe. Lecythis, at least, is said to possess depilatory properties.

Cashew. The cashew or *cajú, Anacardium occidentale*, Anacardiaceae, is native to the sandy dry "taboleiros" of eastern Brazil. The abundant small trees have a gnarled appearance and rather unattractive, small flowers. Yet the pearlike "fruit" (swollen

pedicel or receptacle) offers welcome sustenance to man and beast in the isolated lands of low carrying capacity where the cashew is found wild. The "pear," in addition to being eaten raw, is made into preserves or fermented into cajú wine in Brazil. Temperate Zone peoples scarcely know this delicious fruit, but are familiar with the grotesque seed or nut borne at its summit as if somehow strangely stuck onto the top side of the inverted "pear." This curved seed, the familiar cashew nut, is as astringent when eaten raw as an unripe persimmon. Therefore all cashews entering commerce are roasted before sale.

Typically, ripe pears fallen from the tree are gathered from the ground, the nuts picked free for roasting. The caustic cashew shell oil, today valued industrially, is often extracted with solvents or live steam before the nuts are further processed. In India nuts are mostly cracked and hand shelled by women, who protect their hands by dipping them in lime and linseed or castor oil. The raw kernels are then roasted, and the sheathing testa removed. Annual production is in the neighborhood of 200,000 tons, and there are claims that the cashew is now second only to the almond in world importance among nuts. Immense stands still occur wild through much of tropical America, but selected strains have been introduced into India, Africa, Mexico, Florida, and the Mediterranean area, where the species is now cultivated for the seed. India is the chief center of cashew production, orchards having been extensively planted, some to select cultivars vegetatively propagated. Plantings in Tanzania are mostly on small holdings of only a few acres, but yield upward of 70,000 tons of nuts for export.

Hazelnut or filbert (including Barcelona and Turkish nuts). The hazelnut, *Corylus*, Betulaceae, is known in several species native to North America, Europe, and Asia. The two common wild species in the United States are *C. americana*, the common hazel, and *C. rostrata*, the beaked hazel. Both species are small, hairy-leaved, monoecious trees, often locally abundant, in which the nut occurs surrounded by a leaflike, green involucre. Of greater importance in the Old World and introduced into the United States are several European and Asiatic species. *C. avellana*, a larger European tree, is especially grown.

Hazelnuts are propagated by layering or

A hazelnut orchard in Oregon. (Courtesy USDA.)

from suckers, or in commercial ventures by budding or grafting selected strains onto common seedlings grown from autumn-planted nuts. Hazels bear when 4 to 7 years old, and produce about one-third ton of unshelled nuts per acre. In the United States most commercial growing takes place in Oregon, Washington, and California, producing about 12,000 tons of unshelled nuts annually. Several thousand tons are also imported, chiefly from Turkey and Italy. World production amounts to about 300,000 tons annually, chiefly in Turkey, Italy, Spain, China, and Japan.

Pecan. The pecan, *Carya illinoensis* (*C. pecan*), Juglandaceae, belongs to a genus well known for other types of hickory nuts

Pecan. (Courtesy St. Regis Paper Co.)

Cluster of Schley pecans. (Courtesy USDA.)

in addition to the pecan. None, however, has attained the importance or the perfection as a commercial nut of *C. illinoensis*. The species is native to the middle southern United States and to Mexico, but its importance as a cultivated species has caused it to be selected for strains that can be grown as far north as Indiana and Virginia. The "paper-shell" varieties are today America's premier nut (aside from peanuts), and usually command an excellent market. This, coupled with ready growing and fair yields, has made pecan orchards profitable and popular throughout the South. The invention of suitable shelling machines has further widened the market for this delicious nut, an energy food consisting of 70 per cent or more oil.

Wild pecan trees in the original forest were quite tall and robust, often growing to 100 feet. As orchard plants, however, smaller trees, less than 50 feet high at maturity, are preferred. The species is quite ornamental, with large compound leaves, monoecious flowers in catkins, and a drupaceous fruit, the nut of which forms within a leathery outer husk. Through selection and grafting, thin-shell varieties have been developed that are far superior, for man's purposes, to the native pecans from which

they were originally derived. Propagation is usually by grafting improved stock onto seedlings of wild nuts at about 1 year of age. These are set in the orchard at 2 to 4 years, and are generally clean-cultivated. Yields run about 200 pounds of nuts per acre. Harvest is by hand or sweeping-vacuuming, after the nuts are fully matured and have usually fallen from the husks. Light poles may be used to beat the nuts from the tree to the ground. Some large orchards use mechanical tree shakers. Nuts are "cured" for a few weeks, then are ready for marketing or shelling. If kept over into hot weather they should be refrigerated. Production in good years exceeds 100,000 tons annually, mostly from Texas, Georgia, and Oklahoma, about evenly divided between seedling types and improved cultivars. There is some planting in South Africa and Australia.

Piñon. Piñon nuts, the seeds of a few species of pines, particularly *Pinus edulis*, *P. monophylla*, and *P. cembroides*, Pinaceae, native to the semiarid southwestern United States and Mexico, were an important item in the diet of the Indians in that area.

The species have been cultivated only as

ornamentals, but annual excursions by Indians and other collectors among the distant wild stands provide a considerable supply of nuts for commerce. The nuts are picked from the ground, taken from squirrel caches, or extracted by hand from cones knocked from the tree. About 20 pounds per day constitute good collecting. There is danger that wild stands will be severely decimated, because of land use inimical to piñon existence and propagation. Heavy crops of nuts are borne cyclically, about every 5 years. In years of abundance 3,000 tons of piñon nuts may be gathered. The seed coat or shell is thin, and the kernel is 60 per cent fat.

Pistachio. The pistachio or green almond, *Pistacia vera,* Anacardiaceae, is native to Syria but is cultivated throughout southern Europe, southwestern Asia, and to some extent in the southern United States. It is a small, dioecious, dry-climate tree, with essentially evergreen leaves and small clustered flowers. The fruit is a drupe, of which, as in the almond, the seed constitutes the nut of commerce. This nut is often salted in brine before consumption. Pistachio nuts have a "resinous" flavor and are about 20 per cent protein, 58 per cent oil. Harvest is by knocking the nuts off the tree with poles, or by shaking the tree (sometimes mechanically) onto sheets laid beneath the branches. Hulls of harvested nuts are generally manually removed (although they lend themselves to mechanical processing, practiced to some extent in Turkey and the United States). Production is chiefly from cultivated orchards in the Near East (especially Turkey), and from wild stands in India and Afghanistan. The United States imports nearly 5,000 tons annually. The tree is propagated by budding.

Walnuts. Several types of walnuts, *Juglans,* Juglandaceae, are important as commercial nuts. The black walnut, *J. nigra,* is one of the premier forest trees of the United States in the hardwood belt. When cultivated it is usually planted for the wood rather than for

Pistachio nuts about ready for harvest. (Courtesy USDA.)

Black walnut. (Courtesy St. Regis Paper Co.)

the fruit, but sufficient walnuts are collected from wild and cultivated trees to provide the popular black walnut kernels for ice cream and confections. The nuts are the pits of drupelike fruits and are extremely hard to extract, even after the leathery outer husk has decayed away. Seldom can the kernels be removed entire, so that black walnuts are usually utilized fragmented, as a flavoring, rather than as whole kernel nuts. Production is chiefly by incidental collecting and home shelling in rural areas in the east-central United States, although a cracking machine has been invented, and there is increasing mechanization. The butternut, *J. cinerea*, with the same general range, is similarly collected and locally consumed, although not important commercially.

The English or Persian walnut, *J. regia*, is native to Persia but is extensively cultivated in southern Europe, China, and other parts of Asia. It has been introduced into California, with such success that that state has become the leading producing area of the world. The Persian walnut is a tall, handsome, pinnately leaved tree, ornamental as well as commercially useful. It is ordinarily grafted onto American walnut species stock. The nuts are often marketed with only the outer husks removed, the kernels being much more easily freed from the pericarp than are those of the black walnut. World production averages more than 100,000 tons annually, nearly half (about 50,000 tons) from California and Oregon, and the remainder chiefly from France and Italy.

Nuts of lesser importance. ACORNS. The nuts of many species of oak, particularly white oak, *Quercus alba*, Fagaceae, of the eastern United States, were a staple in the diet of many tribes of American Indians. The acorn was crushed, leached, then cooked in various ways. Today acorns are little consumed by man except perhaps by the poorer rural peoples of Spain and Italy, but over much of the world they constitute prized hog food.

Oak. (Courtesy St. Regis Paper Co.)

BEECH. Seeds from the spiny husk of the American beech, *Fagus grandifolia*, Fagaceae, and the European beech, *F. sylvatica*, serve today, like acorns, mostly as hog or wild animal forage except locally among poorer Europeans.

Beech. (Courtesy St. Regis Paper Co.)

Coconut. (Courtesy St. Regis Paper Co.)

CHESTNUT. The native North American chestnut, *Castanea dentata*, Fagaceae, has been thoroughly eliminated by chestnut blight throughout most of its natural range in the eastern United States, and is no longer a source of market nuts. The European chestnut, *C. sativa*, produces nuts much esteemed in Europe, a few thousand tons of which are imported by the United States annually, mostly from Italy.

The Chinese chestnut, *C. mollissima*, and the Japanese chestnut, *C. crenata*, are important in the Far East, although the nut is inferior to that of the European and American species. The Australian Moreton Bay chestnut, *Castanospermum australe*, which yields even larger nuts than the European chestnut, was introduced into California but has not been commonly cultivated.

COCONUT. *Cocos nucifera*, Palmae, was reviewed as an oil source in Chapter 13. The fruits or nuts are also an important food in southeastern Asia and the western Pacific islands. Dried coconut meat is much imported into the United States as an ingredient in and flavoring for confections.

GINKGO OR SAL-NUTS. The starchy seeds of the Gymnospermous *Ginkgo biloba*, Ginkgoaceae, are used for food in Japan and China, perhaps the world's oldest cultivated nut tree. The seeds are prepared by fermenting off the fleshy covering and then boiling or roasting. Raw they are mildly poisonous, often causing a dermatitis.

HICKORY. The pignut hickory (*Carya cordiformis*, Juglandaceae), the shagbark hickory (*C. ovata*), and other hickories frequent in the hardwood forests of the eastern United States furnish comestible nuts, although none are as esteemed as are the pecans.

OYSTERNUT. *Telfairia pedata*, Cucurbitaceae, yields large, flat seeds, the oysternuts, from a gourdlike fruit of a dioecious, clambering tropical African vine. They may be eaten either raw or roasted. Mature seeds are collected either from the wild or from haphazardly cultivated vines, and the kernel removed by hand from its enclosing integuments, usually with the aid of a penknife.

PIGNOLIA. The seed of the European pine, *Pinus pinea*, Pinaceae, furnishes the pignolia nuts familiar on European markets. These resemble long, narrow piñon nuts.

PILI. The thick-shelled, fatty pili nuts or

The pili nut, *Canarium album*. (Courtesy Paul H. Allen.)

Javanese almonds of the Far East are the seeds of *Canarium ovatum* and other species, Burseraceae. They are consumed raw or roasted, or are sometimes utilized as an oil source. The seed occurs as the pit of an edible plumlike fruit.

QUEENSLAND OR MACADAMIA NUT. The large, smooth, shiny nuts of the evergreen Australian *Macadamia ternifolia*, Proteaceae, are also called Australian Gympie, Bush, or Bopple nuts. The species was introduced from Australia into the Pacific islands less than a century ago, and especially in Hawaii has it gained favor so rapidly as to become one of the most important orchard crops of the islands. Superior cultivars are being selected both in Australia and Hawaii. The seed is derived from a follicle fruit with a fleshy husk. The rather thin shell of the seed must be cracked to retrieve the meaty kernel, now done by machine in Hawaii. After shelling the nuts are dried with warm air, and then boiled in oil for several minutes before salting and vacuum packing for export. Mature macadamia trees are quite tall, and the nuts are harvested where they fall rather continuously for several months. Annual yields of more than 100 pounds per tree have been obtained. Macadamia grows and yields best on well-drained land in climates with abundant rainfall.

SOUARI, SWARRI, BUTTER, PARADISE, OR GUIANA NUTS. Species of *Caryocar*, Caryocaraceae, are abundant tropical trees of northern South America. The nuts somewhat resemble Brazilnuts, but are much larger and richer in taste and have a thick shell

South American souari nuts, *Caryocar glabrum*. (Courtesy Chicago Natural History Museum.)

470

nearly impossible to crack. The woody capsules weigh as much as 25 pounds and resemble rusty cannon balls when fallen to the ground beneath the tree.

TERMINALIA. Species of the tropical genus *Terminalia*, Combretaceae, bear nutlike fruits. *T. catappa*, the tropical almond, has a walnutlike fruit, the flavor of which is said to resemble that of the hazelnut.

WINGNUT. Species of *Pterocarya*, Juglandaceae, may serve as nut trees in the Orient, where there occur about eight species. They are monoecious, deciduous trees, with small, winged fruits that ripen from September to October.

Forage Plants: Legumes and Grasses

It is fitting that this chapter close with a listing of forage plants, which man uses to transform plant carbohydrate and protein into meat and dairy products. About 8 pounds of plant protein are needed to yield 1 pound of meat. Such plants are exceedingly important in the world economy, even though they represent an indirect utilization of vegetation by man. Annual worth of forage in the United States is estimated to exceed $10 billion, and worldwide value is incalculable. Literally, any plant consumed by livestock is a forage, although generally we think of forages as species planted to pasture and meadow, or those harvested for silage (the nutritional value preserved by fermentation). Much of the world's surface is unsuited to growing grains or similar crops, but most of it maintains a green cover of some sort that can be browsed or grazed. Worldwide there is less than 4 billion acres of arable land, but over 6 billion acres are in permanent pasture and meadow. When civilization was young animals were generally turned out to natural pasture, but in an increasingly crowded world amplified use

Natural range. This Idaho grazing land is being sprayed by air to control brush. (Courtesy USDA.)

Most forage today is of cultivated species, a trend destined to increase. These beef cattle are in Texas, once mostly open range. (Courtesy USDA.)

must be made of land by planting it to high-yielding cultivated forages. Well-managed forage is said to yield more TDN (total dry nutrient) per acre than corn or other grains!

All the important cultivated forage plants belong to the grass or legume families, although livestock will of course browse or graze upon plants of many types and of many families. As with cereal grain and legume seeds, energy-rich grass nicely balances legume vegetation relatively concentrated in protein. Grass forage runs in the neighborhood of 15 per cent protein, while legume foliage is generally 20 per cent or better, sometimes as high as 35 per cent (Fig. 17-1). Livestock can subsist indefinitely upon grass and legume forage, although in practice meat animals are "finished" on fattening grain and dairy cows provided additional food supplements. Among the legumes and grasses many wild as well as cultivated species are of importance for hay and pasture.

Hundreds are at least locally of great importance. It is impossible to list all in this section, even were information concerning them available. The grass family contains some 5,000 species, most of them potentially usable as forage, and there are probably 10,000 or more herbaceous legume species.

Which forages were first utilized by man can only be surmised, and doubtless the first domesticated animals (probably goats and sheep) were left more or less to select for themselves. But with man's occupation of colder or more northerly areas, the growing of forage in the abundance of summer to provide winter animal feed became man's responsibility. Hay cut in summer was dried to preserve it, still the general practice. First reliance was probably upon wild hay and hand pulling, but as populations became abundant the planting of hay and pasture crops, and techniques for their curing and storage, entered into early agriculture. The

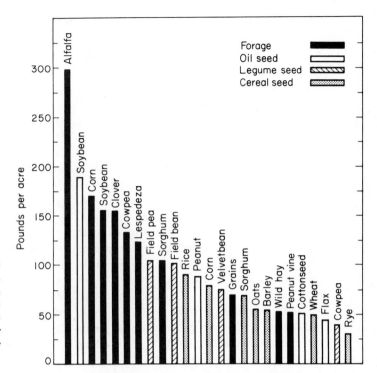

Figure 17-1 Yield of essential amino acids (lysine, methionine, tryptophan, phenylalanine, threonine, valine, leucine, isoleucine) per acre for various crops harvested as forage or as seed. Note the comparative richness of alfalfa. (After Akeson and Stahman, *Economic Botany* **20**, 245 (1966).)

Roman Columella emphasized proper curing of hay about A.D. 50. In those early days, when warfare largely involved the use of horse or camel, hay even became a strategic material. Alfalfa, for example, was carried into Greece by invading Persians five centuries before Christ, possibly as the first cultivated forage plant. Later the Romans developed certain grazing and hay curing practices, and during the Dark Ages the Saxons, at least, recognized clover as a valuable forage plant.

In North America the Indians were little concerned with forages, lacking as they did herbivorous domesticated animals. Colonists early introduced horses, swine, cattle, sheep, goats, and chickens, and at once a winter feeding problem arose. The natural low-lying and swampy meadows of the Atlantic seaboard first provided cut forage, included in which were undoubtedly such grasses as wild rye (*Elymus*), broomsedge (*Andropogon*), beachgrass (*Ammophila*), and switchgrass (*Panicum*), to say nothing of sedges and *Juncus*. When colonization reached first the tallgrass and then the drier shortgrass prairies of the western United States a rich region of native hay grasses, including wheatgrass (*Agropyron*), bluestem (*Andropogon*), grama (*Bouteloua*), needlegrass (*Stipa*), etc. fell before the inexorable tide of agricultural advance. It was upon these prairies that the American "wild West" was nurtured and still today it is from these plains that much American meat comes to market.

Few native forage plants have proven so successful under cultivation as have several introduced species. Early introductions from Europe include several clovers, bluegrass, crabgrass, timothy, orchardgrass, alfalfa, millet, sorghum, Sudangrass, Johnsongrass, etc. Improved strains of these, other clovers, lespedeza, soybean, cowpea, kudzu, peanut, bromegrass, etc. have contributed much to North American agricultural productivity, so that among cultivated crops today hay acreage ranks along with maize and wheat. In total United States production (about 125 million tons annually) alfalfa ranks first (nearly 75 million tons), followed by clover–timothy grown in mixture (approximately 23 million tons). Greatest hay acreage is in the Midwest and the Great Plains states. The alfalfa belt centers chiefly in the northern and western parts of the country, the timothy–clover belt in the northeastern part. The South relies chiefly on annual legumes (e.g., lespedeza, cowpeas, vetches, etc.) and such grasses as Johnson, Sudan, and Bermuda, while the more arid Southwest relies to a great extent upon various sorghums. Wild hay from prairie and marshland still amounts to about 10 million tons annually.

Forage legumes. Among the legumes utilized as forage or green manure plants are many of those discussed for their edible seeds in an earlier section of this chapter. Thus soybeans, cowpeas, velvet beans, cer-

tain vetches (*Vicia*), peas, kudzu, peanuts, lablab beans, and others already treated will not be listed a second time here.

ANTHYLLIS. The drought-resistant kidney vetch, *A. vulneraria,* one of about twenty species of perennial Old World herbs or subshrubs, is indigenous on dry pastures and banks throughout Europe and western Asia. It is used in pasture mixtures, especially for sheep.

CORONILLA. Crownvetch, *C. varia,* is a slow-establishing but self-sufficient Eurasian subshrub with pinnate leaves and attractive blossoms, now widely planted for roadside slopes and spoilbanks in Pennsylvania and neighboring states. It yields hay as abundantly as familiar forage species, and to some extent can be used as cattle browse.

DESMODIUM. This genus furnishes a forage and hay crop in the West Indies from the perennial *D. tortuosum.* The pod is also a

Modern harvest of legume forage. A windrower here prepares alfalfa for pick-up after field curing. (Courtesy Hesston Co.)

good protein supplement in chicken feeds.

INDIGOFERA. *Indigofera* species, from Africa and Asia, are useful as forage plants in the tropics.

LATHYRUS. The sweet pea genus contributes a few forage species of minor importance, particularly the annual *L. sativus*, the grass pea, native to southern Europe and western Asia. *L. tangitanus*, the Tangier pea, finds use as a forage crop in Algeria.

LESPEDEZA. Lespedezas are perennial or annual herbs or undershrubs very valuable for hay and pasturage and for game cover. Asiatic species have found greatest use, especially the perennial sericea, *L. cuneata* (*L. sericea*), and the annuals *L. striata*, common lespedeza, and *L. stipulacea*, Korean lespedeza. The latter two species are slender, prostrate, purple-flowered types that generally are able to maintain themselves through natural seeding. Their ability to grow on poor soils makes them especially valuable in soil conservation. They are hot-weather, short-day plants, grown in the United States from the Ohio valley southward. The winter-hardy *L. cuneata* has been most used as an erosion control and wildlife cover in the southeastern United States, and is often seeded to roadside berms. Nearly 30,000 tons of lespedeza seed are produced in the United States annually, particularly in Missouri, Tennessee, and the Carolinas, in addition to millions of tons of hay.

LOTUS. The perennial *L. corniculatus*, bird's-foot trefoil, is common in meadows and pastures of Europe and is increasingly grown in the cooler sections of the United States. Although slow to establish, it is quite productive and said to hold its own well in competition with grasses. Diploid chromosome number is 12, and there are tetraploid strains as well. Bird's-foot trefoil is native to temperate Eurasia. There are three botanical varieties, including an erect European type, a dwarf form of the British Isles, and a prostrate, winter-hardy variety discovered in New York.

LUPINUS. Various annual lupines are used for soil improvement, particularly in western and central Europe. *L. termis* was grown in ancient Egypt and was given recognition by the Romans. In the United States lupines are often grown as winter annuals in the southeastern states. *L. angustifolius*, *L. albus*, and *L. luteus* are mostly used. Some of the native lupines of the western United States are toxic to livestock.

MEDICAGO. The genus *Medicago* has supplied the foremost forage plants of the world. It is especially noted for perennial alfalfa or lucerne, *M. sativa*. A few annual species, the bur clovers, *M. arabica* (*M. maculata*), *M. hispida*, and to a limited extent other species have attained importance as winter annuals in southern regions, germinating in autumn and growing vigorously during winter to mature in summer. They complement hot-weather pasture grasses such as Bermuda, furnishing winter forage. They are widely used from the Carolinas to Texas, on the Pacific coast, in southern South America, and Australia.

M. sativa, alfalfa, is the most valuable hay plant of the United States, yielding about 75 million tons annually, worth some $1.5 billion. This production is probably almost matched in Argentina, Europe, and Asia. The quality of alfalfa, too, is supreme (see Fig. 17-1), and it has a high carrying capacity as a pasture crop. In most areas it is the highest-yielding legume forage that can be grown. In crop rotation it benefits the soil, and its prolific growth enables it to fight weeds well. Two to six cuttings are made per year. Yield runs as high as 11 tons per acre. Seedings persist as much as a dozen years or more under favorable growing circumstances, although usually after several years grasses gain the upper hand and alfalfa must be reseeded in order to be dominant. The seed is inoculated with the proper nitrifying bacteria for good growth. In northerly areas alfalfa is sometimes winterkilled, and the alfalfa weevil has become a serious pest

Alfalfa or lucerne, the "queen of forages." (Courtesy USDA.)

turies later, when the Spanish colonists brought horses to the New World, they also took along alfalfa with which to feed their horses during the exploration of Central and South America.

Alfalfa has a deep taproot, a characteristic facilitating water absorption during dry times of the year. Many erect, slender stems arise from the crown, and bear trifoliolate, stipulate leaves. Flowers are axillary, and disperse pollen explosively by a "trip mechanism" when the wing petals are separated. Leaf-cutter bees are usually maintained near alfalfa seed-production fields to assure pollination. Alfalfa pods are tightly coiled upon themselves in maturity. Many cultivars have been developed to suit differing climatic zones, and hybrid alfalfa from the crossing of male–sterile lines is becoming commercially available, although costly. Yield increases of 15 per cent or higher are reportedly possible with hybrid plantings. *M. falcata*, Siberian alfalfa, is similar to *M. sativa*, but has yellow flowers. Its germplasm has been used in crossings with *M. sativa*, and *M. media*, sand lucerne, has been considered a natural hybrid between the two species.

Alfalfas are grown in almost all states of the United States, although not widely in the deep South. Total acreage exceeds 33 million acres. Leading producing states are California, Wisconsin, Michigan, and Minnesota. Seed production in the United States amounts to about 70,000 tons annually. Alfalfa seed is best produced in semiarid regions under irrigation, where moisture can be controlled to prevent excessive vegetative growth. Alfalfa is not well adapted to acid soils, and performs best on neutral or alkaline ones. It is a crop of especial importance in India, from central Asia through the Mediterranean area, in temperate parts of South America, as well as in the United States. Although alfalfa sun-dries quickly in the field, in the United States it is increasingly processed into pellets or con-

requiring insecticidal sprayings and other treatments where nonresistant strains are planted. The first plantings of alfalfa in the United States were in Georgia, in 1736. During the gold rush of 1851 the species was taken to California and dispersed rapidly from there throughout the nation. An introduction from Germany by Wendelin Grimm opened the door to hardy cultivars for the more northerly states. Natural selection for survival in Minnesota enabled Grimm to develop hardy strains that became the famous Grimm variety.

Alfalfa is probably native to Persia, and was one of the first domesticated forage plants. Roman records show that alfalfa was introduced into Greece from Persia by the invading Medes about five centuries B.C. (hence the name "Medicago," coined by Linnaeus). It was recognized by the Romans as a superior food for chariot horses. Cen-

White Dutch clover, a long time forage favorite. (Courtesy USDA.)

centrates, in drying facilities inexpensively set up in the growing areas. A wet-fractionation process grinds alfalfa, much like sugar cane, separating the juice from the solids. The latter are dehydrated to become typical feed pellets, but the juice (rich in carotene and xanthophyll) is steam-coagulated to yield a curd containing 40 per cent protein. This is an excellent chick ration, but is suited to human consumption as well if blended into tastier foods.

MELILOTUS. Two species of the genus, *M. alba*, sweet clover, and *M. officinalis*, yellow sweet clover, are fairly important as cultivated and escaped forage plants. Both are annual or biennial erect herbs, with trifoliolate leaves and slender axillary racemes of small, fragrant flowers. They are excellent bee plants in addition to being fair hay plants and fine soil builders. Both are believed native to western Asia, and until recent decades were a weed in grass and alfalfa seedings rather than cultivated plants. The species are most grown in the United States in the Plains states, with Kansas the chief seed-producing state. Total United States production is about 12,000 tons of seed and perhaps 1 million acres planted. *M. indica*, sourclover, is an upright winter annual in the southwestern United States and in Mississippi and Alabama.

ONOBRYCHUS. Sanfoin or espersette, *O. viciaefolia*, is a perennial indigenous to southern Europe and western Asia. It is slower to establish itself than is alfalfa, but is considered of special value on dry, porous soils.

ORNITHOPUS. Serradella, *O. sativus*, is a vetchlike annual native to the Iberian peninsula. It is cultivated as forage and green manure in Europe.

SESBANIA. *S. macrocarpa* is an upright annual of the southern United States and Mexico, somewhat used for soil improvement.

STYLOSANTHES HUMILIS. Townsville lucerne,

S. humilis, is much used for fodder in northern Australia, especially during the dry season. The species is native to tropical America.

TRIFOLIUM. Alfalfa's only rival among leguminous forages is the clover genus, *Trifolium*, which consists of innumerable trifoliolate herbaceous species. Most species are perennial, but the annual crimson clover, *T. incarnatum*, native to southeastern Europe, has long been cultivated there for forage and soil improvement. It was introduced into the United States before 1818, and is grown in the southeastern states as a winter annual. Another annual of comparatively minor importance is *T. alexandrinum*, beerseem or Egyptian clover, native to the Near East.

The biennial red clover, *T. pratense*, probably originated in southwestern Asia and was widely distributed throughout Europe by the Middle Ages. It is planted to over 10 million acres in the United States. It is a humid-region crop, and is mostly grown north and east of Missouri, and in northwestern Europe. Seeding is usually in autumn, the first year's growth used for hay, the second year's for seed. Seed production in the United States runs to about 40,000 tons annually, chiefly from Illinois. Tetraploid red clovers are becoming available.

Alsike clover, *T. hybridum*, native to northern Europe, is utilized for the same purposes as red clover and prefers the same cool, moist climates. There are both diploid and tetraploid strains.

White or Dutch clover, *T. repens*, or ladino clover, *T. repens giganteum*, is another species native to northwestern Europe that has been introduced into and naturalized in all moist, temperate areas of the world. It was brought to North America by the early colonists from England, as a component of "English grass" (the other component was Kentucky bluegrass), and so rapidly did it spread that the Indians called it "white man's foot." White clover is a stoloniferous, abundantly tuberculate perennial found extensively in about 16 million acres of North American pasture. It also volunteers widely in lawns, and is known in many cultivars. Ladino is best adapted to irrigated mountain valleys of the West.

Strawberry clover, *T. fragiferum*, is native to the Mediterranean area and Asia Minor, and comparatively recently has been introduced into the United States. It is adapted to the same conditions and uses as white clover. Subterranean clover, *T. subterraneum*, a perennial of southern Europe, is planted in the coastal counties of northern California. *T. occidentale*, available both as diploids and tetraploids, is used for crosses with both *T. repens* and *T. nigrescens*.

VICIA. Mention was made of *V. faba*, the broadbean, and of several of the forage vetches, in an earlier section of this chapter. The spring vetch, *V. sativa*, and the hairy vetch, *V. villosa*, are perhaps the most widely grown vetches in the United States. The former is a spring or winter annual grown particularly on the Pacific coast. The latter is used mostly as a winter cover crop in the southern United States. *V. villosa* has a diploid chromosome count of 12. Other familiar vetches are purple vetch, *V. atropurpurea*, Hungarian vetch, *V. pannonica*, and *V. angustifolia*.

Forage grasses. Perhaps even more so with grasses than with legumes, the species important for seed are also used as forage plants. Thus oats, barley, rye, wheat, hybrid sorghum, maize, pearl millet, and so on are often used for pasturage or are made into hay and silage. But in addition there are innumerable cultivated and wild grasses, adapted to various local situations, that are of prime importance in maintaining man's herds.

The output from grassland can be enormous, but at present is far from maximum potentiality. One of the problems is re-establishment of grass on the overgrazed prairies and marginally cultivated lands. Soil has eroded as valuable cover was lost in "man-made deserts" of the world, such as the Rajputana area of India and parts of the western United States high plains, to the extent that it is now difficult for good grazing grasses to grow. Also, much once-rich prairie is now clothed in unpalatable brush, the result of careless management. Intensive efforts are being made to reseed ranges, but overgrazing must be stopped, and proper grazing techniques followed, if the world is to approach the potential carrying capacity of its grasslands. Livestock can be expected to depend increasingly upon cultivated pasture rather than natural range.

In addition to the secondary cereals such as the sorghums and millets previously reviewed, the following grass species are important for forage.

WHEATGRASS, *Agropyron.* Several species of this genus are perennial forages of cool, dry regions such as the northern Prairie states of the United States. *A. desertorum* (*A. cristatum*), crested wheatgrass, native to the northern U.S.S.R., has been much employed in reseeding the North American range. Slender wheatgrass, *A. trachycaulon*, is often also used. *A. pauciflorum*, a species native to North America, is not so drought-resistant as crested wheatgrass, but is planted from the Pacific coast to Wisconsin. *A. inerme*, *A. smithii*, and *A. spicatum* are other useful species.

BENTGRASS, *Agrostis. A. alba*, redtop, is the most used species of this genus in North America. *A. stolonifera* (*A. tenuis*) is especially important in the cool, humid parts of Europe. Redtop is grown primarily in the northeastern quarter of the United States, often in combination with clover, and is regarded as the best cultivated wetland grass used in North America.

BLUESTEM, *Andropogon.* Species of blue-stem are important perennial grasses of cool, dry regions of the United States, and are a principle constituent of wild or prairie hay.

TALL MEADOW OATGRASS, *Arrhenatherum elatius.* The species is a hardy perennial bunchgrass a few feet tall, green all winter in the southern part of its range, and giving a heavy yield of hay. It is grown in the mid-eastern and southern United States and on the Pacific coast.

CARPETGRASS, *Axonopus.* Species of this genus are perennial creepgrasses of the United States important for grazing in the hot, humid parts of the deep South, from coastal Texas to the Carolinas.

GRAMAGRASSES, *Bouteloua.* Blue grama, *B. gracilis*, and side-oats grama, *B. curtipendula*, are important prairie species of the northern and middle plains of the United States. Other gramas are equally important in the Southwest.

BROMEGRASS, *Bromus.* Cultivars of brome-grass have become very important as cultivated forage plants for drier, cool areas. Smooth brome, *B. inermis*, a drought-resistant

Bromegrass with alfalfa in Nebraska. The grass is beginning to dominate and will contribute most of the forage. (Courtesy USDA.)

perennial indigenous to Europe and Asia, has been widely planted in the central and northern plains of the United States. It is also extensively utilized in central Europe. Bromegrass is one of the most palatable pasture grasses, and yields as much as 4 tons of hay per acre.

BUFFALOGRASS, *Buchloe dactyloides*. This is an important native grass of the cool, dry prairies of the North American Great Plains, but is low-yielding.

RHODESGRASS, *Chloris gayana*. Rhodesgrass is native to southwestern Africa, and was introduced into the United States at the turn of the century. It is planted also in Australia, South America, northern Africa, the Philippines, and India. It is a sod-forming perennial up to 3 feet tall with spreading stolons that root at the nodes.

BERMUDAGRASS, *Cynodon dactylon*. This hot-weather perennial from India and Africa is the most important pasture species in the warm, humid, southeastern United States. It can stand close grazing, and grows and spreads rapidly at high temperatures, although becoming dormant in cool weather. *Coastal* is a sterile hybrid from a cross of a South African Bermuda with a local Bermuda at the Georgia Coastal Plain Experiment Station. It must be vegetatively planted, but in spite of this is such a high-yielding cultivar that it is now widely used in the southern United States.

ORCHARDGRASS, *Dactylis glomerata*. Orchardgrass or cocksfoot (as it is known in Europe) is another Eurasian species early introduced into America and now naturalized throughout the northeast. It is a long-lived bunchgrass used for permanent pasture, hay, and ensilage. The species is not quite so winter-hardy as timothy.

WILD RYES, *Elymus*. *E. canadensis*, *E. triticoides*, and *E. glaucus* are important native pasture grasses of the Pacific Northwest of the United States.

LOVEGRASS, *Eragrostis*. *E. curvula*, weep-ing lovegrass, is a long-lasting bunchgrass from southern Africa introduced into the United States about 1927. It is one of the best forage species for the southern Great Plains, and is also grown in Brazil, North Africa, and Australia, where it is used for grazing, hay, and soil protection. The genus contains about 250 species, including the sand lovegrass native to the North American mid-continent, and Teff cultivated for grain as well as forage in northern Africa.

TALL FESCUE, *Festuca arundinacea* and MEADOW FESCUE, *F. elatior*. These are hardy perennials from Eurasia adapted to moderately cool, humid regions. In the United States tall fescue is an important forage plant of the border states and the Midwest. It is more tolerant of high summer temperatures than are bluegrass and timothy. Kentucky 31 tall fescue is a selection discovered because of its mid-winter greenness on Kentucky hillsides. Alta fescue is a similar selection from Oregon. Meadow fescue is much planted in New England and California. Tall fescue has a diploid chromosome count of 42, meadow fescue of 14.

RYEGRASS, *Lolium*. Annual or Italian ryegrass, *L. multiflorum*, is much seeded for winter pasture in the southern United States. Perennial ryegrass, *L. perenne*, is widely used in England and northern Europe, and in New Zealand, especially for the grazing of sheep.

PANICUM. Several panicums are important as native pasture plants, adapted to the semiarid southwestern United States. Guinea-grass, *P. maximum*, and paragrass, *P. pur-purascens*, from Africa are commonly cultivated in tropical lowlands for forage.

PASPALUMS. Bahiagrass, *Paspalum notatum*, native to tropical America, is much planted in the deep South of the United States as a hay, pasture, and roadside grass. Dallisgrass, *P. dilatatum*, and vaseygrass, *P. urvillei*, native to southern South America, are other perennial species utilized in subtropical

North America, escapes becoming a pest in the lawn.

BUFFELGRASS, *Pennisetum ciliare*. This species, native to Africa, India, and Indonesia, has been naturalized in northern Australia, and introduced for pasture along the Gulf Coast of the United States. It is adapted to subtropical conditions with long, dry seasons. Pearl millet, in the same genus, previously discussed as a grain plant, is becoming increasingly important as a forage in the southeastern United States.

REED CANARYGRASS, *Phalaris arundinacea*. The species is indigenous to Eurasia, and was brought into Oregon about 1885. It is a clumpy perennial growing to 7 feet tall, spreading by rhizomes, well adapted to swampy habitat. It is especially used in the western United States for pastures, hay, ensilage, and erosion control. Hardinggrass, *P. tuberosa stenoptera*, believed native to North Africa, has been planted in Australia, and introduced from South Africa into California and southern South America.

TIMOTHY, *Phleum pratense*. The perennial timothy or herdgrass has for a long time been the most widely cultivated hay grass in the United States. Timothy is native to Europe and western Asia, and is well adapted to cool, moist climates. It has been grown successfully so far north as the Arctic Circle. It is high-yielding, and one of the most palatable hay species, also used for pasturing in combination with clover (although it does not tolerate close, continuous grazing well). Over 15,000 tons of timothy seed are produced in the United States annually, chiefly from Missouri, Iowa, Minnesota, and Ohio.

During the era of the horse, about 35 million acres were sown to timothy annually.

BLUEGRASS, *Poa*. A number of poa species are important as lawn, pasture, and cover species. Most useful is *P. pratensis*, Kentucky bluegrass, native to Europe but now naturalized throughout the cool, moist portions of the globe. It is one of the most nutritious pasture grasses, but is not very productive during hot summer weather. Canada bluegrass, *P. compressa*, is a rather thin grass, but useful for colonizing poor soil where little else will grow. Annual bluegrass, *P. annua*, is omnipresent wherever cool, moist conditions prevail, too small and erratic to be of much consequence as a pasture species, and quite a pest in cultivated turf.

SLOUGHGRASS OR PRAIRIE CORDGRASS, *Spartina pectinata*. This is an American species found in freshwater marshes and native or prairie hays.

SEDGES. Closely related to grasses, various sedges, of genera such as *Carex*, provide extensive natural hay and pasture lands in certain parts of the world. *C. lyngbyei* (*C. cryptocarpa*), for example, is said to be a superb forage plant in Iceland, cultivated in the bottomlands, and an excellent pasturage in the floodplains of the Alaskan rivers. Sedge meadows also occur in the mountains of North America, and the littoral of the coastal regions. *C. nigra* constitutes a good forage in the high Caucasus, and various species, including *C. halleri* and *C. rariflora*, extend from Finland through Siberia. *Eriophorum angustifolium* is abundant in the fens and moorlands of North Atlantic lands.

SUGGESTED SUPPLEMENTARY REFERENCES

American Seed Trade Association, *Proceedings of Annual Farm Seed Conferences*, Washington, D. C., 1954–present.

Archer, S. G., and C. E. Bunch, *The American Grass Book, A Manual of Pasture and Range Practices*, Univ. of Oklahoma Press, Norman, 1958.

Barnard, C. (ed.), *Grasses and Grasslands*, Macmillan, N. Y., 1964.

Bolton, J. L., *Alfalfa: Botany, Cultivation and Utilization*, Leonard Hill, London, 1962.

Harlan, J. R. *Theory and Dynamics of Grassland Agriculture*, Van Nostrand, Princeton, N. J., 1956.

Harrison, C. M. (ed.), *Forage Economics—Quality*, Amer. Soc. of Agronomy Spec. Pub. 13, Madison, Wisc., 1968.

Howes, F. N., *Nuts: Their Production and Everyday Uses*, Faber & Faber, London, 1948.

Hughes, H. D., M. E. Heath, and D. S. Metcalfe (eds.), *Forages*, 2nd ed., Iowa State Univ. Press, Ames, 1962.

Jaynes, R. A. (ed.), *Handbook of North American Nut Trees*, Northern Nut Growers Assoc., Knoxville, Tenn., 1969.

Myers, R. M., *Forage Plants*, Western Illinois Univ. Biological Science Series, No. 5, Macomb, Ill., 1967.

Stanton, W. R., et al., *Grain Legumes in Africa*, FAO, Rome, 1966.

USDA, *Bibliography of Tree Nut Production and Marketing Research, 1960–65*, Misc. Pub. 1064, Washington, D. C., 1967.

——————, Federal Extension Service, *Trends in Forage Crop Varieties*, Washington, D. C., 1967ff. annually.

USDA Yearbook, *Grass*, Washington, D. C., 1948.

Wheeler, W. A., and D. D. Hill, *Grassland Seeds*, Van Nostrand, Princeton, N. J., 1957.

Whyte, R. O., G. Nilsson-Leissner, and H. C. Trumble, *Legumes in Agriculture*, FAO, Rome, 1953.

——————, T. R. G. Moir, and J. P. Cooper, *Grasses in Agriculture*, FAO, Unipub Inc., P. O. Box 433, N. Y., second printing 1962.

Woodroof, J. G., *Tree Nuts—Production, Processing, Products*, AVI Pub. Co., Westport, Conn., 1967.

CHAPTER 18

"Vegetables": From Root, Stem, and Leaf

CARBOHYDRATES, FATS, AND PROTEINS SUFFI- cient to satisfy the human diet can be largely obtained, directly or indirectly, from the legumes, grasses, and pseudo-cereals dis- cussed in the preceding two chapters. Yet if civilizations were to rely solely upon such seed plants, meals would indeed lack variety, and possibly nutritional balance in vitamins and minerals as well. Enormous numbers of peoples, especially poorer residents of the tropics, exist almost entirely on "rice and beans"—and consider themselves fortunate if they are able to secure adequate quanti- ties of these. But in areas with higher standards-of-living a day seldom passes without one or more "vegetables" being served at the table. These may come from roots (sweet potato), underground stems or rhizomes (Irish potato), stems (asparagus), leaves (cabbage), or inflorescences (cauli- flower). Even some botanical fruits (tomato, squash, and the like), to be considered in a subsequent chapter, are looked upon as "vegetables." The present chapter is devoted to the great variety of vegetable crops exclusive of fruits. First "root crops" are taken up—plants whose underground portions, whether root, rhizome, or corm, are consumed as food. There follows dis- cussion of "green vegetables"—plants in which the aboveground portions are con- sumed as food, usually fresh, canned, or frozen.

Root Crops

A "root crop" in the popular sense is any plant part dug from the soil, whether true root, modified stem, or even leaves condensed into a bulb. Root crops come from many plant families, and the category is quite an artificial one. The underground parts utilized are generally fleshy, the tissues watery, and the concentration of food (energy) less per unit of weight than is the case with such energy-rich foods as cereal and legume seeds. It is not surprising that fleshy root crops generally preserve and handle less satisfac- torily than do dry grains. They are less well adapted to mechanization in harvest and handling, and often require carefully re- gulated storage conditions in order to preserve well for a period of time. Because of their bulk shipping is more costly (in terms of food en- ergy transported), and root crops tend more than grains and pulses to be consumed locally. Most root crops are richer in carbohydrates and poorer in proteins and oils than are grains and pulses, and although tasty are seldom of themselves a balanced diet. Al-

Food products, about half rootcrop, of the San Salvador market. Forty-six of the most popular locally grown items are shown here. Many others are for sale as the season changes. (Courtesy USDA.)

though major root crops such as the white potato, the sweet potato, and cassava are staples in the diet of many peoples, root crops typically eaten as fresh vegetables (e.g., carrots, the onion group, cole plants) are prized more for their flavor and subtle nutrient qualities (such as vitamins) rather than for their ability to satisfy hunger.

All told the tonnage of root crops worldwide must approach that of cereal grains, although, as noted, the dry nutrient weight would be less than the tonnage indicates because of the relatively high moisture content (a potato may be 80 per cent water). Nonetheless, root crops are of great importance for feeding many people, what with annual world production of potatoes reaching

nearly 300 million tons, sweet potatoes and yams over 100 million tons, and cassava (reported) at least 80 million tons. The sugar beet, a root crop, was discussed in Chapter 14; certainly it is a major agricultural commodity, with production of about 165 million tons annually. There are no reliable figures for fresh vegetables, mostly consumed locally, although onions entering commerce apparently total close to 10 million tons annually.

Perhaps more than with most groups of food plants minor root crops are utilized in distant areas removed from the mainstream of world commerce. Many a colonist or explorer dug "roots" as an emergency source of food. Wildlife and primitive peoples have always relied upon native bulbs and tubers, some of which are still gathered today. Even when cultivated many are little changed from the wild form, as is strikingly the case in the Peruvian Andes, where *Arracacia, Lepidium, Oxalis, Pachyrhizus, Tropaeolum, Ullucus,* and other species little known to the world are regularly marketed for the local population. Much the same is true of certain sedges and aroids in tropical Africa and the southeastern Pacific. Given the attention of a modern breeding program some of these could doubtless become important sources of food. Worldwide, however, none of the secondary root crops comes close to rivaling the white potato, the sweet potato, and cassava. These three are to root crops what maize, wheat, and rice are to the cereals.

Irish or white potato, *Solanum tuberosum,* Solanaceae. The common potato is one of the world's most important foods, on a tonnage basis perhaps *the* most important (but note

that the potato tuber is nearly 80 per cent water). Although modern cultivars of the potato are not greatly changed in general appearance from the ancestral wild potato, they have been remarkably increased in size and productivity. Yields of potatoes in favorable growing areas such as the mesas above the Snake River in Idaho average 15 tons per acre, and some fields yield as high as 35 tons per acre. Agronomists feel that 50-ton yields are in sight, and that perhaps 100-ton yields may be a possibility. Few food plants can boast of such productivity.

In industrially advanced parts of the world the trend is toward the processing of potatoes, rather than the cumbersome harvesting and distribution of the tubers themselves. Already the market for potato chips exceeds that for freshly consumed potatoes, and specialty processing for "ready-mix" mashed potatoes and for the ubiquitous French fries is commonplace. With the decline of direct purchase of fresh potatoes by the housewife, breeders

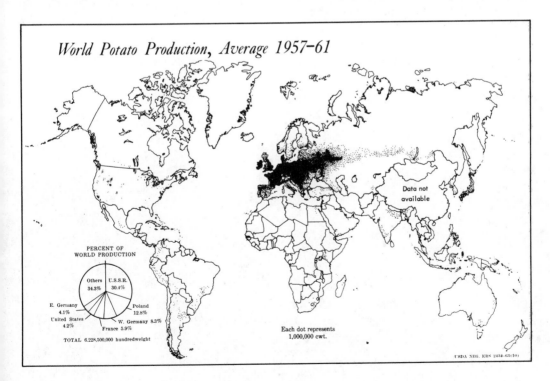

World Potato Production, Average 1957-61

Data not available

PERCENT OF WORLD PRODUCTION

Others 34.3%   U.S.S.R. 30.4%

E. Germany 4.1%

United States 4.2%

Poland 12.8%

France 5.9%   W. Germany 8.3%

TOTAL 6,228,500,000 hundredweight

Each dot represents 1,000,000 cwt.

USDA NEG. ERS 2434-63(10)

are emphasizing yield and high starch content rather than appearance; the commercial cultivars of the future well may be big, solid, and heavy. Planting, care, and harvesting of potatoes is remarkably mechanized in agriculturally advanced areas. But the crop is still hand tended in the technologically less advanced parts of the world and costly in terms of human labor. In the Mexican mountains where potato growing is an ancient art, Ugent (*Economic Botany 22*, 112) describes planting as recently as 1968 as follows:

> ... The soil is cultivated with a wooden-handled iron plowshare, drawn by mules. Furrows are often plowed diagonally up and down inclined slopes, in one direction only, followed by cross-plowing in a horizontal direction. The resulting rhomboid-shaped patterns are about 80–90 cm long, and of equal width. Flat or gently inclined land is plowed in two directions to form squares. Three or four tubers, depending upon their size, are dropped into the intersections formed by the crossing furrows and covered with manure. The soil is then turned over with spades to form hills.

The cultivated potato has a remarkable if not entirely clear history. Its origin was almost certainly from tropical latitudes, but from high elevations that adapt it well as a temperate crop. It is grown worldwide today, but greatest production comes from temperate localities, especially Europe (overwhelmingly the leading continent for potato production). Poland, the U.S.S.R., and Germany are the leading producing countries. The original source of the potato seems very likely to have been the Peruvian Andes, a remarkable sequence of events having transformed this little-known highland plant of the South American Indians into what is today a major staple of the Western world.

Solanum is a large genus of perhaps 2,000 species, many of which are ill-described and poorly understood even today. A number of wild tuber-forming species have had their tubers grubbed from the ground by Indians since time immemorial. Several probably had been casually domesticated in the Andes Mountains before the time of Christ, these cultivars destined to become *S. tubero-*

A field of growing potato plants. (Courtesy Chicago Natural History Museum.)

sum. Correll states that the standard diploid chromosome number in *Solanum* is 24, but that *S. tuberosum* is an autotetraploid with a 2*n* complement of 48. He believes that *S. tuberosum andigena* is the ancestral subspecies. That such a cultivar was long grown in the Peruvian Andes is proven by ancient pottery shards sculptured with figures of the potato. *S. tuberosum* is still an important staple in its Andean homeland, where it is customarily made into "chuño" better to preserve it. Chuño is prepared by trampling and drying the potato during intervals of freezing and thawing. Ugent notes in Mexico even today considerable introgression of wild germplasm (*S.* × *edinense* and other species) into *S. tuberosum*, resulting in a proportion of inedible weed potato plants in the field.

Discovery of the potato by Europeans was probably in 1537, with the Spanish invasion of what is now Colombia. It became familiar to the early explorers, and was sent back to Europe certainly before 1570. From the Iberian peninsula it spread throughout the European continent before 1600, and into the British Isles by 1663. Reintroduction into the New World is said to have been made in 1621, with Bermuda presumably a way station. Nowhere, however, was the potato widely planted until after 1700. European governments, recognizing the potato's food potential, compelled its planting, royal edicts to that effect having been issued in Germany in 1744 and in Sweden in 1764. The potato soon became an important food staple over most of Europe, and especially in Ireland was it a staff of life. During the mid-1840's the late blight disease (caused by *Phytophthora infestans*) wiped out the potato crop two years in a row in Ireland, creating so severe a famine that heavy emigration was undertaken to New England. Even before these migrations the all-important potato was popularly associated with Ireland, hence known in the New World as the "Irish" potato to distinguish it from the sweet potato.

Potatoes dug and windrowed with a rotary-rod digger. (Courtesy USDA.)

The potato is basically a crop of cool environments; it yields poorly when the temperature averages much above 70°F. Although under proper conditions the potato plant may produce true fruit, cultivated potatoes are propagated vegetatively. A "seed" potato is cut into four or more blocky sections, each containing an eye (a dormant bud at the node of this modified stem). About 7 per cent of the crop is saved for propagation. The crop grows best on well-drained soils, with warm days and cool nights. Acid soil helps protect the tubers from scab disease, although many other diseases as well affect the crop. Under intensive production spraying and dusting to control insects and diseases are commonplace. Generally the plant tops are cut off, or killed chemically, just before harvest, to facilitate digging. Maleic hydrazide or a similar growth retardant may be sprayed about 2 weeks before digging of the tubers, too, for inhibition of premature sprouting of the potatoes. Digging machines sift loosened potatoes from the ground, screening away soil and debris. Most of the crop is mechanically harvested in the United States, but where agriculture is less mechanized hand digging still prevails. Harvested potatoes are usually washed to remove soil residues, and then stored in a cool environment (such as a pit house) to inhibit sprouting. It is reported that the cooking quality of a potato is improved if it is held for a few weeks at room temperature before processing. Sugars change to starch, and reducing sugar that causes a dark brown color in potato chips, French fries, and dehydrated potatoes is avoided.

The food value of the potato varies considerably with its growing conditions and handling. The potato is a fairly economical source of starch (nearly 20 per cent), both as food and for industrial extraction. But it is only about 2 per cent protein. In Europe much of the potato crop is fed to livestock, and a sizable portion used also for fermentation and other industrial purposes. Although cornstarch has generally supplanted potato starch in the United States, potatoes do mill more easily than does corn, not requiring so many preliminary soakings and separations. Industrial potato starch goes chiefly into sizings for paper and textiles, and into confections and adhesives.

In the United States potatoes are grown in all states, but especially in Aroostook county, Maine, Kern county, California, Bingham county, Idaho, and Suffolk county, New York. Early potatoes are grown in the South, primarily Florida and California, late spring and summer potatoes in the border states, and autumn potatoes in the northern half of the nation. The potato is responsive to day length, long days stimulating vegetative growth, short days tuberization. The potato is more resistant to late blight disease under a regimen of long days and high light intensity; a cultivar grown in Florida in winter may experience more troublesome blight than the same variety grown in Maine in summer. In recent decades an intensive potato-breeding program has been undertaken by the federal government at Sturgeon Bay, Wisconsin, utilizing as breeding stock collections made from the potato's homeland in South America, as well as cultivars from all over the world.

The potato tuber consists of an outer suberized skin (periderm) varying from red to light brown, a narrow cortex of small starch-rich cells, an adjacent narrow zone of conducting cells (phloem–xylem–internal phloem), and an abundant pith containing most of the starch-bearing cells. Each cell contains greater or lesser quantities of starch grains which swell upon cooking and, according to some authorities, burst the thin cell wall.

Manioc or Cassava, *Manihot esculenta* (*M. utilissima*), Euphorbiaceae. The lowland American tropics have given to the world a counterpart to the Andean highland gift of the potato—manioc (mandioca, cassava, yuca, tapioca, sagu, etc.), probably indigenous to eastern Brazil. Of this plant the root is the part consumed, although young foliage can be consumed as a green and is a possible fodder. Manioc is the staff of life of millions of tropical peoples, frequently the sole food in a village and a mainstay in the diet the year round. It may be consumed boiled, or pulverized and dried into a meal resembling corn meal, termed farinha in Brazil. Manioc is not a balanced food, consisting as it does largely of starch—about 30 per cent by weight—with little protein or fat content. Nonetheless it is a remunerative crop in the hot climates, yielding perhaps more starch per acre (up to 20 tons or better of fresh roots) than any other cultivated crop—and this with a minimum of labor.

When boiled, manioc has a "heavy" consistency, somewhat like that of incompletely cooked macaroni, and is equally tasteless. It is consumed in this form in Paraguay,

Cassava, *Manihot esculenta*, growing on a farm in Tingo Maria, Peru. (Courtesy USDA.)

outlying parts of Brazil, the Andean countries, and parts of Africa and the Far East. In eastern South America cleaned or peeled roots are usually shredded (on hand rasps or in primitive mills), the juices expressed, and the pulpy mass dried over open fires to yield the ubiquitous farinha, available everywhere at a few cents per pound in an area where the daily wage may not permit purchase of any more expensive foodstuff. Farinha may be eaten dry, or, where the income permits, with gravies, sauces, beans, or fats. In the West Indies flat-tasting pancakes, "casabe," are made of manioc meal. Among upper Amazon Indians pulverized manioc may be expressed in locally woven "tipitipis." These gigantic tubes, hung from a tree limb and filled with the manioc pulp, are extended with leverage from a pole; the manioc is thus constricted and the juices expressed. These juices are saved for fermentation to alcoholic beverages, often as important as is the starchy manioc food. The juices also find use in various meat sauces and "West Indian pepper pot." In Africa boiled roots are often pounded into a thick paste called *fufu*.

Manioc is also the source of tapioca starch and tapioca. The former has considerable importance as a size and a raw material for making mucilage. The familiar tapioca "pearls" are manioc starch pellets forced through a mesh and heated at controlled temperatures while being shaken or stirred on a plate, a process that causes swelling, gelatinization, and partial hydrolysis to sugar. Tapioca is mostly prepared (from imported manioc flour) and consumed in North America and Europe. Manioc starch may also be hydrolyzed to simple sugars, syrups, alcohol, acetone, and like products, the residues being used for animal fodder.

The genus *Manihot* embraces multitudes of uncertain species, many of which apparently hybridize with each other (see "Rubber: Maniçoba," Chapter 8). *M. esculenta*, diploid chromosome count of 36, is evidently the only species domesticated as a food plant. Domestication must have been effected in the remote past, by the South American Indians, for the species is not known with certainty in the wild (although numerous escapes are frequent in tropical America). Cassava remains have been found

489

Cassava, roots from a single plant. Thailand. (Courtesy USDA.)

often obscure, but the species is popularly subdivided into sweet and bitter maniocs. The former are edible without preliminary treatment, but the latter contain toxic hydrocyanic principles that must be destroyed (by boiling, expression, or fermentation) prior to consumption. No hard and fast distinction exists between bitter and sweet manioc varieties, and the intermediate condition often occurs. By and large, sweet types are grown in eastern South America and bitter types in the Amazon valley. Both types are grown in Africa and the Far East, apparently first introduced from South America during the seventeenth and eighteenth centuries.

The shrubby manioc plant possesses a ramifying root system, the chief roots of which develop swellings similar to sweet potatoes a short distance from the erect stem. These swellings have a shallow periderm and cortex (in bitter maniocs, most of the poisonous substance occurs here), a modest vascular cylinder, and abundant starch-bearing pith. Leaves are alternate, digitately compound, and the flowers are monoecious and inconspicuous.

in archeological sites in Peru as early as 800 B.C., and indirect evidence shows it to have been cultivated in Colombia and Venezuela at least three millenia ago. Literally hundreds of cultivars are to be had. Differences are

Manioc is usually hand cultured and

Making farinha in a manioc mill, Brazil. (Courtesy Chicago Natural History Museum.)

harvested in the less advanced portions of the globe where it is ordinarily grown. Typically forest is burned over, and stem cuttings of manioc of a few nodes' length inserted into hand dug holes at intervals of about 1 meter. Under experimentation, long sections (about 25 centimeters) from the lower stem, planted horizontally, have given best response. Sometimes the plots are weeded with ponderous hoes; sometimes the manioc is left to fend for itself. Harvestable roots develop in as little as 8 months; maximum yield of quality roots, however, usually takes about 16 months. If not all the roots are dug up, the plant will continue as a "perennial," new stems arising from the roots left in the ground. Roots left in the soil too long become woody and of little value for food.

Production statistics for manioc are all but unobtainable, since the crop is mostly grown and consumed locally, and this to a large extent in tropical regions far off the beaten track. Estimates place tropical American production, chiefly from eastern Brazil, at more than 25 million tons annually; African production probably exceeds 30 million tons, East Indian 20 million tons. In any event, manioc is without doubt one of the great root crops of the world, and is particularly important in hot tropical, often seasonally arid, climates where cereal and potatoes will not grow well.

Sweet potato or Kumara, *Ipomoea batatas*, Convolvulaceae. A second potato gift of the New World is the sweet potato, sometimes erroneously termed "yam" in the southern United States. The true yam is *Dioscorea*, not always distinguished from the sweet potato in statistics. *Ipomoea* is a large genus of about 400 species, noted for many ornamentals ("morning-glories"); but only *I. batatas*, after the Carib name for potato, is of great commercial importance. *I. batatas* is a hexaploid, $2n = 90$, of uncertain ancestry, no longer found in the wild. One suggestion is that the species arose from *I. tiliacea* (*I. fastigiata*) stock through hybridization and ploidy. Modern cultivars are largely self-sterile, and flower poorly except in the tropics. The species is a short-day plant with flowering encouraged at a photoperiod of 11 hours or less. It grows best on well-drained soils, in climates having a lengthy warm sea-

The sweet potato plant. (Courtesy Chicago Natural History Museum.)

son, although in Africa it is often planted at the edge of a swamp to insure survival of planting stock during the dry season.

The sweet potato's swollen underground roots provide a more nutritious food than the Irish potato (about half again as many calories), but unlike the latter it is adapted to hot, moist conditions reflective of tropical lowlands where the average annual temperature is not less than 75°F. The sweet potato was introduced into Spain from Middle America by the early Spanish explorers, probably first by Columbus in 1492. From Spain it spread unobtrusively throughout lower Europe and was carried by navigators of the sixteenth and seventeenth centuries to all parts of the globe. So unspectacular was its spread that, when rediscovered in the East, its American origin was at first overlooked. Evidentally the sweet potato had reached Polynesia in pre-Colombian times, by unknown means (possibly the water-resistant seed pods drifted?), and it was taken into New Zealand by the Maoris before the first visit by Western explorers.

Cultivation of the sweet potato was tried in northern Europe with little success, and even in southern Europe its popularity has been modest. In the southern United States it is in greater favor, both as a stock food and as human food, its use there paralleling in a modest way that of the white potato in Europe. In the Pacific islands, East Indies, China, and India the value of the sweet potato was quickly recognized. In Japan especially it was seized upon as the most productive crop per unit of upland area. It has come to rank as the second most important crop in that overpopulated nation (rice being the first). African production approaches that of the Orient, especially from West Africa, and selected strains of sweet potato are even grown in Australia and southern Soviet Union. Worldwide production probably amounts to more than 100 million tons annually.

The sweet potato plant is a trailing vine with cordate, frequently lobed, alternate leaves and attractive funnelform flowers that bloom frequently in the topics but seldom in temperate climates where the plant is cultivated as an annual. The tops may be used in silage or for animal browse, but the important part commercially is the swollen, tuberous root. In most cultivars this has a white or pale yellow flesh, although in the United States dark yellow to reddish-orange types rich in carotene have become popular. "Dry" types, with a mealy, starchy flesh are usually favored in the North; soft, more gelatinous, and sugary "yam" types in the South. Sloughing of the potato epidermis gives rise to a tough periderm. Within this there occurs a starch-rich but narrow cortex; endodermis; and xylem and phloem radially distorted by secondary growth, the parenchyma of which forms most of the flesh in a mature root. A sweet potato consists of about 70 per cent water, 18 per cent starch, 8 per cent other carbohydrates, 2 per cent protein, 1 per cent mineral, and practically no oil. It is unusually rich in vitamins, iron, and calcium, being about the equal of maize in energy value.

In temperate climates sweet potatoes are planted from shoots taken from "seed" roots kept over from the previous growing season. A potato is planted in a propagation bed, often heated in cooler climates, and the sprouts cut from the end of the potato as they reach several inches in length with basal roots. Three or four sets of slips are obtainable from one potato. In tropical lands where the vines grow continuously, stem cuttings are the usual mode of propagation. Planting is typically in raised hills, sometimes (as in Japan) between maturing rows of cereal crops such as barley. Sweet potatoes are ready for harvest in from 4 to 6 months, the best-keeping potatoes being dug shortly before first killing frost. Yields are abundant—as much as 20 tons per acre under excellent conditions—but most of the cultivation and handling is done by hand, so the crop is costly of labor. It is no wonder, then, that sweet potato growing coincides to a great extent with regions of inexpensive labor, although in the United States mechanization has been initiated.

Harvested sweet potatoes should be kept warm (usually by placing them in the sun) for several days after harvest to encourage development of the protective periderm. They are then best stored in a cool place, perhaps given fungicidal treatment. Even so, storage for protracted periods is difficult, significantly more so than with white potatoes, and losses of 30 per cent are not uncommon during the winter. In Japan, where individual farmers keep and grow sweet potatoes, the danger of insufficient "seed" stock for the next planting because of storage loss frequently arises.

In most countries sweet potatoes are used chiefly as a human food and stock food. In the United States they are usually marketed as dug, and baked or boiled for eating. In Japan potatoes are often cut into slices and dried in the sun until brittle, to improve the keeping quality. The dried potatoes may be pulverized like cornmeal. In Africa sweet potatoes are frequently left in the ground until needed. Industrial uses, especially in Japan, are for production of alcohol by fermentation and as a starch source.

Edible Aroids, **Araceae.** Corms and tubers of several genera of the Araceae have been used for food in the tropics since time immemorial, yet are scarcely known to Temperate Zone people. They go by an exotic assortment of local common names—cocoyam, curcas, dasheen, eddoe, kolokasi, macabo, mafaffa, ocumo, tannia, taro, yautia, and many others that often confuse the same species, and are sometimes the same for different species. The genera *Colocasia* (taros) and *Xanthosoma* (yautias) are the most important. In the family are also many ornamental plants native to the tropics, such as the philodendrons, caladiums, and arum lillies. *Acorus calamus, Calla palustris,* and many other species have been used medicinally since ancient times. Most are adapted to moist, tropical environments, and are watery herbs or trailing vines having characteristic large cordate leaves (popularly termed "elephant ears") and fleshy rhizomes. The family is characterized by a specialized inflorescence consisting of a spadix of inconspicuous flowers subtended by a leafy spathe (which is often colorful and "flowerlike"). Species often have an acrid sap and needlelike crystals of calcium oxalate (raphides). Cultivated species are generally propagated by branch corms or separations, and many do not even produce viable seed.

Most widely utilized is the taro, dasheen, gabi, or eddoe, *Colocasia esculenta*. The taro apparently originated in southeastern Asia, and its cultivation diffused in prehistoric times westward through India and eastward through the Pacific islands as far as Hawaii. It was introduced into the Mediterranean area during the era of the ancient Greeks, and spread southwestward across tropical Africa, and finally to the West Indies and tropical America within recent centuries.

Yautia, tannia, malanga, or macabo (and many other names) is *Xanthosoma* in various species, of which *X. sagittifolium* probably embraces most of the edible cultivars. The species is native to tropical America, and was in cultivation in pre-Columbian times. Yautia was introduced into Africa, Oceania and southeastern Asia at least by the early 1800's, and today is pantropical.

Somewhat less familiar than taros and yautias, but nonetheless often locally important, are the following aroids:

ALOCASIA. These are erect ornamentals of substantial size, of which *A. indica* and *A. macorrhiza* are the principle edible species, used from India to the southeastern Pacific. The stem as well as the cormels are consumed.

A, yam vine, *Dioscorea alata*, trained to a bamboo trellis; B, yam tubers, cultivars of *D. alata*; C, growing yautia; D, edible-leaved yautia, *Xanthosoma brasiliense*; E, corm and branch tubers of taro, variety Trinidad Dasheen; F, tubers developing at the base of yautia plant. (Courtesy USDA.)

494

A Papuan native planting taro with a planting stick. (Courtesy Chicago Natural History Museum.)

AMORPHOPHALLUS. *A. campanulatus*, elephant yam, is grown in India, southeastern Asia, and the Pacific, the large tubers often used for the preparation of a dried flour, as well as eaten fresh and employed for animal food. Other species are grown in Japan and China.

CYRTOSPERMA. *C. edule* (*C. chamissonis*) is the principle edible species. It is a large plant with cylindrical tubers that may weigh as much as 100 pounds after several years' growth. The species is native to southeastern Asia, and is a minor food crop throughout the southeastern Pacific islands.

Under emergency conditions a number of other aroids may be used for food, including in North America such Temperate Zone species as the arrowleaf or arum, *Peltandra*

virginica; the jack-in-the-pulpit, *Arisaema triphyllum*; the skunk cabbage, *Symplocarpus foetidus*; and golden club, *Orontium aquaticum*.

The food value of all aroid corms is similar, like potatoes being rich in carbohydrate (about 20 per cent starch) with minor protein content and little or no oil. Most of the time the corms are boiled like potatoes, or crushed to make native cakes, although roasting, baking, and steaming are also commonplace. Sliced corms fried in oil make good chips. The dasheen (corruption of "eddo de la Chine"), the *antiquorum* variety of taro, is the source of the famed Hawaiian poi, made by steaming the corm, followed by crushing and natural fermentation (initially by bacteria, then by yeast). The taro group is said to contain more calcium oxalate (destroyed in processing) than the yautias, for which reason the latter are often preferred. The starch granules in aroid corms are very small, which contributes to digestibility but makes their extraction for industrial purposes difficult. It is reported that the starch is about 28 per cent amylose, similar to that of the potato.

Cultivation of taros and yautias is mostly by individual peasant farmers in small patches, often as an intercrop with other plantings. Best growth is in moist locations with some shading, but they can be cultivated in uplands if rainfall is adequate. Cormels are usually used as planting material, but sections of a large corm can be used so long as an eye or bud is contained. The plants prefer rich soil, and respond well to mulching and fertilization. Yields of 20 tons per acre have been obtained experimentally. Under favorable conditions a harvestable crop may be obtained in as little as 6 months, although normally a planting is allowed to mature for 9 or 10 months. If corms are kept dry and ventilated they will store well for several months. In larger operations the matured plants may be "plowed out" to reveal the corms, which are gathered by hand and shaken free of soil. In small operations the individual plants are hand dug. Aroid corms are mostly consumed locally, so that there is no reliable estimate of world production; as widely grown as these crops are throughout the tropics, however, total production must reach millions of tons annually, ranking edible aroids among the important tropical crops.

Some other root crops. ACHIRA OR EDIBLE CANNA, *Canna edulis*, Cannaceae. The fleshy rhizome of this species has commerical importance today almost only in the Apurimac valley of Peru, although at times it has served for stock feed and a source of starch in Africa, Asia, and Australia. It was probably domesticated on the eastern slope of

the Andes, and widely distributed by the Incas and other Indian societies of Mesoamerica. The crop is usually cultivated as a casual dooryard plant, planted from small rhizome segments which are kept during the dry season buried in pits to prevent their deterioration. In the Apurimac valley planting is in August, with harvest of mature rhizomes the following May. An achira bake follows, in pits heated with hot stones, producing a light-colored, mucilaginous food with a sweetish taste. In this area this is the only form in which the rhizome is eaten, and large quantities of baked achira (the only Indian cash crop of the area) are taken to Cuzco for sale. Baked achira will not keep more than a few weeks, and sale is linked to the Corpus Christi holidays for which achira is a traditional festival food. Elsewhere achira has lost favor; in the West Indies for example, it has not been able to compete with arrowroot (*Maranta*) which provides a somewhat more digestible starch (see Chapter 14).

AÑU, *Tropaeolum tuberosum*, Tropaeolaceae. This relative of the garden nasturtium is a companion root crop in the original home of the now-widespread white potato. The use of añu is diminishing, although it can still be found in most Andean vegetable markets from Colombia to Bolivia. The raw tubers have a disagreeable odor. They are boiled for consumption. In the Incan Empire they constituted a fairly important food source.

ARRACACHA, *Arracacia xanthorrhiza* (*A. esculenta*), Umbelliferae. The tubers serve as an edible carrot- or parsnip-like vegetable in the Andes from Venezuela to Bolivia; also cultivated to some extent in Central America and the West Indies.

BEET, *Beta vulgaris*, Chenopodiaceae. Mention has already been made of the species in the discussion of sugar (Chapter 14). The species also includes Swiss chard and mangel-wurzels, all of which readily interbreed. The common garden beet is a biennial, storing up food in the fleshy taproot

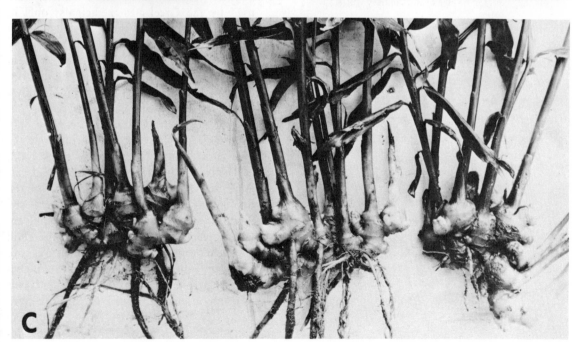

A, apio or arracacha, *Arracacia xanthorrhiza*; B, arrowroot, *Maranta arundinacea*; C, ginger, *Zingiber officinale*. (Courtesy USDA.)

one season to provide for flowering and fruiting the next. Beets grow best in cool weather, and in the tropics are seldom cultivated at elevations below 1,000 meters. As a garden vegetable the beet is grown from seed as an annual, the fleshy root being harvested several weeks after seeding. The modern beet is descended from the wild beet of southern Europe, in which the leaves alone were consumed as a potherb in pre-Christian times, and the species not much cultivated until the third century A.D. Red beets with turnip-like roots seem to have become conspicuous only after the sixteenth century, with improved cultivars developed in France and Germany after 1800. In these the fleshy taproot consists of outer thin layers of periderm and cortex, most of the flesh being formed of phloem and xylem parenchyma. The large, petiolate leaves arise directly from the top of the crown.

Beets are easily grown, and are a favorite

Table beet. (Courtesy USDA.)

in the home garden, being eaten fresh, canned, or pickled. They are tolerant of salt, even responsive to sodium chloride applications. Commercially they are drill-planted in early spring and tended like sugar beets. Mangels or mangel-wurzels, in which the flesh is whitish, were developed from chard-like forms. Today they are mostly grown in Europe for stock feed. Over a half-million tons of beets are grown annually in the United States, mostly for processing. Oregon, New York, and Wisconsin are the leading producing states. The beet root is about nine-tenths water, 8 per cent carbohydrate, and 2 per cent protein plus minerals.

CARROTS, *Daucus carota* var. *sativa*, Umbelliferae. Carrots are believed to have originated in the Near East, and have been food plants since ancient time. A diversity of forms occur wild, naturalized, or cultivated, from the Mediterranean area to the Far East, one of them with a taproot 3 feet long and another purplish-red in color. Carrots were used by the Greeks as medicinals; only in recent centuries have they become a widespread human food, recommended especially for their abundant carotene. The modern fleshy-rooted cultivars are chiefly the result of breeding by the Vilmorin seedsmen of France since the seventeenth century. The custom of eating carrots raw has been accepted only within the present century. Throughout the ages carrots have been used as stock food.

The carrot, like the beet, is a biennial, but is planted from seed and harvested before the first killing frost as an annual. The fleshy taproot possesses wide cortex and phloem layers, where the greater part of the food reserves occur. The compound leaves are borne directly from the crown, with an erect umbel-bearing stem arising the second year to produce flower and seed. Carrots are comparatively expensive to grow, requiring considerable labor even though herbicides are used today for weeding, and planting and harvest are becoming mechanized. In com-

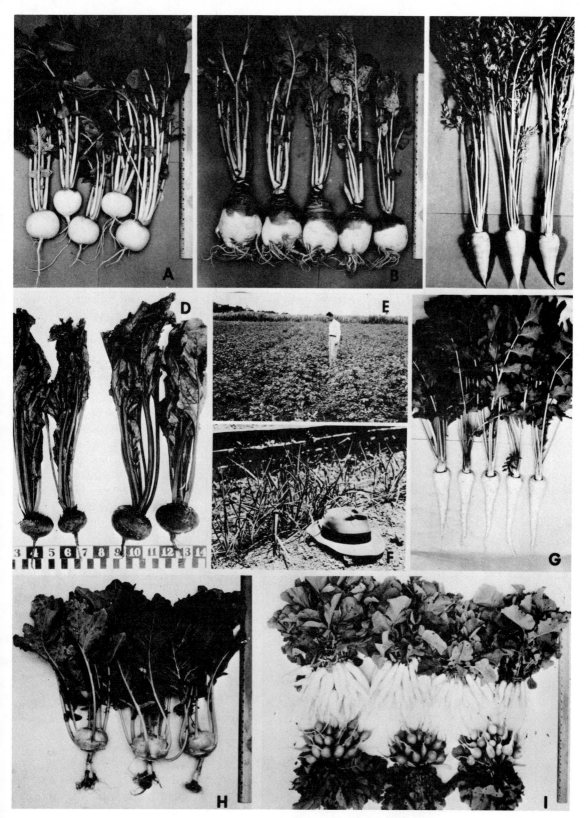

Various rootcrops: A, turnip; B, rutabaga; C, carrot; D, beet; E, field of potatoes; F, salsify; G, parsnip; H. kohlrabi; I, radish cultivars. (Courtesy USDA.)

mercial production yields over 1 ton per acre are average. Production in the United States is nearly 1 million tons annually, chiefly from California, Texas, and the Great Lakes states.

CELERIAC, *Apium graveolens* var. *rapaceum*, Umbelliferae. The large, turnip-like roots of this variety of celery are consumed as human food, especially in soups and stews.

TURNIP-ROOTED CHERVIL, *Chaerophyllum bulbosum*, Umbelliferae. The grayish, carrot-like root of this biennial European and Asiatic species is eaten boiled in Europe.

CHICORY, FRENCH ENDIVE, OR WITLOOF, *Cichorium intybus*, Compositae. This perennial European species, now cultivated and naturalized in North America, has a fleshy

Witloof or French endive. (Courtesy USDA.)

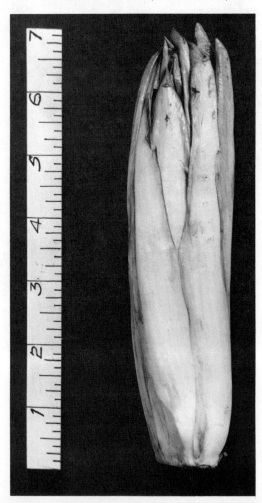

root. Young roots may be boiled and eaten like carrots. The roots are also roasted and pulverized as a coffee additive.

FRA-FRA POTATOES, *Coleus rotundifolius*, Labiatae. This African species, also known as fura-fura potatoes, fabirama, and daso, is cultivated in West Africa for the edible tubers. *C. edulis* is similarly used in northeastern Africa.

GROUNDNUT, *Apios americana* (*A. tuberosa*), Leguminosae. This twining, pinnately leaved herb of northeastern North America, which bears pear-shaped root tubers, was once an important Indian food. This is probably the plant Raleigh reportedly brought to England as the "potato."

HORSERADISH, *Rorippa armoracia* (*Armoracia rusticana*), Cruciferae. This perennial is native to eastern Europe, where it is commonly cultivated, but was introduced into North America about 1806, where it is occasionally found as an escape as well as cultivated. It was probably domesticated about the time of Christ. The roots were employed as a medicinal by ancient and medieval peoples. The species is thought to be an interspecific hybrid of uncertain parentage, highly sterile because of poor chromosome pairing ($2n = 28$ and 32). The fleshy white root is too "strong" for direct consumption, owing to the pungent thiocyanates and the glucoside sinigrin contained, but the grated root finds favor as a condiment and flavor today. The plant is usually propagated by setting small roots trimmed from large ones after harvest. Production in the United States is chiefly in the Mississippi River valley, near St. Louis, and near Eau Claire, Wisconsin, amounting to about 15,000 tons annually.

JERUSALEM ARTICHOKE, *Helianthus tuberosus*, Compositae. The Jerusalem artichoke, a coarse species of sunflower, has been mentioned previously as a source of inulin starch. The name presumably arises from mispronunciation of the Italian *Girasole articiocco* ("sunflower edible"). The potato-like rhizome is consumed directly—boiled, pickled, or

Tubers of Jerusalem artichoke. (Courtesy USDA.)

A

raw. The species is native to North America, and was much cultivated there by the Indians; today it finds more favor as a food plant in Europe, where it was introduced in 1660, and in China. In the United States the tubers once served importantly as hog feed, being dug by the hogs themselves. In food value they are about the equivalent of potatoes. Propagation is chiefly by rhizome sections, and once established plants generally perpetuate themselves from tubers left in the soil.

MACA, *Lepidium meyenii*, Cruciferae. The maca is a little-known food plant of high elevations in the Peruvian Andes, widely used in precolonial times. Macas are cultivated and grow wild in the barren puñas where sheep grazing is the chief occupation. The species is a rosette plant with dissected leaves from a fleshy root resembling a radish. Parenchyma rays extending out from the xylem are rich in starch. Seed is planted to soil cultivated by hand, in September or October, and the crop harvested from May to July. The dug roots can be consumed fresh, or they can be dried (in which condition they last for years). Fresh roots are usually baked in pits, and dried roots often made into a type of porridge.

MATAI OR CHINESE WATERCHESTNUT, *Eleocharis dulcis*, Cyperaceae. The edible corm of this sedge is valued in the Orient as a food delicacy, and as a source of starch. The species is found in the Old World tropics from Madagascar to southeastern Asia and in Polynesia. The plant is seldom cultivated except in China, where improved selections are planted. The plump corms resemble gladiolus bulbs, and remain crisp even after cooking. Like most root crops they are relatively rich in carbohydrate (almost 20 per cent), with a little protein and a minimum of oil. The plants are grown in paddies, dug from the mud by hand after the paddy is drained. Planting is from nursery beds, where select corms are sprouted after danger of frost is past; harvest is in late autumn. It is said that yields up to 15 tons per acre are attainable. Fresh corms are generally served boiled or steamed as a vegetable, and grated corms may be filtered through a fine cloth to yield a starch that is sun-dried.

MELLOCO, *Ullucus tuberosus*, Basellaceae. This species is another north and middle Andean tuber crop, in some districts ranking second in importance only to potatoes. The plant is an upright, succulent herb, with broad, simple cordate leaves and inconspicuous axillary racemes.

OCA, *Oxalis tuberosa*, Oxalidaceae. Another minor tuber crop of the high Andes is the oca. In parts of Colombia and Peru this relative of the wood sorrel is as important as the potato. The herb has the typical shamrock-like trifoliolate leaves of the wood sorrels, but does not seem to bear seeds, being propagated by the tubers. These are cut into sections and cultivated exactly as are potatoes. After harvest the tubers are mellowed (rid of calcium oxalate) by placing them in the sun for several days. They are then consumed raw, boiled, candied, dried (chuño: see discussion of potato), or powdered in soups.

ONIONS (INCLUDING GARLIC, LEEKS, WELSH ONIONS, SHALLOTS, AND CHIVES), *Allium*, Liliaceae. In the genus *Allium* the underground part consumed is the bulb, which consists of a stem core surrounded by specially modified food storage leaves moderately rich in sugar and pungent allyl sulfide. The linear, cylindrical, or flattened photosynthetic leaves arise from the top of the bulb, in the center of which forms

an attractive umbellate inflorescence that makes *Allium* often sought as a garden flower. The familiar onion, *A. cepa*, originating in central Asia and known in a bewildering array of botanical varieties and cultivars, but not in the wild, is cultivated throughout the world. It is known to have been extensively used by the ancient Egyptians. It was later spread by the Spanish colonizers. In the United States most commercial onion growing is done on muck soils of the Great Lakes area, in Texas, and in California–Oregon. Seed is generally drilled directly where the onions are to develop. In the growing of Bermuda onions in Texas seed may be planted in special beds in October and transplanted to the fields weeks later. Home gardeners often prefer to plant onion sets—small, partly grown onions prepared especially for this purpose. Harvesting in all cases involves pulling or plowing, picking up, and drying. United States production amounts to about $1\frac{1}{2}$ million tons annually,

Cultivating onions on irrigated land in Arizona. (Courtesy USDA.)

Field of leeks. (Courtesy USDA.)

and soldiers "to make them strong." It is a perennial with narrow, flat leaves and small bulbs. Much more pungent than the onion, it is used mostly for flavoring, being a favorite of Mediterranean and Far Eastern peoples. Production in the United States amounts to about 20,000 tons annually, mostly from California. The biennial, *A. porrum*, the leek, has the appearance of a large garlic plant, having flat, solid leaves and slender, cylindrical bulb. Native to southern Asia, it has been used as food since ancient times, and is reported to have been a favorite food of Nero's. It is served like asparagus or used as a flavoring.

A. fistulosum, welsh onion, is believed to be of Chinese origin, "welsh" being a corruption of the German "walsch," meaning foreign. This slender but sturdy, tillering, leek-like type is a favorite in the Orient. The perennial *A. ascalonicum*, the shallot, produces clusters of bulbs, and much resembles the common onion. It is especially used for pickling. *A. schoenoprasum*, chives, is an Old World plant, today found wild or escaped in Italy and Greece. It is a hardy perennial, growing in dense clumps and producing very small, clustered bulbs used especially for seasoning. The attractive blue flowers often produce no seed, multiplication then depending upon proliferation of bulblets.

world production to nearly 10 million tons (mostly in Europe). Japan, the United States, the United Arab Republic, Spain, and Turkey are the leading onion-producing countries.

A. sativum, garlic, also native to southern Asia, was fed by the Romans to laborers

Harvested onion field in Egypt. The onions are left in small piles to cure, then tops and roots are clipped off and the onions bagged. (Courtesy USDA.)

In China and Japan *A. tuberosum* is grown as a salad plant, and rakkyo, *A. chinense*, as an important pickling onion much consumed by the Orientals. The onions are processed in brine, then preserved in sweetened vinegar much like the small pickled onions served in martini or gibson cocktails in the United States.

PARSNIPS, *Pastinaca sativa*, Umbelliferae. This relative of the carrot is believed to be native to the Mediterranean region, but it persists as an escape in many parts of the world. Wild plants were consumed in Greek and Roman times. By the sixteenth century the parsnip was widely cultivated in northern Europe. In 1609 it was introduced by the colonists into North America, where even the Indians took to cultivating it. The plant is a hardy biennial cultivated as an annual, with a whitish, carrot-shaped taproot, pinnate basal leaves, and an erect branching stem when in flower. The root consists mostly of phloem cells filled with minute starch grains. The sweet flavor of the roots develops only after exposure to cold. Parsnips are used both for human and for stock food, largely in Europe. They are grown like carrots.

RADISH, *Raphanus sativus*, Cruciferae. Perhaps the most quickly and easily grown vegetable of the home garden is the radish, known in red, black, or white; round or elongate; large or small cultivars. Radishes with taproots weighing 100 pounds were reported in the sixteenth century. The species is presumably native to China, where it has had a long history of cultivation, and had spread to the Mediterranean area before Greek times and to the New World in the early 1500's. Radishes are mostly grown today in Japan and China, where the crop is generally pickled in brine. The species is an annual or biennial herb with a basal rosette of rough, pinnate or deeply lobed leaves.

RUTABAGA, SWEDE, OR SWEDE TURNIP, *Brassica campestris napobrassica*, Cruciferae.

This glabrous-leaved, usually yellow-fleshed, turnip-like plant with 38 chromosomes is believed to have originated as a cross between turnip (20 chromosomes) and (cabbage (18 chromosomes), in Europe some time during the Middle Ages. It was first described by Bauhin in 1620. The rutabaga is a cool-weather plant, and its culture is confined largely to northern Eurasia and northern North America. It is reported to be more nutritious than the turnip, but is extensively grown only in Europe. It is used as human food and for livestock food.

SALSIFY OR OYSTER PLANT, *Tragopogon porrifolius*, Compositae. The long, fleshy, parsnip-like root of this species has much the flavor of oysters when it is boiled. In North America it is of only minor importance, but it is more widely cultivated in Europe. The plant bears a tuft of linear leaves from the crown and purplish flowers maturing seeds similar to those of a dandelion. It is native to the Mediterranean countries, where it can still be found wild.

SCORZONERA OR BLACK SALSIFY, *Scorzonera hispanica*, Compositae. This yellow-flowered Mediterranean plant is similar to other salsifies, but the long taproot is black. The species is little cultivated outside of southern Europe.

SPANISH SALSIFY OR GOLDEN THISTLE, *Scolymus hispanicus*, Compositae. This salsify-like plant of the Mediterranean area has pinnatifid leaves and yellow flowers. It was known to the ancient Greeks, but today it is little cultivated, and it is practically unknown in the United States.

SKIRRET, *Sium sisarum*, Umbelliferae. This is another perennial species grown as an annual, producing bunched, tuberous roots with a grayish flesh. It is eaten like salsify.

TIGERNUT OR EARTH ALMOND, *Cyperus esculentus*, Cyperaceae. The small corms of this species are gathered from the wild in West Africa and eaten raw or roasted. Yields of 3 tons per acre are recorded in north-eastern Nigeria.

TURNIP, *Brassica rapa*, Cruciferae. The turnip is a rough, hairy-leaved biennial, with a large, usually purplish-white and white-fleshed taproot. The root has narrow periderm, cortex, and phloem layers and abundant (parenchymatous) xylem and pith. Many types have been in cultivation since before the Christian era. Probably native to Europe (one kind) and central Asia (another kind), it was spread in ancient times all across Asia to the Pacific. The European turnip was early used for human and stock food, being boiled, baked, or made into kraut for human consumption. The leaves have long served as "greens" as well. The turnip was introduced into North America in 1609, and was spread by colonists and Indians alike. It has since been a fairly common garden vegetable in cooler areas, and is notable for its hardiness and ease of cultivation. Turnips are of comparatively minor importance in the United States, but are widely grown in Europe.

YAMS, *Dioscorea*, Dioscoreaceae. Although moist sweet-potatoes are sometimes called "yams" in the southern United States, the true yams belong to several species of the genus *Dioscorea*. Of these the cultivated ones are indigenous to the tropics. Except for sweet potatoes and manioc they are probably the most widely used of tropical lowland root crops, especially in tropical West Africa from the Ivory Coast to the Cameroons, parts of Middle America, the Pacific islands, and southeastern Asia. Because one must dig deeply for the tubers, yams are an expensive crop in terms of labor, and cannot ordinarily compete with either rice or cassava on a calories-per-man-day basis. Yams are one of the few crops well adapted to tropical rain-forest habitat, however.

Yams exist in a number of species and cultivars, ranging from types with roots no larger than a small potato to those with roots weighing as much as 100 pounds. Coarse, dry, mealy, tender, crisp, mushy, and sweet types are available. Some resemble potatoes, and like the latter are consumed baked, boiled, or fried. Certain kinds are customarily used for soup. Others are boiled and mashed, with perhaps subsequent baking or frying. In West Africa the housewife regularly crushes and pounds yam afresh to prepare the highly prized *fufu* dish. The yam is about 20 per cent starch, and as a food is nearly the equivalent of the potato. Yams also contain tannins, saponins, alkaloids, and other substances; in some cases these may be useful, and in others they are irritants.

The yam plant is a deeply rooted, climbing, dioecious perennial vine with distinctively veined, cordate leaves and inconspicuous, monocotyledonous flowers. The species most commonly cultivated is *D. alata*, probably native to China. It is said that Portuguese slavers provisioned their ships freely with this species for their sixteenth-century world voyaging. The species seldom fruits, and so is propagated chiefly by replanting the tops from dug roots. Stakes are customarily set for the yam plant to climb. In addition to *D. alata* (called water yam), there are cultivated in Africa *D. bulbifera* (air potato), *D. esculenta* (Chinese yam), *D. dumetorum* (three-leaved yam), *D. rotundata* and *D. praehensilis* (white yams), and *D. cayennensis* (yellow or Guinea yam); in the Orient *D. aculeatea*, *D. divaricata*, *D. pentaphylla*, *D. opposita* (*D. batatas*), and several other species under a score of common names; in tropical America *D. trifida* and *D. cayenensis*. An 8- to 10-month growing season is required for a crop to mature, so that little opportunity for cultivation of yams exists outside the tropics. Yields run as high as 10 tons per acre. No reliable production statistics on this locally consumed crop are available, but world production must average several million tons annually.

In recent years yams have been used as the

source for a steroid precursor, diosgenin, used to make birth control pills. Wild, inedible species have mostly been used especially *D. composita*, *D. spiculiflora*, *D. floribunda* of southern Mexico, and other species from China.

YAM BEAN OR MANIOC BEAN, *Pachyrhizus tuberosus*, Leguminosae. This perennial-rooted, twining, trifoliolate species is probably South American in origin, but is now widely cultivated and naturalized throughout the world tropics. *P. erosus*, sincama, native to Central America, has been similarly spread and utilized for its turnip-like root. The crop is propagated from seed, and seed plants are trained to bamboo trellises in the Orient. Plants for root harvest are seeded to light soils, which profit from fertilization and mulching. After about 10 months the roots are plowed or pulled. Yields are about 4 tons per acre. The root is crisp and tender, contains about 10 per cent starch, and is mostly consumed raw as a snack, but may be boiled, fried, or pulverized for a variety of dishes.

MISCELLANEOUS ROOT CROPS OF RESTRICTED IMPORTANCE. These include: arikuma or llacou, *Polymnia sonchifolia*, Compoitae, Peru; arrowhead, *Sagittaria sinensis*, Alismaceae, China; cattail, *Typha*, Typhaceae, eastern North America; chayote *Sechium edule*, Cucurbitaceae, Central America; gamote or wild parsnip, *Phellopterus montanus*, Umbelliferae, northern Mexico; Goa bean, *Psophocarpus tetragonolobus*, Leguminosae, southeastern Asia; groundnut, *Panax triflorus*, Araliaceae, northeastern North America; leren, *Calathea allouia*, Marantaceae, tropical America; yellow pond lily, *Nuphar advena*, Nymphaeaceae, eastern North America; prairie turnip, *Psoralea esculenta*, Leguminosae, western North America; sand root, *Ammobroma sonorae*, Lennoaceae, northern Mexico and western United States; Spanish bayonet, *Yucca spp.*, Liliaceae, western North America; ti, *Cordyline terminalis*, Liliaceae, southeastern

Pacific area; and wild plantain, *Heliconia bihai*, Musaceae, tropical America.

Vegetables

The word "vegetable" is an imprecise term popularly used to include many of the root crops and some of the fruits of the next chapter. Cereals and pulses are sometimes regarded as vegetables, too, especially if consumed in an immature stage (e.g., fresh maize, peas). We are dealing in this section, however, only with those plants whose above-ground parts exclusive of true fruit are eaten. These are often known as "green vegetables" because young stems and leaves are green with chlorophyll.

It is difficult to decide which species should be listed here. Just as multitudes of grass and legume plants make excellent forage for livestock, thousands of plants yield edible shoots that can be used as potherbs. Highly

Above-ground parts of the true yam, *Dioscorea opposita*. (Courtesy Chicago Natural History Museum.)

unlikely plant parts have been so utilized, ranging from young fern fronds ("fiddle-heads"), through many agricultural weeds (dandelions, *Taraxacum;* dock, *Rumex;* lamb's quarters, *Chenopodium;* etc.), to tender tips from aroids of the previous section (though uncooked these may possess irritating oxalate raphides). Even foliage that is toxic before cooking may be consumed. Thus the alphabetical listing which follows is far from inclusive, especially of "wild greens" used in rural areas, distant parts of the world, or under special and emergency conditions.

As foods green vegetables have little energy value. Very few could compete with cereals, pulses, and root crops for sustaining masses of people. But they are excellent sources of supplementary dietary substances, such as vitamins, minerals, and trace materials that round out an otherwise unbalanced diet. This is recognized both in primitive societies where various potherbs are cooked from time to time, and in technologically advanced lands where mothers insist that their children "eat their spinach." Vegetables are also an interest of the gourmet. Seldom do basic foods lend themselves to the creation of delicate shades of flavor so well as do fresh vegetables. Vegetable and salad preparations are the chef's acid test. Taste for many of the vegetables listed must be acquired, and certainly their seasoning is a highly personalized consideration. No wonder that these plants relate more to the artistry of food use than to wholesale sustenance. This is reflected in the lack of production statistics for most vegetables, many of which are consumed locally as fresh produce, or may be seasonally grown in the home garden. Vegetables offer wide choice, but if any one plant family were to be singled out as especially important, it would probably be the Cruciferae, of which the cole plants (the cabbage group) are widely cultivated in all temperate climates. The most widely used single species may be lettuce, the universal salad leaf over much of the world.

Commercial vegetable production in the United States exceeds 20 million tons annually, about three-fifths for the fresh market and two-fifths for processing. California is by far the leading state in commercial production, followed by Florida, Texas, and several of the Great Lakes states.

ARTICHOKE, *Cynara scolymus*, Compositae. The green, French, or globe artichoke is a tall, coarse "thistle," presumably native to the Mediterranean region but domesticated in pre-Christian times and spread throughout most of the Old World. The cardoon, *C. cardunculus*, is similar. Originally the leaves, grown in darkness to be tender and white, were consumed, but since the 1400's the young flower head (base of phyllaries and receptacle) has become the popular part. The plant is a perennial, and is usually propagated by sprouts that arise from the root crowns, since the species seldom comes true from seed. The artichoke is popular chiefly where French or Spanish influence exists. The food value of the artichoke is small. Production amounts to around 30,000 tons annually.

ASPARAGUS, *Asparagus officinalis*, Liliaceae. This species, the most important of several comestible species in the genus, is believed native to the eastern Mediterranean area. It is today escaped and naturalized over much of the world. Originally it was regarded as a cure-all medicinal, but today it is one of the most universally esteemed vegetables. The plant is a dioecious perennial, developing leafless green stems in spring. These are cut by hand when they are only a few centimeters tall. Several cuttings are possible before the spears are allowed to grow to carry on vital activities of the plant. Small flowers and berries mature in late summer on the female plant (which is said to produce larger but fewer shoots than the male), after which the stems die to the ground. Asparagus plants prefer fertile, nitrogenous soils and sunny locations. They may be

Emerging asparagus spears almost ready for harvest. (Courtesy USDA.)

tralian species have been successfully utilized, although more as fodder plants than for human consumption.

BAMBOO. Shoots of several bamboos, such as *Phyllostachys edulis*, *P. mitis*, and species of *Bambusa*, *Dendrocalamus*, and *Gigantochloa* have been used for food since time immemorial in China and Indonesia. Tests in Puerto Rico have shown *Dendrocalamus membranaceus*, *Guadua angustifolia*, and *Bambusa polymorpha*, native to India, to be sweet and tender, the last named especially useful for canning. Typically, bamboo tips are cut as the spreading rhizomes turn up to make an above-ground stem. Outer scales are cut away, and the inner portions boiled for consumption.

BEANS, *Phaseolus* and other genera, Leguminosae. Young pods of most of the pulses reviewed in Chapter 17 are edible, and often consumed—pod and all—as green vegetables. Especially used this way are the so-called green beans or snapbeans, *P. vulgaris*, of which the commercial crop in the United States exceeds a half-million tons annually, mostly used for processing. Oregon, New York, and Wisconsin are the leading producing states. Fresh lima beans, *P. lunatus*, are also of importance as a fresh vegetable, commercial production in the United States running over 100,000 tons annually, chiefly from California.

BROCCOLI, *Brassica oleracea* var. *botrytis* (*B. oleracea italica*), Cruciferae. Presumably broccoli originated about 2,500 years ago from the wild cabbage common in coastal Europe, which had been spread through the Near East to the Orient at a very early date. Ancestral forms of modern varieties seem to have been selected in Italy about the time of Christ. European broccolis are generally different than those grown in North America. "Green-sprouting" varieties are mostly planted in the United States as summer annuals, commercial production there running about 130,000 tons annually, mainly from Califor-

propagated from young rhizomes or from seed (taking 3 years to bear). The cut shoots, asparagus, are largely water (94 per cent), but do contain considerably more protein than most vegetables. Formerly most asparagus was blanched by hilling up soil about the developing shoot, but popular taste has accepted the natural green asparagus. Production in the United States amounts to nearly 200,000 tons annually, most of which is processed, over half grown in California.

ATRIPLEX, *Atriplex sp.*, Chenopodiaceae. The Asiatic orach or mountain spinach, *A. hortensis*, has been long used as a garden green in Europe. Culture and utilization are the same as for spinach. A few Aus-

nia. In Europe, a hardier "cauliflower-heading" type is grown for cutting in winter and spring.

Although the American "sprouting broccoli" traces its ancestry back to the ancient Greeks, it has become popular only within recent decades. Early in this century green vegetables were shown to have more vitamin content than blanched vegetables. This encouraged acceptance of broccoli, already grown to some extent by Italian gardeners on the East coast. Now American selections provide cultivars especially suited to freezing and local markets. The part of the broccoli plant consumed is the fleshy-stemmed flowering head. It is cut before the buds open. When the terminal head is removed, lateral shoots develop, smaller but equally tasty.

BRUSSELS SPROUTS, *Brassica oleracea* var. *gemmifera*, Cruciferae. Brussels sprouts is another of the cabbage clan appearing to have been developed from a primitive cabbage similar to kale or collards. This may have been accomplished in the fourteenth century, near Brussels (hence the name) in northern Europe. In effect the Brussels sprouts plant is a tall, stemmy cabbage, in which many individual "miniature cabbage heads" develop at the stem joints where leaves have been shed, rather than a single terminal head. Sprouts are about the size of a walnut, tender, and flavorful. They are picked individually as they mature (newer ones develop above). In mild climates Brussels sprouts can be planted directly outdoors early in spring, but up to 130 days are required from planting to harvest. Commercial harvest in the United States is almost 40,000 tons annually, chiefly from California.

CABBAGE, *Brassica oleracea* var. *capitata*, Cruciferae. Cabbage is one of the oldest of vegetables. It was in general use by Middle Eastern civilizations 4,000 years ago, already selected for various forms, and possibly had been spread to southern and eastern Asia as far back as eight millenia ago. Today it occurs

The wild cabbage, *Brassica oleracea*, the ancestor of our cabbage, cauliflower, broccoli, brussels sprouts. (Courtesy Chicago Natural History Museum.)

as a widely diversified array of special cultivars in many shapes and hues. The original wild cabbage much resembles kale or collard, being erect and leafy-stemmed rather than fairly squat and forming heads. It is found throughout Europe, but domestication apparently first occurred in the eastern Mediterranean region. Celtic invasions of the Near East in pre-Christian times are said to have brought already domesticated forms of cabbage back into Europe, where at a later date, especially in northern Europe, the heading types today recognized as cabbage are thought to have been developed. Cabbage, along with broccoli and cauliflower, has a diploid chromosome complement of 18, different from the leafy *Brassicas* of the Orient (e.g., *B. campestris*).

Cabbage is an important food plant, rich in vitamins and minerals. It is much grown in Europe, although North America, too, consumes well over 1 million tons annually (nearly a quarter-million for sauerkraut). It is eaten raw, such as for coleslaw, or boiled and stewed, often with meats. Sauerkraut is highly esteemed by Teutonic peoples. Sauerkraut is made by immersing cut cabbage in brine and allowing it to undergo a lactic-acid fermentation under controlled conditions. Leafy, loose-heading cabbages (the Savoy type, sometimes identified as the variety *bullata*) are fairly popular in southern Europe but not much used elsewhere.

Cabbage is a biennial, heading one year and flowering the next. The large, shiny leaves are at first spaced, but grow into a tight ball on compressed stem nodes as the plant ages. If the head is cut for consumption, smaller lateral heads (often no bigger than a Brussels sprout) develop in the exposed leaf axils with most varieties. These (or the original cabbage head), if left unharvested, will bolt into an extended flowering stalk during the second growing season after a cold interlude. In certain climates with hot summers there may be premature bolting instead of head formation, the plant behaving more like an annual than a biennial.

For good head formation cabbages require cool weather. Upstate New York is a well-known cabbage-growing region in the United States, and so of course is northern Europe. In most cases seedlings are started under glass and transplanted to the garden or field in early spring, where they are cultivated and weeded as with any crop. The plants resist light frost well. In warm climates the seed may be sown directly outdoors, as may be done with autumn plantings in the southern United States (Florida and Texas are as important for winter cabbage as New York and Wisconsin are for the late-summer crop). Cabbages take from 2 (early varieties) to 4 (late varieties) months to develop heads after setting out, the large-headed ones (heads weighing 10 pounds or more, often used for kraut) generally being the slower to mature. The heads are cut for market by hand, and may be stored through the winter in cold cellars or pit houses. They hold best under moderate humidity at a constant temperature close to freezing. Home gardeners can store cabbages in trenches covered with mulch. Or the plants can be pulled and hung by the roots in an unheated shed. Gardeners having limited space may wrap the heads in wax

Head of cabbage. (Courtesy USDA.)

paper and store them for many weeks in a cool basement.

Variations in cabbage are legion. Cabbages are often classified according to head shape ("oval," "pointed," "round," "drumhead," etc.), as well as by color and growth cycle. Widespread breeding efforts have resulted in cabbages adapted to almost any climate. In Japan there have been notable breeding programs, with plantings today almost entirely of F_1 hybrids. There are cultivars suited to spring, summer, or autumn sowing, and suited to differing elevations. Types which resist bolting in hot weather have especially been sought, perpetuated for seed production by vegetative cuttings. With adaptation mostly achieved, attention is now directed toward improved nutritional quality.

There has been much selection of disease-resistant cultivars. Serious cabbage diseases include a "yellows" caused by *Fusarium* (which attacks through the roots when soil temperature is above 65°F, blocking sap transfer), a debilitating mosaic caused by a virus (transmitted by aphids), and of course the notorious clubroot caused by the *Plasmodiophora* fungus in the soil.

CABBAGE PALMS, Palmae. The central leaf bud or "heart" of various species of palms is cut out and eaten raw (like celery, or in salads) or cooked (like cabbage). Especially used are the royal palms, *Roystonea*. Cutting out the heart kills the palm; in the cavity left, larvae of certain beetles develop which are considered a delicacy by many tropical peoples. Roasted, these larvae reportedly pop like chestnuts, and are said to have the consistency and flavor of this nut.

CADUSHI, *Cereus repandus*, Cactaceae. In the Antilles of the southern Caribbean, especially on Curaçao, this cactus has overrun land deforested by the early colonists. It is said that stems may be peeled, cut crossways, and fried like potatoes. More often sections of stem have the spines scraped off with a knife, and are then sold as fresh produce; the section is peeled, and strips of the flesh eaten fresh, or dried and pounded into a powdery meal. Mashed pulp is also used to make a mucilaginous soup.

CAULIFLOWER, *Brassica oleracea* var. *botrytis*, Cruciferae. Cauliflower is another of the specialized cole plants. As with broccoli, the thickened fleshy stem that supports the highly modified flowering head is the portion consumed. Unlike broccoli, however, once the head is cut regeneration from shoots does not take place and the usefulness of the plant is ended.

Cauliflower is often regarded as the aristocrat of the cabbage genus, a highly esteemed delicacy both in Europe and North America. Primitive forms of cauliflower were evidently in cultivation in the Near East in pre-Christian times, from which selections adapted to the cooler climate of northern Europe were developed during the Middle Ages that have yielded modern varieties. Cauliflower is a bit more difficult to grow and more troublesome to tend than are other members of the cabbage clan. The head becomes stunted or discolored either by near-

Cauliflower head. (Courtesy USDA.)

freezing or unduly hot weather. Most luscious heads occur on plants which grow vigorously during mildly cool weather. Varieties have been developed that are reasonably heat-tolerant, but stunted heads slowly formed in hot weather may have a bitter flavor.

The white color of cauliflower contributes to its elegance. If the growing head is left exposed the top part tends to bronze or turn purple. To avoid this, one or more of the large upper leaves are customarily tied over the developing head to keep it blanched. This, of course, adds to the labor of tending the crop and its costliness.

Cauliflower is grown as a summer annual. For the "early" crop, seedlings are started under glass and transplanted to the garden after danger of frost has passed. Harvest is generally 60–100 days from planting out. The "late" crop can be seeded directly outdoors in June or July, for autumn harvest. As is true for *Brassicas* in general, cabbage worm may be a pest that requires application of insecticides of a type that will not contaminate the comestible head. The fresh cauliflower head, cut into sections, can be used as a relish; or it may be boiled as are broccoli and cabbage. Commercial production in the United States is about 125,000 tons annually.

CELERY, *Apium graveolens*, Umbelliferae. "Wild" celery, believed to have originated in the Mediterranean area, is found in wet locations throughout Europe and up to the Himalayas. The modern cultivated celery was probably domesticated in Europe from the wild type, first as a medicinal, then (about the close of the Middle Ages) as a flavoring, and finally as a food. Throughout the nineteenth century celery grown in Europe and North America was blanched, usually by banking earth, paper, or planks about the edible portion to keep it white. But more recently green celeries such as "Pascal" have been widely accepted. Celery is an exacting crop, requiring fertile, moist, well-prepared soils and cool climates. It is ordinarily started in hotbeds and transplanted to field or garden, in the United States as a summer crop in the North and a winter crop in the South. Production in the United States amounts to about 750,000 tons annually, largely from California and Florida.

CEYLON OR MALABAR SPINACH, *Basella rubra* (*B. alba*), Basellaceae. *Basella* is one of the better substitutes in the tropics for true spinach, which grows poorly there. Ceylon spinach is widely distributed and cultivated in both the Eastern and Western Hemispheres,

Celery before (left) and after trimming for shipping. (Courtesy USDA.)

This celery harvester in Florida travels five or six feet a minute, covering 24 rows of celery. The celery is cut, trimmed, washed, sorted, and packed in crates before it leaves the field. (Courtesy USDA.)

and seed is being offered for home garden planting in the United States. It is a climbing, annual or biennial herb native to southern Asia. It is much used as a potherb in parts of India (Bengal and Assam), where it is found as a hedgerow in almost every village. The tip growth is consumed boiled like spinach, being rich in minerals and vitamins. The plant grows best during warm, rainy periods. Propagation may be by root or stem cuttings, or by seed. Plants are typically trained to trellises and are ready for picking within 2–3 months.

CHARD OR SWISS CHARD, *Beta vulgaris* var. *cicla*, Chenopodiaceae. This variety is merely a beet without the enlarged, fleshy root. It is similar to the wild beet of the Mediterranean lands, from which it was presumably derived in pre-Christian times. The rosette of leaves with the fleshy petioles serve as the comestible part, and are eaten as greens, like spinach. Chard is grown more widely in Europe than in the United States.

CHINESE CABBAGE, CELERY CABBAGE, OR PETSAI; CHINESE MUSTARD OR PAKCHOI; SPINACH MUSTARD: *Brassica campestris* (*pekinensis*, *chinensis* and *perviridis* groups), Cruciferae. This species is primarily a crop of the Far East where it has been cultivated for more than 2,500 years, and occurs in bewildering variety. It is a mild "mustard" with a celery-like flavor, used mostly for salads in the Western world. Mineral and vitamin content are good. The large, later leaves form a central column or "head" suggestive of cabbage, especially when grown at higher elevations. The species is native to eastern Asia, and has a diploid chromosome complement of 20.

CHINESE SPINACH, EDIBLE AMARANTH OR KULITIS, *Amaranthus tricolor* (*A. viridis*) and other species, Amaranthaceae. The amaranths are an aggressive group of hot-weather weeds, several of which were used for greens by the American Indians, and Chinese spinach long grown in the Orient for the tender tips

513

Climbing malabar spinach withstands hot weather.
(Courtesy Burpee Seeds.)

Heat-resistant tampala, another excellent spinach
substitute. (Courtesy Burpee Seeds.)

consumed like spinach and in stews. Its vitamin A content is quite high. Propagation is generally by seeds broadcast on raised beds. The plant grows remarkably well in weather hot enough to be deleterious to other vegetables. The plants are hand pulled if less than about 20 centimeters tall—or cut at ground level if taller—bundled, and sent to market. *A. hybridus cruentus* is a popular crop in West Africa, and tampala, *A. gangeticus*, said to be the species most grown in the Caribbean area. Tampala has long been grown in India and China, and seed is now being offered for home gardens in the United States.

COLLARDS, *Brassica oleracea* var. *acephala*, Cruciferae. The collard is an unusually large form of kale or borecole, much like the primitive, leafy, nonheading cabbage first domesticated in Europe millenia ago. Collards are extensively grown for greens in the southern United States, where they are quite easily raised and withstand hot weather much better than do most members of the cabbage genus. They are often started in outdoor seedbeds, then transplanted for spacing several feet apart. The leaves are picked as

needed for boiling greens. The plant is a biennial, and in mild climates continues green through winter (indeed, the flavor is said to improve after frost), providing vitamin-rich fare throughout the colder months.

CRESS (GARDEN), *Lepidium sativum*, Cruciferae. This well-known European salad plant is cultivated in cool climates.

CRESS (SPRING), *Barbarea verna*, Cruciferae. This biennial, grown as an annual, is utilized as a salad plant in Europe.

CRESS (WATER), *Nasturtium officinale* (*Rorippa nasturitum-aquaticum*), Cruciferae. This is a well-known perennial salad plant that grows in clear, cold, shallow water.

ENDIVE OR ESCAROLE, *Cichorium endivia*, Compositae. Endive is closely related to chicory. The frilled lower leaves are used for salads or as cooked greens. The species is indigenous to the eastern Mediterranean area, and was cultivated by the ancient Egyptians. It is hardy and more tolerant of summer heat than lettuce is. Most sowing of the plant is for late autumn or winter greens. The commercial crop in the United States runs slightly over 50,000 tons annually, mostly from Florida.

KALE, *Brassica oleracea* var. *acephala*, Cruciferae. This biennial, native to the eastern Mediterranean area and one of the primitive cabbages, is essentially a collard with frilled leaves. In fact no distinction was made between kale and collards in ancient Greece and other early civilizations, and what has been said concerning collards applies equally well to kale. Commercial production in the United States is only about 4,000 tons annually.

KOHLRABI OR CABBAGE TURNIP, *Brassica oleracea*, var. *caulorapa*, Cruciferae. Another member of the cabbage clan, kohlrabi, seems to be of comparatively recent origin. Selections from a cabbage-like ancestral form seem to have been made near the close of the Middle Ages, to give the modern grotesque plant with a few, erect, petiolate leaves and the spherical, thick, fleshy lower stem, the comestible portion. The plant is a hardy, cool weather biennial, cultivated much the same way as cabbage.

LETTUCE, *Lactuca sativa*, Compositae. Lettuce, the world's most popular salad plant, is today grown worldwide. It probably originated in Asia Minor from the wild lettuce, *L. serriola* (*L. scariola*), one of about 100 wild species. Lettuce was popular with pre-Christian peoples, appearing on the royal table of Persian kings as long ago as 550 B.C. It was grown in several forms by the Romans, and had spread to China by the fifth century A.D. Early types were looseleaf, the dense, headed varieties being more recent creations developed during the Middle Ages. Cos or romaine lettuce is a rosette form with erect leaves, comparatively tolerant of heat, probably developed in southern Europe. Various types of lettuce are often grouped under varietal designations, "var. *angustana*" referring to the narrow-leafed asparagus–lettuce or "celtuce" (a name coined by combining syllables from "*cel*ery" and "let*tuce*," implying a flavor and structure combining features of both of these vegetable favorites),

grown for its edible stem; "var. *crispa*," to the broad-leaf, fringed, or curled-leaf lettuce; "var. *longifolia*," to the erect cos types; and "var. *capitata*" to the rounded cabbage or head lettuces. Looseleaf types, because of ease of growing, are common home garden plants, but commercial growing involves chiefly heading types, which keep and market more satisfactorily. Leaf lettuces have higher vitamin content than heading types, although lettuces in general have little nutritional value.

Lettuce is a diploid ($2n = 18$), self-fertilized, herbaceous annual with large, glabrous, crisp leaves, which bolts quickly to dandelion-like flowers and seeds in hot weather, developing at the same time a bitter taste. For spring planting lettuce may be started in hotbeds and transplanted. Some lettuce is grown under glass near urban markets, but most of the crop in the United States is today grown as a winter crop in the irrigated deserts of Arizona and California. Lettuce does best on sandy loams or muck soils. Although the amount of hand labor has been materially reduced in growing lettuce in the United States, it is still not an inexpensive crop to produce. It must be carefully managed to yield marketable, solid heads. Experience has shown growers the exact fertilization practices to achieve this and to delay bolting. Side dressings of nitrogen are usually utilized, and special machines have been developed for

A "crisphead" lettuce. (Courtesy USDA.)

A packing van moves through a lettuce field. The lettuce is wrapped with shrink film, boxed, and dropped off for another truck to take to the sheds. (Courtesy USDA.)

planting, cultivation, and even selective harvesting. Modern equipment simultaneously shapes up the planting bed for irrigation, plants the seed, and distributes starter fertilizer. Gigantic loading machines are driven onto the field for field-packing, the

Two-on-one celtuce has tasty leaves and delicious stalk interiors. (Courtesy Burpee Seeds.)

crop immediately moved to refrigerated cars for shipment. Commercial production of lettuce in the United States exceeds 2 million tons annually. Lettuce growing is practical in the tropics only at high elevations, or where a fairly long cool season prevails. Another lettuce species, *L. indica*, is grown in the oriental tropics as a cooking green.

Sweet Maize or Sweet Corn, *Zea mays* var. *saccharata*, Gramineae. Maize has been discussed along with other cereals, but sweet corn consumed in the immature stage (either fresh as "roasting ear," or canned or frozen) is a "vegetable" used in quantities that dwarf those of most other "fresh vegetables." Commercial production in the United States amounts to about $2\frac{1}{2}$ million tons annually, over three-quarters of which is processed. (See Chapter 16 for a fuller discussion of maize.)

Indian or Leaf Mustard, *Brassica juncea*, Cruciferae. This common annual seems to have evolved by several steps in central Asia. It is an extremely variable, erect, branched, glabrous species, used to some extent for cooked greens. Related *B. carinata*, of Ethiopia, has been introduced as an especially nutritious mustard green in the United States.

New Zealand Spinach, *Tetragonia expansa*, Aizoaceae. This spreading, prostrate, succulent annual of Australia, New Zealand, southern South America, and Japan is used as a spinach substitute. It does not bolt, like spinach, in hot weather, and yields continuously. The tender tip sections of stem are gathered by hand and cooked like spinach. The hard seeds are generally soaked to speed germination.

PARSLEY, *Petroselinum sativum*, Umbelliferae. Parsley has had a long history as a garnish and flavoring, having been utilized since ancient times. One type, Hamburg or "turnip-rooted" parsley, bears a fleshy root which is used like celeriac. It is a biennial species, indigenous to the Mediterranean area.

RHUBARB OR PIEPLANT, *Rheum rhaponticum*, Polygonaceae. This species, one of several that are edible, is believed native to the Asia Minor area. Reportedly a Chinese rhubarb was used as a medicinal as early as 2700 B.C. Rhubarb is one of the few vegetables in which the petiole is the part consumed. The garden type is a coarse perennial with large cordate leaves, adapted to cool, moist climates and cold winters. The fleshy petioles are cut by hand, the leaf trimmed, and the stalks marketed in bundles. The leaf portion is said to contain mildly poisonous substances. Propagation is typically asexual, by dividing clumps.

SPINACH, *Spinacia oleracea*, Chenopodiaceae. This popular herbaceous annual is native to southwestern Asia, and was apparently introduced into Europe and the Far East only after ancient Greek and Roman times. Both smooth-seeded and prickly-seeded types have been developed. The plant, started from seed directly in garden or field, develops a crown of dark green, crisp leaves used as a favorite potherb. The plant bolts quickly in hot weather under long days and is grown chiefly as a spring or autumn crop. It is vigorous and quick-growing, and in well-prepared seedbeds it usually reaches harvestable size ahead of weeds.

More rapid growth and greater succulence are attained with nitrogen fertilization. Commercially, seed is drill-sown, the seedlings mechanically thinned and cultivated as needed. Cutting for harvest is also being mechanized. Production in the United States amounts to about 200,000 tons annually, about one-fourth of which is marketed fresh and the rest processed. California, Texas, and Florida are the leading producing states, harvesting from November until April.

WATER SPINACH, SWAMP CABBAGE, OR KANGKONG, *Ipomoea aquatica*, Convolvulaceae. The species ranges from tropical Africa through southeastern Asia to Australia, and is recorded as a cultivated crop in the records of the Chin dynasty of China about A.D. 300. The plant is an herbaceous annual, semi-aquatic in nature, with trailing shoots and long-petioled, somewhat fleshy leaves. Young stems are usually fried in oil. It is an excellent hot-weather green at times of year when the cabbage group does not flourish, and is said to supply about 15 per cent of the local vegetable output in Hong Kong during the summer season.

MISCELLANEOUS GREENS. The tender shoots of a number of plants serve locally as greens. Among them are: *Portulaca oleracea*, Portulacaceae, circumboreal; *Talinum triangulare*, Portulacaceae, Africa; Sierra Leone bologi, *Gynura biafrae*, Compositae, West Africa; chervil, *Anthriscus cerefolium*, Umbelliferae, Europe and Near East; corn salad, *Valerianella oliteria*, Valerianaceae, Europe and Africa; various *Crotalarias*, Leguminosae, Central America; dandelion, *Taraxacum spp.*, Compositae, circumboreal; jute or bush okra, *Corchorus olitorius*, Tiliaceae, Middle East; ogumoh or efodu, *Solanum nigrum guineense*, Solanaceae, West Africa; poke, *Phytolacca americana*, Phytolaccaceae, eastern North America; roselle, *Hibiscus saddariffa*, Malvaceae, Africa; sea kale, *Crambe maritima*, Cruciferae, Europe; udo, *Aralia cordata*,

Aralaceae, native to Japan; and many others. A large proportion of these are only haphazardly cultivated, or are taken from the wild; all are of restricted importance in the world economy.

SUGGESTED SUPPLEMENTARY REFERENCES

Boswell, V. R., and E. Bostelmann, "Our Vegetable Travelers," *National Geographic* **96**, 145–217, (August, 1949).

Correll, D. S., *The Potato and Its Wild Relatives*, Texas Research Foundation, 1962; Stechert-Hafner, N. Y., 1966.

Coursey, D. G., *Yams*, Humanities Press, N. Y., and Longmans, London, 1968.

Hawkes, J. G., and J. P. Hjerting, *The Potatoes of Argentina, Brazil, Paraguay and Uruguay*, Ann. of Bot. Memoir 3, Oxford Univ. Press, N. Y., 1969.

Huelson, W. A., *Sweet Corn*, World Crops Books, Interscience, N. Y., 1954.

Jones, H. A., and L. K. Mann, *Onions and Their Allies*, World Crops Books, Interscience, N. Y., 1963.

Jones, W. O., *Manioc in Africa*, Stanford Univ. Press, Stanford, Calif., 1959.

Knott, J. E., and J. R. Deanon, Jr., *Vegetable Production in Southeast Asia*, Univ. of The Philippines Press, Laguna, 1967.

Mortensen, E., and E. T. Bullard, *Handbook of Tropical and Sub-tropical Horticulture*, U. S. Dept. of State, Agency for International Development, Washington, D. C., 1964.

Nieuwhof, M., *Cole Crops*, Leonard Hill, London, 1969.

Oomen, H. A. P. C., "Vegetable Greens, A Tropical Undevelopment," from Proceedings of the 7th International Congress on Tropical Medicine and Malaria, Rio de Janeiro, Brazil, 1963, *Chronica Horticulturae* **4**, (1964).

Smith, Ora, *Potatoes: Production, Storing, Processing*, AVI Pub. Co., Westport, Conn., 1968.

Tai, E. A., W. B. Charles, E. F. Iton, P. H. Haynes, and K. A. Leslie (eds.), *Proceedings of the International Symposium on Tropical Root Crops*, Univ. of West Indies, Trinidad, Vols. 1 and 2, 1968.

Thompson, H. C., and W. C. Kelley, *Vegetable Crops*, McGraw-Hill, N. Y., 1957.

Tindall, H. D., *Fruits and Vegetables in West Africa*, FAO, Rome, 1965.

USDA, *Agricultural Statistics*, Washington, D. C., published annually.

Ware, G. W., and J. P. McCollum, *Producing Vegetable Crops*, Interstate Pub., Danville, Ill., 1968.

Winters, H. F., and G. W. Miskimen, *Vegetable Gardening in the Caribbean Area*, USDA Agriculture Handbook 323, Washington, D. C., 1967.

CHAPTER 19

Fruits

THE WORD "FRUIT" EVOKES THE CONCEPT of a pleasant-tasting delicacy plucked from tree or bush, "historically" first accomplished by man in the Garden of Eden. Actually, man's primate ancestors must have subsisted in great measure, at least seasonally, in this very fashion. Even by the time man was finally "out of the trees" and firmly on terra firma, but before agriculture had much evolved, he must have sought out fruits in the wild as a significant part of his hunting–gathering diet. Coprolites confirm that even after man had settled down in communal villages, foraging for wild fruits and game was very much a part of his existence. Just what types of fruit may have been the object of his attention in various parts of the world is indicated by the samplings in Table 19-1. Obviously, this is but a very small representation of the comestible fruits available to man and wildlife worldwide, but it gives some idea of what man has had at his beck and call.

Botanically a fruit is a matured ovary, especially of seed-bearing plants. Popularly, however, this distinction often fails. Thus this chapter, like some before it, has not clear-cut boundaries, but rather is transitional from food plants already discussed as grains, pulses, and "vegetables" to a subsequent chapter on beverages (most of which are derived from fruits). Many nuts are botanically fruits also, or at least the seeds from fruits

freed by removal of hulls consisting of ovarian structures. In this chapter, however, fruits that are more or less fleshy rather than dried (grain, pulses, nuts; even "vegetables" such as sweet corn and snapbeans) are the object of attention. Such fruits vary tremendously in taste and in nutritional value, ranging from such extremes as pump-

One of the largest fruits, a jackfruit, which may weigh nearly 100 lbs. Thailand. (Courtesy USDA.)

519

Table 19-1

Representative Fruits in Three Parts of the World

I. Useful wild fruits of the United States

Amelanchier spp., Shadbush	*Morus rubra*, Red mulberry
Annona glabra, Pond apple	*Nyssa spp.*, Ogeechee lime
Arctostaphylos spp., Bearberry	*Parthenocissus quinquefolia*, Virginia creeper
Ardisia escalloniodes, Marlberry	*Passiflora incarnata*, Maypop
Asimina triloba, Papaw	*Photinia salicifolia*, Christmasberry
Chiogenes hispidula, Creeping snowberry	*Physalis spp.*, Groundcherry
Chrysobalanus icaco, Cocoplum	*Podophyllum peltatum*, May apple
Chrysobalanus pallidus, Gopher apple	*Prosopis spp.*, Mesquite
Chrysophyllum oliviforme, Satinleaf	*Prunus spp.*, Plum
Coccoloba uvifera, Seagrape	*Reynosia septentrionalis*, Darling plum
Cornus spp., Dogwood	*Rhamnus spp.*, Buckthorn
Crataegus spp., Haw	*Rhus spp.*, Sumac
Diospyros virginiana, Persimmon	*Ribes spp.*, Currant
Elaeagnus commutata, Silverberry	*Rosa spp.*, Rose
Empetrum nigrum, Crowberry	*Roystonea elata*, Royal palm
Ficus aurea and *F. laevigata*, Wild fig	*Rubus spp.*, Blackberry
Fragaria spp., Strawberry	*Sambucus spp.*, Elderberry
Gaylussacia spp., Huckleberry	*Shepherdia spp.*, Buffaloberry
Gleditsia triacanthos, Honey locust	*Smilacina spp.*, False Solomon's-seal
Grossularia spp., Gooseberry	*Vaccinium spp.*, Blueberry
Lonicera spp., Honeysuckle	*Viburnum spp.*, Black haw
Lycium spp., Wolfberry	*Vitis spp.*, Grape
Malus spp., Crabapple	*Yucca spp.*, Spanish bayonet
Mitchella repens, Partridgeberry	

II. Familiar village market fruits in central Mexico

Acrocomia mexicana, Palm	*Hylocereus undatus*, Cactus
Ananas comosus, Pineapple	*Mangifera indica*, Mango
Annona cherimola, Cherimoya	*Manilkara zapotilla*, Sapote
Annona reticulata, Custard apple	*Musa paradisiaca sapientum*, Banana
Bumelia laetevirens	*Passiflora ligularis*, Passion fruit
Calocarpum mammosum, Mamey sapote	*Persea americana*, Avocado
Carica papaya, Papaya	*Physalis ixocarpa*, Physalis
Casimiroa edulis, White sapote	*Pouteria campechiano*, Yellow sapote
Citrus aurantifolia, Lime	*Psidium guajava*, Guava
Citrus paradisi, Grapefruit	*Punica granatum*, Pomegranate
Citrus reticulata, Tangerine	*Pyrus communis*, Pear
Citrus sinensis, Orange	*Pyrus malus*, Apple
Diospyros ebenaster, Black sapote	*Tamarindus indica*, Tamarind

Table 19-1 (cont.)

Representative Fruits in Three Parts of the World

III. Familiar cultivated fruits in India

Achras zapota, Sapota	*Litchi chinensis*, Litchi
Ananas comosus, Pineapple	*Malus sylvestris*, Apple
Annona squamosa, Sweetsop	*Mangifera indica*, Mango
Carica papaya, Papaya	*Musa paradisiaca*, Banana
Citrullus vulgaris, Watermelon	*Phoenix dactylifera*, Date
Citrus paradisi, Grapefruit	*Prunus armeniaca*, Apricot
Citrus reticulata, Mandarin	*Prunus domestica*, Plum
Citrus sinensis, Sweet orange	*Prunus persica*, Peach
Cucumis melo, Muskmelon	*Psidium guajava*, Guava
Eriobotrya japonica, Loquat	*Punica granatum*, Pomegranate
Ficus carica, Fig	*Syzygium cumini*, Jambolan
Grewia asiatica, Phalsa	*Vitis vinifera*, Grape

Assembled from listings by J. F. Morton, *Principle Wild Food Plants of the United States;* T. W. Whitaker and H. C. Cutler, *Food Plants in a Mexican Market;* and P. Maheshwari and S. L. Tandon, *Agriculture and Economic Development in India;* respectively in *Economic Botany* **17**, 319–30 (1963); **20**, 6–16 (1966); and **13**, 205–42 (1959).

kin flesh, through Temperate Zone pome and stone fruits like apple and peach, to the delicate tropical tree fruits such as custard apple and mangosteen. That most fruits are watery and contain relatively little energy value is indicated in the data of Table 19-2. The date almost alone is a concentrated source of energy, although the avocado, banana, and fig are certainly as high in energy content as are many vegetables. Almost all of these fruits are good sources of minor minerals and vitamins, however.

In everyday language the distinction between "vegetable" and "fruit" is confused, and botanically quite unacceptable. For example, are tomatoes, rhubarb, and squash vegetables or fruits? Botanically rhubarb is a petiole, and was listed among vegetables in the previous chapter. Tomato and squash are botanically true fruits. Fortunately, most other foods that are botanically fruits are also recognized as fruits by popular definition. As a group the comestible fruit-bearing plants are

more diverse than are root or herbage crops discussed in the preceding chapter. A number have limited or local use only, and a few are still taken from the wild or from semi-cultivated plants. All told world fruit production must run many hundreds of millions of tons annually: figures for commercial production of deciduous fruits (pomes, drupes and grapes) alone approach 100 million tons, mostly from Europe. Throughout the world, in the tropics especially, far more sustenance than this must come from fruits consumed by man and beast, and remains unrecorded.

Not long ago fruit production was rather casual, often from a few trees and bushes planted on the home farm or even urban properties. In most of the tropics fruits are still grown very much as dooryard plants for home consumption. But with the improvements that have come in transportation, refrigeration, freezing, and canning, food industries based upon fruit have frequently become large and businesslike. In keeping

Table 19-2

Constituents of Fruits Grown in Temperate, Subtropical, and Tropical Climates

TEMPERATE ZONE	% Water	% Protein	% Fat	% Carbo-hydrates
APPLE	84	0.3	0.4	12
APRICOT	85	1	0.1	10.4
CHERRY	82	1.2	0.5	10.5
GRAPE	82	0.8	0.4	14.9
PEACH	87	0.5	0.1	8.8
PEAR	83	0.7	0.4	8.9
PLUM	86	0.7	0.2	8.3
STRAWBERRY	90	0.6	0.6	5
SUBTROPICAL				
AVOCADO	68	1.7	20	5
DATE	20	2.2	0.6	75
FIG	78	1.4	0.4	20
GRAPEFRUIT	89	0.5	0.2	10
LEMON	89	0.9	0.6	8
ORANGE	87	0.9	0.2	11
TROPICAL				
BANANA	75	1.2	0.2	23
MANGO	81	0.7	0.2	17
PAPAYA	89	0.6	0.1	10
PINEAPPLE	85	0.4	0.2	14

After J. B. Biale, *Scientific American* **190**, 44 (May, 1954).

with agriculture generally the small orchard on a diversified farm is being replaced by large professionally managed orchards and plantations capable of capitalizing the increasingly complex and expensive equipment now becoming used for harvest. Not only does mechanization bring with it machines that shake fruit loose to fall unbruised upon padded screens, but horticultural manipulations also spell great change. Dwarf stock that can be more easily reached for picking is becoming commonplace, and chemicals are applied not only to control the growth of the tree, but the ripening and fall of the fruit itself. Other chemicals and techniques are much concerned with the keeping quality of the fruits, and their ability to be shipped to distant markets. Unfortunately, the economic considerations involved dictate that only a limited selection of varieties will be marketed; many of the tastier or locally preferred cultivars are of diminishing occurrence except as grown on the home grounds.

In this chapter fruits will be grouped into three categories. First come the fruits of herbaceous annual plants, many of which are popularly recognized as vegetables (e.g., tomato, eggplant, okra, squash). These are botanically true fruits. They are mostly

A powered shaker drops citrus onto a canvas frame in Florida. High labor costs speed the search for mechanization in fruit-picking. (Courtesy USDA.)

grown as summer annuals in temperate climates, although in the tropics they may be short-lived perennials. Next taken up are the perennial fruits of temperate lands, a group recognized popularly as "fruits." A few are herbaceous, such as the strawberry, but most come from woody shrubs (brambles, such as yield the "berry" crops") and trees (typical orchard fruits such as apple, pear, and peach). The final section will be devoted to tropical fruits, a diverse assemblage which is, except for the banana, citrus, and pineapple, generally little known to Temperate Zone peoples. A number are of high importance, however, to primitive peoples in under-developed lands.

Fruits of Herbaceous Annuals

In this group are found a number of the vegetable garden favorites grown as annuals in temperate climates, although sometimes persisting under tropical conditions. The group is dominated by two plant families, the Solanaceae and the Cucurbitaceae. The former consists mainly of rank, erect or sprawling herbs, often strongly pubescent; its most important fruits are the tomato, eggplant, and capsicum peppers. The latter is a family of coarse, trailing, usually tendriled vines; its most important fruits are melons and squashes.

Commercially, these fruits are regarded as "vegetables"; their growing, harvesting, and

Cucurbitaceae fruits. The family is a major source of herbaceous annual fruits. These are different cultivars of the watermelon, *Citrullus.*

handling fits the general scheme for root, stem, and leaf vegetables noted in the previous chapter. A fruit is more perishable than most roots and stems, and hence the crops of this section are more seasonal than items like the potato or even cabbage. Only within recent years has mechanized harvest of such fruits evolved, especially spurred by laws restricting the importation of Mexican stoop labor in the southwestern United States. Over much of the world this group of food fruits is grown in home gardens where labor cost is not a factor.

ACHOCHA, *Cyclanthera pedata edulis,* Cucurbitaceae. This is an annual vine of Peru, the fruits of which are used in pickles and as a vegetable in some parts of South America.

AFRICAN CUCUMBER, BITTER GOURD, OR BALSAM PEAR, *Momordica charantia,* Cucurbitaceae. This is probably an African species, but is now extensively cultivated and escaped in all tropical and subtropical countries. The plant is a rank-smelling, fast-growing, herbaceous vine. The variable, warty pepo may be sliced and dried, and is especially used for curries in India. In the Orient immature fruit (less bitter) is gathered from plants trained to trellises, for cooking with fish or meat, and sometimes for pickling. The species is popular in folk medicine in Africa, and is recommended as an abortifacient in Latin America. Juice from the fruit is a purgative strong enough to be poisonous to children and dogs.

BOTTLE GOURD, *Lagenaria siceraria,* Cucurbitaceae. The species is presumably native to Africa, where "wild" plants (which may be escapes) are still found. Since the bottle gourd occurs in archeological remains in both Mexico and Peru, some 9,000 years old, common ancestry for both Old and New Worlds, or transoceanic transport in prehistoric time must be accepted. The fruit floats well, and, indeed, its greatest economic importance has been as an impervious

container used by primitive peoples. Some authorities view the bottle gourd as being inedible, but others note that young fruits are sold in Latin American markets, and that both the fruits and seeds are used for food in parts of the Orient. In that seeds are infrequently found where gourd remains occur in archeological diggings, it can perhaps be presumed that the seeds were eaten by ancient peoples.

CASABANANA, *Sicana odorifera,* Cucurbitaceae. The fruits are occasionally consumed as a vegetable in Peru and Brazil, and may also be preserved. The pepo is fragrant, and is sometimes used to scent clothing in parts of Latin America.

Cucurbitaceous fruits: A, the chayote; B, a cucumber; C, a loofah, *Luffa cylindrica.* (Courtesy USDA.)

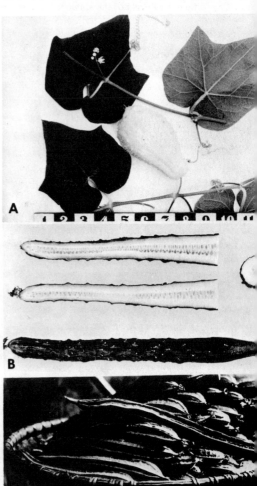

Harvesting cucumbers mechanically in Michigan. (Courtesy USDA.)

CHAYOTE OR CHRISTOPHINE, *Sechium edule*, Cucurbitaceae. This is a common, rapidly-growing, perennial vine native to tropical America. The greenish, soft-spined, one-seeded fruit is cooked like a squash, made into a sort of vegetable pudding, or into sweet-meats. A single plant may grow 100 feet long, bearing several hundred fruits weighing up to 1 pound or more each. The tuberous roots and young shoots occasionally serve as food, also.

CHINESE PRESERVING MELON OR WAX GOURD, *Benincasa hispida*, Cucurbitaceae. This is an annual vine of tropical Asia and Africa. Both young and mature fruits are boiled as a vegetable, pickled, or cut into pieces and candied as a sweetmeat. The species occurs wild in Java, but is most extensively cultivated in southern Japan.

CUCUMBER, *Cucumis sativus*, Cucurbitaceae. The cucumber is another important food plant unknown in the wild, although similar species occur in northern India, the presumed home of the cucumber. Diploid chromosome number is 14, in contrast to the honeydew melon (*C. melo*) in which it is 24. The species

has been cultivated in Asia for at least 3,000 years, being spread widely before Christian times and grown in France by the ninth century. Fruits range from small, stubby types only a few centimeters long to giants a half-meter or more in length, with rough or smooth skin and green or white flesh. They are eaten raw, or are seasoned in vinegar, brine, and spice to become pickles. The plant is a monoecious, rough-stemmed, trailing vine, but more condensed cultivars are being developed better suited to mechanized harvesting. Application of growth-regulator sprays such as maleic hydrazide just prior to flowering is said to increase flower production and fruit set. F_1 hybrid cucumbers are becoming commonplace. United States production of cucumbers approaches a half-million tons annually for pickles, and half this much for the fresh market, from all sections of the nation (but especially North Carolina and Michigan).

EGGPLANT, *Solanum melongena*, Solanaceae. Eggplant, diploid chromosome number 24, is believed to have originated in northern India where wild plants are said

The eggplant, representative of the second great family of herbaceous annual fruits, the Solanaceae. (Courtesy USDA.)

to be found. Small-fruited types were later secondarily developed in China. Apparently the species was spread to Europe by Arabic peoples during the Dark Ages, and yellow- and purple-fruited types were known to both northern Europe and Brazil before the close of the seventeenth century. The United States seldom utilized the eggplant as food before the present century, although it was widely planted for ornament. It is being more widely utilized today although its import- ance in the Western world is not so great as in the Orient. Eggplants are slow- growing and require a long season of warm weather to mature the fruit. Earlier fruit set can be induced with sprays of gibberellate and other chemicals at the beginning of anthesis. Usually eggplants are seeded in hotbeds in temperate climates and transplant- ed when a few centimeters tall. Few crops require so much hand labor. In the United States eggplant is often grown in the home garden, but commercial production is only about 30,000 tons annually, largely from Florida.

GHERKIN, *Cucumis anguria*, Cucurbitaceae. The species is said to be a mutant of the African wild species, *C. longipes*, which has bitter-tasting fruit. It was probably introduced into the New World through the slave trade. It is a slender, trailing, monoecious annual, bearing small, prickly, abundantly-seeded fruits. These are used mostly in pickles, but may be eaten as a cooked vegetable and used in some curries. The familiar "gherkin pickles" of the grocery store are seldom true gherkins, but are made from small cucumbers. The true gherkin is cultivated chiefly in Brazil and the West Indies.

HODGSONIA OR AKAR KAPAJANG, *Hodg- sonia macrocarpa* (*H. heteroclita?*), Cucur- bitaceae. This is a woody vine of the Malaysian area often festooning lofty trees, the fruits and roasted seeds of which are consumed as food in southeastern Asia. The seeds also serve as a source of oil, said to be used for "roasting" opium in the Orient.

HUSK TOMATO OR GROUND CHERRY, *Physalis pubescens* and other species, Solanaceae. A little-grown, decumbent, tomatolike species yielding small fruits with a thin husk. These may be eaten raw, but are ordinarily used for preserves.

MARTYNIA OR UNICORN PLANT, *Martynia proboscidea*, Martyniaceae. The small, curved, beaked pods of this species of the southwestern United States and Mexico are used when young for pickling. The plant is grown to a slight extent in home gardens and is handled much like okra.

MUSKMELON (including Cantaloupe and Honeydew), *Cucumis melo*, Cucurbitaceae. Muskmelons occur in a wide range of forms, including the smooth-skinned fruits known as honeydews, the hard, rough-rind cantaloupes, and various other modifications termed casaba, snake, banana, and other cultivars. Diploid chromosome number is 24. In most cases the fresh flesh is consumed as a delicacy, but cucumberlike types grown in the Orient are sometimes cooked as vegetables (e.g., the Conomon of Japan).

Naudin distinguished the following varieties of *Cucumis*: *reticulatus*, netted melons; *cantalupensis*, cantaloupe melons; *saccharinus*, pineapple melons; *inodorus*, winter or cassaba melons; *flexuosus*, snake melons; *conomon*, oriental melons; *chito*, mango melons; *dudaim*, dudaim melons. Speciation is quite confused, and neither this nor other segregations have been invariably accepted. Seat of origin is not definitely established for the species, some authorities suggesting Africa, others western Asia (Persia) or even India. Early records show muskmelon to have been under cultivation in Egypt 44 centuries ago. Wild plants have never been discovered with certainty anywhere. By the sixteenth century the principle thick-fleshed forms known today had been developed, and the species spread throughout the world.

The muskmelon is a prolific, trailing, annual vine with slender, pubescent stems, angled or lobed leaves, and yellow, monoecious flowers. It is a hot-weather plant, sown outdoors after the soil is warm and all danger of frost is past (or it may be started under glass in cooler climates). Dry climates are preferred for good ripening, and most of the United States production, which runs over 600,000 tons annually, comes from the West, especially California. Since undamaged fruit must be supplied the fresh market, mechanical harvesting has not been widely adopted and labor costs with the crop are high.

OKRA OR GUMBO, *Hibiscus esculentus*, Malvaceae. Okra is believed to have originated in the northeast African center of domestication, and is reportedly found wild in the upper Nile watershed. Used in Egypt for centuries, it spread to the Far East and Europe in Christian times, and was brought to the New World in the 1600's, where it found exceptional popularity in the French cookery of Louisiana. Okra is mainly incorporated in stews, soups, or gravies, or is mixed with other vegetables. It may be used either fresh or dried. It is about 10 per cent carbohydrate, 2 per cent protein; the ripe seeds yield an edible oil.

Common okra, capable of yielding a bast fiber as well as an edible fruit (vegetable). (Courtesy Chicago Natural History Museum.)

The okra plant is a robust annual, with palmate leaves, resembling other mallows. Diploid chromosome range is reported 72-132. The comestible fruit, botanically a capsule, develops as a slender, angled cone that turns mucilaginous upon cooking. Okra is widely grown in the home garden in the tropics, and scatteringly in temperate climates where greatest commercial use is for canned soup. It is a warm-weather plant, tolerant of soil types but demanding full sun.

GARDEN PEPPERS, *Capsicum annuum, C. frutescens, C. chinense*, Solanaceae. The peppers have been an archeological problem as well as a taxonomic puzzle. Fragments recovered from archeological sites are difficult to identify, as are modern taxons based upon the variable fruiting characteristics. Fragments of pepper have been discovered in Mexican caves dating to around 7000 B.C. It is difficult to distinguish wild from cultivated forms, and region of domestication is uncertain. Heiser recognizes five cultivated species in the genus, all with a diploid chromosome number of 24. He regards *C. frutescens* as being of South American origin, with a range extending through Middle America into the southern United States. *C. annuum* is the pepper most widely cultivated throughout the world today; its origin was probably Mexican. It was widely grown in Mexico at the time of the Conquest, and shows great diversity in the region today. A variety of this species, *minimum*, ranges from South America into the southern United States, often as volunteer plants from which small fruits are picked and marketed locally in Latin America. Of course cultivars selected for large fruits with special characteristics are used for the commercial crop. *C. frutescens* exists both as a cultivated and "wild" plant over much of tropical America, and may be conspecific with *C. chinense* (also of South America in spite of the specific name). Heiser summarizes the classification of the cultivated peppers more fully in *Taxon 18*, 36–45, (1969).

Capsicum pepper in a Florida field. (Courtesy USDA.)

The capsicum peppers are, of course, not to be confused with the spice pepper (*Piper*). Capsicum fruits became known as "pepper" because of their pungency, when discovered by Columbus in the New World while seeking true pepper in the Orient. A more appropriate designation would be "chilis" (signifying, in colonial Spain, "from Chile"). Capsicum peppers were brought back to the Old World by the early explorers, and often grown as ornamentals or sometimes "medicinals." They became best known for the "hot" types used to make pungent sauces. As a food capsicum peppers have little energy value, but generous vitamin C content. They are mostly used for salads picked green, but can be allowed to ripen until red; a bright red cultivar known as "pimento" is used for stuffing olives and for dressing up cheese. Paprika and cayenne pepper are derived by grinding dried fruits of certain capsicum cultivars. Chili powders are greatly favored by Mexican peoples, and generally include other spices as well as ground capsicum. The active ingredient of highly pungent chilis is capaicin, an aromatic phenol chemically similar to vanillin, so potent that it is said to be detectible to the taste when present in so diluted a proportion as 1 part per million.

Capsicum pepper production in the United States exceeds 200,000 tons annually, especially from Florida. A good deal of hand labor is required for the growing and harvesting, making it a relatively expensive crop to produce. Peppers thrive best in a warm, moist climate with a long growing season; they are very susceptible to frost. They set fruit poorly early in the season, but parthenocarpy can be induced and fruit set improved with gibberellates and other hormone sprays.

PUMPKIN, *Cucurbita pepo*, Cucurbitaceae. The pumpkins and squashes, along with beans and corn, were a staple in the diet of the ancient civilizations of the New World. Remains have been found in Mexico dating back to nearly 9000 B.C. (which include wild as well as cultivated cucurbits), in South America to 3000 B.C., and in New Mexico to 2500 B.C. It is reasonable to conclude that *Cucurbita* has been under cultivation for at least 9000 years. In that interval the species seems to have changed rather little, with the four most important species similar in appearance but separated by sterility barriers. The pumpkins and summer squashes are *C. pepo* (winter squashes and other pumpkins are regarded as *C. mixta*, *C. moschata*, and *C. maxima*). Many authorities regard *C. pepo* as having originated in northeastern Mexico (the other species in southern Mexico, except *C. maxima* in South America). *C. pepo* prefers cooler, drier habitat than do the other species.

The pumpkin group would have been exceptionally attractive to primitive peoples lacking ready means of food preservation. The fruits, seeds, and even flower can serve as food. The seeds are relatively large, and contain fair amounts of both protein and oil. Fruits gathered in an immature stage serve as a fresh vegetable, or allowed to mature they may be baked, boiled, or roasted. Mature fruits store reasonably well, and if the flesh is cut into strips and dried it will keep almost indefinitely. A chief food in northern Manchuria is dried strips of "vegetable marrow" from *C. pepo*. Late in summer pumpkins are placed on a pointed rod mounted on a bench and cut spirally into long, narrow strings resembling spaghetti. These are put on a rack in the sun for open-air-drying, and later merely hung about the house until used for food. Many ancient civilizations also used the hard rinds of mature pepos for containers after removal of the flesh and seeds. Moreover, *Cucurbitas* are widely adapted and easily cultivated, their needs satisfied by moderate soil moisture and fertility. They are not, however, tolerant of frost.

By the time of New World colonization, the pumpkin ranged from southeastern Canada as far south as Mexico City. It was quickly introduced into Europe, and has been planted extensively from England to Italy. It was taken to Asia Minor during the seventeenth century, where it became highly valued, with selection of many special cultivars. The fruits of *C. pepo* are very variable, and include not only the jack-o-lantern pumpkins recognized in the United States, but various ornamental gourds and squashes (pattypans, marrows, acorns). Diploid chromosome count is 40, as with other *Cucurbita* species. There is speculation that *C. pepo* may have descended from *C. texana* which grows wild in Texas. *Cucurbitas* are harsh, trailing vines, but "bush" forms have been developed for smaller gardens. The crop is cultivated chiefly in drier areas, and is much grown in Africa and the Near East as well as in Europe and America.

SQUASH, *Cucurbita*, Cucurbitaceae. It is

Fruits of squash, *Cucurbita sp.* (Courtesy Chicago Natural History Museum.)

evident from the foregoing discussions that there is no botanical distinction between "pumpkin" and "squash"; pumpkin pie is typically filled with the more richly flavored squashes. Three species are generally grouped as "squashes": *C. mixta*, *C. moschata*, and *C. maxima*. All are typically trailing vines with an extensive root system, with harsh, (often prickly) leaves and stems, and are used for food as was described for the pumpkin. At time of colonization *C. mixta* was found from Guatemala to the southwestern United States. The species is still widely grown in Mexico for the large, tasty seeds, which are roasted as a confection. The fruit, however, is stringy and watery, relatively poor in culinary qualities.

C. moschata occurred abundantly in northern South America and Central America at time of colonization, but only scantily in the southwestern United States. Today it is extensively grown in many parts of the world, especially in Japan. The species is well adapted to the tropical lowlands where high temperatures and high humidity prevail. It is highly esteemed for squash fruits,

including such cultivars as "Butternut," "Kentucky Field", "Dickenson," and "Large Cheese" (the latter three are the chief canning "pumpkins" of the midwestern United States, used for the ubiquitous Thanksgiving pie). *C. moschata* is more resistant to the pesky squash vine borer than is *C. maxima*.

C. maxima was confined to South America at the time of Columbus, but is today widely grown throughout the world. It provides fruit that is generally regarded as the highest quality among the squashes. It is much grown in India and the Philippines as well as the New World. The species is perhaps more tolerant of cool temperatures than are other *Cucurbitas*, and is adapted to growing as far north as New England. "Hubbard," "Buttercup," "Banana," "Boston Marrow," and "Golden Delicious" are familiar *C. maxima* cultivars.

C. ficifolia is of lesser importance than the other species discussed. It is a high-altitude plant from Mexico to the southern Andes, bearing watermelon-like fruits. Within its natural range the species is perennial.

TOMATO, *Lycopersicon esculentum*, Solanaceae. The tomato is native to tropical America, and the large-fruited forms probably arose in the Peru–Ecuador area. A putative ancestral form, *L. esculentum* var. *cerasiforme*, is found wild in the area, and has been proven to cross freely with *L. pimpinellifolium*. There is also the possibility of some introgression with *L. hirsutum glabratum*, *L. peruvianum*, and *L. chilense*, all wild and with at least partly overlapping ranges in the homeland of the cultivated tomato. Diploid chromosome count is 24, and spontaneous tetraploids are not uncommon.

At time of the Conquest the tomato was familiar as a cultivated plant and as an adventive from South America to Mexico. It is said to have been brought from Mexico to Europe in 1523, where by 1600 it had spread through the continent as a curiosity, the "pomme de-amour" or "love apple," rather than being used for food. The large-fruited forms are reported to have been brought directly from Peru to Italy, where the tomato seems first to have gained acceptance as a food plant. It was spread from Europe by early voyagers, across the Pacific into southeastern Asia before 1650, and to North America by the time of the American Revolution. Only relatively recently has the tomato become a major "vegetable," and its acceptance in tropical areas (other than its American homeland) is still not great. World-wide nearly 20 million tons of tomatoes are produced annually for food, mostly in Europe and North America. The United States is the leading consuming country, followed by Italy, Spain, the Arabic nations, Brazil, Japan, and Mexico.

The tomato is a fleshy, trailing annual (sometimes perennial in the tropics) with a characteristically strong scent. It is quite sensitive to frost, and, indeed, will not normally set fruit when night temperatures go below 64°F. However, phthalamic sprays (Duraset) increase early flowering and fruit set. In hot, moist weather, on good soil, the tomato grows rampantly. In the tropics three crops per year can be grown on a single field. Tomatoes are often grown under glass for the winter market in temperate climates, where the fruit brings quite a premium over the abundant summer crop. In the United States most of the crop for the fresh market is grown in Florida and California through the cooler months of the year, while California leads overwhelmingly in the processing crop grown in summer. The tomato is a familiar home garden plant throughout the United States, where tasty fresh-picked fruit is available from a great diversity of cultivars.

The California tomato industry provides an interesting case of rapid mechanization of commercial production. Within just a few years the harvesting of the entire California crop was converted from hand harvesting by Mexican braceros to 95 per cent mechanical picking. By 1968 there were over 1300 mechanical harvesters being used in California; the machines cut the plants near the ground, and shake and sift the vines, while a crew sorts the tomatoes as the machine progresses through the field, throwing back overripe, green, and otherwise undesirable fruit. Up to 5 acres can be harvested per day, yielding up to 100 tons of tomatoes. For this mechanization to be successful plant breeders had to "restructure" the plant from a viney type, with a progressive setting of fruit, to "bushy" types in which the fruits mature early and essentially at the same time. The fruits are of moderate size for good handling,

The most important Solanaceous fruit, the tomato. These smallish, solid types maturing all about the same time are typical of the "new" breeds developed for mechanized harvest. (Courtesy USDA.)

and bruise less than traditional cultivars; the plants can be seeded directly to the field, as much as 50,000 plants to the acre. Techniques for weed control with herbicides were developed, exact fertilization and irrigation schedules worked out to control growth and ripening. Few crops have been so quickly and thoroughly adapted to mechanization.

In the United States a large portion of the tomato crop is processed into juice, catsup, tomato paste, and canned tomatoes. In the tropics tomato cultivars are seldom highly selected, and yield fruit that is smaller than that grown in temperate gardens, and not so uniform. In northern Europe a good deal of greenhouse growing is done, using large-fruited cultivars. In the United States similar cultivars are often started in the greenhouse and sold to the home gardener for planting out in late spring. These plants may be staked to save space and prevent the fruit from becoming "dirty" or rotting in contact with the soil. Such care is of course impossible with the commercial growing. Tomato fruits are fragile, and store for only a comparatively short time; most are picked green for shipping to market, and allowed to ripen at temperatures below 60°F. Tomatoes handled in this fashion can be kept for about 3 weeks. Most consumers feel the flavor of such tomatoes is not the equal of vine-ripened fruit grown at home.

TRICOSANTHES OR SERPENT GOURD, *Tricosanthes anguina* (*T. cucumerina*), Cucurbitaceae. The mature fruit of this sprawling, annual herb is fibrous and bitter, but is eaten boiled when immature. The species is wild from India to Australia, and a cultivated form has long been grown in India. It has been introduced into the West Indies.

WATERMELONS, *Citrullus vulgaris* (*C. lanatus, Colocynthis citrullus*), Cucurbitaceae. One of four species of *Citrullus* ($2n = 22$), the watermelon, has been in cultivation for over 4,000 years. It is thought to be native to Africa, and was in cultivation among the ancient Egyptians. Both yellow-fleshed and red-fleshed, bitter and sweet, types are known. Growing wild and cultivated in Africa, they may frequently be an important water source in times of drought. In modern America watermelons are mostly eaten raw, iced as dessert. The rind is sometimes also made into "pickles." In Russia a fermented beverage is reportedly made of the juice, or the juice is boiled down to a sugary syrup. Orientals preserve chunks of watermelon in brine. In many localities the roasted seeds are eaten, just as are (and were, in pre-Conquest New World cultures) squash seeds. A hard, white-fleshed watermelon, the citron or preserving melon, is used for pickling and jellies.

The watermelon plant is a trailing, tendriled, scabrous, monoecious vine with deeply pinnatifid leaves. Production in the United States amounts to over 3 million tons annually, mostly from Florida, Georgia, Texas, and other southern states.

Perennial Fruits of Temperate Climates

The majority of fruits familiar to the North American palate are included here. Especially prominent are the "stone fruits" in which the fruit is a drupe, like the peach, apricot, cherry, or plum; the pome fruits, notably the apple and pear; and the "berries" (of which the grape, blueberry and persimmon are true berries, but raspberries and strawberries really aggregate or accessory fruits). Almost all are intensively cultivated, in contrast to most tropical fruits (a goodly number of which are semicultivated or even gathered from the wild). A great deal of scientific investigation has been undertaken into the propagation, culture, and handling of Temperate Zone fruits. Today the field of pomology has become one in which trained specialists are on the alert for improved techniques and cultivars, and in which the producer neglecting proper methods has little chance of commercial success. In

A reminder of times past. Children picking wild blackberries in Washington. (Courtesy USDA.)

particular it is necessary to be cognizant of cultivars adapted to the local climate, and to be familiar with growth patterns, planting procedures, spray and fertilizer schedules, pruning techniques, cover cropping, irrigation, and perhaps even budding and grafting.

Most certainly the trend is to mechanized culture and harvest. Mechanical harvesting is not so well perfected as with annual fruit crops of the previous section, partly because a lengthy time is needed for breeders to develop perennial shrubs and trees of suitable stature. In the meanwhile specialized pruning, and widespread use of growth-regulating chemicals, attempts to fit the conventional cultivars to such harvesting machinery as is available today. The trend to dwarfed and special-shaped stock is relatively novel in the United States, but in Europe the limited space available for orchards has long required training and pruning methods to make hedgerows, espaliers, and other forms to fit the demand for space and accessibility. Form control is becoming aided by growth-control chemicals, but especially are these used to regulate fruit fall, substituting for even-maturing growing

Even semi-wild fruits are being subject to mechanized harvest, as with cranberries in this New Jersey bog. (Courtesy USDA.)

stock. Perhaps the most familiar subject of horticultural research in recent years has been the investigation of fruit-thinning compounds, others that induce abscission, still others that restrain fruit fall, induce parthenocarpy, regulate nutrient content of the fruit, overcome inherent tropisms, induce hardiness, and overcome physiological weaknesses.

None of the mechanical devices utilized for harvest do so perfect a job as can the human hand. But modern labor costs being what they are in technologically advanced parts of the world, often as much as one-half the crop can be sacrificed to bruising and other injuries in exchange for the saving of labor. Among the earliest labor-saving devices were positioners of various types—self-propelled machines with platforms elevated or manipulated into positions where fruit could be easily reached. Thus the man on the platform could better spend his time picking fruit rather than maneuvering ladders or climbing trees. In

A mechanical positioner facilitating citrus harvest in California. (Courtesy USDA.)

orchards well pruned to accommodate these machines remarkable economies have been achieved.

A step farther was development of tree shaking devices, with catch frames spread below onto which the loosened fruit falls. These are often most effective in combination with chemical control of fruit drop, and with pruning systems such that interfering branches have been cut away to allow clear fall of the fruit to the catching frame. However, in spite of padding materials used where the shaker bar attaches to the tree, there is often injury to the bark or to the tree itself.

Still another approach has been the use of vacuum devices, which suck the fruit from the tree. This has been somewhat effective with grapes, but is more appropriate to thick-skinned fruit that does not bruise easily, such as citrus. Grapes and blueberries are sometimes mechanically harvested with impactor machines that cause rapid vibration of the plant, shaking the fruits onto a collecting cloth below. Even strawberries are coming to be mechanically harvested, with scoops that comb ripe and green berries alike from the rows (sometimes after defoliation of the strawberry plants; new, even-maturing varieties of strawberries are needed for this method to be fully effective). Advanced harvesting techniques are generally accompanied by improved handling and keeping methods as well, such as the quick cooling of cherries (often by immersion in cold water) to reduce bruising blemishes.

The situation with mechanized fruit harvest as of 1968 is summed up by R. P. Larson of Michigan State University (*HortScience* 4, 233, 1969):

Harvesting machinery has evolved from tractor-mounted boom shakers, hand-carried collecting frames and lug-box containers to highly automated self-propelled units with power-steering, four-wheel drive, various tilt adjustments for uneven ground or different-sized trees, and lights for night harvesting. The frames are usually saran covered, and the conveyer systems have been greatly improved to reduce fruit bruising, The inertia-type branch shaker, mounted on frames, is used by most growers.

Of all plant families, the Rosaceae stands out in importance as a source of temperate fruits. The majority of the leading species are of Old World origin, having arisen especially in the ancient central Asiatic centers of domestication, where prehistoric civilizations may have first selected from wild forms appealing apples, pears, plums, cherries, apricots, strawberries, and bramble fruits—the mainstays in the Temperate Zone fruit diet. Most of these are of very ancient cultivation, and were known in many cultivars by the time of the Greek and Roman civilizations.

Temperate fruits have never been basic to the human diet over any wide area. Nutritionally they provide little food value (some sugar, negligible quantities of fat and protein), consisting as they do largely of water. Yet their vitamin, organic-acid, and mineral contributions earn them an important place in the modern diet. Moreover, they are a prime source of pectins (see discussion under "Gums" in Chapter 9), and are favorite flavors in innumerable confections. Temperate fruits are popular enough worldwide to contribute over 100 million tons of market produce annually; a large additional amount,

consumed locally, never gets into the production statistics. Grapes (including those used for wine) account for the greatest share of market, followed by apples, with pears, peaches, and plums pretty well bunched in third position.

Having high moisture content and delicate texture, fruits in their natural state are unusually perishable. Fresh fruits are thus apt to be seasonal, and generally available only where rapid and efficient transportation prevails. Many fruits can be held for longer or shorter periods of time under modern, improved cold-storage facilities, but much of the fruit crop must be preserved—by canning, drying, sugaring, freezing, or less commonly, by salting, smoking, pickling, or preserving with other chemicals. Some of these practices—drying, for example—date back to prehistoric times, while others—such as freezing—are very modern. Most fruits are consumed raw and are cooked only insofar as that is necessary for their preservation.

In this section the apple, raspberry (brambles), grape, and strawberry will serve as prime examples, followed by briefer comment on other temperate fruits.

APPLE, *Pyrus malus* (*Malus sylvestris, M. communis*) and other species, Rosaceae. No other tree fruit in temperate regions has had the importance of the apple. Adaptability

A modern apple orchard must be meticulously sprayed for quality fruit. (Courtesy Dow Chem. Co.)

of the fruit to handling and marketing, and its excellent keeping qualities (when properly stored) permit sale throughout most of the year. Moreover there are a wealth of cultivars bearing fruit in summer, autumn, and with some harvesting even as late as December. Thousands of tons of apples are processed annually (especially for apple sauce and dried apples), or are converted into cider and vinegar.

The apple evidently originated in the Caucasus Mountains of western Asia, where it has been under cultivation for thousands of years and where wild forms still exist today. At least 2,000 cultivars have been selected and named, all except a few being chance seedlings or mutants rather than the end results of a directed breeding program (obviously, it is difficult with a tree crop having a relatively long generation cycle to produce inbred lines and develop tailor-made crosses). A few species other than *P. malus* also bear large-fruited modern apples, such as the "*Malus pumila*" types of China.

The Temperate Zones favorite tree fruit, the apple. This is a Golden Delicious tree in West Virginia. (Courtesy USDA.)

Most of the world's commercial apple crop comes from European and American cultivars now spread widely throughout the world. These are mainly diploids ($2n = 34$) and triploids, the former the better pollinators, the latter best interplanted with suitable diploids. Almost 70 per cent of world apple production centers in Europe, much of the remainder in North America. Apple cultivation was introduced by the English into Himalayan India, where over 100,000 acres is devoted to growing cultivars with such exotic names (for India) as "Beauty of Bath," "Cox's Orange Pippin," "Lady Sudeley," as well as North American varieties such as "Delicious" and "Jonathan." World production of apples amounts to about 20 million tons annually. France, Italy, Germany, the United States, and Japan are the leading apple-producing countries.

In the United States only a score or so of apple cultivars are much grown. Most of the best-liked cultivars originated from unknown sources. One only, the Cortland, results from a breeding program. The Rhode Island Greening apple is a typical case. A seedling was raised by Mr. Green at his tavern near Newport, Rhode Island, from which visitors obtained scions after tasting the delicious fruit that this tree bore. Delicious, Golden Delicious, Jonathan, York Imperial, and so on are other horticultural varieties having experienced similar chance beginnings.

The apple tree is a spreading, long-lived perennial able to bear for as long as a century, although most orchards are renewed much sooner with younger trees of improved variety. The trend in recent years has been toward dwarf trees conveniently spaced for mechanized tending and simplified harvesting. Commercial propagation is by budding or grafting onto proven rootstock. In general, the apple flowers and sets fruit abundantly only in alternate years, pollination being chiefly by bees. Sprays are routinely used in technologically advanced parts of the world not only to protect the apple from numerous insect and fungus pests, but to control set of fruit, its retention on the tree, or its fall where thinning is needed. In some areas sprays are applied by airplane.

The apple is quite hardy in cold climates, withstanding subfreezing temperature well. It is not adapted to tropical conditions, requiring a winter conditioning in order to fruit. Apple orcharding is favored by steady spring temperatures, such as are found leeward from large bodies of water. These prevent early flowering and subsequent loss of blossom to frost. Orchards are mostly located in rolling country above low-lying frost pockets. Climate in almost all of Europe is excellent for apple growing. In the United States the Pacific slope of Washington, Oregon, and northern California is the leading apple-growing area, with Michigan, New York, and Virginia important in the East.

The apple fruit is a pome, having an inner cartilaginous core surrounded by fleshy tissues. About half of the apple production is consumed as fresh fruit, with much of the remainder processed into juice (cider). A controlled bacterial inoculation of cider results in an acetic-acid fermentation to yield vinegar, or sweet cider may undergo a natural alcoholic yeast fermentation to become "hard" cider. Fruit residues at the cider mills are an important source of pectin gum. A ripe apple fruit is approximately 84 per cent water, 11 per cent sugar, and with inconsequential amounts of protein and fat; minerals and fruit acids are, of course, well represented. Overripe apples turn mealy because of decomposition of the pectic materials in the cell walls.

BRAMBLES, *Rubus spp.*, Rosaceae. Under the general heading of brambles or bush berries are included red raspberries (some with yellow fruit), *R. strigosus* (the hardy American types), and *R. idaeus* (the large-fruited European types) and their hybrids; black raspberries or blackcaps, *R. occidentalis* (American); purple raspberries, hybrids between reds and blackcaps; blackberries, a highly variable group of about thirty-five species including *R. procerus*, and *R. laciniatus* (introduced), *R. allegheniensis*, and *R. argutus*, and a host of other American species; dewberries, trailing blackberries; youngberries, loganberries, and boysenberries, *R. ursinus* hybrids; and other species. The genus is large and complex, including aneuploids and a polyploid series of blackberries up to $12n$. There is much hybridization and self-sterility in wild blackberries, and intentional crossings with introductions from Asia and elsewhere to yield modern cultivars such as Cascade ($12x$ Zielinski $\times$ $6x$ Logan). Commercial raspberry cultivars are almost exclusively diploid ($2n = 14$), although a few triploids and tetraploids have been tried in Europe.

Rubus plants are mostly prickly shrubs (a few thornless cultivars have recently been developed), some as tall as 7 meters, with stems or canes commonly biennial and underground parts perennial. They range from sub-

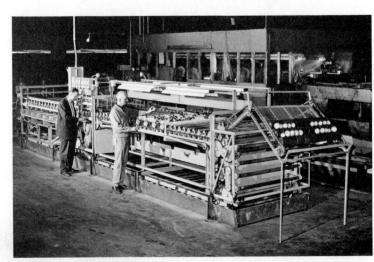

Mechanized sorting and packing of apples in Washington. (Courtesy USDA.)

tropical types such as *R. trivialis* to hardy boreal types like *R. hispidus*. Brambles typically flower on second-year canes (floricanes) only. Flowers are perfect, and develop into an aggregate of druplets, the "berries." In the raspberry group the prominent receptacle separates readily from the aggregate, producing a hollow berry, but with blackberries the core or receptacle adheres to the druplets.

Black raspberries are usually propagated by layering, red raspberries from underground stems called "suckers," and blackberries from root cuttings. Newly propagated brambles are dug during the dormant season, usually for spring sale and transplanting. These may be clean-cultivated, mulched, or treated with herbicides to control weeds. Canes that have fruited are generally pruned after fruiting. Topping and pruning are usually necessary for increasing fruit size and for facilitating harvest. There is some mechanized picking, and care is taken to avoid bruising and excessive handling of the fruits. Bramble fruits are sometimes "conditioned" before shipping in a controlled atmosphere to enhance keeping quality.

By and large, red raspberries find "apple climates" congenial, while black raspberries

Typical blackberry fruits. (Courtesy USDA.)

A red raspberry, "London." (Courtesy USDA.)

and blackberries are a little more tolerant of heat and less hardy, preferring "peach climates." All are mesophytic in water requirements. Brambles are comparatively recent as domesticated plants. Red raspberries (*R. idaeus*) were brought into cultivation in Europe at the close of the Middle Ages, and a few cultivars introduced into North America by 1800. A few years later the American red raspberry (*R. strigosus*) was domesticated, and hybridized with the European species. The first main variety of black raspberry was developed in Cincinnati, Ohio, in 1832. Wild blackberries were long esteemed by the Indians and early colonists, but were not brought into cultivation until the nineteenth century. The cutleaf blackberry (*R. laciniatus*) originated during this period from a European selection. Since then many exotics, mutants, and hybrids have been intermixed, until today blackberry types (including the dewberry, loganberry, laxtonberry, nessberry, veitchberry, and so on) are legion and include a thorough mixture of many botanical species themselves not clearly delimited. Well-known commercial varieties among red raspberries are Willamette, Canby, September, Chief, and Latham; among black raspberries Cumberland, Dundee, Black Hawk, Bristol, and Black Pearl; among blackberries Eldorado, Cascade, Himalaya, Lucretia, and many others. Production of bramble fruits in the United States approaches 100,000 tons annually, with appreciable amounts of wild fruits still harvested locally. Raspberries account for about half the commercial acreage, and are produced chiefly in Oregon–Washington. Blackberries come from many regions, including the south-central states as well as the Pacific Northwest. In season much of the crop is consumed fresh, but additional large amounts are preserved (especially as jellies and fruit sauces), and frozen.

The northern fox grape, *Vitis labrusca*, part parent of some modern grape varieties. (Courtesy Chicago Natural History Museum.)

GRAPE, *Vitis spp.*, Vitaceae. Few plants have been accorded so much esteem as has the grape. The wine grape, *V. vinifera*, is intimately associated with the very beginning of Western civilization, a source of fruit and drink throughout Biblical times. Fossil leaf prints from the Quaternary are found in France and Italy, and the species still grows wild in Europe, the Near East, and northern India. It was apparently domesticated in southwestern Asia somewhere between India and southeastern Europe, well before 5000 B.C. (at which time it was brought into Palestine from the north). Inscriptions show wine grapes to have been grown in Egypt as early as 2375 B.C. Today *V. vinifera* constitutes the major bloodline for most of the world's grape crop.

In North America before the coming of the white man, the Indians were already utilizing several native grape species, mostly with a diploid chromosome count of 40, compared to *V. vinifera* with 38. Included are the fox grape, *V. labrusca;* the riverbank grape, *V. riparia;* and the muscadine grapes of the southeast, *V. rotundifolia* (the famous Scuppernong), *V. munsoniana*, and *V. popenoei*. Indeed, the early explorers of the New World were impressed by the profusion of wild grapes found there. It is perhaps indicative that the ancient Norse name for Viking discoveries in the West was Vineland. Size and quality of the New World wild grapes, however, were not equal to *V. vinifera*. As early as 1619 Lord Delaware had French stock introduced into the Virginia colony. For over 200 years thereafter *V. vinifera* was to be introduced into eastern North America periodically with only scant success, largely because of susceptibility to the phylloxera root louse (today most *vinifera* grapes are grafted onto phylloxera-resistant American rootstock). There was probably some crossing between the introduced species and American grapes, which were often planted side by side in colonial gardens. And before 1800 American grape varieties had been taken to Europe, where they probably contributed to some of the bloodlines.

Catawba was the first American cultivar to achieve much fame; it originated in 1819 as a cutting from a vine found growing in Maryland. Then in 1852, E. W. Bull of Concord, Massachusetts, raised a seedling from fruit borne on a *V. labrusca* vine that had been growing near a Catawba in his garden. This he named "Concord," a cultivar destined to dominate grape growing in the northeastern United States up to the present time (over 80 per cent of grape production in

Abundantly fruiting grapes. (Courtesy Chicago Natural History Museum.)

New York state is still this variety). Meanwhile, in the southeastern United States, improved clones of muscadine grapes were selected, a type of grape used only locally because of rapid deterioration that prevents extensive shipping. The oldest and best known, still widely planted, is the Scuppernong, but many newer cultivars have been bred, including the self-pollinating Southland, Bountiful, and Chief.

Grape culture in North America gained greatest importance, however, in California. The Jesuit fathers had taken Spanish grapes to Mexico, and about 1635 established a vineyard at what is now Socorro, New Mexico, the earliest planting in the western United States. During the century following, the missions in California acquired similar cuttings, and a mission vineyard at San Diego is thought to be the source of the Mission cultivar (believed to be a seedling out of *vinifera* stock) that is still grown in the area. After California achieved statehood in 1850, widespread introduction of French and Near Eastern cultivars took place. The dry central valleys of California were free from phylloxera and mildew, and under irrigation lent themselves to efficient growing of grapes having excellent quality. Today about 90 per cent of the United States' domestic crop originates there, from select cultivars grown on resistant rootstock. The California vineyards are becoming highly mechanized. The vines, 400–600 to the acre, are trained to standardized trellises and provided chemical weed control as well as pest protection. The fruit is often sprayed with growth-control chemicals to increase size. Even grape picking, traditionally a job for hand labor, is becoming mechanized as earlier described.

California production contrasts with grape growing in the Near East. Although select cultivars grafted to resistant rootstocks are becoming more used, the vineyards are largely hand tended. In Greece, for example, the grapes are generally planted in holes punched

Sun drying grapes for raisins. (Courtesy USDA.)

into the rocky soil, about 1,200 vines to the acre, the holes laboriously filled with water before settling the soil about the roots. The vines are sometimes left to trail on the ground, but are usually staked (rather than trellised as in California). Hand girdling of the bark below a fruit cluster to increase fruit size is not uncommon, contrasted with the gibberellin sprays used in California. Cultivation and weed control are still by hand, as is harvest.

In Europe grape growing is rapidly mechanizing, but not to the extent possible in California because of the prevalence of small vineyards on hillside terraces difficult to service en masse. Strangely enough, in the famed Rhine and Moselle valleys the vines are generally planted in rows up the slope rather than across, which encourages extravagant soil erosion under clean cultivation. Many of the vineyards are tended with small tractors that fit the narrow avenues between vine rows; the tractors often have fogging or spraying attachments for convenient pesticide application.

Efficient grape growing requires a climate having long, sunny summers (the vintage years are those with bright, dry weather that produces a high sugar content in the grapes), and winters mild enough to avoid winterkill. The grape profits from a well-drained soil, and is propagated by cuttings (usually with scions grafted onto phylloxera-resistant rootstock). Grapes flower and fruit on the new wood, and in modern cultivars seldom need cross-pollination. The vines are generally pruned back to a few paired branches that are trained to wire trellises, from which at most a few dozen buds are let to mature into grape-bearing shoots.

World grape production runs in the neighborhood of 50 million tons annually, of which two-thirds is European, especially from France, Italy, and Spain. In many localities as much as 90 per cent of the crop is used for making wine, a subject for further discussion in the chapter on beverages. In the Near East and California a significant portion of the crop is dried for raisins; the United States produces over 200,000 tons annually and Greece nearly as much (on a fresh-fruit basis these figures would be nearly 5 times as great). California production of raisin-grape cultivars runs nearly 3 times that for wine cultivars and about 4 times that for table grapes. Production from the Great Lakes area, principally New York, is almost a quarter-million tons annually, largely Concords for juice and preserving, but with some introduction of *vinifera* cultivars for the fresh-fruit market. More will be said about grapes in Chapter 20, where wine making is discussed.

STRAWBERRY, *Fragaria spp.*, Rosaceae. Perhaps the favorite "berry" among Temperate Zone peoples of both hemispheres is the strawberry, embracing especially *F. chiloensis*, probably indigenous to Chile and the North American mountains from California to Alaska. Bailey regarded its var. *ananassa* as representing most of the common cultivated strawberries. *F. virginiana*, the wild strawberry of eastern North America, was domesticated in early colonial times, and has probably entered into the evolution of the garden strawberry. The European *F. vesca* and its varieties is the source of the everbearing strawberries. All species are herbaceous perennials adapted to temperate climates, with a condensed stem (crown) bearing a rosette of long-petiolate, stipulate, trifoliolate, dentate leaves, and most types propagating readily by stolons which root at intervals. The white flowers are borne in short racemelike clusters, and may be either perfect or pistillate. Cultivars of the latter kind require pollen-bearing types in association with them to effect set of fruit. The fruit is not a berry botanically but an aggregate of achenes (the small black dots on a strawberry) on a fleshy receptacle.

The diploid everbearing strawberry, *F. vesca*, has been under sporadic cultivation

The strawberry, a favorite perennial fruit of the Temperate Zone. (Courtesy USDA.)

in Europe ever since pre-Christian times. It was mentioned by Pliny and by various herbalists thereafter. It was apparently not seriously cultivated, however, until after the fifteenth century, and has never been so widely popular as the common strawberry. The latter seems to have originated in Europe in the nineteenth century, from the hybridization of the two introduced octoploid species, *F. chiloensis* and *F. virginiana*. *F. virginiana* had been introduced into Europe from North America in the early 1600's, with only modest success because of its small fruit. *F. chiloensis* was introduced into France from Chile (four plants only surviving the voyage, one of which alone was to sire the now-important French strawberry industry), and was grown, at least in England later, alongside of *F. virginiana*. Hybridization apparently occurred, to produce large-fruited strains whose true value was not realized for a century. Several cultivars selected from these strains were introduced into the United States during the 1800's and have given rise to the modern industry.

Strawberries are available today that suit almost any well-drained soil. The plants are propagated by runners. The runners bear fruit the year following transplanting so that in commercial production the strawberry is commonly treated as a biennial. The plants are reset every second or third year for best yields. In the home garden, strawberries will continue to bear for several years, but the berries become progressively smaller and less abundant. The first year the flower buds are best removed so that all vigor may be directed toward more robust growth. Commercially this may be accomplished to an extent with hormone sprays. During this first year careful cultivation and weed control must be practiced, and in cooler climates protective mulching during the winter may be needed. The second year the strawberry requires little attention other than harvesting of the abundant fruits.

Strawberries are mostly picked by hand, although scooplike mechanical harvesters are being developed useful for cultivars in which fruit matures all about the same time. Strawberries bruise readily and keep poorly, so that great care must be taken in picking them. The labor required for the crop makes the rather high market price of strawberries no surprise. Strawberries are consumed fresh, made into preserves, or frozen.

A number of large-fruited commercial cultivars have been developed, unfortunately seldom with the flavor of the "old-fashioned" homegrown types. Commercial production in the United States exceeds 200,000 tons annually, slightly over half for the fresh market. Limited winter strawberry production comes from Florida; early spring production primarily from Louisiana and California; mid-spring production largely from Arkansas, North Carolina and Tennessee; and late spring production mostly from Michigan, Oregon, and Washington. California and Oregon are the leading producing states.

Other Temperate Fruits. APRICOT, *Prunus armeniaca*, Rosaceae. The apricot is probably native to China, but was early introduced into the Near East, where it likely escaped in ancient Greek and Roman times. Apricot trees were brought into California during the eighteenth century, and several cultivars have been developed there. Manchurian and Russian selections have yielded late-blooming, hardier strains. Apricot culture is essentially the same as for the peach. Apricots are cultivated only in warmer climates because early bloom is susceptible to frost. World production is a little over 1 million tons annually, mostly from southern Europe, Iran, Australia, and California. California production runs about 200,000 tons annually, much of this sun-dried.

BARBERRY, *Berberis vulgaris*, Berberidaceae. This and other species of the genus have at times been cultivated for fruit used to make jellies and preserves. This species is objectionable insofar as it is the alternate host for wheat rust.

BLUEBERRY, BILBERRY, HUCKLEBERRY, OR WHORTLEBERRY, *Vaccinium spp.*, Ericaceae. The blueberry group is a variable one, involving many species and presumed hybrids, important mainly in North America. There is still much harvest from wild stands, especially in Maine and eastern Canada, and to some extent in the Pacific Northwest. Wild blueberries combined with venison were an important food of the American Indians—"pemmican." *V. corymbosum* is a "highbush" species of eastern North American swampland, from which several notable selections have been domesticated, the cultivars ranging up to the hexaploid. *V. myrtilloides* (a diploid) and *V. angustifolium* (a tetraploid) are "lowbush" species constituting much of the wild crop. The small lowbush types are seldom cultivated; however, wild stands are biennially burned-over in early spring, to control pests (especially weeds) and rejuvenate growth. Bloom is in spring, the flowers cross-pollinated by insects (fruit set seldom exceeds 20 per cent of the flowers), and the fruit is gathered in late summer.

Blueberries are small shrubs adapted to acid, generally infertile barrens, where they spread by rhizomes. Commercial plantings are becoming more common in the above-mentioned regions, and in Michigan; they are quite responsive to fertilization. Most cultivars have been selected from wild stock only within the last half-century (the first was in 1908). The berries have been traditionally hand picked, although mechanized harvest is increasingly used with commercial plantings. Wild berries from burned-over stands are commonly "raked up" with toothed scoops, along with foliage and other detritus; these are used chiefly for processing, after being winnowed free of debris. Cultivated plants are the source of most fresh fruit, and blueberry

"Earliblue" cultivar of the blueberry(right): note how fruit ripens without dropping compared to wild type (left). (Courtesy USDA.)

Harvesting lowbush blueberries for the canneries in eastern Maine with a toothed rake. (Courtesy USDA.)

orcharding has become a small but thriving industry in boreal North America. Total world annual production is probably about 50,000 tons, almost exclusively from North America (there is some growing of the bilberry, *V. myrtillus*, and limited highbush plantings in Europe, a little local use of tropical species in Hawaii and the Andean countries).

BUFFALOBERRY, *Shepherdia argentea*, Eleagnaceae. The fruit of this frequent shrub of the Plains region and westward in the United States is used mostly to make jellies.

CHERRY, *Prunus spp.*, Rosaceae. The diploid sweet cherry, *P. avium*, and the tetraploid sour cherry, *P. cerasus*, are believed to be native to Asia Minor, and were mentioned in the writings of the ancients. Duke cherries, hybrid types from a cross between these species (but tetraploid, from an unreduced egg nucleus), give a somewhat more sour "sweet cherry." Wild *P. avium*, the mazzard cherry, serves as propagative grafting stock for horticultural varieties, as does also the generally less satisfactory Eurasian *P. mahaleb* and occasionally other species.

The sour cherry, *Prunus cerasus*. (Courtesy Chicago Natural History Museum.)

The sour cherry is one of the most widely distributed tree fruits, having few prejudices regarding soil and climate. Montmorency is the most popular cultivar. Sweet cherries are less hardy, and prefer peach rather than apple climates. Napoleon, developed in Europe in the eighteenth century, is a favorite, as is a seedling from it, Lambert, much planted in the northwestern United States. Orchard practices for both types are much like those described for apples. Like the peach, apricot, and plum, the cherry fruit is a drupe.

World production of cherries is $1\frac{1}{2}$ million tons annually, mostly from Europe. Annual United States production usually exceeds 200,000 tons, almost equally divided between sour and sweet types. Sweet cherries are produced mostly in the Pacific states, sour cherries especially in Michigan and New York.

CRABAPPLE, *Pyrus* (*Malus*) *spp.*, Rosaceae. There are a great many cultivated or semi-cultivated species of crabapple, including those native to the United States (*P. coronaria*, *P. ioensis*, etc.) and many others that have been introduced (*P. baccata*, etc.). Crabapples are prized mostly as ornamen-tals, but the small, applelike fruit does make a superb jelly.

CRANBERRY, *Vaccinium macrocarpon*, Ericaceae. This creeping, vinelike relative of the blueberry is cultivated in the low, acid bogs of New England, Wisconsin, and the Pacific Northwest. Wild as well as cultivated stands are employed. Erect flowering branches bear small oval leaves and round berries that turn red toward autumn. The latter are gathered by specially designed, toothed cranberry scoops by a combing action. Water-raking is sometimes practiced: the fields are flooded, and when the berries float to the surface they are rapidly raked free from the vines. The berries are extremely tart, and are cooked with sugar to become cranberry sauce or jelly. Cranberry plants are propagated by cuttings. Construction of a cranberry bog is expensive, but once started a bog should yield for a half-century. A mycorrhizal association seems essential for successful growing. The cranberry is almost exclusively an American product, first cultivated in New England in the early nineteenth century. It remains today the leading export crop of Massachusetts, where nearly half of

A cranberry plant in fruit. (Courtesy USDA.)

Harvesting cranberries by machine in Massachusetts. (Courtesy USDA.)

the world's annual production of $1\frac{1}{2}$ million barrels is grown. A European cranberry, *V. oxycoccus*, is also grown.

CURRANTS, *Ribes sativum, R. nigrum, R. rubrum*, and other species, Saxifragaceae. Currants, along with gooseberries, are regarded by pomologists as "bush fruits." They are cool-climate plants, hardy far north in temperate regions but not adapt-

able to the southern regions. Currants are generally propagated by winter stem cuttings, planted in autumn in permanent locations and in succeeding years pruned free of old wood (3-year canes). The fruits, true berries, are borne on spurs near the base of second-year wood. Currants are of rather minor importance in North America but do find moderate use in jellies and occasionally

Currants, *Ribes sp.* (Courtesy Chicago Natural History Museum.).

dried as a seasoning. They are also delicious served fresh with sugar and cream. The species listed are indigenous to Europe, where they are quite popular. *R. nigrum* has been discouraged in the United States because it is the alternate host for the white pine blister-rust. Currants are modestly grown commercially, some in New York state and about 2000 tons annually in Washington. Native species furnish fruit from the wild over all the northern half of the country.

ELDERBERRY, *Sambucus canadensis*, Caprifoliaceae. This pithy, compound-leaved North American shrub occurs wild over the eastern half of the United States. The small fruits, borne in large cymes, are locally collected, especially for the making of wines.

GOOSEBERRY, *Ribes* (*Grossularia*) *hirtellum* and *R. grossularia*, Saxifragaceae. This bush fruit, like its relative the currant, is restricted to cool, moist climates. In the United States it is grown commercially only in Oregon, although varieties and other species are not uncommon in home gardens in many states. *R. grossularia* has been in cultivation in Europe at least since the early Middle Ages; there gooseberries find more favor than in North America. The American *R. hirtellum* has been developed as a cultivated plant mostly since the turn of the century, and has perhaps been instrumental in formation of better varieties through hybridization with *R. grossularia*. Like currants, gooseberries may serve as alternate hosts for white pine blister-rust.

HAWTHORN, *Crataegus spp.*, Rosaceae. This large genus of confusing species yields a crabapple-like fruit more frequently consumed by wildlife than by man. Burbank introduced a variety useful for jellies.

CALIFORNIA HOLLY, *Photinia arbutifolia*, Rosaceae. The small, red, berrylike fruit of this evergreen shrub is reported relished by the American Indians. Today it serves more as Christmas decoration in California than as food.

HUCKLEBERRY OR DANGLEBERRY, *Gaylussacia spp.*, Ericaceae. Although some species of *Vaccinium* (blueberry) are termed huckleberry, the name usually refers to this related genus of a similar nature whose boney-seeded fruit is similarly utlized.

JUNEBERRY OR SERVICEBERRY, *Amelanchier spp.*, Rosaceae. As with hawthorn, species of *Amelanchier* yield a small fruit mostly of value to wildlife, but sometimes collected by man.

MEDLAR, *Mespilus germanica*, Rosaceae. This small tree or shrub, similar to the hawthorn and cultivated in Europe, yields brownish fruits that are consumed after frost or made into preserves.

MULBERRY, *Morus spp.*, Moraceae. Various species of this worldwide genus are cultivated for their multiple fruit, although in the United States the trees serve chiefly to provide food for wildlife. In Europe and southern South America mulberries are highly esteemed as a table "berry," and also are made into wine and feed.

NECTARINE, *Prunus persica* var. *nectarina*, Rosaceae. The nectarine is merely a fuzzless type of peach, with perhaps a slightly richer flavor. The tree is identical with the peach, and its culture and care are the

Nectarines growing in California. (Courtesy USDA.)

same. Nectarines are little grown outside of the Pacific coast area, where cultivars of the two botanical forms (clingstone and freestone types) are grown.

PAWPAW, *Asimina triloba*, Annonaceae. This small tree of shaded, lowland woods in the eastern United States produces the cucumber-shaped, pulpy fruit unfortunately known but to few. No effort has been made to select superior fruit types, and pawpaws remain an occasional rich but variable delicacy of the eastern woods.

Fruiting branch of the North American pawpaw, *Asimina triloba*. (Courtesy Chicago Natural History Museum.)

PEACH, *Prunus persica*, in several varieties, Rosaceae. The peach appears to have originated in China, where it was mentioned in literature several centuries before Christ. It was introduced into Persia before Christian times, and was spread by the Romans throughout Europe. Many cultivars were developed there, some of which were introduced by the Spanish into Florida and spread throughout the United States by the Indians during the 1600's. The peach had been naturalized as far north as Pennsylvania by the time the English colonists landed in Virginia. Commercial peach growing was not undertaken in North America until the nineteenth century. Some types were then introduced from China, from which leading cultivars were developed. In 1870, two seeds taken from a "Chinese cling" peach planted among other peach trees on the farm of Samuel Rumph of Georgia yielded the famed freestone Elberta cultivar, the leading commercial peach of the eastern United States, and the Belle of Georgia, said by many to be the best-flavored peach ever discovered (but of little commercial interest since it ships and keeps poorly).

The peach tree is similar in general appearance to the apple, but has longer, tapered leaves. The pink flowers appear before the leaves, and develop into the drupaceous fruit characteristic of the Prunoideae. Compared to the apple, the peach is not hardy. In the United States it can be grown well only in the warm, sunny parts of the country. Propagation is typically by peach-on-peach (stocks from wild or cannery pits) or peach-on-almond budding of named varieties. Most varieties are self-fertile.

Varieties of peaches are broadly grouped as freestone or clingstone, depending upon whether the flesh of the drupe separates readily from the pit. Some of the favorite named varieties (all freestone), are Halehaven, Valiant, Belle, Oriole, and of course the famous Elberta. World production is

The peach in blossom. (Courtesy USDA.)

about 5 million tons annually, equally divided between Europe (especially Italy) and North America, with significant production from Japan, South Africa, and Australia. United States production is mostly from California and the South Atlantic states. More than half of the crop is consumed fresh.

PEAR, *Pyrus communis* (in a number of varieties) and a few other species, Rosaceae. The common pear, *P. communis*, embraces

Close-up of a Kiefer pear. (Courtesy USDA.)

many cultivated strains and hybrids with other species. Indigenous to western Asia, it has long been cultivated there and in Europe, where it is represented by a number of naturalized as well as cultivated forms. Extensive selection of cultivars was made in northern Europe in the eighteenth and nineteenth centuries, and almost all of the successful cultivars in the United States (such as Bartlett and Seckel) were brought directly from Europe. The inferior-fruited Chinese sand pear, *P. serotina*, has been utilized in the United States for crossing with the European pear, to give such cultivars as Kieffer, Orient, and Waite. Other exotic species (e.g., *P. ussuriensis*, *P. calleryana*, and *P. betulaefolia*) have been utilized for crossing and as fruit stock, as has also the quince (*Cydonia oblonga*). An extensive pear selection and breeding program was undertaken by the USDA in Beltsville, Maryland, and at Wooster, Ohio, during the 1960's.

The pear tree, a close relative of the apple, is very similar to the latter in general appearance, although the fruit is readily distinguishable in shape and texture. A distinctive feature of the pear is the grit in the flesh due to the presence of stone-cell clusters. The care and culture of the pear are about the same as for the apple, but the pear is less tolerant of either extreme heat or cold.

Chilling requirement for flowering is similar. Silvex sprays are often used to increase fruit set; hormone sprays prevent fruit drop. Fertile soils and good stock for the budded or grafted scion are essential to prolific bearing of quality fruit. Fireblight disease is a serious threat to pear growing, especially with susceptible cultivars such as Bartlett. Pears are likely to be self-sterile, so that it usually pays to plant more than one cultivar in the orchard. World pear production exceeds 5 million tons annually, mostly from Europe (Italy especially). Much of the crop in Europe is used to make cider (perry) and wine. United States production generally runs about 750,000 tons annually, mostly from Pacific coast states where the Bartlett and Anjou cultivars make up most of the crop.

PERSIMMON, *Diospyros spp.*, Ebenaceae. The oriental persimmons of eastern Asia, *D. kaki* and *D. lotus*, have been introduced into the subtropical highlands throughout the world, and are somewhat cultivated commercially in California, but the persimmon best known to millions of North Americans is the wild *D. virginiana* omnipresent in the hardwood forests from Texas to New Jersey. The fruit, a true berry, has an aromatic flavor superior to that of the oriental species, but an extreme astringency when eaten unripe. It is often supposed that ripeness develops only after a frost, but actually the finest persimmons are usually those that ripen early.

PLUM, *Prunus domestica* and other species, Rosaceae. The various types of plums include species from many parts of the world and hybrids from these species. Over 2,000 cultivars are grown. The common European plum, *P. domestica* (Bradshaw, Reine, Claude, and other Greengage types; various prune plums; and others), is thought to have originated as a natural hybrid between *P. cerasifera* and *P. spinosa* in the Asia Minor area about the time of Christ. At least plums were not known in Italy much before the Christian era. The small-fruited Damson, Mirabelle, or Bullace plums (var. *instititia*) of the same region are probably of like antiquity. *P. salicina*, the Japanese plum (Abundance, Beauty, Burbank, Formosa) originated in the Orient and was introduced into the United States by Burbank in 1870. It has a very low chilling requirement. *P. mira* and *P. holosericea* were collected on extensive expeditions into Tibet. Several American species (*P. nigra*, *P. americana*, *P. hortulana*, *P. munsoniana*, and others) were doubtless known to the Indians, and have yielded a number of cultivars (DeSoto, Hawkeye, Downing, Miner, Cheney, and so on).

With such a variety of ancestry, it is no wonder that strains of plums can be found for almost any site. Often more than a single cultivar must be planted, since many varieties are self-sterile. Perhaps more

Preparing plums for drying to prunes, California. (Courtesy USDA.)

than any other Temperate Zone fruit tree, plums are resistant to disease and insect troubles. Plum fruits, botanically drupes, are less perishable than pears or peaches, and with refrigeration may be stored for considerable periods. Plums are widely grown throughout the world, for table fruit, cooking, canning, jelly making, and drying to prunes. The prune industry is quite important on the Pacific coast of the United States. Thoroughly ripe plums of the *domestica* type are dried in the sun or by artificial heat. They are aged or "sweated" a few weeks and then "glossed" with a steam, glycerine, or fruit-juice bath to produce a sterile, glossy skin. Nearly 200,000 tons of fresh plums are produced in the United States annually, and over 150,000 tons of prunes (dry basis), chiefly from California. World plum output is about 5 million tons annually, mainly from southeastern Europe.

QUINCE, *Cydonia oblonga*, Rosaceae. This Asiatic species is grown in temperate climates primarily for a preserving fruit and as dwarfing stock on which to bud or graft pears. It is a small, unarmed, crooked tree bearing solitary flowers at the ends of new shoots that mature astringent, pear-shaped fruits. The quince is susceptible to fireblight and fruit moth damage. Commercial growing in the United States is confined almost entirely to California and New York. Crossed with *Pyrus*, the quince has yielded the "genus" *Pyronia*.

Tropical Fruits

The diversity of tropical fruits reflects the diversity of tropical vegetation in general. Scores of fruits from a wide assortment of families are at least modestly known and utilized in the tropics, and hundreds more are potentially available for use although still coming from jungle plants. With three exceptions—the banana, citrus fruits, and the pineapple—tropical fruits are not as a rule so widely or so scientifically cultivated and marketed as are the familiar Temperate Zone fruits. Some, such as the avocado, guava, annonaceous fruits, papaya, mango, and mangosteen, are today becoming better known. With more rapid air shipment, they may one day become as familiar items in the larger Temperate Zone fruit markets as bananas and coconuts. Others, such as the tamarind, sapodilla, loquat, breadfruit, jackfruit, and durian, have widespread but modest use in the tropics, and many more are utilized only where they grow.

In this section major attention is given to the important banana, citrus fruits, and pineapple. An alphabetical listing of other

Tropical fruits of a Colombian market. Pictured from left to right are mamoncillo, plantains, oranges (below cassava root), papaya (above bananas), cherimoya (above limes), and tomatoes (above small bananas). (Courtesy USDA.)

New small-farm banana planting in Puerto Rico. (Courtesy USDA.)

tropical fruits follows, with brief comment concerning those of greater importance or interest. A list of southern Florida fruits, compiled by R. Bruce Ledin (*Economic Botany* **11**, 372–4, 1957), embraces thirty plant families, and provides a good general summary of tropical fruits.

Banana, Musa spp., Musaceae. Cultivated bananas are considered to be *M. paradisiaca sapientum*, *M. sapientum*, or *M. cavendishii*. Cultivars include the rather tasteless "plantains," large fruits with greenish or reddish skins, usually cooked rather than eaten raw, in addition to the tasteful banana. The fiber-yielding *M. textilis*, abacá, was reviewed in Chapter 7.

The banana is probably native to tropical Asia, where it has long been domesticated. Alexander the Great encountered it there on his expedition to India. The humid Asiatic tropics from India to Indonesia still yield "wild" (escaped?) seedless bananas. Polynesians evidently spread the banana throughout most of the Pacific, as did the Arabian merchants across Africa, where the name "banana" originated in Sierra Leone. Portuguese and Spanish colonizers completed the establishment of the banana throughout the

warm regions of the globe. Yet as late as 1880 the banana was little known in the United States, for refrigerated ships were not yet bringing perishable tropical fruits to the Temperate Zone.

Most bananas marketed in North America before 1964 were the Gros Michel cultivar, the "Poya banana" (named for its finder, Jean Francois Pouyat, a Jamaican who discovered the cultivar while strolling through a banana plantation on Martinique in 1836). The Gros Michel is triploid, of uncertain origin. It is of large size, and is especially durable for shipping; Temperate Zone peoples know the banana almost alone by this cultivar, although those having lived in the tropics generally find the thin-skinned "ladyfinger" types much more flavorful. The Gros Michel was, however, vulnerable to the Panama disease and other ailments, so that during the 1960's expeditions scoured the oriental tropics for disease-resistant types that might be used in the banana-breeding program. Scientists have learned to stimulate seed production in normally seedless strains with certain sprays, a promising development for a breeding program. Meanwhile the disease-resistant Lacatan and Valery strains of the Gros Michel became available in the early

Bananas growing on United Fruit Co. property. Palma Sur, Costa Rica. (Courtesy Paul H. Allen.)

leaves quickly, and in less than a year a central inflorescence which terminates the plant's life upon fruiting. The banana is adapted to the humid tropical lowlands, and is best protected from persistent winds which quickly shred the large leaves. Under favorable growing conditions the banana yields 10 or more tons of fruit to the acre annually, with reasonably little attention.

Bananas destined for market are cut green, to be allowed to ripen after delivery; they are quickly perishable once ripe (yellow). Experience has shown just the proper stage for cutting fruit for shipment to the distant

A typical pendulous "stem" of developing bananas. Peru. (Courtesy USDA.)

1960's, and have been the dominant cultivars marketed since 1964. Spraying programs are often undertaken to control leafspot.

Tabulated world production of the banana runs about 23 million tons annually, and undoubtedly there is much additional growing as a home garden plant in the tropics, where the banana is often a staple in the diet. Tropical America is responsible for about three-fourths of the recorded world production, followed by Africa, India, Thailand, and the Philippines. Brazil is the world's leading banana-producing country, followed by Ecuador, Venezuela, Honduras, and Panama.

Although the banana grows to heights several times that of a man, botanically it is still an herb lacking woody tissues. Being mostly sterile and seedless, it is propagated from rhizomes or "root separations," which when planted alone yield a cluster of gigantic

Temperate Zone markets, so that the green bananas may be loaded according to carefully scheduled arrival of refrigerated ships. The temperature in the ship's hold is maintained at about 57° F, which prevents ripening. After delivery, a few days at room temperature causes the banana to turn golden, accompanied by hydrolysis of starch to fruit sugars and development of flavor. The fruit on the whole is a fairly well-balanced source of nutrients, containing minerals and vitamins as well as abundant carbohydrate, with some oil and a little protein. Bananas are increasingly moved to market prepackaged in containers in the tropics, helpful in preventing the bruising of fruit and an aid to its distribution in those market cities with high labor costs.

During the colonial era, the banana underwrote some of the best-managed tropical plantations the world has seen. Even though these foreign-owned and -managed operations were generally beneficent to the workers—providing improved living, educational and sanitation facilities—the rising nationalism and anti-colonialism of the mid-twentieth century has witnessed something of an abandonment of the plantation and a return to private growing, with the corporation merely purchasing and delivering the fruit. Well-capitalized plantations were quite efficient in clearing the lowland forests, preparing the planting sites, and instituting pest control until the plantings were established. Once the bananas are flourishing they make such a tight canopy that little further attention is needed.

Harvest of the fruit has always been a hand operation. The pendulous cluster of fruit, weighing as much as 100 pounds, is typically cut with a knife on a long pole, in such a way that it settles directly onto the shoulder of a worker who is to carry it to mule, railcar, or canoe, for transportation to a central gathering point. There the bananas are sorted and packed into individual

Chaco Indians loading bananas on a coastal boat. Yaviza, Darien, Panama. (Courtesy Paul H. Allen.)

cartons, or are moved in bunches on conveyer belts directly into the air-conditioned ship for transportation to distant markets. The banana is also an ubiquitous fruit of the village markets everywhere in the lowland tropics, individual clusters from home plantings being regularly offered.

Citrus fruits, *Citrus spp.,* Rutaceae. About sixteen species of citrus are recognized, some of which were mentioned in earlier chapters as source of essential oils. Most familiar for fruit are sweet oranges, *C. sinensis;* lemons, *C. limon;* limes, *C. aurantifolia;* mandarin oranges or tangerines, *C. reticulata;* and grapefruits (pomelos), *C.*

paradisi. Less extensively used are the citrons, *C. medica;* pummelos or "shaddocks" (after a Captain Shaddock, reputed to have brought the first seed to the West Indies), *C. grandis;* sour oranges, *C. aurantium;* and calamondins, *C. mitis.* The chironja, a putative natural cross between the sweet orange and the grapefruit, was discovered and named in Puerto Rico only in 1956. Scores of lesser-known fruits of related genera (e.g., kumquats, *Fortunella japonica;* Indian baels, *Aegle marmelos; Clausenia dentata; Feronia limonia;* and others mainly of ornamental value) frequently enter commerce as well. The vast majority of species, of both *Citrus* and related genera, wild and domesticated, are native to southeastern Asia. There the finer *Citrus* types were selected by primitive man, and there innumerable interspecific and intergeneric hybridizations undoubtedly took place. All species of *Citrus* have a diploid chromosome complement of 18.

The group of species embracing the commonly cultivated citrus fruits consists of small trees with mycorrhizal roots, often bearing single spines in the axils of the younger leaves. Leaves are unifoliolate, glandular-punctuate, aromatic, with a prominently winged petiole. Flowers are axillary, solitary, or in short racemes, usually perfect, fragrant, and provided with a disc. They mature into the familiar hesperidium berry (orange-type fruit), filled with watery pulp vesicles, the rind containing numerous oil glands. The fruit is approximately 40–45 per cent juice, 20–40 per cent rind, 20–35 per cent pulp and seeds. Chemically it is 86–92 per cent water, 5–8 per cent sugar, 1–2 per cent pectin, with small amounts of fruit acids (citric, especially), protein, essential oils, and minerals.

The attractiveness and utility of citrus species caused them to be mentioned in the accounts of travelers of the earliest ages. Curiously enough, only the citron from among the well-known oriental selections was introduced into the Mediterranean area before Christian times—to be followed centuries later by the sour orange, lemon, lime, and sweet orange. After the decline of the Roman civilization, the diffusion of the "new" citrus fruits—sour orange, lemon, and lime—through the Mediterranean area was largely carried out by the Arabic Empire of the Middle Ages. Later the Crusades extended the knowledge and culture of citrus in Europe, and finally the Portuguese and Spanish explorers introduced *Citrus* into the New World. About the time of Portuguese and Spanish domination of the seas the sweet orange first reached Europe.

A citrus grove in Florida. (Courtesy USDA.)

Grapefruit in a Florida orchard. (Courtesy USDA.)

presumably originated from seed spread and dropped by the Indians. After Florida was ceded to the United States in 1821, orange growing was greatly stimulated, and almost all the wild groves were top-worked with buds of superior strains before the turn of the century. Late in the 1800's grapefruit growing was started in Florida.

In 1862 William Hancock purchased a farm near Lakeland, Florida, where a Mrs. Rushing had set out three seedling grapefruit trees. One of these happened to bear seedless fruit, and became the "Marsh seedless strain," named after C. M. Marsh, a nurseryman who eventually purchased budwood from this cultivar. Also arising by chance in Florida were the pink-meat grapefruits. A bud sport was found on a Marsh tree growing near Oneco, Florida, in 1913. This became the Thompson seedless grapefruit, named for the Thompson Groves where the sport was found. A bud mutation out of this stock was discovered at McAllen, Texas, in 1929, later to be dubbed "Ruby," the first patented grapefruit cultivar and the one most grown in Texas today. The temple orange does not keep well, but is one of the more flavorful types. It originated at Winter Park, Florida, from trees budded with scions obtained in Jamaica. The cultivar may represent a cross between a tangerine and a sweet orange (a tangor).

Citrus was introduced into California during the eighteenth century, by way of the Spanish missions. The first sizable orange grove was set out in 1804, but only with the great influx of gold rush 49'ers was any considerable demand for the fruit created. In 1871 the famed navel orange, introduced from Bahía, Brazil, gave another lift to the youthful citrus industry in California, and with the advent of transcontinental shipment toward the close of the nineteenth century, orange growing became firmly established. The United States (mainly Florida, California, Arizona, and Texas) produces 200 million boxes (nearly 10 million

Possibly it came from India in a ship of the early Portuguese voyagers, possibly overland by the caravan routes, as the citron had come centuries before. The pummelo had reached Europe during the era of Muslim hegemony, as had the more important fruits mentioned above, but it was probably introduced into Barbados and Jamaica directly from the East by Captain Shaddock. There it is thought to have given rise to the modern grapefruit.

From the West Indies the orange and other citrus fruits soon spread to continental America. In the United States they were known in Florida before 1565, and only slightly later in Georgia and South Carolina. By the eighteenth century there were many "wild" orange groves (mostly of the sour type, but some of the sweet) in Florida; these

The navel orange, a gift from Brazil. (Courtesy Chicago Natural History Museum.)

reduced old orchards to root sprouts. It took them 15 years to recover, by which time California had assumed production leadership. In 1950–51 many orchards in Texas were similarly hit. Because of the danger of frost, smudge pots and ventilation fans are kept ever-ready in the citrus groves.

Oranges and grapefruits are usually allowed to nearly ripen on the tree. Lemons and limes are picked green. Many oranges on the market are artificially colored, or are treated with gas or chemicals to destroy the green chlorophyll and thereby allow other pigments to show through. Traditionally fruit is gathered by hand, providing seasonal employment for unskilled labor, but mechanized harvesting is fast developing in the United States. In other parts of the world growing is less mechanized. Orcharding is intensive in Japan, where the mandarin or tangerine is much esteemed, but ample labor is available for harvesting. In less advanced parts of the

tons) of citrus fruits annually, sufficient, if piled one upon another, to reach one-third of the distance to the moon. Recorded annual world production of *Citrus* is about 22 million tons, 18 million of them the popular oranges and tangerines. Citrus orcharding is still increasing, growing rapidly where formerly of little importance, such as in Australia, South Africa, and Israel.

Establishment and care of citrus groves involve many of the problems inherent in any orchard management. There is budding of selected varieties to be done—usually on sour orange stock started from seed. Spraying against insects and disease, and irrigation at proper season, are ordinarily essential. At one time the fruit fly seriously threatened the industry in Florida, but has been brought under control. Virus "decline" is still a problem in some areas, as are trace-element imbalances. Citrus growers must also guard against frost. In 1894–95 the great freeze in Florida destroyed nearly the whole crop and

The pummelo. (Courtesy Chicago Natural History Museum.)

tropics citrus growing is rather casual, with occasional trees of doubtful origin planted in the home garden. In western Africa, where *Citrus* should do well (it is the leading agricultural industry in Ghana, for example), harvest may be only from chance trees neither consciously planted nor given any attention.

The primary use of citrus fruits is for fresh consumption, but there is an ever-ready demand for juices. Citrus juice is obtained commercially either by burring or by pressing cut halves of the fruit. Juice obtained by the former method contains little of the oils and need not be centrifuged, as must that produced by the latter method, to remove the essential oils. Concentrates of juice for beverage flavoring are made by boiling juice in vacuum pans. Citric acid may be extracted from lemon juice. Mention has been made in previous chapters of extraction of pectic compounds and essential oils from citrus rinds. Residues are shredded in hammermills, limed to precipitate the pectin, cured, and the liquid portion expressed. The liquid is evaporated to yield oils, and a molasses (about 28 per cent water, nearly 50 per cent sugar). The press cake is dried separately to about 8 per cent moisture content and used for cattle feed. The oils are primarily terpenes, and are used for flavoring, in perfumery, with pharmaceuticals and soap. The molasses can be fermented to yield alcohol, or as a medium for yeast growth that can be incorporated into the livestock feed. An oil resembling olive oil can be extracted from the seeds.

A few words concerning each of the chief *Citrus* species follows:

Citron—China and India southward; cultivated especially in Italy; introduced to Mediterranean area about 300 B.C. probably by Alexander the Great, and popular in Roman times; fruit oblong, large, fragrant, yellow, with a very thick rind that is frequently candied; species more sensitive to cold than other commercial citrus fruits.

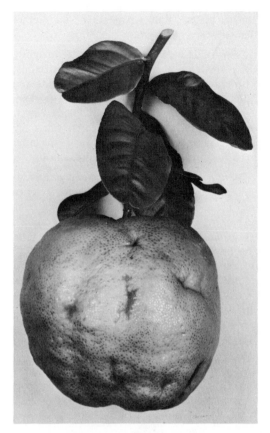

The citron, *Citrus medica*. (Courtesy Chicago Natural History Museum.)

Grapefruit—Originated in the West Indies, where it was first known about 1700; introduced into Tampa, Florida, in 1823, by Count Philippe, seedlings from which yielded the famed Duncan cultivar; possibly a hybrid of the pummelo and a sweet orange, or a sport; fruits large, subglobose, yellowish; has found great favor in the United States (where 90 per cent of the world's grapefruits are grown).

Lemon—southeastern Asia; introduced to the Mediterranean region about 1000 A.D. by the Arabs; fruit oval, with broad apical papilla, yellow; first used in the Western world medicinally, now also as flavor, garnish, and vitamin source; hybridizes with other species; in numerous cultivars (Eureka, Lisbon, Villa Franca, and others); chief producing areas are the

United States, Italy, Argentina, and other Mediterranean countries.

Lime—East Indies; fruit small, oval, greenish, with a small apical papilla; used as are lemons; hybridizes freely; one of the chief Florida and California cultivars is the seedless triploid, "Persian" or "Tahiti" lime; small, seedy Key or Mexican limes a specialty for pies; limes are generally grouped as "acid" or "sweet," the latter perhaps representing crosses with *C. medica*.

Mandarin orange—Philippines and southeastern Asia; did not reach Europe and America until about 1800; fruits subglobose, with thin, loose peel, bright yellow to scarlet orange; hybridizes freely (e.g., with grapefruit to yield tangelos, with orange to give tangors, and so on); especially prominent in Japan and grown to a considerable extent on the Gulf Coast of the United States; known in King, Satsuma, Tangerine, Mandarin, and Mitis varietal groupings.

The king mandarin, *Citrus reticulata*. (Courtesy Chicago Natural History Museum.)

Orange—China and Indochina, but so long domesticated that it is no longer known in the wild; not introduced to the Mediterranean region until about 1500 A.D.; fruits subglobose, with a solid core at maturity; dominant in the citrus industry; used primarily for fresh fruit and for juice; hybridizes freely; known in "normal," "navel," and "blood" varietal groupings; the Bahía or Washington navel the chief winter orange in California, the Valencia providing summer harvest; Florida varieties generally thinner-skinned and less highly colored.

Pummelo or shaddock—southeastern Asia and East Indies; fruit very large, subglobose, easily peeled; mostly consumed in the East, where the pulp is shelled out into dishes; does not squirt like grapefruit; probable ancestor of grapefruit.

Sour orange, seville orange, or bigarade—southeastern Asia; widely cultivated; introduced to the Mediterranean region by Arabs about 1000 A.D.; fruit hollow in maturity, subglobose, the thick peel a brilliant orange, pulp sharply acid; used medicinally, for flavoring ("ades" and curaçao liqueur), for essential oil (bergamot or bigarade), and for preserves (especially marmalade)—the flowers too for perfume (neroli); generally used as rootstock for lemons, etc.; known in many varieties (*amara, bergamia, myrtifolia*, etc.), hybrids, and horticultural forms.

Tangerine—Deep orange or scarlet-fruited *mandarin oranges*—which see.

Pineapple, ananas, piña, or abacaxi, *Ananas comosus* (*A. sativus*), Bromeliaceae. The pineapple was domesticated by South American Indians long before the coming of the white man to the New World. Three wild species of the genus, and another closely related genus (*Pseudananas*), are found in various parts of South America from northern Argentina to the Caribbean coast. *A. comosus* occurs frequently as an escape, but is not known certainly from the wild. Wild fruit of the other species is occasionally harvested, and in rare instances the plants even cultivated. The diploid chromosome count is 50 for all species, but diploids and tetraploid plants are not uncommon. Crosses are possible between most of the species, but *A. comosus* is sterile to its own pollen and does not set seed when self-pollinated.

Pineapple was apparently first domesticated by the Guarani Indians of what is now northern Paraguay. The species spread throughout lowland South America east of the Andes, and was taken throughout the Caribbean area by the Caribs. Columbus received the pineapple in barter with the Indians when visiting the island of Guadeloupe in 1493. The new fruit was quickly appreciated, and within decades the pineapple had been spread throughout the Spanish and Portuguese colonies. It arrived in India by 1548, and in the East Indies soon after. The fruit became highly esteemed as a delicacy in Europe during the seventeenth and eighteenth centuries, grown in the greenhouse. Selection and refinement occurred under glass, leading to some of the modern cultivars. Pineapple stock improved in England was sent to Australia, and thence imported into Hawaii in 1896. Hawaii has become the leading pineapple-producing area, accounting for about a third of the world's total production of around 3 million tons annually. Other prominent producing areas are Brazil, Thailand, Malaysia, Mexico, Taiwan, the Philippines, South Africa, West Africa, and much of Latin America.

The pineapple plant is a bristly, agave-like biennial bearing numerous overlapping leaves with sharp points. During the first year of growth the plant stores food in the central axis, after which (in response to slightly lower temperature) an inflorescence is stimulated that matures in about 6 months. The pineapple fruit bears a crown of miniature leaves (which is sometimes used to propagate the plant), and slips (just below) and suckers (farther down the stem) below the fruit. The latter are mainly used for propagation; suckers are quicker to yield a new harvest than are slips and crowns. After the fruit is

Maturing pineapple fruit showing "slips" at its base which can be used for new plantings. (Courtesy USDA.

cut two fruit-bearing side sprouts develop bearing slightly smaller fruit, and these again will bear a "second ratoon crop" if treated in the same fashion. In commercial fruit production seldom is the process carried beyond the first ratoon stage before replanting the field. The pineapple itself is a multiple fruit from the partial fusing of numerous berrylike fruitlets, in which the hardened sepals make something of a continous rind over the outside. The fruit is normally seedless due to self-incompatibility, and triploid cultivars such as the Cabezona of the West Indies do not produce functional reproductive cells at all.

A number of cultivars of the pineapple have been selected, but relatively few have attained commercial importance. The smooth Cayenne cultivar is responsible for most Hawaiian and African production (there is a spiny, recessive mutation that must be continuously rogued), while Red Spanish is the dominant variety in Latin America, and Queen of considerable commercial importance in Australia, Malaysia and South Africa. The Cayenne cultivar probably traces back to the original hothouse selections made in England, and Queen was noted as early as 1658 by the Frenchman de Rochefort. Cayenne is large-fruited, and the most popular canning cultivar. Queen is more frequently used for the fresh fruit, as is almost entirely Pernambuco of Brazil and Monte Lirio of Central America.

In the mechanized plantations of Hawaii, suckers are usually planted to a well-prepared soilbed, through slits punched into a paper or plastic "mulch." Planting is still mostly a hand operation, as is picking of the fruit (although tremendous harvesting units enter the fields to pick up the cut fruits on a conveyer belt). Soil fumigation is generally practiced, and fertilizer applied abundantly. A planting reaches fruiting maturity within 15 months to 2 years, and a ratoon crop about 1 year after the "first-fruit" harvest. On good pineapple land yield should exceed 30 tons of fruit to the acre with the first crop, 20 to 25 with the first ratoon crop. The paper or plastic mulch between the rows helps to control weeds, raises soil temperature, slows down evaporation of moisture, and increases nitrogen mineralization. In Hawaii it is often necessary to spray trace elements, especially iron and sometimes manganese. Hormone sprays are effective in initiating fruit production, enabling better timing of the crop. Other sprays may be necessary for weed, disease, and insect control.

The major portion of the pineapple fruit crop is canned. Pineapple juice is becoming more important, from which sugar syrup, alcohol, or citric acid can be recovered as occasion warrants. Bromelin, a protein-digesting enzyme, has been extracted (but is expensive), and, of course, the juice can be fermented into wine, brandies, and liqueurs. By-products include a pineapple bran used for cattle feed, and fiber can be extracted from the leaves to make "piña cloth." The fruit develops full flavor only when ripened on the plant, and then contains about 15 per cent sugar, plus fruit acids, vitamins, and minerals. The pineapple is moderately perishable, but can be held for as long as a month under refrigeration.

Other tropical fruits. ABIU, LUCUMA, OR CAIMITO, *Pouteria* (*Lucuma*) *caimito*, Sapotaceae. This North Andean tree bears ovoid, yellow, two- or three-seeded fruits used especially for "geladas" and "refrescos" (sherbets and "ades").

AKEE, *Blighia* (*Cupania*) *sapida*, Sapindaceae. This West African tree has been introduced into tropical America. Each locule of its reddish trilocular capsule bears a black seed having a large, white, fleshy, edible aril that is usually consumed stewed or browned.

Akee, *Blighia sapida*. (Courtesy Paul H. Allen.)

AMBARELLA OR GOLDEN APPLE, *Spondias dulcis* (*S. cytherea*), Anacardiaceae. This Polynesian species, scatteringly cultivated in all tropical regions, bears fruit resembling a small mango. It is eaten fresh, stewed, or in "refrescos."

ASSAI, *Euterpe oleracea*, Palmae. This lowland species of northern South America bears small, bluish fruits about the size of a cherry that are kneaded in water to give a mash commonly consumed with manioc.

AVOCADO, AGUACATE, OR ALLIGATOR PEAR, *Persea americana* and other species, Lauraceae. The avocado is native to the tropical American lowlands, and has been introduced into the Mediterranean area, western Africa, and various parts of the Orient. It has been cultivated since time immemorial by the Indians of tropical America, and archeological diggings prove it to have been introduced into the Tehuacan area of south-central Mexico before 7000 B.C., from a more humid habitat. Today it remains a common and important food plant from Mexico to southern Brazil. Three races are recognized: the West Indian (Simmonds, Pollock, Catalina cultivars, planted mostly in Florida), the Guatamalan (Hass, Nabal, Itzamna, Taylor cultivars, much planted both in Florida and California), and the Mexican (Puebla, somewhat grown in Texas), as well as hybrids such as Lula and Booth 8, widely planted. The greenish one-seeded drupe is an especially rich source of digestible fat (up to 30 per cent) and protein (about 2 per cent). Pound for pound the avocado has more energy value than meat, and it is also rich in vitamins and iron. No wonder that the avocado has assumed much importance in the tropics! It should become a familiar table item in the United Sates, although its perishable nature makes it costly to market at the peak of taste and in the more flavorful cultivars.

Considerable attention has been given to the culture, care, and propagation of the

The "fuerte" cultivar of the avocado, *Persea americana,* in Mexico. (Courtesy USDA.)

avocado, so that orcharding and budding techniques have been developed to nearly as fine a point as with *Citrus.* Many of the older strains in Florida and California have been top-worked with buds of choice cultivars. The trees are evergreen, producing in autumn or spring terminal racemes of small, apetalous flowers that mature the female and male parts at different times. Some strains bear fruit in as little as 2 years, and at maturity yield several hundred avocados per tree annually. The bearing season north of the equator is usually summer and autumn, although the California crop is marketed chiefly in winter and spring. Fruits are gathered by hand when nearly ripe, the pedicel usually being severed with an orange clipper. Once ripe, avocados are quickly perishable. Avocado production in the United States runs about 80,000 tons annually, and a few thousand additional tons are imported. World production is many times this, but no statistics are available for the largely local consumption. Avocado oil extraction has been considered for less developed countries such as Ghana, where facilities for handling fresh fruit are minimal.

BACUPARI, *Rheedia brasiliensis* and other species, Guttiferae. Wild trees of southern Brazil yield a yellowish subacid fruit with a leathery rind, used chiefly for making "doce" (dessert jam).

BACURI, *Platonia insignis*, Guttiferae. The bacuri is a northern South American tree similar to the bacupari but yielding larger, orange-size fruits with an agreeably flavored pulp. They are used for "doces" and "refrescos."

BARBADOS CHERRY OR ACEROLA, *Malpighia glabra*, Malpighiaceae. The small, three-lobed, acid fruits of this tropical American shrub are used for "doces" and "refrescos." It is said to have the highest ascorbic acid content of any fruit.

BILIMBI, *Averrhoa bilimbi*, Oxaliadaceae. This widely cultivated Malayan tree is similar to the carambola and has similar uses.

BOROJOA, *Borojoa patinoi*, Rubiaceae. The genipa-like fruit of this Colombian genus is used for "refrescos."

BREADFRUIT, *Artocarpus altilis* (*A. communis; A. incisa*), Moraceae. The breadfruit, one of several useful species of the genus, is native to the East Indies but spread throughout the tropical lowlands of the world. It is famed for the drama of its introduction from Tahiti to the West Indies. Probably upon suggestion of Captain Cook, a special British expedition was sent to Tahiti under the command of Captain William Bligh, of the "Bounty." Enchanted with the Tahitian way of life, the crew mutinied on the return voyage, putting Bligh off at sea in a small boat with eighteen loyal followers. Bligh and his men survived a 41-day run to the East Indies, the captain living to command a second expedition, successful in introducing breadfruit to the West Indies. The mutinous sailors with a number of Tahitians migrated to Pitcairn Island to establish a utopian colony.

The breadfruit never proved to be the expected bonanza in the West Indies, the residents there preferring the banana. Breadfruit is still, however, a staple in the diet of Pacific island peoples and is scat-

The bilimbi, *Averrhoa bilimbi*. (Courtesy Chicago Natural History Museum.)

Tip foliage and young fruit of the breadfruit. (Courtesy USDA.)

teringly cultivated in the New World tropics. Handsome, monoecious trees with leathery, lobed leaves produce the large, rough, multiple (ovaries plus receptacle) fruits resembling North American "hedge-apples" (*Maclura*). The fruit is starchy, normally 30 to 40 per cent carbohydrate, and is usually eaten baked, boiled, or fried. In taste and consistency it resembles potatoes. Both seedless and seeded strains are known. Seeded forms are of little use, although the seeds, boiled or fried, are edible. Propagation of seedless types is usually by root sprouts.

CANISTEL, *Pouteria campechiana* (*Lucuma nervosa, L. salicifolia*), Sapotaceae. This Central American tree bears orange-fleshed fruit that is very sweet and cloying. It is usually consumed fresh.

CAPULIN. This is the common name variously given fruits of *Prunus spp.*, Rosaceae, of the tropical American highlands; *Eugenia acapulcensis*, Myrtaceae, of Mexico; and *Muntingia calabura*, Tilaceae, of tropical America.

CARAMBOLA, *Averrhoa carambola*, Oxalidaceae. Native to Malaya and now widely distributed, this pinnately leaved tree bears curious five-winged acidulous, fleshy, yellow fruits from terminal inflorescences. The fruit is usually consumed cooked.

CARISSA OR NATAL PLUM, *Carissa grandiflora*, Apocynaceae. The plum-size, brilliant red, ovoid fruit of this spiny, ornamental, South African shrub is eaten fresh, stewed, or preserved; in the preserved form it resembles cranberry jelly.

CASHEW, *Anacardium occidentale*, Anacardiaceae. This fleshy receptacle (pear) below the nut of this Brazilian tree is consumed as a fresh or preserved fruit.

CERIMAN, *Monstera deliciosa*, Araceae. This Mexican and Central American climbing vine, with large, perforate leaves, produces conelike inflorescences with a "pineapple" flavor, edible when fully ripe.

CHERIMOYA, CUSTARD APPLE, PINHA,

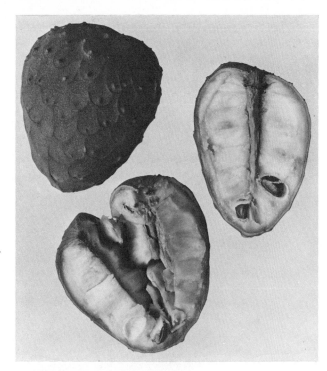

Fruit of the delicious cherimoya, *Annona*. (Courtesy Chicago Natural History Museum.)

ILAMA AND SIMILAR ANNONACEOUS FRUITS, especially *Annona*, Annonaceae. Many of the tropical representatives of the Annonaceae, the family of the Temperate Zone pawpaw (*Asimina*), produce an aggregate–accessory fruit of creamy consistency with a delightful sweetish, "fruity" flavor. They are small trees with alternate simple leaves and multipistillate flowers. The cherimoya, *Annona cherimola*, is native to the northern Andes and is now grown scatteringly throughout the tropics. The fruits attain weights of up to 15 pounds. The custard apple or bullock's heart, *A. reticulata*, is native to tropical America. The fruit is often chilled and the ice-cream-like pulp eaten with a spoon. The pinha (or ata, fruta de conde, sugar apple, or sweetsop), *A. squamosa*, of the same area, bears smaller, cone-shaped fruits, which are eaten fresh. The soursop or guanabana, *A. muricata*, of tropical America, bears large fruits similar to those of the pinha but with a more acid flavor. It is most used for "ades" and sherbets. The related biriba, *Rollinia*

The soursop, *Annona muricata*. (Courtesy Chicago Natural History Museum.)

pulchrinerva, of Brazil, is similar to the soursop. The ilama is *A. diversifolia*, native to Central America, its fruit similar to the cherimoya. A hybrid of the cherimoya and *A. squamosa* is known as the atemoya. Several lesser-known species of *Annona* serve as wild and infrequently cultivated fruit trees. As a group the annonaceous fruits are among the most exquisite of tropical fruits, in taste and texture ranking with the mangosteen in this respect. In general, however, they are neither scientifically selected nor cultivated. The fruits are seldom shipped since the soft, perishable, ripe pulp ferments quickly, and so are scarcely known to Temperate Zone peoples.

CHINESE GOOSEBERRY, *Actinidia chinensis*, Actinidiaceae. This is a climbing shrub native to China, producing fruits about the size of a hen's egg that may be eaten fresh when fully ripe; it is much esteemed in Australia and New Zealand.

CHUPANDILLA, *Cyrtocarpa procera*, Anacardiaceae. This is a Mexican fruit little grown elsewhere, from a small thorn–scrub tree of a semidesert area. The fruit is known to have been consumed by ancient peoples as early as 6500 B.C.

COCONUT, *Cocos nucifera*, Palmae. "Man's most useful tree," the coconut, was discussed in some detail in the chapter on oils. Coconut palms dot the shores of all tropical lands, and the fruit is often a main source of food among island peoples. The "milk" (liquid endosperm) furnishes a cool drink at any time; and the "meat," when eaten young and thin, is soft and tender.

COSAHUICO, *Sideroxylon spp.*, Sapotaceae. This relative of the sapote is little known outside of the area to which it is native in south-central Mexico. Today a small tree of irrigation ditches and dooryard gardens, the cosahuico is known to have been consumed by primitive peoples as long ago as 8,000 years, and was improved by selection through several millenia before Christ.

CUPU-ASSU, *Theobroma grandiflorum*, Sterculiaceae. This Amazonian relative of cacao bears large fruits, the fleshy pulp of which is used for "refrescos" and "doces."

DATE, *Phoenix dactylifera*, Palmae. The familiar date of commerce is the fruit of this dioecious, subtropical feather-palm native to dry areas of southwestern Asia. It has been under cultivation in the Holy Land for at least 8,000 years. According to Muslim tradition the date palm was made from dust left over after the creation of Adam, and is thus considered "the Tree of Life." The species was widely spread in the Mediterranean area by the Arabs, and its fruit is often a staple in the diet of these nomadic peoples. It was introduced into California by the Spanish in 1765, where it naturalized in the arroyos near the mission San Ignacio. The date is an excellent food (as found on the market it consists of about 70 per cent carbohydrate, 2 per cent protein, and 2.5

Date palms in the Near East. (Courtesy Chicago Natural History Museum.)

per cent fat). Dates may be eaten fresh, dried, or pounded into pastes. In the Arabic world they are commonly consumed with milk products that bolster protein content. A fermented maceration of dates yields one form of the famous "arrak," which Pedro Texeira, a sixteenth-century traveler, described as "the strongest and most dreadful drink ever invented."

The techniques of date selection, culture, care, and marketing have perhaps been improved to as great a degree as for any tropical fruit. Propagation is sometimes by seed but more commonly by sucker (offshoot from base of trunk) to assure continuing quality and female types. The offshoots are chiseled from the parent plant and usually "seasoned" to prevent rotting. Then they are planted in warm, humid sheds, where a year or more may elapse before rooting is complete. The palms begin to flower about the fourth year. Flowering occurs from February to June, during which time matured male inflorescences are cut and hand dusted on emerging female inflorescences from which the spathe has been split back. This necessity had been discovered by the ancient Hebrews and Babylonians, even though incompletely understood and liberally sprinkled with magic ritual. The pollen is viable for a few years, and a reserve supply may be obtained by dusting inflorescences into a paper bag. The fertilized female inflorescence is carefully watched, the number of fruiting branches being limited by trimming to that which can be best brought to perfection without undue drain on the vitality of the palm. Maximum yield is about 100 pounds of dates per palm per year, under a system of soil fertilization and careful culture. Ripening occurs from June to December, during which time it may be necessary to bag the inflorescence in cheesecloth or paper to prevent bird and insect damage. Picking is by hand, and the fruit is commonly pasteurized or sterilized with carbon bisulfide fumes. In commercial production dates are usually artificially ripened by "sweating" in incubators, during

A fruiting date palm. (Courtesy Chicago Natural History Museum.)

which sugars develop and astringent tannin becomes insoluble.

The date is cultivated quite intensively in California and Arizona, where production amounts to perhaps 22,000 tons annually. The eastern Mediterranean area continues to be the center of the date industry, with Egypt, Iraq, Iran, and Saudi Arabia accounting for over two-thirds of world production that reaches nearly 2 million tons annually. Subsisting as it does with little care on saline subsurface water in hot locations, the date has been replanted after centuries absence in parts of Israel and the Jordan valley where other crops fit less well.

DURIAN, *Durio zibethinus*, Bombacaceae. The durian has for centuries been a favorite fruit source in the Malay area and the East Indies, but has mostly been cultivated only haphazardly. The immense, spiny fruits have a thick rind and are quite malodorous, but nonetheless they are a great attraction to wildlife, from ant to elephant—and a hazard to whoever may pass beneath the trees at ripening time. The comestible portion is a

A sectioned durian, one of the prominent tropical tree fruits. (Courtesy Paul H. Allen.)

fleshy aril that surrounds the seeds. Human reaction to the durian varies—those unaccustomed to it finding its foul effluvium unbearable and others actually arranging home leaves so as not to miss the durian season. In his classic, *The Malay Archipelago,* Alfred R. Wallace penned a description of the durian fruit which, after a century, is still unsurpassed:

> . . . a rich butter-like custard highly flavored with almonds, but intermingled with it come wafts of flavor that call to mind creamcheese, onion-sauce, brown sherry, and other incongruities. Then there is a rich glutinous smoothness in the pulp which nothing else possesses, but which adds to its delicacy. It is neither acid nor sweet, nor juicy, yet one feels the want of none of these qualities, for it is perfect as it is.

One analysis of the fruit shows it to be nearly 58 per cent water, 34 per cent carbohydrate, 4 per cent fat, and 3 per cent protein, with a sprinkling of minerals, vitamins, and perhaps butyrate. In recent decades the durian has been introduced into the New World, where it has as yet few enthusiasts. *D. kutejensis* of Indonesia, and occasionally both *D. oxleyanus* and *D. dulcis* of Borneo, are also utilized in the same fashion as the durian.

EMBLIC, *Phyllanthus emblica,* Euphorbiaceae. Also known as the emblic myrobalan, the emblic was introduced into Florida from southeastern Asia as a possible source of tannin. It is a large tree, native from India to Polynesia, usually monoecious, bearing nearly sessile, yellowish fruits an inch or two in size that are astringent and extremely acid. The pulp is reported to contain very little protein and fat, but about 14 per cent carbohydrate, relatively abundant minerals, vitamins (ascorbic acid), and pectin. The fruit is consumed raw or preserved in India and southeastern Asia, and is used for bakery goods.

EUGENIA AND MINOR MYRTACEOUS FRUITS, *Eugenia* and other genera, Myrtaceae. The South American Surinam cherry or pitanga, *Eugenia uniflora,* yields a small ribbed fruit eaten fresh or cooked. Several pitombas, pera do compos, and others, *E. spp.,* of Brazil, serve in the same way, including *E. aggregata* and *E. luschnathiana* introduced into *Florida. E. (Syzygium) jambos,* the roseapple, is a Malayan species with larger fruit, introduced into the New World and now escaped, as is also the Malay-apple or ohia, *E. (S.) malaccensis,* which yields a pearshaped fruit. The Curaçao-apple, *E. (S.) javanica,* is a small ornamental of Malaya; it also bears clusters of small pear-shaped fruits. The Java-plum, jambu or jambolan, *E. (S.) cuminii,* indigenous to Java but now widely spread, bears clusters of juicy, purplish, plumlike fruits from the older branchlets. Species of *Feijoa* and *Abbevillea* (Brazil) and *Rhodomyrtus* (Orient) yield fruits similar to those of *Eugenia,* and are similarly consumed. All are small and of decidedly secondary importance in the tropical diet.

FIG, *Ficus carica,* Moraceae. The large

The malay apple, *Eugenia malaccensis.* (Courtesy Paul H. Allen.)

The cultivated fig. (Courtesy Chicago Natural History Museum.)

tropical and subtropical genus *Ficus* contains many incompletely understood wild, domesticated, and semicultivated species. These are woody shrubs, climbers, epiphytes, or trees, characterized by a specialized fruit (syconium) in which the perfect or unisexual flowers are borne inside a globose or pyriform receptacle (Fig. 19-1). *F.*

carica is the common fig native to the Mediterranean region, cultivated in the Holy Land for over 5,000 years and found abundantly as an escape or wilding there. It is frequently mentioned in the Bible, and recorded in Egyptian documents of the fourth dynasty (about 2700 B.C.).

The species is a small tree with broad, palmately veined, lobed leaves. The fruit is self-pollinated or parthenogenetic. A nearly continous sequence of summer crop is borne, the early fruits ("brebas") from the old wood, juicy and often consumed fresh, the later "second crop" from new wood being smaller and being generally dried or preserved. Caprifigs (var. *sylvestris*) are wild and semicultivated figs of the Mediterranean area, probably ancestors of the other types but of no commercial value for fruit. They are, however, host for a fig wasp that is essential to pollination of Smyrna figs (var. *smyrniaca*). The Smyrna-type figs (called Calimyrna in California) are very important commercially; however, they bear only pistillate flowers and so are dependent for fertilization upon wasps that bring pollen from caprifig plants (a process called caprification). For successful production Smyrna figs must be planted in association with caprifigs. Smyrna figs, when first introduced in California, dropped their fruit prematurely and lacked the characteristic "nutty" flavor until caprifigs and fig

Figs grown in Guatemala. (Courtesy USDA.)

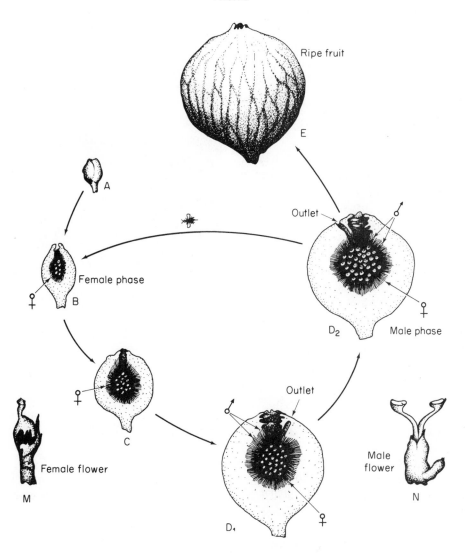

Figure 19-1 Developmental cycle of the syconium, the fig fruit of *Ficus sycomorus*. At phase B fertile female wasps enter syconium which has mature ovaries but immature pollen, oviposits; by phase D a new generation of wasps matures, emerges coated with new mature pollen which is carried to younger syconia in the B phase. (After Galil and Eirsckowitch, *Ecology* **49**, 263 (1968).)

wasps were introduced in 1899 from the Mediterranean area. Caprification, although poorly understood in ancient times, seems to have been first adopted by the Greeks centuries before the Christian Era, accompanied by suitable ritual and tribute to the Gods.

Commercial figs are subtropical, deciduous plants propagated by cuttings. They can be grown, with winter protection (such as straw), as far north as St. Louis and Philadelphia. In California most figs are sun-dried in the semiarid San Joaquin Valley, where the trees are cultivated under irrigation.

About 70,000 tons of figs are produced annually in California, and are consumed dried, fresh, and preserved. World production runs about 1½ million tons annually, mostly from Portugal, Italy, Greece, and Turkey. Modernized fig growing is being given emphasis in Israel. In the Old World dried figs find considerable use as stock feed.

GENIPAPO, *Genipa americana*, Rubiaceae. This little-known fruit from large trees of eastern South America and the West Indies may be eaten fresh or in marmalade form, but is more commonly made into an "ade" or wine.

GRANADILLA, MARACUJÁ, OR PASSION FRUIT, *Passiflora spp.*, Passifloraceae. Several passion · flower vines, especially *P. edulis flavicarpa*, cultivated and wild, yield a useful, juicy, aromatic fruit similar to a small gourd. The acidulous arils of this fruit are used for "refrescos." The species are native to the Americas, but have found greatest popularity in Australia, Hawaii, and South Africa. Attempt has been made to establish a passion fruit juice industry in Kenya.

GROSELHA OR OTAHEITE GOOSEBERRY, *Phyllanthus distichus* (*P. acidus*), Euphorbiaceae. This oriental tree, introduced into tropical America, bears on the stems abundant clusters of small, greenish, acid fruits which are tasty when cooked with sugar, or made into pies and preserves.

GUAVA, *Psidium guajava*, Myrtaceae. The guava is native to lowland South America and was apparently semidomesticated more than 2,000 years ago. It has been distributed to practically all tropical lands, where it often becomes naturalized and may even become a pest. To yield quality fruits the plants must be given some care. The species is a small tree with coarse, opposite leaves bearing raised veins below. The white axillary flowers arise on the new wood, and develop into the globose, gritty, somewhat acid, applelike fruits most used to make "goiabada," a dessert marmalade or paste, or guava jelly.

Ripe guava fruits, Cuba. (Courtesy USDA.)

Guavas can also be eaten fresh or in pie, or a juice may be extracted from them for jelly or drink. Guavas are a good source of minerals and vitamins.

In many tropical areas the guavas are given little cultural attention, but in Florida, California, Cuba, and India managed orchards have been established and cultivars (e.g., Supreme, Red Indian, Rolfs, Ruby, Florida) developed by selection and hybridization. Most commercial planting is still done from seed, though propagation of named types may be accomplished by air-layering, budding, or grafting. Fruits are picked from the trees by hand, or the trees are shaken to release the fruits. World production is difficult to estimate but must certainly amount to several hundred thousand tons annually, most of which is consumed locally. Another species, the Cattley or strawberry guava, *P. cattleianum*, is sometimes planted in the West Indies.

ICACO OR COCO-PLUM, *Chrysobalanus icaco*, Rosaceae. This coastal species occurs generally throughout the American tropics. The plumlike fruit is frequently made into a dessert paste.

IMBÚ OR UMBÚ, *Spondias tuberosa*, Anacardiaceae. This species is a spreading, usually wild tree of northeastern Brazil, bearing abundant small, yellow-green, acidulous fruits very refreshing to travelers in the semiarid climate.

INGA, *Inga spp.*, Leguminosae. Several species of *Inga* yield fleshy pods consumed locally in tropical America.

JABOTICABA, *Myrciaria* (*Eugenia*) *cauliflora* and other species, Myrtaceae. These are evergreen trees native to southern Brazil, bearing on the trunk juicy, grapelike fruits consumed fresh, made into jelly or wine, or frozen.

JACKFRUIT, *Artocarpus heterophylla* (*A. integrifolia*), Moraceae. This handsome, evergreen brother to the breadfruit tree bears huge, sweetly acid, multiple fruits from the trunk and lower branches. It is a lowland Malayan species, introduced into Brazil and Jamaica. In Brazil it serves as a "poor man's fruit," but is much esteemed by workers accustomed to its heavy, aromatic sweetness. Both pulp and seeds (roasted) may be eaten, or unripe fruit used as a cooked vegetable or in soup. The fruit is about 24 per cent carbohydrate, with little protein or fat. The Marang, *A. odoratissima*, of the Philippines is similar to the jackfruit.

JUJUBE, *Zizyphus jujuba* (*Z. vulgaris*), Rhamnaceae. This is a small Chinese species adapted to dry, hot weather, long cultivated in China, the Mediterranean area, and to some extent in the southwestern United States. The olive-size fruit is consumed fresh, dried, candied, or preserved, and the pit can serve as a "nut." Dried fruits resemble dates. The Indian jujube, *Z. mauritiana*, is similarly utilized.

KUMQUAT, *Fortunella spp.*, Rutaceae. These are bushy trees with small fruits having a mild, edible rind, used in salads (often for decoration).

LANGSAT, *Lansium domesticum*, Meliaceae. The langsat is one of the better Malaysian fruit species. It bears clusters of globose, velvety fruits with a bitter rind but a delicate, flavorful, sweetish pulp somewhat resembling the mangosteen. Well known in the East, it has been little cultivated in the New World.

LITCHI OR LYCHEE, *Litchi chinensis*, Sapindaceae. This famous and prolific tree of southern China has been in cultivation in the Far East since time immemorial, and since 1906 in Florida where fruit of the Brewster cultivar is marketed fresh. Hundreds of thousands of tons of the fresh and dried fruits are produced annually in China, part of these exported to Chinese living abroad. The species is also important in India, South Africa, and elsewhere. The tree is usually propagated by air-layering, and takes 5 years to bear. The warty-shelled, white-fleshed, viniferous-flavored fruits are borne in loose, terminal clusters. Similar Sapindaceous relatives of the litchi, similarly used, are the longan, *Euphoria* (*Nephelium*) *longana*; the rambutan, *Nephelium lappaceum*, with a larger, spinier fruit; and the pulasan, *N. mutabile*, all of eastern Asia.

Rambutan, *Nephelium lappaceum*. (Courtesy Paul H. Allen.)

The pulasan, *Nephelium mutabile*, considered in Singapore to be a variety of rambutan. (Courtesy Paul H. Allen.)

Mamee, *Mammea americana*. (Courtesy Paul H. Allen.)

LOQUAT OR JAPANESE MEDLAR, *Eriobotrya* (*Photinia*) *japonica*, Rosaceae. This tomentose Chinese tree bears plum-size fruits reminiscent in flavor of apples or pears. It is widely grown in Japan, China, India, and the Mediterranean area, and to some extent in the New World subtropics. It is perhaps the most important of the limited selection of Rosaceous fruits for the tropics, and cultivars of it have been selected (in California, Advance, Champagne, Premier, Thales, etc.; in Japan, Tanaka; in Algeria, Olivier; in Florida, Pineapple). The fruit is eaten like apples.

Loquat fruits, *Eriobotrya japonica*. (Courtesy Chicago Natural History Museum.)

MAMMEE, MAMEY, OR ABRICO DO PARÁ, *Mammea americana*, Guttiferae. The mamey is a beautiful and frequent dooryard tree of the West Indies, Central, and South America resembling an evergreen magnolia, first reported by Columbus in 1502. The thick, firm flesh of the fruit is consumed fresh or stewed with sugar.

MAMONCILLO, *Melicocca bijuga*, Sapindaceae. The species is a tropical American relative of the litchi; the greenish, plumlike fruit is popular among the poorer classes.

MANGABA OR MANGABEIRA, *Hancornia speciosa*, Apocynaceae. This latex-yielding species of eastern Brazil (see "Rubber, in Chapter 8) bears tasty persimmonlike fruits which are a welcome addition to the diet on the sandy taboleiros of its native area.

MANGO, *Mangifera indica*, and possibly other species, Anacardiaceae. This large eastern Asiatic tree is today one of the better-known tropical fruits and is cultivated throughout the tropics in numerous cultivars. It is certainly one of the most popular fruits of the Orient, and holds an honored place in Hindu culture and ceremonies. Wild trees are found from India through the Malay archipelago. The genus contains about forty species, with a basic diploid chromosome

Fruiting mango tree. (Courtesy USDA.)

introduced into Brazil by the Portuguese colonists before the early 1700's. There and elsewhere in the New World it has become naturalized and frequent in inhabited areas. It is a magnificent tree with abundant, tapered, glossy, dark-green leaves and large, terminal polygamo-monoecious panicles of small flowers. Fruiting is best in humid climates having a dry season. The fruit, in spite of being fleshy, keeps fairly well, and carefully packed shipments of select types have been made from India to Europe and elsewhere. Kept at a temperature of 45–50°F the fruit keeps as long as 1 month, and it may be picked before it is fully ripe, like the banana. Alphonso and Mulgoa cultivars are much planted in India; a seedling from the latter in Florida became the well-known Haden. Trees have a tendency toward biennial bearing, and a mature tree may produce 600 fruits in an "on" year. There are no reliable figures on world mango production, but India alone is said to produce over 5 million tons annually.

Sandersha variety of cooking mango. (Courtesy Paul H. Allen.)

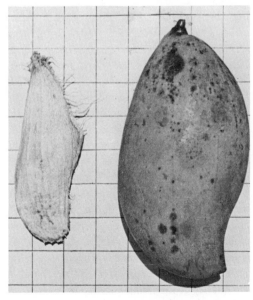

complement of 40 (although some cultivars are polyploid and apomictic). The large drupe has a juicy, usually fibrous, aromatic pulp much esteemed by some but considered disagreeable by others. The mango is of the same family as poison ivy, and some people develop a rash on the lips similar to mild poison ivy irritation from eating it. The mango is usually consumed fresh, but may be canned or made into jam, pickles, or chutney. Only the less fibrous and fine-flavored grafted cultivars merit planting today, although anywhere in tropical countries seedling mangos spring up.

The species has been in domestication since time immemorial in the East, and was

The mangosteen, *Garcinia mangostana*, "queen of tropical fruits." (Courtesy Chicago Natural History Museum.)

MANGOSTEEN, *Garcinia mangostana*, Guttiferae. The mangosteen is a small tree native to the Malayan area, bearing purplish fruits with a thick rind and a center containing four to six seeds with delicate white arils of exquisite flavor. It has long been termed the "queen of tropical fruits," but its establishment in the New World has met with little success because of ecological difficulties. In the East Indies and Malaysia marketed fruit usually comes from scattered dooryard trees, although a few plantations have been established. Up to 2,000 fruits may be obtained annually from a single tree under ideal conditions. The species is extremely susceptible to cold, and even grafts on hardy stock have little chance of surviving in the United States. Little selection or varietal improvement has been accomplished. *G. dulcis*, *G. livingstonei*, and *G. xanthochymus* are other species having an edible fruit, although not of the quality of the mangosteen.

MANZANILLA, *Crataegus spp.*, Rosaceae. Mexican and Central American species of this large genus yield ornamental and edible fruits in the highland areas.

MOMBIN (RED), *Spondias purpurea*, Anacardiaceae. This small Mexican tree bears cauline fruits that are much consumed in Central America, either fresh or boiled and dried.

MOMBIN (YELLOW) OR CAJA, *Spondias mombin* (*S. lutea*), Anacardiaceae. This is a tall, spreading tree indigenous to tropical America, bearing ovoid, yellow fruits over 1 inch long, with a large central seed and a pungent pulp. Propagation is usually by cutting.

NARANJILLA, *Solanum quitoense*, Solanaceae. This gigantic, perennial herb of the northern Andes, 6 to 10 feet tall and with leaves more than 1 foot long, produces small, orange, tomatolike fruits during every season of the year. The plants are started in seedbeds and transplanted. Plantings are maintained for about 3 years, after which new plantings are generally made in virgin soil as production drops off (probably due to nematode buildup). The fruits are crushed to produce a refreshing juice.

Naranjilla fruits, Ecuador. (Courtesy USDA.)

Fruiting branch of olive. (Courtesy USDA.)

OLIVE, *Olea europea*, Oleaceae. The olive is primarily a crop of the Mediterranean area, where as much as 90 per cent of the production is expressed for olive oil. The olive was intimately associated with ancient civilizations in the Near East, and is referred to frequently in the Bible. Olives were introduced into California through the Mission San Diego in 1769, and "Mission" is still a leading cultivar. The olive is killed by severe temperatures too much below freezing, but needs some winter cold for floral induction. It is prone to biennial bearing, which is overcome in California to some extent by hormonal spraying. Propagation is chiefly by layered cuttings, or budding of select cultivars on seedlings. The olive is best adapted to semi-arid regions, grown under irrigation. Established orchards are said to yield a half-ton of olives per acre. Harvesting is increasingly accomplished with mechanical shakers.

Fruit for the familiar pickled olive is carefully picked to avoid bruising, at "straw-color" stage for green olives and at "black" stage for ripe olives. The processing of olives neutralizes the bitter glucoside, oleuropein, present in fresh olives. With black ripe olives, the fruit is stored several months in a weak salt solution, where a lactic fermentation takes place, and is then treated with sodium hydroxide, washed, and finally stored in dilute brine and sterilized. Green olives are treated similarly, except that the lactic fermentation is omitted. In Spain olives are soaked in caustic soda for 14 hours, then held in casks of brine for a 30-day fermentation. Stuffed olives are pitted one at a time with a hand plunger, then filled by hand with brine-preserved pimento, etc. World production of olives reaches about 10 million tons annually, mostly from Spain, Italy, Greece, Turkey,

Pruning olive orchard in Greece. (Courtesy USDA.)

and Morocco. California produces over 50,000 tons of olives annually. Olive oil production is reviewed in Chapter 13.

PALM NUTS, *Elaeis guineensis*, Palmae. Palm nuts are chiefly important as a source of vegetable oil, discussed in Chapter 13, but in Africa the boiled pericarp is pulped and combined with other vegetables in soups and stews. It is said to be eaten as often as twice weekly in the forested and coastal zones of west Africa. The palm pericarp provides both carbohydrate and fat, and it is rich in vitamin A.

PAPAYA OR MAMAO, *Carica papaya*, Caricaceae. The papaya is an excellent fruit of a small, unbranched, soft-stemmed, usually dioecious tree found wild and naturalized though most of tropical America. As early as 1598 it had been introduced into India, and soon after it spread through the Pacific islands, where today it is common. In as little as 10 months it may bear ripe fruit. Plantations are established by clearing the land of all vegetation, planting seeds in a

"Colombia" cultivar of papaya, in Puerto Rico. (Courtesy USDA.)

nursery (germination is in 3–4 weeks), transplanting the young plants to the field, and keeping weeds in check. With dioecious cultivars all but a few staminate trees are destroyed; some cultivars bear perfect flowers, partly as a response to environment. Named cultivars include the Hawaiian Solo, perhaps the nearest thing to a "pure line," the Florida Betty, and the African Hortus Gold, as well as recent cultivars such as Bluestem, Graham, and Kissimee. In the course of its short life (commonly only 3 to 4 years) the papaya may produce in continuous succession 100 or more immense, hollow-centered, pyriform fruits, in appearance much like honeydew melons. The fruit is usually consumed fresh, but may be made into juice, pickles, preserves, jellies, or sherbets, or may be served cooked like squash. It is the source of the digestive enzyme papain, which finds some commercial use as a meat-tenderizing sauce. The enzyme is obtained by scoring the rind of maturing fruits and collecting the latex which exudes in nonmetallic containers. It is said that tough, unaged meats in tropical regions where refrigeration is unavailable can be readily tenderized by wrapping them for short periods in papaya leaves. If picked unripe, papayas may be shipped with moderate success. They are little known in temperate regions, but in the tropics they constitute an esteemed fruit staple.

PEJIBAYE OR PEACH PALM, *Guilielma gasipaes* (*G. speciosa*), Palmae. A spiny, slender New World feather-palm which produces orangish fruits about the size of a hen's egg which are mealy and nutty in flavor when boiled in salted water. These drupes are highly nutritious, but a taste for them must be acquired. Yields are said to be as remunerative on a per acre basis as is maize in temperate climate agriculture. This would seem to augur well for widespread utilization of pejibaye with undernourished peoples in many parts of the tropics, but in general this "poor people's food" has not caught on.

A peach palm, *Guilielma gasipaes*. (Courtesy Paul H. Allen.)

The fruit of the peach palm, *Guilielma gasipaes*. (Courtesy Paul H. Allen.)

POMEGRANATE, *Punica granatum*, Punicaceae. This species is a small shrubby tree growing in dense wild stands in eastern Asia and naturalized in the Mediterranean area. It was cultivated in Israel more than 5,000 years ago, and was regarded by the ancients as a symbol of the prolific. The fruit bears many seeds, each with a juicy red aril, inside a tough rind. It has excellent keeping qualities and will hold for up to 6 months in cold storage. The pomegranate is grown to a limited extent from California southward in the American tropics but is more popular in the Old World than anywhere else. Propagation is by seeds, cuttings, or layering. Favorite cultivars are Wonderful, Paper Shell, and Spanish Ruby.

Pomegranate growing in California. (Courtesy USDA.)

RAMONTCHI OR GOVERNOR'S PLUM, *Flacourtia indica* (*F. ramontchi*), Flacourtiaceae. The ramontchi is a small, easy-to-grow, dioecious tree or shrub of southern Asia, introduced elsewhere, bearing berries 2–3 centimeters in diameter which are eaten fresh or as preserves. *F. rukam* and *F. cataphracta* (*F. jangomas*) are similar to the ramontchi although less used.

SAPODILLA, *Achras sapota* (*Manilkara sapotilla*), Sapotaceae. The species, the latex of which yields chewing gum, also furnishes man with a thin-skinned, sweet-fleshed, dessert fruit (berry) slightly larger than a plum. It is a large tree native to the forests of northern Central America, but today planted throughout the world tropics in numerous cultivars perpetuated by budding, grafting, and marcottage.

SAPOTE OR MAMEY, *Calocarpum sapota* (*C. mammosum*), Sapotaceae. This Central American lowland tree is cultivated from seed through most of the Caribbean area. It bears moderately large drupes with a thick, firm, spicy, nonacid flesh that is eaten fresh or made into conserves. The fruit is frequently an important item of the diet among hinterland Indians.

Sapote or mamey, *Calocarpum mammosum*. (Courtesy Paul H. Allen.)

SAPOTE (WHITE) OR CASIMIROA, *Casimiroa edulis*, Rutaceae. This species is native to the highlands of Mexico and Central America and has been introduced throughout the Caribbean area. The thin-skinned, sweet, orange-size fruits are usually eaten fresh. A few cultivars have been selected in Florida and California, and *C. tetrameria* is also planted.

STAR-APPLE OR CAINITO, *Chrysophyllum cainito*, Sapotaceae. This tropical American species is distinguished by leaves that are glossy green and velvety brownish-pubescent below. The applelike, purplish fruit possesses a star-shaped, acidulous pulp that is usually eaten fresh. The trees propagate by seed and are dooryard counterparts of the sapodilla and sapote over most of the Caribbean area.

TAMARIND, *Tamarindus indica*, Leguminosae. This ornamental Indian species is now firmly established throughout the world tropics. The pulpy legume is reputed to contain more sugar and fruit acid per unit volume than any other fruit. In the New World, the fruit finds most use for "refrescos" and as a flavor in guava jellies, but in Arabic countries and the East Indies it is an important market item in many forms. Propagation is commonly by seed. The related carob or St. John's bread, *Ceratonia siliqua*, was mentioned both as a source of gum and as a food seed; the pods are rich in protein and sugar, sometimes ground into a flour used as a health food, or consumed dry by man and beast in the Mediterranean lands.

TREE TOMATO, *Cyphomandra betacea*, Solanaceae. Native to the Peruvian area, this small tree has been spread to various parts of the world, where the small, egg-shaped, succulent fruits are eaten fresh or, more commonly, stewed.

TUNA, PITAYA, AND OTHER CACTACEOUS FRUITS, Cactaceae. "Tuna" and "pitaya" are names applied to fruit of several species of *Opuntia, Cereus,* and related genera. This fruit is usually seedy, sometimes spiny. It is widely cultivated but not of great importance. *Pereskia aculeata,* Barbados gooseberry, yields fruit that bears leaves; it is very acid and is usually cooked with sugar.

UMKOKOLA OR KEI-APPLE, *Dovyalis (Aberia) caffra,* Bixaceae. This dioecious, African species is little cultivated for fruit outside of its native area, although it is known elsewhere as a hedge plant. The fruit is used for preserves and jams. The similar ketembilla or Ceylon gooseberry, *D. hebe-*

carpa, occurs in Ceylon; it yields a pubescent very acid fruit. *D. abyssinica* has a milder fruit eaten fresh in Ethiopia.

OTHER FRUITS. It is obviously impossible even to list all tropical fruits in a volume of this kind. For additional information the reader may turn to E. O. Fenzi, *Frutti Tropicali e Semitropicali* (Firenze, Italy: Instituto Agricola Coloniale Italiano, 1915), which gives an alphabetical list of 727 species, with native name, Italian, French, and English indices; and to more recent indices listed among the Supplementary References.

SUGGESTED SUPPLEMENTARY REFERENCES

Anderson, W., *The Strawberry: A World Bibliography 1920–1966,* Scarecrow Press, Metuchen, N. J., 1969.

Batjer, L. P., H. A. Schomer, E. J. Newcomer, and D. L. Coyier, *Commercial Pear Growing,* Agriculture Handbook 330, USDA, ARS, Washington, D. C., 1967.

Chandler, W. H., *Deciduous Orchards,* Lea and Febiger, Philadelphia, 1959.

——————, *Evergreen Orchards,* Lea and Febiger, Philadelphia, 1958.

Collins, J. L., *The Pineapple: Botany, Cultivation and Utilization,* 2nd ed., Leonard Hill, London, 1967.

Condit, I. J., *Ficus: The Exotic Species,* Univ. of California Division Agricultural Science, Berkeley, 1969.

——————, *The Fig,* Chronica Botanica, Waltham, Mass., 1941: Ronald Press, N. Y., 1947.

Darrow, G. M., "The Cultivated Raspberry and Blackberry in North America—Breeding and Improvement," *American Horticultural Magazine* **46,** 203–218 (October, 1967).

——————, *The Strawberry: History, Breeding and Physiology,* Holt, N. Y., 1966.

Dawson, V. H. W., and A. Aten, *Dates—Handling, Processing and Packing,* FAO, Rome, 1962.

Duckworth, R. B., *Fruit and Vegetables,* Pergamon Press, Elmsford, N. Y., 1966.

Eck, P., and N. F. Childers (eds.), *Blueberry Culture,* Rutgers Univ. Press, New Brunswick, N. J., 1967.

Heiser, C. B., *Nightshades, The Paradoxical Plants,* Freeman, San Francisco, 1969.

Indian Council Agricultural Research, *The Mango: A Handbook,* New Delhi, 1968.

McPhee, J., *Oranges,* Farrar, Straus and Giroux, N. Y., 1967.

Mortensen, E., and E. T. Bullard, *Handbook of Tropical and Subtropical Horticulture,* U. S. Dept. of State, A. I. D., Washington, D. C., 1966.

Mowry, H., L. R. Toy, and H. S. Wolfe, *Miscellaneous Tropical and Subtropical Florida Fruits* (rev. G. D. Ruehle), Univ. of Florida Agricultural Extension Bul. 156, Gainesville, 1953.

Popenoe, W., *Manual of Tropical and Subtropical Fruits*, Macmillan, N. Y., 1920.

Shoemaker, J. S., *Small Fruit Culture*, Blakiston, Philadelphia, 1948.

——————, and B. J. E. Teskey, *Tree Fruit Production*, Wiley, N. Y., 1959.

Simmonds, N. W., *Bananas*, 2nd ed., Longmans, London, 1966.

Singh, L. B., *The Mango: Botany, Cultivation and Utilization*, Leonard Hill, London, 1960.

Smith, J. R., *Tree Crops*, Devin-Adair, N. Y., 1950.

Smock, R. M., and A. M. Neubert, *Apples and Apple Products*, Interscience Publishers Inc., N. Y., 1950.

Tindall, H. D., *Fruits and Vegetables in West Africa*, FAO, Rome, 1965.

Tressler, D. K., and M. A. Joslyn, *Fruit and Vegetable Juice Processing Technology*, AVI Pub. Co., Westport, Conn., 2nd ed. 1971.

Upshall, W. H., (ed.), *North American Apples: Varieties, Rootstocks, Outlook*, Michigan State Univ. Press, E. Lansing, 1970.

USDA, ARS, *Muscadine Grapes*, Farm. Bul. 1785, Washington, D. C., 1965.

Webber, J. H., and L. D. Batchelor (eds.), *The Citrus Industry*, Univ. of Calif. Press, 1943; 2nd ed., W. Ruether, et al. (eds.), Vols. I and II, 1967–68.

Whitaker, T. W., and G. N. Davis, *Cucurbits: Botany, Cultivation and Utilization*, Leonard Hill, London, 1962.

Ziegler, L. W., and H. S. Wolfe, *Citrus Growing in Florida*, Univ. of Florida Press, Gainesville, 1961.

CHAPTER 20

Beverage Plants

DRINK IS A MORE PRESSING NEED FOR MAN even than food. Water, of course, is the great quench for thirst, essential to any life. Water is so plentiful that alternative beverages can hardly be called essential. Nevertheless, from time immemorial man has searched for ways to make his drinks more flavorful and zestful than is water alone. An obvious alternative to plain water is the juice of fruits, such as were discussed in the preceding chapter. The apple, the cherry, and other temperate tree fruits have always had their juices expressed to make ciders, and the chief use of the grape has always been for wine. In temperate lands the breakfast table is as likely as not to offer citrus or pineapple juice in these days of facile transportation. Less familiar tropical fruits such as naranjilla, granadilla, and species of *Spondias* are used in their homelands chiefly for making "refrescos" or fruit drinks. Modern "soft drinks" imitate fruit juices, being compounded from sugar, fruit acids, and other flavors. They have become the hallmark of youth, and are consumed by the hundreds of millions of bottles per day.

A step beyond expressed fruit juice are diffusions and decoctions from processed plant parts, such as coffee, tea, and chocolate (cacao). Mostly these are made by steeping specially prepared materials in water; various substances diffuse into the beverage including stimulating caffeine (found in most nonalcoholic drinks). Cacao could as well be regarded a food as a beverage, for today it is consumed largely as a flavor or component of confections rather than for the chocolatl-type drink once of such great importance in Aztec ceremonies. But people still drink hot chocolate and relish chocolate sodas. Beer and malt beverages are derived from cereal grains, and, indeed, almost any carbohydrate extract can be fermented to make a potable drink. Beverages classify rather neatly into nonalcoholic sorts such as coffee, tea, and chocolate that usually contain caffeine or caffeinelike derivatives; and alcoholic ones like beer, wine, and liquors derived by alcoholic fermentation. Beverage fermentation is an art as old as man's alliance with plants, and an important feature of all civilizations from earliest times to the present day.

Nonalcoholic Beverages

Apart from plain water and milk, most familiar nonalcoholic beverages are derived from plants. Tea is consumed by at least half the world's people, and coffee by a third or more. In the United States coffee and tea are served in almost all homes. Cacao provides food and drink for millions more. Needless to say, the production, compounding, and

French coffee buyers in Yemen. (Courtesy Chicago Natural History Museum.)

marketing of these beverages are enterprises of great importance to the economic life of many nations and peoples. Coffee and tea are among the foremost items of international commerce. Two-thirds of the world's coffee comes from tropical America, and Brazil alone accounts for almost half of world production. Consumption is largely in North America and Europe. Tea is produced chiefly in the Orient, but it is consumed widely in the West as well as there. Chocolate is a major export from tropical Africa to temperate lands. Other nonalcoholic beverage plants, such as maté, guaraná, and so on, although much used locally, are of comparatively minor importance in world trade.

Coffee, *Coffea arabica* and other species, Rubiaceae. Unlike most cultivars, coffee is a relatively new domesticate, popular as a beverage only in recent centuries. It may be that the coffee berry served as a stimulant and "medicinal" long before that, among tribes of the Near East. In some places the flesh of the berry rather than the seed is consumed, and dried coffee leaves are sometimes used like tea. As for the coffee berry or "bean," one legend has it that in a monastery near the Red Sea goats tended by the Monks returned one night capering and sleepless. A shepherd traced their visit to a remote wadi where they had chewed upon "a shrub with leaves like laurel, flowers like jasmine ... and little dark berries the chief object of the animals' nibbling." Berries brought to the monastery were pulverized by the imam and dropped into boiling water, from which "a most exhilarating odor arose ... and a warm comforting glow ran through his holy limbs." When offered to other brothers in the monastery each "leaped alertly to his feet ready to praise Allah." They called the drink "Kahveh," meaning "stimulating" in the language of the times.

Coffee was certainly cultivated in Yemen before 1500, and known to Europeans by then. It was carried from Yemen to the East Indies and other parts of the world, and was introduced into the New World in 1720 from the progeny of a single plant that had been brought to Holland in 1706 from Java. *C. arabica*, source of three-fourths of the world's commercial coffee, was named by Linnaeus from cultivated specimens only. He presumed the species to be native to Felix Arabia (the ill-known southern coast near the Gulf of Aden) and only recently have wild stands been confirmed in the southwestern highlands of Ethiopia, from which coffee was doubtless first brought. Few plants have risen to popularity so quickly as has coffee. Little used two centuries ago, the coffee crop today exceeds 4 million tons annually. Nearly half of world production comes from Brazil, with Colombia the next most important source; various nations in tropical Africa and Central America, Mexico, and Indonesia follow.

Edicts closing the "kaffa" houses in the Near East were frequent during the 1600's. But in spite of intermittent official and religious opposition to coffee houses for fear that they might breed disrespect of authority, the custom of coffee drinking spread widely. It reached Europe in the sixteenth and seventeenth centuries through the pilgrimages and caravan trade of the times. In Europe the first cafés were opened in Paris in the early 1600's, and they soon became frequented by the most illustrious literary and political figures of the day. Marseilles, at that time far outranking Paris as a commercial center, accepted the custom of coffee drinking, but quickly abandoned it upon a rumor that coffee was an anaphrodisiac. Lloyds of London, the world-famed underwriting establishment, originated as a coffee house of the late 1600's given to posting sailing data for the benefit of its patrons. The first coffee house of North America was opened in Boston in 1669, but it and its kind never attained in the New World the intellectual importance of the European coffee houses.

Mocha, the city whose name a type of coffee still bears, remained for two centuries the center of the coffee trade, an Arabian monopoly until coffee growing became widespread in the late 1700's. Coffee was first carried to India, Ceylon, and the

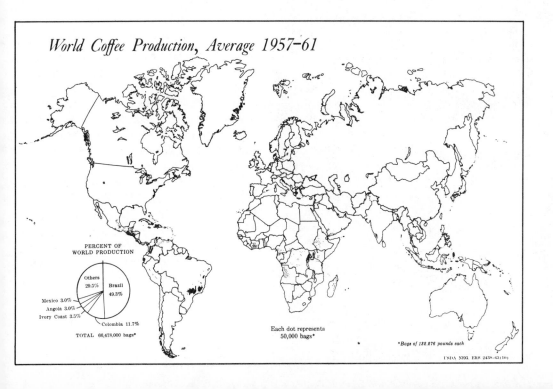

World Coffee Production, Average 1957–61

PERCENT OF WORLD PRODUCTION

Others 29.5%
Brazil 49.3%
Mexico 3.0%
Angola 3.0%
Ivory Coast 3.5%
Colombia 11.7%

TOTAL 66,478,000 bags*

Each dot represents 50,000 bags*

*Bags of 132.276 pounds each

USDA NEG. ERS 2438-63(10)

East Indies during the sixteenth and seventeenth centuries, and from there, via Amsterdam, to the West Indies and South America. It was once the premier crop of Ceylon, but disease destroyed the plantings there and through much of the East in the late 1800's, and tea became dominant among beverage plants. In the New World, French Guiana plants were introduced into Amazonia Brazil in 1727, and nearly half a century later reached Rio de Janeiro and southern Brazil. There, in the highlands of São Paulo, the coffee plant found soils and climate perhaps more favorable than anywhere else in the world.

The genus *Coffea* may include as many as seventy species ranging from eastern Africa to southeastern Asia. Only *C. arabica* and *C. canephora* have achieved commercial importance. The former is the milder, more delicately flavored species, and accounts for most coffee grown in tropical America. It is the cultivar typically used for brewed

Picking coffee berries by hand in Costa Rica. (Courtesy USDA.)

coffee. *C. canephora* (*C. robusta*) yields Robusta coffee, especially suitable for the "instant" drink; it is produced chiefly in Africa, and now accounts for about one-quarter of world production. *C. arabica* is an allotetraploid having 44 chromosomes while all other species of the genus are diploids. Species other than *C. arabica* are said to be self-incompatible, but the Arabica species is facultatively self-fertilized and accumulative of homozygous mutations making it less variable than its wild progenitors. There is now little doubt that *C. arabica* originated in mountainous southwestern Ethiopia, where wild plants are still found (although sometimes difficult to tell from escapes). Attempts are being made to discover strains there resistant to the *Hemileia* rust disease so destructive in Africa (and unfortunately gaining a toehold in Brazil by 1970 as well). Strangely, the rust is not too troublesome on Robusta coffees, provided the crop is grown in partial shade such as from an overstory of *Albizzia*. Portugal has undertaken an extensive breeding program for coffee, since the crop is especially important to commerce of the Angola colony.

Basically coffee is a tropical rainforest species. Rainfall in its Ethiopian homeland exceeds 60 inches per year, much of it from April to September. Coffee prefers the cool temperatures of the highlands, but cannot stand freezing. The crop grows best where temperature varies little from 70°F. Well-drained volcanic soils are very productive in Africa, as are the red soils rich in iron and potassium in southern Brazil. Mulching is effective for increasing yields; typically alternate rows are mulched one year, the process reversed the next. The quality of coffee seems much influenced by growing conditions. Highly prized Blue Mountain coffee of Jamaica proved no more tasty when grown in Africa than typical African varieties such as Nyasa, or Brazilian cultivars such as Nacionale. Central American, Colombian, and

Jamaican production of Arabica coffee is generally considered superior to Brazilian, and Brazilian superior to African, even though the same cultivars may be planted. Mundo Novo is the high-yielding Brazilian cultivar responsible for most commerical coffee.

The coffee plant is a shrub or small tree with glossy, deep-green, opposite leaves. The leaves, as well as the fruit, contain caffeine. Fragrant, white axillary flowers are borne two or three times a year, in one principal and one or two secondary flowering seasons that correspond to times of dryness. The inferior ovary of the flower matures into a reddish, two-seeded drupe, the coffee "berry" or "cherry." Each seed bears a delicate, glistening membranous seed coat, the "silver skin," and lies within the parchmentlike endocarp and fleshy mesocarp. One tree bears up to 6 pounds of berries per year, and yields per acre generally run $1\frac{1}{2} - 2\frac{1}{2}$ tons of clean coffee beans.

Preparation of berries for market follows one of two methods. In the older "dry" method, utilized in the Near East and wherever water is scarce, the gathered berries are dried in the open, cover being kept handy to protect the hand turned piles in case of rain. Once they are sufficiently dry to prevent further fermentation of the pericarp, the berries may be stored in bins. At a later time they are mechanically separated from the pericarp.

In the "wet" process used for higher-quality coffee, water flotation first separates defective fruit and debris. The berries are then conducted as soon as possible to pulping machines, where the seeds (still in the endocarp) are freed from the mesocarp. Fragments of remaining pulp and mucilage are removed by controlled fermentation over a 12- to 24-hour period. Sometimes chemical or mechanical methods are used instead of fermentation. Then follows a thorough washing in running water and drying in trays, on paved areas, or on mats, much as in the dry method. Dry depulped fruits are mechanically freed of their parchment (endocarp) in special mills, after which the silver skin is removed by polishing. Sieving to size and grading follow. The entire process reduces 5 or 6 pounds of fresh coffee berries to 1 pound of "beans" ready for roasting.

The practice of roasting the coffee bean further affects its flavor. The degree of roasting (according to "secret" but well-regulated procedures—in modern roasters about 5 minutes at 500°F) varies from locality to locality and is dictated by popular taste. The roasting develops certain aromas (essential oils and caffeol) that give coffee its appetizing flavor. The process also helps break down the cellulosic cell walls, facilitating grinding. Once roasted and ground, coffee beans lose their flavor in a matter of weeks and the oils gradually tend to turn rancid. Therefore coffee is generally fresh-roasted, and some Near Eastern peoples will not accept coffee unless it has been roasted immediately before its consumption. During roasting the beans lose 12 to 18 per cent of their weight but gain about 30 per cent in volume.

The composition of green coffee is roughly 34 per cent cellulose, 10 to 13 per cent oil, 7 per cent sugar, 14 per cent protein, 12 per cent residual water, and minor amounts of other components. The soluble content of coffee—i.e., the difference between dry weight of coffee grounds before and after brewing— is about 25 per cent. The chief stimulative constituent of coffee is of course caffeine ($C_8H_{10}N_4O_2$). The quantity of this alkaloid supplied by a few cups of coffee is enough to promote a feeling of well-being and to relieve fatigue. There seems to be no habit-forming or depressing after-effect, such as is found with many other stimulants. The caffeine content of coffee berries from differing species and cultivars ranges from 0 to 3 per cent by weight. An "average" caffeine content for coffee is possibly about 1.5 per cent. Coffee has been used medicinally for centuries, having an effect on blood pressure, pulse, sensory perception, and other functions.

The best and most productive coffee is grown in highland habitats, where constant, moderate temperature and frequent mists seem to benefit the plants. Coffee trees demand fertile, well-drained soils and moist climates for best growth. Generally they are planted in rows about 5 feet apart, from seedlings started in seed-beds, and come to bearing age in about 5 years. In Brazil an economical method of field planting is increasingly used, consisting of setting about four seeds in a hole that is covered with a palm leaf. One or more of the seedlings is let mature to a bearing plant. In Java, some planting is of selected stock grafted onto robusta-type rootstock. Shade is usually provided for coffee, frequently in the form of leguminous trees (such as *Gliricidia*) that also benefit the soil. In many less progressive areas the coffee is intercropped

with bananas or figs; and in southern Brazil no shading is used after the first year. In Haiti coffee is planted by primitive methods and allowed to grow practically wild under shade of the natural cover of the mountain slopes. The fruit is harvested by peasants when they need money, and dried and milled according to centuries-old procedures.

In Brazil and Central America the plants are mostly cultivated by hand to control weeds, from seedling stage on, and in managed plantings are given fertilization every few years if possible. Coffee trees may live a half-century or more, but are generally productive only for about 25 years. Unless proper conservation measures are taken, coffee plantings are destructive of soil. It is reported that many of Brazil's finest soils in São Paulo are on the verge of exhaustion because of prolonged cropping with coffee.

A good deal of hand labor is generally needed for coffee growing. It is estimated that one man per hectare is required, although most plantations manage with a smaller

Coffee beans drying at a *beneficio* near San Jose in the Costa Rican highlands. (Courtesy Paul H. Allen.)

proportion. Planting, pruning or training, and picking of the berries are all hand operations. The fruit should be picked selectively as it ripens to avoid expense and waste for removal of useless berries during wet-processing. Thus several pickings must ordinarily be made from the same tree, increasing labor costs. In Brazil little care is taken in picking: usually the whole branch, including leaves, green fruits, and buds is stripped between thumb and forefinger. Strippings are simply dropped to the ground to be swept up later, perhaps winnowed by throwing them in the air, and taken to the flotation tanks, where the ripe berries float off and green berries and heavy debris sink. Mechanical vibrators that shake the fruits onto cloths spread beneath the trees are coming to be used in various parts of the world, but even so coffee growing remains a cheap-labor industry. In some parts of Africa portable equipment moves from village to village to process the crop.

The coffee industry, although it has become important only within the last few centuries, has had many problems. The susceptibility of the plants to disease is of course one: coffee was eliminated by rust in Ceylon at a time when the island was the foremost producer of the berry, and *Hemileia* has restricted African production mostly to the Robusta type (and has recently been discovered in Brazil). In Brazil plantings often suffer cicada attack, and a July freeze in 1969 set back production nearly 40 per cent. From time to time surpluses plague the industry, and the Brazilian government has been known to purchase and destroy millions of sacks of coffee beans to protect the price. Because the economies of many newly independent tropical countries rely heavily on exports, overproduction rather than underproduction seems likely. In times of scarcity, such as during war, adulterants and coffee substitutes have been used—even roasted tulip bulbs in the Netherlands. Parched-cereal substitutes for coffee find a minor market in the United States, and in some localities a modest percentage of chicory root is mixed into coffee. What with the "coffee break" having supplanted the coffee house as a social amenity, there seems no likelihood of coffee failing to remain one of the world's important beverages.

Tea, *Camellia* (*Thea*) *sinensis*, Theaceae. Tea or cha consists of the dried tip leaves of *C. sinensis*, native from southwestern China and northeastern India to Cambodia. Three varieties are often recognized: low-growing *sinensis*, high-yielding *assamica*, and the less important *cambodiensis*. Selected cultivars are legion. Diploid chromosome number is 30, and triploid and tetraploid variants have been discovered. Tea has long been used in the Orient, at first probably medicinally. Its precise mode and date of origin are uncertain, but it was an item of commerce with the Mongols thousands of years ago. Tea was introduced to Europe in 1610 by the Dutch, and it reached the tea-capital-to-be,[*] London, in 1664, Boston in 1714. In Salem, Massachusetts, as in parts of China and Mongolia, not only was the infusion drunk, but the extracted leaves, mixed with butter and salt, were eaten.

The tea plant, left unpruned and unplucked, grows as high as 10 meters. It bears thick, alternate, elliptic, serrate leaves, which possess numerous oil glands containing an essential oil. Whitish axillary flowers appear singly or in clusters of a few. These

---

[*] Even more than spice, tea was to form the backbone of the British East India Company, the famed empire within an empire. For a most interesting account of the "China trade," "the great tea races," and other historical events, the reader is referred to W. H. Ukers, *All About Tea* (New York: Tea and Coffee Trade Journal Co., 1935). Tea gradually replaced coffee as the favorite beverage in England during the eighteenth century, and the habit of tea drinking was spread throughout the British sphere of influence in subsequent centuries.

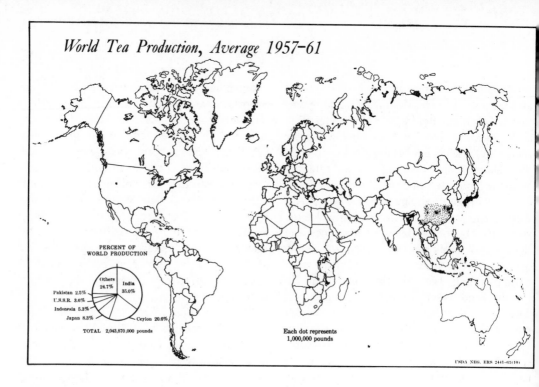

World Tea Production, Average 1957–61

PERCENT OF
WORLD PRODUCTION

Others 24.7%
India 35.0%
Pakistan 2.5%
U.S.S.R. 3.6%
Indonesia 5.3%
Japan 8.3%
Ceylon 20.6%

TOTAL 2,043,870,000 pounds

Each dot represents
1,000,000 pounds

USDA NEG. ERS 2441–63(10)

have the many stamens characteristic of the family, and are mildly fragrant. The fruit is a three-celled capsule, each compartment ordinarily yielding a solitary seed that is used for propagation.

The tea plant thrives best in protected localities, on well-drained, acid soil. It is grown from sea level to several-thousand-foot elevations, but requires warm climate with ample rainfall (rainfall ranges from 60 to 200 inches in the Assam growing areas). Tea has traditionally been planted under shade of sheltering trees, and no doubt the protection afforded has some advantage, especially in drier climates. But tests in Africa show tea there to be more productive and of higher quality if left unshaded. Seeds are usually started in nursery beds as forest is cleared for a plantation, although sometimes they are planted directly in the plantation row. When about 1 foot high the seedlings are transplanted to permanent locations. Select clones, vegetatively propagated, are becoming more extensively used, especially in Assam and Africa. Some of these are frost-tolerant. Since tea is cross-fertilized, seedlings do not come

true-to-type. Weak seedlings are replaced annually from the nursery. Cultivation, weeding, spraying against disease, cover cropping, etc. are generally practiced. Nitrogen fertilizer is used liberally to force leaf growth, usually ammonium sulfate, which has an acidifying influence on the soil. Trace elements are generally needed as well.

A portion of the plantation is usually pruned back each year, heavily the third year from seed, to low, flat-topped bushes that can be readily reached for plucking. After 10 years the plants are often cut back to the ground and suckers allowed to replace the old bush. Economic life of a tea plant is seldom more than 40 years. Plucking starts when a tea bush is 4 or 5 years old. For best-quality tea this must be a hand operation to select just the right leaves, although mechanical mass-shearing is sometimes practiced during rush periods. Laborers, mostly women and children, move agilely among the bushes, a basket suspended from the waist or back, leaving the hands free, and break the twig tips off with skilled fingers. Only the terminal two leaves (or occasionally in poorer-quality teas the terminal three or four leaves) are plucked. Plucked plants are stimulated to new growth, permitting another plucking in about 2 weeks. A worker can pluck 40 to 80 pounds of leaves per day, an amount producing 20 pounds of dried tea.

Harvested tea leaves may be treated in either of two ways—to produce black or green tea. The former, a fermented type, is the preferred tea in the United States. In preparing black tea, fresh leaves, after stalk and debris removal, are withered for about 24 hours on well-ventilated racks, usually indoors. "Tannins" (polyphenols, caffeine, and certain essences) are more readily extracted after this process. Withered leaves

are routed to roller machines, which, in several rolling and breaking operations, rupture the cells. Some rolling is by hand—in primitive areas even between the palms. More recently chopping machines such as are used for processing tobacco have increasingly replaced rolling, and withering may be dispensed with entirely. Such processing frees juices and enzymes, permitting oxidation of the polyphenols. Processed leaves are spread in cool, moist fermentation rooms for several hours; here the brown color and pleasing aroma develop. Finally the fermented leaves are "fired"—i.e., conveyed through hot-air chambers, a process which reduces moisture content to about 3 per cent in about

Tea plantation in South America. (Courtesy USDA.)

20 minutes, and turns the product black. The tea is then sifted into grades such as Orange Pekoe, and so on, and packed for marketing.

Green-tea production is less elaborate: the leaves are dried immediately, rolled, then further dried for marketing as Gunpowder, Imperials, and Hyson teas. A tea of Taiwan known as Oolong is a semifermented type produced by preliminary sun-drying with accompanying rolling. A similar semifermented type is jasmine tea, prepared by spreading jasmine flowers over the dried leaf, which absorbs their fragrance. Brick tea is a cheap, twiggy, coarse, fermented type of central Asia that is pressed into briquets. Tablet tea is made from the fine residual dust left from the screening of better grades. In addition to designations of grade, tea labels usually indicate the locale of production as well.

The quality of the tea leaf is dependent not only upon care taken in cultivation and processing, but also upon the elevation at which the tea is grown and the rainfall. The flavor of tea is due to essential oils of the leaf, but its stimulative properties stem from the alkaloid theine (identical with caffeine, $C_8H_{10}N_4O_2$) which occurs in the leaf in a concentration of 2 to 5 per cent. Commercial caffeine is, in fact, obtained from damaged tea and tea wastes. High tannin content, such as occurs in Brazilian teas (15 per cent or more) is considered desirable, to give "body" (color, pungency) to the beverage. Tea ready for market contains 38 to 45 per cent soluble matter, including dextrins, pectins, and like substances furnishing some consistency to the liquor. As in the coffee industry, a small group of professional tasters determine quality and uniformity in tea. These men are said to develop an amazing ability to recognize from taste alone not only country of origin but even district, and picking season as well.

China leads the world in tea acreage, most tea there being grown on small home plots

and consumed domestically. *Sinensis* cultivars are planted almost exclusively. Tea growing in India, Ceylon, and the East Indies is typically on large estates, utilizing *Assamica* cultivars. India is the greatest exporter of tea, followed closely by Ceylon, where the interesting switch from coffee to tea mentioned in the discussion of coffee was made only in 1867. Mixed Assam and Chinese cultivars are grown in Brazil, eastern Africa, and elsewhere. World production of tea amounts to over 1 million tons annually, more than half of the commercial crop coming from India and Ceylon. Interest in growing tea is increasing in both highland Africa and Brazil. Yields of 1 ton or more of tea per acre are not uncommon.

Cacao, *Theobroma cacao,* Sterculiaceae. Cacao (cocoa is the name for its products) is a small tropical tree, the source of chocolate. The name "cacao" is derived from the

Bearing cacao tree in Costa Rica. (Courtesy USDA.)

Mayan Nahuatl dialect, where the plant was known as "cacahuatl" or "cacahoatl," and the chocolate drink prepared from the fruit as "xocoatl" or "chocoatl." *Theobroma*, so named by Linnaeus, translates as "food of the Gods," in reference to the Indian belief of divine origin of cacao. The tree had apparently been cultivated in Central America from southern Mexico into northern South America for centuries before the Spanish conquest. The domesticate likely originated in the upper Amazon basin, with a secondary center of speciation on the Pacific slopes of the Andes. Escaped cacao is frequent throughout tropical America, even in remote locations.

Cacao is a small tree of the understory in tropical forests, often found along river banks where its roots are periodically under water. Terminal growth separates into several meristems giving a low fan of spreading branches, on plantations the tree usually restricted to a few such whorls by pruning. Leaves are simple, alternate, seasonally deciduous, the younger ones usually hanging vertically. Flowers are cauliforus, inconspicuous, apparently mostly pollinated by species of midges. Cross-compatible, cross-incompatible, self-compatible, and self-incompatible clones are known. *T. cacao* has a diploid chromosome count of 20. Seldom more than 8 meters tall, cacao naturally grows in the shade of taller trees. Thus plantings are usually protected with a scattering of taller shade trees planted along with the cacao, or left standing as forest is cleared. Some experiments in Africa, however, have shown yields to be materially increased by elimination of the shading overstory. Other tests indicate that cacao flourishes best in 50 per cent shade, but associated factors are involved. For one thing cacao has many feeder roots near the soil surface; under normal forest conditions these draw nutrients from the decomposing leaves that fall to the forest floor. Shade trees supply something of this natural condition, and many lateritic soils lose their organic content, baking hard like bricks, when exposed to full sun. Research also shows that certain overstory trees inhibit the growth of cacao, while others encourage it.

Cacao is adapted to regions of heavy rainfall and equable temperature. For successful growing rainfall should be 80 inches

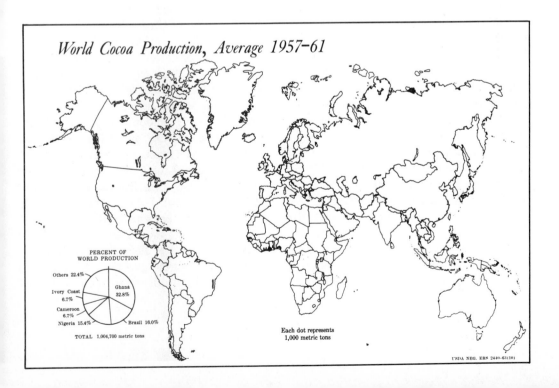

World Cocoa Production, Average 1957–61

PERCENT OF
WORLD PRODUCTION

Others 22.4%
Ivory Coast 6.7%
Cameroon 6.7%
Nigeria 15.4%
Ghana 32.8%
Brazil 16.0%

TOTAL 1,004,700 metric tons

Each dot represents
1,000 metric tons

USDA NEG. ERS 2440-63(10)

annually or more, and temperature variations no more than about 65–95°F both daily and seasonally. In other words cacao grows best in an environment much like that of a warm and humid greenhouse. Propagation is usually by seed, three seeds to a hole directly in the field in West Africa (later thinned to one seedling), or more often by transplanting from a nursery. As select cultivars are developed, rooted cuttings are coming more to be used even though this form of propagation is far more costly. Several diseases attack cacao, mostly combatted by cultural practices and selection of resistant cultivars, although on plantations some fungicidal spraying is undertaken.

The cultivated races of cacao are not well understood, but generally two subspecies are recognized: the *cacao* group, including most of the original high-quality "criollo" varieties; and the *sphaerocarpum* group, containing the more vigorous but often less distinctively flavored ("forasteiro") types that have gradually dominated cacao plantings because of their greater vigor and higher yield. "Criollo" signified cacao originally grown in western Venezuela, and when later introductions were made from Trinidad (from uncertain sources), these were called "forasteiro" (foreign). These designations persist and others have been added; chiefly the forasteiro types have been spread throughout the world and today constitute the main source of commercial chocolate. A forasteiro known as Amelonado predominates. Trinitario is a name given a heterogeneous assemblage of hybrids generally centering in Trinidad and Venezuela.

Cacao is important for its seeds, thirty to fifty of which occur in football-shaped pods, weighing about a quarter-pound, that grow from leafless "cushions" on the lower bare branches and trunks. These are produced continuously through most of the year. The seeds are surrounded by mucilaginous pulp, and in turn by a fairly hard outer shell something like that of a melon. Technically the pod is a berry. Vigorous trees yield many dozens of pods annually. Acre yields may run as much as 2 tons of dry cocoa under good conditions.

The seeds or "beans" are a rich source of nutriment and flavor. The kernels are especially rich in oil (cacao butter), some 50 per cent of the bean. Starch and protein content are each about 15 per cent. There is up to 3 per cent of the alkaloid theobromine, small quantities of caffeine, and traces of various aromatic oils that contribute to the flavor. Theobromine, a stimulant of commercial value extracted from the bean residues, is rather easily transformed to caffeine (much used in "colas").

Spanish explorers of the early 1500's found cacao highly esteemed by the indigenous Central American cultures. Cortez, when he invaded the Aztec Empire, was honored by Montezuma through the serving of a chocolate drink. Tribute to the rulers was often in cacao beans rather than gold. The Spanish, and later the Portuguese, soon spread cacao widely throughout their colonies. Cacao was introduced to Trinidad and other Caribbean islands very early, and not much later to islands in the Gulf of Guinea off the west coast of Africa. Eventually it was taken to mainland Africa and the Far East.

The pulp of cacao is often sucked by the native peoples as a sweet. But cacao is mainly important for a tremendous world commerce that brings temperate peoples this tropical delicacy. Over 1 million tons of cacao beans are exported annually, about 80 per cent from Africa (mainly Ghana and Nigeria), most of the remainder from Latin America (mainly Brazil, but also from Bolivia northward to Mexico and Cuba). They are imported chiefly into western Europe and North America. At one time cacao was mostly produced on plantations. The trend more recently has been toward small-farm growing, especially in Africa, which affords some greater flexibility in adjusting to the market (although relaxing control over quality).

The familiar chocolate taste and aroma is not found in the fresh bean; it must be developed by a processing that involves fermentation. Ripening pods can be discerned by their color, in harvestable condition for about 2 weeks. They are cut by hand from the trunks and branches of the cacao tree, piled in convenient locations for subsequent removal of the seeds. The pods are cut open with a bush knife (which may damage a few of the seeds), or cracked with a mallet. Women typically scoop the pulpy seeds from the husks. Under primitive circumstances the fresh seeds are piled in small mounds atop banana leaves, covered with more leaves, and left to ferment. About 1 week is needed to develop the flavor and aroma, during which time the kernel changes from purple to brown. Larger farms and plantations have a series of fermentation boxes, insulated to retain the heat. The seeds are mixed each 48 hours, for about three turnings. Yeasts and bacteria attack the sweet pulp surrounding the seed, heating the slimy mass and causing unpleasant odors. Fermentation, of course, kills the seed.

After fermentation the beans must be dried, sorted, and cleaned. Usually drying is by exposure to the sun over a period of several weeks. Drying beds may be equipped with a portable roof in case of rain. In some cases the beans are artificially heated. Further processing takes place at commercial centers: the shell of the bean is cracked and removed, the cacao butter expressed from the kernels, and the remaining "cake" treated to retrieve the theobromine. Cacao butter is an especially valuable fat, long used in the finest chocolate confections. It may be incompletely extracted, and the kernels ground and compounded with milk, sugar, and other ingredients to yield milk chocolate, cocoa and other familiar items of the grocery shelf.

Minor nonalcoholic beverages. CASSINE, DAHOON, YAUPON, OR CAROLINA TEA, *Ilex vomitoria*, Aquifoliaceae. This relative of maté, growing wild in the southeastern United States, has been used as a source of cassina tea, drunk locally, and at times used for soft drink and ice cream flavorings.

COLA, *Cola nitida*, Sterculiaceae. The cola tree, relative of cacao, is native to tropical western Africa but has been introduced into various parts of the world. There are about sixty species in the genus, several of them chewed as narcotic stimulants (e.g., *C. acuminata*, *C. anomala*, *C. verticillata*). Seeds of *C. nitida* especially have long been used as a masticatory and as a source of flavor

Fruiting branch of the cola tree, *Cola acuminata*. (Courtesy Paul H. Allen.)

and caffeine in "cola" drinks. The species, diploid chromosome count of 40, is a moderate forest tree with simple, glossy leaves and small axillary cymes of flowers. Cultural requirements are much as for cacao. The star-shaped fruits, 5 to 10 centimeters long, contain eight reddish or white seeds (color depending upon the cultivar). In the African hinterland these are sometimes chewed fresh as a stimulant, the 2 per cent of caffeine plus certain essential oils and alkaloids serving to inhibit fatigue and hunger. Pulverized and boiled in water, the seeds make a beverage. Propagation is from seed, the trees bearing when 5 to 6 years old and continuing to bear profitably for about 50 years. Fruiting usually takes place twice a year. Orchards are generally intercropped with food plants in the early years, and provided with an overstory of shade trees later. Fruits are hand gathered, with knives on long poles as the trees grow taller. The seeds are removed by hand, freed from seed coats, "sweated" for a few days, and sun-dried for shipment. Production is chiefly African, from the tropical belt extending from Sierra Leone to Ghana, running about 150,000 tons annually, shipped mainly to Europe.

GUARANÁ, *Paullinia cupana* var. *sorbilis,* Sapindaceae. Guaraná, the "cola" of Brazil, contains more caffeine (5 per cent) than do coffee and tea, to say nothing of a good proportion of tannin (5 per cent). Reputedly a single cup is sufficiently stimulating to counteract feelings of extreme fatigue. *Paullinia cupana* is a trailing shrub or vine found cultivated and wild in the northern part of South America. It has pinnate leaves, inconspicuous flowers, and hard, small, triangular fruits the black seeds from which when pulverized (often mixed with cassava flour) are the source of guaraná paste. Plantings are made from seed and are desultorily tended. The fruits are picked by hand from October to December and are pulverized to form a paste that is usually shaped into bars or cylinders before drying. These are later used in solutions to provide the beverage. Production is chiefly from the Brazilian state of Amazonas.

GUAYUSA, *Ilex guayusa*, Aquifoliaceae. Guayusa is another relative of maté, native to the eastern Andean foothills. A letter to the viceroy of Peru in 1683, from the explorer Lucero, noted the use of guayusa by the savage Jivaro Indians, especially as an antisomniferant when remaining alert for battle. Later the Jesuits sold guayusa as a general remedy in Quito, a practice abandoned when the Jesuits were expelled about 1766. Guayusa is still drunk in parts of eastern Ecuador and Peru, the foliage harvested from wild plants and those planted around the missions. Like maté an infusion from the dried leaves is

Branch of the guaraná plant, *Paullinia cupana*. (Courtesy Chicago Natural History Museum.)

the customary way of taking the beverage.

KAVAKAVA, *Piper methysticum*, Piperaceae. A beverage containing sleep-producing alkaloids is prepared from the masticated roots of this species, widely grown in the Pacific islands and an integral part of the religious and social life there.

KHAT, *Catha edulis*, Celastraceae. Leaves of this species of northeastern Africa have long served in Arabia to make a tea and masticatory. They contain a stimulating alkaloid similar to caffeine.

MATÉ, *Ilex paraguariensis*, Aquifoliaceae. Maté, also known as Paraguay tea, Jesuit tea, or matté, is a moderate-sized tree of southern Brazil, Paraguay, and northern Argentina, and is found both wild and extensively cultivated. The species was in use for beverages among the Guaraní Indians long before the Conquest, and is today one of the chief beverages of southern South America. Modest quantities are exported to Europe and North America, but cannot compete there with tea and coffee.

Maté is made by trimming most of the terminal branches from either wild or plantation trees. These are commonly "toasted" over open fires to reduce moisture content and to coagulate or vaporize resins or other substances. The smaller twigs and leaves are next removed, dried by any of several methods (commonly by a stream of hot air or on platforms over open fires), and the leaves threshed free of twigs and debris (in machines or by beating over canvas). A final aging that helps develop flavor and aroma precedes blending, packaging, and possibly redrying.

The beverage is prepared much as is tea or, in the hinterland, by pouring cold water over coarse leaf. Every gaucho or traveler carries a gourd or hollow horn filled with maté leaf to provide drink wherever water is encountered. A "bombilla," a strawlike tube with a perforated bowl, is introduced into the newly made infusion, and the gourd and bombilla passed from hand to hand "peace-pipe fashion" among members of a party. The gourd is filled several times without renewing the leaf.

Production of maté amounts to a few hundreds of thousands of tons annually, of which almost half is produced (largely on plantations) and two-thirds consumed in Argentina. Brazil is the largest exporter. United States importers have occasionally used maté as a base for soft drinks, and for extraction of caffeine (over 1 per cent of the leaf), vitamins, and chlorophyll.

NEW JERSEY TEA, *Ceanothus americanus*, Rhamnaceae. Leaves of this shrub of the eastern United States were used as a tea substitute in the early years of the Republic.

ROOIBOS TEA, *Aspalathus contaminatus*, Leguminosae. This species, along with related *Cyclopedia*, is a beverage plant of minor importance in southern Africa. The plant is a semiprostrate shrub with fragile branches and terete, needlelike leaves. The plant is typically propagated by seed sowed in shaded nurseries, and transplanted to the fields. It grows best on well-drained soils at fairly high elevations (about 1,500 feet). Plantings are sheared to stimulate branching, and yields are not profitable until about the third year, continuing until about the tenth year, when the planting plays out. Harvest in South Africa is generally with scythes, and the cut foliage immediately chopped with a tobacco cutting machine. Cut foliage is further bruised with wooden hammers or by rolling, and as with tea fermented for up to 24 hours to develop the characteristic flavor. It is then dried, marketed, and utilized much as is tea. Production may run 100,000 tons annually, for export as well as being "South Africa's national drink."

SARSAPARILLA, *Smilax spp.*, Liliaceae. The roots of these tropical American vines serve as medicinals and beverage flavor. Production is principally from wild or cultivated vines of Mexico, Honduras, Costa Rica, Ecuador, Peru, and Jamaica.

SASSAFRAS, *Sassafras albidum* (*S. variifolium*), Lauraceae. Aromatic bark of the sassafras was an important export from colonial America to Europe for use as a medicinal tea. In rural areas of the eastern United States it is still so used.

MISCELLANEOUS PLANTS. Many plants in addition to those mentioned have at one time or another yielded nonalcoholic beverages consumed locally. To list only a few: *Athrisca phylicoides* leaves and *Treculia africana* seeds in Africa; *Anthemis nobilis, Sideritis spp.,* and *Tilia spp.* leaves and *Cyperus esculentus* tubers in Europe; soybean milk in Manchuria; *Colubrina reclinata* bark in Haiti; *Lippia citriodora* leaves in Chile; *Dictamnus albus, Fragaria spp., Gaultheria procumbens, Lindera benzoin, Mentha piperata, Micromeria douglasii, Monarda didyma, Rubus occidentalis, Solidago odora,* and *Thymus serpyllum* leaves in the United States; and so on. Parched flour of cereals, nuts, legume seeds, and the like serves to make many beverages of local importance. Locally, by-products of other crops serve as beverages. For example, juice expressed from sugar cane is often drunk, as are various nonalcoholic derivatives from chicha making.

The place of natural fruit juices has been usurped to a large extent by synthetic soft drinks. Especially in the United States have carbonated beverages, consisting of water charged with carbon dioxide, sweetening, and flavor, overrun the country. "Ginger ale" is flavored with ginger–capsicum extract; "root beer" with caramel, sarsaparilla, wintergreen, and other aromatics; "colas" with secret formulas utilizing caffeine or similar alkaloids and sometimes extracted coca leaf; and so on. Most cherry, strawberry, grape, cream, or other types of soda are synthetically flavored and artificially colored.

Alcoholic Beverages

Few plant products are more involved in the customs and traditions of mankind than

Wine making in the Tyrol. (Courtesy Chicago Natural History Museum.)

alcoholic beverages. They have been regarded as everything from blessing to scourge. In modern America, a country in which alcohol as a beverage is perhaps used with dubious discretion, there is something mildly wicked yet fascinating about it. In the older cultures of Europe alcoholic beverages are a usual and accepted drink, particularly with meals; who can imagine the Frenchman or Italian without his table wine or the northern European without beer? Per capita consumption of beer in parts of northern Europe averages more than a pint per day, and in Elizabethan England a population of 5 million is said to have consumed 13 million barrels of beer annually.

In other parts of the world, too, fermented beverages are basic in the life and customs of the people. Amazonian Indians brew ceremonial beverages from semipoisonous manioc juice; Andean workers seek out the ubiquitous chicha shops; Pacific peoples partake of kavakava in rituals fundamental to the race; Orientals prepare a toddy from the palm tree, as well as sake from rice. Among early peoples, before the advent of selected seed plants, fermentation may even have been essential for "predigestion" of the hard, low-quality food seed collected in the wild. Yeast, man's favorite agent of fermentation, may be the first domesticated plant! It was carried by the Polynesians, for example, from island to island for use in making pulverized seeds more readily digestible.

It was not until the nineteenth century, however, that the art of fermentation became a science. Pasteur's fundamental research established the importance of microorganisms in transformation of organic matter, and since his epochal work industrial fermentations have become better understood and of increasing importance in the economy. Production of beers, wines, and spirits has become a highly regulated procedure, capable of yielding constant-quality products. Their flavor usually stems from minor quanti-

ties of aromatics and other rather complex organic substances. Their stimulative properties, of course, come from ethyl alcohol, C_2H_5OH. Alcohol in excess first causes cerebral excitation, followed by depression and dulling of the senses. It is thus not a "stimulant" in the same sense as caffeine. Ethyl alcohol can be readily derived from carbohydrates, by anaerobic respiration of the yeast plant. It is of great importance not only as a constituent of alcoholic beverages but also as an industrial raw material.

Beers and beerlike beverages. Undistilled alcoholic beverages from grains or other nongrape sources have been brewed for thousands of years. Clay tablets from Sumerian and Assyrian times more than 6,000 years old depict beer brewing, and Egyptian documents from 2600 B.C. describe barley malting and beer fermentation. When Columbus reached the New World he found the Indians drinking a beer made from maize (chicha). In the Middle Ages beer brewing was a household task the same as baking bread. Hops seem to have been somewhat used in the beers of the ancient Finns, but not until the ninth century were they utilized as a regular constituent, a practice begun in Germany. Without refrigeration, beer was often stored in caves for lagering. The first commercial brewery in North America was constructed in what is now New York, in 1623. Refrigeration (allowing for year-round lagering) was introduced about the time of the Civil War, and pasteurization in 1873.

BEER. Beer as commonly produced is a malt beverage—i.e., brewed from malt, the germinated seeds especially of barley. To the malt are generally added adjuncts—usually rice, maize, other of the cereal grains, potatoes, or manioc—and hops for flavoring. The cereals have been discussed at length in another chapter and hops reviewed in Chapter 11. Thus most plant products entering directly

Steaming rice for fermentation in a small brewery of Thailand. (Courtesy USDA.)

into the making of beer have already received attention, with the exception of the essential ferment, yeast. A flow sheet for the brewing process is given as Fig. 20-1.

Successful beer making involves considerable experience and skill, especially in the malting and fermenting. In brief the procedure is as follows. Quality barley in the United States, most of it specially grown in the Midwest from two-row cultivars rich in carbohydrate, is thoroughly cleaned in blowers and sieves. It is then washed and steeped in huge vats of water for about 2 days. Germination of the barley grain starts and proceeds for 4 to 6 days thereafter in special germinating drums where humidity and temperature can be controlled. After the primary root has developed but before the coleoptile has been ruptured, germination is cut short by a carefully regulated drying sequence in special kilns, at final temperatures of about 180°F. During germination the barley sprouts develop enzymes, especially diastase, capable of hydrolizing starch or other complex foods to simpler products such as sugar. Most of the starch of the original barley has been changed into sugar during malting, and the hydrolizing enzymes are retained in the quickly dried malt. Higher temperatures during the kiln-drying pro-

duce a darker, caramelized malt used for bock-type beers.

Malt is next sent to the brew house, where up to 35 per cent of adjuncts (rice, corn, or other carbohydrate source) may be added in making mash. The measured quantity of pure water in which the pulverized mash is soaked at moderately high temperatures (about 170°F) becomes wort, containing many soluble substances from malt and adjunct that will later afford food for yeast. The malt enzymes are, of course, instrumental in making insoluble materials in the adjunct soluble. Spent grains left after the wort is drawn off (in "lauter tubs") can be used in stock feeds. The wort is boiled with hops for 2 to 3 hours, to give the bitter flavor characteristic of beer, to clarify the liquor (coagulate high-molecular nitrogenous substances), and to increase its keeping qualities. Where liquid-hop concentrates are used, this step is eliminated and the hop flavor generally added after fermentation.

Cooled wort is conducted under sterile conditions to starting cellars, where measured quantities of yeast, *Saccharomyces cerevisiae*, Ascomycetes, from a previous brew are introduced in large tanks at cool temperatures. There fermentation begins, the yeasts acting upon sugars of the wort to

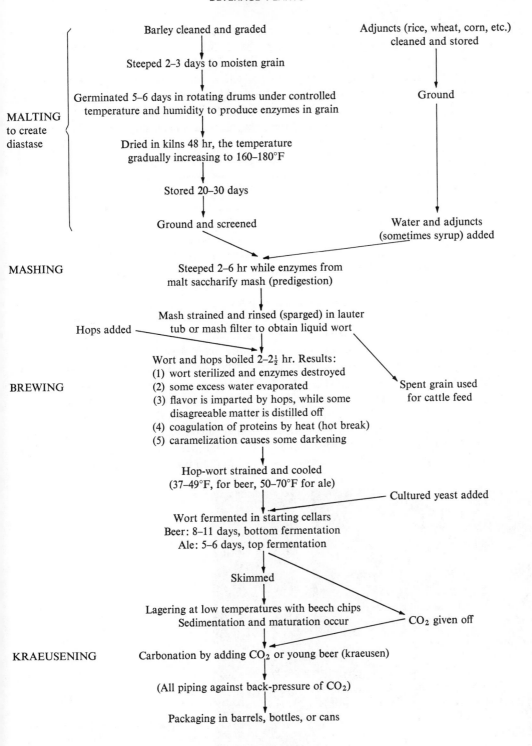

MALTING
to create
diastase

Barley cleaned and graded

Steeped 2–3 days to moisten grain

Germinated 5–6 days in rotating drums under controlled
temperature and humidity to produce enzymes in grain

Dried in kilns 48 hr, the temperature
gradually increasing to 160–180°F

Stored 20–30 days

Ground and screened

Adjuncts (rice, wheat, corn, etc.)
cleaned and stored

Ground

Water and adjuncts
(sometimes syrup) added

MASHING

Steeped 2–6 hr while enzymes from
malt saccharify mash (predigestion)

Hops added —

Mash strained and rinsed (sparged) in lauter
tub or mash filter to obtain liquid wort

BREWING

Wort and hops boiled 2–2½ hr. Results:
(1) wort sterilized and enzymes destroyed
(2) some excess water evaporated
(3) flavor is imparted by hops, while some
    disagreeable matter is distilled off
(4) coagulation of proteins by heat (hot break)
(5) caramelization causes some darkening

Spent grain used
for cattle feed

Hop-wort strained and cooled
(37–49°F, for beer, 50–70°F for ale)

Cultured yeast added

Wort fermented in starting cellars
Beer: 8–11 days, bottom fermentation
Ale: 5–6 days, top fermentation

Skimmed

Lagering at low temperatures with beech chips
Sedimentation and maturation occur

CO_2 given off

KRAEUSENING

Carbonation by adding CO_2 or young beer (kraeusen)

(All piping against back-pressure of CO_2)

Packaging in barrels, bottles, or cans

Figure 20-1. Brewing flow sheet.

produce carbon dioxide, alcohol, and other organic compounds having a characteristic flavor. Differing strains of yeast vary in the flavors produced, and precisely on this account selected strains are highly prized. The carbon dioxide may be saved for later carbonation or for other purposes. After about a day in the starting celler the green beer is transferred to lagering cellars, where secondary fermentation may proceed for several days. Various suspensions coagulate and float off as a scum, along with spent yeast cells. The young beer is then usually aged for a matter of weeks or months. Generally, just before bottling or kegging, a small quantity of green beer is introduced along with beechwood chips (nucleus for yeast growth) to effect a final rapid fermentation and resultant carbonation, or carbon dioxide may be introduced under pressure to provide the same effect. All filtering and piping proceeds against a back-pressure of carbon dioxide, so that the beer cannot undergo oxidation and always remains sufficiently carbonated to give a good head of foam. Bottled beers are generally pasteurized for better keeping, but draft beers traditionally were not and had to remain re-frigerated. Recently unpasteurized (draft) beers have been marketed after especially thorough filtration and treatment with a calcium diethylate sterilant (which shortly breaks down into alcohol and a calcium precipitate); such beer can be held without re-frigeration. Chemically beer is largely water (about 90 per cent), plus about 5 per cent ethyl alcohol and smaller quantities of malt-ose, nitrogenous substances, gums, dextrins, and so on.

Over 100 million barrels (31 gallons to a barrel) of beer are produced in the United States annually and world production is several times this amount. American beers are usually clearer and lighter than European beers, with less hopping (often less than 1 pound of hops per barrel in the United States beers compared to as much as 4 pounds in Europe).

Other malt beers

ALE. Ale is a beer produced by "top fermentation"—i.e., by yeasts floating on the surface of the wort rather than working at the bottom of the tank. It is usually of higher alcoholic content, paler and more tart, and possessing a higher hop concentration than beer.

Brewery in Thailand. Yeast is mixed with steamed rice for fermentation. (Courtesy USDA.)

KVASS OR QUASS. Kvass is a Russian beer made of barley and rye, with peppermint flavoring.

POMBE OR BOUSA. Pombe results from spontaneous fermentation of sprouted millet grain, especially in Africa.

PORTER. Porter is a dark, sweetish ale produced, as is bock beer, from caramelized malt.

STOUT. Stout is a porter of somewhat higher alcoholic content and stronger hop flavor.

WEISS. Weiss is a light, malty ale made mostly from wheat.

Usually unmalted beers

CHANG. Chang is a beverage of Tibet prepared from unmalted barley. The grain is boiled, then dried, and yeast added. Added to water in sealed pots it yields a yellow, oily brew within 10 days. The beer may be decanted, and more water added for second and third (progressively weaker) brewings. Thereafter the spent grain is given to livestock. Chang can be distilled to make the potent arak. About two-fifths of the Tibetan barley crop is said to be used for chang.

CHICHA. A number of South American beverages, a few of them nonalcoholic (from toasted corn, peanuts, quinoa, and so on), are termed *chicha*. Alcoholic chichas may be fermented from maize, manioc, plantain, potatoes, palm fruits, and so on. Malt enzymes are not generally available, and saliva instead has traditionally supplied the necessary diastase for conversion of starches to sugar. The former custom, still much used in Bolivia, is to masticate materials intended to become chicha and permit semispontaneous fermentation to follow. This method has been termed the "chew-and-spit" process of beer production. In Peru it is rapidly becoming supplanted by manufacture of chicha from germinated maize. Cultivars of maize especially suited to chicha making have long been selected in the Andean countries.

In traditional chicha production maize is pulverized to flour and then worked in the mouth with little chewing. The salivated flour, usually with some "adjunct," is emptied into earthen pots. Water is added and the mixture warmed. After a day the liquid above the denser layers of sediment is removed, and on the third day this liquid is boiled. The jellified portion of sediment is caramelized and added to the liquor, initiating rather violent fermentation. After 6 more days of diminishing fermentation the liquor is considered ready for consumption, the superficial froth (of yeast, oils, and other substances) being pushed aside before drinking. This froth may serve as a starter for fermentation of green chicha, but ordinarily the earthen pots, seldom cleaned, contain sufficient yeast to make this unnecessary. More modern methods of chicha production vary from region to region, but essentially involve the soaking of a specially selected maize in pots full of water for 12 to 18 hours, followed by spreading the soaked grain in layers in the dark for about 3 days to effect germination. Once sprouted, the maize is gathered into a heap and covered for a fermentation heating that leaves it whitish as though thoroughly parched. This is then dried in the sun for a few days, and ground to become "pachucho" from which chicha is fermented. The pachucho is boiled in water, cooled, the solids grated, and the material boiled a second time before straining. The strained liquid is allowed to ferment in pots already inoculated with yeast from past fermentations until judged ready, then decanted into an upper, more potent layer, and a sediment-containing bottom layer. Sometimes sugar is added at the second boiling to increase fermentable carbohydrates and potency of the eventual chicha. Generally 100 pounds of maize yields approximately 15 gallons of chicha.

Alcoholic content of chicha may run as high as 12 per cent, although it usually averages about 5 per cent, the same as for beer. In many ways production of this important South American drink, the annual consumption of which must be reckoned as many millions of gallons, strikingly parallels the basic but refined techniques of beer making previously discussed.

HARD CIDER. The juice of apples allowed to ferment spontaneously develops a modest alcohol content, to become hard cider. Further controlled bacterial fermentation produces vinegar. Many fruits the world over are utilized to make hard ciders of one kind or another.

GINGER BEER. The acid ginger beer results from a bacterial and yeast fermentation of a sugar solution containing pieces of ginger rhizome.

MARWA. Marwa is a beer made from millet in the Himalayan kingdoms of Sikkim and Bhutan. When millet is unobtainable almost any other grain will serve, and even the pith of a tree fern may be used. Production is essentially the same as was described for chang.

MEAD. Mead is only indirectly a plant product, being fermented from a honey solution. It is an ancient beverage used especially in Africa and Scandinavia.

PALM TODDY. Sugary sap from the cut inflorescences of a number of palms is gathered, especially in the Orient and Africa, to be spontaneously fermented to toddy.

PEPPER-TREE. A chichalike beverage is made from fruit of the pepper tree, *Schinus molle*, Anacardiaceae, ubiquitous in the Peruvian highlands.

PULQUE. Pulque, common in Mexico, is made by spontaneous fermentation of the sweetish juice from crushed *Agave* leaves.

SAKE. Sake is a Japanese beverage prepared by fermentation of steamed rice. The starch of rice is usually hydrolized to sugar for subsequent fungus fermentation by inoculation with *Aspergillus oryzae*. A similar beverage is made from rice in China.

SORGHO. Sorgho results from alcoholic fermentation of sorghum grain.

Wines. These are alcoholic beverages obtained from fermentation of grapes, which were discussed in the preceding chapter. As mentioned there, grapes are of greater economic importance fermented to wine than as a table fruit, and have probably been so since ancient times. Evidence of wine making is revealed by artifacts of prehistoric man. Certainly wines were well known to the ancients; and wine is cherished in the ritual of the Christian Church. A tremendous industry has grown up about the science and art of wine making, attaining greatest magnitude in Europe but also important in North Africa, South Africa, Argentina, Australia, and in the United States especially in California and New York. Europe produces over 3 billion gallons of wine yearly, mostly from France and Italy (almost 1 billion gallons each). Spain, Portugal, and Algeria produce about a half-billion gallons each. United States production, over 80 per cent from California, amounts to nearly 200 million gallons annually. World production exceeds 7 billion gallons annually. In France per capita consumption of wine is 33 gallons annually, but in the United States it is less than 1 gallon.

The world's greatest wines are made from *Vitis vinifera* grapes, of which new, hopefully better wine cultivars are continuously being bred. There has been some crossing with *V. labrusca* and *V. riparia*, and parentage from at least six species is said to enter the labrusca hybrids grown in New York where the climate is unsuited to *V. vinifera*. Labrusca grapes contain methyl anthranilate, which gives a "foxy" taste to wine that is objectionable especially to Europeans accustomed to vinifera wines. Yet Concord, Catawba, Delaware, Niagara, and other cultivars of the northeastern United States are being specially processed to yield not only traditional

A heavily bearing grape vine. Major use of grapes is for wine. (Courtesy USDA.)

soil and growing conditions. Such factors enter into the final "orchestration" of a good wine, composed as it is of ethyl alcohol, other alcohols, sugars and other carbohydrates, polyphenols, aldehydes, ketones, enzymes, pigments, a half-dozen or so vitamins, a couple of dozen minerals and organic acids, and traces of other ingredients. The possible variation in all these is limitless, and it is no wonder that wine making remains as much an art as a science. Certainly human judgment enters strongly when deciding that the grape is just properly ripe for harvest, and how long the wine should be racked and aged.

Wines may be designated as either fortified or unfortified. Fortified wines are "cocktail" or "dessert" wines such as port, sherry, muscatel, and vermouth, in which the alcoholic content is supplemented by the addition of brandy (distilled alcohol from fruits) to a strength unobtainable by simple fermentation. At about 14 per cent alcohol content the yeast responsible for fermentation is destroyed by its own by-product. Whether fortified or natural, wines may be further classified as dry or sweet (on the basis of acid–sugar percentages—largely a reflection of grape variety and prolongation of fermentation), white or red (depending upon pigmentation and anthocyanin extraction from the grape skins), still or sparkling (referring to carbonation of the beverage—fortified wines are always still or uncarbonated). Further classification of wines may be made on the basis of country of origin, grape variety, vintage, and vinification practices. Well-known white wines, unfortified, include Riesling (Germany); Tokay (Hungary); Chablis, Grenache, Champagne, Sauterne, and white Burgundy (France); Chardonnay, Sauvignon Blanc, and so on (California). Widely known red table wines are Burgundy and Claret (France); Chianti (Italy); Cabernet and Zinfandel (California).

wines of the area, but acceptable baked sherry (even Flor Sherry, from trickling wine over a substrate of *Saccharomyces oviformis* obtained from the Arbois region of France). The long history of wine production in Europe has served empirically to prove out certain cultivars, even yeasts, suited to a particular soil and climate; regions there tend to specialize in certain types of wine. For economic reasons California vintners produce a broad line of wines, and have less opportunity to specialize in a single type. Other than this European wines have no advantage over those from California. In general grapes and the wines they yield are more acid and less sugary when grown in more northerly areas (compared to those farther south). Two plants of given cultivar grown only a few miles apart may show differences due to

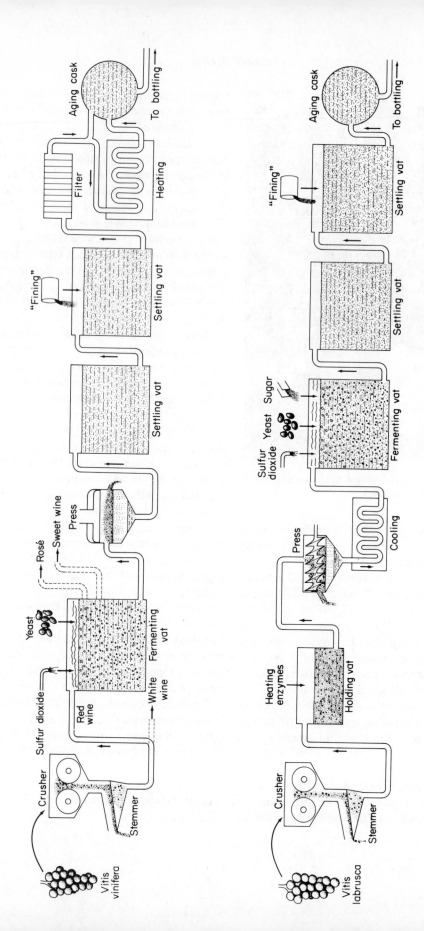

As in making beer, yeast is the plant responsible for fermentation in wine making. Sugars in the grape juice are changed to alcohol, and desirable flavor overtones are developed, through activity of strains of *Saccharomyces ellipsoideus*, the wine yeast. Also as with beer, undesirable strains of yeast, bacteria, and molds must be kept from the fermentation to insure uniformity and quality of the product. Mechanically expressed and settled grape juice (termed "must") is generally treated with small quantities of sulphur dioxide to check growth of undesirable organisms, then inoculated with pure-culture yeast for fermentation. Fermentation proceeds for several days in huge vats, during which time occasional agitation may be necessary to assure complete extraction of tannins and pigments from the pomace. Red wines are fermented with the skins included in the must (at least in the beginning) to give maximum color and aroma. White wines are fermented from free-run must only. After preliminary fermentation the must is transferred to closed vats where fermentation usually proceeds at rather low temperatures in the presence of carbon dioxide. Diagrams of the steps in wine making are given in Fig. 20-2.

Sparkling wines such as champagne are formed by addition of sugar to a suitable fermented and clarified table wine, initiating a new fermentation in a tightly closed bottle or container. Since the carbon dioxide cannot escape, the wine becomes carbonated —i.e., "sparkling." Sediment caused by this second fermentation is worked onto the cork by turning the bottles and is removed before marketing. Vermouth is a special wine that has been blended with aromatic and bitter flavors, such as coriander, bitter orange, wormwood, quinine, calamus, gentian, cloves, and the like.

Newly fermented wine, drawn off from the pomace or lees (racking), is usually aged in wooden casks for periods sometimes lasting many years, depending upon the type of wine. Some sedimentation occurs; in consequence at least one and sometimes more than one new racking is advisable. Complex chemical changes occurring during aging improve the bouquet of the wine. If

Figure 20-2 (on facing page) *Wine making.* The first step in wine making is to crush the grape in order to extract the juice. Once done by stomping the fruit with hobbled boots (so as to not crush the seed, giving bitter flavors), this is now accomplished mechanically between rollers. The sugar in grape juice is comprised of dextrose and levulose in almost equal parts (the latter is about twice as sweet as the former), and cultivars rich in levulose are prized for sweeter wines. Other prominent ingredients are malic and tartaric acids (responsible for the agreeable tartness of wine), and pigments and tannins occurring largely in the grape skins. The grape skin typically has a whitish bloom containing yeast capable of fermenting the juice, and many other microorganisms. But for controlled fermentation these are usually killed by sterilization, such as bubbling sulfur dioxide through the must. Then the must is inoculated with a starter yeast of known pedigree. After fermentation the incipient wine is decanted or finally expressed from the residues. Then begin thousands of subtle chemical reactions, especially bacterial change of malic to lactic acid which mellows the young wine. This secondary fermentation is best conducted in sealed containers away from oxygen, so that *Acetobacter* fermentation to vinegar does not proceed (vinegar is from "vin aigre" meaning sour wine). Cream of tartar and other sediments precipitate out during this stage. Finally the wine is aged in oak casks during which time there is some oxidation and esterification which refines the wine and develops its bouquet. Temperature and the length of aging are controlled according to the kind of wine. Contary to popular opinion, length of aging is not necessarily indicative of quality; some wines are best when only a year old, others not until they have aged 5 to 30 years. Diagrams after M. A. Amerine, *Scientific American* 211 : 48–49, 1964.

aging is insufficient to clarify the wine by causing complete deposition of suspensions, the wine is "fined" (treated with gelatin or a similar substance) and filtered. By-products of wine making include pomace (for fertilizers), and cream of tartar and alcohol (both obtained from the lees).

Chemically wines are largely water (70 per cent or more). Table wines contain about 14 per cent alcohol; dessert wines about 20 per cent. Dry wines contain less than 0.2 per cent sugar, barely perceptible to taste; sweet table wines may contain up to 6 per cent sugar, and sweet dessert wines as much as 18 per cent. Acidity of wines ranges from practically nothing to about 0.65 per cent. Various organic compounds responsible for bouquet are found as traces, as are minerals that may be of some value in promoting good health. Indeed, earliest use of wine was as a medicinal, and many modern physicians still recommend it.

Using the term broadly, fermented juices of fruits other than grapes are sometimes called wine: thus one hears of blackberry, elderberry, cherry, and other wines. Technically, however, these are considered hard ciders.

Distilled spirits. Distilled spirits are the many potable alcoholic beverages whose alcohol content is increased by distillation beyond the percentage obtainable through natural fermentation. Distillation involves boiling a fermented mash or "beer" at regulated temperatures, and condensing the fractionated alcohol. Commonly the beer is fed into the top of a columnar still, and flows across a series of perforated plates. Steam entering below boils and vaporizes alcoholic portions of the beer, which are condensed and usually redistilled to a higher proof. Below 160 proof the distillates may be designated bourbon, rye, etc.; between 160 and 190 proof it may be termed whisky only; and above 190 proof it is "grain neutral spirits." "Proof," incidentally, is double the alcohol percentage figure: thus 100 proof whisky contains 50 per cent alcohol.

In distillation alcohol is accompanied by various volatile substances, which have similar boiling points, and these impart the distinctive character to whiskies, rums, brandies, and similar spirits. The process sounds simple, as indeed it is if used simply to derive alcohol. Skill and experience are needed, however, to produce a standard, well-flavored beverage, in which minute traces of substances other than alcohol must be correctly blended and aged to give the desired flavor. Such skill is not a secret solely of the modern age; for centuries before Christ, in Egypt, China, Arabia, and elsewhere, beverage distillation was a prized art. By the tenth century A.D. efficient stills had been developed in Arabia, and in 1826 Robert Stein of Scotch whisky fame invented the patent still. Alcohol has often been condemned, but in itself it seems to be in no way harmful. In small quantities it is relaxing; in excess, however, it narcotizes parts of the brain and thus may lead to many social problems. More than 300 million gallons of distilled spirits are produced annually in the United States; almost half of this is in the form of whisky.

BRANDY. Brandies are obtained by distillation of wine or fermented mash of fruits, the distillate being suitably aged in wood. They usually contain 40 to 50 per cent ethyl alcohol. Legend has it that an enterprising Dutch shipmaster invented brandy in an attempt to make a wine concentrate for easier shipment to Holland. In Holland the product was reportedly so esteemed that it was decided not to dilute the distillate back to "wine." Cognac is perhaps the most famed brandy; it is made in the ancient city of Cognac on the Charente River, in France, from fine wines of the region. Other well-known brandies include armagnac; Spanish, Greek, and American brandies;

kirsch (cherry) brandy; calvados or apple-jack; and slivovitz (plum) brandy. Brandy is usually consumed as an after-dinner drink.

WHISKIES. Just as it was natural that brandies should be made in grape-growing areas, so whiskies are found in cereal-consuming regions, for whiskies are the distillates of grain fermentation suitably aged in (white oak) wood. Whiskies usually contain about 50 per cent alcohol. Scotch whisky is prepared largely from barley that has been malted, mashed, fermented (processes already described in principle for beer), distilled, matured, and blended. The malt is usually smoke-cured over open peat fires. Irish whisky is almost identical with Scotch, but the malt is kiln-dried.

American whisky was early made in the colonies, and a tax imposed upon it in the time of George Washington led to the "Whisky Rebellion." Whisky makers moved into Indian territory to escape taxation and in so doing came upon the finest whisky-making areas in the country, in western Pennsylvania, Indiana, and Bourbon county, Kentucky. These localities purportedly owe their pre-eminence largely to the quality of the water used, water that has percolated through limestone rock. Maize is the chief grain utilized, and only a small amount of malt is added to the pulverized corn, the mash being cooked during conversion of starches. Fermentation is by *Saccharomyces cerevisiae*, the beer yeast, although bacteria play a part in developing the flavor. This distillate is aged in charred white oak barrels to give color and better flavor to the product. Bourbon or corn whisky contains not less than 51 per cent corn grain in the mash, rye whisky not less than 51 per cent rye, wheat whisky not less than 51 per cent wheat, and so on. Any of these may be blended with others, or with "grain neutral spirits" (ethyl alcohol). "Bottled in bond" is no guarantee of quality,

but simply indicates that the whisky has been aged at least 4 years in a bonded warehouse and is 100 proof.

GINS. Gins differ from whiskies and brandies in that the flavor is not carried from the fermented mash but is added from aromatic sources, principally *Juniperus* "berries," before or after distillation of the alcoholic fraction. The mash is usually a maize–malt–rye mixture similar to that used for whisky, but may occasionally be molasses or another carbohydrate source. Nor is aging necessary to produce flavor in gin. Flavoring materials (or essential oils derived from them) often added, in addition to the juniper cones or "berries" gathered principally from *Juniperus communis*, include coriander, cardamon, angelica, anise, bitter almonds, caraway, calamus, cassia, fennel, orris, licorice, sweet and bitter orange peel, and so on. Gins contain not less than 40 per cent alcohol. Dry gins are unsweetened (London dry gin); Old Tom gins have syrup added. In Holland gins the aromatics are included with the mash; in United States gin the distilled alcoholic vapors pass through a "gin head" containing the flavoring agents. Sloe gin is secondarily flavored with blackthorn fruit (*Prunus spinosa*) instead of juniper, and is thus really a liqueur.

RUMS (RON, RHUM). Rum is a potable beverage distilled from fermented sugar cane juice or molasses and suitably aged in wood. It probably originated in the oriental tropics, but mention of rum especially brings to mind the buccaneering of the Caribbean and the Yankee Clipper trade. Rums of Cuba, Haiti, the Dominican Republic, Puerto Rico, and the like are generally light-bodied and distilled at high proof, and frequently have caramel, fruit flavor, or perhaps cognac added before marketing to give color and added flavor. Rums of Jamaica, Trinidad, Martinique, Barbados, and the like are generally heavy-bodied; they are made from spontaneous fermentation rather than with cultured yeasts, and the molasses bouquet and "impurities" are carried over in distillation. As with other rums, caramel is added for color. Raw, unaged rum is known as aguardiente or caxaça (Brazil), and is commonly a drink of the poorer classes. Batavian Arak is a rum made by special treatment (such as addition of Javanese rice cakes) from molasses of the East Indies sugar mills. Other araks are the distillates of local beers, such as chang and marwa in the Tibetan region. Bacardi is perhaps Cuba's best-known rum.

LIQUEURS OR CORDIALS. Liqueurs were the invention of the alchemists, before the age of chemistry, and like gin were believed to have magical elixir and love-potion properties. They are made by combining brandy-like spirits with any of many flavorings (from fruit to medicinals) and sweetened by the addition of 2.5 per cent or more sugar. Fruit liqueurs are made by steeping fresh or dried fruits in brandy for several months, after which the product is then strained and usually aged. Thus are made apricot, peach, cherry, strawberry, and other fruit liqueurs. Liqueurs made with various aromatics are usually produced by macerating the aromatics in the brandy and then distilling the latter. Additional flavorings and colorings may be added after distillation, as is done in making crème de menthe.

crème de cacao, and the like. Benedictine is a liqueur compounded (according to a secret formula handed down since 1510) on a cognac base. Chartreuse is another liqueur, originally made by French monks of a convent at Grenoble, but after their expulsion produced at Tarragona, Spain. Scores of other distinctive liqueurs are likewise marketed, including cointreau, kummel, curaçao, triple sec, and many others.

Miscellaneous other spirits

Absinthe—a strongly alcoholic (nearly 70 per cent) liqueur-type beverage made in Spain from high-proof brandy, wormwood, and other aromatics such as fennel, artemisia, and the like.

Akvavit—an unaged Scandinavian beverage distilled from grain or potato fermentation and flavored with caraway.

Bitters—tonics containing about 40 per cent alcohol, made principally in the West Indies from herbs (including gentian) and usually used to flavor mixed drinks.

Okolehao or Oke—a Hawaiian distillate from a molasses–rice–dasheen ferment, aged in charred barrels.

Tequila—the distillate from certain species of Mexican *Agave*, first fermented much as is pulque.

Vodka—an unaged, unflavored Russian distillate made from a wheat–malt mash.

SUGGESTED SUPPLEMENTARY REFERENCES

Amerine, M. A., H. W. Berg, and W. V. Cruess, *Technology of Wine Making*, 2nd ed., AVI Pub. Co., Westport, Conn., 1967.

Cook, A. H. (ed.), *Barley and Malt: Biology, Biochemistry and Technology*, Academic Press, N. Y., 1962.

Cuatrecasas, Jose, *Cacao and its Allies: A Taxonomic Revision of Theobroma*, Contributions U. S. National Herbarium, U. S. National Museum, **35⁹**, 1964.

Eden, T., *Tea*, 2nd ed., Longmans, London, 1958.

FAO, *World Cocoa Survey*, Rome, 1964.

Haarer, A. E., *Modern Coffee Production*, Leonard Hill, London, 1962.

Harler, C. R., *The Culture and Marketing of Tea*, 3rd ed., Oxford Univ. Press, London, 1964.

Joslyn, M. A., and M. A. Amerine, *Dessert, Appetizer and Related Flavored Wines*, Univ. of California Press, Berkley, Calif., 1964.

Singleton, V. L. and P. Esau, *Phenolic Substances in Grapes and Wine, and their Significance*, Academic Press, N. Y., 1969.

Sivetz, M., and H. E. Foote, *Coffee Processing Technology*, AVI Pub. Co., Westport, Conn., Vols. I and II, 1963.

Urquhart, D. H., *Cocoa*, 2nd ed., Wiley, N. Y., 1961.

Vogel, E. H., F. H. Schwaiger, H. G. Leonhardt, and J. A. Merten, *The Practical Brewer*, Master Brewers Assoc., Von Hoffmann Press, St. Louis, 1947.

Wellman, F. L., *Coffee: Botany, Cultivation and Utilization*, Leonard Hill, London, 1961.

Winkler, A. J., *General Viticulture*, Univ. of California Press, Berkeley, Calif., 1962.

Younger, W., *Gods, Men and Wine*, World, Cleveland, 1966.

CHAPTER 21

From Microorganisms to Miscellanea

OUR FINAL CHAPTER IS CONCERNED WITH diverse and often minor uses of plants—subjects that do not fit neatly into the previous chapters. Industrial microbiology, aquatic plants, mushrooms, ornamentals, hayfever and toxic plants, weeds, wild herbs, and several other topics fall here. Much more might be included—for example, plants useful in an ecological sense. But obviously neither every economic plant nor every obscure relationship between plants and man can be mentioned in any one book. Even the miscellanea of this chapter can be touched upon only lightly.

Lower Plants

Of microorganisms and industrial fermentations. Fungi are among the more important of the lower plants, both directly and for their indirect influence. Beneficial activities include the decomposition of organic wastes, nitrogen fixation, and mycorrhizal aid to the roots of higher plants. Fungi serve as food, as agents in food production, and for medicinals. Industrial fermentations yield alcohol; gallic, citric, gluconic, itaconic, kojic and other acids; various enzymes; glycerol and fats. Fungi serve as biological controls, and are used in

certain assays. On the negative side they cause disease, rot materials, spoil foods, and in some cases toxify the environment.

From time immemorial microorganisms have been the agents of man. No doubt before the dawn of the first civilization man had learned to let his meat stand a few days before eating it. It developed, then, a more pleasing taste and consistency because of preliminary digestion by microorganisms. Modern-day meat is still tenderized for market by aging in cool chambers. Certainly primitive man knew how to ferment drinks from grain and

Germinating mold spores highly magnified. (Courtesy Chas. Pfizer & Co., Inc.)

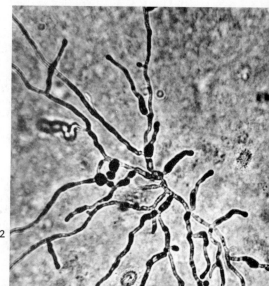

612

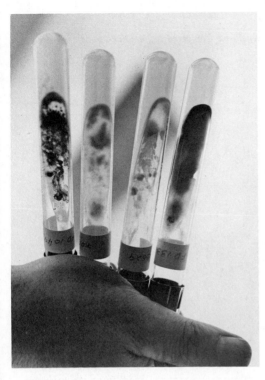

Test tubes in which microorganisms of potential usefulness are cultured. (Courtesy Chas. Pfizer & Co., Inc.)

fruit, a topic of the preceding chapter. Of course primitive man was not conscious that bacteria, molds, and yeasts engineered the response he empirically noted. More likely he credited this to magic.

Science has since revealed much about microorganisms. Today they are bred and cultivated as intensively as are conventional crop plants, domesticated animals, and garden flowers. Highly purified cultivars provide tasty foods and beverages, and yield metabolites that have become man's most potent medicines. The arts of cheese and bread making have become sciences as select molds and yeast have been harnessed, their performance related to substrate, pH, and temperature. *Penicillium*, Ascomycetae, grown upon corn steep liquor, yields penicillin; various Actinomycetes give other medicinals

such as tetracyclene and aureomycin; bacteria are controlled with bacitracin and polymyxin (both of which, incidentally, contain polypeptides that sometimes cause allergic reactions limiting their usefulness). The proteolytic enzyme maxatase is derived from *Bacillus subtilis*, a bacterium, for use in washing compounds. *Aspergillus niger*, Ascomycetae, grown on sugar residues, yields commercial citric acid, and other species kojic and itaconic acids much used in plastics. *Mucor rouxii* and *Rhizopus nigricans*, Phycomycetae, provide important diastatic enzymes substituting nicely for malt (such as are used to make sake in Japan). Throughout earlier chapters the importance of fermentation has been mentioned, in preparation of such foods as pickles, sauerkraut and olives, and of various beverages such as coffee and cocoa.

Because microorganisms can use hydrocarbons as sole source of energy, they offer an unusual opportunity for creating useful products from inexpensive raw materials. It should be possible to develop microbial protein feeds from a wide variety of carbon compounds, ranging from wood-scrap cellulose to inexpensive petroleum fractions. Although feeding upon carbohydrate, the microorganisms become themselves rich sources of amino acids and vitamins. With the population facing malnutrition and starvation in many corners of the world, certainly it seems wise to take as great advantage as possible of this little-used source for animal and even human proteinaceous foodstuffs. The amino acid content of several representative microorganisms compared to conventional food is given in Table 21-1.

Bacteria are normally less economically collected than the larger-celled yeasts, Ascomycetae. Harvest is typically by centrifugation. On the other hand bacteria grow on a wider range of substrates, including ubiquitous cellulose. Mixed cultures of bacteria should yield acetic, butyric, lactic, and for-

Table 21-1

Amino Acid Content of Representative Microorganisms Compared to Familiar Foods

	Content (%)	
Sample	Essential amino acids (dry wt)	Nitrogen
Bacteria		
Staphylococcus aureus	21.6	10.75
Escherichia coli	33.1	13.19
Bacillus subtilis	23.8	10.07
Yeasts		
Saccharomyces cerevisiae, av.	17.1–23.8	5.9–8.2
Saccharomyces cerevisiae	23.1	8.94
Torula yeast	29.5	8.35
Torula yeast	24.4	7.47
Molds		
Aspergillus niger	9.2	5.21
Penicillium notatum	12.8	6.13
Rhizopus nigricans	9.6	5.80
Mushroom		
Tricholoma nudum	20.8	8.64
Nonmicrobial samples		
Animal muscle, av.	48.1	15.4
Fish meal, av.	32.1	9.8
Alfalfa meal	6.9	2.72

From M. J. Johnston, *Science* **155,** 1516 (1967).

Soil from the desert near Timbuktu, Mali, is collected in the search for new antibiotic-producing microbes. (Courtesy Chas. Pfizer & Co., Inc.)

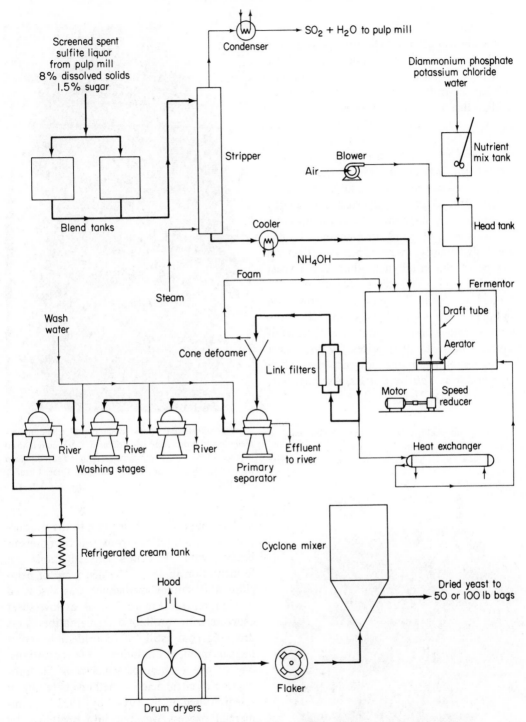

Figure 21-1 Flow sheet for production of food yeast from sulfite liquors at the Rhinelander, Wisconsin, plant of Lake States Yeast Corp. (After *Industrial and Engineering Chemistry* **43**, 1704 (1951).)

mic acids, and ethanol and glucose economically from the fermentation of cellulosic substance. Yeasts and molds can release similar substances when an appropriate cultivar is grown on properly maintained media. But the ability of yeasts to provide protein more economically than bacteria would seem to destine them rather than bacteria for direct consumption as food. Except for wastes, of which the cost of disposal constitutes a subsidy, the most economical source of hydrocarbon energy for microbes is methane; cost is often only a cent or two per pound. Propane and butane are only slightly more expensive. Bacteria can be grown on all such alkanes, but yeast probably not. The least expensive yeast protein is from Torula yeast grown on sulfite pulp waste liquor. This runs little more than 10 cents a pound, about twice the cost of soybean meal at 1970 prices. A flow sheet showing production of food yeast from sulfite pulp liquor by *Torulopsis utilis*, Ascomycetae, is given as Fig. 21-1.

Aspergillus niger. Black "bread mold" used commercially in the production of citric acid. Citric acid is used in thousands of ways from giving "tang" to soft drinks to recovering oil from oil wells. (Courtesy Chas. Pfizer & Co., Inc.)

This green mold is *Penicillium chrysogenum*, a mutant form of which now produces almost all of the world's commercial penicillin. (Courtesy Chas. Pfizer & Co., Inc.)

The chain of events leading to the production of antibiotics from microorganisms began with Pasteur's investigations of fermentation in 1857. The discovery of penicillin and similar medicinals was discussed in Chapter 12. Figure 21-2 is a flow sheet characterizing antibiotic fermentation. Less dramatic but still of considerable commercial moment are industrial fermentations, one of the earliest of which was the production of citric acid by methods based upon research begun in 1914. By 1923 a commercial process was put into operation by Chas. Pfizer and Co., in which *Aspergillus niger* was used to transform sugars into citric

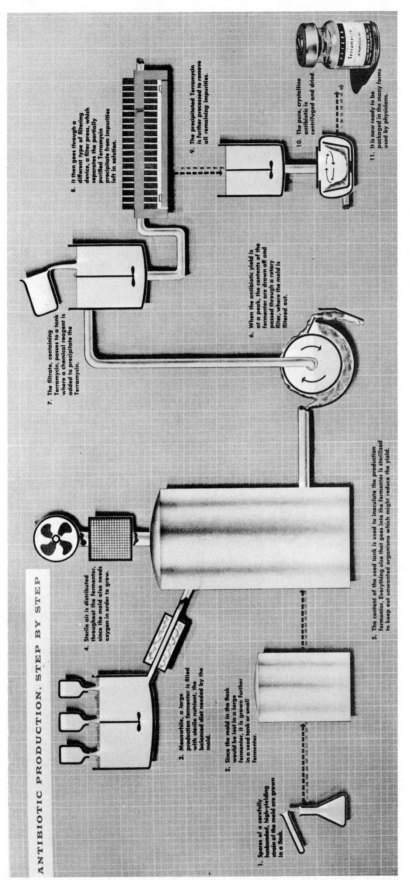

Figure 21-2 A typical antibiotic fermentation flow sheet. (Courtesy Chas. Pfizer & Co., Inc.)

acid. This cut the price of the commercial acid 75 per cent, the mold yielding citric acid so much more economically than could the citrus tree. Although no longer of commercial moment, Weizmann's process for deriving acetone from corn during a World War I shortage of this product (essential for explosives) is another of the classic industrial fermentations.

At one time riboflavin, a member of the vitamin B complex, was produced commercially by fermentation with *Ashbya gossypii* or *Eremothecium ashbyii*, Ascomycetae. This has been largely supplanted by more economical synthetic production, although ascorbic acid, a raw material in the synthesis,

is a fermentation product. Riboflavin is produced by many other fungi as well, but less economically. One such is *Aspergillus flavus*, long known to damage seed during storage. Investigations showed *A. flavus* to yield kojic acid abundantly, and this product, important to the plastics industry, is now derived from a cultivar cultured on a peanut oil medium. A partial list of chemical substances derivable from molds is given in Table 21-2. Other potentially useful employments of fungi include extraction from *Poria obliqua*, Basidiomycetae, of a material possibly useful in treating cancer, and use of a *Fusarium* species (that causes canker on southeastern yellow pines) to increase flow of

Table 21-2

Partial List of Chemical Substances Derivable from Molds

Acids	Alcohols	Enzymes	Polysaccharides	Sterols	Miscellaneous
Acetic	Ethyl	Amidase	Capreolinose	Cholesterol	Aldehyde
Aconitic	Glycerol	Amylase	Galactocarolose	Ergosterol	Antibiotics
Allantoic	Mannitol	Catalyse	Glycogen	Fungisterol	Ergot
Carlic		Cytase	Gums	Phytosterol	Ethylacetate
Citric		Dextrinase	Luteic acid		Lipins
Formic		Dipeptase	Mannocarolose		Vitamins
Fulvic		Enulsin	Mycodextrin		
Fumaric		Eripsin	Polygalactose		
Fusarinic		Inulase	Polymannase		
Gallic		Invertase	Rugulose		
Glaucic		Lactase	Selerotiose		
Glauconic		Lecithinase	Starch		
d-Gluconic		Lipase	Trehalose		
Glycolic		Maltase	Varianose		
Itatartaric		Nuclease			
Itatonic		Protease			
Kojic		Raffinase			
Lactic		Rennet			
Luteic		Sulfatase			
Malic		Tannase			
Malonic		Urease			
Oxalic		Zymase			
Penicillic					
Pyruvic					
Stipitatic					
Succinic					
Terrestric					

From F. A. Gilbert and R. F. Robinson, *Economic Botany* **11**, 142 (1957).

Complex equipment used in industrial microbiological operations. (Courtesy Chas. Pfizer & Co., Inc.)

naval stores from tapped pine trees of the southeast. Species of *Peziza* and *Lactarius*, Basidiomycetae, yield nearly 2 per cent of rubber, although extensive commercial production has not yet come from lower plants. Dextran, a polysaccharide substitute for blood in transfusions and extender of blood plasma, is obtained by fermentation of sugar solutions with a special strain of the bacterium, *Leuconostoc mesenteroides*. Certain yeasts, *Saccharomyces*, Ascomycetae, can be used for commercial production of glycerol, and the genus is well known for providing baker's yeast and ergosterol. Glutathione, coenzymes I and II, ribonucleic acid, trehalose, adenosine triphosphate, invertase, lactase, and many other chemical intermediates are also obtainable from various species of yeast.

Many common molds, including *Aspergillus flavus* and a number of species of *Penicillium*, produce mycotoxins (aflatoxin) that are responsible for poisoning of food and feed. This is not unlike the side-effects with antibiotic medicinals, toxic to persons allergic to particular associated substances. This is not to condemn molds in food, however. *Penicillium roqueforti* produces the proteolytic and lipolytic enzymes essential for the development of the characteristic flavors in Roquefort, Gorgonzola, Stilton, and blue cheeses. The mold is sufficiently abundant in such cheeses to create the characteristic soft texture (aided in partial digestion by abundant *Streptococcus lactis*, a characteristic bacterium of milk). In a like manner *P. camemberti* is responsible for Camembert and Brie cheeses. The process for manufacturing blue cheese typically involves blending skim milk and homogenized cream in a vat that is inoculated with *Streptococcus lactis* for a "ripening" of about

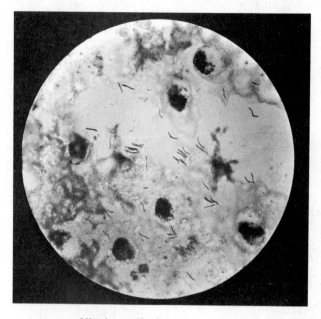

Microbes affecting man—tuberculosis bacteria.
(Courtesy Chicago Natural History Museum.)

1 hour at a warm temperature. Then the enzyme rennin is added to effect coagulation of a curd. Liquid whey is drained from the curd, a little salt added, and *Penicillium roqueforti* powder mixed in. This inoculum is prepared separately by culturing it in the interior of a loaf of bread, started with spores from an agar slant. The inoculated curd is shaped in perforated containers until firm, then removed to a cool, humid room where salt is applied to the surface. The fresh cheese is then set aside for curing for about 3 months.

Miso is a familiar fermented food of Japan, prepared from soybeans, salt, and rice, inoculated with *Aspergillus oryzae*. Nearly 1 million tons of this and similar fermented dishes are consumed annually in Japan. Typically whole soybeans are soaked, cooked, salted, ground, then inoculated with the mold (cultured separately on trays of rice). Several other microorganisms play a part,

Restrictive influence on pathogen growth of differing antibiotics is tested by drops on agar culture in petri dish. (Courtesy Chas. Pfizer & Co., Inc.)

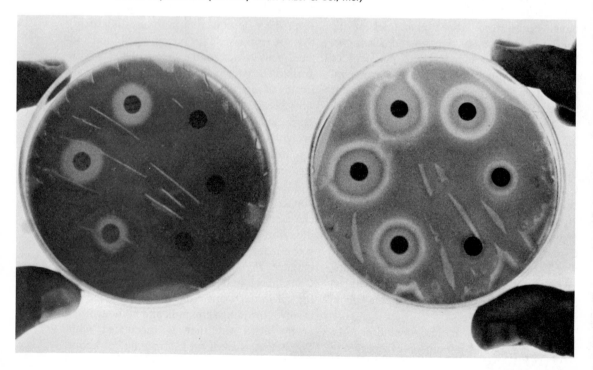

including species of *Zygosaccharomyces*, Ascomycetae, and several bacteria. Fermentation may proceed for 6 months or more, during which time the soybeans and rice are partially digested. The fermented meal is then aged at room temperature for a few weeks, ground into a paste, and packaged for distribution.

Mushrooms. Sporophores of many of the Basidiomycetes (and a few Ascomycetes) are harvested as edible mushrooms. Species of *Agaricus, Morchella, Pholiota, Cortinellus, Tricholoma, Collybia, Armellaieu,* and *Pleurotus* have been cultivated in various parts of the world. Poisonous mushrooms are generally regarded as "toadstools." Some mushrooms, such as *Coprinus*, may produce toxic reactions when ingested along with alcohol. Since ancient times wild mushrooms have been a table delicacy, and few foods command more respect than the truffles (*Tuberales*) of France and the morels (*Morchella*) of North America. Wild *Agaricus* is still sought by collectors in the cool of the spring, while oyster mushrooms (*Pleurotus*) and similar less frequently eaten types are a prize for the connoisseur. Mushroom cultivation has become a sizable industry in many parts of the world, and no longer are fresh mushrooms a delicacy reserved for the wealthy. Mushrooms are not an important food, however, containing little carbohydrate, fat, and protein, although there are some minerals, vitamins, and amino acids.

Contrary to popular belief, cultivars of *Agaricus bisporus* (cf. *A. campestris*), Basidiomycetae, the chief commercial mushroom, do not need dark, dank habitats. Light is of no consequence one way or another, and excessive moisture actually gives an inferior, diseased crop. Nor does production require cool temperatures, although mushrooms cannot be grown in the heat of summer nor in warmer climates without artificial temperature control. The first step in mushroom growing is preparation of a composted growth medium, once made chiefly from horse manure. Grain, legume fodder, chicken manure, chemical nutrients, etc. may be included. Mushrooms can be grown on artificial media, but nothing has proved more satisfactory than compost. Actinomycetes are chiefly responsible for reducing manure to an earthy humus, during which several turnings and waterings are necessary. A final fermentation, in the tiers of planting trays in closed sheds where the mushrooms are to be grown, raises compost temperatures as high as 140°F, effecting pasteurization. Meanwhile mushroom spores have been artificially cultured under sterile conditions, then grown in jars on a medium often made from tobacco stems, humus, and added minerals to produce "spawn." This spawn may be planted in the beds within 10 days, or be gradually dried until dormant, in which condition it will keep for many months. Chunks of spawn are set about 2 inches deep and 10 inches apart. Temperature is first held about 70°F, then dropped to around 62°F as growth proceeds during the next 4 weeks. A shallow layer of slightly alkaline loam is spread over the compost and its ramifying mycelial network, and in about 10 days the temperature dropped to 50°F. Beds are watered lightly every other day. A week or two later mushroom fructifications appear. The young "buttons" are picked by a twisting motion at the stage when the veil distends. Waterings and harvest continue until the bed is exhausted, usually in 3 to 7 months. Nearly 2 pounds of mushrooms are obtained from each square foot of bed space. For maximum year-round efficiency, shed and beds must be equipped for winter heating and temperature control.

In the United States, Pennsylvania has long been an important source of mushrooms, at times providing over half of domestic production estimated to reach a few tens of thousands of tons annually. The region

Table 21-3

Checklist of Mushrooms Which May Be Sold in the Markets of the State of Geneva

Amanita caesarea	C. cornucopioides	Mitrophora sp.
Boletus aereus	Fistulina hepatica	Pholiota aegerita
B. appendiculatus	Gomphidius glutinosus	Polyporus confluens
B. aurantiacus	Hydnum imbricatum	P. frondosus
B. edulis	H. repandum	P. umbellatus
B. granulatus	Hygrophorus caprinus	Psalliota arvensis
B. luteus	H. marzuolus	P. augusta
Cantharellus cibarius	H. obrusseus	P. campestris
C. lutescens	H. puniceus	P. silvatica
Chaeromyces meandriformis	H. virgineus	Tricholoma aggregatum
Clitocybe cyathiformis	Laccaria laccata	T. colombetta
C. geotropa	Lactarius deliciosus	T. Georgii
C. gigantea	L. sanguifluus	T. irinum
Clitopilus prunulus	Lepiota excoriata	T. nudum
C. orcella	L. procera	T. paneolum
Coprinus comatus	L. rhacodes	T. personatum
Cortinarius praestans	Marasmius oreades	Tuber sp.
Craterellus clavatus	Morchella sp.	

From C. Weber, *Economic Botany* **18**, 255, (1964).

surrounding Paris, France, is the center for the canned-mushroom industry. A special market for wild mushrooms has been created at Geneva, Switzerland, under supervision of a state mycologist; a checklist of species which may be sold there is given as Table 21-3.

Algae. Plants discussed in this book have been overwhelmingly land plants. Yet there is much more sea than land to this world, of which most of the arable soil has already been brought into cultivation. The aquatic vastness is domain for algae. Photosynthesis by these lower plants provides the energy for the "pastures of the sea," upon which marine animal life from bottom worm to baleen whale is dependent. With burgeoning human populations tending to overtax the carrying capacity of the land, attention has turned to cultivating the sea. Harvest of fish from the sea is as old as civilization itself, and in times of emergency there have been attempts at "sea gardening." Only in Japan and China, where abundant hand labor is available, has extensive cultivation of algae for food been

undertaken. In many parts of the world, of course, wild algae are regularly harvested— for food, agar (see Chapter 9), and fertilizer. Jacques Zaneveld has reviewed the seaweed resources of the Orient in *Economic Botany*, giving vernacular names, general distribution, and uses, as well as photographs or drawings of many of the species most utilized. The genera involved are listed in Table 21-4. But the growing of kelp, especially *Laminaria japonica, Phaeophyceae*, by the Japanese and Chinese, shows greatest parallel with land agriculture.

Two Colleges of Aquatic Products Studies, concerned with the culture of kelp and other marine organisms, had been opened on mainland China by 1965. Many tens of thousands of tons of kelp are grown annually, worth commercially, it is reported, $120,000 as of 1965. Dried kelp is nearly 60 per cent carbohydrate, about 8 per cent protein, with very little fat. It is rich in inorganic salts (nearly 13 per cent), and contains several vitamins. In addition to being used for food, kelp is a source of iodine, mannitol, and alginates (important in textiles, printing inks, medicinals, emulsifiers, ice cream, rubber, etc.). Chinese scientists have learned to selectively breed cultivars of *Laminaria japonica*, and young sporophytes are hand fastened to floating bamboo rafts or other anchored devices in the shallow sea off the China coast. Interestingly, growth is much more satisfactory if nitrogen fertilization is provided, usually accomplished by fastening porous earthenware containers filled with soluble nitrates in the vicinity of the growing algae. There is some culture upon rocks placed on the sea floor, where the algae benefit from the excrement of various sea creatures. Strains of kelp have been bred suited to prevailing water temperatures and other environmental conditions.

Floating algae—plankton—are an early step in the food chain that leads to fish and other seafood. Various diatoms are of special importance. They are seasonally abundant, especially after rainy seasons when rivers bring minerals leached from the land to the ocean. Much the same thing occurs when water upwells from the deep, as off the coast of Peru where one of the world's richest fishing grounds is found. The food chain from unicellular algae to fish prevails in rivers, lakes, and farm ponds, too. It is reported that, pound for pound, more fish can be produced on some of the worn-out farmlands (by damming streams and impounding water) than traditional crops grown. Sometimes fertilization of the pond is practiced to insure adequate production of algae, and eventual fish harvest.

There is growing concern about pollution of lakes and watercourses by speeded-up eutrophication; "scummy" algal growth is encouraged by minerals and pollutants dumped into the water. Gradual eutrophication is natural and to be expected, but man's activities are greatly shortening the life of clear-water streams and lakes for recreation and water supplies. Regulated growth of algae on lagoons of waste can yield appreciable quantities of plant material useful for stock feed, and in some cases even for human consumption. *Chlorella*, a unicellular member of the *Chlorophyta*, has been a favorite source of so-called SCP, "single-cell protein." It can be successfully grown on properly prepared waste materials from sewage-treatment plants, utilizing the energy of sunlight several times more efficiently than does the average crop plant. Indicated yields as high as 15 tons of dry material per acre per year have been obtained. Unfortunately the cost of growing an alga such as *Chlorella* in shallow layers of nutrient to take maximum

Table 21-4

Marine Algae of the Oriental Tropics Frequently Harvested by Man

Division	Genus
CYANOPHYTA (MYXOPHYCOPHYTA)	*Nostoc*
CHLOROPHYTA	*Enteromorpha, Ulva, Chaetomorpha, Caulerpa, Acetabularia, Codium*
PHAEOPHYTA	*Chnoospora, Hydroclathrus, Dictyota, Padina, Sargassum, Turbinaria*
RHODOPHYTA	*Porphyra, Liagora, Gelidiella, Gelidium, Grateloupia, Halymenia, Agardhiella, Eucheuma, Catenella, Hypnea, Gymnogongrus, Gracilaria, Corallopsis, Gelidiopsis, Sarcodia, Rhodymenia, Caloglossa, Bostrychia, Acanthophora, Laurencia*

Adapted from J. S. Zaneveld, *Economic Botany* **13**, 90 (1959).

effect of the sunlight has seldom proved economical. Moreover, *Chlorella* has a sharp taste, making it unpleasant to the human palate. However, *Chlorella* has been extensively cultivated in China for use as hog fodder and fertilizer, and included in baked goods and sweets. *Scenedesmus* is another green alga sometimes cultured in Japan, and *Spirulina maxima*, of the *Cyanophyta*, a palatable species consumed for years by inhabitants of the Lake Chad region of central Africa.

Lichens and Bryophytes. Lichens constitute almost the only plant life able to grow under rigorous Arctic and Antarctic conditions. There they provide sustenance for reindeer upon which the Eskimos, Lapps, and others are so dependent. Some lichens are used directly for industrial purposes, such as raw material for dyes and cosmetics, medicinals, and even for brewing. Prominent genera are *Alectoria*, *Umbilicaria*, *Cladonia*, *Cetraria*, and *Parmelia*. *Evernia furfuracea* of the Nile valley yields an extract that is antibiotic against several disease-causing bacteria; fragments have been found in an Egyptian vase from the seventeenth century B.C., apparently used medicinally. Extracts of other lichens exhibit fungicidal antagonisms, and some have been considered useful for treatment of tuberculosis. The Bryophytes— mosses and liverworts—are responsible for extensive peat deposits, especially the genus *Sphagnum*. Peat is used for fuel, in the curing of malt for Scotch whiskey, and as the world's most important source of horticultural humus. The leaf cells of *Sphagnum* contain large, hollow chambers capable of absorbing liquid much as does a sponge. Thus *Sphagnum* peat serves well as a packing material about the roots of living plants, preventing desiccation. In times of emergency it has even been used as a bandaging material to staunch blood flow. Medicinals have been extracted from the liverwort, *Marchantia*, and several

mosses are consumed along with lichens by wildlife in subarctic regions. In hostile environments lichens and bryophytes are often pioneer plants, helpful in building soil on raw rock and barren habitat.

Petroleum and coal. Petroleum and coal are derived from plant residues trapped beneath sediments in ages past. Civilization has been built upon the energy accumulated through past photosynthesis, mainly of plants long extinct. Most of the coal and oil no doubt derives from "lower plants" (although of tree size) that were ancestral to the *Spermatophyta*, including especially giant forms of the *Lycopodinae* and *Equisetinae*. This subject was touched upon in the opening chapter.

Higher Plants

Hayfever plants. Many people react to breathing plant pollen with allergic symptoms similar to those of a severe cold. Some people are sensitive only to certain species, although any abundant pollen protein would seem potentially an allergin. Over much of temperate North America hayfever plants may be grouped into three more-or-less overlapping categories. The early spring hayfever season is largely due to pollen from trees. Relatively few people are sensitive to tree pollen. In late spring and summer grass and herb pollen is abundant, again affecting a relatively limited group of hayfever sufferers. The most troublesome hayfever season is autumn, when the predominating pollen in the air is usually ragweed (*Ambrosia trifida* and *A. elatior*) and other species of the Compositae family. Goldenrod, *Solidago*, once had a notorious reputation as the prime cause of hayfever, but the pollen has been shown to be too heavy to carry widely in the air.

Poisonous plants. Many substances found in plants are toxic, especially when con-

Poison ivy vine. (Courtesy USDA.)

sumed in considerable quantities. Toadstools (poisonous mushrooms) are a case in point. Even goldenrod contains a number of biologically active substances, mainly alkaloids that can have strong physiological influence. Many seeds are highly toxic. The jequerity bean, *Abrus precatorius*, Leguminosae, received notoriety in 1968, when the attractive red-black seeds, used as eyes for figurines brought into the United States by tourists from the Caribbean area, were the object of a nationwide recovery campaign last they be inadvertently eaten by some child. Many of the extractions from plants mentioned in Chapter 12 could prove deadly if ingested in sufficient quantity, as might even juice from the ubiquitous cassava root.

More an irritant than a deadly poison is poison ivy, *Rhus toxicodendron*, Anacardiaceae. An essential oil found in all parts of the plant causes skin allergy with many people. Very sensitive persons may even be affected by small concentrations of the oil that vaporize into the air near poison ivy plants. Vaporization can be quite pronounced if the plant is burned.

Livestock loss frequently occurs from consumption of poisonous plants. Animals may die even from an imbalance of forage, suffering occasionally from such afflictions as saponin bloat, nitrogen toxicity, or illness caused by associated fungal toxins such as ergot and fescue-foot. On the western ranges of the United States locoweeds (species of *Lupinus* and *Astragalus*, Leguminosae, and of *Delphinium*, Ranunculaceae) may be a problem, and some plants absorb sufficient selenium from the soil to render them deadly. Especially troublesome plants are listed in the USDA bulletin, *22 Plants Poisonous to Livestock in the Western States.* (USDA Agric. Inf. Bull. 327, 1968). Included, in addition to the locoweeds, are: *Triglochin maritima* and *T. palustris* of the Alismataceae, bracken fern (*Pteridium aquilinum*), *Prunus virginiana* of the Rosaceae, *Oxytenia acerosa* of the Compositae, *Zigadenus spp.* of the Liliaceae, *Sarcobatus vermiculatus* of the Chenopodiaceae, *Halogeton glomeratus* of the Chenopodiaceae, *Tetradymia spp.* of the Compositae, *Apocynum cannabinum* of the Apocynaceae, *Oxytropis lambertii* of the Leguminosae, *Asclepias spp.* of the Asclepiadaceae, several species of *Quercus* of the Fagaceae, *Conium maculatum* of the Umbelliferae, *Hymenoxys spp.* of the Compositae,

Hypericum perforatum of the Hypericaceae, *Helenium hoopesii* of the Compositae, *Cymopterus watsonii* of the Umbelliferae, *Veratrum californicum* of the Liliaceae, and *Cicuta douglasii* of the Umbelliferae. Toxic alkaloids are especially frequent in the Leguminosae, Liliaceae, Ranunculaceae, and Solanaceae families; glucosides (some of which hydrolyze to hydrogen cyanide) in the Rosaceae; resinlike products in the Umbelliferae and Ericaceae; and proteinaceous materials, oxalates, nitrates, higher alcohols, and coumarols in many families.

Another form of plant toxicity has only recently received much attention—interaction between plants themselves. Metabolites from one plant may severely repress growth of others, or even the sprouting of seed in the vicinity of the mother plant (a survival factor of special importance in desert climates where moisture is limited). Such ecological interplay has only been partially investigated, although it has long been recognized that juglone, for example, secreted by the black walnut tree, suppresses growth of many species (especially in the Rosaceae), and that quackgrass, *Agropyron repens*, imparts toxic qualities to soil that repress growth of other plants even after the quackgrass itself has been eliminated. Most living organisms probably interact in this fashion to a greater or lesser degree, sometimes with benefit (certainly *Trifolium* aids *Poa pratensis*), and often with repression of competition. Weed seeds may be kept dormant in the soil by interaction with other seeds and plant residues, and many weeds seem to restrict crop performance more than can be accounted for solely by the competition for growth essentials. Germination of *Abutilon* seeds has been shown to be especially repressive to other species. Volatile phytotoxins from *Salvia leucophylla* and *Artemisia californica* inhibit invasion of the chaparral by natural grasslands in California. *Catabrosa aquatica* has been shown to have persistent influence in preventing incursion of other vegetation in the marshy polders of Holland.

Weeds. Weeds are plants growing where they are not wanted. By this definition almost any plant could qualify as a weed, depending upon the time, place, climate, and human preference. Take, for example, the common white clover (*Trifolium repens*, Leguminosae), introduced from Europe into North America. In carefully tended bluegrass lawns, golf courses, or chat driveways, this plant is a troublesome weed; yet in an average lawn it may be tolerated or encouraged, and in a pasture it is certainly a welcome member of the plant community. So it is with any plant. Where overabundant or unwanted, plants become weeds; elsewhere they may be respected members of the flora, directly useful or ornamental, and not competitive with crops, lawns, or gardens.

Control of weeds has always consumed much effort and income. The ancient Mayan abandoned overgrown fields and burned new forests; the farmer laboriously cultivates his land. Both practices are only differing aspects of man's continuing fight against weeds that compete with crops. In the late twentieth

The dandelion, *Taraxacum*, familiar weed of lawns here shows typical curl of pedicels after treatment with 2,4-D herbicide. (Courtesy The Lawn Institute.)

century the control of weeds takes many forms. Over much of the earth it is still a direct process of pulling up or hoeing out weeds by hand. But these practices can rarely be adopted in areas of labor scarcity or high costs. Thus man uses the plow, the disk, the harrow, and other mechanical devices to duplicate the work of the hand wielded hoe. In the United States today almost all cultivation to destroy weeds is done by machines. Prevention of weeds by the use of "clean" seed and avoidance of weed dissemination is often effective. Sterilization of seedbed by heat or chemicals may help to a limited extent. Burning is used to destroy both weed seeds and growth that harbors disease. Mulching is useful for control of weeds. The use of large-scale paper mulching was mentioned in the discussion of pineapple growing. Selective sprays used in chemical weed control are spread by devices ranging from sprinkling can to airplane. Many older chemicals such as ammate (sodium sulfamate), chlorates, arsenicals, copper sulfate, oil, and the like are still used in the field, but newer herbicides are continually being discovered and developed. The phenoxys are widely used to control dicotyledonous weeds among monocotyledonous crop, triazines for general weed control in corn. Biological control is a bit more difficult. It pits one organism against another, from the use of mere "smother crops" to the introduction of weed diseases.

Apart from their direct competition with crops, weeds may have other detrimental effects. Various algae help foul ship bottoms, necessitating costly expenditure of fuel, drydocking, and painting. Water weeds may clog drainage or irrigation ditches, or even navigation channels; among such weeds is the water hyacinth, *Eichhornia crassipes*, Pontederiaceae. Other weeds may serve as host or alternate host for crop diseases. Some weeds, such as wild garlic (*Allium spp.*, Liliaceae), give an objectionable odor and flavor to the milk produced by cows which have eaten them. Some weeds have spines, thorns, or burs that cause injury to domestic animals and discomfort to gardener, farmer, and woodsman. Of course weeds are not entirely harmful. They provide cover to otherwise unprotected soil; plowed under, they provide humus for soil building; their seeds furnish food for birds and wildlife; some are of occasional medicinal or food value.

Edible wild plants. Edible wild plants have more to offer modern man as a hobby and for sympathetic understanding of nature than as an economic reward. Yet in times of emergency edible wild plants could pay dividends. Few plants are poisonous, although may are unpalatable. Little danger will be experienced from tasting small quantities of plant parts presumed to be edible. Edible wild plants are discussed at length in several books, a few of which are listed among the Supplementary References at the end of this chapter. Wild rice, nuts, fruits, and Indian foods were mentioned in earlier chapters. Some of the wild plants of eastern North America deemed useful for food are listed in Table 21-5.

Edible wild plants are useful not only for direct human consumption, but also as food and protection for wildlife. Marsh grass seeds in large measure determine the carrying capacity of waterfowl preserves, and wild plants provide quail with food and cover. Bee pasture is of direct economic importance less for yields of honey and wax than for plant pollination effected by bees. Entomophilous plants depend increasingly upon domesticated bees, as natural habitat is destroyed. It has been estimated that the economic value of bees as pollinators runs more than 50 times the value of all honey produced in the United States (over 100,000 tons annually).

The flavor of honey is determined largely by the nectar the bee secures (e.g., clover gives a light, mild honey; buckwheat a dark,

Table 21-5

Uses of Edible Wild Plants of Eastern North America

PURÉES AND SOUPS

Typha (young flowering spike), *Smilax* (young leaves and sprouts), *Oxyria* and *Rumex* (leaves), *Silene* (young leaves), *Epilobium* (young shoots), *Cryptotaenia* (leaves), *Sambucus* (pith), certain grasses (seeds), meats of nuts, *Hemerocallis* (buds and flowers), *Sassafras* (shoots or pulverized leaves), seeds of *Helianthus* and several mallows, roots of *Viola* and *Arctium*, etc.

COOKED STARCHY VEGETABLES AND ROOT VEGETABLES

Tubers or rootstock of *Typha, Sparganium, Potamogeton, Sagittaria, Butomus, Phragmites, Cyperus, Scirpus, Peltandra, Orontium, Commelina, Claytonia, Polygonum, Nuphar, Nymphaea, Nelumbo, Ranunculus, Potentilla, Glycyrrhiza, Hedysarum, Lathyrus, Apios, Oenothera, Aralia, Erigenia, Cyptotaenia, Carum, Sium, Pastinaca, Heracleum, Daucus, Ipomoea, Stachys, Lycopus, Valeriana, Campanula, Helianthus, Arctium, Tragopogon*, etc.

Seeds of *Pinus, Pontederia*, various nut trees, *Cannabis, Nuphar, Nelumbo, Lathyrus, Gymnocladus, Trapa, Staphylea*, etc.

BREADSTUFFS

Inner bark or cambium of *Pinus, Tsuga, Acer*

Underground portions of *Butomus*, grasses, *Arisaema, Peltandra, Calla, Symplocarpus, Orontium, Polygonatum, Smilax, Pueraria, Symphoricarpos, Valeriana*

Seeds or fruits of many grasses, *Scirpus, Symplocarpus, Orontium, Pontederia*, many nut trees, *Rumex, Polygonum, Kochia, Amaranthus, Spergula, Portulaca, Nuphar, Nelumbo, Capsella, Sorbus, Amelanchier, Fragaria, Prunus, Trifolium, Aesculus, Diospyros, Verbena, Xanthium, Helianthus, Madia*, etc.

Flowers of *Cercis, Sambucus*

Pollen of *Typha, Scirpus*

COOKED GREENS OR POTHERBS

Young shoots of many grasses, *Commelina, Aneilema, Tradescantia, Pontederia, Eichhornia, Clintonia, Trillium, Fagus, Humulus, Urtica, Rumex, Polygonum, Chenopodium, Kochia, Amaranthus, Sesuvium, Stellaria, Claytonia, Nelumbo, Capsella, Brassica, Arabis, Sedum, Cassia*, various mallows, *Epilobium, Aralia, Hydrocotyle, Cryptotaenia, Clethra, Anagallis, Hydrophyllum, Lamium, Lycium, Veronica, Dicliptera, Plantage, Galium, Eclipta, Cosmos, Galinsoga, Petasites, Erechites, Tragopogon, Lactuca*, and many others that usually have a stronger or more bitter flavor

Asparaguslike stems of several ferns ("fiddleheads") and grasses, *Typha, Uvularia, Allium, Smilacina, Polygonatum, Smilax, Humulus, Polygonum, Phytolacca, Nelumbo, Epilobium, Aralia, Monotropa, Asclepias, Sambucus, Carduus, Tragopogon*, etc.

SALAD AND RELISH PLANTS

Very young leaves of grasses, *Acorus, Tradescantia, Oxyria, Arenaria, Portulaca*, various cresses, *Lepidium, Capsella, Cochlearia, Cakile, Brassica, Reseda, Sedum, Saxifraga, Sanguisorba, Oxalis, Erodium, Cryptotaenia, Carum, Anagallis, Plantago, Valerianella. Campanula, Helianthus, Cosmos, Arctium, Cichorium, Tragopogon, Taraxacum, Sonchus*, etc.

Underground parts of *Medeola, Dentaria, Glycerrhiza, Stachys, Lycopus*

Stems of *Smilax, Lathyrus, Dolichos, Angelica, Ligusticum*

Flowers of *Yucca, Rosa, Cercis*

PICKLES AND CONDIMENTS

Allium, Smilacina, Polygonatum, Medeola, Myrica, Asarum, Oxalis, Sassafras, Rorippa, Sedum, Osmorrhiza, Carum, Foeniculum, Gaultheria, various mints, *Tanacetum* and many other genera

Table 21-5 (cont.)

Uses of Edible Wild Plants of Eastern North America

BEVERAGES

Underground parts of *Cyperus, Geum, Cichorium, Taraxacum*

Twig tips of *Tsuga, Thuja, Lindera*

Bark of *Betula, Ulmus, Sassafras*

Leaves of *Myrica, Betula, Chenopodium, Hamamelis, Fragaria, Potentilla, Rubus, Ilex, Ceanothus, Epilobium, Aralia, Dirca, Gaultheria, Chiogenes,* various mints, *Veronica*

Flowers of *Trifolium, Tilia, Sambucus, Solidago*

Fruits and seeds (usually parched) of *Juniperus, Triglochin, Castanea, Gymnocladus, Cassia, Cytisus, Tilia, Diospyros, Galium, Triosteum, Helianthus*

Many wild fruits or fleshy underground parts may furnish the basis for fermentation of alcoholic beverages (e.g., *Morus, Podophyllum, Berberis, Rubus, Prunus, Rhus, Vitis, Diospyros, Sambucus,* etc.)

SWEETS

Sugars and syrups from *Juglans, Carya, Platanus, Acer* and *Tilia* sap, *Castanea* and *Gleditsia* fruit, *Scirpus* rootstock

Confections from shoots or roots of *Pinus, Phragmites, Acorus, Asarum, Ligusticum, Angelica, Inula, Tussilago* and *Arctium;* and reproductive structures or juices of *Castanea,* several mallows, *Viola, Shepherdia, Trapa,* and *Marrubium*

Fruits and jellies from many species such as were listed useful for beverages

MISCELLANEOUS

Oils from various nuts and seeds

Masticatories from *Pinus, Picea, Rumex, Berberis, Prunus, Oxalis, Vitis, Nyssa, Symplocos, Asclepias,* etc.

Adapted from M. L. Fernald and A. C. Kinsey, *Edible Wild Plants of Eastern North America,* Cornwall-on-Hudson, N. Y., Idlewild Press, 1943.

strong product). Among the important bee plants are maples (*Acer*), willows (*Salix*), and dandelions (*Taraxacum*) in the early spring; clovers (*Trifolium* and *Melilotus*), alfalfa (*Medicago*), fruit trees (orange, apple, pear, etc.), and brambles (*Rubus*) for later major honey flow; and Polygonaceous plants, goldenrod (*Solidago*), *Eupatorium,* various mints, and so on for summer or autumn flow. *Tilia, Liriodendron, Oxydendrum, Diospyros, Melaleuca, Sabal, Prosopis, Gleditsia,* and *Robinia* are noted honey trees, and *Acacia, Clethra, Baccharis, Ilex, Rhus,* etc. locally important shrubs. Herbs of considerable value include melons, cabbage, chives, *Echinops, Phacelia, Epilobium, Asclepias, Cleome, Lathyrus, Lotus,* and many other genera.

A few curiosities. Tetrapanax papyriferum, Araliaceae, is the source of rice paper, raw material for making realistic-looking artificial flowers. The spongy pith of this large-leaved shrub or small tree of China and Taiwan is utilized. Stems 2 or 3 years old are cut, soaked for several days, sectioned, and the pith forced out of the woody cylinder with a peglike stick. Pith is sold to small factories and "peeled" much as rotary veneer is made from a log, by dextrous women wielding a constantly sharpened knife. The soft, velvety pith ribbons so produced are sold as rice paper, sometimes used for paintings in the Orient, but mostly imported into Japan and the United States for making artificial flowers.

A curious use of yareta, *Azorella yareta,*

Umbelliferae, is for fuel above the timberline in the puña of Peru and Chile. The low cushions of this condensed "shrub" look much like colonies of coral or a large cauliflower head. Stems and foliage contain combustible resin, and dried plants yield an unexpectedly large amount of heat when burned. Yareta is collected chiefly by Indian herdsmen, stacked like cordwood, and sold as a commercial fuel as well as being used in the home.

The calabash tree, *Crescentia cujete*, Bignoniaceae, grows both wild and planted throughout tropical America. The large, gourdlike fruits have a hard, durable shell, and are much used to make the ubiquitous "maracas" or rattles used musically. Highly ornate, carved designs are typically made on the shell, and wooden handles affixed. Dried (and often smoked) calabash shells serve well as containers, pails, even cooking pots and occasionally as a "hat." Seeds and pulp are reputed to have medicinal properties.

Leaf protein is of increasing interest as a supplementary foodstuff for people, especially in the tropics. Cassava leaf has been suggested as a raw material, but almost any abundant, inexpensively gathered, nontoxic foliage would serve. Leaves are comminuted into a pulp by special machines, from which 50 per cent or more protein can be expressed, coagulated by heating, and formed into blocks resembling cheese (see the discussion of alfalfa in Chapter 17).

The emphasis on esthetic use of plants for landscaping increases with affluence.

Producing lawnseed in Oregon. A field of fine fescue just heading, being rogued by chemical spraying. (Courtesy Union Pacific Railroad.)

Horticulture and ornamentals. Many plants grown in croplike quantities are used for esthetic enjoyment rather than substance. Some home garden plants serve both purposes, such as nut and fruit trees. As a society becomes affluent, interest seems to increase in esthetic use of plants. Thus in economically advanced lands, especially, horticultural specialities have sparked important industries. In the aggregate (including equipment), a market worth several billion dollars annually has been generated in the United States alone. There over 50 million home gardeners tend more than 10 million acres, and consume 200 million pounds of seed, over $1 billion worth of "green goods" (bedding plants, bulbs, roses, shrubs, evergreens, perennials, shade trees, etc.), and voluminous quantities of power equipment, outdoor furniture, fertilizers, and pesticides for landscaping and beautification. Important floriculture items include cut flowers, house plants, bedding plants, and greenery such as ferns. Food plants such as tomatoes and lettuce may be grown in the greenhouse for off-season sale, but will not be discussed here since they were dealt with in earlier chapters.

SEED OF ORNAMENTALS. A specialized seed industry has developed only during the twentieth century to supply pedigreed weed-free flower, vegetable, and lawn seed. The mid-century migration of Americans to the suburbs intensified interest in attractive home grounds, and created new demand for select lawngrass and garden cultivars. The most important lawngrasses planted from seed of the United States are the Kentucky bluegrasses (*Poa pratensis*), the fine or red fescues (*Festuca rubra*), the bentgrasses (*Agrostis spp.*), and Bermudagrass (*Cynodon dactylon*). Bluegrass, fescue, and bentgrass are adapted to the northern two-thirds of the United States, Bermudagrass to the area from the border states southward. Several coarser species are used for lawns in the deep South, and are often vegetatively planted. Seed of bahíagrass (*Paspalum notatum*), carpetgrass (*Axonopus spp.*), centipedegrass (*Eremochloa ophiuroides*), and unselected *Zoysia* is available. Special cultivars of zoysia and Bermuda must be propagated by plugs, sprigs, or sod, since they do not come true-to-type from seed. St. augustinegrass, *Stenotaphrum secundatum*, is marketed entirely as vegetative material, since it yields seed poorly.

Traditionally grass seed was harvested from pastures and meadows, and to a certain extent still is with Canadian fescues, coarse fescues in the Midwest, and bahíagrass in Florida and Texas. Seed production of the elite bluegrasses, fine fescues, and bentgrasses now centers in Oregon and Washington in the United States, where special trade channels have evolved and special equipment has been developed for efficient growing and processing of grass solely for its seed. Along with the specialized growing have come standards assuring clean seed of known genetic identity.

Garden seed is likewise produced mainly in the western states, most flower seed in California, most vegetable seed in Idaho. More so than with lawnseed, elite cultivars must be maintained under isolation to preserve pure lines of distinctive characteristics. For the same economic reasons that made hybrid corn so successful, the trend is toward hybrid proprietary cultivars that become the exclusive property of the originator. Both pure-line and hybrid strains of annual flowers and vegetables are intensively merchandised, with colorful catalogues and promotional schemes (such as the "all-American" selections) intended to direct attention to newly bred cultivars. A great deal of technical skill undergirds the industry, including special hybridizing techniques,

controlled pollination, customized harvesting, and so on. Most firms maintain widespread trial grounds, and corps of experts visit test locations to keep abreast of performance and new entrees.

NURSERY PLANTS. Live plants of some maturity used in landscaping are the basis of an extensive nursery industry. Growers usually concentrate upon a particular specialty. Holland is world renowned for its bulb production, and one of the world's largest lily bulb farms is located in Gresham, Oregon. Production of herbaceous perennials (often propagated vegetatively) and shrubbery for landscaping is more scattered. Gaining impetus from the landscaping of the interstate roadway system in the United States, the growing of ornamental trees has enjoyed rapid expansion.

Commercially a "bulb" may be a corm, tuber, rhizome, or fleshy root, as well as a true bulb. Thus *Iris, Hemerocallis,* and

Planting lilies in Oregon for propagation as ornamentals. (Courtesy Jagra, Oregon Bulb Farms.)

Young lily seedlings are protected under lathe shelter. (Courtesy Jagra, Oregon Bulb Farms.)

Young lilies first year from seed. (Courtesy Jagra, Oregon Bulb Farms.)

Field of new trumpet hybrid lilies being checked for commercial possibilities. Part of the extensive operations in satisfying demand for ornamentals. (Courtesy Jagra, Oregon Bulb Farms.)

Harvesting lily bulbs for marketing, reflecting the burgeoning interest in ornamentals. (Courtesy Jagra, Oregon Bulb Farms.)

Dahlia join *Tulipa*, *Hyacinthus*, and *Narcissus* as "bulbs." Grown indoors, or seasonally outdoors, are *Hippeastrum*, *Anemone*, tuberous *Begonias*, *Caladium*, *Canna*, *Clivia*, *Freesia*, *Gladiolus*, *Hymenocallis*, *Montbretia*, *Ranunculus*, *Tigridia*, *Polianthes*, *Zephyranthes*, and several others. Generally hardy are the crocuses (*Colchicum* and *Crocus*), *Fritillaria*, *Galanthus*, *Lycoris*, *Scilla*, and the familiar bulbs earlier cited.

Most of the bulb species were long ago domesticated, improved by unrecorded selection and crossing through the years. The majority are of Old World origin. Tulips (*Tulipa*, Liliaceae) are available as botanical species, but prized garden tulips result from innumerable crosses involving several species (e.g., "early tulips," "breeders tulips," "cottage tulips," "darwin tulips," "lily-flowered tulips," "triumph tulips," "mendel tulips," "parrot tulips," and so on). The earliest tulip cultivars seem to have originated in Turkey, but were soon spread throughout Europe and especially adopted by the Dutch. During the "tulip craze" of the eighteenth century, single select bulbs sold for thousands of dollars. Holland exports to the United States over a quarter-million bulbs annually, mostly tulips. Patient, skillful, intensive gardening is required to nurse a seed or bulblet to commercial size over a period of several years.

Garden perennials are admired seasonally for their flowers, or as a ground cover. Typical species are the hollyhock (*Althea rosea*), the columbine (*Aquilegia*), the bellflower (*Campanula*), *Chrysanthemum*, *Delphinium*, *Dianthus*, coral bells (*Heuchera sanguinea*), *Phlox*, poppies (*Papaver*), primroses (*Primula*), and speedwells (*Verconica*). Their production and sale is usually a sideline to other nursery activities.

In the United States ornamental shrubbery is grown especially in the Great Lakes area, the Middle Atlantic states, Florida, and California. Prominent coniferous species used for landscaping are *Taxus*, *Juniperus*, *Picea*, and *Pinus*; broadleaf evergreens include *Rhododendron*, *Camellia*, *Ilex*, and *Buxus*; typical deciduous shrubs are *Forsythia*, *Viburnum*, *Berberis*, *Ligustrum*, and *Rosa*. Most are propagated as rooted cuttings. "Lining-out" stock is started by specialists and sold to other nurserymen for field planting. Grown to marketable size, the plants are sold wholesale to retail nurseries and garden centers. Rose growing is highly specialized. Select cultivars are bud-grafted onto mass-produced rootstocks, usually *Rosa multiflora*. This is the only means for achieving rapid increase of new selections. Rose growing has tended to center in the southwestern United States, where advantage can be taken of the nearly year-round growing

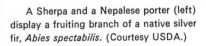

A Sherpa and a Nepalese porter (left) display a fruiting branch of a native silver fir, *Abies spectabilis*. (Courtesy USDA.)

weather, and where the dry atmosphere limits problems from disease. The budding operation requires skillful labor, and is not inexpensive.

Shade trees are grown and handled much as are shrubs. Many native species are still used, but opportunity for transplanting a tree from some woodlot has largely disappeared in the United States with the swelling of the population. The tendency is toward development of select cultivars propagated in the nursery. These are usually grafted, budded, or propagated by cuttings. Especially esteemed for flower or fruit are the crabapples (*Pyrus*), redbud (*Cercis canadensis*), dogwoods (*Cornus*), hawthorns (*Crataegus*),

Cut lilies are a favorite floriculture crop. (Courtesy Jagra, Oregon Bulb Farms.)

and *Magnolias*. Larger trees planted chiefly for shape and foliage are maples (*Acer*), ash (*Fraxinus*), honey locust (*Gleditzia*), *Tilia*, etc. among Angiosperms. Gymnosperms include *Pinus*, *Thuja*, *Picea*, and most other genera mentioned in the chapter on forests; somewhat rare are the "living fossil" trees, *Ginkgo biloba* and *Metasequoia glyptostroboides*.

FLORICULTURE. A wide array of plants are grown in the greenhouse for greenery, cut flowers, and potted plants. Successful operation requires heavy capitalization and well-planned procedures. Timing of the crop must be exact, to hit seasonal demands (for example, poinsettias at Christmas, lilies at Easter, chrysanthemums, carnations, orchids, camellias, etc. for special occasions). Familiar potted plants include *Cyclamen*, *Saintpaulia*, *Poinsettia*, *Lilium*, *Chrysanthemum*, *Hydrangea*, and so on. Much-used foliage plants are various ferns, palms, ivy, *Philodendron*, and succulents. *Begonia*, *Pelargonium*, and *Vinca* are typical bedding plants started under glass, and garden annuals often started in the greenhouse include *Petunia*, *Tagetes*, *Salvia*, and so on. To compete successfully in floriculture a greenhouse must be highly automated, and constantly supervised to assure sterile growing conditions (including use of pesticides for insect and disease control). Special techniques have been developed to grow plants hydroponically where warranted, to provide special misting machines for the rooting of cuttings, and for the use of chemical growth substances to regulate shape and size and to effect disbudding.

* * *

Thus are plants involved in most of man's activities. Forest products, extractives, food plants, and esthetics all tie him to the thin layer of green that calcimines this earth. Hopefully the reader has gained an appreciation of this dependence from the discussion of the many topics in this and earlier chapters.

SUGGESTED SUPPLEMENTARY REFERENCES

Coats, A. M., *The Plant Hunters*, McGraw-Hill, N. Y., 1970.

Culberson, C. F., *Chemical and Botanical Guide to Lichen Products*, Univ. of North Carolina Press, Chapel Hill, 1969.

Fernald, M. L., and A. C. Kinsey, *Edible Wild Plants of Eastern North America*, Idlewild Press, N. Y., 1943.

Firth, F. E., (ed.), *The Encyclopedia of Marine Resources*, Van Nostrand Reinhold Co., N. Y., 1969.

Frazier, W. C., *Food Microbiology*, 2nd ed., McGraw-Hill N. Y., 1967.

Gillespie, W. H., *Edible Wild Plants of West Virginia*, Scholar's Library, N. Y., 196?.

Goldberg, H. S. (ed.), *Antibiotics, Their Chemistry and Non-Medical Uses*, Van Nostrand, Princeton, N. J., 1959.

Gray, W. D., *The Relation of Fungi to Human Affairs*, Holt, N. Y., 1959.

Gray, W. D., *The Use of Fungi as Food and in Food Processing*, Chemical Rubber Co., Cleveland, 1970.

Hanson, A. A., and F. V. Juska (eds.), *Turfgrass Science*, American Society of Agronomy, Madison, Wisc., 1969.

Harrington, H. D., *Edible Native Plants of the Rocky Mountains*, Univ. of New Mexico Press, Albuquerque, N. Mex., 1967.

Harshbarger, G. F., *McCall's Garden Book*, Simon and Schuster, N. Y., 1968.

Hartman, H. T., and D. E. Kester, *Plant Propagation*, 2nd ed., Prentice-Hall, Englewood Cliffs, N. J., 1968.

Havas, N., *Graphic View of the Retail Florist Industry*, USDA Marketing Research Report 788, Washington, D. C., 1967.

Hawthorn, L. R., and L. H. Pollard, *Vegetable and Flower Seed Production*, Blakiston, N. Y., 1954.

Isely, D., *Weed Identification and Control*, 2nd ed., Iowa State Univ. Press, Ames, 1960.

Jackson, D. F. (ed.), *Algae, Man and the Environment*, Syracuse Univ. Press, Syracuse, N. Y., 1968.

——————, (ed.), *Algae and Man*, Plenum Press, N. Y., 1964.

King, L. J., *Weeds of the World*, Leonard Hill, London, and Interscience, N. Y., 1966.

Kingsbury, J. M., *Poisonous Plants of the United States and Canada*, Prentice-Hall Inc., Englewood Cliffs, N. J., 1964.

Laurie, A., D. C. Kiplinger, and K. S. Nelson, *Commercial Flower Forcing*, McGraw-Hill, N. Y., 1968.

Levring, T., H. A. Hoppe, and O. J. Schmid, *Marine Algae*, Botanica Marina Handbooks, Vol. I, Cram, de Gruyter and Co., Berlin, 1969.

Medsger, O. P., *Edible Wild Plants*, Macmillan, N. Y., 1939.

Muenscher, W. C., *Weeds*, 2nd ed., Macmillan, N. Y., 1955.

Chas. Pfizer & Co., *Our Smallest Servants*, N. Y., 1955 ff.

Rose, A. H., and J. S. Harrison (eds.), *The Yeasts*, Academic Press, N. Y., Vol. I, *Biology of the Yeasts*, 1969.

Salisbury, Sir Edward, *Weeds and Aliens*, Collins, London, 1961.

Schery, R. W., *Householder's Guide to Outdoor Beauty*, Pocket Books, N. Y., 1963.

Schery, R. W., *The Lawn Book*, Macmillan, N. Y., 1973 (in press).

Singer, R., *Mushrooms and Truffles*, Leonard Hill, London, and Interscience, N. Y., 1961.

Singh, R. N., *Role of Blue-Green Algae in Nitrogen Economy of Indian Agriculture*, Indian Council of Agricultural Research, New Delhi, 1962.

Smith, A. H., *The Mushroom Hunters Field Guide*, Univ. of Michigan Press, Ann Arbor, 1958.

Umezawa, H. (ed.), *Index of Antibiotics from Actinomycetes*, Univ. Park Press, Baltimore, 1968.

USDA Yearbook, *Science for Better Living*, Washington, D. C., 1968.

——————, *Seeds*, Washington, D. C., 1961.

Weiser, H. H., *Practical Food Microbiology and Technology*, AVI Pub. Co., Westport, Conn., 1962.

Wodehouse, R. P., *Hayfever Plants*, Chronica Botanica, Waltham, Mass., 1945.

Wyman, D., *Shrubs and Vines for American Gardens*, Macmillan, N. Y., 1953.

——————, *Trees for American Gardens*, Macmillan, N. Y., 1951.

Zajic, J. E., *Properties and Products of Algae*, Plenum Press, N. Y. 1970.

Index